Tissue engineering using ceramics and polymers

Related titles:

Artificial cells, cell engineering and therapy
(ISBN 978-1-84569-036-6)
This authoritative reference work provides a detailed study of the most up-to-date developments in artificial cells, cell engineering and cell therapy. It covers design, engineering and uses of artificial cells. The clinical relevance of artificial cells and cell engineering is assessed. This is a highly dynamic sector with growing interest in the use of engineered cells in controlled release of active drugs and other therapeutic agents.

Biomedical polymers
(ISBN 978-1-84569-070-0)
This book presents a detailed study of the structure, processing and properties of biomedical polymers and the relationship between them. Emphasis is placed on the various groups of biopolymers including natural polymers, synthetic biodegradable and non-biodegradable polymers.

Molecular interfacial phenomena of polymers and biopolymers
(ISBN 978-1-85573-928-4)
This book combines three fundamental areas of interest to the science and engineering community, these being material science, nanotechnology and molecular engineering. Although there have been various results published in this field, there has yet to be a fully comprehensive review. This book covers key research on molecular mechanisms and thermodynamic behaviour of (bio)polymer surfaces and interfaces, from theoretical and experimental perspectives.

Details of these and other Woodhead Publishing materials books, as well as materials books from Maney Publishing, can be obtained by:

- visiting www.woodheadpublishing.com
- contacting Customer Services (e-mail: sales@woodhead-publishing.com; fax: +44 (0) 1223 893694; tel.: +44 (0) 1223 891358 ext. 130; address: Woodhead Publishing Limited, Abington Hall, Abington, Cambridge CB21 6AH, England)

Maney currently publishes 16 peer-reviewed materials science and engineering journals. For further information visit www.maney.co.uk/journals.

Tissue engineering using ceramics and polymers

Edited by Aldo R. Boccaccini and Julie E. Gough

Woodhead Publishing and Maney Publishing
on behalf of
The Institute of Materials, Minerals & Mining

CRC Press
Boca Raton Boston New York Washington, DC

WOODHEAD PUBLISHING LIMITED
Cambridge England

Woodhead Publishing Limited and Maney Publishing Limited on behalf of
The Institute of Materials, Minerals & Mining

Published by Woodhead Publishing Limited, Abington Hall, Abington,
Cambridge CB21 6AH, England
www.woodheadpublishing.com

Published in North America by CRC Press LLC, 6000 Broken Sound Parkway, NW,
Suite 300, Boca Raton, FL 33487, USA

First published 2007, Woodhead Publishing Limited and CRC Press LLC

British Library Cataloguing in Publication Data
A catalogue record for this book is available from the British Library.

Library of Congress Cataloging in Publication Data
A catalog record for this book is available from the Library of Congress.

Woodhead Publishing Limited ISBN 978-1-84569-176-9 (book)
Woodhead Publishing Limited ISBN 978-1-84569-381-7 (e-book)
CRC Press ISBN 978-1-4200-4454-6
CRC Press order number WP4454

The publishers' policy is to use permanent paper from mills that operate a sustainable forestry policy, and which has been manufactured from pulp which is processed using acid-free and elementary chlorine-free practices. Furthermore, the publishers ensure that the text paper and cover board used have met acceptable environmental accreditation standards.

Project managed by Macfarlane Production Services, Dunstable, Bedfordshire, England (macfarl@aol.com)
Typeset by Godiva Publishing Services Limited, Coventry, West Midlands, England
Printed by TJ International Limited, Padstow, Cornwall, England

Contents

Contributor contact details

(* = main contact)

Editors
Aldo R. Boccaccini
Department of Materials
Imperial College London
Prince Consort Road
London SW7 2BP
UK
E-mail: a.boccaccini@imperial.ac.uk

Julie E. Gough
Materials Science Centre
School of Materials
University of Manchester
Grosvenor Street
Manchester M1 7HS
UK
E-mail: j.gough@manchester.ac.uk

Chapter 1
J. Huang*
Department of Mechanical Engineering
University College London
Torrington Place
London WC1E 7JE
UK
E-mail: jie.huang@ucl.ac.uk

S. M. Best
Department of Materials Science and Metallurgy
University of Cambridge
Pembroke Street
Cambridge CB2 3QZ
UK
E-mail: smb51@cam.ac.uk

Chapter 2
Guobao Wei and Peter X. Ma*
Department of Biologic and Materials Sciences
Department of Biomedical Engineering
Macromolecular Science and Engineering Center
The University of Michigan
Ann Arbor
Michigan 48109-1078
USA
E-mail: mapx@umich.edu

Chapter 3
J. R. Jones
Department of Materials
Imperial College London
South Kensington Campus
London SW7 2AZ
UK
E-mail: julian.r.jones@imperial.ac.uk

Chapter 4
S. K. Misra and A. R. Boccaccini*
Department of Materials
Imperial College London
Prince Consort Road
London SW7 2BP
UK
E-mail: a.boccaccini@imperial.ac.uk

Chapter 5
J. Mansbridge
Tecellact LLC
1685 Calle Camille
La Jolla
USA
E-mail: Verajonath@aol.com
JonathanMansbridge@yahoo.com

Chapter 6
K. Shakesheff* and G. Tsourpas
Division of Drug Delivery and Tissue Engineering
School of Pharmacy
University of Nottingham
Nottingham NG7 2RD
UK
E-mail:
Kevin.shakesheff@nottingham.ac.uk

Chapter 7
N. Tirelli* and F. Cellesi
School of Pharmacy
University of Manchester
Oxford Road
Manchester M13 9PL
UK
E-mail:
Nicola.tirelli@manchester.ac.uk
f.cellesi@manchester.ac.uk

Chapter 8
F. Davis and S. P. J. Higson*
Cranfield Health
Cranfield University
Barton Rd
Silsoe MK45 4DT
UK
E-mail: f.davis@cranfield.ac.uk
s.p.j.higson@cranfield.ac.uk

Chapter 9
Andrew J. Urquhart and Morgan R. Alexander*
School of Pharmacy
University of Nottingham
University Park
Nottingham NG7 2RD
UK
E-mail:
Morgan.Alexander@nottingham.ac.uk

Chapter 10
A. M. Donald
Department of Physics
Cavendish Laboratory
University of Cambridge
JJ Thomson Avenue
Cambridge CB3 0HE
UK
E-mail: amd3@cam.ac.uk

Chapter 11
S. I. Anderson
Advanced Microscopy Unit
School of Biomedical Science
E Floor Medical School
Queens Medical Centre
Clifton Boulevard
Nottingham NG7 2UH
UK
E-mail:
susan.anderson@nottingham.ac.uk

Chapter 12
I. Notingher
School of Physics and Astronomy
University of Nottingham
University Park
Nottingham NG7 2RD
UK
E-mail:
ioan.notingher@nottingham.ac.uk

Chapter 13
A. Atala
Wake Forest Institute for Regenerative Medicine
Wake Forest University Health Sciences
Medical Center Boulevard
Winston-Salem
NC 27157
USA
E-mail: aatala@wfubmc.edu

Chapter 14
P. Woźniak*
Department of Biophysics and Human Physiology
Medical University of Warsaw
Ul. Chalubinskiego 5
02-004 Warsaw
Poland
E-mail: pwozniak@amwaw.edu.pl
wozniak_piotr@yahoo.com

A. J. El Haj
Institute for Science & Technology in Medicine
Keele University
Thornburrow Drive
Hartshill
Stoke-on-Trent ST4 7QB
UK
E-mail: a.j.el.haj@bemp.keele.ac.uk

Chapter 15
L. Di Silvio
Senior Lecturer in Biomaterials & Biomimetics
King's College London Dental Institute
Biomaterials Science
Floor 17, Guy's Tower
Guy's Campus
St Thomas' Street
London SE1 9RT
UK
E-mail: lucy.di_silvio@kcl.ac.uk

Chapter 16
Q. Z. Chen, H. Jawad, A. R. Boccaccini*
Department of Materials
Imperial College London
Prince Consort Road
London SW7 2BP
UK
E-mail: a.boccaccini@imperial.ac.uk

S. E. Harding and N. N. Ali
National Heart and Lung Institute
Imperial College London
Dovehouse Street
London SW3 6LY
UK

Chapter 17
J. Hoyland* and T. Freemont
Tissue Injury and Repair Group
School of Medicine
Stopford Building
The University of Manchester
Oxford Road
Manchester M13 9PT
UK
E-mail:
judith.hoyland@manchester.ac.uk
tony.freemont@manchester.ac.uk

Chapter 18
S. MacNeil
Tissue Engineering Group
Kroto Research Institute
University of Sheffield North Campus
Broad Lane
Sheffield S3 7HQ
UK
E-mail: s.macneil@sheffield.ac.uk

Chapter 19
K. Shakesheff
Division of Drug Delivery and Tissue Engineering
School of Pharmacy
University of Nottingham
Nottingham NG7 2RD
UK
E-mail: kevin.shakesheff@nottingham.ac.uk

Chapter 20
A. Saito
Division of Nephrology and Metabolism
Department of Medicine
Tokai University School of Medicine
Bohseidai
Isehara
Kanagawa 259-1193
Japan
E-mail: asait@is.icc.u-tokai.ac.jp

Chapter 21
A. M. Turner and J. Southgate*
Jack Birch Unit of Molecular Carcinogenesis
Department of Biology
University of York
York YO10 5YW
UK
E-mail: js35@york.ac.uk
alexturner64@doctors.org.uk

R. Subramaniam and D. F. M. Thomas
Department of Paediatric Urology
St James's University Hospital
Leeds LS9 7TF
UK
E-mail: Ramnath.Subramaniam@leedsth.nhs.uk
D.F.M.Thomas@leeds.ac.uk

Chapter 22
P. Kingham* and G. Terenghi
Blond McIndoe Laboratories
Tissue Injury and Repair Group
3.106 Stopford Building
The University of Manchester School of Medicine
Oxford Road
Manchester M13 9PT
UK
E-mail:
paul.j.kingham@manchester.ac.uk
giorgio.terenghi@manchester.ac.uk

Chapter 23
A. E. Bishop* and H. J. Rippon
Stem Cells & Regenerative Medicine
Section on Experimental Medicine & Toxicology
Imperial College Faculty of Medicine
Hammersmith Campus
Du Cane Road
London W12 ONN
UK
E-mail: a.e.bishop@imperial.ac.uk
hj.rippon@imperial.ac.uk

Chapter 24
D. A. J. Lloyd* and S. M. Gabe
St Mark's Hospital
Northwick Park
Harrow HA1 3UJ
UK
E-mail: dajl@btinternet.com
s.gabe@imperial.ac.uk

Chapter 25
A. Fritsch and L. Dormieux
Laboratory for Materials and Structures
Ecole Nationale des Ponts et Chaussées (LMSGC-ENPC)
Marne-la-Vallee
France

C. Hellmich*
Vienna University of Technology
Karlsplatz 13/E 202
1040 Wien
Austria
E-mail:
christian.hellmich@tuwien.ac.at

J. Sanahuja
Lafarge Research Center
San-Quentin Fallavier
France

Chapter 26
J. E. Gough
Materials Science Centre
School of Materials
University of Manchester
Grosvenor Street
Manchester M1 7HS
UK
E-mail: j.gough@manchester.ac.uk

Introduction

J E GOUGH, University of Manchester, UK and
A R BOCCACCINI, Imperial College London, UK

The field of tissue engineering has advanced dramatically in the last 10 years, offering the potential for regenerating almost every tissue and organ of the human body. Tissue engineering and the related discipline of regenerative medicine remain a flourishing area of research with potential new treatments for many more disease states. The advances involve researchers in a multitude of disciplines, including cell biology, biomaterials science, imaging and characterisation of surfaces and cell–material interactions. In the field of biomaterials, for example, many additional and novel processing techniques have been developed for the construction of improved porous scaffolds and matrices, including a much wider variety of polymers, ceramics and their composites with tailored micro- and nanostructure, surface topography and chemistry as well as optimised properties for the intended application in tissue engineering strategies. There are in fact many more materials being investigated for potential scaffolds, including novel uses of natural materials, combinations of natural and synthetic materials and new structures designed to mimic extracellular matrix at all relevant scales (macro, micro and nano), which are intended to provide scaffolds that are closer to the *in vivo* cellular environment. There is also strong emphasis on biomechanics and the effects of mechanical forces on the cell response and subsequent tissue formation, again, aiming to mimic the conditions *in vivo*. Along with major advances in the design of materials and control of their properties, there have been huge advances in understanding the cell biology of the cell response to artificial environments, substrates and scaffolds, with recent emphasis on the suitability of stem cells for tissue regeneration and the restoration of function.

This book aims to combine some of the most recent innovative research with reviews of specific aspects of tissue engineering to provide an up-to-date source for undergraduate, postgraduate, academic and industrial readers. Advances in the related areas of biomaterial science, cell biology and characterisation techniques are reported, and the relevance of the chosen biomaterial (polymer, ceramic) is highlighted.

Part I of this book contains chapters concerning the materials of choice for tissue engineering – polymers, ceramics, bioactive ceramics and glasses and composites. It contains some general issues related to novel tissue engineering strategies with emphasis on transplantation of engineered tissues, surface modification of scaffolds and combined tissue engineering and drug delivery approaches. Specific chapters on state-of-the-art characterisation techniques are also included with particular reference to those techniques of major relevance to tissue engineering: X-ray photoelectron spectroscopy and secondary ion mass spectrometry, environmental scanning electron microscopy, confocal microscopy and Raman spectroscopy.

Part II of this book focuses on some specific examples of organ and tissue regeneration. It covers some of the highly challenging, more recent advances in tissue engineering such as liver, kidney, bladder, intervertebral disc, lung, cardiac tissue and intestine and also recent advances in nerve bioengineering. Some of the largest, longer-standing areas of tissue engineering research (including bone, skin and cartilage) which require new innovative approaches are covered. There has been a great deal of activity in determining and optimising new cell sources for tissue engineering which is mentioned in various areas of the book.

We hope this book will be a valuable source of information and an inspiration to new and existing researchers in the field.

Part I

General issues

1
Ceramic biomaterials

J HUANG, University College London, UK and
S M BEST, University of Cambridge, UK

1.1 Introduction

Ceramic materials have been part of everyday life for thousands of years, and traditional ceramics include porcelain, refractories, cements and glass. More recently, advanced ceramics have been developed, including ferroelectrics, structural oxides, borides, carbides and nitrides and glass-ceramics, and these find applications in telecommunications, the environment, energy, transportation and health. There is no complete definition for a ceramic; but they are generally solid materials composed of inorganic, non-metallic substances. They exist as both crystalline and non-crystalline (amorphous) compounds, and glasses and glass-ceramics (partially crystallised glasses) are subclasses of ceramics.

A biomaterial is a non-viable material used in a medical device, intended to interact with biological systems (Williams, 1987). Various engineering materials, including ceramics, metal (alloys), polymer and composites, have been developed to replace the function of the biological materials. The focus of this chapter is to consider ceramics used in biological applications, now generally referred to as bioceramics, and their applications in implants and in the repair and reconstruction of diseased or damaged body parts. Most clinical applications of bioceramics relate to the repair of the skeletal system, comprising bone, joints and teeth, and to augment both hard and soft tissue. According to the types of bioceramics and host tissue interactions, they can be categorised as either bioinert or bioactive, the bioactive ceramics may be resorbable or non-resorbable, and all these may be manufactured either in porous or dense bulk form, or granules or coatings.

The chapter begins by introducing various ceramics used in medical applications, including bioinert ceramics (i.e. alumina and zirconia) and bioactive ceramics (i.e. calcium phosphates, bioactive glasses and glass-ceramics). To understand the nature and formation of ceramic structures, it is essential to have an understanding of the atomic arrangements, the forces between atoms and the location of atoms in a crystalline lattice. The difference between crystalline and non-crystalline materials with the examples of hydroxyapatite ceramics and

Bioglass® and apatite-wollastonite glass-ceramic, the most widely applied bioceramics, is discussed in Section 1.2. The properties of a ceramic are determined by its microstructure (e.g. grain size and porosity). A brief summary of the common techniques for characterisation of the microstructure of ceramics is included in Section 1.3. This is followed by a review of the properties of ceramics, particularly mechanical properties, surface properties, biocompatibility and bioactivity, which are crucial for the biological application of the ceramics. Alumina and zirconia have excellent mechanical properties for load-bearing applications, while the bioactivity of glass and ceramics leads to the potential for osteoconduction. A brief review of the processing of ceramics with an example of hydroxyapatite is presented in Section 1.5. The processing of porous ceramics scaffolds and surface modification using coating and thin film deposition is also discussed. The chapter finishes with a summary, highlighting the importance of understanding of the clinical requirement and relationships between processing, microstructure and properties, which will help to develop better ceramic materials for tissue engineering.

1.1.1 Bioinert ceramics

Alumina and zirconia have been used as an important alternative to surgical metal alloys in total hip prostheses and as tooth implants. The main advantages of using ceramics over the traditional metal and polymer devices are lower wear rates at the articulating surfaces and the release of very low concentrations of 'inert' wear particles. For example, using femoral heads of alumina ceramic bearing against alumina cup sockets significantly reduces wear debris when against ultra-high molecular weight polyethylene cups. Excessive wear rates can contribute to loosening and eventual implantation failure. Zirconia ceramics have the advantages over alumina ceramics of higher fracture toughness and higher flexural strength, and relatively lower Young's modulus, although after a number of implant failures the choice of zirconia components in hip procedures suffered a significant setback. There are two types of zirconia ceramics for surgical implants, yttria-stabilised tetragonal zirconia (Y-TZP) and magnesium oxide partially stabilised zirconia (Mg-PSZ). Typical mechanical properties of these ceramics are listed in Table 1.1.

To improve the fracture toughness of alumina ceramics, nanophase alumina with grain size of 23 nm was synthesised. The modulus of elasticity of nanophase alumina decreased by 70% (Webster *et al.*, 1999). The fracture toughness of alumina can then be controlled through the use of nanophase formulations; furthermore, enhanced biological responses of osteoblast cells to the nanophase materials were found, indicating the improved osseointegration potential for nanophase alumina (Webster *et al.*, 2000). However, the flexure strength of nanophase alumina was unclear, which is a critical parameter for its clinical application.

Table 1.1 A summary of mechanical properties of various biomaterials (Hench and Andersson, 1993; Holand and Vogel, 1993; Hulbert, 1993; Hench and Best, 2004)

Materials	Density ($g\,cm^{-3}$)	Hardness (Vickers, HV)	Young's modulus (GPa)	Bending strength (MPa)	Compressive strength (MPa)	Fracture toughness K_{IC} ($MPa\,m^{1/2}$)
Bioglass® 45S5	2.66	458	35	40–60		0.4–0.6
A-W glass-ceramic	3.07	680	118	215	1080	2.0
Bioverit glass-ceramic	2.8	5000	70–88	140–180	500	1.2–2.1
Sintered HA	3.156	500–800	70–120	20–80	100–900	0.9–1.3
Alumina	3.98	2400	380–420	595	4000–4500	4–6
Zirconia (TZP)	6.05	1200	150	1000	2000	7
Zirconia (Mg-PSZ)	5.72	1120	208	800	1850	8
316 stainless steel	8		200	540–1000*		~100

* = tensile strength.

Alumina and zirconia have good biocompatibility, adequate mechanical strength, but are relatively biologically inactive (nearly inert) and lack direct bonding with host tissue. Bioactive materials are conceptually different from bioinert materials in that chemical reactivity is essential. A series of bioactive ceramics, glasses and glass-ceramics are capable of promoting the formation of bone at their surface and of creating an interface, which contributes to the functional longevity of tissue.

1.1.2 Bioactive ceramics

Bioactive ceramics include several major groups, such as calcium phosphate ceramics, bioactive glasses and glass-ceramics.

Calcium phosphate ceramics

Calcium phosphates are the major constituents of bone mineral. Table 1.2 lists several calcium phosphates with their chemical formula and Ca/P ratio (from 0.5 to 2). These calcium phosphates can be synthesised by mixing calcium and phosphate solution under acid or alkaline conditions. Only certain compounds are useful for implantation in the body, compounds with a Ca/P ratio less than 1 are not suitable for biological implantation because of their high solubility.

The most extensively used synthetic calcium phosphate ceramic for bone replacement is hydroxyapatite (HA) because of its chemical similarities to the inorganic component of bone and teeth. HA with a chemical formula of $Ca_{10}(PO_4)_6(OH)_2$, has a theoretical composition of 39.68 wt% Ca, 18.45 wt% P; Ca/P wt ratio of 2.151 and Ca/P molar ratio of 1.667. It is much more stable than other calcium phosphate ceramics within a pH range of 4.2–8.0.

Table 1.2 Ca/P ratio of various calcium phosphates (Aoki, 1991)

Name	Abbreviation	Formula	Ca/P ratio
Tetracalcium phosphate	TTCP	$Ca_4O(PO_4)_2$	2.0
Hydroxyapatite	HA	$Ca_{10}(PO_4)_6(OH)_2$	1.67
Tricalcium phosphate ($\alpha,\alpha',\beta,\gamma$)	TCP	$Ca_3(PO_4)_2$	1.50
Octacalcium phosphate	OCP	$Ca_8H_2(PO_4)_6 \cdot 5H_2O$	1.33
Dicalcium phosphate dihydrate (brushite)	DCPD	$CaHPO_4 \cdot 2H_2O$	1.0
Dicalcium phosphate (montite)	DCP	$CaHPO_4$	1.0
Calcium pyrophosphate (α,β,γ)	CPP	$Ca_2P_2O_7$	1.0
Calcium pyrophosphate dihydrate	CPPD	$Ca_2P_2O_7 \cdot 2H_2O$	1.0
Heptacalcium phosphate	HCP	$Ca_7(P_5O_{16})_2$	0.7
Tetracalcium dihydrogen phosphate	TDHP	$Ca_4H_2P_6O_{20}$	0.67
Calcium phosphate monohydrate	CPM	$Ca(H_2PO_4)_2H_2O$	0.5

The stoichiometry of HA is highly significant where thermal processing of the material is required. Slight imbalances in the ratio of Ca/P can lead to the appearance of extraneous phases. If the Ca/P is lower than 1.67, β-tricalcium phosphate (β-TCP) and other phases, such as tetracalcium phosphate (TTCP), will be present with HA. If the Ca/P is higher than 1.67, calcium oxide (CaO) will be present with the HA phase. The extraneous phases may adversely affect of the biological responses of the implants. TCP is a biodegradable bioceramic with the chemical formula of $Ca_3(PO_4)_2$. It dissolves in wet media and can be replaced by bone during implantation. TCP has four polymorphs, the most common being α and β.

In an ideal situation, a biodegrable implant material is slowly resorbed and replaced by natural tissue. However, to match the rate of resorption with that of the expected bone tissue regeneration for a biodegradable material is a great challenge. When the solubility of calcium phosphate is higher than the rate of tissue regeneration, it will not be of use in cavity filling. TCP with Ca/P ratio of 1.5 is more rapidly resorbed than HA. The use of a mixture of HA and β-TCP, known as biphasic calcium phosphate (BCP), has been attempted as bone substitutes (Daculsi *et al.*, 2003). It has the advantage of tailor making its chemical properties, such as varying the ratio of HA/β-TCP. The higher the TCP content in BCP, the higher the dissolution rate. The resorption rate of BCP can then be monitored and controlled.

Calcium phosphate cement is another important type of bioceramic (Chow and Tagaki, 2001; Fernandez *et al.*, 1998). By mixing with various calcium phosphates, an injectable paste was formed, which will be cured over time and the final product is a carbonate apatite. The cements cure *in situ*, and are gradually resorbed and replaced by the newly formed bone. However, their properties, such as mechanical toughness and curing time, are still being fully optimised.

Bioactive glasses and glass-ceramics

The concept of a bioactive glass was initiated by Hench and colleagues (Hench *et al.*, 1971). The composition of Bioglass® is a series of special designed glasses, consisting of a Na_2O–CaO–SiO_2 glass with the addition of P_2O_5, B_2O_3 and CaF_2 (Table 1.3). A biologically active hydroxy-carbonate apatite (HCA) layer was formed on the surface of bioactive glasses *in vitro* and *in vivo*. This HCA phase is chemically and structurally equivalent to the mineral phase in bone, so it provides direct bonding by bridging host tissue with implants. It is possible to control a range of chemical properties in bioactive glasses and the rate of bonding to tissue. Some specialised compositions of Bioglass® (i.e. 45S5) can bond to soft tissue as well as bone, in either bulk or particulate form (Hench, 1991).

Apatite–wollastonite (A-W) glass-ceramic, with an assembly of small apatite particles effectively reinforced by β-wollastonite, exhibits not only bioactivity, but also a fairly high mechanical strength (Kokubo *et al.*, 1986). The bending

Table 1.3 Composition of bioactive glasses and glass-ceramics (wt%) (Gross *et al.*, 1993; Hench and Andersson, 1993; Holand and Vogel, 1993)

	45S5 Bioglass	52S4.6 Bioglass	55S4.3 Bioglass	58S Sol–gel glass	A-W GC	KG Cera GC	Kgy 213 GC	Bioverit GC
SiO_2	45	52	55	60	34.2	46.2	38	29.5–50
P_2O_5	6	6	6	4	16.3			8–18
CaO	24.5	21	19.5	36	44.9	20.2	31	13–28
CaF_2					0.5			2.5–7
$Ca(PO_3)_2$						25.5	13.5	
MgO					4.6	2.9		6–28
Na_2O	24.5	21	19.5			4.8	4	3–5
K_2O						0.4		3–5
Al_2O_3							7	0–19.5
Ta_2O_5							5.5	
TiO_2							1	

strength, fracture toughness and Young's modulus of A-W glass-ceramic are the highest among bioactive glasses and glass-ceramics (Table 1.1), enabling it to be used in some compression load-bearing applications, such as vertebral prostheses and iliac crest replacement (Kokubo, 1991).

The invention of Bioglass® encouraged the design of new glass and glass-ceramic compositions, of which Ceravital-type glass-(ceramic) was one. The term 'Ceravital' means a number of different compositions of glasses and glass-ceramics and not only one product (Gross *et al.*, 1993). One of the advantages of Ceravital® glass is that the solubility of the material can be adjusted by the addition of metal oxides (Table 1.3), but with negative influence on the cellular function and the development and maturation of the extracellular matrix.

The machinable bioactive Bioverit® glass-ceramics have been successfully applied in the middle ear, nose and jaw and in the general region of the head and neck or as thoracic vertebra substitutes in several patients suffering from tumours (Vogel and Holand, 1990). The presence of the mica phase in the glass ceramic enables it to be machined, with standard metal working tools by the surgeon, if necessary during an operation.

In general, the advantages of bioactive glasses are the speed of their surface reactivity and the ability to alter the chemical composition, thus enabling bonding with a variety of tissues. A disadvantage is their mechanical properties, as these materials have relatively low bending strength and Young's modulus. Many attempts have been made to improve the poor fracture toughness of bioactive glasses and ceramics, such as stainless steel fibre or titanium fibre-reinforced Bioglass® composites (Ducheyne and Hench, 1982; Ducheyne *et al.*, 1993) and hydroxyapatite particles reinforced polymer composites (Bonfield *et al.*, 1981; Bonfield, 1988).

1.2 Characteristics of ceramics

The major characteristics of ceramics are their brittleness, high hardness, thermal and electrical insulation, and corrosion resistance. The chemical inertness fits in the initial criterion for selection of suitable materials in the biological applications, as the human body is a hostile environment for any material. Fundamentally, it is the type of bonding between atoms that controls the properties of a material. There are three primary interatomic bonds: metallic, ionic and covalent; and secondary bonds such as van der Waals and hydrogen. Metallic bonding is the predominant bond mechanism for metals. Ionic bonding occurs when one atom gives up one or more electrons, and another atom or atoms accept these electrons, such that electrical neutrality is maintained. Covalent bonding occurs when two or more atoms share electrons. Atomic bonding in ceramics is mainly ionic or covalent, or a combination of two (Kingery, 1976).

The crystal structure of a material is the periodic arrangement of atoms in the crystal, while a lattice is an infinite array of points in space, in which each point has identical surroundings to all others. A basic concept in crystal structures is the unit cell. It is the smallest unit of volume that permits identical cells to be stacked together to fill all space. By repeating the pattern of the unit cell over and over in all directions, the entire crystal lattice can be constructed. There are seven crystal classes: cubic, tetragonal, orthorhombic, hexagonal, triagonal, monoclinic and triclinic, with four possible unit cell types of symmetry, thus giving 14 Bravais lattices. The letters a, b and c are used to define the unit cell edge length, while α, β and γ are used for the angles. The most common crystal structures in ceramics are simple cubic, closed packed cubic and closed packed hexagonal (Kingery, 1976). The spatial arrangement of individual atoms in a ceramic depends on the type of bonding, the relative sizes of the atoms and the need to balance the electrostatic charges. The brittle nature of ceramic material stems from its crystal structure.

1.2.1 Hydroxyapatite and substituted hydroxyapatite

Hydroxyapatite possesses a hexagonal lattice and a $P6_3/m$ space group. This space group is characterised by a six-fold c-axis perpendicular to three equivalent a-axes (a_1, a_2, a_3) at angles 120° to each other, with cell dimensions of $a = b = 0.9418$ nm and $c = 0.6884$ nm (Hench and Best, 2004).

The mineral phase of bone, biological apatite, is not stoichiometric hydroxyapatite. The apatite is hospitable to a variety of cationic and anionic substitutions, and the type and amount of these ionic substitutions in the apatite phase vary from the wt% level (e.g. 3–8 wt% CO_3) to the ppm–ppb level (e.g. Mg^{2+} or Sr^{2+}). Substitution in the apatite structure for (Ca), (PO_4) or (OH) groups result in changes in properties, such as lattice parameters, morphology, solubility without significantly changing the hexagonal symmetry.

The fluoride substitution (F^- for OH^-) has the consequence of increasing the crystallinity, crystal size and the stability of the apatite, which in turn reduces solubility. Fluoride substitution has been implicated in caries prevention, where its presence in enamel crystals increases stability, which helps to resist dissolution in the acidic oral environment (LeGeros and LeGeros, 1993).

Carbonate, CO_3, can substitute for either the hydroxyl (OH) groups or the phosphate groups, and the resulting apatite is designated as Type A or Type B respectively. An important effect of carbonate substitution in HA is on crystal size and morphology. An increase in carbonate content leads to changes in the size and shape of apatite crystal (LeGeros *et al.*, 1967) and the carbonate-substituted apatites are more soluble than carbonate-free synthetic apatites.

Although silicon has been found in only trace quantities in bone mineral (up to a maximum level of ~0.5 wt%), it has been shown to have a crucial role in bone mineralisation, and is believed to be essential in skeletal development (Carlisle, 1970, 1972). *In vitro* and *in vivo* bioactivities are enhanced with the incorporation of silicon into HA lattice (Gibson *et al.*, 1999a; Patel *et al.*, 2002). The Si substitution in HA inhibited densification and grain growth at higher sintering temperatures (Gibson *et al.*, 1999b, 2002), thus increasing the total surface area/volume ratio of grain boundaries (Porter *et al.*, 2004a), this may also have an effect on the *in vivo* responses. The surface charge of silicon-substituted HA (SiHA) was significantly more negative compared with pure HA (Botelho *et al.*, 2002), it may attribute to the faster bone-like apatite formation *in vitro* induced by SiHA. Silicon-substituted HA as bone graft has been used successfully for spinal fusion.

In addition to the above-mentioned substitutions, there are other substitutions, including strontium (Sr), magnesium (Mg), barium (Ba), lead (Pb), etc. for calcium, vanadates, borates, manganates, etc. for phosphates.

1.2.2 Bioactive glasses

Polycrystalline ceramics are solids in which the atoms or ions are arranged in regular array. In contrast, the regularity (order) is only short range in glass (amorphous materials), because a glass is formed when a molten ceramic composition is rapidly cooled, the atoms do not have time to arrange themselves in a periodic structure. For example, the structure of Bioglass® is regarded as a three-dimensional SiO_2 network, modified by incorporation of other oxides. A number of bioactive glasses have been developed and investigated for tissue engineering and probably the best known of these is Bioglass® (Hench, 2006). Three key compositional features distinguish them from traditional soda-lime–silica glasses: (a) less than 60% of SiO_2, (b) high Na_2O and CaO content and (c) high $CaO:P_2O_5$ ratio. These compositional features make the surface highly reactive when exposed to an aqueous medium and therefore lead to *in vitro* and *in vivo* bioactivity (Hench, 1991).

Sol–gel-derived bioactive glasses have a porous texture in the nanometre range, giving them a surface area of 150–600 m^2g^{-1}, which is two orders of magnitude higher than that of melt-derived glasses. Dissolution is therefore more rapid for sol–gel glasses of similar composition, and more silanol groups are on the sol–gel glass surfaces to act as nucleation sites for the formation of the apatite layer.

1.2.3 Bioactive glass-ceramics

A glass-ceramic is polycrystalline solid prepared by the controlled crystallisation or devitrification of a parent glass. It generally consists of fine grain (with crystal sizes ranging from 0.1 to 10 μm) and has a small volume of residual glass sited at the grain boundary. One advantage of glass-ceramics is that the crystallisation and the formation of the crystal phases can be controlled to develop materials with a combination of special properties, such as bioactivity, machineability and improved mechanical properties. A-W glass-ceramic is the most extensively studied glass-ceramic for use as a bone substitute (Kokubo, 1991). A dense and homogeneous composite was obtained after heat treatment of parent glass, in which 38 wt% was oxyfluorapatite ($Ca_{10}(PO_4)_6(O,F)_2$) and 34 wt% β-wollastonite ($CaOSiO_2$), both grain-like particles, 50–100 nm in size in a MgO–CaO–SiO_2 glassy matrix. It combines high bioactivity with desirable mechanical properties and has been successful in the load-bearing spinal area of the body.

1.3 Microstucture of ceramics

The microstructure of ceramics determines their mechanical and biological properties. Ceramics are commonly polycrystalline, with phases that are physically or chemically distinguishable from each other, and may vary in the crystal structures. The arrangement of crystals (or grains) and phases constitutes the microstructure of the ceramics. Various grain sizes are observed, which depend on the manufacture method, raw materials and grain growth during sintering, a glassy phase, grain boundaries and gas-filled pores may also exist. Typical microstructures of hydroxyaptite and zirconia ceramics are shown in Fig. 1.1: various sized grains and pores were observed on HA while a relatively uniform small grain size (about 1 μm) was found in yttria-stabilised zirconia, which partially determines the excellent mechanical properties (strength and toughness) of zirconia ceramics.

Medical grade alumina with an average grain size of less than 4 μm and 99.7% purity exhibits good flexural and compressive strength (Table 1.1). A very small amount of magnesia (<0.5%) is used as an sintering aid to limit grain growth, as the strength, fatigue resistance and fracture toughness of alumina are a function of grain size and percentage of sintering aid (Hulbert, 1993). High concentrations of sintering aids must be avoided because they remain in the grain boundaries and degrade fatigue resistance, especially in a corrosive physiological environment. This is particularly important for orthopaedic prostheses to be used in younger patients.

The grain structure of ceramics can be observed by optical and electron microscopy after polishing and etching. Glass-ceramics microstructures are characterised by a dispersion of crystals dispersed in a continuous glassy matrix. In contrast, no microstructural features can be observed in glasses. In addition to microscopy, other analytical methods, such as X-ray diffraction, infrared spectroscopy and spectrochemical analysis for detecting impurities are equally important for understanding the microstructure of ceramics.

X-ray diffraction (XRD) is a common technique for structure determination, phase analysis, detection of preferred orientation and order/disorder and determination of crystal size. Rietveld refinement can be carried out after the collection of X-ray diffraction data. It involves comparing the experimental data with data derived from a theoretical model, and the lattice parameters are allowed to vary and are refined to match the experimental data. The substitution of electronegative

(a)

(b)

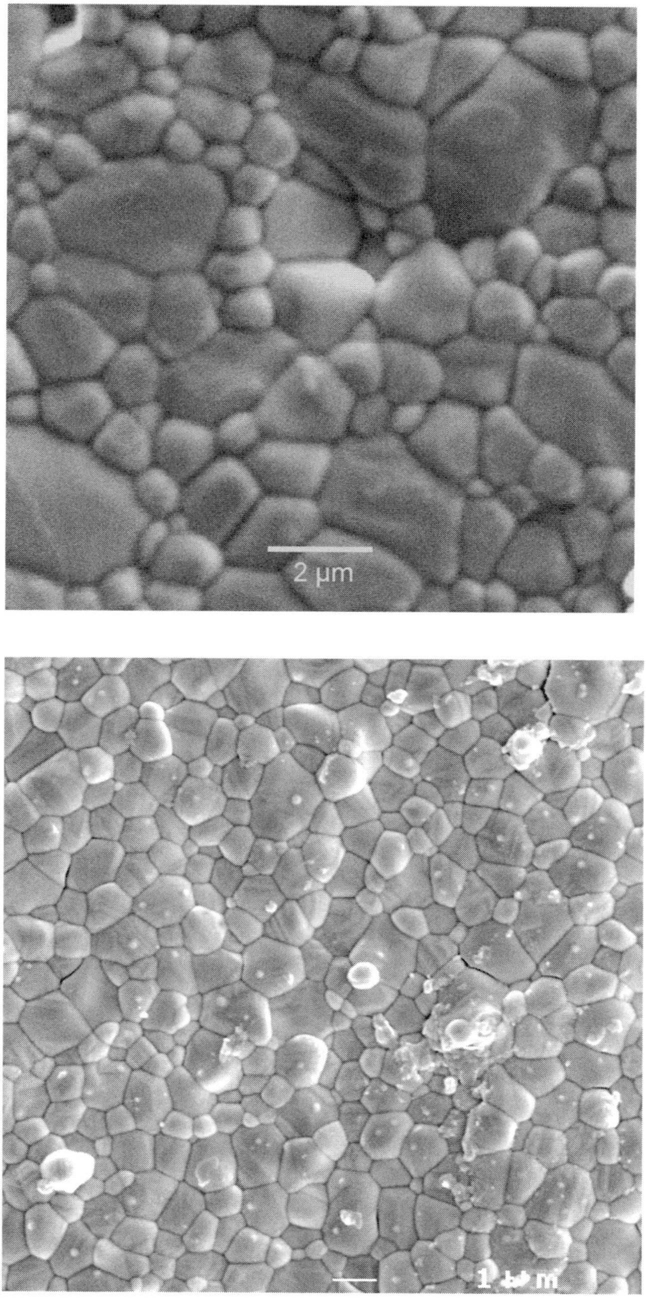

1.1 Comparison of the microstructure of hydroxyapatite (a) and zirconia (b) ceramics, various sized grains and pores were observed on hydroxyapatite while a relatively uniform grain size was found in zirconia.

anions, such as fluorine and chlorine for OH in HA, has been reported to alter the lattice parameters of the material, causing a decrease in the *a*-axis while leaving the *c*-axis unchanged. In carbonated substitute HA, the substitution of the larger planar CO_3 group by the smaller linear OH group causes an increase in the *a*-axis and a decrease in the *c*-axis in the Type A substitution; for Type B, the substitution of a smaller planar CO_3 group for a larger tetrahedral PO_4 group, causes a decrease in the *a*-axis and an increase in the *c*-axis (LeGeros and LeGeros, 1993). Substituting the larger silicate ions (SiO_4^{4-}) for the smaller phosphate (PO_4^{3-}) group causes the decrease in the *a*-axis and an increase in the *c*-axis with increasing silicate content (Gibson *et al.*, 1999b).

Thin film XRD, fixing the sample at small angle (i.e. 1°) to the incident beam, has been used to analyse surface structure, which enables the changes in the surface of materials after immersion in simulated body fluid to be determined. This technique is a useful tool for analysing the formation of surface apatite layers on bioactive glass, ceramic, polymer and composite (Kokubo *et al.*, 1990; Huang *et al.*, 1997).

Infrared spectroscopy is a non-destructive technique enabling the presence of certain bonds in a material to be established. It has been used to study reactions between aqueous solutions and the surfaces of bioactive glasses and glass-ceramics. The reaction stages occurring on the material side can be clearly delineated by changes in the vibrational modes of the chemical species in the surface (Hench, 1991). The presence of carbonate substitution in HA can be observed directly in infrared spectra in the form of weak peaks at between 870 cm^{-1} and a stronger doublet at between 1470 cm^{-1} (Rehman and Bonfield, 1997).

Raman spectroscopy is another analytical technique that can detect chemical bonds present in a material. It has been used to compare the bone-like apatite formed on Bioglass® surfaces *in vitro* with biological apatite (Rehman *et al.*, 1994), and the crystal imperfection in silicon-substituted HA (Zou *et al.*, 2005).

Scanning electron microscopy (SEM) has the advantages of high resolution and deep focal length. The fractography of ceramics is important in understanding the performance of materials. The size, shape and connectivity of pores of a HA ceramic scaffold can be revealed (Fig. 1.2). The selection or optimisation of suitable material for required application can be carried out accordingly. Combining SEM with X-ray microanalysis makes it possible to examine the changes of structure and composition on material surface and implant–bone interface, i.e. the thickness of each layer in the reaction zone on the surface of implant *in vivo*. The difference in the reaction zones between A-W glass-ceramic and Bioglass® was that no Si-rich layer was observed on A-W glass-ceramics (Hench and Ethridge, 1982; Kitsugi *et al.*, 1987). Such quantitative analysis is useful in designing a new material.

Complementary to SEM, laser scanning confocal microscopy, with the advantage of non-destructive optical sectioning, has recently been used to

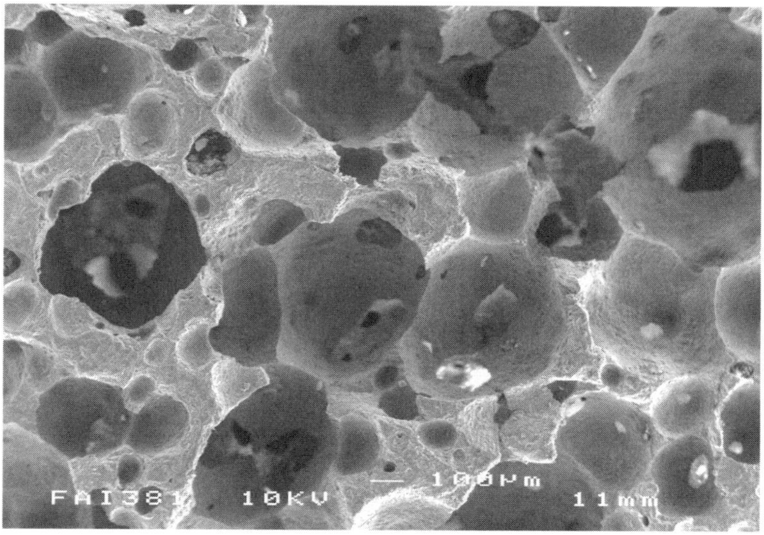

1.2 Fracture surface of porous HA ceramics.

generate 3D surface structure of A-W glass ceramics *in vitro* (Akhshi *et al.*, 2005). The detailed structure in the *Z* direction will provide further understanding of the *in vitro* responses of the material.

Transmission electron microscopy (TEM) is another technique to investigate the biomaterial–tissue interface. TEM with its high resolution and selected area electron diffraction, is a powerful tool for ultrastructural analysis. The incorporation of silicate ions into HA has been shown to increase the rate of bone apposition to HA bioceramic implants. Using high-resolution TEM, dislocations and grain boundaries have been characterised in HA and SiHA with a significant increase in density of triple junctions per unit area in SiHA over HA (Porter *et al.*, 2003, 2004b). Dissolution was observed to follow the order 1.5 wt% SiHA > 0.8 wt% SiHA > HA, and was prevalent at grain boundaries and triple junctions, which suggested that an increased number of defects in SiHA lead to an increased rate of dissolution of SiHA. The findings will help to understand the mechanisms by which silicate ions increase the *in vivo* bioactivity.

Porosity and pore size, distribution and interconnectivity are important parameters for porous glasses and ceramics materials, which determine the mechanical properties and biocompatibility. For a porous material, the pore sizes and volume can be measured using a mercury intrusion porosimeter. X-ray microtomography (XMT) is another recently developed technique to characterise material structure three dimensionally and non-destructively. For a tissue engineering scaffold, it is important to obtain pore size and distribution and interconnectivity in three dimensions. Figure 1.3 shows the pore size and

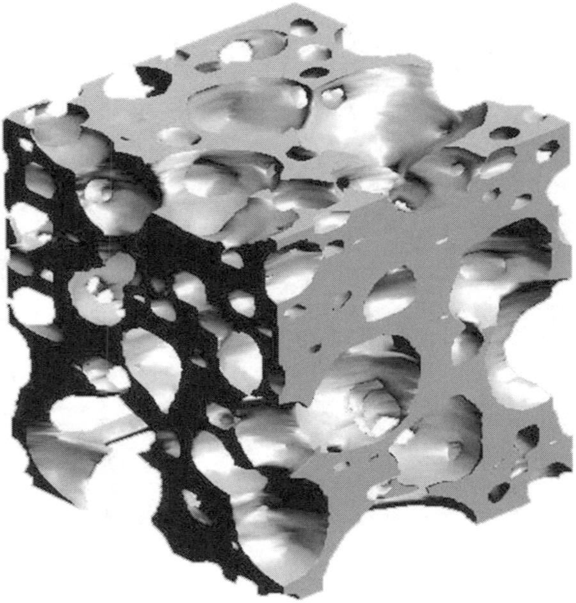

1.3 Three-dimensional structure of a hydroxyapatite scaffold obtained by X-ray microtomography ($2 \times 2 \times 2\,mm^3$). Courtesy of Dr X Fu, University of Cambridge.

connectivity of a HA scaffold obtained from XMT: it can be seen that XMT is a valuable tool for studying the 3D structure of porous scaffolds non-destructively.

In general, none of these analysis techniques may adequately be used in isolation to characterise ceramics, but they complement each other and reveal the various characteristics of tissue engineering materials *in vitro* and *in vivo*.

1.4 Properties of ceramics

1.4.1 Mechanical properties

The mechanical properties of a material are important as they determine its structural applications. Table 1.1 summarises some of the mechanical properties of bioactive and bioinert ceramics. Strength can be measured in a number of different ways, such as uniaxial tensile, three-point bending, four-point bending and uniaxial compression. Tensile strength testing is typically used for characterising ductile metals. Ceramics are not normally characterised by tensile testing, owing to the high cost of test specimen fabrication and the requirement for extremely good alignment of the load during testing. Compressive strength is the crushing strength of a material, and is commonly measured for ceramics, especially those that must support loads. The strength of ceramics materials is generally characterised by bending testing, also referred to as flexure testing. Apart from understanding these uniaxial stresses of a material, biaxial strength provides the data under a biaxial stress condition (e.g. under both tensile and shear stresses) as many applications for materials impose multiaxial stress fields. The strength value is dependent on the type of test conducted, flaw size distribution of the material and the stress distribution in the test specimen. Based on the concept of the failure of the weakest link, the strength distribution of ceramic materials can be described as Weibull modulus, a dimensionless number used to characterise the variability in measured strength of brittle materials (ceramics) which arises from the presence of flaws having a distribution in size and orientation. The higher the Weibull modulus, the more consistent the material, which means uniform defects are evenly distributed throughout the entire volume.

The fracture mechanics approach considers a fracture in terms of crack surface displacement and the stresses at the tip of the crack. The stress concentration at a crack tip is denoted in terms of the stress intensity factors K_I. The subscripts refer to the direction of load application with respect to the position of the crack. In tensile and bend tests the load is perpendicular to crack, and Mode I is most frequently encountered for ceramic materials. K_{IC} is the stress intensity factor at which the crack will propagate and lead to fracture, also referred to as fracture toughness, and is considered a basic property of the material. The higher the fracture toughness, the more difficult it is to initiate and propagate a crack.

The hardness of a material is a measure of its resistance to localised deformation by indentation or scratching. A small indentor is pressed into the surface of the material and the size of the indent is measured to calculate a hardness value. The hardness and fracture toughness of a ceramic influence its wear behaviour. The wear property of bioceramics is important for their applications in joint replacement, as wear debris can trigger inflammatory responses in the surrounding tissue and eventually lead to failure of the implant.

1.4.2 Surface properties

The physicochemical properties of implant materials, such as surface energy and charge, hydrophilicity or hydrophobicity, tend to affect cellular response by influencing protein absorption and cell attachment. Therefore, a material's surface properties can have a great impact on cellular responses. The nature and development of a stable interface between an implanted bioceramic and bone, which is crucial for the clinical success of the implant, are affected by many factors. It is becoming recognised that a key aspect is the surface modification of the bioceramic, which occurs due to interaction with the local environment. For materials such as bioactive glasses, a series of surface reactions occur when immersed in a physiological solution. Exchanges of Na^+ and K^+ with H^+ and

H_3O^+ ions from solution at the glass surface lead to the loss of soluble silica and the formation of SiOH (silanols) at the glass solution interface. This stage is followed by the migration of calcium and phosphate ions through the silica-rich layer to the surface and the formation of an amorphous calcium phosphate layer, which can then crystallise by the incorporation of hydroxyl, carbonate and fluoride ions to create an apatite layer. In the presence of osteogenic precursors, bioactive glasses favour the formation of osteoblasts which govern the further steps of bone development (Hench and Andersson, 1993).

Enhancing the interaction between tissue and biomaterials is another way to achieve a desirable tissue response to implant materials. Cells recognise surface features and react to them, resulting in contact guidance (Curtis and Wilkinson, 1997). A key design parameter for achieving maximal cell responses is the material topography. There is increasing evidence that surface topography both on the micro- and nanoscale are important in determining the cell response to biomaterials (Gray *et al.*, 1996; Dalby *et al.*, 2002). The creation of micro- and nanoscaled surface topography using HA has recently been attempted by electrohydrodynamic spraying and print-patterning with the aim of up-regulating cell activity (Huang *et al.*, 2004; Ahmad *et al.*, 2006).

In order to understand the mechanism of bonding of a bioactive material with host tissue, it is necessary to characterise the surface of a material *in vitro* (in physiological solution initially and then the cell and tissue responses), as the specific tissue compatibility of a material is highly dependent on the composition and structure of surface layers. Such *in vitro* analysis will be an aid to understanding the potential *in vivo* host tissue responses and to provide the information of material characteristics for developing new tissue engineering scaffold materials.

There are two approaches, solution analysis and surface analysis, for studying the mechanism and reaction at material surface. In solution analysis, the constituents released into the surrounding environment (which will potentially be released into the tissue) are examined. The leaching of alkali and alkaline earth ions from bioactive glass and ceramic surfaces, the dissolution of silica and the precipitation of calcium phosphates were revealed by this method. The dissolution products of bioactive glasses were found to have a positive effect on the expression of genes regulating osteogenesis (Xynos *et al.*, 2000, 2001). Therefore, the quantification of the release products from bioactive glasses and ceramics is important for their application as scaffolds for the formation of bioengineered bone tissue.

Surface analysis is used to examine the surface of the material using instrumental tools such as atomic force microscopy, Auger electron spectroscopy, infrared reflection spectroscopy, thin-film X-ray diffraction and scanning electron microscopy with X-ray analysis. Solution analysis can give the information about the total depths of reaction on the surface of a material; however, if precipitation occurs in the system (on the material surface or elsewhere), it will yield misleading information about the surface reactions of a material. With the aid of instrumental tools, the surface structure regarding either the top 1.5 μm or only the top 0.5–5 nm of the sampling depth can be obtained. Using these two techniques, the characteristics of a reactive material though its thickness can be understood.

1.4.3 Bioactivity and biocompatibility

Bioactivity

To assess the bioactivity of a material, Hench proposed an *in vivo* bioactivity index I_B, which is defined as $I_B = 100/t_{50bb}$, where t_{50bb} is the time required for more than 50% of the interface to be bonded (Hench, 1991). The rate of bone bonding to implant and the strength and stability of the bond vary with the composition and microstructure of the bioactive materials. Bioactive glasses with 42–53% SiO_2 form a bond to bone very rapidly, within days, and also form an adherent bond with soft tissues. Bioactive glasses with 54–60% of SiO_2 require 2–4 weeks to bond with bone, but do not bond with soft tissues. Bioactive glasses with more than 60% of SiO_2 do not have the ability to bond to any living tissue. However, using the sol–gel process, the compositional range of bioactivity was extended to 100% SiO_2. Generally, the higher the value of bioactivity index of the material, the faster rate of apatite formation on its surface and the better bonding with bone.

Deionised water is the simplest solution for studying the behaviour of a material *in vitro*, but it is unlikely to reflect the *in vitro* and *in vivo* situation, since there is a lack of buffer capacity to the changes in the pH of the solution. Various kinds of buffers, such as phosphate-, carbonate-, tris- and HEPES, have been introduced to maintain the pH of the solution in the physiological range (7.2–7.4) and form physiological solutions or pseudo-extracellular fluids. Kokubo developed and named the solutions as simulated body fluid (SBF) K1 to K9 in ascending order of ion content and concentration (Kokubo *et al.*, 1990). SBF K1 is equivalent to the tris buffer solution, containing no physiological cations, SBF K2 is physiological saline. It was suggested that SBF K9 was a suitable medium for initial *in vitro* study, as its ion concentrations is close to those of human blood plasma (Table 1.4). The presence of Ca and P ions in SBF accelerated the repolymerisation of silica on the glass, the formation of an amorphous Ca, P layer and the crystallisation to hydroxycarbonate apatite (HCA). Mg ion in SBF slowed down both the formation of the amorphous Ca,P phase and the crystallisation of HCA (Kokubo *et al.*, 1990).

The interaction of Bioglass® with the solution was modified by adsorption protein from the medium, which influenced the development of the surface structure. The crystallisation of HCA on the surface of Bioglass® became more complex or delayed in the solutions containing proteins. In contrast, the lag time

Table 1.4 Comparison of the ion concentrations of SBF K9 with those of blood plasma (Kokubo *et al.*, 1990)

	Concentration (mM)	
	SBF K9	Blood plasma
Na^+	142.0	142.0
K^+	5.0	5.0
Mg^{2+}	1.5	1.5
Ca^{2+}	2.5	2.5
Cl^-	147.8	103.0
$(HCO_3)^-$	4.2	27.0
$(HPO_4)^{2-}$	1.0	1.0
SO_4^{2-}	0.5	0.5

needed for the nucleation and formation of apatite on HA in normal physiological solution was significantly retarded, or even blocked, in media with serum proteins and other organic molecules (Radin and Ducheyne, 1996). The role of various proteins and enzymes on the bonding of implant materials with host tissue and bone mineralisation is still unclear.

Biocompatibility

The mechanism of tissue attachment is directly related to the type of tissue response at the interface. There are four types of materials according to the type of tissue response at the material–tissue interface (Table 1.5).

Biocompatibility has been defined as 'the ability of a material to perform with an appropriate host response in a specific application' (Williams, 1989), and is a critical property for a biomaterial. Biological assessment of a biomaterial usually involves two stages, *in vitro* and *in vivo*. Although a direct study of the host tissue response to a scaffold material *in vivo* would be ideal, the high complexity of the *in vivo* processes could cause difficulty in understanding a specific cellular response. Therefore, instead of studying the complex *in vivo* response initially, the *in vitro* assessment of the responses of isolated cell lines to a biomaterial is performed, which allows a controlled study of a specific cellular response to a test material. The knowledge obtained *in vitro* is useful for tailoring of the material to the host site, which certainly aids the screening of a new biomaterial, but it cannot replace the *in vivo* evaluation completely. Especially for a bioactive system, a controversial result exists for a bioactive glass tested *in vitro* and *in vivo*.

A wide range of cell lines are available for use in the *in vitro* modelling of biological responses. They are derived either from animal (mouse, rat) or human, which can be primary cells, transformed sarcoma cells or, more recently, stem cells. The *in vitro* biological responses of a biomaterial can be assessed

Table 1.5 Various types of materials and tissue response at material–tissue interface

	Type of material	Tissue response
Toxic		Death of the surrounding tissue
Non-toxic	Biologically inactive (bioinert)	Formation of a fibrous tissue
	Biologically active (bioactive)	Formation of an interfacial bond
	Dissolves or degrades (biodegradable)	Replacement of the surrounding tissue

qualitatively, i.e. assessment of the organisation of cytoskeletal proteins, and quantitatively, i.e. cytotoxicity testing, measurement of the growth, proliferation and differentiation, phenotype and gene expression of cells.

The function of cells grown in contact with a material is affected by the physicochemical characteristics of the material, such as crystallinity and surface roughness. It was also found that micromolar concentration of inorganic ions, such as Si could stimulate osteoblast proliferation, differentiation and gene expression (Xynos *et al.*, 2001; Reffitt *et al.*, 2003).

Simulating the complexities of the host–scaffold interface poses considerable challenges. The local *in vivo* conditions of cellular activity, pH, ionic concentration and the transportation of soluble products away from the scaffold site are difficult to replicate *in vitro*. To assess the biocompatibility/bioactivity of scaffold materials, suitable *in vivo* studies are required. There are difficulties in selecting appropriate animal models for evaluation, as tissue responses to scaffold materials may differ with species and anatomical location. Large animals show bone growth and remodelling similar to that observed in human; smaller animals (mouse or rat) rarely exhibit lamellar cortical bone remodelling. Rats have accelerated bone metabolism and are able to spontaneously regenerate proportionally greater bone defects than humans. Choosing the appropriate animal model for an *in vivo* study needs careful consideration. Smaller animals have been used for studying initial bone formation (Patel *et al.*, 2002; Hing *et al.*, 2006), while larger animals are preferable for studies of bone growth and remodelling. Sheep are often ideal candidates for the study of load-bearing effects on bone healing, and a larger defect size can be created. The model has been used successfully to evaluate the bioactivity of calcium phosphate implants (Patel *et al.*, 2005).

In addition to histological examination of implant–tissue interface, quantitative histomorphometry is also used to estimate the percentage of bone ingrowth and bone coverage within the implant, which is performed on toluidine blue stained section using point counting and linear intercept. Mineral apposition rates can be calculated by dividing the distance between two time-spaced fluorochrome labels with the time between the administration of the labels. The commonly used fluorochromes for bone mineral are tetracycline, alizarin red and calcein green. Statistical analysis or ranking of test materials can then be performed based on these quantitative measurements.

1.5 Processing of ceramics

The objective of ceramic processing is to make a form of the material that will perform a specific function, such as space-filling, tissue bonding or replacement. This requires the production of a solid object, a coating or particulates. There are various ways of making a specific shape including casting from the liquid state or pre-forming the shape from fine-grained particulates followed by consolidation. When a shape is made from powders it is called forming. The powders are usually mixed with water and an organic binder to achieve a plastic mass that can be cast, injected, extruded or pressed into a mould of the desired shape. The green body formed is subsequently subjected to a rising temperature to increase its density. After cooling and finishing steps (e.g. grinding and polishing), a product with desired properties is obtained.

An example of preparation of HA ceramics is given in the following. It consists of HA powder preparation, consolidation and densification (sintering).

1.5.1 Preparation of HA ceramics

There are numerous methods for the preparation of synthetic apatites, which can be grouped as aqueous reactions, solid state reactions and hydrothermal reactions. The aqueous reactions may be divided into chemical precipitation and hydrolysis methods. Chemical precipitation is the commonly used method, because of its simplicity and the ability of producing a wide variety of particle sizes and morphologies.

Methods based on those described by Akao *et al.* (1981) (equation 1.1) and Hayek and Newesely (1963) (equation 1.2) are the most frequently used. They consist of dropwise addition of phosphate solution into a stirred calcium solution. Addition of ammonium hydroxide is needed to keep the pH of the reaction alkaline to ensure the formation of HA after sintering the precipitate.

$$10Ca(OH)_2 + 6H_3(PO_4) \rightarrow Ca_{10}(PO_4)_6(OH)_2 + 18H_2O \tag{1.1}$$

$$10Ca(NO_3)_2 + 6(NH_4)_2H(PO_4) + 2H_2O \rightarrow Ca_{10}(PO_4)_6(OH)_2 + 12NH_4NO_3 + 8HNO_3 \tag{1.2}$$

The concentrations of reagents need to keep the Ca/P molar ratio of 1.67 for the stoichiometric HA. The concentration of calcium can be adjusted if substitution for calcium (e.g. strontium, magnesium) is required. Similarly, the phosphate concentration can be adjusted and replaced with required amount of carbonate or silicate when carbonate or silicon substitution is desired. Fluoride or chloride substitute apatite can be prepared by addition of fluoride or chloride ions in the reactions.

The next step of ceramic processing is to break down the materials received from chemical synthesis, which is a solid aggregation of particles in a dried, filtered precipitate. The agglomerates have a deleterious effect on the microstructure of the ceramic, and therefore needs to be broken down by crushing and grinding. Milling is then used to further reduce the particle size. The most widely used millings include ball and attrition mills. The particle size reduction is vital for good sinterability, but the ability to handle the powder is equally important, as it ensures the powder flows properly and can be compacted practicably. Spray drying generates agglomerates that will allow handling and forming of the powder. Calcination, a heat treatment, is another means of improving powder handling.

The HA powder obtained can then be made into dense or macroporous products using compaction (die pressing, isostatic pressing, slip casting, etc.) followed by solid state sintering. The properties of powder, such as morphology, surface area, mean particle size and particle size distribution, need to be adequately characterised, as this will greatly influence the handling and processing.

In die pressing, the load is applied uniaxially, which invariably leads to density variations throughout the compact. Isostatic pressing eliminates these density variations.

The compacted or compressed body can then be sintered at the temperature of 950–1300 °C. The processing of densifying a powder compact without the presence of a liquid phase is called solid state sintering. Solid material is moved to areas of contact between particles. The driving force is reduction of surface energy gradients by the mechanisms of grain boundary diffusion, volume diffusion and surface diffusion. During solid state sintering the material moves to eliminate the pores and open channels that exist between the grains of the compact, the crystals become tightly bonded together at their grain boundaries and the density, strength, toughness and corrosion resistance of sintered material increase greatly.

1.5.2 Porous ceramics

Porous ceramics have attracted great interest as scaffolds for tissue engineering, particularly bioactive ceramics and glasses, as they are able to bond to the host tissues. To be able to regenerate a tissue, a scaffold should act as a template for tissue to grow in three dimensions. The template must be a network of large pores (macropores, at least 100 μm in size) and the pores must be connected to each other, thus allowing essential nutrients to reach the whole network and stimulate blood vessels to grow inside the pore network.

Macroporosity can be introduced by mixing the powder with a volatile component, e.g. hydrogen peroxide or naphthalene, or adding poly(methyl methacrylate) (PMMA) beads to the powder slurry, using porous polymer (polyurethane foam) as a template for impregnating with ceramic suspensions, the volatile components or polymer phases with low decomposition temperatures are removed during the evaporating and sintering process. The ceramics component with a porous structure remains.

Ceramic slurries can also be foamed to obtain a porous structure. The incorporation of bubbles is achieved by injection of gases though the fluid medium, mechanical agitation, blowing agents and evaporation of compounds. A surfactant is generally used to stabilise bubbles formed in the liquid phase by reducing the surface tension of the gas–liquid interface. The gel-casting method has been used to produce macroporous HA with interconnected pores, the compressive strengths of HA foams were above 10 MPa, which is similar to that of trabecular bone (Sepulveda *et al.*, 2005).

Porous HA has also been produced by hydrothermal transformation from reef-building corals. These methods employ the use of elevated temperatures, pressures and controlled atmospheres to convert the calcium carbonate skeleton into HA. The route has the benefit of preserving the original architecture, since corals serve as a template to make a porous structure (Roy and Linnehan, 1974).

The design and control of the internal architectural of the porous HA structure influence the tissue regeneration (Chu *et al.*, 2002). Rapid prototyping (RP) has emerged as a new processing technique for making a scaffold, which allows highly complex structures to be built as a series of thin 2D slices using computer-aided design (CAD) and computer-aided manufacturing (CAM) programs. This technique allows properties such as porosity, interconnectivity and pore size to be predefined. Rapid prototyping, especially 3D printing, has been developed for making custom-made 3D porous HA scaffolds for bone replacement, and the repair of osseous defects from trauma or disease. A 3D HA structure with controlled patterns or porosity has been built by direct-write assembly (Michna *et al.*, 2005). A complex-shaped porous HA ceramic with fully interconnected channel was generated from HA powder (Seitz *et al.*, 2005), which was printed with a binder solution layer by layer. Unglued powder was removed and the ceramic green body obtained was consolidated by sintering at high temperature. It is possible to design and manufacture parts according to an individual patient's anatomy. Patient-derived cells can then be seeded onto the scaffolds for tailor-making tissue engineering implants

1.5.3 Glass

Melting and sol–gel processing are two well-known methods for producing glasses. Traditional glass synthesis consists of melting the precursor mixture and quenching. New components can be added to the system to tailor the glass composition for clinical applications, i.e. increase or decrease bioactivity, decreasing glass-forming temperatures, etc. An alternative approach is the use of sol–gel techniques to prepare glasses, and this route can produce high-purity glasses, which are more homogeneous than those obtained by melting, and require relatively low processing temperatures. Three methods can be used to make sol–gel materials: gelation of colloidal powders, hypercritical drying, controlled hydrolysis and condensation of metal alkoxide precursors followed by drying at ambient pressure. They involve transformation of a sol (suspension of colloidal particles) to a gel (3D, interconnected network). Colloids are solid particles with a diameter less than 100 nm. Thermal treatment increases the density, strength and hardness of the gels and converts the network to a glass with properties similar to melt-derived glass. An advantage of the sol–gel process is the ability to control the surface chemistry of the material by the thermal treatment, so making it possible to expand the bioactive compositional ranges studied in the phase diagram of melted glasses. The glasses obtained exhibit higher surface area and porosity, which are the critical factors in their bioactivity.

Glass in the fibre form can be produced by the melt-spinning approach. The fibre diameter is limited to micrometre-scale (e.g. tens to hundreds of micrometres). Electrospinning can produce various polymer fibres in the range of 10–1000 nm, and has recently been applied to produce bioactive glass fibres in the micro-/nanoscale using the sol–gel glass. Nanoglass fibres in the forms of bundled filament, fibrous membranes and 3D scaffold have been produced (Kim *et al.*, 2006).

Porous bioactive glasses have been produced by foaming of melt-derived and sol–gel-derived bioactive glasses (Chen *et al.*, 2006; Sepulveda *et al.*, 2002; Jones and Hench, 2003). During the foaming process of sol–gel glass, air was entrapped in the sol under vigorous agitation as viscosity increased and the silica network formed. As the porous foam became a gel, the bubbles were stabilised. The gel was then subjected to controlled thermal processing of ageing (60 °C), drying (130 °C) and sintering to remove organic species (500–800 °C). The resulting bioactive glass foam scaffolds had macropores up to 600 μm in diameter and compressive strength up to 2.5 MPa (Jones, 2005). An overview of structural characteristics and mechanical properties of highly porous bioactive ceramic or glass forms for bone tissue engineering was summarised by Chen *et al.* (2006).

1.5.4 Glass-ceramics

Glass-ceramics are produced by the transformation of the glass into a ceramic. The glass is firstly heated at the temperature of 450–700 °C to produce a large number of nuclei, then the temperature is increased to 600–900 °C to promote crystal growth. The resulting microstructure is fine-grained with a uniform size distribution.

1.5.5 Coating processing

Although the relatively poor mechanical properties of calcium phosphate ceramics limit their clinical applications to non-major load-bearing parts of the skeleton, calcium phosphate-coated metallic implants are used in the load-bearing parts. The most popular commercial routes of calcium phosphate coatings are based on plasma spraying (de Groot *et al.*, 1987). The ceramic powder is suspended in the carrier gas and fed into the plasma where it can be fired at a substrate. However, the high temperature involved in the processing can lead to changes in phase purity and crystallinity of calcium phosphate ceramics. Recently, a variety of thin film and low-temperature techniques have been developed to deposit a bioactive layer on the surface of bioinert materials, such as electrostatic atomisation spray deposition, sol–gel deposition, biomimetic deposition and magnetron sputtering (Abe *et al.*, 1990; Jansen *et al.*, 1993; Gross *et al.*, 1998; Habibovic *et al.*, 2002; Huang *et al.*, 2005; Leeuwenburgh *et al.*, 2005; Thian *et al.*, 2006a). Condensation from a vapour by sputtering has been used to produce nano-sized coatings, which are uniform, non-porous and very fine grains. Nano-crystalline SiHA coating by RF magnetron sputtering has been found to promote better biological responses (Thian *et al.*, 2006b).

1.6 Conclusions

Ceramic biomaterials have been widely used in biological applications as orthopaedic and dental implants and porous scaffolds for tissue engineering. Bioinert ceramics, such as alumina and zirconia, have excellent mechanical properties for load-bearing applications, while bioactive glasses and ceramics have the potential for osteoconduction. Therefore, it is of great importance to understand the clinical and materials requirements to allow the production of tailor-made scaffolds. This chapter has described the microstructure and properties of ceramics with intention of developing an understanding of the relationships between processing, microstructure and properties of ceramics. The most widely applied orthopaedic bioceramics, such as hydroxyapatite ceramics, Bioglass® and apatite-wollastonite glass-ceramic were discussed from the aspects of microstructure, processing, mechanical properties, surface properties, biocompatibility and bioactivity. An understanding of their properties and behaviour will help in developing better ceramic materials for tissue engineering.

1.7 Future trends

In comparison with the usage of autogenous bone repair, there are some drawbacks to the majority of current biomaterials, such as their inability to self-repair, to maintain a blood supply and to modify their structures and properties in response to the physiological and mechanical environment. As such, tissue engineering offers the potential for production of tissue-like constructs in the laboratory. A patient's own cells can then be isolated and seeded onto a scaffold with the desired architecture *in vitro*, and with the appropriate biological stimuli, these cells will proliferate and differentiate, and finally, be ready for implantation into a human body to regenerate a diseased or damaged tissue.

Considerable efforts have been made towards developing and engineering structures and surfaces that could elicit rapid and desired reactions with cells and proteins for specific applications in addition to the activities of materials synthesis, optimisation, characterisation and the biological testing of host–material interactions. However, our knowledge on the physical and chemical functioning of biomaterials, the response of these materials on human, and the interaction mechanism between the materials and the biological systems still need to be further understood (Anderson, 2006). The search for suitable materials with the desired degradation rates, products and mechanical properties for the desired tissue engineering scaffold is still ongoing, so as to achieve the architectures with the desired pore size, morphology, surface topography and bioactivity. The advances in materials science, engineering, cell and molecular biology, and medicine will be able to offer new solutions. The incorporation of advanced fabrication technology and the synthesis of new materials will enhance the complexity and bioactivity of tissue engineering constructs. The emergence of biotechnology and nanotechnology will also offer great potential for calcium phosphate ceramics, bioactive glasses and glass-ceramics to be further developed in regenerative medicine and tissue engineering (Hench and Polak, 2002; Hollister, 2005).

1.8 References

Abe Y, Kokubo T and Yamamuro T (1990), 'Apatite coating on ceramics, metals and polymers utilising a biological process', *J Mater Sci Mater Med*, **1**, 233–238.

Ahmad Z, Huang J, Edirisinghe M J, Jayasinghe S N, Best S M, Bonfield W, Brooks R A and Rushton N (2006), 'Electrohydrodynamic print-patterning of nano-hydroxyapatite', *J Biomed Nanotech*, **2**, 201–207.

Akao M, Aoki H and Kato K (1981), 'Mechanical properties of sintered hydroxyapatite for prosthetic applications', *J Mater Sci*, **28**, 809.

Akhshi M, Huang J, Best S M, Farrar D, Rose J and Bonfield W (2005), 'Study of *in situ* apatite formation on bioactive substrates using confocal microscopy', *Key Eng Mater*, 284–286, 457–460.

Anderson J M (2006), 'The future of biomedical materials', *J Mater Sci: Mater Med*, **17**, 1025–1028.

Aoki H (1991), *Science and Medical Applications of Hydroxyapatite*, JAAS, Tokyo.

Bonfield W (1988), 'Composites for bone replacement', *J Biomed Eng*, **10**, 522–526.

Bonfield W, Grynpas M D, Tully A E, Bowman J and Abram J (1981), 'Hydroxyapatite reinforced polyethylene: a mechanically compatible implant material for bone replacement', *Biomaterials*, **2**, 185–186.

Botelho C M, Lopes M A, Gibson I R, Best S M, Santos J D (2002), 'Structural analysis of Si-substituted hydroxyapatite: zeta potential and X-ray photoelectron spectroscopy (XPS)', *J Mater Sci: Mater Med*, **13**, 1123–1127.

Carlisle E M (1970), 'Silicon: a possible factor in bone calcification', *Science*, **167**, 279–280.

Carlisle E M (1972), 'Silicon: an essential element for the chick', *Science*, **178**, 619–621.

Chen Q Z, Thompson I D, Boccaccini A R (2006), '45S5 Bioglass-derived glass-ceramic scaffolds for bone tissue engineering', *Biomaterials*, **27**, 2414–2425.

Chow L C and Takagi S (2001), 'A natural bone cement – A laboratory novelty led to the development of revolutionary new biomaterials', *J Res Natl Inst Stand Technol*, **106**, 1029–1033.

Chu T M G, Orton D G, Hollister S J, Feinberg S E and Halloran J W (2002), 'Mechanical and *in vivo* performance of hydroxyapatite implants with controlled architectures', *Biomaterials*, **23**(5), 1283–1293.

Curtis A and Wilkinson C (1997), 'Topographical control of cells', *Biomaterials*, **18**, 1573–1583.

Dalby M J, Kayser M V, Bonfield W and DiSilvio L (2002), 'Initial attachment of osteoblasts to an optimised HAPEX^{TM} topography', *Biomaterials*, **23**, 681–690.

Daculsi G, Laboux O, Malard O, Weiss P (2003), 'Current state of the art of biphasic calcium phosphate bioceramics', *J Mater Sci: Mater Med*, **14**, 195–200.

de Groot K, Geesink R, Klein C P A T and Serekian P (1987), 'Plasma sprayed coatings of hydroxyapatite', *J Biomed Mater Res*, **21**, 1375–1381.

Ducheyne P and Hench L L (1982), 'The processing and static mechanical properties of metal fibre reinforced Bioglass', *J Mater Sci*, **17**, 595–606.

Ducheyne P, Marcolongo M and Schepers E (1993), 'Bioceramic composites', in Hench L L and Wilson J, *An Introduction to Bioceramics*, World Scientific, Singapore, 281–297.

Fernandez E, Gil F J, Best S M, Ginebra M P, Driessens F C M and Planell J A (1998), 'Improvement of the mechanical properties of new calcium phosphate bone cements in the $CaHPO_4$-alpha-Ca-3(PO_4)(2) system: compressive strength and microstructural development', *J Biomed Mater Res*, **41**, 560–567.

Gibson I R, Huang J, Best S M and Bonfield W (1999a), 'Enhanced *in vitro* cell activity and surface apatite layer formation on novel silicon-substituted hydroxyapatites', in Ohgushi H, Hastings G W and Yoshikawa T, *Bioceramics 12* (Proceedings of the 12th International Symposium on Ceramics in Medicine), 191–194.

Gibson I R, Best S M and Bonfield W (1999b), 'Chemical characterization of silicon-substituted hydroxyapatite', *J Biomed Mater Res*, **44**, 422–428.

Gibson I R, Best S M and Bonfield W (2002), 'Effect of silicon substitution on the sintering and microstructure of hydroxyapatite', *J Am Ceram Soc*, **85**(11), 2771–2777.

Gray C, Boyde A and Jones S J (1996), 'Topographically induced bone formation *in vitro*: implications for bone implants and bone grafts', *Bone*, **18**, 115–123.

Gross K A, Chai C S, Kannangara G S K, Ben-Nissan B and Hanley L (1998), 'Thin hydroxyapatitie coatings via sol–gel synthesis', *J Mater Sci: Mater Med*, **9**, 839–843.

Gross U M, Muller-Mai C and Viogt C (1993), 'Ceravital bioactive glass-ceramics', in Hench L L and Wilson J, *An Introduction to Bioceramics*, World Scientific, Singapore, 105–123.

Habibovic P, Barrere F, van Blitterswijk C F, de Groot K and Layrolle P (2002), 'Biomimetic hydroxyapatite coating on metal implants', *J Am Ceram Soc*, **85**, 517–522.

Hayek E and Newesely H (1963), 'Pentacalcium monohydroxyorthophosphate', *Inorganic Syntheses*, **7**, 63–65.
Hench L L (1991), 'Bioceramics: from concept to clinic', *J Am Ceram Soc*, **74**, 1487–1510.
Hench L L (2006), The story of Bioglass®, *J Mater Sci: Mater Med*, **17**, 967–978.
Hench L L and Andersson O H (1993), 'Bioactive glasses', in Hench L L and Wilson J, *An Introduction to Bioceramics*, World Scientific, Singapore, 41–62.
Hench L L and Best S M (2004), 'Ceramics, glasses and glass-ceramics', in Ratner B D, Schoen F J, Hoffman A S and Lemons J E, *Biomaterials Science: An Introduction to Materials in Medicine*, 2nd edition, Elsevier Academic Press, San Diego, 153–170.
Hench L L and Ethridge E C (1982), *Biomaterials: An Interfacial Approach*, Biophysics and Bioengineering Series, Vol. 4, Academic Press, New York.
Hench L L and Polak J M (2002), 'Third generation biomaterials', *Science*, **295**, (5557), 1014–1017.
Hench L L, Splinter R J, Allen W C and Greenlee T K (1971), 'Bonding mechanisms at the interface of ceramic prosthetic materials', *J Biomed Mater Res Symp*, **2**, 117–141.
Hing K A, Revell P A, Smith N and Buckland T (2006), 'Effect of silicon level on rate, quality and progression of bone healing within silicate-substituted porous hydroxyapatite scaffolds', *Biomaterials*, **27**(29), 5014–5026.
Holand W and Vogel W (1993), 'Machineable and phosphate glass-ceramics, in Hench L L and Wilson J, *An Introduction to Bioceramics*, World Scientific, Singapore, 125–137.
Hollister S J (2005), 'Porous scaffold design for tissue engineering', *Nat Mater*, **4**, 518–524.
Huang J, Di Silvio L, Wang M, Rehman I and Bonfield W (1997), 'Evaluation of *in vitro* biocompatibility of Bioglass® reinforced polyethylene composites', *J Mater Sci: Mater Med*, **8**, 809–813.
Huang J, Jayasinghe S N, Best S M, Edirisinghe M J, Brooks R A and Bonfield W (2004), Electrospraying of a nano-hydroxyapatite suspension, *J Mater Sci*, **39**, 1029–1032.
Huang J, Jayasinghe S N, Best S M, Edirisinghe M J, Brooks R A, Rushton N and Bonfield W (2005), 'Novel deposition of nano-sized silicon substituted hydroxyapatite by electrostatic spraying', *J Mater Sci: Mater Med*, **16**, 1137–1142.
Hulbert S (1993), The use of alumina and zirconia in surgical implants, in Hench L L and Wilson J, *An Introduction to Bioceramics*, World Scientific, Singapore, 25–40.
Jansen J A, Wolke J G, Swann S, Van der Waerden J P and de Groot K (1993), 'Application of magnetron sputtering for producing ceramic coatings on implant materials', *Clin Oral Implants Res*, **4**(1), 28–34.
Jones J R (2005), 'Scaffolds for tissue engineering', in Hench L L and Jones J R, *Biomaterials, Artificial Organs and Tissue Engineering*, Woodhead Publishing, Cambridge, 201–214.
Jones J R and Hench L L (2003), 'Effect of surfactant concentration and composition on the structure and properties of sol–gel-derived bioactive glass foam scaffolds for tissue engineering', *J Mater Sci*, **38**, 1–8.
Kim H W, Kim H E and Knowles J C (2006), 'Production and potential of bioactive glass nanofibers as a next-generation biomaterial', *Adv Funct Mater*, **16**(12), 1529–1535.
Kingery W D (1976), *Introduction to Ceramics*, J Wiley & Sons, New York.
Kitsugi T, Nakamura T, Yamamuro T, Kokubo T, Shibuya and Takagi M (1987), 'SEM-EPMA observation of three types of apatite-containing glass-ceramics implanted in

bone: the variance of a Ca-P-rich layer', *J Biomed Mater Res*, **21**, 1255–1271.

Kokubo T (1991), 'Bioactive glass ceramics: properties and applications', *Biomaterials*, **12**, 155–163.

Kokubo T, Ito S, Sakka S and Yamamuro T (1986), 'Formation of a high-strength bioactive glass-ceramic in the system MgO-CaO-SiO_2-P_2O_5', *J Mater Sci*, **21**, 536–540.

Kokubo T, Kushitani H, Sakka S, Kitsugi T and Yamamuro T (1990), 'Solutions able to reproduce *in vivo* surface-structure changes in bioactive glass-ceramic A-W', *J Biomed Mater Res*, **24**, 721–734.

Leeuwenburgh S C G, Wolke J G C, Schoonman J and Jansen J A (2005), 'Influence of deposition parameters on chemical properties of calcium phosphate coatings prepared by using electrostatic spray deposition', *J Biomed Mater Res*, **74**, 275–284.

LeGeros R Z and LeGeros J P (1993), 'Dense hydroxyapatite', in Hench L L and Wilson J, *An Introduction to Bioceramics*, World Scientific, Singapore, 139–180.

LeGeros R Z, Trautz O R, LeGeros J P and Shirra W P (1967), 'Apatite crystallites: effect of carbonate on morphology', *Science*, **155**, 1409–1411.

Michna S, Wu W and Lewis J A (2005), 'Concentrated hydroxyapatite inks for direct-write assembly of 3-D periodic scaffolds', *Biomaterials*, **26**, 5632–5639.

Patel N, Best S M, Bonfield W, Gibson I R, Hing K A, Damien E and Revell P A (2002), 'A comparative study on the *in vivo* behavior of hydroxyapatite and silicon substituted hydroxyapatite granules', *J Mater Sci: Mater Med*, **13**, 1199–1206.

Patel N, Brooks R A, Clarke M T, Lee P M T, Rushton N, Gibson I R, Best S M, Bonfield W (2005), '*In vivo* assessment of hydroxyapatite and silicate-substituted hydroxyapatite granules using an ovine defect model', *J Mater Sci: Mater Med*, **16**, 429–440.

Porter A E, Patel N, Skepper J N, Best S M and Bonfield W (2003), 'Comparison of in vivo dissolution processes in hydroxyapatite and silicon-substituted hydroxyapatite bioceramics', *Biomaterials*, **24**, 4609–4620.

Porter A E, Best S M and Bonfield W (2004a), 'Ultrastructural comparison of hydroxyapatite and silicon-substituted hydroxyapatite for biomedical applications', *J Biomed Mater Res*, **68A**, 133–141.

Porter A E, Patel N, Skepper J N, Best S M and Bonfield W (2004b), 'Effect of sintered silicate-substituted hydroxyapatite on remodelling processes at the bone-implant interface', *Biomaterials*, **25**, 3303–3314.

Radin S and Ducheyne P (1996), 'Effect of serum proteins on solution-induced surface transformations of bioactive ceramics', *J Biomed Mater Res*, **30**(3), 273–279.

Reffitt D M, Ogston N, Jugdaohsingh R, Cheung H F J, Evans B A J, Thompson R P H, Powell J J and Hampson G N (2003), 'Orthosilicic acid stimulates collagen type 1 synthesis and osteoblastic differentiation in human osteoblast-like cells *in vitro*', *Bone*, **32**(2), 127–135.

Rehman I and Bonfield W (1997), 'Characterisation of hydroxyapatite and carbonated apatite by photo acoustic FTIR spectroscopy', *J Mater Sci: Mater Med*, **8**, 1–4.

Rehman I, Hench L L, Bonfield W and Smith R (1994), 'Analysis of surface layers on bioactive glasses', *Biomaterials*, **15**, 865–870.

Roy D M and Linnehan S K (1974), 'Hydroxyapatite formed from coral skeletal carbonate by hydrothermal exchange', *Nature*, **247**, 220–222.

Seitz H, Rieder W, Irsen S, Leukers B and Tille C (2005), 'Three-dimensional printing of porous ceramic scaffolds for bone tissue engineering', *J Biomed Mater Res Part B: Appl Biomater*, **74B**, 782–788.

Sepulveda P, Jones J R and Hench L L (2002), 'Bioactive sol–gel foams for tissue repair', *J Biomed Mater Res*, **59**A, 340–348.

Sepulveda P, Binner J G, Rogero S O, Higa O Z and Bressiani J C (2005), 'Production of porous hydroxyapatite by the gel-casting of foams and cytotoxic evaluation', *J Biomed Mater Res*, **50**(1), 27–34.

Thian E S, Huang J, Best S M, Barber Z H, Vickers M E and Bonfield (2006a), 'Silicon-substituted hydroxyapatite (SiHA): a novel calcium phosphate coating for biomedical applications', *J Mater Sci,* **41**, 709–717.

Thian E S, Huang J, Best S M, Barber Z H, Brooks R A, Rushton N and Bonfield W (2006b), 'The response of osteoblasts to nanocrystalline silicon-substituted hydroxyapatite thin films', *Biomaterials*, **27**, 2692–2698.

Vogel W and Holand W (1990), 'Development, structure, properties and application of glass-ceramics for medicine', *J Non-Cryst Solid*, **123**, 349–353.

Webster T J, Siegel R W and Bizios R (1999), 'Design and evaluation of nanophase alumina for orthopaedic/dental applications', *Nanostruct Mater*, **12**, 983–986.

Webster T J, Ergun C, Doremus R H, Siegel R W and Bizios R (2000), 'Enhanced functions of osteoblasts on nanophase ceramics', *Biomaterial*, **21**, 1803–1810.

Williams D F (1987), 'Definitions in biomaterials', *Proceedings of a Consensus Conference on the European Society of Biomaterials*, Elsevier, New York.

Williams D F (1989), 'A model for biocompatibility and its evaluation', *J Biomed Eng*, **11**, 185–191.

Xynos I D, Edgar A J, Buttery L D, Hench L L and Polak J M (2000), 'Ionic products of bioactive glass dissolution increase proliferation of human osteoblasts and induce insulin-like growth factor II mRNA expression and protein synthesis', *Biochem Biophys Res Commun*, **276**, 461–465.

Xynos I D, Edgar A J, Buttery L D K, Hench L L and Polak J M (2001), 'Gene expression profiling of human osteoblasts following treatment with the ionic dissolution products of Bioglass® 45S5 dissolution', *J Biomed Mater Res*, **55**, 151–157.

Zou S, Huang J, Best S M and Bonfield W (2005), 'Crystal imperfection studies of pure and substituted hydroxyapatite using Raman and XRD', *J Mater Sci: Mater Med*, **16**, 1143–1148.

2
Polymeric biomaterials

G WEI and P X MA, The University of Michigan, USA

2.1 Introduction

Biomaterials are materials used in therapeutic or diagnostic systems that are in contact with tissue or biological fluids. The categories of biomaterials include metals, ceramics, carbons, glasses, modified natural biomolecules, synthetic polymers and composites consisting of various combinations of these material types (Dumitriu, 1996). Being utilized to manufacture various medical devices, diagnostic products and pharmaceutical preparations, biomaterials provide solutions to medical and healthcare problems.

Advances in medicine have changed the concept from surgery (the deletion of damaged tissues for the preservation of remaining healthy tissues) to transplantation and to regenerative medicine (the repair and replacement of lost tissues by initiating the natural regeneration process). Most approaches currently pursued or contemplated within the framework of regenerative medicine, including cell-based therapies and engineered living tissues, are dependent on the ability to synthesize or otherwise generate novel materials, fabricate or assemble materials into appropriate 2D and 3D forms, and precisely tailor physical, chemical, structural and biological properties to achieve desired clinical responses. In these aspects, biodegradable polymeric biomaterials offer the advantages of being able to be eliminated from the body after fulfilling their intended purposes. It is not surprising that biodegradable polymers are becoming the biomaterials of choice in tissue engineering and drug delivery areas (Langer, 1990; Langer and Vacanti, 1993; Liu and Ma, 2004; Saito *et al.*, 2005). They play a pivotal role in reparative and regenerative medicine to treat damaged or diseased tissues.

This chapter will review recent advances in biodegradable polymer scaffolds for tissue engineering and drug delivery applications. The first part will discuss polymer scaffold fabrication, 3D porous structures, surface modification, and their effects on tissue regeneration. The second part will highlight drug delivering polymer scaffolds and their applications in tissue engineering.

2.2 Polymeric scaffolds for tissue engineering

In a tissue engineering strategy, cells are seeded on a scaffold that acts as a template to guide cell growth and to facilitate the formation of functional new tissues and organs. Scaffolds promote new tissue formation by providing an appropriate surface and adequate spaces (volumes) to foster and direct cellular attachment, migration, proliferation, and desired differentiation to specific cells in three dimensions. The design of a scaffold is critical because it affects the formation and ultimate function of neo tissues. There are many general well-accepted criteria for ideal scaffolds in tissue engineering applications although they could vary in some degree among tissue types (Ma, 2004a,b). Critical variables in scaffold design and function include the bulk material, the mechanical properties, the 3D architecture, the surface morphology and chemistry, and the scaffold environment during and after fulfilling the function, which is affected by degradation characteristics. Generally, a tissue engineering scaffold should be: (1) biocompatible, that is, non-immunogenic and non-toxic to living cells and tissues; (2) biodegradable or capable of being remodeled in tune with regeneration or repair process; (3) porous to provide a suitable 3D environment for cell and tissue penetration as well as nutrients and wastes transportation; (4) surface conductive to facilitate cellular functions; (5) mechanically stable for surgical handling; and (6) easy to manufacture and sterilize. In addition, the scaffolds also should possess the ability to carry biological signals such as growth factors, cytokines and to deliver them in a controlled manner. In this aspect, the scaffold serves as a drug delivery system as well. This is particularly important for large tissue defect repair and functional tissue regeneration.

2.2.1 Polymeric scaffold fabrication

The search for ideal scaffolding materials and appropriate scaffolding structure that fulfill the above design criteria continues to be important and challenging in tissue engineering (Burg *et al.*, 2000; Liu and Ma, 2004; Ma, 2004a). Polymeric biomaterials, particularly biodegradable polymers such as the family of poly(α-hydroxy esters) including poly(lactic acid) (PLA), poly(glycolic acid) (PGA) and their copolymers (PLGA), have been used extensively in medical and surgical applications. These biodegradable polymers have been of interest primarily because of their biocompatibility and biodegradability, their established safety as suture materials (approval from the US Food and Drug Administration, FDA), and the versatility and flexibility that they offer for producing well-defined highly porous scaffolds with different geometry and structures to meet the needs of specific tissue engineering applications (Mikos *et al.*, 1994; Whang *et al.*, 1995; Harris *et al.*, 1998; Giannobile *et al.*, 2001; Borden *et al.*, 2002; Liu and Ma, 2004). Because degradation occurs mainly by hydrolysis, the degradation rate can be modulated over a wide range by tailoring the composition, molecular weights, end groups and geometry of the device (Hollinger and Leong, 1996; Shive and Anderson, 1997).

A number of techniques have been explored to fabricate biodegradable polymers into 3D porous scaffolds with different porosities, pore architectures, pore orientations, pore sizes, inter-pore connections, and pore wall surface morphologies (Mikos *et al.*, 1994; Zhang and Ma, 2000; Ma and Choi, 2001; Chen and Ma, 2004; Wei and Ma, 2006). Solvent-casting/particulate leaching is a conventional technique that has been widely used to fabricate porous scaffolds for tissue engineering applications (Mikos *et al.*, 1994). It is technically simple and easy to carry out. The pore size can be controlled by the size of salt particles and the porosity by the polymer/particulate ratio. However, this technique has limited control over pore shape and inter-pore connectivity. To prepare scaffolds with well-controlled interconnected porous structures, paraffin spheres or sugar spheres are used as alternative porogen materials to salt particles. Because paraffin or sugar spheres can be heat treated to form a bound template, interconnected spherical pore structures are created after the removal of the bound porogen template (Fig. 2.1(a), (b)). The heat treatment time and temperature of porogen materials control the inter-pore connectivity (or interpore opening area) in the scaffold (Chen and Ma, 2004; Wei and Ma, 2006).

Porous structures can also be introduced into a material using a phase separation technique without using a porogen material. A thermally induced solid–liquid phase separation is achieved by lowering the temperature of a homogeneous polymer solution to induce solvent crystallization (Ma and Zhang, 1999, 2001). Subsequent removal of solvent crystals results in porous polymer scaffolds (Fig. 2.1(c)). The characteristics of pores are varied with polymer material, concentration, solvent, phase separation temperature and thermal transfer direction. For example, manipulation of thermal transfer direction controls the direction of solvent crystal growth during the phase separation, which results in a scaffold with anisotropic microtubular structure (Fig. 2.1(d)). Such a parallel array of microtubules facilitates the organization and regeneration of certain tissues (nerve, muscle, tendon, ligament, dentin, and so on) which naturally have oriented tubular or fibrous bundle architectures (Fig. 2.2).

Phase separation can be used in combination with other scaffold fabrication techniques such as porogen leaching and solid freeform fabrication (SFF). The combined technique provides broader control over porous architectures in size from macro, micro to nano levels (Chen and Ma, 2004; Chen *et al.*, 2006; Wei and Ma, 2006). The hierarchical pore structure mimics natural extracellular matrix (ECM) more closely and has been shown to promote tissue regeneration (Chen *et al.*, 2006).

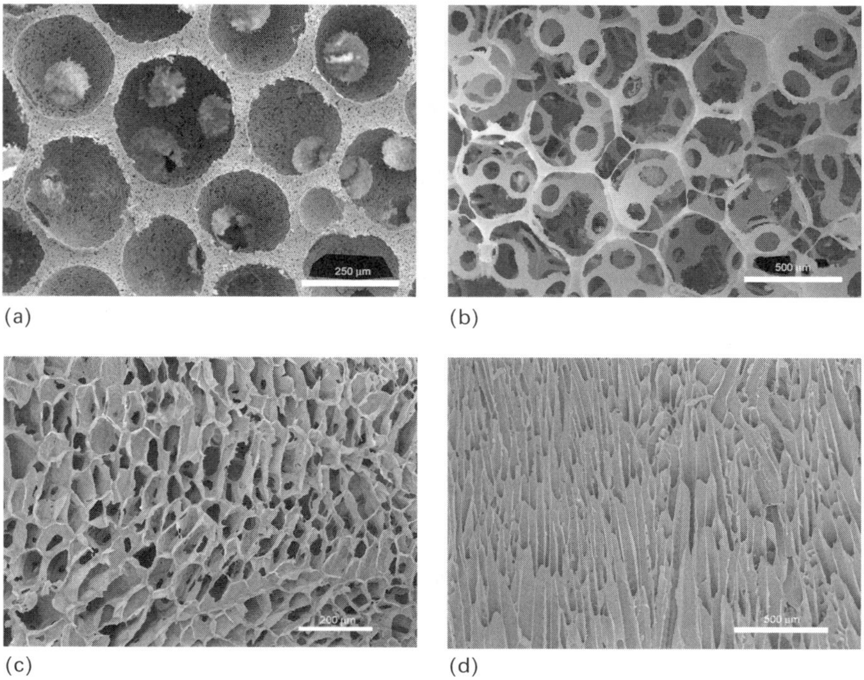

2.1 (a, b) Scanning electron micrographs (SEM) of scaffolds fabricated using a spherical porogen (paraffin sphere) leaching technique: (a) PLLA; (b) PLGA (85/15); (c, d) PLLA scaffolds fabricated using a thermally induced solid–liquid phase separation technique in dioxane (c) or benzene (d). (a, b) From Ma and Choi (2001), Copyright © 2001 by Elsevier. Reprinted with permission of Elsevier. (c, d) From Ma and Zhang (2001) and Ma *et al.* (2001), Copyright © 2001 by John Wiley & Sons. Reprinted with permission of John Wiley & Sons.

2.2.2 Three-dimensional porous architectures

In the body, nearly all cells are embedded or exposed to a 3D microenvironment that is tightly regulated by interactions with the surrounding cells, soluble factors, and ECM molecules. Chondrocytes in monolayer culture lose their phenotypic properties, while cultures in 3D agarose gels lead to re-expression of chondrocyte phenotype (Benya and Shaffer, 1982). The *in vitro* results demonstrate that cells show very different phenotypic characteristics when cultured on a 3D scaffold versus a 2D substrate. A 3D scaffold offers a local micro-environment that mimics natural ECM–cell context more closely and where the functional properties of cells can be manipulated and optimized (Ma, 2004a,b; Zhang, 2004).

Consequently, the degree of success of various tissue engineering strategies depends significantly on 3D architectures of the polymeric scaffold (3D

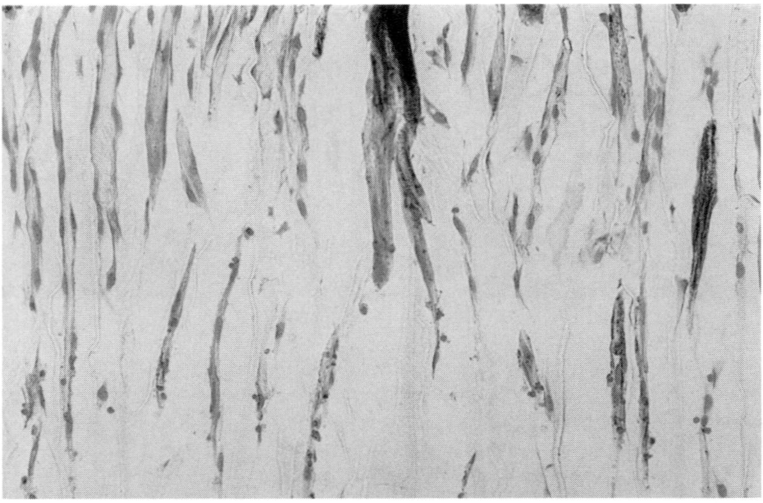

2.2 MC3T3-E1 cell growth on the oriented microtubular PLLA scaffold (4 weeks *in vitro* cell culture, von Kossa's silver nitrate staining). From Ma and Zhang (2001), Copyright © 2001 by John Wiley & Sons. Reprinted with permission of John Wiley & Sons.

environment for cells). Three-dimensional scaffold architecture refers to the way in which a bulk material is distributed in space from the macro, micro to nano scales (corresponding to tissue, cellular and molecular scales in a specific tissue, respectively) (Muschler *et al.*, 2004). Such hierarchical porous properties affect not only cell seeding, survival, migration, proliferation, and organization but also their gene expression and phenotypic characteristics (Chen *et al.*, 2006). They also define the mechanical structure of the scaffold and the initial void space that is available for progenitor cells to form new tissues, including new blood vessels, as well as the pathways for mass transport via diffusion and/or convection.

The importance of macroporosity (>100 μm) on neo tissue formation has been investigated (Tsuruga *et al.*, 1997; Kuboki *et al.*, 2001; Roy *et al.*, 2003). PLGA scaffolds with a higher porosity (>80%) promote more tissue ingrowth and new tissue formation (Roy *et al.*, 2003). The interconnection between macropores (interpore opening size, density, and pathway) is important for cellular activity and bone tissue formation. Scaffolds with open pore structures favor cell and tissue penetration, blood vessel invasion and new bone formation (Lu *et al.*, 1999). Failures of cell/tissue ingrowth often resulted from insufficient inter-pore connection where cell colonization was limited to the very peripheral and superficial layers (e.g. about 240 μm from the surface of a 1.9 mm thick scaffold made from traditional solvent-casting/particulate leaching technique) (Ishaug-Riley *et al.*, 1998). Increase in interconnected macroporosity, however, may adversely affect mechanical properties of a scaffold. More advanced scaffold design and fabrication techniques are desired. For example, scaffolds prepared by thermally induced phase separation offer both higher porosity (up to 98%) and improved mechanical properties over scaffolds produced by traditional salt-leaching technique (Zhang and Ma, 1999a). The oriented microtubular scaffolds have shown anisotropic mechanical properties similar to some fibrillar and tubular tissues, and has been demonstrated to facilitate cell organization into oriented tissues (Ma and Zhang, 2001) (Fig. 2.2).

2.2.3 Nanofibrous scaffolds

While interconnected macroporosity of a scaffold is important to provide sufficient space for cellular activity, interactions between cells and biomaterials occur at the interface, i.e. the entire internal pore walls of the 3D scaffold. Manipulation of the morphology or topography of the macropore walls directly and significantly affects cell–scaffold interactions and eventually tissue formation and function (Chen *et al.*, 2006; Woo *et al.*, 2007a).

Collagen is the major ECM component of many tissues and has been actively investigated as a substrate or scaffold for cell attachment, proliferation and differentiation (Elsdale and Bard, 1972; Strom and Michalopoulos, 1982). Importantly, the nanoscaled collagen fibrillar structure has been recognized to enhance cell–matrix interaction (Grinnell and Bennett, 1982; Kuntz and Saltzman, 1997). To mimic collagen fiber bundles in nano size (50–500 nm) and to eliminate possible immunogenicity brought by collagen, nanofibrous features have been introduced into synthetic biodegradable polymer scaffolds (Ma and Zhang, 1999; Zhang and Ma, 2000; Woo *et al.*, 2003; Chen and Ma, 2004; Chen *et al.*, 2006; Wei and Ma, 2006). A combined technique of sugar sphere template leaching and phase separation has been developed to prepare spherical macropores in nanofibrous scaffolds (Wei and Ma, 2006) (Fig. 2.3(a), (b)). The resulting polymer nanofibers have a diameter between 50 and 500 nm, which is similar in size to collagen fibers. The nanofibrous scaffold has high surface area of about 100 m^2/g, which is more than 100 times higher than that of a non-fibrous (solid-walled) scaffold with the same macroporosity and pore structures. Alternatively, a sugar fiber template or solid freeform fabricated wax mold have been used to prepare nanofibrous scaffolds with interconnected macro-tubular or channeled structures, respectively (Zhang and Ma, 2000; Chen *et al.*, 2006) (Fig. 2.3(c), (d)).

The macroporous and nanofibrous scaffolds enhance protein adsorption and positive cellular response as compared to solid-walled scaffolds without nanofibrous structures (Woo *et al.*, 2003, 2007a; Chen *et al.*, 2006). Such scaffolds also allow for more uniform matrix and mineral production throughout (Fig. 2.4). Cells in nanofibrous scaffolds showed significantly higher expression of osteocalcin and bone sialoprotein mRNAs. The high surface area, the microporosity between nanofibers (several μms), and selective adsorption of ECM

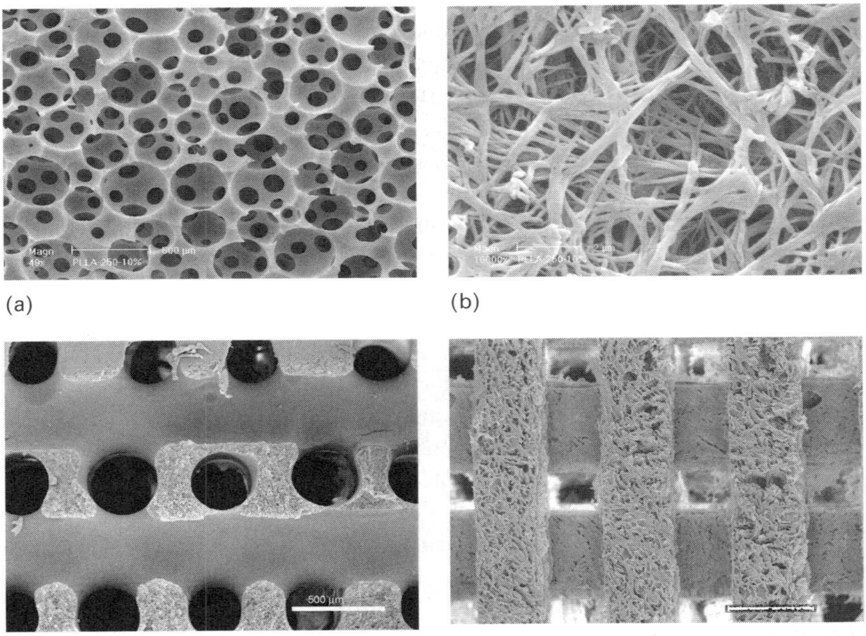

(a) (b) (c) (d)

2.3 SEM micrographs of 3D nanofibrous PLLA scaffolds: (a, b) prepared from sugar sphere template leaching and phase separation; (c) prepared from sugar fiber template leaching and phase separation; (d) prepared from solid freeform fabrication and phase separation. (a, b) From Wei and Ma (2006), Copyright © 2006 by John Wiley & Sons; (c) From Zhang and Ma (2000), Copyright © 2000 by John Wiley & Sons. Reprinted with permission of John Wiley & Sons. (d) From Chen *et al.* (2006), Copyright 2006 by Elsevier. Reprinted with permission of Elsevier.

proteins in nanofibrous scaffold may all contribute significantly to the enhanced tissue regeneration.

Other techniques that have the capability of producing nanofibers are electrospinning (Huang *et al.*, 2003; Xu *et al.*, 2004) and self-assembly (Whitesides and Grzybowski, 2002). However, designed 3D macroporosity is difficult to incorporate into electrospun scaffolds and the thickness of the scaffold is limited to 1 mm (Nair *et al.*, 2004). Self-assembly has also drawn attention in developing nanofibrous materials with the potential to be tissue engineering scaffolds. This far, the self-assembly techniques are limited to small molecules such as peptides and amphiphiles in a form of hydrogels (Whitesides *et al.*, 1991; Whitesides and Grzybowski, 2002; Zhang, 2004; Beniash *et al.*, 2005). The control of biodegradability and 3D pore structure has not been demonstrated.

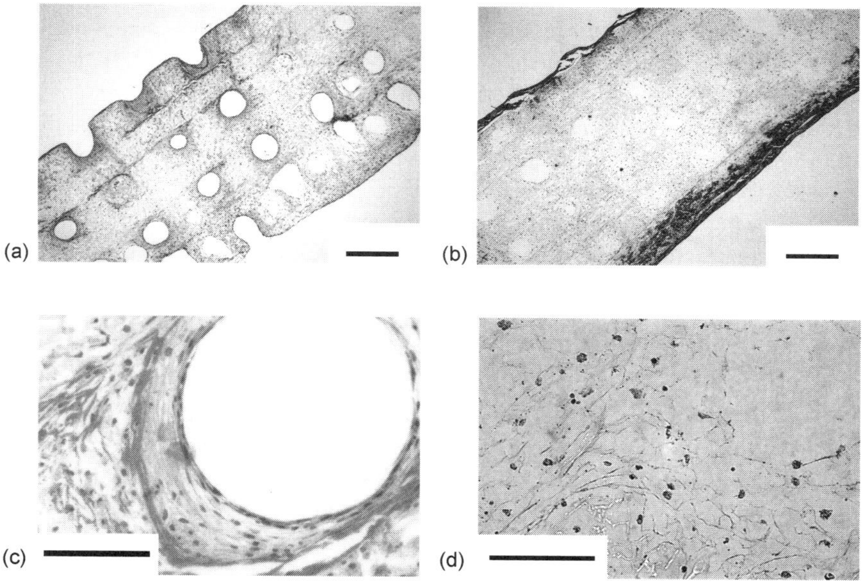

2.4 Responses of MC3T3-E1 cells on nanofibrous (a, c) and non-fibrous solid-walled (b, d) PLLA scaffolds (hematoxylin and eosin (H & E) staining). Scaffolds were prepared using a combined solid freeform fabrication and phase separation techniques. Cells were cultured *in vitro* for 6 weeks. From Chen *et al.* (2006), Copyright © 2006 by Elsevier. Reprinted with permission of Elsevier.

2.2.4 Surface modification of polymeric scaffolds

Besides the manipulation of physical structures such as morphology and topography, the scaffold surface can also be chemically modified to improve its physiological functions. Plasma exposure is a widely used approach to modify the hydrophilicity of a surface or induce certain functional groups on the surface of a substrate for further modification in combination with other chemical methods (Chim *et al.*, 2003; Yamaguchi *et al.*, 2004). Because of its limited penetration depth, plasma treatment is effective on 2D substrates but is difficult to modify a 3D porous scaffold.

In a complex 3D porous scaffold, the cell– or tissue–material interface is not just the outside surface of the scaffold, but also the entire internal 3D pore wall surfaces. Surface modification of a 3D scaffold requires the modification to be uniform. Dip coating is a technically simple approach to adsorb required surface-modifying species from a solution onto prefabricated 3D scaffold surface. However, the modified surface is unstable because of the weak interactions between bulk materials and surface-modifying species (Wu *et al.*, 2006). Another strategy for surface modification is to introduce functional groups in polymer chains before scaffold fabrication, which allows for later coupling with certain peptides (Yoon *et al.*, 2004). PLGA porous scaffold was functionalized with a primary amine group to which Gly-Arg-Gly-Asp-Tyr (GRGDY) was immobilized. Bone marrow cell adhesion was substantially enhanced by the GRGDY modification of porous scaffold surfaces. The GRGDY-immobilized PLGA scaffold also showed enhanced osteoblastic differentiation as determined by alkaline phosphatase activity (Yoon *et al.*, 2004).

To mimic both the nanofibrous structure and chemical composition of collagen fibers, a nanofibrous scaffold was surface modified with gelatin, an ECM-like biomacromolecule derived from collagen (Liu *et al.*, 2005a,b, 2006). In a recent study, a new entrapment method was developed to effectively incorporate gelatin onto both interior and exterior surfaces of the nanofibrous pore walls (Liu *et al.*, 2005b). The entrapment modification method can be used for various geometries, morphologies, and thicknesses of 3D polymer scaffolds without affecting the architecture and properties of the scaffolds. In a further study, gelatin spheres acted as both porogen for scaffold fabrication and surface-modification agent, where pore generation and surface modification were accomplished in a simple one-step process (Liu *et al.*, 2006). The gelatin-modified porous nanofibrous scaffold was demonstrated to significantly improve initial osteoblast cell adhesion and proliferation throughout the scaffold (Liu *et al.*, 2005b, 2006). Since most biomacromolecules are charged cationic or anionic polyelectrolytes, a layer-by-layer self-assembly process has also been introduced for surface modification of a 3D scaffold (Liu *et al.*, 2005a). One advantage of the self-assembly approach is that biomolecules can be immobilized on 3D scaffold surfaces under mild conditions.

Three-dimensional polymer scaffolds can also be modified with a layer of bone-like apatite via a biomimetic process in a simulated body fluid (SBF) or with growth factors. The former modification is specifically designed to increase osteoconductivity and bone bonding properties in bone tissue regeneration. The latter, growth factor immobilization/release, belongs to the category of controlled drug delivery, which has been an exciting area in inductive tissue regeneration. We will discuss these two modification approaches separately in sections of composite scaffolds (Section 2.2.5) and polymer scaffolds with controlled release capacity (Section 2.3).

2.2.5 Polymer/apatite composite scaffolds for bone regeneration

Polymer/apatite composite materials have been developed for mineralized tissue engineering applications such as bone tissue regeneration. Being similar to the major inorganic component of natural bone, the inorganic component such as hydroxyapatite (HA) in the composite scaffold provides good osteoconductivity while the polymer component provides the continuous structure and design flexibility to achieve the high porosity and high surface area necessary for

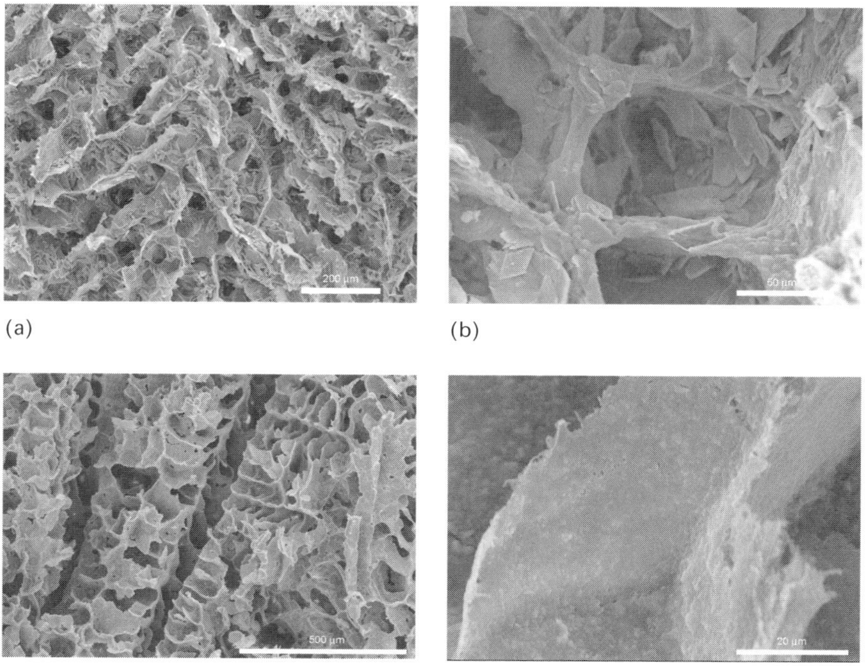

(a) (b) (c) (d)

2.5 SEM micrographs of PLLA/mHAP (a, b) and PLLA/nHAP composite scaffolds fabricated using phase separation. (a, b) From Zhang and Ma (1999a), Copyright © 1999 by John Wiley & Sons. Reprinted with permission of John Wiley & Sons; (c, d) From Wei and Ma (2004), Copyright © 2004 by Elsevier. Reprinted with permission of Elsevier.

anchorage-dependent cells such as bone cells to survive and differentiate. By blending and phase separation techniques, polymer/hydroxyapatite composite scaffolds have been developed with improved mechanical properties and osteoconductivity (Zhang and Ma, 1999a; Ma *et al.*, 2001) (Fig. 2.5(a), (b)). The HA-containing scaffolds improve osteoblastic cell seeding uniformity and show significantly enhanced expression of osteocalcin and bone sialoprotein over plain polymer scaffolds. Bone tissue formation throughout the scaffold has been demonstrated (Ma *et al.*, 2001). The nano-HA/polymer composite scaffolds not only improved the mechanical properties, but also significantly enhanced protein absorption over micro-sized HA/polymer scaffolds (Wei and Ma, 2004) (Fig. 2.5(c), (d)). The enhanced protein adsorption improves cell adhesion and function (Ma *et al.*, 2001; Woo *et al.*, 2007b).

To efficiently modify the internal pore wall surfaces with bone-like apatite without altering the bulk structures and properties of the scaffolds, a

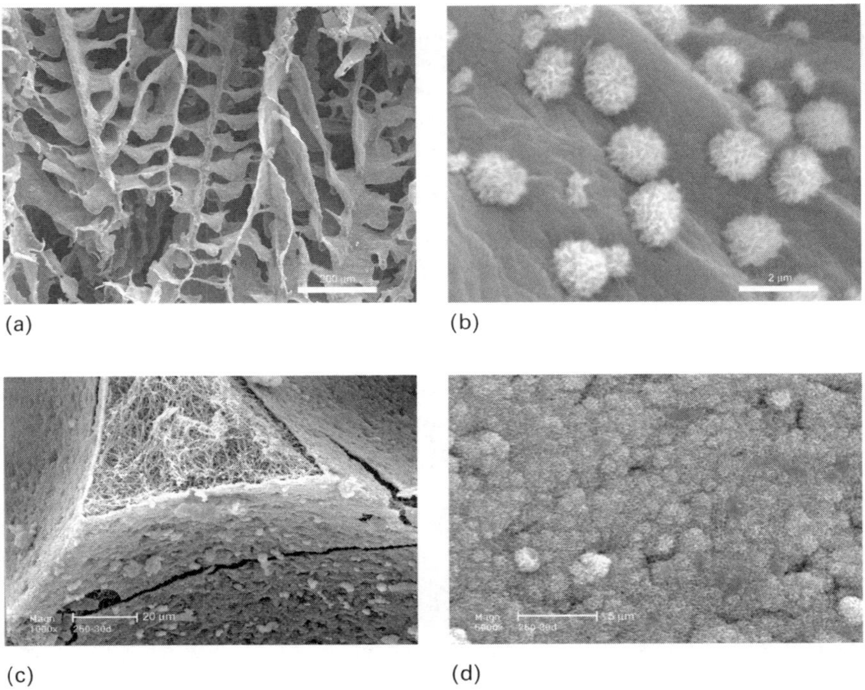

2.6 SEM micrographs of PLLA/apatite composite scaffold prepared by a biomimetic approach in a simulated body fluid (SBF). (a, b) PLLA scaffolds phase separated in dioxane; (c, d) PLLA nanofibrous scaffolds prepared by sugar template leaching and phase separation in THF. Scaffolds were incubated in 1.5 × SBF at 37 °C for 30 days. (a, b) From Zhang and Ma (2004), Copyright © 1999 by John Wiley & Sons; (c, d) From Wei and Ma (2006), Copyright © 2006 by John Wiley & Sons. Reprinted with permission of John Wiley & Sons.

biomimetic approach has been developed to grow bone-like apatite particles on prefabricated porous polymer scaffolds in an SBF (Zhang and Ma, 1999b, 2004; Wei and Ma, 2006). It has been observed that the growth of apatite crystals was significantly affected by the polymer materials, porous structure, ionic concentration of SBF as well as the pH value (Zhang and Ma, 2004). When macroporous nanofibrous scaffolds were investigated for bone-like apatite deposition, a uniform and dense layer of nano-apatite was found to cover the entire internal pore wall surface (Fig. 2.6). It was also found that the existence of nano-HA in a composite scaffold had the ability to promote new apatite deposition, and therefore greater amounts of apatite particles were formed on nanocomposite scaffolds than on plain polymer scaffolds (Wei and Ma, 2006).

2.3 Polymeric scaffolds with controlled release capacity

Owing to the rapid advance in recombinant technology and the availability of a large amount of purified recombinant polypeptides and proteins, protein drugs such as growth factors have been widely used in recent tissue engineering studies to stimulate cellular activity and regulate tissue regeneration (Reddi, 1998; Babensee *et al.*, 2000; Lee *et al.*, 2002). However, protein and peptide drugs in general have short plasma half-lives, are unstable in the gastrointestinal tract and also have low bioavailability due to their relatively large molecular weight and high aqueous solubility. These properties have limited their effective clinical applications (Lee, 1988). Carriers are needed to achieve high therapeutic efficacy of these peptides and proteins (Langer, 1990; Ferrara and Alitalo, 1999; Morley *et al.*, 2001).

2.3.1 Micro-/nano-encapsulation

Polymeric particulate carriers (micro- and nano-spheres) have been demonstrated to be effective to controlled release substances and to protect unstable biologically active molecules from denature and degradation after administration (Langer, 1990). Among the natural or synthetic polymers used for particulate carrier fabrication, PLLA and poly(lactic-co-glycolic acid) (PLGA) were found to be remarkable for their application in drug delivery due to their excellent biocompatibility and biodegradability through natural pathways (Uludag *et al.*, 2000; Wei *et al.*, 2004). Most importantly, the released proteins were able to maintain a high level of biological activity with desired prolonged duration (Oldham *et al.*, 2000; Wei *et al.*, 2004). Microspheres and nanospheres can be obtained using a double emulsion technique (Fig. 2.7). The release of

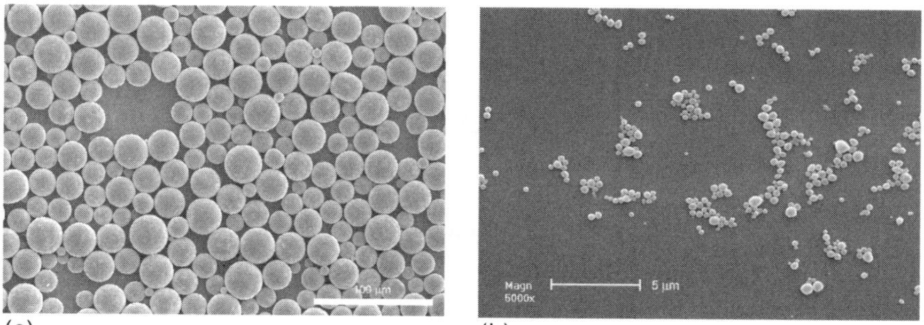

(a) (b)

2.7 SEM micrographs of PLGA50-74K microspheres (a) and PLGA50-6.5K nanospheres (b). (a) From Wei *et al.* (2004), Copyright © 2004 by Elsevier; (b) From Wei *et al.* (2006), Copyright © 2006 by Elsevier. Reprinted with permission of Elsevier.

proteins was controlled in the first stage by diffusion and in the second stage by the degradation of polymer micro- or nanospheres. By varying the molecular weight and the ratio of LA/GA in PLGA copolymers, sustained protein release over days to months can be achieved (Wei *et al.*, 2004). Both *in vitro* and *in vivo* assays demonstrated that the bioactivity of a biomolecule parathyroid hormone (PTH) was retained during the fabrication of PLGA microspheres and upon release. Consistently, the released platelet-derived growth factor (PDGF) from PLGA nanospheres was biologically active and was able to stimulate the proliferation of human gingival fibroblasts (Wei *et al.*, 2006). These studies illustrate the feasibility of achieving local delivery of bioactive factors by a microsphere/nanosphere encapsulation technique to induce cellular responses.

2.3.2 Factor-releasing scaffolds

As discussed above, biomimetic scaffolds are able to mimic natural ECM for optimal 3D structures, surface morphology and chemistry to achieve improved cell–matrix interactions and tissue regeneration from the materials perspective. Growth factors, employed properly, are able to directly stimulate cellular activities. Delivery of growth factors from a 3D scaffold, therefore, is an attractive strategy for tissue regeneration. In this way, the 3D scaffold serves both as a temporary substrate for cell functions and as a delivery carrier for the controlled release of growth factors.

A novel immobilization technology has been developed to incorporate growth factor-encapsulated nanospheres into prefabricated porous polymer scaffolds (Wei *et al.*, 2006, 2007). The nanospheres were uniformly distributed throughout the macroporous and nanofibrous scaffold without interfering the macro-, micro-, and nanostructures of the scaffold (Fig. 2.8). Immobilization onto a scaffold significantly reduced the initial burst release of the growth factors. Various release profiles were achieved through the use of nanospheres with different degradation rates (Wei *et al.*, 2006). Recombinant bone morphogenetic protein (rhBMP-7) delivered from nanosphere-immobilized scaffolds induced significant ectopic bone formation throughout the scaffold while simple adsorption of the same amount of rhBMP-7 onto the scaffold failed to induce bone formation due to either the loss of rhBMP-7 biological function or insufficient duration of the factor present in the scaffold (Wei *et al.*, 2007) (Fig. 2.9). Clearly, the new nanosphere-immobilization technique protects growth factors from denaturation as compared with simple adsorption of growth factors onto a scaffold, which has been reported to result in complete degradation of many growth factors (such as rhVEGF, BMP-4, and bFGF) during a very short release time of 3 days (Ziegler *et al.*, 2002). In addition, the burst release is very high and temporal control over release kinetics is very limited when a simple adsorption method is used (Miyamoto *et al.*, 1992; Tamura *et al.*, 2001). Growth factors can also be incorporated into a scaffold during emulsion freeze drying

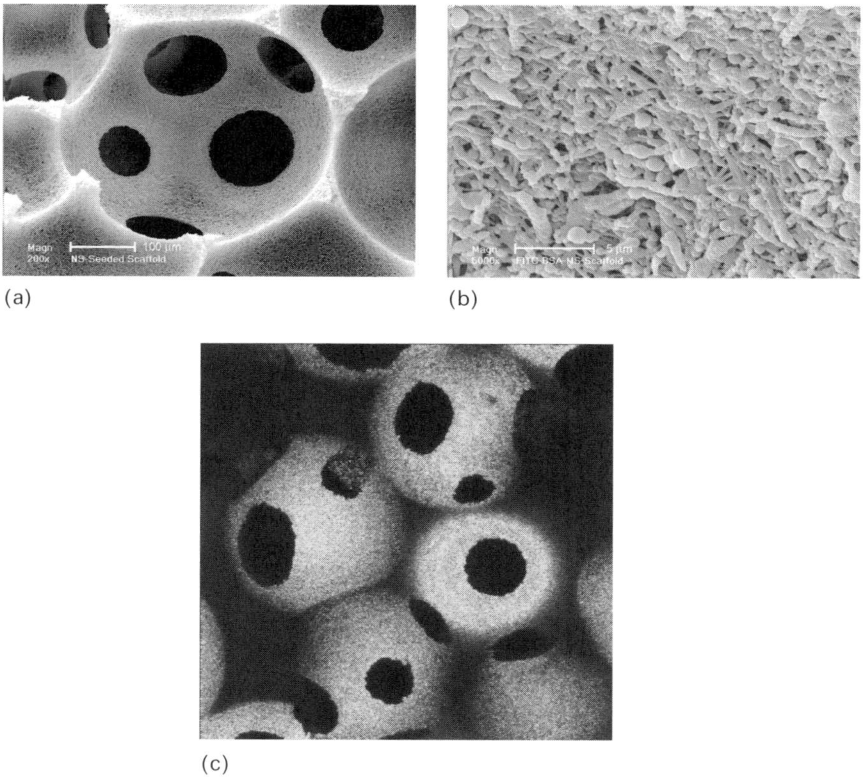

2.8 Scanning electron micrographs ((a) low magnification, (b) high magnification) and laser scanning confocal micrograph (c) of PLGA nanosphere-immobilized PLLA nanofibrous scaffolds. Fluorescein isothiocyanate (FITC)-labeled bovine serum albumin was encapsulated in the PLGA nanospheres (the areas around the black circles), shown as green emission under confocal microscopy (c). From Wei *et al.* (2006), Copyright © 2006 by Elsevier. Reprinted with permission of Elsevier.

(Whang *et al.*, 1998) or gas foaming (Sheridan *et al.*, 2000) scaffold fabrications. The resulting rhBMP-2 or VEGF-containing scaffolds were reported to release the growth factors to induce bone formation and angiogenesis, respectively. One disadvantage associated with these two incorporation techniques is the difficulty of achieving sufficient macroporosity and open pore structures in the scaffold. As an improvement, particulate leaching was combined with gas foaming to obtain an open porous structure of the scaffold (Murphy *et al.*, 2000). However, significant loss of growth factors during the leaching process is a major concern.

Many current growth factor delivery approaches for tissue engineering have focused on delivering single growth factors to trigger specific phenotypic repair processes. This approach is very limited considering the highly and precisely

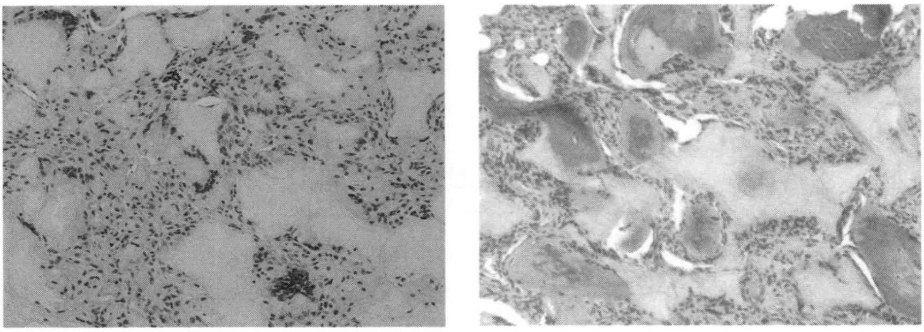

(a) (b)

2.9 New bone formation in rhBMP-7 incorporated PLLA nanofibrous scaffolds retrieved 6 weeks after subcutaneous implantation in rats. (a) 5 μg rhBMP-7 adsorbed to scaffold; (b) 5 μg rhBMP-7 incorporated in nanosphere immobilized scaffold. H & E staining with original magnification of 100×. From Wei *et al.* (2007), Copyright © 2007 by Elsevier. Reprinted with permission of Elsevier.

regulated signaling molecules involved in natural tissue regeneration (Rasubala *et al.*, 2003; Franceschi, 2005). It has been suggested that specific growth factors and cytokines are required to regulate tissue regeneration during different stages. Therefore, dual or multiple factor delivery systems may be advantageous in emulating the natural repair process.

A PLGA scaffold has been prepared to deliver two factors (VEGF and PDGF) using a gas foaming method (Richardson *et al.*, 2001). An improved angiogenic effect was reported as compared with VEGF or PDGF delivered alone. However, the scaffolds prepared from gas foaming have uncontrolled pore structure and poor mechanical properties. Most problematically, the imbedded particles in the matrix have uncontrollably altered the release profiles from the original particles. The delivery of more than two factors with predetermined release profiles is difficult to achieve due to the inherent design limitations of the system.

Multiple factor delivery, each with an individualized release profile, has been achieved by immobilizing multiple types of micro-/nanospheres (MNS) onto an advanced macroporous nanofibrous scaffold (Wei, 2006). Different biodegradable polymer nanospheres are utilized to individually control the release profiles of different biological molecules. Multiple factors with distinctly different release periods from days to weeks, months, and years are achievable from one single scaffold. The macro-, micro- and nano-features of the scaffolds were retained after MNS incorporation. Such novel scaffolds, while supporting 3D neo tissue genesis by cells, allow for the release of multiple biological factors in a spatially and temporally controlled fashion. The novel technology provides a platform to program biological signals into a 3D scaffold to regulate the cellular activities and to orchestrate predictable tissue regeneration.

2.4 Conclusions

Significant progress has been made in biomaterials from synthesis, characterization, processing to applications in tissue engineering, and drug delivery. Tissue engineering entails the successful interplay between cells, biomaterials, and bioactive factors, which needs to be further understood both from biology and materials perspectives. Advances in 3D scaffold fabrication technologies have provided a platform for delivering both cells and biological signals to the site that tissue regeneration is needed. Combining the multiple types of information (material, physical, chemical, structural, and biological) into one single biomaterial system will more closely mimic the multifunctional features of the native ECM and will be a powerful approach to regulate tissue regeneration.

2.5 References

Babensee J E, McIntire L V and Mikos A G (2000), 'Growth factor delivery for tissue engineering', *Pharm Res*, **17**(5), 497–504.

Beniash E, Hartgerink J D, Storrie H, Stendahl J C and Stupp S I (2005), 'Self-assembling peptide amphiphile nanofiber matrices for cell entrapment', *Acta Biomater*, **1**(4), 387–97.

Benya P D and Shaffer J D (1982), 'Dedifferentiated chondrocytes reexpress the differentiated collagen phenotype when cultured in agarose gels', *Cell*, **30**(1), 215–24.

Borden M, Attawia M and Laurencin C T (2002), 'The sintered microsphere matrix for bone tissue engineering: *in vitro* osteoconductivity studies', *J Biomed Mater Res*, **61**, 421–29.

Burg K J, Porter S and Kellam J F (2000), 'Biomaterial developments for bone tissue engineering', *Biomaterials*, **21**(23), 2347–59.

Chen V J and Ma P X (2004), 'Nano-fibrous poly(L-lactic acid) scaffolds with interconnected spherical macropores', *Biomaterials*, **25**(11), 2065–73.

Chen V J, Smith L A and Ma P X (2006), 'Bone regeneration on computer-designed nano-fibrous scaffolds', *Biomaterials*, **27**(21), 3973–79.

Chim H, Ong J L, Schantz J T, Hutmacher D W and Agrawal C M (2003), 'Efficacy of glow discharge gas plasma treatment as a surface modification process for three-dimensional poly (D,L-lactide) scaffolds', *J Biomed Mater Res A*, **65**(3), 327–35.

Dumitriu S (1996), *Polymeric Biomaterials*, New York, Marcel Dekker, Inc.

Elsdale T and Bard J (1972), 'Collagen substrata for studies on cell behavior', *J Cell Biol*, **54**(3), 626–37.

Ferrara N and Alitalo K (1999), 'Clinical applications of angiogenic growth factors and their inhibitors', Nature Medicine, **5**(12), 1359–64.

Franceschi R T (2005), 'Biological approaches to bone regeneration by gene therapy', *J Dent Res*, **84**(12), 1093–103.

Giannobile W V, Lee C S, Tomala M P, Tejeda K M and Zhu Z (2001), 'Platelet-derived growth factor (PDGF) gene delivery for application in periodontal tissue engineering', *J Periodontol*, **72**(6), 815–23.

Grinnell F and Bennett M H (1982), 'Ultrastructural studies of cell–collagen interactions', *Methods Enzymol*, **82** (Pt A), 535–44.

Harris L D, Kim B S and Mooney D J (1998), 'Open pore biodegradable matrices formed with gas foaming', *J Biomed Mater Res*, **42**(3), 396–402.

Hollinger J O and Leong K (1996), 'Poly(alpha-hydroxy acids): carriers for bone morphogenetic proteins', *Biomaterials*, **17**(2), 187–94.

Huang Z M, Zhang Y Z, Kotaki M and Ramakrishna S (2003), 'A review on polymer nanofibers by electrospinning and their applications in nanocomposites', *Compos Sci Technol*, **63**(15), 2223–53.

Ishaug-Riley S L, Crane-Kruger G M, Yaszemski M J and Mikos A G (1998), 'Three-dimensional culture of rat calvarial osteoblasts in porous biodegradable polymers', *Biomaterials*, **19**(15), 1405–12.

Kuboki Y, Jin Q and Takita H (2001), 'Geometry of carriers controlling phenotypic expression in BMP-induced osteogenesis and chondrogenesis', *J Bone Joint Surg Am*, **83-A** Suppl 1(Pt 2), S105–15.

Kuntz R M and Saltzman W M (1997), 'Neutrophil motility in extracellular matrix gels: mesh size and adhesion affect speed of migration', *Biophys J*, **72**(3), 1472–80.

Langer R (1990), 'New methods of drug delivery', *Science*, **249**(4976), 1527–33.

Langer R and Vacanti J P (1993), 'Tissue engineering', *Science*, **260**(5110), 920–26.

Lee H, Cusick R A, Browne F, Kim T H, Ma P X, Utsunomiya H, Langer R and Vacanti J P (2002), 'Local delivery of basic fibroblast growth factor increases both angiogenesis and engraftment of hepatocytes in tissue-engineered polymer devices', *Transplantation*, **73**(10), 1589–93.

Lee V H (1988), 'Enzymatic barriers to peptide and protein absorption', *Crit Rev Ther Drug Carrier Syst*, **5**(2), 69–97.

Liu X and Ma P X (2004), 'Polymeric scaffolds for bone tissue engineering', *Ann Biomed Eng*, **32**(3), 477–86.

Liu X, Smith L A, Wei G B, Won Y and Ma P X (2005a), 'Surface engineering of nano-fibrous poly(L-lactic acid) scaffolds via self-assembly technique for bone tissue engineering', *J Biomed Nanotech*, **1**(1), 54–60.

Liu X, Won Y and Ma P X (2005b), 'Surface modification of interconnected porous scaffolds', *J Biomed Mater Res A*, **74**(1), 84–91.

Liu X, Won Y and Ma P X (2006), 'Porogen-induced surface modification of nano-fibrous poly(L-lactic acid) scaffolds for tissue engineering', *Biomaterials*, **27**(21), 3980–87.

Lu J X, Flautre B, Anselme K, Hardouin P, Gallur A, Descamps M and Thierry B (1999), 'Role of interconnections in porous bioceramics on bone recolonization *in vitro* and *in vivo*', *J Mater Sci Mater Med*, **10**(2), 111–20.

Ma P X (2004a), 'Scaffolds for tissue fabrication', *Mater Today*, **7**(5), 30–40.

Ma P X (2004b), 'Tissue Engineering', in Kroschwitz, J. I., *Encyclopedia of Polymer Science and Technology*, Hoboken, NJ, John Wiley & Sons.

Ma P X and Choi J W (2001), 'Biodegradable polymer scaffolds with well-defined interconnected spherical pore network', *Tissue Eng*, **7**(1), 23–33.

Ma P X and Zhang R (1999), 'Synthetic nano-scale fibrous extracellular matrix', *J Biomed Mater Res*, **46**(1), 60–72.

Ma P X and Zhang R (2001), 'Microtubular architecture of biodegradable polymer scaffolds', *J Biomed Mater Res*, **56**(4), 469–77.

Ma P X, Zhang R, Xiao G and Franceschi R (2001), 'Engineering new bone tissue *in vitro* on highly porous poly(alpha-hydroxyl acids)/hydroxyapatite composite scaffolds', *J Biomed Mater Res A*, **54**(2), 284–93.

Mikos A G, Thorsen A J, Czerwonka L A, Bao Y, Langer R, Winslow D N and Vacanti J P (1994), 'Preparation and characterization of poly(L-lactic acid) foams', *Polymer*, **35**(5), 1068–77.

Miyamoto S, Takaoka K, Okada T, Yoshikawa H, Hashimoto J, Suzuki S and Ono K (1992), 'Evaluation of polylactic acid homopolymers as carriers for bone morphogenetic protein', *Clin Orthop Relat Res*, **278**, 274–85.

Morley P, Whitfield J F and Willick G E (2001), 'Parathyroid hormone: an anabolic treatment for osteoporosis', *Curr Pharm Des*, **7**(8), 671–87.

Murphy W L, Peters M C, Kohn D H and Mooney D J (2000), 'Sustained release of vascular endothelial growth factor from mineralized poly(lactide-co-glycolide) scaffolds for tissue engineering', *Biomaterials*, **21**(24), 2521–27.

Muschler G F, Nakamoto C and Griffith L G (2004), 'Engineering principles of clinical cell-based tissue engineering', *J Bone Joint Surg Am*, **86-A**(7), 1541–58.

Nair L S, Bhattacharyya S and Laurencin C T (2004), 'Development of novel tissue engineering scaffolds via electrospinning', *Expert Opin Biol Ther*, **4**(5), 659–68.

Oldham J B, Lu L, Zhu X, Porter B D, Hefferan T E, Larson D R, Currier B L, Mikos A G and Yaszemski M J (2000), 'Biological activity of rhBMP-2 released from PLGA microspheres', *J Biomech Eng*, **122**(3), 289–92.

Rasubala L, Yoshikawa H, Nagata K, Iijima T and Ohishi M (2003), 'Platelet-derived growth factor and bone morphogenetic protein in the healing of mandibular fractures in rats', *Br J Oral Maxillofac Surg*, **41**(3), 173–78.

Reddi A H (1998), 'Role of morphogenetic proteins in skeletal tissue engineering and regeneration', *Nature Biotechnol*, **16**(3), 247–52.

Richardson T P, Peters M C, Ennett A B and Mooney D J (2001), 'Polymeric system for dual growth factor delivery', *Nat Biotech*, **19**(11), 1029–34.

Roy T D, Simon J L, Ricci J L, Rekow E D, Thompson V P and Parsons J R (2003), 'Performance of degradable composite bone repair products made via three-dimensional fabrication techniques', *J Biomed Mater Res A*, **66**(2), 283–91.

Saito N, Murakami N, Takahashi J, Horiuchi H, Ota H, Kato H, Okada T, Nozaki K and Takaoka K (2005), 'Synthetic biodegradable polymers as drug delivery systems for bone morphogenetic proteins', *Adv Drug Deliv Rev*, **57**(7), 1037–48.

Sheridan M H, Shea L D, Peters M C and Mooney D J (2000), 'Bioabsorbable polymer scaffolds for tissue engineering capable of sustained growth factor delivery', *J Control Release*, **64**(1–3), 91–102.

Shive M S and Anderson J M (1997), 'Biodegradation and biocompatibility of PLA and PLGA microspheres', *Adv Drug Deliv Rev*, **28**(1), 5–24.

Strom S C and Michalopoulos G (1982), 'Collagen as a substrate for cell growth and differentiation', *Methods Enzymol*, **82** (Pt A), 544–55.

Tamura S, Kataoka H, Matsui Y, Shionoya Y, Ohno K, Michi K I, Takahashi K and Yamaguchi A (2001), 'The effects of transplantation of osteoblastic cells with bone morphogenetic protein (BMP)/carrier complex on bone repair', *Bone*, **29**(2), 169–75.

Tsuruga E, Takita H, Itoh H, Wakisaka Y and Kuboki Y (1997), 'Pore size of porous hydroxyapatite as the cell-substratum controls BMP-induced osteogenesis', *J Biochem (Tokyo)*, **121**(2), 317–24.

Uludag H, D'Augusta D, Golden J, Li J, Timony G, Riedel R and Wozney J M (2000), 'Implantation of recombinant human bone morphogenetic proteins with biomaterial carriers: a correlation between protein pharmacokinetics and osteoinduction in the rat ectopic model', *J Biomed Mater Res*, **50**(2), 227–38.

Wei G, Jin Q, Giannobile W V and Ma P X (2006), 'Nano-fibrous scaffold for controlled delivery of recombinant human PDGF-BB', *J Control Rel*, **112**(1), 103–10.

Wei G, Jin Q, Giannobile W V and Ma P X (2007), 'The enhancement of osteogenesis by nano-fibrous scaffolds incorporating rhBMP-7 nanospheres', *Biomaterials*, **28**(12), 2087–96.

Wei G B (2006), 'Growth factor-delivering nano-fibrous scaffolds for bone tissue regeneration', PhD Thesis, Biomedical Engineering, The University of Michigan, Ann Arbor, MI.

Wei G B and Ma P X (2004), 'Structure and properties of nano-hydroxyapatite/polymer composite scaffolds for bone tissue engineering', *Biomaterials*, **25**(19), 4749–57.

Wei G B and Ma P X (2006), 'Macro-porous and nano-fibrous polymer scaffolds and polymer/bone-like apatite composite scaffolds generated by sugar spheres', *J Biomed Mater Res A*, **78**(2), 306–15.

Wei G B, Pettway G J, McCauley L K and Ma P X (2004), 'The release profiles and bioactivity of parathyroid hormone from poly(lactic-co-glycolic acid) microspheres', *Biomaterials*, **25**(2), 345–52.

Whang K, Thomas C H, Healy K E and Nuber G (1995), 'A novel method to fabricate bioabsorbable scaffolds', *Polymer*, **36**(4), 837–42.

Whang K, Tsai D C, Nam E K, Aitken M, Sprague S M, Patel P K and Healy K E (1998), 'Ectopic bone formation via rhBMP-2 delivery from porous bioabsorbable polymer scaffolds', *J Biomed Mater Res*, **42**(4), 491–99.

Whitesides G M and Grzybowski B (2002), 'Self-assembly at all scales', *Science*, **295**(5564), 2418–21.

Whitesides G M, Mathias J P and Seto C T (1991), 'Molecular self-assembly and nanochemistry – a chemical strategy for the synthesis of nanostructures', *Science*, **254**(5036), 1312–19.

Woo K M, Chen V J and Ma P X (2003), 'Nano-fibrous scaffolding architecture selectively enhances protein adsorption contributing to cell attachment', *J Biomed Mater Res A*, **67**(2), 531–37.

Woo K M, Jun J H, Chen V J, Seo J, Baek J H, Ryoo H M, Kim G S, Somerman M J and Ma P X (2007a), 'Nano-fibrous scaffolding promotes osteoblast differentiation and biomineralization', *Biomaterials*, **28**(2), 335–43.

Woo K M, Seo J, Zhang R and Ma P X (2007b), 'Suppression of osteoblast apoptosis by enhanced protein adsorption on polymer/hydroxyapatite composite scaffolds', *Biomaterials*, **28**(16), 2622–40.

Wu Y C, Shaw S Y, Lin H R, Lee T M and Yang C Y (2006), 'Bone tissue engineering evaluation based on rat calvaria stromal cells cultured on modified PLGA scaffolds', *Biomaterials*, **27**(6), 896–904.

Xu C Y, Inai R, Kotaki M and Ramakrishna S (2004), 'Aligned biodegradable nanofibrous structure: a potential scaffold for blood vessel engineering', *Biomaterials*, **25**(5), 877–86.

Yamaguchi M, Shinbo T, Kanamori T, Wang P C, Niwa M, Kawakami H, Nagaoka S, Hirakawa K and Kamiya M (2004), 'Surface modification of poly(L-lactic acid) affects initial cell attachment, cell morphology, and cell growth', *J Artif Organs*, **7**(4), 187–93.

Yoon J J, Song S H, Lee D S and Park T G (2004), 'Immobilization of cell adhesive RGD peptide onto the surface of highly porous biodegradable polymer scaffolds fabricated by a gas foaming/salt leaching method', *Biomaterials*, **25**(25), 5613–20.

Zhang R and Ma P X (1999a), 'Poly(alpha-hydroxyl acids)/hydroxyapatite porous composites for bone-tissue engineering. I. Preparation and morphology', *J Biomed Mater Res*, **44**(4), 446–55.

Zhang R and Ma P X (1999b), 'Porous poly(L-lactic acid)/apatite composites created by biomimetic process', *J Biomed Mater Res*, **45**(4), 285–93.

Zhang R and Ma P X (2000), 'Synthetic nano-fibrillar extracellular matrices with predesigned macroporous architectures', *J Biomed Mater Res*, **52**(2), 430–38.

Zhang R and Ma P X (2004), 'Biomimetic polymer/apatite composite scaffolds for mineralized tissue engineering', *Macromol Biosci*, **4**(2), 100–11.
Zhang S (2004), 'Beyond the Petri dish', *Nat Biotechnol*, **22**(2), 151–52.
Ziegler J, Mayr-Wohlfart U, Kessler S, Breitig D and Gunther K P (2002), 'Adsorption and release properties of growth factors from biodegradable implants', *J Biomed Mater Res*, **59**(3), 422–28.

3
Bioactive ceramics and glasses

J R JONES, Imperial College London, UK

3.1 Introduction

One aspect of tissue engineering is the use of materials as templates (scaffolds) for tissue growth in 3D. Here bioactive ceramics and glasses will be introduced, their properties and current uses explained and their use as potential bone scaffold materials assessed from the materials chemistry perspective. An ideal scaffold must be a template for tissue growth in three dimensions, i.e. it should have an interconnected porous network that mimics that of porous bone, and it should stimulate bone regeneration while safely dissolving at a controlled rate (Jones *et al.*, 2006).

3.1.1 Bioactive materials

What is a bioactive material? There are several uses for the word bioactive. In its most general use it means that the material stimulates an advantageous biological response from the body on implantation. The termed was coined by Larry Hench in 1971, when he and his colleagues, at the University of Florida, invented Bioglass®, the first material that formed a strong bond to bone. This discovery not only launched the field of bioactive glasses but bioactive ceramics in general. Initially bioactivity meant materials that could bond to bone, but Bioglass® was later found to also stimulate new bone growth (osteogenesis) and to bond to soft tissues. Other materials, usually polymers, have also been developed that can release biological stimulants such as bone morphogenic proteins (BMP), which can stimulate bone growth and therefore can also be considered bioactive. In this chapter, bioactivity will be considered, the ability to bond to tissue and to stimulate bone growth without drugs or biological agents incorporated into the material.

Bioglass® was designed as a glass that would contain large amounts of calcium and phosphorous, as they are found in bone. However, as bone mineral is carbonated hydroxyapatite (HCA), a more obvious choice of material for a bone-repairing implant was a synthetically produced apatite. Sintered hydroxyapatite (sHA) has become a very popular bioactive implant material and has many more clinical products than Bioglass®. Bioglass® and sHA have spawned several new materials, such as sol–gel-derived bioactive glasses and substituted HA, which have not yet had widespread clinical use but are of particular interest in tissue engineering applications and will be discussed later. Aside from glasses and ceramics, a third class of material has been developed, which is in widespread use in Japan: glass-ceramics, particularly the apatite–wollastonite glass-ceramics that originated from Bioglass®. This chapter will introduce each of these materials, their clinical products and indicate their future potential in tissue engineering applications.

3.1.2 Classifying bioactivity

Large differences in rates of *in vivo* bone growth, between bioactive glasses and sHA were observed by Oonishi *et al.* (1999, 2000), indicating that there are two classes of bioactive materials. The level of bioactivity of a specific material has been related to the time taken for more than 50% of the interface to bond to bone ($t_{0.5bb}$) by the bioactivity index (I_B):

$$I_B = 100/t_{0.5bb} \tag{3.1}$$

Materials exhibiting an I_B value greater than 8 can be considered Class A. Class A materials are also expected to bond to soft tissue. Bioglass® is a Class A bioactive material. Materials with an I_B value less than 8 but greater than 0 are considered Class B bioactive materials, e.g. sHA, which will bond only to hard tissue. However, perhaps a more important distinction between the two classes is that Class A bioactive materials stimulate *both* osteoconduction and osteoinduction (Hench, 1998; Wilson and Low, 1992). Osteoconduction is the growth of bone along the implant material surface from the implant/bone interface. Osteoinduction is the generation of new bone on the implant, but away from the implant/interface, therefore bone cells or their progenitors must be recruited by the material and the signals provided for them to produce bone (Hench and Polak, 2002). The mechanisms will be discussed later.

3.1.3 Clinical use

All bioactive ceramics are used in dental applications, maxillofacial restoration and bone defect fillers in powder and moulded forms. HA and A-W glass-ceramics have been used in vertebral disc replacements and other bone defect replacements. Although these materials have been used to repair bone and show regenerative potential, none of them has been used in clinical tissue engineering applications. There are two main reasons for this; first, there are not suitable regulatory procedures for such constructs, and secondly their mechanical properties are not ideal for all defect sites, especially those under tensile load. That said, their bioactive properties are unparalleled by other materials, so there is potential for the tissue engineering strategy to work around these disadvantages. This chapter will review the properties, history and applications of bioactive glasses and ceramics and discuss their potential use in bone tissue engineering.

3.2 Synthetic hydroxyapatite

Synthetic hydroxyapatite (HA), chemical formula $Ca_{10}(PO_4)_6OH_2$, is an attractive bone repair material because it is chemically similar to bone mineral (Posner, 1969). However, bone mineral is a carbonatehydroxyapatite, approximated by $(Ca,X)_{10}(PO_4,HPO_4,CO_3)_6(OH,Y)_2$, where X is a cation, either a magnesium, sodium or strontium ion that can substitute for the calcium ions, and Y is an anion (a chloride or fluoride ion) that can substitute for the hydroxyl group (LeGeros, 2002). Synthetic HA can be created by solution chemistry or derived from natural materials. It is a Class B bioactive material that bonds to bone and is osteoconductive. The mechanism for bone bonding is thought to be cellular activity on the implant surface that causes partial dissolution and liberation of Ca and P ions. In addition to released Ca and P ions, Ca, P and other ions (Mg, CO_3) are recruited from the body fluid and combine to form HCA microcrystals on the material surface, which bond to the host bone. HA is osteoconductive because it encourages bone cell attachment, proliferation, migration and phenotypic expression, leading to formation of new bone in direct apposition to the material (Oonishi *et al.*, 1999, 2000).

3.2.1 Sintered HA

Conventional synthetic HA (sHA) is produced by creating an HA powder, which is sintered. It is a crystalline ceramic and stoichiometric HA (Ca/P ratio of 1.67) is commonly produced by precipitation by reacting calcium hydroxide with orthophosphoric acid solution (at $pH > 9$) at temperatures of between 25 and 90 °C. Another faster precipitation method involves calcium nitrate, diammonium hydrogen phosphate and ammonium hydroxide. However, this approach requires washing of the precipitate to remove nitrates and ammonium hydroxide, which is used to maintain constant pH. The production rate of these two processes is similar. Continued stirring and ageing are usually carried out after the reactants have been combined as the calcium is slowly incorporated into the apatite structure. This process also helps the material to approach stoichiometric Ca/P ratios. During a thermal maturation process, needle-like crystals change to crystals of blocky morphology. In both cases, the powders are sintered to produce an implant (Kato *et al.*, 1979). Processing at pH of less than 9 can also result in the production of a calcium-deficient hydroxyapatite.

HA can also be synthesised by the water extraction variant of the sol–gel process (Deptula *et al.*, 1992). A solution of calcium acetate and phosphoric acid (molar ratio Ca/P = 1.67) is emulsified in dehydrated 2-ethyl-1-hexanol. Drops of the sol are solidified by extraction of water with the solvent, which are then calcined to spherical sHA particles. HA formation begins above 400 °C and carbonate HA forms above 580 °C, which is decomposed to stoichiometric sHA above 750 °C.

All these processes produce sHA powders. The first suggestion of using sHA as a bone or tooth implant was in 1969 (Levitt *et al.*, 1969). Attempts were made to take it into commercial production, but it was not used clinically until 1978 when dense sintered HA cylinders were used as immediate dental root implants after tooth extraction (Denissen *et al.*, 1989).

Porous sHA

In a macroporous form, HA ceramics can be colonised by bone tissue (Passuti *et al.*, 1989). Several methods have been used to produce porous sHA. The simplest way to generate porous scaffolds from ceramics such as HA is to sinter spherical particles. Porosity is often increased by adding sacrificial porogens, which are often particles that will be removed after compaction or during sintering. The sacrificial particles can either be soluble, e.g. salt or sucrose (Andrade *et al.*, 2002) or combustible, e.g. poly(methyl methacrylate) (PMMA) microbeads. However, these methods give rise to a heterogeneous pore distribution and the interconnectivity of the pores is low. To improve interconnectivity, open-celled polyurethane foams can be immersed in slurries of sHA under vacuum to allow the slurry to penetrate into the pores of the foam. The foams are then heated at 250 °C to burn out the organic components (pyrolysis) and sintered at 1350 °C for 3 h, producing a scaffold with 300 μm interconnected pore diameters (Zhang *et al.*, 2001). This method is successful, but leaves the foam struts hollow. Perhaps the most successful method for creating interconnected sHA is the gel-casting method (Sepulveda *et al.*, 2000), in which suspensions of sHA particles and organic monomers are foamed with the aid of a surfactant. Once the foam is formed the monomers are polymerised and the porous network is set. The polymer gelling agent is burnt out (during sintering) before casting. The materials produced exhibited pores of maximum diameter of 100–200 μm suggested to be necessary for tissue engineering applications. Porous HA networks can therefore be produced with a variety of methods.

Clinical products

One of the most common clinical applications for sHA is the coating of orthopaedic implants. The aim is to form a bond between the host bone and the metal alloy (usually titanium or cobalt-chrome alloy) and negate the need for bone cement. The plasma-spray method is used for depositing the coating. This process uses a source material of sHA, the plasma-sprayed coatings are not pure

HA but a mixture of crystalline calcium phosphates, of which approximately 95% is sHA, and an amorphous calcium phosphate (Tisdel *et al.*, 1994).The crystalline to amorphous ratio in the coating can vary from 30 : 70 to 70 : 30 and this ratio is higher in the layer closest to the metal compared with the outermost layer. The presence of the amorphous phase increases the degradation rate of the material (Gross *et al.*, 1998).

Although sHA has excellent biocompatibility and osteoconductivity, it does not resorb in the body and therefore will not fulfil the criteria of an ideal tissue scaffold. The rate of dissolution depends on porosity and crystallinity (Koerten and van der Muelen, 1999). Therefore it is not an ideal material for regenerative applications, but is an excellent material as a permanent bone defect filler or permanent grafting material (bone substitute).

A successful porous sHA product is ApaPore® (Apatech Ltd., Elstree, UK), which is a porous HA that has an interconnected macroporosity and some microporosity. It has been successfully used in impaction grafting for the cemented revision of failed total joint arthroplasties, spinal fusions and bone defect treatment. Hi-Por Ceramics Ltd (Sheffield, UK) have released a foamed competitor and Ossatura™ (Isotis Orthobiologics, US) is a granular macroporous (75%) material composed of approximately 80% sHA and 20% β-tricalcium phosphate (TCP).

3.2.2 Natural source HA

Hydroxyapatite has been derived from special species of corals (*Porites*) (Roy and Linnehan, 1974) and from bovine bone (Valentini *et al.*, 2000). These HA are not pure but contain some of the minor and trace elements originally present in the coral or in the bone. Coralline HA contains traces of Mg, Sr, CO_3 and F. Bovine bone-derived apatite contains Mg, Na, CO_3 that were originally present in the bone (LeGeros, 2002). The great advantage of these materials is that they have an interconnected macroporous network that is conserved from the coral and bone.

Hydroxyapatite grafts derived from coral or coralline HA are prepared by hydrothermal converting of coral, which is often $CaCO_3$, at 260 °C and 15 000 psi in the presence of ammonium phosphate (Roy and Linnehan, 1974). Ions such as F, Sr, and CO_3 present in the coral become incorporated in the resulting HA. A secondary phase of β-TCP also forms during the hydrothermal conversion.

Bovine-derived apatites are produced by removal of the organic matrix. The resulting material is then either left unsintered or sintered above 1000 °C. The unsintered bone mineral consists of small crystals of HCA, whereas the sintered bone mineral consists of much larger apatite crystals without CO_3 (LeGeros, 2002). The small crystals in the unsintered HA mean that it will slowly resorb in the presence of body fluid, and therefore this material may be suitable for tissue engineering applications.

Clinical products

HA has become accepted as a commercial material (LeGeros, 2002). A porous bovine bone-derived HA graft is marketed under the name Bio-Oss® (Osteohealth, Shirley, New York), which is an unsintered material and therefore has some resorbability *in vivo* (Valentini *et al.*, 2000) and Osteograf-N™ (CeraMed Co, Denver, CO) and Endobon™ (Merck Co, Darmstadt, Germany), which is a sintered material. Porous coralline HA is sold as Interpore™ and Pro-Osteon™ (Roy and Linnehan, 1974), which are manufactured by Interpore International, Inc, Irvine, CA.

3.2.3 Substituted hydroxyapatite

Although stoichiometric sHA has very low degradation rates, degradation can be increased by substituting in other components that are found in biological apatites. The type and extent of substitution affect the rate of dissolution. Carbonate substitution contributes to the most soluble apatite (HCA), however F substitution (for OH) decreases the solubility to lower than sHA. Substituted apatite ceramics are also of interest because they offer the potential to improve the bioactive properties of implants.

Carbonate substituted apatites

Bone is a carbonate substituted apatite, with 5–8 wt% CO_3^{2-} (DeGroot, 1983) and therefore the synthesis of a carbonated apatite (CHA) may provide benefits over sHA. Calcium nitrate tetrahydrate was reacted dropwise to a solution of diammonium hydrogen orthophosphate solutions with the addition of 0.2–0.4 mol sodium hydrogen carbonate at pH > 9 (Merry *et al.*, 1998). The precipitate was aged, washed and sintered in a CO_2 atmosphere to prevent carbonate loss. The aim was to substitute CO_3^{2-} for the PO_4, however, the Ca/P ratio measured by XRF was higher than that of sHA which indicates that some CO_3^{2-} may also have substituted for OH instead of PO_4. Increasing carbonate content was shown to reduce the temperature at which decomposition occurred, to phases of CaO and β-TCP. The CHA specimens had strengths similar to sHA. This process was then improved to produce a high-purity single phase mixed AB-type carbonate-substituted hydroxyapatite (CHA), i.e. no sodium was present in the final material (Gibson and Bonfield, 2002).

Silicon substituted apatites

Silicon has been substituted into the structure of sHA, $Ca_{10}(PO_4)_6(OH)_2$, via aqueous precipitation. The silicate ion was proposed to substitute for the phosphate (Gibson *et al.*, 1999). A single-phase Si-HA was obtained by calcining the precipitated material to temperatures above 700 °C. Small structural differences were observed between sHA and Si-HA, with Si-HA having a unit cell with a shorter *a*-axis and longer *c*-axis and a lower number of OH groups. The incorporation of silicon in the HA lattice also caused an increase in the distortion of the PO_4 tetrahedra.

In vitro tests using human osteoblasts cultured on sHA and Si-HA discs with 0.8 wt% and 1.5 wt% Si substitutions have shown that cell response and gene expression are enhanced by the presence of the Si and are dependent on Si content. Bone nodules were present after 21 days of culture when β-glycerophosphate and hydrocortisone were added and were not related to the presence or level of silicon in the substrate (Botelho *et al.*, 2006). This dose-dependent effect was also observed *in vivo* for porous Si-HA. Bone response to porous Si-HA was investigated by implanting 4.6 mm diameter cylinders with various silicon levels (0–1.5 wt% Si) in rabbits for up to 12 weeks. Improved bone ingrowth and mineral apposition rate in Si-HA compared with sHA were observed within a week. The effect appeared to be dose-dependent in that an Si content of 0.8 wt% seemed to stimulate optimal bone ingrowth at 3 weeks. At 12 weeks, bone ingrowth continued to occur, apparently as a result of partial resorption (thought to be a cellular process) of the scaffold but ingrowth levels remained highest in the 0.8 wt% Si group (Hing *et al.*, 2006). The porous Si-HA scaffolds were produced using a slip foaming technique, where Si-HA particles in solution was agitated in the presence of a low molecular weight polymer, and sintered at 1250 °C. Total porosities of 70% were achieved.

Clinical products

Apatech Ltd. (Elstree, UK) have recently released Actifuse®, an 80% porous Si-HA (0.8 wt%), produced by the slip foaming method. Applications are small void filling, e.g. after removal of a small bone tumour, plastic surgery or spinal fusion, a cage or screw fixation device is used to relieve the graft site from physiological loads. It is applied as a granular structure that is combined with localised blood or with blood marrow aspirate, in a ratio of 1–1.5:1 (blood/BMA to Actifuse) to allow for optimal handling and placement.

It is not intended to be used in place of cortical strut allograft bone where high tensile, torsion and/or bending strength are required. Commercial sHA is used as bone substitutes in several orthopaedic and dental applications but they generally differ from bone in mechanical strength and physicochemical properties. They cannot be used in load bearing applications because of their low fracture strength (DeGroot, 1983).

3.3 Bioactive glass

Bioactive glasses are amorphous silicate-based materials which bond to bone and can stimulate new bone growth while dissolving over time, making them candidate materials for tissue engineering. The first bioactive glass (45S5 Bioglass®, 46.1% SiO_2, 24.4% NaO, 26.9% CaO and 2.6% P_2O_5, in mol%) was the first material seen to form an interfacial bond with host tissue after implantation (Hench *et al.*, 1971). The strength of the interfacial bond between Bioglass® and cortical bone was equal to or greater than the strength of the host bone (Weinstein *et al.*, 1980). Bioglass® particulate has been in clinical use since 1985 (USBiomaterials Corp., Alachua, Florida). Other compositions and glass types have also been found to be bioactive. Phosphate-based glasses have also been developed, which have a high rate of resorption. There are two processing routes to produce bioactive glasses; the traditional melt-derived approach and the sol–gel process, each yielding very different glasses. The interest in bioactive glasses now focuses on their osteogenic (osteoinductive) potential and applications in tissue engineering.

3.3.1 Mechanism of bioactivity

The bonding of bioactive glasses to bone has been attributed to the formation of an HCA layer on the glass surface on contact with body fluid. The HCA is similar to bone mineral and therefore forms a bond (Hench *et al.*, 1971). The HCA layer forms as a result of a sequence of chemical reactions on the surface of the implant when exposed to body fluid (Pantano *et al.*, 1974; Hench, 1998). There are five proposed reaction stages that lead to the rapid release of soluble ionic species, essentially glass corrosion, and the formation of a high surface area hydrated silica and polycrystalline HCA bi-layer on the glass surface.

- *Stage 1*: Rapid exchange of Na^+ and Ca^{2+} with H^+ or H_3O^+ from solution, causing hydrolysis of the silica groups, which creates silanols (Si–OH), e.g.:

  $$\text{Si–O–Na}^+ + \text{H}^+ + \text{OH}^- \rightarrow \text{Si–OH}^+ + \text{Na}^+_{(aq)} + \text{OH}^-$$

  Ion exchange is diffusion controlled with a $t^{1/2}$ dependence. The pH of the solution increases as a result of H^+ ions in the solution being replaced by cations.
- *Stage 2*: Stage 1 increases the hydoxyl concentration of the solution, which leads to attack of the silica glass network. Soluble silica is lost in the form of $Si(OH)_4$ to the solution, resulting from the breaking of Si–O–Si bonds and the continued formation of Si–OH (silanols) at the glass solution interface:

  $$\text{Si–O–Si} + \text{H}_2\text{O} \rightarrow \text{Si–OH} + \text{OH–Si}$$

  This stage is an interface-controlled reaction with a $t^{1.0}$ dependence.
- *Stages 3–5*: Condensation and repolymerisation of the Si–OH groups is then thought to occur, leaving a silica-rich layer on the surface, depleted in alkalis and alkali-earth cations (stage 3). Ca^{2+} and PO_4^{3-} groups then migrate to the surface through the silica-rich layer and from the surrounding fluid, forming a

film rich in $CaO–P_2O_5$ on top of the silica-rich layer (stage 4). The $CaO–P_2O_5$ film crystallises as it incorporates OH^- and CO_3^{2-} anions from solution to form a mixed HCA layer.

This mechanism was based on soda-lime–silica glass corrosion mechanisms and completes within hours. The kinetics of these two stages implied that stage 1 was the rate-determining step for bone bonding.

The biological mechanisms of bonding that follow HCA layer formation are suggested to involve the adsorption of growth factors, followed by the attachment, proliferation and differentiation of osteoprogenitor cells (Hench and Polak, 2002). Osteoblasts (bone growing cells) lay down extracellular matrix (collagen-based matrix), which will mineralise to create a nanocomposite of mineral and collagen on the surface of the bioactive glass implant while the dissolution of the glass continues over time (Ducheyne and Qin, 1999).

Bioactive glasses have been found to bond more rapidly to bone than sHA, and to be osteoinductive, i.e. they stimulate new bone growth on the implant away from the bone/implant interface. sHA is classified as osteoconductive, i.e. it encourages bone to grow along the implant from the bone/implant interface (Oonishi *et al.*, 1999).

The reasons for bioactive glasses being a Class A bioactive and osteo-inductive material and sHA being a Class B osteoconductive material have long been linked to the rate of formation of the HCA surface layer, allowing bone bonding to occur more rapidly. However, this would only explain improved osteoconductivity of Bioglass® over HA, not the cause of osteoinduction. Therefore the mechanism for osteoinduction is more complicated. It is vital to understand the biological response to bioactive materials, i.e. what signals do the osteogenic cells such as osteoblasts receive from the material? As Bioglass® degrades it releases silica, calcium, sodium and phosphate species into solution. It is thought that the combination of some of these ions triggers the cells to produce new bone, especially critical concentrations of soluble silicon and calcium ions (Hench and Polak, 2002). Molecular biology studies have shown that seven families of genes involved in osteogenesis have been stimulated by bioactive glass dissolution products, including *IGF-II* with IGF binding proteins and proteases that cleave *IGF-II* from their binding proteins (Xynos *et al.*, 2001). It is thought that bioactive glasses determine gene expression by the rate and type of dissolution ions released. The intracellular signalling pathways, however, remain uncertain.

It is important to ascertain which ions cause osteoinduction via gene expression, and at what concentrations. The effect has been seen to be concentration dependent (Xynos *et al.*, 2001), with approximately 17–20 ppm of soluble Si and 88–100 ppm of soluble Ca ions required. Sodium ions are not thought to be beneficial to cells, and the phosphate content of the glass is not thought to affect gene expression, although it may be needed in the body fluid for the extracellular matrix to mineralise and form HCA. Recent studies have shown that phosphate is not required to be released from the glass for extracellular matrix production and bone cells can mineralise, as long as it is present in the solution (Jones *et al.*, 2007). This has been supported by work by Reffitt *et al.* (2003), who showed enhanced differentiation of osteoblastic cell lines when exposed to soluble silica (orthosilicic acid) and demonstrated that the collagen extracellular matrix production increased in all cells treated with orthosilicic acid. In fact, dietary silica supplements have long been associated with increased bone mineral density (Jugdaohsingh *et al.*, 2004), and Si-doped HA materials have recently showed enhanced bone bonding compared with conventional synthetic HA (Porter *et al.*, 2004).

3.3.2 Melt-derived bioactive glasses

The original Bioglass® was produced by melt-processing, which involves melting high-purity oxides (SiO_2, Na_2CO_3, $CaCO_3$ and P_2O_5) in a crucible in a furnace at 1370 °C (Hench *et al.*, 1971). Platinum crucibles must be used to ensure there is no contamination of the glass. Bioglass® particulate is made by pouring the melt into water to quench, creating a frit. The frit is then dried and ground to the desired particle size range. Bioglass® can also be poured into pre-heated (350 °C) moulds (e.g. graphite) to produce rods or as-cast components.

The compositional range for bonding of bone to bioactive glasses and glass-ceramics is illustrated in Fig. 3.1. The most bioactive glasses lie in the middle (region S) of the Na_2O–CaO–SiO_2 diagram (assuming a constant 6 wt% of P_2O_5) (Hench, 1998). Compositions that exhibit slower rates of bonding lie between 52 and 60% by weight of SiO_2 in the glass. Compositions with greater than 60% SiO_2 (region B) are bio-inert. Adding multivalent cations, such as Al^{3+}, Ti^{4+} or Ta^{5+} to the glass shrinks the bone bonding boundary (Greenspan and Hench,

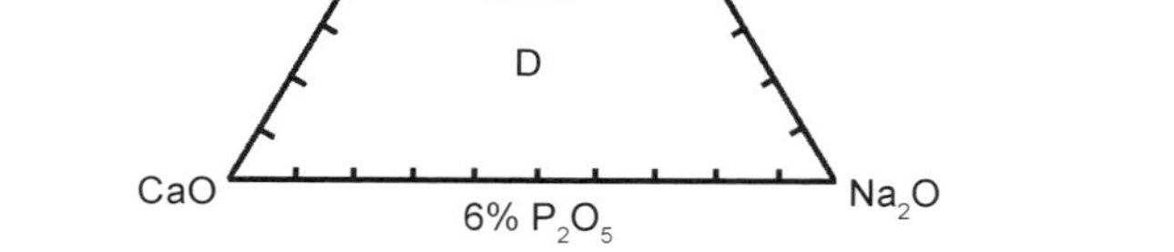

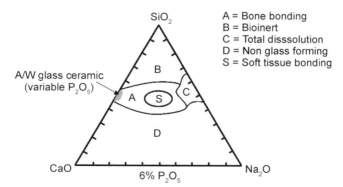

3.1 Compositional diagram for bioactivity of melt-derived silicate glass. Region S is a region of Class A bioactivity where bioactive glasses bond to both bone and soft tissues and are gene activating.

1976). The scientific basis for the compositional boundaries is associated with the dissolution rate of the glasses.

Andersson *et al.* (1990) modified the 45S5 composition and implanted 16 different compositions of the SiO_2–Na_2O–CaO–P_2O_5–Al_2O_3–B_2O_3 system into rabbit tibia. Bone bonding only occurred for glasses that could form an HCA layer when tested in Tris buffer solution *in vitro*. The presence of alumina in the composition inhibited bone bonding by slowing HCA formation rate by stabilising the silica structure enough to prevent calcium phosphate build-up within the layer. Up to about 1.5 wt% Al_2O_3 can be included in the glass without destroying the bioactivity. Compositions within boundaries similar to those in Fig. 3.1 bonded to bone; glasses outside the bioactive boundary did not bond.

3.3.3 Sol–gel derived bioactive glasses

For a melt-derived glass to bond to bone, the silica content has to be 60 mol% or lower. However, HCA layer formation and bone bonding can be achieved for glasses with up to 90 mol% silica if the glass is sol–gel derived (Li *et al.*, 1991). The first sol–gel derived bioactive glasses were developed in the early 1990s (Li *et al.*, 1991; Pereira *et al.*, 1994). The 58S (60 mol% SiO_2, 36 mol% CaO and 4 mol% P_2O_5) composition, which was a close resemblance to the melt-derived compositions developed previously, was found to form the HCA surface layer more rapidly than any melt-derived glass (Sepulveda *et al.*, 2002b).

The sol–gel process is shown schematically in Fig. 3.2 and involves the hydrolysis of alkoxide precursors to create a sol. In the case of silicate-based bioactive glasses, the silicate precursor would be an alkoxide such as tetraethyl orthosilicate (TEOS) or similar. If other components apart from silica are required in the glass composition they are added to the sol either as other

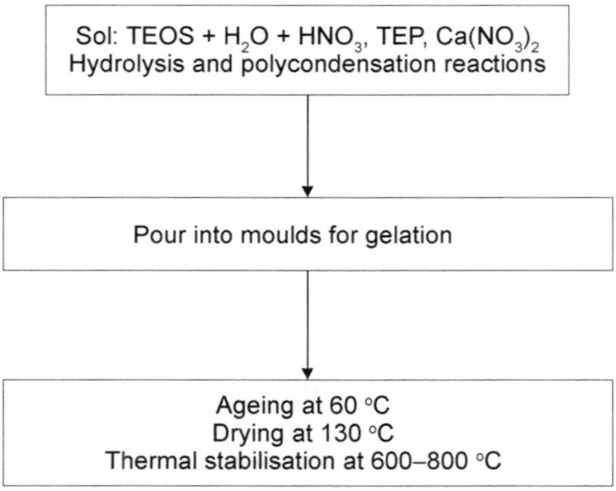

3.2 Flow diagram of the process for sol–gel bioactive glass synthesis.

alkoxides or as salts. In the case of 58S, phosphate is incorporated by adding triethyl phosphate (TEP) and calcium nitrate tetrahydrate. The sol can be considered as a solution of silica species that can undergo polycondensation to form the silica network of Si–O–Si bonds (Hench and West, 1990). A gel forms within 3 days at ambient temperature. Water and ethanol are by-products of the condensation reaction, which must be evaporated by carefully using carefully controlled low heating rates. The final step is to heat the dried gel to at least 600 °C in order to remove organic by-products and especially nitrates from the calcium nitrate, which burn off at approximately 560 °C (Saravanapavan and Hench, 2003).

Sol–gel derived glasses have a specific surface area two orders of magnitude higher than melt-derived glasses (Sepulveda *et al.*, 2001). This is because gel-glasses contain a nanoporous network that is inherent to the sol–gel process, whereas melt-derived glasses are fully dense. The nanopores are usually in the range of 1–30 nm diameter. The nanopore size can be tailored during processing, by controlling the pH of the catalyst (Brinker and Scherer, 1990), the nominal composition (Arcos *et al.*, 2002) and the final temperature (Jones *et al.*, 2006). It is, however, difficult to produce large, crack-free monoliths (greater than 10 mm thickness) because the driving off of water, organics and nitrates can create capillary stresses that cause cracking.

Advantages of sol–gel-derived bioactive glasses over melt-derived glasses of similar compositions are that they are generally more bioactive and they can be bioactive while having silica contents of up to 80 mol%. The enhanced bioactivity is due to the nanoporosity and enhanced surface area (Sepulveda *et al.*, 2001), which causes increased rates of dissolution, accelerating stages 1 and 2 of the bioactivity mechanism. Because of this, sol–gel-derived glasses can be considered truly bioresorbable. The compositions of gel-glasses can also be bioactive while containing fewer components, e.g. glasses of the 70 mol% SiO_2, 30 mol% CaO (70S30C) composition forms an HCA layer as rapidly as the 58S composition (Saravanapavan *et al.*, 2003).

One reason for this is that one part of the bioactivity mechanism is the formation of Si-OH groups, which are thought to play a role in HCA layer nucleation. Apatite layers can be nucleated on various materials that have a high concentration of surface OH groups when they are placed in solutions that contain high concentrations of calcium and phosphate ions (Li *et al.*, 1994). This has even been demonstrated on polymers (Miyazaki *et al.*, 2003). Sol–gel-derived glasses inherently contain a substantial number of OH groups in the glass network. This is because even in glasses produced with only silica in the nominal composition, OH groups become network modifiers and break up the network. Hence, several sol–gel-derived silicas have nanoporosity, often microporosity, and also dissolve in solution.

3.3.4 Bioactive glass scaffolds

Perhaps the biggest benefit of using the sol–gel route over melt-derived is that porous scaffolds with interconnected macropores suitable for tissue engineering applications have been developed (Sepulveda *et al.*, 2002a). Although Bioglass® was invented in 1971, no successful porous scaffolds have been synthesised from it because a sintering process is employed in all known methods for the processing of glass powder into porous structures. Sintering requires glasses to be heated above their glass transition temperature in order to initiate localised flow and the Bioglass® composition crystallises immediately above its glass transition. Glass-ceramic scaffolds have been developed from the Bioglass® composition (Chen *et al.*, 2006).

By foaming sol–gel-derived bioactive glasses, porous glass scaffolds can be created. The sol–gel foaming process involves the foaming of the sol by vigorous agitation with the aid of a surfactant, as the viscosity rapidly increases. On gelation, the spherical bubbles become permanent in the gel and as drainage occurs in the foam struts, the gel shrinks and the bubbles merge, interconnections open up at the point of contact between neighbouring bubbles, creating a highly connected network similar to trabecular bone. Figure 3.3 shows an X-ray micro-computer tomography (μCT) image of a bioactive glass foam scaffold. The image shows that the macropores are very interconnected. In fact, the pore structure is hierarchical because the nanoporosity inherent to the sol–gel process is maintained (Jones *et al.*, 2007). Compressive strengths of 2.4 MPa

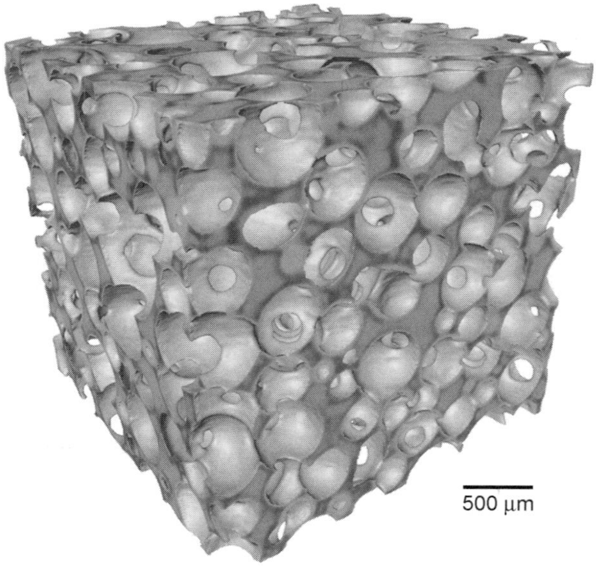

3.3 X-ray microtomography (μCT) image of a typical bioactive glass scaffold produced by the sol–gel foaming process. Image: Gowsihan Poologasundarampillai.

while maintaining modal interconnect diameters above 100 μm have been achieved by tailoring the nanoporosity during processing (Jones *et al.*, 2006). The strength values are similar to those of clinically used porous HA (Valentini *et al.*, 2000) and are continually rising as the process is improved. Although their compressive strength may be suitable, bioactive scaffolds suffer similar problems to sHA for brittleness. The only way a porous scaffold could be developed that will have the properties of a bioactive glass and the toughness of a composite would be to create an inorganic/inorganic nanocomposite by incorporating biodegradable polymers into the sol–gel process (Pereira *et al.*, 2005; Vallet-Regi *et al.*, 2006).

Whatever material is used as a scaffold, it is vital to be able to characterise the pore networks of scaffolds in 3D to ensure the scaffold has the potential to allow 3D bone growth throughout. Many authors have used μCT images to display the pore networks of their scaffolds, but little has been done to quantify the images. Now, novel 3D image analysis methods have been developed to quantify the structure by applying combinations of computer algorithms to μCT images (Jones *et al.*, 2007). The μCT data can also be input into finite element models to predict mechanical properties and permeability as a function of specific pore networks. For *in vivo* tests or clinical trials it is imperative to know the exact structure of the pore network before and after implantation, and therefore a non-destructive technique is required for imaging and quantification of the scaffold.

Cell response studies on the bioactive glass foam scaffolds have found that primary human osteoblasts lay down mineralised immature bone tissue, without additional signalling species. This occurred in scaffolds of both the 58S composition (Gough *et al.*, 2004) and the 70S30C composition (Jones *et al.*, 2007), which showed that phosphate is not required in the glass composition for bone matrix production and mineralisation to occur.

3.3.5 Clinical products

Despite all these developments, bioactive glass clinical products in use remain few, especially when compared with the less bioactive synthetic HA. This may be due to regulatory issues in that genetic stimulation cannot be claimed under the current FDA regulations; there are regulatory procedures for a medical device and for osteogenic drugs, but not for something that is implanted and dissolves over time, releasing osteogenic agents. To date, all Bioglass® devices placed into use in the USA have been cleared via the 510[k] process. Thus, the Bioglass® devices have been able to demonstrate equivalence in safety and efficacy to devices that were already in commerce prior to 1976 (Warren *et al.*, 1989).

The first Bioglass® device was approved in the USA in 1985 and was used to treat conductive hearing loss by replacing the bones of the middle ear. The device was called the 'Bioglass® Ossicular Reconstruction Prosthesis' (MEP®) and was a solid, cast Bioglass® structure that acted to conduct sound from the tympanic membrane to the cochlea. The advantage of the MEP® over other devices in use at the time was its ability to bond with soft tissue (tympanic membrane) as well as bone tissue. A modification of the MEP design was made to improve handling in the surgery and it is used clinically with the trademark name of the DOUEK MED.

The first particulate material cleared for sale in the USA was PerioGlas, which was cleared via the 510[k] process in December, 1993 and produced by USBiomaterials, Alachua, Florida. In 1995, PerioGlas® obtained a CE Mark and marketing of the product began in Europe. The initial indication for the product was to restore bone loss resulting from periodontal disease in infrabony defects (Karatzas *et al.*, 1999). During a 10-year clinical history, PerioGlas® has demonstrated excellent clinical results with virtually no adverse reactions to the product and is sold in over 35 countries. Bioactive glass particles have also been shown to undergo continual dissolution after the formation of the HCA layer in mandible sites, to such an extent that a hollow shell of HCA remained, in which new bone formed (Radin *et al.*, 2000). This led to the product Biogran®, which is Bioglass® granules, marketed by Orthovita Corp.

Building on the successes of PerioGlas® in the market, a Bioglass® particulate for orthopaedic bone grafting was introduced into the European market in 1999, under the trade name NovaBone®. The product was cleared for general orthopaedic bone grafting in non-load-bearing sites in February, 2000. To date, NovaBone® is being sold in the USA and Europe, as well as in China and a number of other countries.

Bioglass® particulate is also used for the treatment of dentinal hypersensitivity. The Bioglass® material used in this application is a very fine particulate that is incorporated into toothpaste, or used with an aqueous vehicle and applied to the tooth surface around exposed root dentin. When Bioglass® particles are put in contact with dentin, they adhere to the surface, rapidly form an HCA layer and occlude exposed tubules, thereby relieving pain. Studies have shown that the Bioglass® particulate performs better than current therapies in relatively low concentrations (Gillam *et al.*, 2002). Early in 2004, FDA cleared two products for sale through the 510[k] process, and product sales began in mid-2004 by Novamin Corp., Alachua, Florida.

Products involving sol–gel-derived bioactive glasses have only just started to appear. Novabone Corp. recently modified its product by introducing sol–gel-derived glass particles of the 58S composition. The aim is to that the 58S particles will dissolve more rapidly than the Bioglass® particles, creating spaces that will encourage bone ingrowth between the Bioglass® particles. The new Novabone® product was FDA approved in 2005 and was the first product containing sol–gel-derived bioactive glasses to be approved, which paves the way for bringing the scaffolds to the clinic.

Novathera Ltd, Cambridge, UK, have developed a wound healing gel that incorporates particles of the 70S30C gel-glass composition, which is modified by 2 mol% of silver ions, termed Theraglass®. Low concentrations of silver ions have been found to be bactericidal without killing useful cells (Bellantone *et al.*, 2002).

3.4 Glass-ceramics

Glass-ceramics are crystalline materials produced by heating a parent glass above its crystallisation temperature. A glass-ceramic was developed using a parent glass with composition similar to Bioglass® (SiO_2 40–50%, P_2O_5 10–15%, Na_2O 5–10%, CaO 30–35%, K_2O 0.5–3.0% MgO 2.5–5.0%, all wt%), which was later marketed as Ceravital® (Xomed, Jacksonville, FL) (Bromer *et al.*, 1977), which been shown to bond to bone via the formation of the HCA layer, although the formation of this layer was in several days, compared to a few hours for bioactive glass. Ceravital® has been used successfully in radical mastoidectomy (canal reconstruction) in the middle ear and ear canal for over 20 years (Della Santina and Lee, 2006); however, there are concerns about its long-term stability.

The most important modification of bioactive glasses was the development of A/W (apatite/wollastonite) bioactive glass-ceramic by Yamamuro, Kokubo and colleagues at Kyoto University, Japan. Composition modification and higher processing temperatures lead to a very fine-grained glass-ceramic composed of very small apatite (A) and wollastonite (W = $CaSiO_3$) crystals bonded by a bioactive glass interface (Nakamura *et al.*, 1985). Mechanical strength, toughness and stability of AW glass-ceramics (AW-GC) in physiological environments are excellent and bone bonded to A/W-GC implants with high interfacial bond strengths. The bioactivity of this glass-ceramic was attributed to apatite formation on its surface in the body, brought about by the dissolution of calcium and silicate ions from the glass-ceramic (Kokubo *et al.*, 1990). The AW-GC material was approved for orthopaedic applications in Japan with particular success in vertebral replacement and spinal repair (Neo *et al.*, 1992). It showed bioactivity and a high compressive strength (80 MPa). It is used clinically as artificial vertebrae and in iliac crest reconstruction. Bioactive glass-ceramics are reviewed in detail in Kokubo (1991). No A/W scaffolds have been developed.

3.5 Conclusions

Since the invention of Bioglass®, the first bioactive ceramic, several bioactive ceramics and glass-ceramics have been developed. At the time of writing, the bioactive materials with the most clinical applications are sintered hydroxy-apatite and apatite–wollastonite glass-ceramics. Bioactive glasses have been found to release ions that stimulate bone cells at the genetic level, causing osteoinduction. This leads to the development of silicon-substituted hydroxyapatite. It is this biological mechanism that must be fully understood if bioactive materials are to be fully optimised. Porous scaffolds can then be developed that will take these mechanisms into account. In this way, materials can be optimised from the atomic to the macro level with respect to cell response.

3.6 References

Andersson, O. H., Liu, G. Z., Karlsson, K. H., Niemi, L., Miettinen, J., Juhanoja, J. (1990) *In vivo* behavior of glasses in the SiO_2-Na_2O-CaO-P_2O_5-Al_2O_3-B_2O_3 system. *J. Mater. Sci. Mater. Med.*, **1**, 219–227.

Andrade, J. C. T., Camilli, J. A., Kawachi, E. Y., Bertran, C. A. (2002) Behavior of dense and porous hydroxyapatite implants and tissue response in rat femoral defects. *J. Biomed. Mater. Res.*, **62**, 30–36.

Arcos, D., Greenspan, D. C., Vallet-Regi, M. (2002) Influence of the stabilization temperature on textural and structural features and ion release in SiO_2–CaO–P_2O_5 sol–gel glasses. *Chem. Mater.*, **14**, 1515–1522.

Bellantone, M., Williams, H. D., Hench, L. L. (2002) Broad-spectrum bactericidal activity of Ag_2O-doped bioactive glass. *Antimicrob. Agents Chemother.*, **46**, 1940–1945.

Botelho, C. M., Brooks, R. A., Best, S. M., Lopes, M. A., Santos, J. D., Rushton, N., Bonfield, W. (2006) Human osteoblast response to silicon-substituted hydroxyapatite. *J. Biomed. Mater. Res. A*, **79A**, 723–730.

Brinker, J., Scherer, G. W. (1990) *Sol–Gel Science: The Physics and Chemistry of Sol–Gel Processing*, Boston, Academic Press.

Bromer, H., Deutscher, K., Blencke, B., Pfeil, E., Strunz, V. (1977) Properties of the bioactive implant material Ceravital. *Sci. Ceram.*, **9**, 219–225.

Chen, Q. Z., Thompson, I. D., Boccaccini, A. R. (2006) 45S5 Bioglass®-derived glass-ceramic scaffolds for bone tissue engineering, *Biomaterials*, **27**, 2414–2425.

DeGroot, K. (1983) *Bioceramics of CaP*, Boca Raton, FL, CRC Press.

Della Santina, C. C., Lee, S. C. (2006) Ceravital reconstruction of canal wall down mastoidectomy. *Arch. Otolaryngol. Head. Neck Surg*, **132**, 617–623.

Denissen, H. W., Kalk, W., Veldhuis, A. A. H., Vandenhooff, A. (1989) 11-Year study of hydroxyapatite implants. *J. Prosthetic Dentistry*, **61**, 706–712.

Deptula, A., Lada, W., Olczak, T., Borello, A., Alvani, C., Dibartolomeo, A. (1992) Preparation of spherical powders of hydroxyapatite by sol–gel process. *J. Non-Cryst. Sol.*, **147**, 537–541.

Ducheyne, P., Qiu, Q. (1999) Bioactive ceramics: the effect of surface reactivity on bone formation and bone cell function. *Biomaterials*, **20**, 2287–2303.

Gibson, I. R., Bonfield, W. (2002) Novel synthesis and characterization of an AB-type carbonate-substituted hydroxyapatite. *J. Biomed. Mater. Res*, **59**, 697–708.

Gibson, I. R., Best, S. M., Bonfield, W. (1999) Chemical characterization of silicon-substituted hydroxyapatite. *J. Biomed. Mater. Res.*, **44**, 422–428.

Gillam, D. G., Tang, J. Y., Mordan, N. J., Newman, H. N. (2002) The effects of a novel Bioglass (R) dentifrice on dentine sensitivity: a scanning electron microscopy investigation. *J. Oral Rehab.*, **29**, 305–313.

Gough, J. E., Jones, J. R., Hench, L. L. (2004) Nodule formation and mineralisation of

human primary osteoblasts cultured on a porous bioactive glass scaffold. *Biomaterials*, **25**, 2039–2046.
Greenspan, D. C., Hench, L. L. (1976) Chemical and mechanical-behavior of Bioglass-coated alumina. *J. Biomed. Mater. Res*, **10**, 503–509.
Gross, K. A., Berndt, C. C., Herman, H. (1998) Amorphous phase formation in plasma-sprayed hydroxyapatite coatings. *J. Biomed. Mater. Res.*, **39**, 407–414.
Hench, L. L. (1998) Bioceramics. *J. Am. Ceram. Soc.* **81**, 1705–1728.
Hench, L. L., Polak, J. M. (2002) Third-generation biomedical materials. *Science*, **295**, 1014–1017.
Hench, L. L., West, J. K. (1990) The sol–gel process. *Chem. Rev.*, **90**, 33–72.
Hench, L. L., Splinter, R. J., Allen, W. C., Greenlee, T. K. (1971) Bonding mechanisms at the interface of ceramic prosthetic materials. *J. Biomed. Mater. Res.*, **2**, 117–141.
Hing, K. A., Revell, P. A., Smith, N., Buckland, T. (2006) Effect of silicon level on rate, quality and progression of bone healing within silicate-substituted porous hydroxyapatite scaffolds. *Biomaterials*, **27**, 5014–5026.
Jones, J. R., Lee, P. D., Hench, L. L. (2006) Hierarchical porous materials for tissue engineering. *Phil. Trans R. Soc.*, **364**, 263–281.
Jones, J. R., Poologasundarampillai, G., Atwood, R. C., Bernard, D., Lee, P. D. (2007) Non-destructive quantitative 3D analysis for the optimisation of tissue scaffolds. *Biomaterials*, **28**, 1404–1413.
Jugdaohsingh, R., Tucker, K. L., Qiao, N., Cupples, L. A., Kiel, D. P., Powell, J. J. (2004) Dietary silicon intake is positively associated with bone mineral density in men and premenopausal women of the Framingham Offspring cohort. *J. Bone. Min. Res.*, **19**, 297–307.
Karatzas, S., Zavras, A., Greenspan, D., Amar, S. (1999) Histologic observations of periodontal wound healing after treatment with PerioGlas in nonhuman primates. *Int. J. Periodon. Rest. Dent.*, **19**, 489–499.
Kato, K., Aoki, H., Tabata, T., Ogiso, M. (1979) Biocompatibility of apatite ceramics in mandibles. *Biomater. Med. Dev.*, **7**(2), 291–297.
Koerten, H. K., van der Meulen, J. (1999) Degradation of calcium phosphate ceramics. *J. Biomed. Mater. Res*, **44**, 78–86.
Kokubo, T. (1991) Bioactive glass-ceramics – properties and applications. *Biomaterials*, **12**, 155–163.
Kokubo, T., Ito, S., Huang, Z. T., Hayashi, T., Sakka, S., Kitsugi, T., Yamamuro, T. (1990) Ca, P-Rich layer formed on high-strength bioactive glass-ceramic A-W. *J. Biomed. Mater. Res.*, **24**, 331–343.
LeGeros, R. Z. (2002) Properties of osteoconductive biomaterials: calcium phosphates. *Clin. Orthop. Rel. Res.*, **395**, 81–98.
Levitt, G. E., Crayton, P. H., Monroe, E. A., Condrate, R. A. (1969) Forming methods for apatite prosthesis. *J. Biomed. Mater. Res*, **3**, 683–685.
Li, P. J., Ohtsuki, C., Kokubo, T., Nakanishi, K., Soga, N., Degroot, K. (1994) The role of hydrated silica, titania, and alumina in inducing apatite on implants. *J. Biomed. Mater. Res.*, **28**, 7–15.
Li, R., Clark, A. E., Hench, L. L. (1991) An investigation of bioactive glass powders by sol–gel processing. *J. Appl. Biomater.*, **2**, 231–239.
Merry, J. C., Gibson, I. R., Best, S. M., Bonfield, W. (1998) Synthesis and characterization of carbonate hydroxyapatite. *J. Mater. Sci. Mater. Med.*, **9**, 779–783.
Miyazaki, T., Ohtsuki, C., Akioka, Y., Tanihara, M., Nakao, J., Sakaguchi, Y., Konagaya, S. (2003) Apatite deposition on polyamide films containing carboxyl group in a biomimetic solution. *J. Mater. Sci. Mater. Med.*, **14**, 569–574.

Nakamura, T., Yamamuro, T., Higashi, S., Kokubo, T., Itoo, S. (1985) A new glass-ceramic for bone-replacement – evaluation of its bonding to bone tissue. *J. Biomed. Mater. Res.*, **19**, 685–698.

Neo, M., Kotani, S., Nakamura, T., Yamamuro, T., Ohtsuki, C., Kokubo, T., Bando, Y. (1992) A comparative study of ultrastructures of the interfaces between 4 kinds of surface-active ceramic and bone. *J. Biomed. Mater. Res.*, **26**, 1419–1432.

Oonishi, H., Hench, L. L., Wilson, J., Sugihara, F., Tsuji, E., Kushitani, S., Iwaki, H. (1999) Comparative bone growth behavior in granules of bioceramic materials of various sizes. *J. Biomed. Mater. Res.*, **44**, 31–43.

Oonishi, H., Hench, L. L., Wilson, J., Sugihara, F., Tsuji, E., Matsuura, M., Kin, S., Yamamoto, T., Mizokawa, S. (2000) Quantitative comparison of bone growth behavior in granules of Bioglass (R), A-W glass-ceramic, and hydroxyapatite. *J. Biomed. Mater. Res.*, **51**, 37–46.

Pantano, C. G., Clark, A. E.,Hench, L. L. (1974) Multilayer Corrosion Films on Bioglass Surfaces. *J. Am. Ceram. Soc.*, **57**, 412–413.

Passuti, N., Daculsi, G., Rogez, J. M., Martin, S., Bainvel, J. V. (1989) Macroporous calcium-phosphate ceramic performance in human spine fusion. *Clin. Orthop. Rel. Res.*, **248**, 169–176.

Pereira, M. M., Clark, A. E., Hench, L. L. (1994) Calcium-phosphate formation on sol–gel-derived bioactive glasses *in-vitro*. *J. Biomed. Mater. Res.*, **28**, 693–698.

Pereira, M. M., Jones, J. R., Orefice, R. L., Hench, L. L. (2005) Preparation of bioactive glass-polyvinyl alcohol hybrid foams by the sol–gel method. *J. Mater. Sci. Mater. Med.*, **16**, 1045–1050.

Porter, A. E., Patel, N., Skepper, J. N., Best, S. M., Bonfield, W. (2004) Effect of sintered silicate-substituted hydroxyapatite on remodelling processes at the bone-implant interface. *Biomaterials*, **25**, 3303–3314.

Posner, A. S. (1969) Crystal chemistry of bone mineral. *Phys. Rev.*, **49**, 760–792.

Radin, S., Ducheyne, P., Falaize, S., Hammond, A. (2000) *In vitro* transformation of bioactive glass granules into Ca-P shells. *J. Biomed. Mater. Res.*, **49**, 264–272.

Reffitt, D. M., Ogston, N., Jugdaohsingh, R., Cheung, H. F. J., Evans, B. A. J., Thompson, R. P. H., Powell, J. J., Hampson, G. N. (2003) Orthosilicic acid stimulates collagen type 1 synthesis and osteoblastic differentiation in human osteoblast-like cells *in vitro*. *Bone*, **32**, 127–135.

Roy, D. M., Linnehan, S. K. (1974) Hydroxyapatite formed from coral skeletal carbonate by hydrothermal exchange. *Nature*, **247**, 220–222.

Saravanapavan, P., Hench, L. L. (2003) Mesoporous calcium silicate glasses. I. Synthesis. *J. Non-Cryst. Solids*, **318**, 1–13.

Saravanapavan, P., Jones, J. R., Pryce, R. S., Hench, L. L. (2003) Bioactivity of gel–glass powders in the CaO-SiO_2 system: a comparison with ternary (CaO-P_2O_5-SiO_2) and quaternary glasses (SiO_2-CaO-P_2O_5-Na_2O). *J. Biomed. Mater. Res. A*, **66A**, 110–119.

Sepulveda, P., Binner, J. G. P., Rogero, S. O., Higa, O. Z., Bressiani, J. C. (2000) Production of porous hydroxyapatite by the gel-casting of foams and cytotoxic evaluation. *J. Biomed. Mater. Res.*, **50**, 27–34.

Sepulveda, P., Jones, J. R., Hench, L. L. (2001) Characterization of melt-derived 45S5 and sol-gel-derived 58S bioactive glasses. *J. Biomed. Mater. Res.*, **58**, 734–740.

Sepulveda, P., Jones, J. R., Hench, L. L. (2002a) Bioactive sol–gel foams for tissue repair. *J. Biomed. Mater. Res.*, **59**, 340–348.

Sepulveda, P., Jones, J. R., Hench, L. L. (2002b) *In vitro* dissolution of melt-derived 45S5 and sol–gel derived 58S bioactive glasses. *J. Biomed. Mater. Res.*, **61**, 301–311.

Tisdel, C. L., Goldberg, V. M., Parr, J. A., Bensusan, J. S., Staikoff, L. S., Stevenson, S. (1994) The influence of a hydroxyapatite and tricalcium-phosphate coating on bone-growth into titanium fiber-metal implants. *J. Bone Joint Surg.*, **76A**, 159–171.

Valentini, P., Abensur, D., Wenz, B., Peetz, M., Schenk, R. (2000) Sinus grafting with porous bone mineral (Bio-Oss) for implant placement: a 5-year study on 15 patients. *Int. J. Periodon. Rest. Dent.*, **20**, 245–254.

Vallet-Regi, M., Salinas, A. J., Arcos, D. (2006) From the bioactive glasses to the star gels. *J. Mater. Sci. Mater. Med.*, **17**, 1011–1017.

Warren, L. D., Clark, A. E., Hench, L. L. (1989) An investigation of Bioglass powders – quality assurance test procedure and test criteria. *J. Biomed. Mater. Res.*, **23**, 201–209.

Weinstein, A. M., Klawitter, J. J., Cook, S. D. (1980) Implant-bone interface characteristics of Bioglass dental implants. *J. Biomed. Mater. Res.*, **14**, 23–29.

Wilson, J., Low, S. B. (1992) Bioactive ceramics for periodontal treatment – comparative studies in the patus monkey. *J. Appl. Biomater.*, **3**, 123–129.

Xynos, I. D., Edgar, A. J., Buttery, L. D. K., Hench, L. L., Polak, J. M. (2001) Gene-expression profiling of human osteoblasts following treatment with the ionic products of Bioglass (R) 45S5 dissolution. *J. Biomed. Mater. Res.*, **55**, 151–157.

Zhang, C., Wang, J. X., Feng, H. H., Lu, B., Song, Z. M., Zhang, X. D. (2001) Replacement of segmental bone defects using porous bioceramic cylinders: a biomechanical and X-ray diffraction study. *J. Biomed. Mater. Res.*, **54**, 407–411.

4
Biodegradable and bioactive polymer/ceramic composite scaffolds

S K MISRA and A R BOCCACCINI, Imperial College London, UK

4.1 Introduction

Bone tissue regeneration is one of the areas within tissue engineering that has gained considerable attention by the research community (Vacanti and Langer, 1999; Hench and Polak, 2002; Rose and Oreffo, 2002; Gomes and Reis, 2004). Critical size bone defects due to trauma or disease are very difficult to repair via the natural growth of host tissue. Therefore, there exists a need to fill these defects with a bridging (usually porous) material (termed scaffold), which should also, in combination with relevant cells and signalling molecules, promote the regeneration of new bone tissue. The biomaterials of choice for the development of bone tissue engineering scaffolds are those exhibiting bioactive properties (Hench and Polak, 2002). Bioactive materials react with physiological fluids and form tenacious bonds to bone through the biological interaction of collagen fibres with the material surface and thus they can transfer loads to and from living bone. Prominent bioactive materials are inorganic compounds such as bioceramics, including selected compositions of silicate glasses and glass-ceramics, as well as hydroxyapatite (HA) and related amorphous or crystalline calcium phosphates (Hench, 1998). Like most ceramic materials, the major disadvantage of bioactive ceramics is their low fracture toughness (i.e. brittleness). They are thus often used combined with biopolymers, both stable polymers, e.g. poly(methyl methacrylate) (PMMA), polyethylene (PE) (Bonfield *et al.*, 1981; Ramakrishna *et al.*, 2001), and, specially for tissue engineering applications, biodegradable polymers, e.g. aliphatic polyesters (Rezwan *et al.*, 2006). This chapter will discuss specifically the materials science and technology of composites based on the combination of biodegradable polymers and bioactive inorganic particles, added as filler or coating to the polymer matrix (Fig. 4.1), for development of bone tissue engineering scaffolds. Since the requirements for optimal scaffolds are manifold (Fig. 4.2) (Hutmacher, 2001), it is often considered that combination of degradable polymers and inorganic bioactive particles represents the best approach in terms of achievable mechanical and biological performance (Roether *et al.*, 2002; Lu *et al.*, 2003;

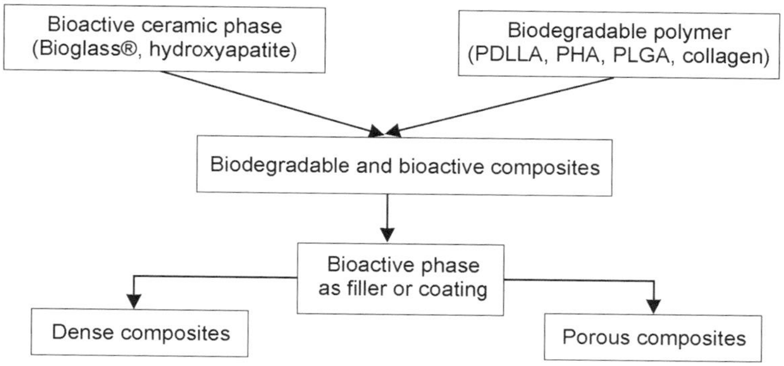

4.1 Schematic diagram showing the different combinations of biodegradable polymers and bioactive inorganic particles to form composite materials. PDLLA = poly(D,L-lactic acid), PHA = polyhydroxyalkanoate, PLGA = poly(lactic-co-glycolide).

Rezwan *et al.*, 2006). Not only the combination of the 'right' biomaterials but also the structure and morphology of the scaffold, characterised by a highly interconnected, three-dimensional (3D) pore network as well as tailored surface characteristics, determine the suitability of the scaffold for a given application. For bone tissue engineering, porosity of 90% and pore size >100 μm are desirable, as well as high pore interconnectivity, in order to facilitate the attachment and proliferation of cells and the ingrowth of new tissue into the scaffold, as well as to enable mass transport of oxygen, nutrition and waste products (Hutmacher, 2001; Ma, 2004).

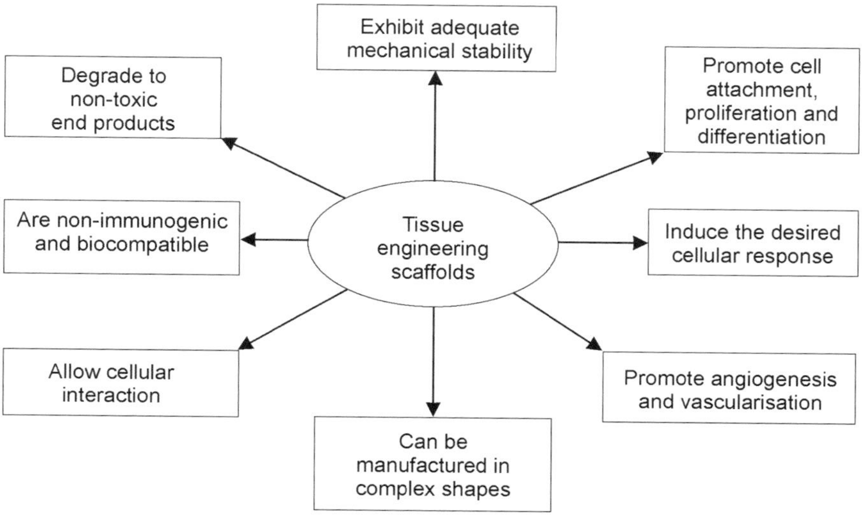

4.2 Requirements for bone tissue engineering scaffolds (Hutmacher, 2001).

The chapter will thus focus on the composite material strategy for developing scaffolds, including aspects of materials selection (Sections 4.2 and 4.3) and processing technologies for composite scaffold fabrication (Section 4.4) intended for bone tissue engineering. In Section 4.5 selected composite systems are described in detail with emphasis on their structure, properties and applications in bone tissue engineering. Finally Section 4.6 presents a summary and the scope for future developments in the field. This chapter complements Chapters 14 and 15 which focus on bone regeneration and biomineralisation strategies.

4.2 Biodegradable polymers and bioactive ceramics

4.2.1 Biodegradable polymers

In conventional tissue engineering strategies, the scaffold must provide a suitable environment for inducing tissue regeneration at the defect site, acting as a transient substrate, mimicking the extracellular matrix (ECM) to temporarily support cell attachment and subsequent proliferation and differentiation. Cells are then expected to infiltrate the scaffold pore structure and to proliferate and differentiate therein, provided the scaffold, acting effectively as the ECM, is biologically compatible with the cells. Once new tissue has been regenerated, the scaffold should have degraded completely. Remaining scaffold material (debris) could cause a physical hindrance against effective tissue regeneration, hence biopolymers used for tissue engineering scaffolds must be biodegradable. While the polymer matrix is degrading, the physical support provided by the 3D scaffold must be maintained until the engineered tissue has sufficient mechanical integrity to support itself. It is thus mandatory for successful tissue regeneration to control the degradation time profile of the scaffold at the defect site. The above-mentioned strategy works well in cases where the tissue around the defect has inherent potential for regeneration. However, for the contrasting situation where tissue lacks regenerative ability, tissue regeneration cannot always be expected to occur only by supplying the scaffold. In such cases the scaffold is used in combination with relevant cells, usually also adding growth factors, which have the potential to accelerate tissue regeneration. In these cases combinations of scaffolds and drug delivery systems can be developed to create such an environment, which will have the additional function of minimising microbial infections and maximising tissue regeneration (Pitt, 1990; Gittens and Uludag, 2001).

Several biodegradable polymers (natural and synthetic) are being considered for tissue engineering scaffolds, as briefly discussed next. More detailed treatments of the synthesis and properties of degradable polymers for biomedical applications are available in the specialised literature (Jagur-Grodzinski, 2007; Pachence and Kohn, 2000).

Natural polymers

Natural polymers can be classified as proteins (silk, collagen, gelatin, fibrinogen, elastin, keratin, actin and myosin), polysaccharides (cellulose, amylose, dextran, chitin and glycosaminoglycans) or polynucleotides (DNA, RNA) (Yannas, 2004). Natural polymers, because of their similar macromolecular structure to tissues, offer the convenience of recognition from the biological system. This further leads to the avoidance of issues related to toxicity and stimulation of a chronic inflammatory reaction, as well as lack of recognition by cells, which are frequently provoked by many synthetic polymers. Natural polymers are known to degrade by the effect of naturally occurring enzymes. It is further possible to control the degradation rate of the implanted polymer by chemical crosslinking or other chemical modifications. On the other hand, natural polymers are frequently known to trigger immunogenic responses and have inadequate mechanical properties for load-bearing applications (Yannas, 2004). Also because of natural polymers being structurally more complex than synthetic polymers, it is often complicated to identify the most suitable manufacturing techniques (compression moulding, extrusion) to be used on them (Brown *et al.*, 2005). The natural variability in the structure of the macromolecular substances further complicates the problem, as each of these polymers appears as a chemically distinct entity from one species (species specificity) to another and from one tissue to the following (tissue specificity) (Yannas, 2004). Nevertheless, owing to their proven biocompatibility, natural polymers are being used for various biomedical applications such as cardiovascular applications, peripheral nerve regeneration, surgical sutures and drug delivery systems (Dang and Leong, 2006). In particular for bone tissue engineering applications, collagen gel is being increasingly considered (Brown *et al.*, 2005; Kim, H.W. *et al.*, 2006; Sachlos *et al.*, 2006).

Synthetic polymers

Synthetic polymers represent the largest group of biodegradable polymers and they can be produced under controlled conditions. They exhibit, in general, predictable and reproducible mechanical and physical properties such as tensile strength, elastic modulus and degradation rate (Pachence and Kohn, 2000; Gunatillake *et al.*, 2006). Possible risks such as toxicity, immunogenicity and favouring of infections are lower for pure synthetic polymers with constituent monomeric units having a well-known and simple structure. Moreover synthetic polymers provide the freedom to tailor the properties for specific applications. Some synthetic polymers are hydrolytically unstable and degrade in the body while others may remain essentially unchanged for the lifetime of the patient. Biodegradable polymers have been used to repair nerves, skin, vascular system and bone and a large proportion of the currently investigated synthetic degradable polymers for tissue engineering scaffolds are polyesters, as shown in

Table 4.1 Classification of aliphatic polyesters into two groups with regard to the mode of bonding of constituent monomers, i.e. poly(hydroxyl acid)s and poly(alkylene dicarboxylate)s

Chemical structure	Polyester	Examples
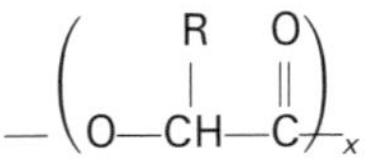	Poly(α-hydroxy acid)	Poly(glycolic acid) Poly(L-lactic acid)
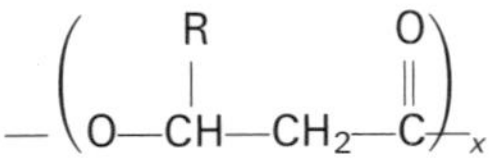	Poly(β-hydroxyalkanoate)	Poly(β-hydroxybutyrate) Poly(β-hydroxybutyrate-co-β-hydroxyvalerate)
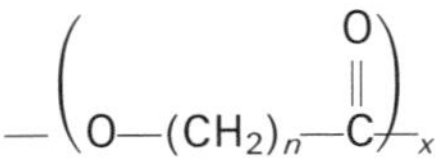	Poly(ω-hydroxybutyrate)	Poly(β-propiolactone) Poly(ϵ-caprolactone)
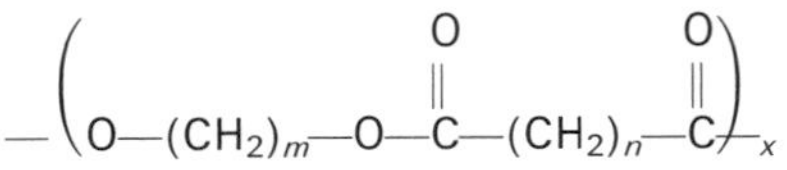	Poly(alkylene dicarboxylate)	Poly(ethylene succinate) Poly(butylene succinate) Poly(butylene succinate-co-butylene adipate)

Table 4.1. A wide range of physical properties and degradation times can be achieved by varying the monomer ratios when using copolymers such as lactide/glycolide copolymers, or hydroxybutyrate–hydroxyhexanoate copolymers. As mentioned above, since tissue engineering scaffolds do not have to be removed surgically once they are no longer needed, degradable polymers are of value in short-term applications and can circumvent some of the problems related to the long-term safety of permanently implanted devices. However, because scaffold degradation products are released in the body, considerable attention is needed towards designing the composition of the scaffold (composite strategy, see below) and the testing for potential toxicity or inflammatory reactions as consequence of polymer degradation is of paramount importance.

Poly(lactic acid) (PLA), poly(glycolic acid) (PGA) and poly(lactic-co-glycolide) (PLGA) copolymers (Table 4.1) are among the most commonly used synthetic polymers in tissue engineering (Pachence and Kohn, 2000; Ma, 2004). PLA exists in three forms: L-PLA (PLLA), D-PLA (PDLA) and racemic mixture of D,L-PLA (PDLLA). The chemistry of these polymers allows hydrolytic degradation through de-esterification. The human body already contains highly regulated mechanisms for completely removing monomeric components of lactic and glycolic acids. Owing to these properties PLA and PGA have been used in numerous biomedical products and devices, such as degradable sutures and fixation plates, which have approval by the US Food and Drug Administration (FDA). Moreover, PLA and PGA can be processed easily and their degradation rates, physical and mechanical properties are adjustable over a

Table 4.2 Physical properties of synthetic, biocompatible and biodegradable polymers used as scaffold materials

	P3HB	PGA	PDLLA	PLLA	PLGA	PCL
Melting temperature (°C)	170–175	225–230	nd	173–178	nd	58
Glass transition temperature (°C)	−4–10	35–40	55–60	60–65	45–55	−72
Young's modulus (GPa)	1.1–3.5	7–10	1.9–2.4	1.2–3.0	1.4–2.8	0.4
Elongation (%)	2–6	15–20	3–10	5–10	3–10	>70
Water contact angle (°)	70–80		60–70	70–80		66
Crystallinity (%)	55–80	55–56	0	37	0	
Degradation period (months)	>18	6–12	12–16	>24	1–12	>24

P3HB, poly(3-hydroxybutyrate); PGA, poly(glycolic acid); PDLLA, poly(DL-lactic acid); PLLA, poly(L-lactic acid); PLGA, poly(lactic-co-glycolide); PCL, poly(ε-caprolactone).

wide range by using various molecular weights and copolymers (Mano *et al.*, 2004). However, these polymers undergo a bulk erosion process and they can cause scaffolds to fail prematurely. In addition, abrupt release of acidic degradation products during degradation can cause a strong inflammatory response. In general, PGA degrades faster than PLA.

Polyhydroxyalkanoates (PHA) belong to a class of microbial polyesters and are being increasingly considered for applications in tissue engineering (Chen and Wang, 2002; Valappil *et al.*, 2006). Members of the PHA family can exist as homopolymers of hydroxyalkanoic acids, as well as copolymers of two or more hydroxyalkanoic acids. The chemically different structure of these polymers allows the development of polymers with different properties that greatly affect the mode/rate of degradation in biological media as well as their mechanical properties. Applications of PHA in medicine are being expanded to include wound management, vascular system devices, orthopaedics and drug delivery systems (Freier *et al.*, 2002; Valappil *et al.*, 2006). However, a major obstacle in using polymers from the PHA family is their availability, as only two of the PHA polymers are commercially available; i.e. poly(3-hydroxybutyrate) and poly(3-hydroxybutyrate-co-hydroxyvalerate).

Some of the common properties of the synthetic polymers most widely used in tissue engineering are listed in Table 4.2. For applications in hard tissue repair and load-bearing sites these polymers on their own are mechanically inadequate, this being one of the reasons for their combination with stiff inorganic phases in composites, as discussed in Section 4.3.

4.2.2 Bioactive ceramics

Bioactive ceramics, such as hydroxyapatite (HA), tricalcium phosphate (TCP) and certain compositions of silicate and phosphate glasses (bioactive glasses) and glass-ceramics (such as apatite–wollastonite) react with physiological fluids and through cellular activity form tenacious bonds to hard (and in some cases soft) tissue (Hench, 1998). As discussed in detail in Chapter 3, a characteristic of many bioactive materials is the formation of a biologically active HA surface layer in the presence of body fluids *in vitro* or *in vivo*. The materials with the highest levels of bioactivity develop a silica gel layer that promotes HA formation. Thus HA formation on material surfaces upon immersion in acellular simulated body fluid (SBF) is considered a (qualitative) measure of the material bioactivity. Bioactive glasses (e.g. 45S5 Bioglass®) with compositions in the system SiO_2–Na_2O–CaO–P_2O_5, having <55% SiO_2, exhibit high bioactivity index (Class A), and bond to both soft and hard connective tissues (Hench, 1998). The class of bioactivity depends on the rate and type of tissue response. Class A bioactive materials are osteogenetic and osteoconductive while Class B bioactive materials such as HA exhibit only osteoconductivity (Hench, 1998). It has been also found that reactions on bioactive glass surfaces release critical concentrations of soluble Si, Ca, P and Na ions, which induce intracellular and extracellular responses (Xynos *et al.*, 2001). For example, a synchronised sequence of genes is activated in osteoblasts that undergo cell division and synthesise an extracellular matrix, which mineralises to become bone (Xynos *et al.*, 2001). In addition, bioactive glass compositions doped with AgO_2 have been shown to elicit antibactericidal properties while maintaining their bioactive function (Bellantone *et al.*, 2002). More recently, 45S5 Bioglass® has also been shown to increase secretion of vascular endothelial growth factor (VEGF) *in vitro* and to enhance vascularisation *in vivo*, suggesting scaffolds containing controlled concentrations of Bioglass® might stimulate neo-vascularisation which is beneficial to large tissue engineered constructs (Day *et al.*, 2004).

The excellent properties of bioactive glasses and their long history of applications in biomedical implants (Hench, 1998) has recently prompted extensive research regarding their use in bone engineering and regeneration strategies, mainly in the form of powder aggregates, porous substrates (foam scaffolds) (Chen and Boccaccini, 2006) and also as a particulate addition (as filler or coating) to biodegradable polymers (Roether *et al.*, 2002; Maquet *et al.* 2003; Lu *et al.*, 2003), which is the focus of the present chapter, as discussed next.

4.3 Composite material approach

A composite material consists of two or more chemically distinct phases (metallic, ceramic or polymeric) which are separated by an interface. The classification of engineering composite materials is based on the matrix materials (metals, ceramics or polymers) or on the reinforcement dimensions and morphology (particulates, short fibres, continuous fibres, nanofillers) (Matthews and Rawlings, 1994). Biodegradable composites for tissue engineering applications must exhibit specific properties such as high initial strength and tailored initial elastic modulus close to the elastic modulus of bone. In addition, they must have controlled strength and modulus retention *in vivo* so that they can provide the necessary support for cell attachment and proliferation as well as augment the tissue capacity to regenerate. Polymers by themselves are generally flexible and exhibit a lack of mechanical strength and stiffness, whereas inorganic materials such as ceramics and glasses are known to be too stiff and brittle. Moreover, polymers can be easily fabricated to form complex shapes and structures, yet, in general, they lack a bioactive function (e.g. strong bonding to bone), being too flexible and weak to meet the mechanical demands in surgery and in the physiological environment. There are thus several reasons for the combination of biodegradable polymers and bioactive ceramics and glasses for tissue engineering applications (Maquet *et al.*, 2003; Lu *et al.*, 2003; Blaker *et al.*, 2005; Rezwan *et al.*, 2006). Firstly, the combination of polymers and inorganic phases leads to composite materials with improved mechanical properties due to the inherent higher stiffness and strength of the inorganic material. Secondly, addition of bioactive phases to bioresorbable polymers can alter the polymer degradation behaviour, by buffering the pH of the nearby solution and hence controlling the fast acidic degradation of the polymer, in particular in case of polylactic acid. Inorganic filler materials have been shown to influence the degradation mechanism of polymers by preventing the autocatalytic effect of the acidic end groups resulting from hydrolysis of the polymer chains. Moreover, incorporation of a bioactive phase in the polymer matrix helps to absorb water due to the internal interfaces formed between the polymer and the more hydrophilic bioactive phases, hence providing a means of controlling the degradation kinetics of the scaffolds (Blaker *et al.*, 2003; Kim *et al.*, 2005). The incorporation of a bioactive inorganic phase such as HA, Bioglass® or TCP has an extra function: it induces a surface topography (nano or micro roughness) and it also allows the composite to interact with the surrounding bone tissue by forming a tenacious bond via the growth of a carbonate HA layer, as mentioned above. Therefore, development of composite materials for tissue engineering is attractive since their properties can be engineered to suit the mechanical and physiological demands of the host tissue by controlling the volume fraction, morphology and arrangement of the reinforcing phase (Rezwan *et al.*, 2006).

Two types of reinforcements are normally used for biomedical composites: fibres and particulates (Ramakrishna *et al.*, 2001). It has been shown that increased volume fraction and higher surface area to volume ratio of inclusions favour bioactivity (Rezwan *et al.*, 2006). Therefore for certain applications incorporation of fibres is preferred instead of particles (Kim *et al.*, 2005). In addition, the mechanical properties are influenced by the reinforcement shape and size as well as by the distribution of the reinforcement in the matrix and the reinforcement–matrix interfacial bonding.

The major factor affecting the mechanical properties and structural integrity of scaffolds, however, is their porosity, e.g. pore volume, size, shape, orientation and connectivity, hence it is very important to consider the suitability of fabrication technologies developed for production of 3D scaffolds with controlled porosity, as discussed next.

4.4 Materials processing strategies for composite scaffolds

Tissue engineering scaffolds must mimic the numerous functions of the natural extracellular matrix, including providing support for cell adhesion and migration, and organising cells into 3D structures. Several fundamental scaffold design requirements have been identified and they are summarised in Fig. 4.2 (Murphy and Mooney, 1999; Hutmacher, 2001; Boccaccini *et al.*, 2002).

Porosity plays an important role in determining the characteristics of a scaffold. Average pore size, pore size distribution, pore volume, pore interconnectivity, pore shape, pore throat size and pore wall roughness are important parameters to consider while designing a scaffold. A minimum pore size is required for tissue ingrowth and high 3D interconnectivity is necessary for access of nutrients, transport of waste products, better cell spreading and vascularisation (Yang *et al.*, 2001; Maquet *et al.*, 2003). It is thus very important to quantify and optimise the porosity and to understand the relationship between scaffold properties (mechanical properties, permeability) and pore structure.

It is recognised that pore structure and properties of the scaffolds are dictated by the choice of the manufacturing process. The range of scaffold manufacturing techniques developed in recent years is immense, covering distinct methods such as the use of porogens, chemical segregation and rapid prototyping. Techniques such as solvent casting, particulate leaching, 3D printing, thermally induced phase separation (TIPS) and fused deposition modelling are among the most used for fabricating 3D structures with variable porosity. Though each of these techniques has the ability to produce scaffolds with different architecture, they also have limitations with respect to various properties achievable as shown in Table 4.3 (Hutmacher, 2001; Liu and Ma, 2004; Rezwan *et al.*, 2006). Four of the most advanced techniques developed for composite scaffold production are discussed next.

4.4.1 Solvent casting and particulate leaching

Solvent casting in combination with particulate leaching is one of the simplest and more common methods used for scaffold preparation. Solvent casting involves the dissolution of the polymer in an organic solvent, mixing with ceramic granules, and casting the solution into a predefined 3D mould. The solvent is subsequently allowed to evaporate. The main advantage of this technique is the ease of manufacturing and ability to incorporate drugs and chemicals within the scaffold. However the method has significant dis-

Table 4.3 Typical composite scaffold fabrication processes and their advantages and disadvantages

Processing	Advantage	Disadvantages
Thermally induced phase separation (TIPS)	High porosities Control of pore structure and pore sizes Interconnected pores	Shrinkage issues Long time to sublime solvents Small-scale production
Solvent casting and particulate leaching	Controlled porosity Faster and relatively inexpensive approach	Pore interconnectivity Limited to thin membranes Solvent residue
Microsphere sintering	Controllable and graded porosity Fabricated in complex shapes	Pore interconnectivity Mechanical properties
Scaffold coating	Rapid and simple approach Applied to all types of composites	Lack of interfacial strength Clogging of pores
Fibrous composites	Superior compressive strength Independent control of porosity and pore sizes	Solvent residue Oriented pore structures

advantages. Firstly, only simple shapes, e.g. flat sheets and tubes, can be formed. Secondly, pore interconnectivity is very low and usually unsuitable for tissue engineering applications. Finally, there are chances of residual solvent to remain trapped which would reduce the activity of bioinductive molecules (e.g. protein), if incorporated.

4.4.2 Microspheres sintering

In this process, ceramic/polymer composite microspheres are synthesised first, using an emulsion/solvent evaporation technique. Sintering the composite microspheres yields a 3D porous scaffold. Lu *et al.* (2003) have worked on this technique using PLGA and Bioglass® as the starting materials. The composite spheres were prepared using a water–oil–water emulsion method. The microspheres obtained were then sintered and a well-integrated microstructure was obtained, with a porosity of 40%. The mechanical properties of these composites were found to be similar to those of cancellous bone.

4.4.3 Thermally induced phase separation (TIPS)

The TIPS method can produce homogeneous and highly porous (~95%) scaffolds with highly anisotropic tubular morphology and extensive pore

(a) (b)

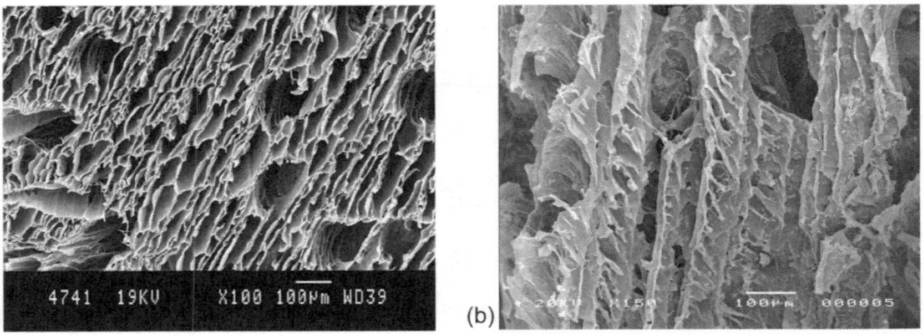

4.3 SEM micrographs showing the microstructure of PDLLA/Bioglass®-filled composite foams (5 wt% Bioglass®): (a) orthogonal to the pore direction and (b) parallel to the pore direction (Blaker *et al.*, 2005).

interconnectivity (Blaker *et al.*, 2003; Maquet *et al.*, 2003). This technique allows controlling the macro- and microstructures of the scaffolds. The morphology of the porous membrane varies depending on the polymer, solvent, concentration of the polymer solution and phase separation temperature. Pore morphology, in turn, affects the mechanical properties of the scaffold. The controlled phase separation of the polymer solution can be induced in several ways, including non-solvent-induced phase separation, chemically induced phase separation and thermally induced phase separation. The membranes obtained from this process usually exhibit oriented tubular pores of diameters of several hundred micrometres (>100 μm) and an isotropic pore network of smaller pore size (~10 μm) connecting the large tubular pores, as Fig. 4.3 shows (Blaker *et al.*, 2005). Foams are prepared by rapidly cooling the polymer solution to a lower temperature, in order to solidify the solvent and induce solid–liquid phase separation. The samples are then freeze dried for a longer duration to ensure complete removal of solvent.

4.4.4 Solid freeform fabrication techniques (SFFT)

SFFT, such as fused deposition modelling, have been employed to fabricate highly reproducible scaffolds with fully interconnected porous networks (Hutmacher *et al.*, 2001; Ramakrishna *et al.*, 2001; Yang *et al.*, 2002). SFFT refers to computer-aided manufacture (CAD/CAM) methodologies such as stereolithography, selective laser sintering, ballistic particle manufacturing and 3D printing. SFFT offer the possibility to fabricate polymer scaffolds with well-defined architecture because local composition, macrostructure and microstructure can be specified and controlled at high resolution in the interior of the components. This technique has been applied for composites containing calcium phosphates as the bioactive phase. For example, Xiong *et al.* (2002) fabricated PLLA/TCP composites with porosities of up to 90% and mechanical properties close to human cortical bone by using low-temperature deposition based on a layer-by-layer manufacturing method. Taboas *et al.* (2003) produced PLA scaffolds with computationally designed pores (500–800 μm wide channels) and solvent-derived local pores (50–100 μm). However, a shortcoming of this route is increased scaffold fabrication time and complex equipment requirement compared with direct methods.

4.5 Case studies

There have been numerous combinations of polymers and bioactive ceramics to achieve a range of properties in 3D porous scaffolds for bone tissue engineering, as reviewed in the recent literature (Rezwan *et al.*, 2006). Some representative examples will be highlighted in this section to provide an understanding of the 'composite strategy' used and to show how the addition of inorganic phases can affect the properties of the composite.

4.5.1 PDLLA/Bioglass® scaffolds

Blaker *et al.* (2003) and Boccaccini and Blaker (2005) have worked extensively on the PDLLA/Bioglass® composite system. The TIPS method was used to prepare PDLLA foams with different concentrations of Bioglass® particles (<40 wt%) as filler. Porosity >90% was obtained and the density of the composite foams was found to increase on addition of Bioglass®. Figure 4.3 shows the typical microstructure of a PDLLA/Bioglass® foam (containing 5 wt% Bioglass®). Compressive strength analysis carried out on the samples showed mechanical anisotropy in the axial and transverse directions. The transverse mechanical response to compression displayed a different response from that in the axial direction (Blaker *et al.*, 2005). There appeared to be no obvious micro-failure response due to buckling of the walls of tubular macropores, indeed the behaviour of all foams was dominated by the densification of the foams under uniaxial loads. The axial modulus and transverse modulus increased due to the addition of Bioglass® particles, as expected. Owing to the potential advantages the system offers, PDLLA/Bioglass® composites have been further investigated by other authors (Zhang *et al.*, 2004; Helen *et al.*, 2006, 2007; Yang *et al.*, 2006; Tsigkou *et al.*, 2007) in terms of its *in vitro* and *in vivo* response. For example, the adhesion, growth and differentiation of human bone marrow mesenchymal stem cells on composites made of PDLLA foams and Bioglass® particles was investigated *in vitro* and *in vivo* and the potential for *in vivo* bone formation on the composites was analysed in immunocompromised animals (Yang *et al.*, 2006). Moreover PDLLA/Bioglass® films were demonstrated to enhance bone nodule formation and displayed enhanced alkaline phosphatase activity of primary human fetal osteoblasts in the absence of osteogenic supplements (Tsigkou *et al.*, 2007). The attachment and spreading of osteoblast cells onto PDLLA/Bioglass® 3D

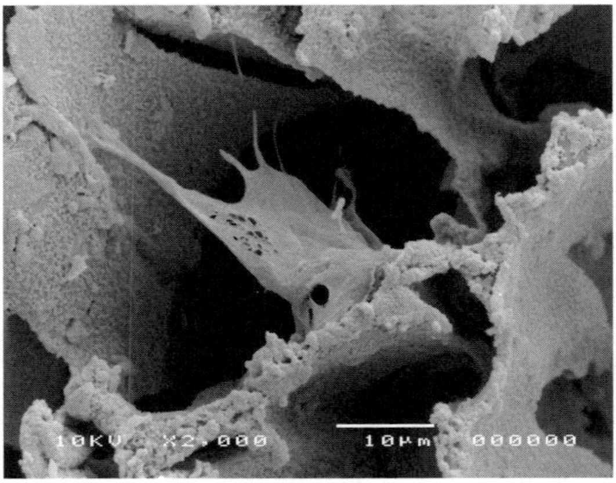

4.4 SEM micrograph showing the well-spread and flattened morphology of an MG-63 cell adhering to PDLLA/40 wt% Bioglass® foam after 24 h in culture. (Reproduced from: Blaker, J. J. *et al.*, *J. Biomed. Mater. Res.* 2003 **67A**, 1401–1411.)

composite foams was investigated by Blaker *et al.* (2003) and a typical micrograph showing the well-spread and flattened morphology of an MG-63 cell adhering to PDLLA/40 wt% Bioglass® foam after 24 h in culture is presented in Fig. 4.4. The results achieved so far have demonstrated that the regulatory role on cell differentiation and mineralisation of the Bioglass® containing PDLLA composites is likely to be a combination of both the cell–scaffold interaction (including topographic contributions) and the ionic release of Bioglass® dissolution products discussed above (Xynos *et al.*, 2001). In more recent developments, Helen *et al.* (2007) have shown that composite PDLLA/Bioglass® films are an appropriate substrate for the culture of annulus fibrous cells *in vitro* and have proposed the composite as a suitable material for intervertebral disc tissue repair.

4.5.2 PLGA/hydroxyapatite scaffolds

As mentioned in Section 4.2, PLGA is a commonly used polymer for preparing tissue engineering scaffolds. Kim, S.S. *et al.* (2006) have used a combination of gas forming/particulate leaching method to prepare PLGA/HA (75:25) scaffolds. This method was shown to efficiently expose the bioactive ceramic particles compared with a normal particulate leaching method. Pores with diameters in the range 100–200 μm along with porosity of 90% were achieved with this technique. Analysis also showed that the gas forming/particulate leaching technique leads to scaffolds with better mechanical properties than those obtained by the solvent casting/particulate leaching method. The average compressive modulus for the scaffolds was 4.5 MPa and the tensile modulus 26.9 MPa. These scaffolds showed enhanced osteogenic potential resulting from the exposure of HA particles on the scaffold surface. Another novel technique used to prepare such composite system is colloidal non-aqueous chemical precipitation technique, as demonstrated recently (Petricca *et al.*, 2006). The ultimate tensile strength and Young's modulus of the PLGA/HA composite system was found to decrease on increasing the HA content from 10 to 30 wt%, most probably due to uncontrolled agglomeration of HA particles. On the other hand, the level of cell attachment was higher for the composites containing 30 wt% HA particles, confirming the composite bioactivity. In order to broaden the application of tissue engineering scaffolds, drug delivery is usually incorporated as a vital function for tissue regeneration. Various drugs and growth factors can be used. One such example is the use of an antibiotic (gentamicin) within the PLGA composite system (Schnieders *et al.*, 2006).

4.5.3 Collagen/hydroxyapatite scaffolds

Collagen proteins represent the main organic part of bone. This molecule spontaneously forms triple helices fibrils of great tensile strength and its 3D structure provides a template for hydroxyapatite nanocrystals to grow (Sachlos *et al.*, 2006). Itoh *et al.* (2002) synthesised collagen/HA implants using a combination of precipitation, uni-axial compression and dehydration methods. The strength of the composites (collagen/HA = 20/80) was 39.5 MPa, with a Young's modulus of 2.5 GPa. The increase in the mechanical properties of the composite was shown to be due to the incorporation of stiff HA particles. Collagen/HA microspheres have also been prepared using water-in-oil emulsion method as a grafting material for clinical applications (Wu *et al.*, 2004). The bioactivity of such composite systems has been confirmed using Fourier transform infrared (FTIR) and Raman spectroscopies as well as X-ray diffraction (XRD). Various biological analyses on collagen/HA systems have shown that these composites are compatible with osteoblast cells. In a comparative study between collagen gel matrix and collagen/HA microspheres, it was shown that the growth of osteoblast cells (e.g. increase in cell numbers) was higher for the composite spheres than for pure collagen (Hsu *et al.*, 1999). Kikuchi *et al.* (2001) have also demonstrated the possibility of using HA/collagen nanocomposite as a biodegradable artificial bone. Attempts have been recently directed towards using magnetic fields to create composites with unidirectionally oriented HA crystals within the collagen matrix (Wu *et al.*, 2007). Moreover it has been shown that incorporating HA in the collagen matrix reduces the piezoelectricity of the collagen/HA system compared with collagen on its own (Silva *et al.* 2001). Although the combination of collagen and HA closely mimics the composition of natural bone and it could be considered the ideal composite to fabricate scaffolds, preparing such composites in controlled 3D porous structures with given pore size and orientation is extremely difficult. The method presented recently by

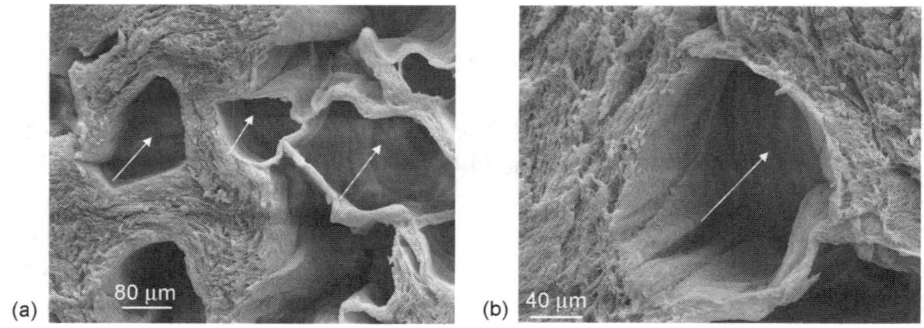

4.5 Channel formation via the degradation of phosphate glass fibres in fibre-collagen scaffolds: (a) SEM image of the cross-section through a collagen scaffold showing a cluster of channels formed as the fibres degraded and (b) higher magnification SEM image of a channel. (Reprinted with permission from: Nazhat *et al.*, *Biomacromolecules* 2007 **8** 543–551. Copyright (2007) American Chemical Society.)

Brown *et al.* (2005) based on a novel plastic compression technique of hyperhydrated collagen gel could be a suitable way towards the development of strong collagen structures with controllable nano- and microscale features. More recently, Nazhat *et al.* (2007) developed a technique to introduce controlled microchannels in dense collagen scaffolds by incorporating soluble phosphate glass fibres. Figure 4.5 shows the channel formation obtained by the degradation of a phosphate glass fibre in a compacted collagen gel-based scaffold. It was demonstrated (Nazhat *et al.*, 2007) that the process is scalable to produce multilayered dense sheets incorporating fibres of different orientations, chemistry and size as well as particulates and short fibre reinforced matrices.

4.5.4 Poly(hydroxybutyrate-co-hydroxyvalerate)/inorganic phase composites

Poly(hydroxybutyrate-co-hydroxyvalerate) (PHBV) is the most extensively studied polymer from the PHA family for biocomposite development. Experiments have been carried out to form PHBV composites using tricalcium phosphate, HA, wollastonite and sol–gel bioactive glasses as fillers (Misra *et al.*, 2006). Chen and Wang (2002), for example, demonstrated that the addition of HA particles reduced the degradation temperature and crystallinity of PHBV composites. Increasing the concentration of HA particles also increased the microhardness and Young's modulus of PHBV/HA composites. A similar result was recorded by Li and Chang (2004), wherein the compressive strength of the composite increased from 0.16 to 0.28 MPa by increasing the content of wollastonite. However, a study carried out using HA showed that increasing the HA content above 40 wt% resulted in a sharp decrease in the elastic modulus of the composite (Galego *et al.*, 2000). It has also been shown that addition of 10 and 20 wt% of bioactive glass caused an increase in the compressive yield strength of the scaffolds (Li *et al.*, 2005). A detailed *in vitro* analysis using wollastonite and bioactive glass showed that the presence of wollastonite helps in neutralising the acidic by-products and aids in stabilising the pH of the region around the scaffold (Li *et al.*, 2005). Incorporation of wollastonite and bioactive glasses also resulted in a decrease of the water contact angle from 66° to 16° and 65° to 32°, respectively, indicating the possibility of tailoring the hydrophobicity of the composite by addition of the inorganic phase. Moreover, the average molecular weight (M_w) of the polymer was found to decrease more than that of the composite upon immersion in SBF (9 weeks) (Li *et al.*, 2005). From the available studies, it can be concluded that the tailored incorporation of bioceramic particles in PHBV matrices can enhance the applicability of this polymer in tissue engineering by increasing its mechanical competence and bioactivity. The novel composites based on PHBV matrices should be compared in their biological performance (*in vivo*) with those described above, in particular PDLLA/Bioglass® and PLGA/HA systems.

4.6 Conclusions and future trends

Biodegradable, polymer/inorganic bioactive phase composites considered in this chapter are particularly attractive as bone tissue engineering scaffolds due to their bioactivity and adjustable biodegradation kinetics. From the materials science perspective, the present challenge in tissue engineering is to design and fabricate reproducible bioactive and bioresorbable 3D scaffolds of tailored porosity and pore structure, which are able to maintain their structure and integrity for predictable times, even under load-bearing conditions.

Achieving the mechanical properties of bone might also allow the replacement of bigger parts of damaged bone tissue than is possible today. Stronger composite scaffolds might be achievable by increasing the organic/inorganic interfacial bonding by using for example surface functionalised particles. The inclusion of nanoparticles into the biopolymer matrix with the dual objective of improving the mechanical properties as well as of incorporating nanotopographic features that mimic the nanostructure of natural bone is currently an area of intensive research (Liu and Webster, 2007; Torres *et al.*, 2007). The role of the scaffolds is being extended from being a mere mechanical support to include intelligent surfaces capable of providing both chemical and physical signals to guide cell attachment and spreading, possibly influencing also cell differentiation (Berry *et al.*, 2006). In addition, the tailoring of roughness and topography of the pore surfaces is being explored due to the profound effect that surface roughness has on early cell attachment behaviour as well as possibly on subsequent cell adhesion, cytoskeletal organisation and gene expression (Berry *et al.*, 2006).

It is now accepted that the response of host tissue at the protein and cellular level to nanostructured surfaces is different from that observed to conventional

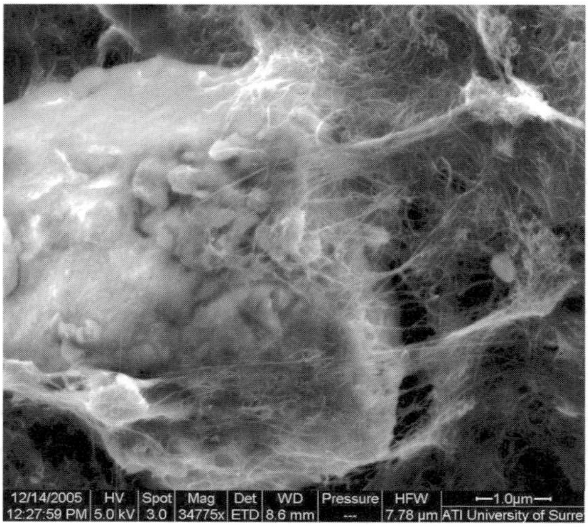

4.6 Scanning electron micrograph showing the fracture surface of a P(3HB) composite material incorporating bioactive glass particles and CNT with potential to be used as scaffold with sensing function (Reproduced from: Misra, S. K., *et al.*, *Nanotechnology* 2007 **18** 075701 (7pp.), with permission of IoP Publishing Limited.)

(μm) surfaces (Liu and Webster, 2007). Scaffold enhancement by incorporation of carbon nanotubes (CNT) is another area attracting research efforts (Misra *et al.*, 2007; Harrison and Atala, 2007). CNT have the potential for providing enhanced structural reinforcement in a polymer matrix at very small concentrations (to counterbalance the fact they are non-degradable). Figure 4.6 shows the microstructure of a P(3HB) composite incorporating both bioactive glass particles and CNT, recently developed by Misra *et al.* (2007). The addition of CNT is expected to add extra functionalities to the scaffold; for example a sensing function exploiting the electrical conductivity of CNT would be possible or the release of bioactive factors by CNT functionalisation (Harrison and Atala, 2007). However, issues related to the cytotoxicity of CNT (and of nanoparticles in general) remain unresolved and they should be investigated in parallel as there are controversial reports and debate in the literature.

The incorporation of biomolecules such as growth factors in the scaffold with the aim of accelerating local bone healing is promising and currently under extensive research as well. Moreover there is significant scope in the application of surface modification, through the use of protein adsorption or plasma treatment, to provide more cues to cell attachment and response, thus making the scaffold more biocompatible.

There is limited understanding regarding the long-term *in vitro* and *in vivo* characterisation of porous 3D composite scaffolds, specifically regarding the long-term effect of the incorporation of inorganic bioactive phases on the degradation and ion release kinetics of these highly porous systems. In this regard, the development of appropriate characterisation techniques coupled with predictive analytical models is mandatory in order to be able to comprehensively assess the degradation of these systems with respect to pore structure, scaffold geometry, fluid flow and the influence of the bioactive additions. Here, the use of X-ray microtomography as a reliable tool for 3D pore structure quantification is likely to gain increased impetus. Finally, in order to target clinical applications, *in vitro* and *in vivo* studies are inevitable and the need for more studies on composite scaffolds in biological systems is imperative. This includes also research directed at assessing the suitability of bioactive composite scaffolds for enhancing the angiogenesis and vascularisation of tissue/scaffold constructs.

4.7 References and further reading

Bellantone M, Williams HD, Hench LL. 'Broad-spectrum bactericidal activity of AgO_2-doped bioactive glass', *Antimicrob. Agents Chemother.* 2002 **46**, 1940–1945.

Berry CC, Dalby MJ, Oreffo ROC, McCloy D, Affrosman S, 'The interaction of human bone marrow cells with nanotopographical features in three dimensional constructs', *J. Biomed. Mater. Res.* 2006 **79A** 431–439.

Blaker JJ, Gough JE, Maquet V, Notingher I., Boccaccini AR, '*In vitro* evaluation of novel bioactive composites based on Bioglass®-filled polylactide foams for bone tissue engineering scaffolds', *J. Biomed. Mater. Res.* 2003 **67A** 1401–1411.

Blaker JJ, Maquet V, Jérôme R, Boccaccini AR, Nazhat SN, 'Mechanical properties of highly porous PDLLA/Bioglass® composite foams as scaffolds for bone tissue engineering', *Acta Biomater.* 2005 **1** 643–652.

Boccaccini AR, Blaker JJ, 'Bioactive composite material for tissue engineering scaffolds', *Expert Rev. Medical Devices* 2005 **2** 303–317.

Boccaccini AR, Roether JA, Hench LL, Maquet V, Jerome R, 'A composite approach to tissue engineering', *Ceramic Eng. Sci. Proc.* 2002 **23**(4) 805–816.

Bonfield W, Grynpas, MD, Tully AE, Bowman J, Abram J, 'Hydroxyapatite reinforced polyethylene. A mechanically compatible material for bone replacement', *Biomaterials* 1981 **2** 185–186.

Brown RA, Wiseman M, Chuo C-B, Cheema U, Nazhat SN, 'Ultrarapid engineering of biomimetic materials and tissues: fabrication of nano- and microstructures by plastic compression', *Adv. Funct. Mater.* 2005 **15** 1762–1770.

Chen LJ, Wang M, 'Production and evaluation of biodegradable composites based on PHB-PHV copolymer', *Biomaterials* 2002 **23** 2631.

Chen QZ, Boccaccini AR, 'Poly(DL-lactic acid) coated 45S5 Bioglass-based scaffolds: Processing and characterisation', *J. Biomed. Mater. Res. A* 2006 **77A** 445–457.

Dang JM, Leong KW, 'Natural polymers for gene delivery and tissue engineering', *Adv. Drug Deliv. Rev.* 2006 **58** 487–499.

Day RM, Boccaccini AR, Shurey S, Roether JA, Forbes A, Hench LL, Gabe SM, 'Assessment of polyglycolic acid mesh and bioactive glass for soft-tissue engineering scaffolds', *Biomaterials* 2004 **25** 5857–5866.

Freier T, Sternberg K, Behrend D, Scmitz KP, 'Health issues of biopolymers: polyhydroxybutyrate' in: *Biopolymers Polyesters III*, Vol. 4, Doi Y, Steinbüchel A (eds), Wiley-VCH, Weinheim (2002).

Galego N, Rozsa C, Sanchez R, Fung J, Vazquez A, Tomas JS, 'Characterisation and application of poly(β-hydroxyalkanoates) family as composite biomaterials', *Polym. Test.* 2000 **19** 485.

Gittens SA, Uludag H, 'Growth factor delivery for bone tissue engineering', *J. Drug Target* 2001 **9** 407–429.

Gomes ME, Reis RL, 'Tissue engineering: some key elements and some trends', *Macromol. Biosci.* 2004 **4** 737–742.

Gunatillake P, Mayadunne R, Adhikari R, 'Recent developments in biodegradable synthetic polymers', *Biotechnol. Ann. Rev.* 2006 **12** 301–347.

Harrison BS, Atala A, 'Carbon nanotube applications for tissue engineering', *Biomaterials* 2007 **28** 344–353.

Helen W, Gough JE, '*In vitro* studies of annulus fibrosus disc cell attachment, differentiation and matrix production on PDLLA/45S5 Bioglass® composite films', *Biomaterials* 2006 **27** 5220–5229.

Helen W, Merry CLR, Blaker JJ, Gough JE, 'Three-dimensional culture of annulus fibrosus cells within PDLLA/Bioglass® composite foam scaffolds: assessment of cell attachment, proliferation and extracellular matrix production', *Biomaterials* 2007 **28**, 2010–2020.

Hench LL, 'Bioceramics', *J. Am. Ceram. Soc.* 1998 **81** 1705–1728.

Hench LL, Polak JM, 'Third-generation biomedical materials', *Science* 2002 **295** 1014–1017.

Hsu FY, Chhueh SC, Wang YJ, 'Microspheres of hydroxyapatite/reconstituted collagen as supports for osteoblast cell growth', *Biomaterials* 1999 **20** 1931–1936.

Hutmacher DW, 'Scaffolds in tissue engineering bone and cartilage', *Biomaterials* 2001 **21** 2529–2543.

Hutmacher DW, Schantz T, Zein I, Ng KW, Teoh SH, Tan KC, 'Mechanical properties an cell cultural response of polycaprolactone scaffolds designed and fabricated via fused deposition modeling', *J. Biomed. Mater. Res.* 2001 **55**(2) 203–216.

Itoh S, Kikuchi M, Koyama Y, Takakuda K, Shinomiya K, Tanaka J, 'Development of an artificial vertebral body using a novel biomaterial, hydroxyapatite/collagen composite', *Biomaterials* 2002 **23** 3919–3926.

Jagur-Grodzinski J, 'Polymers for tissue engineering, medical devices and regenerative medicine. Concise general review of recent studies', *Polym. Adv. Technol.* 2007 **17** 305–418.

Kikuchi M, Itoh S, Ichinose S, Shinomiya K, Tanaka J, 'Self organisation mechanism in a bone like hydroxyapatite/collagen nanocomposites synthesised *in vitro* and its biological reaction *in vivo*', *Biomaterials* 2001 **22** 1705–1711.

Kim H-W, Lee EJ, Jun IK, Kim HE, Knowles JC , 'Degradation and drug release of phosphate glass/polycaprolactone biological composites for hard-tissue regeneration', *J. Biomed. Mater. Res.* 2005 **75B** 34–41.

Kim H-W, Song J-H, Kim H-E, 'Bioactive glass nanofibre–collagen nanocomposite as a novel bone regeneration matrix', *J. Biomed. Mater. Res.* 2006 **79A** 698–705.

Kim SS, Park MS, Jeon O, Choi CY, Kim BS, 'Poly(lactide-co-glycolide)/hydroxyapatite composite scaffolds for bone tissue engineering', *Biomaterials* 2006 **27** 1399–1409.

Li H, Chang J, 'Fabrication and characterisation of bioactive wollastonite/PHBV composite scaffolds', *Biomaterials* 2004 **25** 5473.

Li H, Du R, Chang J, 'Fabrication, characterization, and *in vitro* degradation of composite scaffolds based on PHBV and bioactive glass', *J. Biomater. Appl.* 2005 **20** 137.

Liu H, Webster TJ, 'Nanomedicine for implants: a review of studies and necessary experimental tools', *Biomaterials* 2007 **28** 354–369.

Liu X, Ma PX, 'Polymeric scaffolds for bone tissue engineering', *Ann. Biomed. Eng.* 2004 **32**(3) 477–486.

Lu HH, El Amin SF, Scott KD, Laurencin CT, 'Three dimensional bioactive, biodegradable, polymer-bioactive glass composite scaffolds with improved mechanical properties support collagen synthesis and mineralization of human osteoblasts-like cells *in vitro*', *J. Biomed. Mater. Res. A* 2003 **64A** 465–474.

Ma PX, 'Scaffolds for tissue fabrication', *Materials Today* 2004 (May) 30–40.

Mano JF, Sousa RA, Boesel LF, Neves NM, Reis RL, 'Bioinert, biodegradable and injectable polymeric matrix composites for hard tissue replacement: state of the art and recent developments', *Compos. Sci. Technol.* 2004 **64** 789–817.

Maquet V, Boccaccini AR, Prayata L, Notingher I, Jerome R, 'Preparation and characterisation, and in vitro degradation of bioresorbable and bioactive composites based on Bioglass-filled polylactide foams', *J. Biomed. Mater. Res. A* 2003 **66A**(2) 335–346.

Matthews FL, Rawlings RD, *Composite Materials: Engineering and Science*, Chapman and Hall, London (1994).

Misra SK, Vallappil SP, Roy I, Boccaccini AR, 'Polyhydroxalkanoate/inorganic phase composites for tissue engineering applications', *Biomacromolecules* 2006 **7** 2249.

Misra SK, Watts PCP, Valappil SP, Silva SRP, Roy I and Boccaccini AR, 'Poly(3-hydroxybutyrate)/Bioglass® composite films containing carbon nanotubes', *Nanotechnology* 2007 **18** 075701 (7pp).

Murphy WL, Mooney DJ, 'Controlled delivery of inductive proteins, plasmid DNA and cells from tissue engineering matrices', *J. Period Res.* 1999 **34**(7) 413–419.

Nazhat SN, Abou Neel EA, Kidane A, Ahmed I, Hope C, Kershaw M, Lee PD, Stride E, Saffari N, Knowles JC, Brown RA, 'Controlled microchannelling in dense collagen scaffolds by soluble phosphate glass fibres', *Biomacromolecules* 2007 **8** 543–551.

Pachence JM, Kohn J, 'Biodegradable polymers', in: *Principles of Tissue Engineering*, 2nd edition, Lanza RP *et al.* (eds) Academic Press, San Diego (2000) p. 263.

Petricca SE, Marra KG, Kumta PN, 'Chemical synthesis of poly(lactic-co-glycolic acid)/hydroxyapatite composites for orthopaedic applications', *Acta Biomater.* 2006 **2** 277–286.

Pitt CJ, 'Poly(ϵ-caprolactone) (PCL) and its copolymers', in: *Biodegradable polymers as drug delivery systems*, Chasin M, Langer R (eds), New York, Marcel Dekker (1990), pp. 71–120.

Ramakrishna S, Mayer J, Wintermantel E, Leong KW, 'Biomedical applications of polymer-composite materials: a review', *Compos. Sci. Technol.* 2001 **61** 1189–1224.

Rezwan K, Chen QZ, Blaker JJ, Boccaccini AR, 'Biodegradable and bioactive porous polymer/inorganic composite scaffolds for bone tissue engineering', *Biomaterials* 2006 **27** 3413–3431.

Roether JA, Boccaccini AR, Hench LL, Maquet V, Gautier S, Jérôme R, 'Development and *in-vitro* characterisation of novel bioresorbable and bioactive composite materials based on polylactide foams and Bioglass® for tissue engineering applications', *Biomaterials* 2002 **23** 3871–3878.

Rose, FRAJ, Oreffo, ROC, Bone tissue engineering: hope vs hype, *Biochem. Biophys. Res. Commun.* 2002 **292** 1–7.

Sachlos E, Gotora D, Czernuszka JT, 'Collage scaffolds reinforced with biomimetic composite nano-sized carbonate-substituted hydroxyapatite crystals and shaped by rapid prototyping to contain internal microchannels', *Tissue Eng.* 2006 **12** 2479–2487.

Schnieders J, Gbureck U, Thull R, Kissel T, 'Controlled release of gentamicin from calcium phosphate/poly(lactic-co-glycolic acid) composite bone cement', *Biomaterials* 2006 **27** 4239–4249.

Silva CC, Thomazini D, Pinheiro AG, Aranha N, Figueiro SD, Goes JC, Sombra ASB, 'Collagen-hydroxyapatite films: piezoelectric properties', *Mater. Sci. Eng. B* 2001 **B86** 210–218.

Taboas JM, Maddox RD, Krebsbach PH, Hollister SJ, 'Indirect solid freeform fabrication of local and global porous, biomimetic and composite 3D polymer–ceramic scaffolds', *Biomaterials* 2003 **24** 281–294.

Torres FG, Nazhat SN, Sheikh Md Fadzullah SH, Maquet V, Boccaccini AR, 'Mechanical properties and bioactivity of porous PLGA/TiO2 nanoparticle-filled composites for tissue engineering scaffolds', *Comp. Sci. Technol.* 2007 **67** 1139–1147.

Tsigkou O, Hench LL, Boccaccini AR, Polak JM, Stevens MM, 'Enhanced differentiation and mineralization of human fetal osteoblasts on PDLLA containing Bioglass® composite films in the absence of osteogenic supplements', *J. Biomed. Mater. Res.* 2007 **80A** 837–851.

Vacanti JP, Langer R, 'Tissue engineering: the design and fabrication of living replacement devices for surgical reconstruction and transplantation', *Lancet* 1999 **354** 32–34.

Valappil SP, Misra SK, Boccaccini AR, Roy I, Biomedical applications of polyhydroxyalkanoates, an overview of animal testing and *in vivo* responses, *Expert Rev. Medical Devices* 2006 **3** 853–868.

Wu C, Sassa K, Iwai K, Asai S, 'Unidirectionally oriented hydroxyapatite/collagen composite fabricated by using a high magnetic field', *Materials Lett.* 2007 **61**(7) 1567–1571.

Wu TJ, Huang HH, Lan CW, Lin CH, Hsu FY, Wang YJ, 'Studies on the microspheres comprised of reconstituted collagen and hydroxyapatite', *Biomaterials* 2004 **25** 651–658.

Xiong Z, Yan YN, Wang SG, Zhang RJ, 'Fabrication of porous scaffolds for bone tissue engineering via low temperature deposition', *Sci. Mater.* 2002 **46** 771–776.

Xynos ID, Edgar AJ, Buttery LDK, Hench LL, Polak M, 'Gene expression profiling of human osteoblasts following treatment with the ionic products of Bioglass® 45S5 dissolution', *J. Biomed. Mater. Res.* 2001 **55** 151–157.

Yang S, Leong K, Du Z, Chua C, 'The design of scaffolds for use in tissue engineering. Part I Traditional factors', *Tissue Eng.* 2001 **7**(6) 679–689.

Yang S, Leong K, Du Z, Chua C, 'The design of scaffolds for use in tissue engineering. Part II Rapid prototyping techniques', *Tissue Eng.* 2002 **8**(1) 1–11.

Yang XBB, Webb D, Blaker J, Boccaccini AR, Maquet V, Cooper C, Oreffo ROC, 'Evaluation of human bone marrow stromal cell growth on biodegradable polymer/Bioglass® composites', *Biochem. Biophys. Res. Commun.* 2006 **342** 1098–1107.

Yannas IV, 'Classes of materials used in medicine: natural materials', in: *Biomaterials Science – An introduction to materials in medicine*, Ratner BD, Hoffman AS, Schoen FJ, Lemons J (eds), California, Elsevier Academic Press (2004).

Zhang K, Wang Y, Hillmayer MA, Francis LF, 'Processing and properties of porous poly(L-lactide)/bioactive glass composites', *Biomaterials* 2004 **25** 2489–2500.

5
Transplantation of engineered cells and tissues

J MANSBRIDGE, Tecellact LLC, USA

5.1 Introduction

For more than 50 years it has been clear that transplantation of organs from one animal to another of the same species would cause, with very few exceptions, a vigorous immune response that would lead to the rejection of the transplanted organ (Medawar, 1957). The concept of the impossibility of organ transplantation, without the use of immunosuppressants or other means to control the immune system, has reached general acceptance and is regarded as dogma. However, in recent years, since the development of tissue constructs grown *in vitro*, it has been found that such structures, when implanted, cause little immunological reaction and have never been clinically rejected, even when allogeneic. At this point, more than 50 000 patients have been treated without observable reaction, placing an upper limit to the incidence of the phenomenon. A major reason for this is the lack of professional antigen-presenting cells (macrophages, B lymphocytes, dendritic cells, endothelial cell), which provoke the rejection of transplanted organs, in the implanted constructs so far used.

Two major pathways of organ rejection are considered in this chapter – the direct and the indirect (Fig. 5.1). The direct pathway involves activation of host T-lymphocytes. Acute rejection by the direct pathway involves an initial interaction between a T-cell receptor (TCR) on a host lymphocyte and a surface molecule of the major histocompatibility complex (MHC) on a donor antigen-presenting cell (APC). This is usually a class II molecule, such as HLA-DR (Fig. 5.1(a)). The lymphocyte may secrete γ-interferon which induces HLA-DR and CD40 in many cells. In the second reaction (Fig. 5.1(b)), CD40 on the donor cell binds with CD154 on the lymphocyte. This induces expression of CD80 and CD86 on the transplant that interacts with CD28 on the host lymphocyte. These co-stimulatory reactions, along with the specific initial reaction of the host TCR with the foreign MHC, make up an 'immune synapse' that activates the lymphocyte (Fig. 5.1(d)). There are several such lymphocyte-activating pathways but this appears to be the most important in acute transplant rejection. In the chronic, or indirect, pathway, antigens shed by the allogeneic transplant

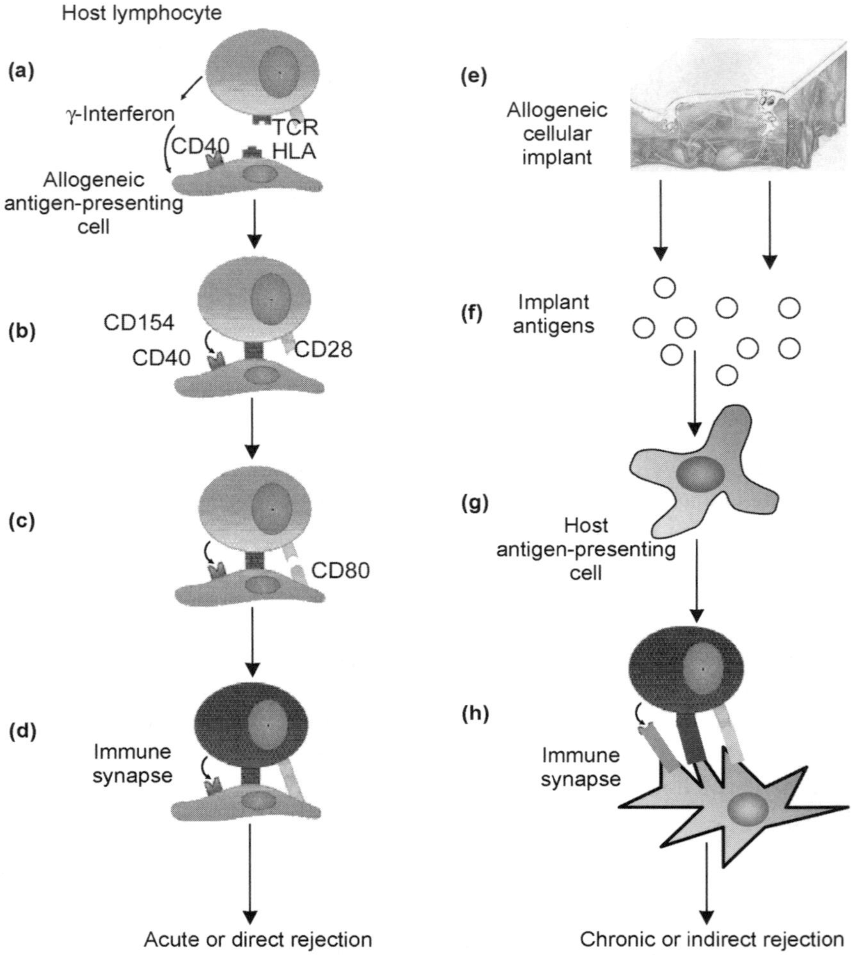

5.1 Acute and chronic transplant rejection of allogeneic transplants.

(Fig. 5.1(e)) are ingested by antigen presenting cells, such as the macrophage (Fig. 5.1(f)), which then matures to a dendritic cell (Fig. 5.1(g)). The dendritic cell then exposes these peptides on MHC Class II molecules, such as HLA-DR. If these peptides are detected as foreign by a TCR on a lymphocyte, the same set of co-stimulatory reactions as occur in the acute pathway lead to activation of the lymphocyte. Activation of the lymphocytes, which are of the T-helper class, leads to both cell-mediated and humoral responses. The response develops more gradually than the acute pathway and persists as a chronic condition, as the release of antigens and the early parts of the pathway are slow processes.

Most of this discussion will be concerned with the first mechanism. It has been observed that fibroblasts, used in scaffold-based, three-dimensional culture, lack co-stimulatory CD80 and CD86. It has also been found that they respond selectively to γ-interferon, failing to induce HLA Class II molecules (e.g. HLA-DR) that are frequently involved in initiating rejection. Fibroblasts in this type of culture fail to induce either humoral or cell-mediated immunity and fail to activate lymphocytes in a mixed-cell reaction.

This discussion will centre on experience with Dermagraft. After a brief description of the product and clinical experience with it, evidence will be presented for its persistence in wounds. Dermagraft is made by seeding neonatal, human fibroblasts to a degradable, biocompatible scaffold made of lactate-glycolate copolymer. Under these conditions, the fibroblasts proliferate and lay down a large amount of extracellular matrix, much as the capsule made *in vivo* in response to a foreign body. This material is implanted into chronic wounds and aids their healing.

The consequences for the commercialization of tissue-engineered products of the use of allogeneic rather than autologous cells are large. While the testing required by the regulatory authorities for an allogeneic product may be more extensive, the advantages in manufacturing, including bioreactor design, automation and the advantages of scale-up, more than outweigh the disadvantages. This is true even if some parts of the construct, such as the vascular system, have to be autologous.

5.2 Rejection of tissue-engineered products

Experience with tissue-engineered skin products designed for treating chronic wounds has shown no evidence for rejection. To date, about 90 000 pieces of Dermagraft have been implanted (about 25 000 patients), over 100 000 pieces of Apligraf and smaller numbers of other allogeneic, live cell implants, without a single report of immunological rejection. This provides the best evidence for lack of rejection, but the question arises as to why this should be so.

As an FDA requirement, in later clinical trials of Dermagraft, and for post-market surveillance, immunological responses to the allogeneic implant were specifically investigated. Neither humoral nor cell-mediated responses were detected. Briscoe *et al.* (1999) made a direct comparison between the responses of SCID-Hu mice to allogeneic human skin and to a bilayered human allogeneic tissue skin construct (Apligraf), which included both keratinocytes and fibroblasts. While they could quite clearly show rejection of the skin, the tissue-engineered construct was not rejected.

In several xenogeneic implants using Dermagraft performed at Advanced Tissue Sciences in intact animals (pigs, dogs, rats), without immunosuppression, that lasted for 2–4 weeks, no rejection was observed. These results all show that immune-based rejection has not been a clinical problem in the use of allogeneic implants for the treatment of wounds.

5.2.1 Reasons for the lack of rejection of tissue-engineered implants

One possible explanation for the lack of immune rejection of tissue-engineered implants is that the cells do not survive. In the case of the bilayered construct Apligraf, the keratinocyte layer is lost within a couple of weeks. This, however, cannot be taken as evidence for immune rejection as the keratinocytes would be expected to be lost in any case. During expansion culture to obtain sufficient cells from the initial biopsy to be commercially viable, stem cells are diluted or lose their phenotype and essentially all the keratinocytes in the implant are transit amplifying cells that may be expected to proceed to terminal differentiation and cornify.

5.2.2 Persistence of implanted allogeneic fibroblasts

The persistence of fibroblasts in these systems has been studied by implanting the construct, which is derived from a male donor, into female patients and examining extracted DNA for the presence of sequences unique to the Y chromosome. This approach has been criticized on the grounds that it is insufficient to demonstrate the presence of cells (Epstein, 2003). However, it has been used in several diverse systems (Rooney *et al.*, 1998; Tisdale *et al.*, 1998; Shapero *et al.*, 2000) and shows great sensitivity. In the case of Apligraf, used for treating venous ulcers, male DNA has been found within the first 4 weeks (Phillips *et al.*, 2002), occasionally at 6 weeks (Griffiths *et al.*, 2004) and, in the case of the treatment of epidermolysis bullosa, 28 weeks (Fivenson *et al.*, 2003).

In the case of Dermagraft, studies were conducted in conjunction with a venous ulcer trial in which 10 female patients received a single implant. Biopsies were taken at 1–2 weeks and at 6 months or when the ulcer healed. DNA was extracted and the male-specific SRY sequence was amplified by a two-stage nested technique (Fig. 5.2a). Samples were extracted from histological sections. Dewaxed paraffin sections, from which the epidermis was removed, or PBML were incubated for 3 h at 55 °C followed by 15 min at 99 °C in 50 mM tris pH 8.5, 0.045% Tween 20, 0.045% NP40, 1 mM EDTA, 1 mg/ml Proteinase K and stored at −20 °C (Leach, 1993, personal communication). Two sets of primers that can be nested were used for polymerase chain reaction (PCR) amplification of SRY. The use of nested primers permits very high PCR amplification without incurring problems of overamplification (appearance of artifactual bands) to the extent seen with comparable amplification with a single set of primers. Thus SRY sequences amplified with SRY1 and SRY2 (Calzolari *et al.*, 1993) for 30 cycles followed by SRY4 and SRY5 (Petrovic *et al.*, 1992) gave clean 290 bp bands without background with male DNA (Fig. 5.2b) and no band with female whereas 60 cycles using female DNA with either primer pair

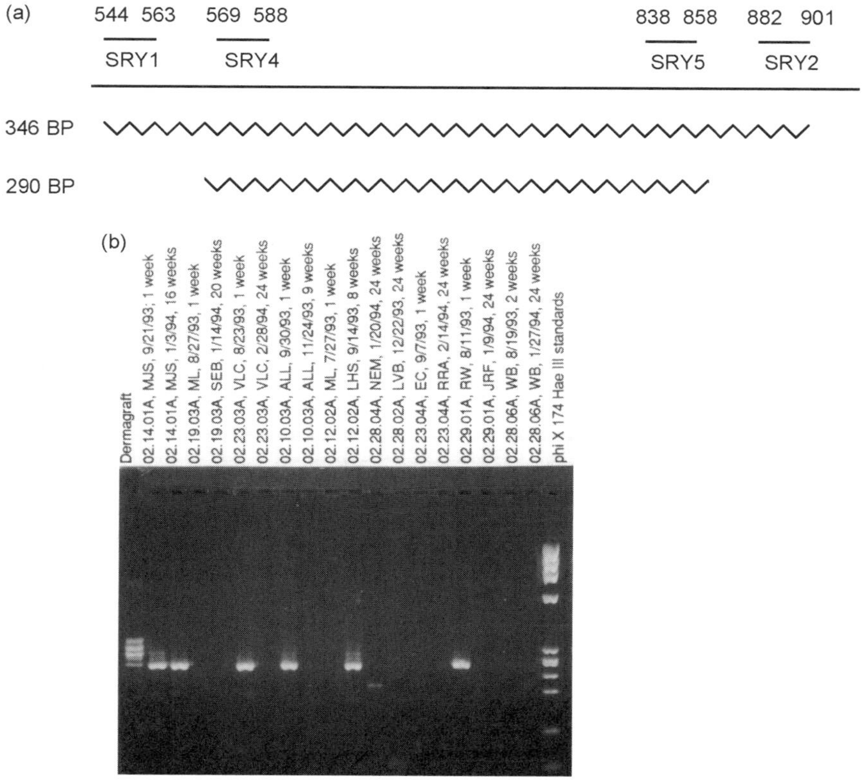

5.2 (a) Amplification of SRY sequences gave clean 290 bp bands; (b) positive or negative signals.

alone gave variable artifactual bands, some of which showed mobilities comparable to those obtained from male DNA.

This technique was capable of detecting single molecules, giving easily distinguishable positive or negative signals (Fig. 5.2b). DNA samples were split into aliquots predicted to contain one DNA molecule in every 10 reactions to allow counting of molecules. In this experiment, cells were invariably detected in the first biopsy from all the patients. However, Dermagraft DNA was also detected in six of nine of the later biopsies, which were taken at 2–6 months (Table 5.1). One problem with this very sensitive type of determination is the possibility of male cells circulating and persisting in the mother following a pregnancy with a male fetus (Bianchi *et al.*, 1996). At least one of the patients who showed persistence of Y-chromosome-detectable sequences had not had a male child.

Thus, fibroblasts from tissue-engineered skin implants colonize the ulcer bed and may persist for up to 6 months. The number of cells detected was very small in any individual biopsy (<10); however, this is not surprising considering the

Table 5.1 Detection of male DNA in biopsies from sites implanted with Dermagraft

	Early point		Late point	
Patient	Week	Allograft cell density (cells/mm^3)	Week	Allograft cell density (cells/mm^3)
1	1	22	9	51
2	1	109	8	54
3	1	53	16	168
4	1	156	20	0
5	1	200	24	68
6	1	47	24	0
7			24	2912
8	2	495	24	0
9	2	512	24	224
10	1	99	24	0

volume of the biopsy used (about 1 nl) and the frequency of apoptosis at late stages of normal wound healing (Greenhalgh, 1998).

5.2.3 Humoral immune responses to Dermagraft

In a series of 50 sera from patients treated with at least three pieces of Dermagraft, the presence or absence of antibodies reactive with Dermagraft components was assessed by staining immunoblots of proteins from detergent-lysed Dermagraft. No increase in antibodies to human proteins was observed in comparing blots stained with sera taken after treatment compared with sera taken before. The only reaction consistently seen was to bovine serum albumin, However, no consistent change was seen in this protein following Dermagraft implantation.

5.2.4 Cellular immune response to Dermagraft

Immune rejection of transplants is largely through a cell-mediated T-lymphocyte response. The activation of a cell-mediated response to Dermagraft implantation was investigated using the secretion of cytokines by activated T-cells. This was chosen in preference to determination of the blast response by tritiated thymidine incorporation because of the difficulty of estimating cellular tritium uptake in the heterogeneous lymphocyte/matrix/attached cell system. The secretion of γ-interferon by patient-derived lymphocytes in response to Dermagraft was determined by enzyme-linked immunosorbent assay (ELISA). Blood samples were taken before treatment and at the end of treatment, following at least three Dermagraft implants. The samples were transported to the Mayo clinic within 24 h, lymphocytes isolated and incubated with

Dermagraft for 3 days. The supernatant was harvested and stored at −70 °C until sufficient had been collected for economical analysis by ELISA.

In all patients, peripheral blood mononuclear lymphocytes isolated and tested in this way showed substantial stimulation by phytohemagglutinin, the positive control. However, the results showed no activation following Dermagraft treatment in any of the 25 patients. In all the data, a single case of lymphocyte activation was observed, but this occurred in a patient prior to treatment with Dermagraft. It thus appeared to be unrelated to Dermagraft and was interpreted as a non-specific result.

A similar study was performed using samples from 10 patients implanted intra-orally with a single piece of Dermagraft in conjunction with a clinical trial for the application of Dermagraft for gingival reconstruction. In this case, all positive controls showed reaction, but none responded to Dermagraft. Taking the results together, no case of implant-stimulated cell-mediated immunological response was observed in 35 patients.

These results provide no evidence for an immunological response to Dermagraft, and, with the lack of clinical evidence for rejection, such a response does not occur, or is, at most, a very rare event. These data are consistent with numerous reports of variable survival of allograft cells in a host without immunosuppression (Hefton *et al.*, 1983; Hull *et al.*, 1983, Sher *et al.*, 1983, Thivolet *et al.*, 1986, Palmer *et al.*, 1991, Otto *et al.*, 1995).

5.2.5 Selective response to γ-interferon by fibroblasts in scaffold-based three-dimensional culture

While the lack of antigen-presenting cells in tissue-engineered constructs may be the major factor in the lack of immune response to such implants, 3D culture also modifies cells in a manner that may contribute. In the presence of γ-interferon, which is likely in a wound site, cells induce MHC Class II molecules, such as HLADR, and co-stimulatory molecules, such as CD40. Such a reaction might be expected to precipitate an immune response. However, we have found that fibroblasts in scaffold-based 3D culture respond selectively, the majority of cells failing to induce these molecules (Fig. 5.3). Gene expression analysis shows a large number of genes induced in fibroblasts by γ-interferon and a spectrum of differential responses between monolayer and 3D cultures (Table 5.2). The set of genes markedly less induced in 3D than monolayer culture include MHC Class II and molecules involved in the antigen-presenting pathway, while other genes, such as HLA Class I, are comparatively unaffected. Three-dimensional fibroblast cultures, therefore, show a selective response.

In an investigation of the reason for the selective response, we have examined parts of the signal transduction pathway from γ-interferon. In particular, we have determined the phosphorylation of Signal Transducer and Activator of Transcription-1 (STAT-1), which transmits a signal from the γ-interferon

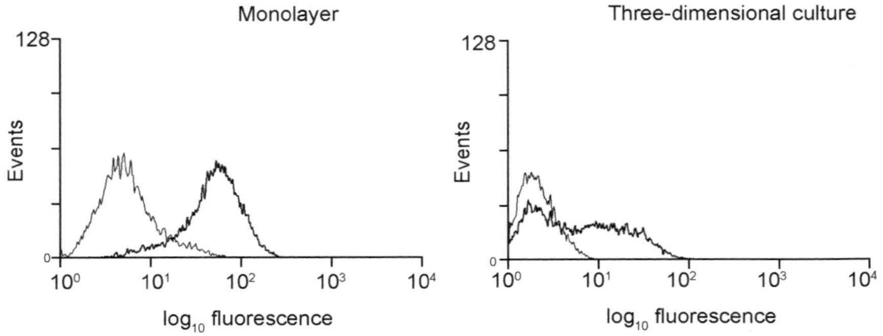

5.3 Comparison of the induction of HLA Class I and HLA-DR in monolayer and three-dimensional cultures by γ-interferon. Fibroblasts, cultured in monolayer, in scaffold-based 3D cultures were suspended by collagenase digestion stained for HLA-DR and analyzed by flow cytometry. In each case the culture without γ-interferon and the irrelevant IgG control (fine line —) were identical. The bold (——) line represents the culture following treatment for 3 days with 500 U/ml γ-interferon. The ordinates indicate the frequency of events and the abscissa the intensity of fluorescence of a 4 decade log scale.

receptor to the nucleus, and the Class II Transactivator (CIITA) which activates the promoter for HLA Class II genes such as HLA-DR. We observed that the initial γ-interferon receptor-mediated phosphorylation of STAT-1 on tyrosine 701 is very similar in monolayer and 3D culture. Phosphorylated STAT movement into a detergent insoluble fraction, presumably the nucleus, although slightly slower in 3D than in monolayer culture, is also little different (Fig.

Table 5.2 Comparison between monolayer and scaffold-based three-dimensional culture of genes induced by γ-interferon

	Induced within 85% in monolayer and 3D cultures	Induced 3–6× more in monolayer than 3D culture
	HLA-class I, β2 microglobulin 9-27, 1-8U, C1 inhibitor	
HLA Class II		HLA-DR, HLA-DMB, HLA-DP, IP-30, butyrophilin (BTF5), BiP
Virus-associated		HPB, HPC, CMV associated, IDO, 2′-5′ polyA synthetase
Transcription factors		IRF-9, IFP-35, IRF-1

Comparisons were made between median values from three independent determinations using Affymetrix U95A expression arrays, Genes are selected from a total of 13 genes showing similar induction in monolayer and 3D culture and 43 that showed a difference similar to those discussed. Genes were selected on their relevance to the expression of MHC Class II genes and antiviral properties.

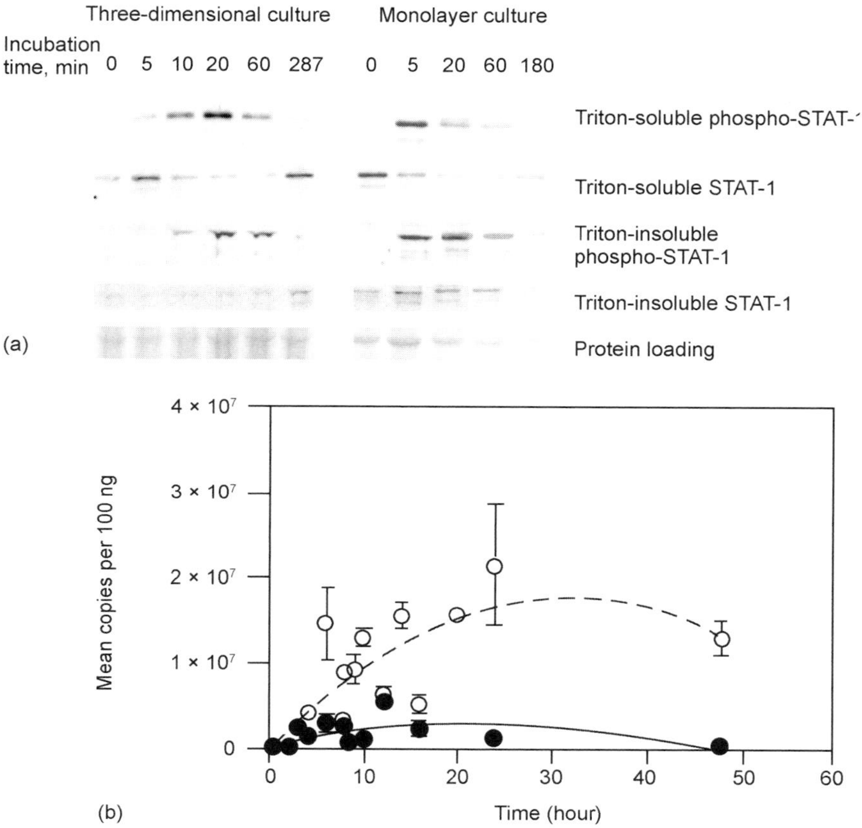

5.4 (a) Phosphorylation of STAT-1 on tyrosine-701. Cultures were incubated with 500 U/ml γ-interferon for times up to 3 h and harvested. The Triton-soluble and Triton-insoluble fractions were isolated and phosphorylation of tyrosine-701 was detected by immunoblots using an anti-phospho-STAT-1 antibody. (b) Comparison of the induction of CIITA by γ-interferon in fibroblasts in monolayer and three-dimensional culture. Cultures were incubated for times up to 48 h with 500 U/ml γ-interferon and RNA was prepared. Messenger RNA for CIITA was determined using real-time PCR. All values were normalized to total RNA: ○ monolayer culture; ● 3D culture. Lines mark trend lines, the broken line (-----) follows monolayer culture and the solid line (——), 3D culture. The bars are mean $\pm$SEM.

5.4a). The difference in kinetics may be largely explained by differences in diffusion. However, expression of CTIIA, is markedly different. The CTIIA gene has a complex control region, comprising four promoters, some of which are important in particular cells, such as B-lymphocytes, and others are affected by agents such as γ-interferon (O'Keefe *et al.*, 2001) (Fig. 5.4b). CIITA is much more highly induced by γ-interferon in monolayer than in 3D culture. Other factors also involved in this pathway, such as Interferon Response Factor-1

(IRF-1), is also less induced in 3D culture than monolayer. While this phenomenon does not appear to apply in collagen-based 3D culture (Kern *et al.*, 2001), it may be expected to occur as the cells migrate into the wound bed.

The general conclusion is that in the absence of cells involved in antigen presentation, the cells generally used for tissue engineering may well not be immunogenic. The one immunogenic cell important for tissue engineering is the vascular endothelial cell. Barring a method to render them non-immunogenic, they will have to be obtained from an autologous source (Williams, 1995; Williams *et al.*, 1989, 1994). It may be noted in passing that one organ that is regularly transplanted without major immune consequences in the majority of cases is the cornea, which contains no blood vessels.

5.3 Testing and regulatory consequences

At this stage, regulatory authorities are requesting testing of new, allogeneic products for immunogenicity. This places an appreciable resource burden on companies developing products, both in terms of collection of samples, effort and money, so the numbers requested have not been large. The results demonstrate the general lack of humoral or cell-mediated response to the allogeneic implants, but cannot exclude rare or idiosyncratic events. At this stage, the lack of clinically observed rejection of marketed skin products places the upper limit for such events at about 1 in 50 000 patients.

5.4 Generality of the resistance of tissue-engineered products to immune rejection

The scope of immunity from rejection among the allogeneic cells used in tissue engineering has not been determined. It is true for fibroblasts, almost certainly for keratinocytes and smooth muscle cells, possibly for chondrocytes, certainly not for endothelial cells and hemopoietic cells, but for other cells it is unknown. As discussed above, the loss of keratinocytes from bilayered tissue-engineered skin implants, such as Apligraf, may be explained without involving immune rejection. Indeed, it is likely that activation of the immune system by keratinocytes would lead to complete destruction of the implant as occurs following activation by endothelial or dendritic cells or B-lymphocytes. No clinical symptoms of rejection have been observed during or after implantation of such bilayered devices.

In unpublished experiments performed at Advanced Tissue Sciences, small diameter vascular implants were tested in dogs. These constructs comprised a smooth muscle tube lined with endothelial cells. Endothelial cells are antigen-presenting cells and immunogenic, so autologous cells were used (Williams *et al.*, 1994). However, the smooth muscle cells were of allogeneic canine origin. The constructs were implanted for up to 12 weeks without any evidence for immunological rejection. These experiments need to be extended; however, it is likely that cells other than fibroblasts are tolerated, provided that antigen-presenting cells, such as endothelial cells, B-lymphocytes, macrophages and dendritic cells are carefully excluded.

5.5 Manufacturing consequences

The contrast between approaches to autologous and allogeneic tissue engineering is striking. Autologous tissue engineering is a service industry, requiring heavy investment in plant (individual tissue culture suites, or burdensome cleaning, are required for each patient) and is usually performed by expensive, skilled and highly trained personnel. Economies of scale are not possible. However, safety testing is much reduced. In allogeneic approaches, the burden of testing is much greater. In recent experience, the cost of a new master cell bank was placed at approaching a million dollars overall. However, with judicious use, such a cell bank may last 10–15 years, so the cost is averaged over a considerable time. The advantage of allogeneic approaches is that manufacturing is much simplified, can be performed by less expensive, although trained, personnel and shows economies of scale. This approach is more suitable for large-scale production and can incorporate more easily specialized bioreactor design, automation and so forth. The cost of the product is much reduced. While these advantages are more difficult to achieve with autologous systems, ingenious design, such as development of completely automatic machines that will accept a fresh biopsy and produce the final product, all within disposable containers and without human intervention, is possible.

Although it may be feasible to make the bulk of a tissue-engineered construct from allogeneic cells, certain components, and particularly the vascular system, will have to be autologous. At this stage, few tissue-engineered vascular organs have been implanted, as problems associated with the development of a practical vascular system have not been solved. Thus far, implants, such as skin constructs, which do become vascularized, have been vascularized directly from the host following implantation, an approach that may well have limited their size.

One case in which constructs containing vascular endothelial cells have been implanted is small diameter blood vessels. Implanted artificial blood vessels fail to be endothelialized by the host for more than a few millimetres. In the case of small diameter blood vessels, lack of endothelialization leads to blood clotting and rapid failure. In human trials of such constructs, the entire device is of autologous origin (L'Heureux *et al.*, 1993, 2006). In the canine experiment, mentioned above, although the smooth muscle cells were allogeneic, the endothelial cells had to be autologous. In that case, endothelial cells were isolated from liposuction fat (Williams *et al.*, 1994). Although such cells might not be identical to small vessel aortic endothelial cells, they were found, after conditioning to the fluid shear to be expected after implantation, to perform satisfactorily.

Another method for vascularizing an organ in an autologous manner is implantation of the construct in a host location rich in blood supply and inducing vascular development in the implant by using growth factors, followed by transfer to its ultimate site. To do this, the organ would probably have to start as a primordial structure, which could survive as vascularization was established, and then grow within the host. While this approach will provide an autologous vascular bed, it is likely to involve multiple connections to the host. If an organ is to be transplanted, it is desirable to limit the number of inlets and outlets to minimize microvascular surgery during transplantation.

5.6 Conclusions and future trends

Despite experience with organ transplantation, implantation of tissue-engineered products containing live cells has thus far shown no clinical evidence for immune rejection. Limited testing for specific mechanisms of rejection have shown no evidence of humoral or cell-mediated reaction to the implants. The major reason for the lack of immunogenicity lies in the lack of antigen-presenting cells in the implants, and inclusion of such cells has been found experimentally to confer rejectability (Rouabhia, 1996). In a transplanted organ, endothelial cells are a major source of antigen-presenting cells and acute rejection may be seen as an attack on the vascular system that rapidly extends to other cells. It has also been found that fibroblasts in scaffold-based 3D culture show a selective response to γ-interferon, which, although it induces molecules associated with antigen presentation in many cell types under suitable culture conditions, does not do so in this case.

The use of allogeneic cells has many advantages for manufacturing, even if it can be applied only to part of the final construct, so it would be valuable to ascertain the scope of cells types immune to rejection. At present, it applies to fibroblasts and smooth muscle cells, but not to endothelial cells or hemopoietic cells, although claims have been made for bone marrow stem cells (Aggarwal and Pittenger, 2005) including their ability to suppress responses to allogeneic cells (Klyushnenkova *et al.*, 2005). In the case of stem cells, it may be true that undifferentiated cells are non-immunogenic but that they may become immunogenic if they differentiate into antigen-presenting cells such as macrophages, dendritic cells and endothelial cells. Determination of the range of cells showing minimal immunogenicity would be a valuable contribution to tissue engineering.

Ultimately, the development of organs requiring development of a vascular system will entail the inclusion of endothelial cells that will have to be autologous. Methods are available now to obtain such cells comparatively easily, but methods will have to be devised to establish a vascular network rapidly. One possible method is inosculation (Black *et al.*, 1998), in which vessels preformed by allogeneic cells are populated from the host. The allogeneic endothelial cells would have to be destroyed before implantation, leaving only basement membrane channels. Another approach is growing the organ within its ultimate host, using the host circulatory system to supply nutrition, in much the same way as in some early experiments, where organs (among them an ear) were grown in the mouse (Cao *et al.*, 1997). A third is to develop machines that would grow the vascular system within a preformed tissue-engineered implant. However, our current understanding of the mechanisms involved in the development of blood vessels, particularly large vessels, is inadequate to allow us to do this.

5.7 Sources of further information and advice

Much of the literature in this field is primary, the remainder being generated within companies. A series of early articles indicating that skin substitutes were not immunogenic include Hefton *et al.* (1983), Hull *et al.* (1983), Sher *et al.* (1983), Thivolet *et al.* (1986), Palmer *et al.* (1991) and Otto *et al.* (1995). The demonstration that adding antigen-presenting cells to constructs confers rejectability is described in Rouabhia (1996) and Briscoe *et al.* (1999). Methods for obtaining, comparatively conveniently, autologous endothelial cells are described by Williams *et al.* (1995) and a company attempting to develop automatic devices for growing tissue-engineered products may be found at www.millenium-biologix.com.

5.8 Acknowledgements

PCR amplification of DNA to determine the persistence of implanted fibroblasts was performed by Kathryn MacKenzie. Determination of lymphocyte activation by γ-interferon production was performed by Dr L. Oliver at the Mayo Clinic. Figures 5.3 and 5.4 were previously published by the *Journal of Investigative Dermatology*. Immunoblots to determine humoral responses to Dermagraft were performed by E. Pinney. RNA for expression array analysis was prepared by Dr K. Liu.

5.9 References

Aggarwal, S. & Pittenger, M. F. (2005) Human mesenchymal stem cells modulate allogeneic immune cell responses. *Blood*, **105**, 1815–22.

Bianchi, D. W., Zickwolf, G. K., Weil, G. J., Sylvester, S. & Demaria, M. A. (1996) Male fetal progenitor cells persist in maternal blood for as long as 27 years postpartum. *Proc Natl Acad Sci USA*, **93**, 705–8.

Black, A. F., Berthod, F., L'Heureux, N., Germain, L. & Auger, F. A. (1998) In vitro reconstruction of a human capillary-like network in a tissue- engineered skin equivalent. *Faseb J*, **12**, 1331–40.

Briscoe, D. M., Dharnidharka, V. R., Isaacs, C., Downing, G., Prosky, S., Shaw, P.,

Patenteau, N. L. & Hardin-Young, J. (1999) The allogeneic response to cultured human skin equivalent in the hu-PBL-SCID mouse model of skin rejection. *Transplantation*, **67**, 1590–99.

Calzolari, E., Patracchini, P., Palazzi, P., Aiello, V., Ferlini, A., Trasforini, G., Uberti, E. D. & Bernadi, F. (1993) Characterization of a deleted Y-chromosome in a male with Turner stigmata. *Clin Genet*, **43**, 16–22.

Cao, Y., Vacanti, J. P., Paige, K. T., Upton, J. & Vacanti, C. A. (1997) Transplantation of chondrocytes utilizing a polymer-cell construct to produce tissue-engineered cartilage in the shape of a human ear. *Plast Reconstr Surg*, **100**, 297–302; discussion 303–4.

Epstein, E. (2003) Evidence for living cells: DNA fragments are not enough. *Arch Dermatol*, **139**, 541; author reply 541.

Fivenson, D. P., Scherschun, L., Choucair, M., Kukuruga, D., Young, J. & Shwayder, T. (2003) Graftskin therapy in epidermolysis bullosa. *J Am Acad Dermatol*, **48**, 886–92.

Greenhalgh, D. G. (1998) The role of apoptosis in wound healing. *Int J Biochem Cell Biol*, **30**, 1019–30.

Griffiths, M., Ojeh, N., Livingstone, R., Price, R. & Navsaria, H. (2004) Survival of Apligraf in acute human wounds. *Tissue Eng*, **10**, 1180–95.

Hefton, J. M., Madden, M. R., Finkelstein, J. L. & Shires, G. T. (1983) Grafting of burn patients with allografts of cultured epidermal cells. *Lancet*, **ii**, 428–30.

Hull, B. E., Sher, S. E., Rosen, S., Church, D. & Bell, E. (1983) Fibroblasts in isogeneic skin equivalents persist for long periods after grafting. *J Invest Dermatol*, **81**, 436–8.

Kern, A., Liu, K. & Mansbridge, J. N. (2001) Expression HLADR and CD40 in three-dimensional fibroblast culture and the persistence of allogeneic fibroblasts. *J Invest Dermatol*, **117**, 112–18.

Klyushnenkova, E., Mosca, J. D., Zernetkina, V., Majumdar, M. K., Beggs, K. J., Simonetti, D. W., Deans, R. J. & McIntosh, K. R. (2005) T cell responses to allogeneic human mesenchymal stem cells: immunogenicity, tolerance, and suppression. *J Biomed Sci*, **12**, 47–57.

L'Heureux, N., Germain, L., Labbe, R. & Auger, F. A. (1993) In vitro construction of a human blood vessel from cultured vascular cells: a morphologic study. *J Vasc Surg*, **17**, 499–509.

L'Heureux, N., Dusserre, N., Konig, G., Victor, B., Keire, P., Wight, T. N., Chronos, N. A., Kyles, A. E., Gregory, C. R., Hoyt, G., Robbins, R. C. & McAllister, T. N. (2006) Human tissue-engineered blood vessels for adult arterial revascularization. *Nat Med*, **12**, 361–5.

Medawar, P. B. (1957) *The Uniqueness of the Individual*, Basic Books.

O'Keefe, G. M., Nguyen, V. T., Ping Tang, L. L. & Benveniste, E. N. (2001) IFN-gamma regulation of class II transactivator promoter IV in macrophages and microglia: involvement of the suppressors of cytokine signaling-1 protein. *J Immunol*, **166**, 2260–69.

Otto, W. R., Nanchahal, J., Lu, Q. L., Boddy, N. & Dover, R. (1995) Survival of allogeneic cells in cultured organotypic skin grafts. *Plast Reconstr Surg*, **96**, 166–76.

Palmer, T. D., Rosman, G. J., Osborne, W. R. A. & Miller, A. D. (1991) Genetically modified skin fibroblasts persist long after transplantation but gradually inactivate introduced genes. *Proc Natl Acad Sci USA*, **88**, 1330–34.

Petrovic, V., Nasioulas, S., Chow, C. W., Voullaire, L., Schmidt, M. & Dahl, H. (1992)

Minute Y chromosome derived marker in a child with gonadoblastoma: cytogenetic and DNA studies. *J Med Genet*, **29**, 542–46.
Phillips, T. J., Manzoor, J., Rojas, A., Isaacs, C., Carson, P., Sabolinski, M., Young, J. & Falanga, V. (2002) The longevity of a bilayered skin substitute after application to venous ulcers. *Arch Dermatol*, **138**, 1079–81.
Rooney, C. M., Heslop, H. E. & Brenner, M. K. (1998) EBV specific CTL: a model for immune therapy. *Vox Sang*, **74** Suppl 2, 497–8.
Rouabhia, M. (1996) *In vitro* production and transplantation of immunologically active skin equivalents. *Lab Invest*, **75**, 503–17.
Shapero, M. H., Kundu, S. K., Engleman, E., Laus, R., Van Schooten, W. C. & Merigan, T. C. (2000) *In vivo* persistence of donor cells following adoptive transfer of allogeneic dendritic cells in HIV-infected patients. *Cell Transplant*, **9**, 307–17.
Sher, A. E., Hull, B. E., Rosen, S., Church, D., Friedman, L. & Bell, E. (1983) Acceptance of allogeneic fibroblasts in skin equivalent transplants. *Transplantation*, **36**, 552–7.
Thivolet, J., Faure, M., Demidem, A. & Mauduit, G. (1986) Long-term survival and immunological tolerance of human epidermal allografts produced in culture. *Transplantation*, **42**, 274–80.
Tisdale, J. F., Hanazono, Y., Sellers, S. E., Agricola, B. A., Metzger, M. E., Donahue, R. E. & Dunbar, C. E. (1998) Ex vivo expansion of genetically marked rhesus peripheral blood progenitor cells results in diminished long-term repopulating ability. *Blood*, **92**, 1131–41.
Williams, S. K. (1995) Endothelial cell transplantation. *Cell Transplant*, **4**, 401–10.
Williams, S. K., Jarrell, B. E., Rose, D. G., Pontell, J., Kapelan, B. A., Park, P. K. & Carter, T. L. (1989) Human microvessel endothelial cell isolation and vascular graft sodding in the operating room. *Ann Vasc Surg*, **3**, 146–52.
Williams, S. K., Rose, D. G. & Jarrell, B. E. (1994) Microvascular endothelial cell sodding of ePTFE vascular grafts: improved patency and stability of the cellular lining. *J Biomed Mater Res*, **28**, 203–12.
Williams, S. K., McKenney, S. & Jarrell, B. E. (1995) Collagenase lot selection and purification for adipose tissue digestion. *Cell Transplant*, **4**, 281-9.

6

Surface modification to tailor the biological response

K SHAKESHEFF and G TSOURPAS, University of Nottingham, UK

6.1 Introduction

A cell interrogates and signals to its local environment using surface interactions. Within any tissue the massive combined surface area of component cells results in an immense number of specific molecular interactions between cell membrane components and the extracellular matrix (ECM). One of the central functions of a tissue engineering scaffold is to mimic the ECM. Therefore, the nature of the interfacial relationship between cells and scaffolds is of great importance.

This chapter begins with an overview of the biochemistry of cell interactions with the ECM to establish the general principles of natural cell surface interactions. Then the broad aims of surface modification of tissue engineering scaffolds are reviewed. Tissue engineering scaffolds are a complex material type to successfully surface modify and, therefore, the next section reviews general strategies to achieve surface engineering. Finally, specific examples of surface engineering techniques that have been used with scaffolds are presented.

6.2 The biochemistry of cell interactions with the ECM

6.2.1 The extracellular matrix

Native ECM physically holds cells together, influences the tissue structure and regulates the cell phenotype. There are different types of ECMs within different types of tissues. ECM interactions sustain specialised cell functions such as strength (in tendon), filtration (in the kidney glomerulus), adhesion (through adhesion molecules), signalling (through growth factors and hormones), development (in morphogenesis) as well as wound healing (degradation and re-synthesis of matrix components to regenerate tissue).

ECM is a gel of relatively insoluble ligands containing collagens and other glycoproteins (such as fibronectin (FN), vitronectin (VN) and laminin (LN)), proteoglycans, glycosaminoglycans, elastins and hyaluronic acid. These compounds provide natural scaffolding for cells and tissues and help to provide a permeability barrier between tissue compartments (Pierschbacher and Ruoslahti, 1987). ECM also accommodates molecules such as growth factors, cytokines, matrix-degrading enzymes and their inhibitors. Many ECM molecules are large (M_W of more than 100 000), containing several functional domains or regions involved in interactions with other molecules or regions for receptor mediated cell adhesion (Ruoslahti and Obrink, 1996).

Cell adhesion molecules of ECM bind to other cell matrix components, imparting further strength and rigidity to the ECM complex. The first adhesion domain was found by Pierschbacher in 1984, and consisted of the three amino acids Arg-Gly-Asp (RGD) (Pierschbacher and Ruoslahti, 1984). RGD ($M_W \approx 400$ Da) was identified as the minimal sequence required to mediate cell adhesion to fibronectin (FN) ($M_W \approx 450$ kDa). Only one year after the discovery of RGD, a family of receptors that promotes cell adhesion, migration and differentiation was found, and named integrins (Pytela *et al.*, 1985).

6.2.2 The integrin family of receptors

Since the discovery of integrins and the recognition of their role (Hynes 1987) in cell–cell and cell–ECM interactions, their functionalities have been studied by several groups. Integrins have been found to play an important role in:

- activating intracellular signalling molecules important for transmembrane signal transduction to the cytoskeleton;
- the maximal activation of some growth factor receptors; for example an insulin receptor undergoes maximal activation only following integrin-mediated cell attachment;
- mediating the functional pre- and post-synaptic maturation at synapses; and
- mediating virus penetration into host cells.

The integrin receptors are composed of two subunits α and β which associate non-covalently, in a calcium-dependent manner (Hynes, 1992). According to Hynes within the complete mammalian set of integrin family there are at least 24 distinct integrins consisting of 8β subunits and 18α subunits. Each subunit (>1600 amino acids) crosses the cell membrane with most of the sequence in the ECM domain and only 20–50 amino acids in the cytoplasm (Fig. 6.1(a)). Binding occurs at the β subunit but the affinity is regulated by the α subunit. In resting phase they are stabilised by divalent cations such as Ca^{2+} (low affinity) and when active (high affinity) are regulated by the presence of Mn^{2+} or Mg^{2+} (Fig. 6.1(b)). While most transmembrane connections regulate the actin-based microfilament system, the β_4 subunit interacts with intermediate filaments instead, utilising its large cytoplasmic domain (Hynes, 2002). Either transmembrane interaction confirms the significant role of integrins in the cytoskeleton organisation.

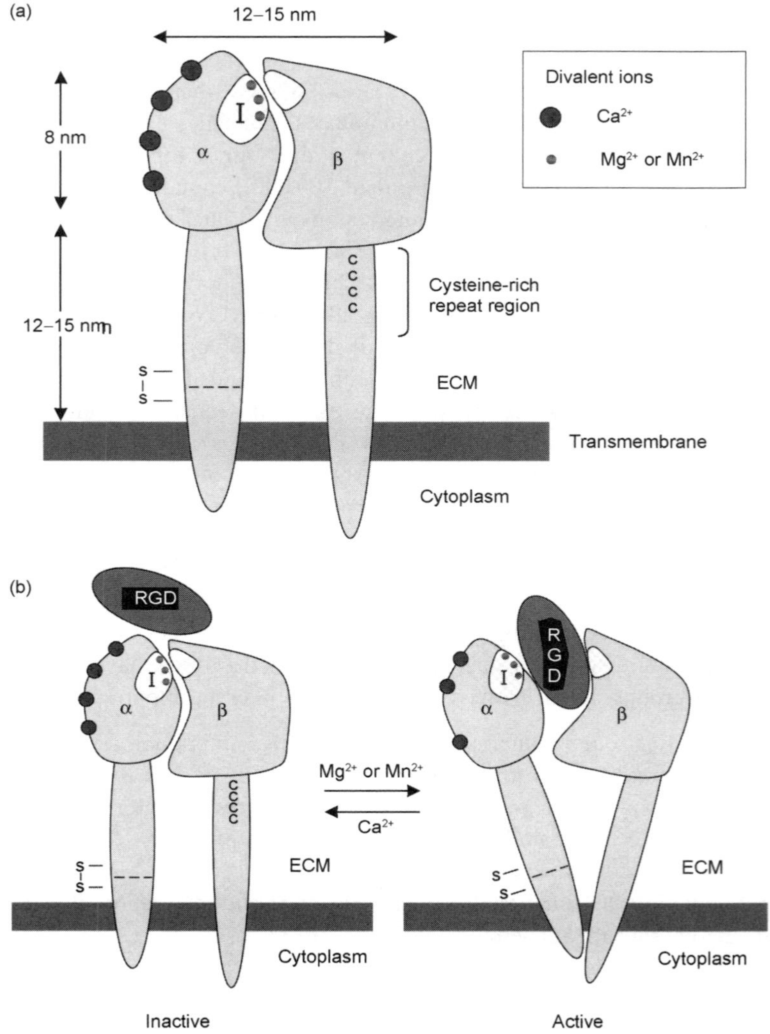

6.1 (a) Schematic representation of the structure of integrin molecule, and (b) its active and inactive conformation.

Integrins connect ECM with the cytoskeleton at sites known as focal contacts. Focal adhesion sites contain proteins (talin, paxillin, vicullin and tensin) along with other signalling molecules stimulating pathways such as tyrosine phosphorylation, mitogen protein kinases, calcium influx and pH changes (Longhurst and Jennings, 1998) (Fig. 6.2). For the purpose of signalling, as the integrins bind to the ECM ligands, they come together at the cell membrane, triggering the assembly of actin filaments in the cytoplasm. The rearrangement

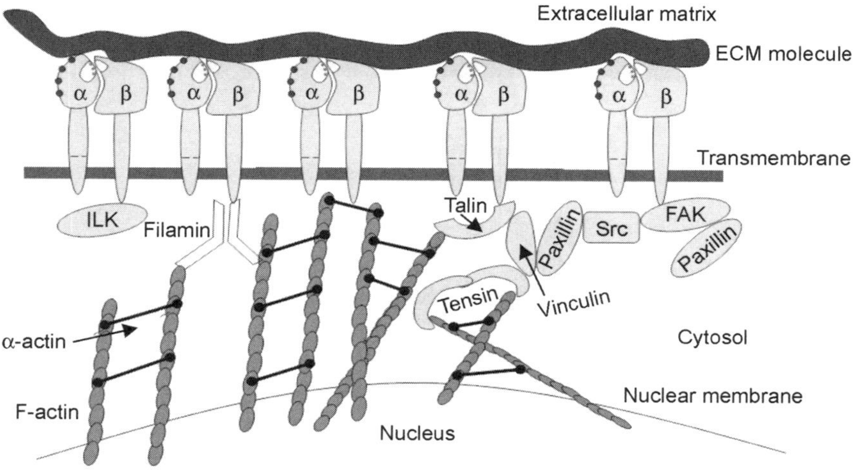

6.2 Schematic representations of focal adhesion sites. Transmembrane signal transduction pathways regulated by integrins.

of actins creates a positive feedback, causing a more profound clustering effect of integrins on the cell surface.

Integrins are involved in a bi-directional signal transduction pathway:

- Outside-in signalling where ECM binding initiates a number of signalling events to the interior of the cell such as focal adhesion kinase (FAK) activation, integrin-linked kinase (ILK) and others.
- Inside-out signalling where the extracellular binding affinity (i.e. cell adhesion) is regulated by the cytoplasmic portion of integrins.

6.2.3 RGD and other cell adhesion peptides

The diversity of integrin function and their vital role in mediating cellular attachment and spreading ensures that generating appropriate scaffold-to-integrin interactions is of central importance to many tissue engineering strategies. The RGD–integrin breakthrough initiated a cascade of discoveries of other cell adhesion motifs and their receptors. Many short adhesion sequences, peptides, have been identified and characterised within ECM molecules, indicating that intact adhesion molecules interact with a wide range of cell types with a high degree of specificity. Table 6.1 contains examples of cell adhesion peptides identified within protein molecules important for various functions.

RGD binding sites have been identified in the majority of ECM proteins including fibronectin, vitronectin, fibrinogen, von Willebrand factor, collagen, laminin family, osteopontin and tenascin. The affinity and specificity of the RGD binding domains vary for each protein due to different conformation of the RGD binding loop and its adjacent amino acids (Table 6.2).

Table 6.1 Peptide sequences identified within ECM proteins responsible for cell adhesion. T3 peptide is the motif from tumstatin (69–88 amino acids). Region II plus sequence is EWSPCSVTCGNGIQVRIK

Protein	Adhesion motif	Role	Reference
Fibronectin	LDV	Adhesion	Wayner and Kovach (1992)
	REDV	Adhesion of HUVEC cells	Hubbell *et al.* (1991)
	GREDVY		
	KQAGDV	Adhesion of vascular smooth muscle cells	Mann and West (2002)
	PHSRN	Adhesion of baby hamster kidney or HT-1080 cells	Aota *et al.* (1994)
	PHSRN	Macrophage adhesion and FBGC formation	Kao *et al.* (2001)
	(synergistically with RGD)	Human primary blood-derived macrophages	Liu and Kao (2002)
	PRRARV	Macrophage adhesion	Kao *et al.* (1999)
	KNEED	Spreading of fibroblasts	Wong *et al.* (2002)
HBFN-f	WQPPRARI	Human articular cartilage adhesion	Yasuda *et al.* (2003)
HEP II FN	HBP/III5 (with RGD induced adhesion)	Cytoskeletal response of melanoma cells (via chondroitin sulphate proteoglycan receptor)	Moyano *et al.* (2003)
	KRSR	Osteoblast adhesion	Hasenbein *et al.* (2002)

Elastin	VAPG	Human primary blood-derived macrophages	Gobin and West (2003)
Laminin $\alpha1$	RNIAEIIKDI	Adhesion of nerve cells	Schense *et al.* (2000)
	IKVAV	Neurite extension	Tashiro *et al.* (1989)
	CQAASIKVAV	Strong adhesion of nerve cells	Shaw and Schoichet (2003)
	IKLLIS	Heparin binding and cell adhesion	Tashiro *et al.* (1999)
Laminin $\beta1$	YIGSR	Adhesion of variety of cells (via 67LR) including neurite extension	Hirano *et al.* (1993)
	CDPGYIGSR	Very good neurite extension	Shaw and Shoichet (2003)
Collagen I	DGEA	Neurite outgrowth	Schense *et al.* (2000)
		Osteoblastic differentiation	Mizuno *et al.* (2000)
	KDGEA	Adhesion of platelets	Yamamoto *et al.* (1993)
	GPP	Platelet adhesion (via Gp VI receptor)	Knight *et al.* (1999)
	GFP*GER	Platelet adhesion (via Gp VI receptor)	Farndale *et al.* (2003)
Collagen IV	T3 Peptide	Adhesion of endothelial cells	Maeshima *et al.* (2001)

Table 6.2 Specificity of RGD sequences responsible for integrin mediated adhesion of various cell lines

RGD sequence			Polymer	Cell line	Reference
	RGD		PLLA	3T3 FB (mouse)	Yamaoka *et al.* (1999)
	RGD		PET	Hepatocytes	Krijgsman *et al.* (2002)
G	RGD		PHPMA	Rat spinal cord	Woerly *et al.* (2001)
	RGD	**T**	PEA	CHO-K1, NRK	Hirano *et al.* (1993)
G	RGD	**S**	P(D,L)LA/PLL	BAE-1	Quirk *et al.* (2001a)
G	RGD	**Y**	P(LA-co-lysine)	BAE	Cook *et al.* (1997)
G	RGD	**F**	PAN	Human platelets	Beer *et al.* (1992)
YG	RGD		PEG star	WTNR 6 fibroblasts	Maheshwari *et al.* (2000)
	RGD	**S**	PEG copolymer	Human dermal fibroblast	Mann and West (2002)
G	RGD	**SP**	PVA	BAE	Sugawara and Matsuda (1995)
G	RGD	**SPK**	PTFE/PLL	Adult HSVEC	Walluscheck *et al.* (1996)
G	RGD	**SY**	PU copolymer	HUVEC	Rouhi (1999)

6.3 The need for surface modification of scaffolds

Direct mimicking of the ECM provides the strongest need for surface modification of scaffolds. Some natural scaffolding materials will be derived from the ECM – the most common example is collagen – and will possess inherent cell adhesion properties. However, many scaffolds are produced from synthetic or natural polymers that lack ECM structural properties. Surface modification provides a method, often simple to achieve, to maintain desirable bulk properties of scaffolds (e.g. mechanical strength, controlled growth factor release and batch reproducibility) while radically changing the interaction of the scaffold with cells.

In addition to direct ECM mimicking, there are a number of other motivations for surface modification of scaffolds.

6.3.1 Improving wettability

Wettability issues are common with scaffolds composed of poly(lactic acid) (PLA), poly(caprolactone) (PCL) and others. Simple modifications to increase surface hydrophilicity are effective in ensuring a rapid wetting of the pore structure of the scaffold and enhanced cell seeding/migration. This is an important issue in both *in vitro* tissue engineering, where the interaction occurs between tissue culture medium and the scaffold, and *in vivo* tissue engineering, where the interaction is with extracellular fluid and blood.

6.3.2 To block detrimental protein conditioning

This is mainly an issue for *in vivo* tissue engineering strategies. The body's defence against foreign materials involves a protein coating processes (Vroman effect) in which a material can be tagged for exclusion (e.g. production of a thick collagen layer isolated the biomaterial from neighbouring cells) or attack (a cellular response to remove the material). Surface modification, for example to present high densities of poly(ethylene glycol) (PEG), can block or modify the pattern of initial protein conditioning of scaffold surfaces.

6.3.3 To present other signalling molecules

Non-ECM molecules can be attached to scaffold surfaces to stimulate specific biological responses. For example, growth factors and DNA have been immobilised and can generate potent cellular responses due to the high local concentration at the site of cell adhesion.

6.3.4 To spatially pattern the response of cells to a scaffold

Here the scaffold is surface engineering to mimic the ECM but the modification is restricted to a zones or patterns within the 3D structure. This strategy offers a method to spatially pattern cell responses.

6.4 General strategies for surface modification

The most appropriate strategy for surface modification will be determined by the chemical nature of the bulk material of the scaffold, the nature of chemical change required at the scaffold surface and the architecture of the scaffold.

The major strategies to be explored further in this chapter are as follows:

- *Surface coating from contacting solution.* The most common method of surface modification relies on an electrostatically or surface energy driven adsorption of a molecule from a contacting solution. Tissue culture media containing serum or defined ECM protein additives provide a source of cell adhesion molecules for adsoption. Fibroblasts and related cell types will also secrete more ECM molecules over time in culture.
- *Chemical functionalisation.* A common method of increasing hydrophilicity. A reactive species is exposed to the scaffold surface for sufficient time to generate more reactive or hydrophilic end groups and side groups.
- *Entrapment and blending.* A surface-restricted interpenetrating network of the scaffold bulk polymer and the surface modifying species is generated.

- *Thin layer deposition.* A nanometre thick layer of a polymer is deposited on the scaffold surface. Plasma polymerisation is a convenient method of achieving a controlled deposition and has recently been proven to work on highly porous 3D scaffolds.
- *Covalent attachment.* This method relies on the presence of sufficient density of reactive groups on the scaffold surface. Often covalent attachment is preceded by one of the above strategies to increase reactive group density.

6.5 Examples from the literature

6.5.1 Surface coating from contacting solution

The examples of surface coating are too numerous to list in a review. Most examples of scaffold use involve a modification of the surface chemistry by adsorbed species. A number of groups have attempted to refine the adsorption process using electrostatic interactions. Electrostatic self-assembly has been employed by Zhu *et al.* (2004) to build up layers of charges protein and polycationic species on the surfaces of PLA microparticles and scaffolds. The polycation poly(ethylene imine) (PEI) was deposited on to the PLA surface exploiting the charge of PLA end groups. Sequential addition of gelatin allowed multiple layers to grow from the PLA surface. Cell population viability on PLA surfaces approximately doubled on scaffolds with ESA engineering compared to controls.

Electrostatically driven adsorption has also been employed to surface engineer PLGA scaffolds with DNA (Jang *et al.*, 2006). In the work described by Jang *et al.* the polyanionic DNA was first complexed with PEI and then exposed to the polymer scaffold. The authors note that a relatively low concentration of DNA generates a high transfection efficiency and propose that this is achieved because the cells adhere to the surface that directly presents the DNA.

Yang *et al.* (2001) have used surface coating to present either the ECM protein fibronectin or a synthetic mimic consisting of poly(L-lysine) with coupled GRGDS (PLL-RGD) on the surfaces of poly(lactic *acid-co-*glycolic acid) *PLGA* scaffolds. Successful expansion, migration and differentiation of human osteoprogenitors on foam scaffolds were achieved at optimal concentrations of both the natural protein and the mimic molecule. Figure 6.3 shows an example of work from this paper in which the cell adhesion and specific functions are compared on the different surfaces.

Liu *et al.* (2006) report that surface engineering can be achieved by the retention of small quantities of porogen during leaching to generate macroporosity in scaffolds. Gelatin particles were used with a porogen and combined with a solution of P_LLA in a water/THF mixture. On leaching at 40 °C in distilled water sufficient gelatine remained coated to the P_LLA pore walls significantly enhance osteoblast adhesion and proliferation.

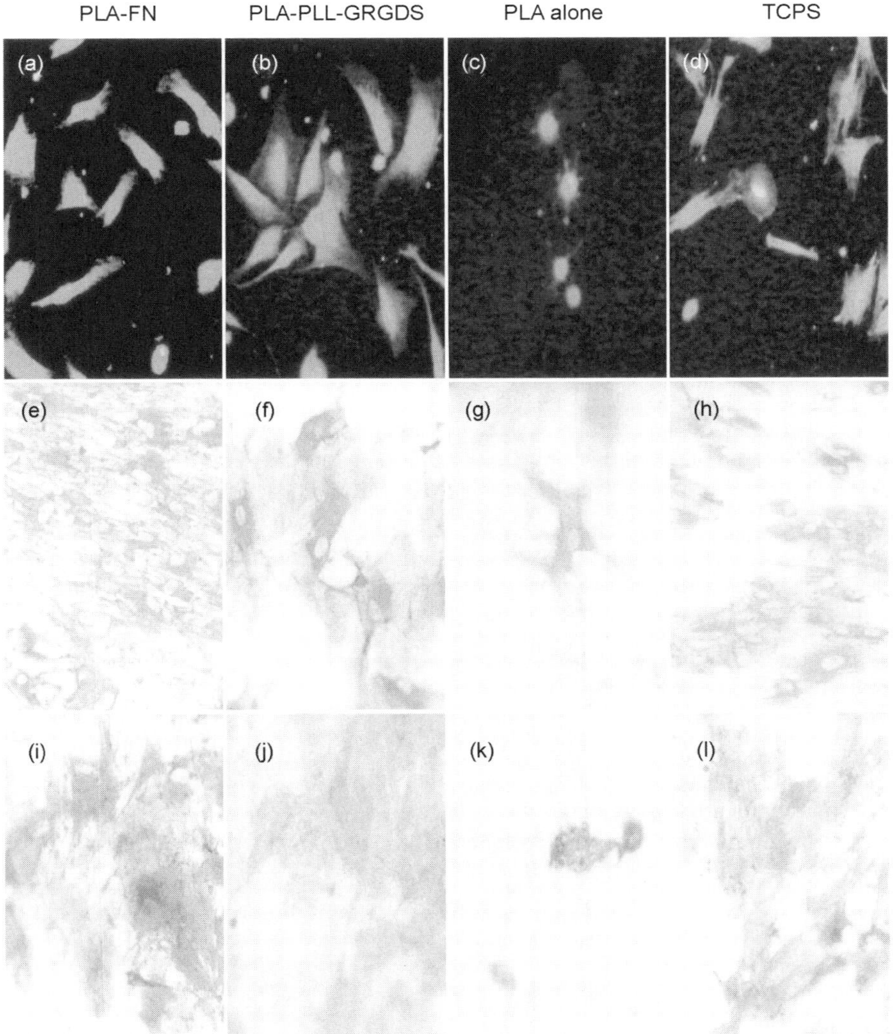

6.3 Surface coating to present fibronectin or an RGD containing molecule. (a–d) Cell attachment and spreading on PLA films (a–c) and tissue culture plastic coated with serum (TCPS) (d) after 5 h as detected by fluorescence microscopy. Magnification ×200. (e–h) Type I collagen immunohistochemistry using LF 67 on PLA films and TCPS. Note enhanced adhesion and expression of type I collagen on modified PLA films. Original magnification ×200. (i–l) Comparable alkaline phosphatase activity detected histochemically on PLA films and TCPS. Negligible cell adhesion was observed on PLA alone. Original magnification ×200. Reproduced from Yang *et al.*, *Bone* **29**, 523, 2001 with permission.

6.5.2 Chemical functionalisation

As discussed in the previous section, chemical functionalisation is important as a method of modifying wettability and subsequent protein adsorption. The use of chemical functionalisation as a preparatory step for covalent attachment is discussed in more detail in Section 6.5.5. Functionalisation is generally achieved by degrading polymer chains in the surface region of the scaffold to create more end groups or by radical reaction with side groups.

Gao *et al.* (1998) have described a broadly applicable method of surface hydrolysis of polyester scaffolds to increase the density of hydroxyl and carboxylic acid groups. 1*N* NaOH was exposed to poly(glycolic acid) (PGA) fibres causing fibre diameter to decrease at a rate of 0.65 μm/min. A serum-dependent enhancement in the adhesion of vascular smooth muscle cells was observed. The method can be applied to many other polymer types with minor tailoring of the exposure time to account for changes in the polymer chemical structure.

Plasma etching provides a rapid method of functionalisation for the enrichment of oxygen and nitrogen-containing species. Chim *et al.* (2003) provide an example of the ability to use gas plasma to surface engineer poly(DL-lactic acid) ($P_{DL}LA$) scaffolds. A glow discharge gas plasma was formed within a pure O_2 environment. Measurement of contact angle in 2D films of the polymer confirmed an 18° drop in water contact angle. Cell proliferation was slightly faster on the treated scaffolds and a marker of cell proliferation was also increased. The authors noted that better cell penetration was observed in the centre of the treated scaffolds.

Wan *et al.* (2006) used an ammonia plasma treatment of P_LLA to improve wettability and cell penetration into 4 mm thick scaffolds. A 30 min exposure time was required for full thickness surface engineering and the authors note a partial degradation of the polymer bulk.

In an approach specifically designed to stimulate interactions with bone cells, Oyane *et al.* (2005) used an O_2 plasma-treated PCL scaffold to reduce the time period required to stimulate the formation of apatite layers.

6.5.3 Entrapment and blending

The concept of surface entrapment engineering (SEE) is that a preformed scaffold can be surface modified by exposing it to a solution of a surface-modifying molecule if the solution acts as a partial solvent for the scaffold material. This approach was first described as the formation of a surface physical interpenetrating network by Desai and Hubbell (1992) and used to entrap PEG into poly(ethylene terephthalate). Our group used a related approach to modify PLA surfaces using a mixture of 2,2,2-trifluoroethanol (TFE) and water containing the surface modifier (Quirk *et al.*, 2000). The composition of the TFE/water mixture must be tailored to gel the PLA surface without causing dissolution of the scaffold material. Figure 6.4 shows an example taken from the work of Quirk *et al.* in which X-ray photoelectron spectroscopy (XPS) is used to measure a high density of entrapment PEG. Next, the approach was extended to simultaneously entrap PEG and PLL-RGD (Quirk *et al.*, 2001b). The resulting

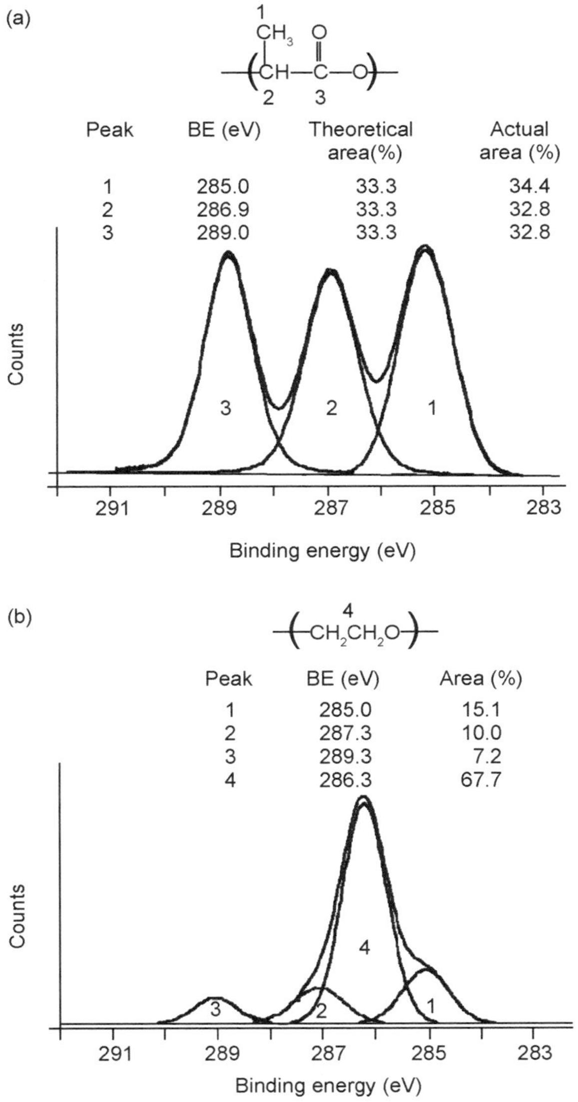

6.4 XPS analysis of surface entrapment PEG in PLA. XPS scans of C 1s regions: (a) PLA before treatment; (b) PLA following PEG modification (10% TFE, 50% w/v PEG, 24 h immersion time. Reproduced with permission from *Macromolecules* **33**, 258, 2000.

surface was both repellant to serum proteins and adhesive to 3T3 fibroblasts. As a result this surface could generate a specific cell adhesion response to a single peptide type and block non-specific protein events.

A further example and extension of the concept of surface entrapment has been described by Liu *et al.* (2005). Using mixtures of dioxane and water as the solvent/non-solvent miscible carrier and gelatine as the surface modifier, this team surface engineered poly(L-lactic acid) (P_LLA) and PLGA. An additional step in the process over the approach of Quirk *et al.* was to chemical crosslink the gelatin to enhance retention during the non-solvent wash stage. Statistically significant improvements in osteoblast cell attachment were measured for the entrapped surfaces with a small additional benefit to the chemical crosslinking. Surface entrapment engineering has also been developed for postfabrication modification of alginate fibres by Hou *et al.* (2005).

Although not a surface engineering approach, the work of Bhattarai *et al.* (2006) shows how surface characteristics can drive the design of the bulk of a scaffold. In this work blends of P_LLA and PEG were electrospun into nano-fibrous mats. At a ratio of P_LLA : PEG of 80 : 20 the highest cell adhesion and proliferation were achieved. Excess PEG reduced cell adhesion owing to excessive leaching.

6.5.4 Thin layer deposition

The major work in this area has been on the use of plasma polymerisation to deposit a thin layer of a functionalised polymer onto the walls of pores within the scaffold. Barry *et al.* (2005) described the use of allylamine deposits within P_{DL}LA foam scaffolds. Surface analysis revealed that deposition of allylamine was more efficient from the plasma than via simple surface grafting. The penetration of the surface coating into the 3D core of the scaffold was restricted but with sufficient exposure time the entire volume of a 10 mm diameter × 4 mm high disc could be surface modified.

In a further study Barry *et al.* (2006) used plasma polymerisation to coat two polymer types and to zone the location and density of each type. Initially, a complete poly(allyl amine) coating was applied throughout the scaffold. This promoted cell penetration into the core of the scaffold but cells still preferentially adhered to the outer regions of the scaffold. Next a sheath region of plasma polymerised hexane was applied to the outer region of the scaffold. Using the slow penetration of the plasma-generated radicals into the scaffold it was possible to restrict the hexane-derived species to the outer region and maintain the cell attractive allyl amine region in the centre. Figure 6.5 provides an example of the work in this paper in which cell population in the centre of the material is favoured over the outer regions.

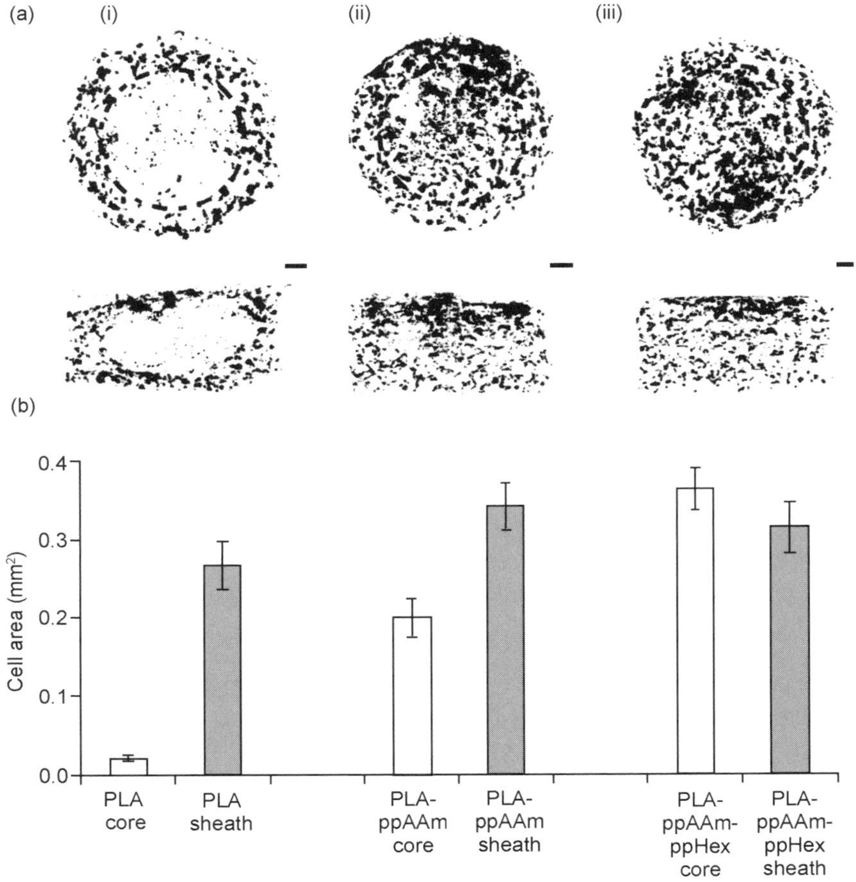

6.5 (a) Scaffolds cultured with 3T3 fibroblasts for 24 h with mild agitation (the scale bars are 1 mm and cells are colour-coded in red): (i) PDLLA, (ii) PDLLA/ppAAm and (iii) PDLLA/ppAAm/ppHex. The lower images show X-ray μCT images from approximately 2 mm slices through the centres of the scaffolds. (b) Cumulative cell area in the 0.01 mm slices through the centres of the scaffold within the core and the sheath denoted by the black dotted lines in (a). The error bars are the standard error in the mean where $n = 20$. Reproduced with permission from Barry *et al.* (*Advanced Materials* **15**, 1143, 2005).

6.5.5 Covalent attachment

Covalent attachment strategies have been widely employed. For many ECM-derived materials or other natural glycoproteins, covalent attachment is readily achieved with a water-soluble carbodiimide reaction. For example, Chou *et al.* (2006) showed that a growth factor, TGF-β1, can be immobilised using 1-ethyl-

3-(3-dimethylaminopropyl)carbodiimide as a crosslinking agent onto a scaffold composed of gelatine, hyaluronic acid and chondroitin-6-sulphate.

For many synthetic polymer scaffolds a barrier to the use of covalent attachment is the lack of reactive groups available. This is often the case for high molecular weight polyesters where end groups are sparse. However, it is realively simple to introduce more reactive groups using any of the methods described throughout this review.

In a complex but very effective example of covalent attachment strategies, Yoo *et al.* (2005) have immobilised hyaluronic acid on the surface of PLGA scaffolds. The initial step in the surface engineering involves including an amine terminated PLGA–PEG–NH_2 diblock copolymer in a blend formulation of the bulk scaffold with the PLGA. This blending ensured a high density of primary amines at the surface of the scaffold for subsequent crosslinking with carboxylic acid groups of the HA.

Yoon *et al.* (2004) have used a simpler blending formulation in which the carboxyl terminal groups of PLGA are aminated by conjugating with hexa-ethylene glycol-diamine. The amine groups were then used to couple the cell adhesion peptide GRGDY and the non-adhesive control peptide GRGEY to films and scaffolds. Cell attachment efficiency of rat bone marrow cells was controlled by increasing the ratio of aminated PLGA versus normal PLGA and by the use of the specific GRGDY sequence from their paper. This strategy has also been employed by Santiago *et al.* (2006) to surface engineer RGD, YIGSR and IKVAV peptides to polycaprolactone scaffolds. Adipose-derived stem cells demonstrated the highest attachment efficiency to the IKVAV presenting scaffolds.

Zhu *et al.* (2006) used the amination strategy to modify flexible films of poly(L-lactide-*co*-caprolactone*)* and then covalently attach collagen and fibronectin via the glutaldehyde crossliking reaction. This approach generated a more physically consistent coating. The fibronectin-coated materials enhanced the culture of oesophageal smooth muscle cells.

For applications of scaffolds in which a high density of biological molecules are required, the use of highly branched PEG star polymers has been employed. Groll *et al.* (2005) showed that a reactive six-armed star PEG could react with glass, silicon or titanium surfaces and present controlled concentrations of the cell adhesion peptide RGD. Cell adhesion was controlled by the number of RGD motifs per PEG star and human mesenchymal stem cells displayed differentiation ability on the surfaces.

Ma *et al.* (2005) have surface engineering P_LLA scaffolds to present both collagen and basic fibroblast growth factor. Their approach, termed 'grafting and coating', begins with the formation of carboxylic acid groups on the polymer surface by exposing the scaffold to H_2O_2 under UV light at 50 °C for 40 min. The exposed carboxylic acid groups are then used as initiation points for the polymerisation of poly(methyl methacrylate) in the form of a surface coating with high densities of further carboxylic acid groups. Covalent coupling of collagen is then achieved using a water-soluble carbodiimide strategy. Finally bFGF was added as an aqueous solution to the collagen phase before covalent anchorage for samples that required a depot of the growth factor within the collagen surface region. Initial studies of chondrocyte viability showed significant increases with collagen coating and a further boost with bFGF.

Yang *et al.* (2002) have targeted the asialoglycoprotein receptor of hepatocytes using an alginate scaffold with galactose covalently attached via aqueous carbodiimide chemistry. The efficient capture of hepatocytes by surface-presented galactose groups resulted in higher cell viability and spheroid formation leading to an improvement in a number of markers of the liver cell functionality.

6.6 Future trends

The techniques described in the previous section provide an excellent toolbox for the modification of conventional tissue engineering scaffold. Current techniques can cover a wide range of scaffold types and surface modifier types. However, it should be noted that these modified surfaces are much simpler in design and biological response than the ECM. The future for this area of scaffold design is to consider how to mimic more complex aspects of ECM functionality while contributing to the simplification of clinical procedures involving tissue engineering and regenerative medicine. Key advances in surface modification will focus on the following.

6.6.1 Spatial control of surface modification to induce tissue architecture

Early examples of this work used 2D patterns of cell adhesion peptides to guide neuronal cells and endothelial cells to form patterned structures (Patel *et al.*, 1998). Recently, Yu and Shoichet (2005) have provided an elegant example of combining architecture control and peptide immobilisation to promote neurite extension in multichannel nerve guide scaffolds. In a further example of spatial control, Koegler and Griffith (2004) have demonstrated the ability to create surface engineered zones within 3D scaffolds using surfactant coating combined with 3D printing. In the 3D printing technique a PLGA 2D layer can be printed from a concentrated collagen solution of the polymer. Sequential layers printed on top of each other generate a 3D structure with spatial control of pore location at a resolution of approximately 100 μm. The 3D printing method was adapted to deposit stripes of the PEG containing surfactant Pluronic F-127. After exposure of the scaffold, a collagen solution, rat osteoblasts were seeded and their distribution, proliferation and functionality recorded. The PEG-rich regions attracted fewer cells but also restricted cell spreading and increased certain bone tissue related functions.

6.6.2 Surfaces that respond to cell requirements

The work of Lutolf *et al.* (2003a,b) highlights the great potential for systems that possess some of the responsivity of the ECM to cells. This work has described a hydrogel system that presents cell adhesion peptides and possesses chemical linkers that are suspectible to cleavage when a cell secretes a matrix metallo-protease. As a result cells can adhere and migrate through the hydrogel, creating new surfaces as they decided to enter further into the scaffold. This concept addresses the issue that surface interactions occur throughout tissue formation and the type and number of interactions is driven by cell demand.

6.6.3 Bespoke materials that self-assemble to present tailored surfaces

To date many surface modification strategies have been developed to address fundamental shortcomings in available materials for scaffold manufacture. Increasingly, new materials are being designed specifically for tissue engineer-ing and a key feature of these materials is the ability to spontaneous form highly porous materials with ideal surface properties. For example, Jayawarna *et al.* (2006) have decribed a peptide that self-assembles into a nanostructured hydrogel with the potential to ensure high presentation densities of cell adhesion molecules.

6.7 References

Aota, S., M. Nomizu, *et al.*, (1994). 'The short amino-acid-sequence Pro-His-Ser-Arg-Asn in human fibronectin enhances cell-adhesive function.' *Journal of Biological Chemistry* **269**(40): 24756–24761.

Barry, J. J. A., M. Silva, *et al.* (2005). 'Using plasma deposits to promote cell population of the porous interior of three-dimensional poly(D,L-lactic acid) tissue-engineering scaffolds.' *Advanced Functional Materials* **15**(7): 1134–1140.

Barry, J. J. A., D. Howard, *et al.* (2006). 'Using a core-sheath distribution of surface chemistry through 3D tissue engineering scaffolds to control cell ingress.' *Advanced Materials* **18**(11): 1406–1410.

Beer, J. H., K. T. Springer, *et al.* (1992). 'Immobilized Arg-Gly-Asp (Rgd) peptides of varying lengths as structural probes of the platelet glycoprotein-Iib/Iiia receptor.' *Blood* **79**(1): 117–128.

Bhattarai, S. R., N. Bhattarai, *et al.* (2006). 'Hydrophilic nanofibrous structure of polylactide; fabrication and cell affinity.' *Journal of Biomedical Materials Research Part A* **78A**(2): 247-257.

Chim, H., J. L. Ong, *et al.* (2003). 'Efficacy of glow discharge gas plasma treatment as a surface modification process for three-dimensional poly (D,L-lactide) scaffolds.' *Journal of Biomedical Materials Research Part A* **65A**(3): 327–335.

Chou, C. H., W. T. K. Cheng, *et al.* (2006). 'TGF-beta 1 immobilized tri-co-polymer for articular cartilage tissue engineering.' *Journal of Biomedical Materials Research Part B – Applied Biomaterials* **77B**(2): 338–348.

Cook, A. D., J. S. Hrkach, *et al.* (1997). 'Characterization and development of RGD-peptide-modified poly(lactic acid-co-lysine) as an interactive, resorbable biomaterial.' *Journal of Biomedical Materials Research* **35**(4): 513–523.

Desai, N. P. and J. A. Hubbell (1992). 'Surface physical interpenetrating networks of poly(ethylene-terephthalate) and poly(ethylene oxide) with biomedical applications.' *Macromolecules* **25**(1): 226–232.

Farndale, R. W., P. R. M. Silijander, *et al.* (2003). Collagen–platelet interactions: recognition and signalling. *Proteases and the Regulation of Biological Processes.* **70:** 81–94.

Gao, J. M., L. Niklason, *et al.* (1998). 'Surface hydrolysis of poly(glycolic acid) meshes increases the seeding density of vascular smooth muscle cells.' *Journal of Biomedical Materials Research* **42**(3): 417-424.

Gobin, A. S. and J. L. West (2003). 'Val-Ala-Pro-Gly, an elastin-derived non-integrin ligand: Smooth muscle cell adhesion and specificity.' *Journal of Biomedical Materials Research Part A* **67A**(1): 255–259.

Groll, J., J. Fiedler, *et al.* (2005). 'A novel star PEG-derived surface coating for specific cell adhesion.' *Journal of Biomedical Materials Research Part A* **74A**(4): 607–617.

Hasenbein, M. E., T. T. Andersen, *et al.* (2002). 'Micropatterned surfaces modified with select peptides promote exclusive interactions with osteoblasts.' *Biomaterials* **23**(19): 3937–3942.

Hirano, Y., M. Okuno, *et al.* (1993). 'Cell-attachment activities of surface immobilized oligopeptides Rgd, Rgds, Rgdv, Rgdt, and Yigsr toward 5 cell-lines.' *Journal of Biomaterials Science-Polymer Edition* **4**(3): 235–243.

Hou, Q. P., R. Freeman, *et al.* (2005). 'Novel surface entrapment process for the incorporation of bioactive molecules within preformed alginate fibers.' *Biomacromolecules* **6**(2): 734–740.

Hubbell, J. A., S. P. Massia, *et al.* (1991). 'Endothelial cell-selective materials for tissue engineering in the vascular graft via a new receptor.' *Bio-Technology* **9**(6): 568–572.

Hynes, R. O. (1987). 'Integrins – a family of cell-surface receptors.' *Cell* **48**(4): 549–554.

Hynes, R. O. (1992). 'Integrins – versatility, modulation, and signaling in cell-adhesion.' *Cell* **69**(1): 11–25.

Hynes, R. O. (2002). 'Integrins: bidirectional, allosteric signaling machines.' *Cell* **110**(6): 673–687.

Jang, J. H., Z. Bengali, *et al.* (2006). 'Surface adsorption of DNA to tissue engineering scaffolds for efficient gene delivery.' *Journal of Biomedical Materials Research Part A* **77A**(1): 50–58.

Jayawarna, V., M. Ali, *et al.* (2006). 'Nanostructured hydrogels for three-dimensional cell culture through self-assembly of fluorenylmethoxycarbonyl-dipeptides.' *Advanced Materials* **18**(5): 611–614.

Kao, W. J., J. A. Hubbell, *et al.* (1999). 'Protein-mediated macrophage adhesion and activation on biomaterials: a model for modulating cell behavior.' *Journal of Materials Science – Materials in Medicine* **10**(10–11): 601–605.

Kao, W. J., D. Lee, *et al.* (2001). 'Fibronectin modulates macrophage adhesion and FBGC formation: the pole of RGD, PHSRN, and PRRARV domains.' *Journal of Biomedical Materials Research* **55**(1): 79–88.

Knight, C. G., L. F. Morton, *et al.*, (1999). 'Collagen–platelet interaction: Gly-Pro-Hyp is uniquely specific for platelet Gp VI and mediates platelet activation by collagen.' *Cardiovascular Research* **41**(2): 450–457.

Koegler, W. S. and L. G. Griffith (2004). 'Osteoblast response to PLGA tissue engineering scaffolds with PEO modified surface chemistries and demonstration of patterned cell response.' *Biomaterials* **25**(14): 2819–2830.

Krijgsman, B., A. M. Seifalian, *et al.* (2002). 'An assessment of covalent grafting of RGD peptides to the surface of a compliant poly(carbonate-urea)urethane vascular conduit versus conventional biological coatings: its role in enhancing cellular retention.' *Tissue Engineering* **8**(4): 673-680.

Liu, X. H., Y. J. Won, *et al.* (2005). 'Surface modification of interconnected porous scaffolds.' *Journal of Biomedical Materials Research Part A* **74A**(1): 84-91.

Liu, X. H., Y. J. Won, *et al.* (2006). 'Porogen-induced surface modification of nano-fibrous poly(L-lactic acid) scaffolds for tissue engineering.' *Biomaterials* **27**(21): 3980–3987.

Liu, Y. P. and W. Y. J. Kao (2002). 'Human macrophage adhesion on fibronectin: The role of substratum and intracellular signalling kinases.' *Cellular Signalling* **14**(2): 145–152.

Longhurst, C. M. and L. K. Jennings (1998). 'Integrin-mediated signal transduction.' *Cellular and Molecular Life Sciences* **54**(6): 514–526.

Lutolf, M. P., J. L. Lauer-Fields, *et al.* (2003a). 'Synthetic matrix metalloproteinase-sensitive hydrogels for the conduction of tissue regeneration: engineering cell-invasion characteristics.' *Proceedings of the National Academy of Sciences of the United States of America* **100**(9): 5413–5418.

Lutolf, M. R., F. E. Weber, *et al.* (2003b). 'Repair of bone defects using synthetic mimetics of collagenous extracellular matrices.' *Nature Biotechnology* **21**(5): 513–518.

Ma, Z. W., C. Y. Gao, *et al.* (2005). 'Cartilage tissue engineering PLLA scaffold with surface immobilized collagen and basic fibroblast growth factor.' *Biomaterials* **26**(11): 1253–1259.

Maeshima, Y., U. L. Yerramalla, *et al.* (2001). 'Extracellular matrix-derived peptide binds to alpha(v)beta(3) integrin and inhibits angiogenesis.' *Journal of Biological Chemistry* **276**(34): 31959–31968.

Maheshwari, G., G. Brown, *et al.* (2000). 'Cell adhesion and motility depend on nanoscale RGD clustering.' *Journal of Cell Science* **113**(10): 1677–1686.

Mann, B. K. and J. L. West (2002). 'Cell adhesion peptides alter smooth muscle cell adhesion, proliferation, migration, and matrix protein synthesis on modified surfaces and in polymer scaffolds.' *Journal of Biomedical Materials Research* **60**(1): 86–93.

Mizuno, M., R. Fujisawa, *et al.* (2000). 'Type I collagen-induced osteoblastic differentiation of bone-marrow cells mediated by collagen-alpha 2 beta 1 integrin interaction.' *Journal of Cellular Physiology* **184**(2): 207–213.

Moyano, J. V., A. Maqueda, *et al.* (2003). 'A synthetic peptide from the heparin-binding domain III (repeats III4-5) of fibronectin promotes stress-fibre and focal-adhesion formation in melanoma cells.' *Biochemical Journal* **371**: 565–571.

Oyane, A., M. Uchida, *et al.* (2005). 'Simple surface modification of poly(epsilon-caprolactone) to induce its apatite-forming ability.' *Journal of Biomedical Materials Research Part A* **75A**(1): 138–145.

Patel, N., R. Padera, *et al.* (1998). 'Spatially controlled cell engineering on biodegradable polymer surfaces.' *Faseb Journal* **12**(14): 1447–1454.

Pierschbacher, M. D. and E. Ruoslahti (1984). 'Cell attachment activity of Fibronectin can be duplicated by small synthetic fragments of the molecule.' *Nature* **309**(5963): 30–33.

Pierschbacher, M. D. and E. Ruoslahti (1987). 'Influence of stereochemistry of the sequence Arg-Gly-Asp-Xaa on binding-specificity in cell-adhesion.' *Journal of Biological Chemistry* **262**(36): 17294–17298.

Pytela, R., M. D. Pierschbacher, *et al.* (1985). 'Identification and isolation of a 140-Kd cell-surface glycoprotein with properties expected of a fibronectin receptor.' *Cell* **40**(1): 191–198.

Quirk, R. A., M. C. Davies, *et al.* (2000). 'Surface engineering of poly(lactic acid) by entrapment of modifying species.' *Macromolecules* **33**(2): 258–260.

Quirk, R. A., W. C. Chan, *et al.* (2001a). 'Poly(L-lysine)-GRGDS as a biomimetic surface modifier for poly(lactic acid).' *Biomaterials* **22**(8): 865–872.

Quirk, R. A., M. C. Davies, *et al.* (2001b). 'Controlling biological interactions with poly(lactic acid) by surface entrapment modification.' *Langmuir* **17**(9): 2817–2820.

Rouhi, A. M. (1999). 'Contemporary biomaterials.' *Chemical & Engineering News* **77**(3): 51–52.

Ruoslahti, E. and B. Obrink (1996). 'Common principles in cell adhesion.' *Experimental Cell Research* **227**(1): 1–11.

Santiago, L. Y., R. W. Nowak, *et al.* (2006). 'Peptide-surface modification of poly(caprolactone) with laminin-derived sequences for adipose-derived stem cell applications.' *Biomaterials* **27**(15): 2962–2969.

Schense, J. C., J. Bloch, *et al.* (2000). 'Enzymatic incorporation of bioactive peptides into fibrin matrices enhances neurite extension.' *Nature Biotechnology* **18**(4): 415–419.

Shaw, D. and M. S. Shoichet (2003). 'Toward spinal cord injury repair strategies: Peptide surface modification of expanded poly(tetrafluoroethylene) fibers for guided neurite outgrowth *in vitro*.' *Journal of Craniofacial Surgery* **14**(3): 308–316.

Sugawara, T. and T. Matsuda (1995). 'Photochemical surface derivatization of a peptide-containing Arg-Gly-Asp (Rgd).' *Journal of Biomedical Materials Research* **29**(9): 1047–1052.

Tashiro, K., G. C. Sephel, *et al.* (1989). 'A synthetic peptide containing the Ikvav sequence from the A-chain of laminin mediates cell attachment, migration, and neurite outgrowth.' *Journal of Biological Chemistry* **264**(27): 16174–16182.

Tashiro, K., A. Monji, *et al.*, (1999). 'An IKLLI-containing peptide derived from the laminin alpha 1 chain mediating heparin-binding, cell adhesion, neurite outgrowth and proliferation, represents a binding site for integrin alpha 3 beta 1 and heparan sulphate proteoglycan.' *Biochemical Journal* **340**: 119–126.

Walluscheck, K. P., G. Steinhoff, *et al.* (1996). 'Improved endothelial cell attachment on ePTFE vascular grafts pretreated with synthetic RGD-containing peptides.' *European Journal of Vascular and Endovascular Surgery* **12**(3): 321–330.

Wan, Y. Q., C. F. Tu, *et al.* (2006). 'Influences of ammonia plasma treatment on modifying depth and degradation of poly(L-lactide) scaffolds.' *Biomaterials* **27**(13): 2699–2704.

Wayner, E. A. and N. L. Kovach (1992). 'Activation-dependent recognition by hematopoietic-cells of the Ldv sequence in the V-region of fibronectin.' *Journal of Cell Biology* **116**(2): 489–497.

Woerly, S., V. Doan, *et al.* (2001). 'Spinal cord reconstruction using NeuroGel (TM) implants and functional recovery after chronic injury.' *Journal of Neuroscience Research* **66**(6): 1187–1197.

Wong, J. Y., Z. P. Weng, *et al.* (2002). 'Identification and validation of a novel cell-recognition site (KNEED) on the 8th type III domain of fibronectin.' *Biomaterials* **23**(18): 3865–3870.

Yamamoto, M., K. Yamamoto, *et al.* (1993). 'Type-I collagen promotes modulation of cultured rabbit arterial smooth-muscle cells from a contractile to a synthetic

phenotype.' *Experimental Cell Research* **204**(1): 121–129.
Yamaoka, T., Y. Hotta, *et al.* (1999). 'Synthesis and properties of malic acid-containing functional polymers.' *International Journal of Biological Macromolecules* **25**(1–3): 265–271.
Yang, J., M. Goto, *et al.* (2002). 'Galactosylated alginate as a scaffold for hepatocytes entrapment.' *Biomaterials* **23**(2): 471–479.
Yang, X. B., H. I. Roach, *et al.* (2001). 'Human osteoprogenitor growth and differentiation on synthetic biodegradable structures after surface modification.' *Bone* **29**(6): 523–531.
Yasuda, T., A. R. Poole, *et al.* (2003). 'Involvement of CD44 in induction of matrix metalloproteinases by COOH-terminal heparin-binding fragment of fibronectin in human articular cartilage in culture.' *Arthritis and Rheumatism* **48**(5): 1271–1280.
Yoo, H. S., E. A. Lee, *et al.* (2005). 'Hyaluronic acid modified biodegradable scaffolds for cartilage tissue engineering.' *Biomaterials* **26**(14): 1925–1933.
Yoon, J. J., S. H. Song, *et al.* (2004). 'Immobilization of cell adhesive RGD peptide onto the surface of highly porous biodegradable polymer scaffolds fabricated by a gas foaming/salt leaching method.' *Biomaterials* **25**(25): 5613–5620.
Yu, T. T. and M. S. Shoichet (2005). 'Guided cell adhesion and outgrowth in peptide-modified channels for neural tissue engineering.' *Biomaterials* **26**(13): 1507–1514.
Zhu, H. G., J. Ji, *et al.*, (2004). 'Biomacromolecules electrostatic self-assembly on 3-dimensional tissue engineering scaffold.' *Biomacromolecules* **5**(5): 1933–1939.
Zhu, Y. B., K. S. Chian, *et al.*, (2006). 'Protein bonding on biodegradable poly(L-lactide-co-caprolactone) membrane for esophageal tissue engineering.' *Biomaterials* **27**(1): 68-78.

7
Combining tissue engineering and drug delivery

N TIRELLI and F CELLESI, University of Manchester, UK

7.1 Introduction

The tasks that tissue engineering aims at accomplishing (tissue replacement and regeneration) are characterised by the challenging need to control a hierarchically complex system, which spans from the level of individual cells, through that of their collective behaviour, to that of complex tissue architectures. Furthermore, this system evolves, and the real hurdle is therefore to control it in a dynamic fashion during its development, i.e. during the morphogenesis of the tissue.

Many, probably most, tissue engineering solutions make use of biological signals, which are naturally involved in modulating this morphogenetic process. Implantable tissue engineering constructs could thus be seen not only as physical, possibly remodellable supports for cell adhesion and migration, but also as sophisticated delivery systems for the molecules (often, but not necessarily, macromolecules) that bear the signals.

The kind of signal, its intensity and the possibility to modulate it, would all depend on physical state (immobilised vs. free to diffuse), identity and physical properties of the molecule (size, diffusion coefficient, interactions), and on its concentration and spatial localisation (gradients). The situation is therefore complex; I will try to reduce it to more elementary problems narrowing it down where possible and introducing *caveats* and *notes* to point out interesting spin-outs.

7.1.1 Signalling molecules: physical state (*where?*)

We may individuate three limiting cases: the signalling molecules can be:

- *In solution*; they could therefore diffuse over long distances, possibly also affecting cells far from the target site. Soluble molecules are at the basis of paracrine or endocrine intercellular communication, and trigger the vast majority of systemic reactions, as well as a number of localised ones (inflammation, foreign body reactions, tissue decay, etc.).

- *Aggregated/complexed*; they would therefore produce, possibly together with other molecules, objects sized from a few nanometres to 1 μm (colloids). Owing to their small size, these objects substantially retain good diffusion properties (see above): on the other hand, the increased local concentration (clustering) of the assembled molecules could dramatically change their mode of action.
- *Immobilised* in/on a solid-like phase; their influence on cells would therefore depend on their accessibility: for example, they can be accessed only through cell-mediated degradation of a matrix, or when cells migrate within a porous scaffold. The communication is established through direct contact of cells with a solid support and mostly influences cell differentiation, adhesion and motility.

7.1.2 Signalling molecules: identity (*who?*)

In the overwhelming majority of cases the signalling compounds used in tissue engineering fall in the categories of cytokines, chemokines, growth factors or immunomodulators; the nomenclature of these classes of compounds has often generated confusion, it is therefore useful to indicate how these terms are used here.

- The terms cytokines, chemokines and growth factors are always referred to as regulatory proteins, while immunomodulators may have a non-proteic composition.
- Most authors agree in considering chemokines (whose specific effects are in the activation and mobilisation of leukocytes) a subclass of cytokines (more generally defined as mediators of immune responses).
- Under the term immunomodulators we can also comfortably consider both cytokines and a number of other often non-proteic and also non-macromolecular compounds, with the common characteristics of acting on immune responses.
- The inclusion of growth factors in the cytokine class is more debated, and the majority of researchers probably prefer to consider them separately, highlighting their function in cell proliferation, differentiation, migration, motility and adhesion, although they often have immunomodulating functions.

7.1.3 Aims and objectives

This chapter will only marginally review the effects of signalling molecules in tissue engineering. We will focus on how, when externally provided, they are made available to target biological systems. In order to narrow down this complex landscape, we will introduce two limitations: (1) since the very concept of delivery is intimately related to diffusion, we will hereafter concentrate only on the use of *diffusible signalling molecules*, i.e. those in either a completely soluble or in an aggregated/complexed form (the first two bullet points in Section 7.1.1), and (2) owing to the broader spectrum of inducible responses (they can be rightly defined as *morphogens*), growth factors have always been the most employed signalling molecules in tissue engineering. We will hereafter specifically focus our attention on them.

7.2 Growth factor (GF) delivery

7.2.1 A first caveat

Each GF can be produced by several, not necessarily related cell lines and can have different effects on different cell types. In a very generic sense, they up- or down-regulate the synthesis of receptors, proteins and even other growth factors in a concentration-dependent manner. This interconnection makes it almost impossible to single out the effect of a single GF, above all in the complex and continuously evolving phenomena of tissue differentiation and regeneration. In addition, the action of GFs can also depend on cell location within a tissue and on the cell cycle state. Owing to these limitations, the use of GFs in tissue engineering has always been largely phenomenological, with limited possibilities to achieve an even qualitative predictability.

7.2.2 Strategies for GF release

The use of GF action in tissue engineering follows two general strategies, which exploit external bodies either for a direct GF release (strategy 1), or for stimulating an endogenous GF release (strategy 2); these approaches are graphically summarised in Fig. 7.1.

1. The *GFs diffuse from the site of application* (e.g. an implanted material, a blood vessel) and influence cells both at the site and in the surrounding tissues. The GFs may:
    - be originally present in an implant (*growth factor therapy*). Although this approach has been widespread (down to extensive clinical trials, a quick summary of the most common systems is provided in Table 7.1), its success has been fairly limited to date. A major drawback is that unequivocal effects are often observed only at high dosages, which correspond to concentrations up to several orders of magnitude higher than the naturally occurring ones, e.g. up to three orders of magnitude for bone morphogenetic proteins (BMPs) (Giannoudis and Tzioupis, 2005; Luhmann *et al.*, 2005). These high doses originate from the need to have a sustained presence of GFs for prolonged periods, which is hampered by the fact that simple routes of administration offer (1) negligible protection and thus poor stability: the released GFs are very sensitive e.g. to the action of proteases, and (2) non-sustained delivery: the GFs are all delivered at the beginning (see Section 7.3.1).

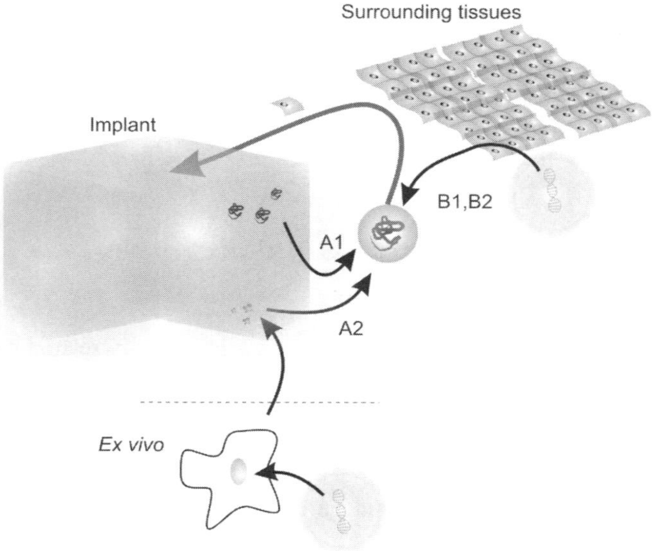

7.1 A tissue engineering construct may exploit GF action by releasing pre-loaded GFs (A1) or producing them *in situ* through encapsulated cells (A2), which will possibly recruit cells from the surrounding tissues. Alternatively, these cells may be forced to (over)express GFs acting therefore as a non-immunogenic and renewable source of them (B1 and B2).

- Be synthesised there by appropriate, encapsulated cells (*cell therapy*); this approach, however, requires standardised and effective procedures for encapsulation and reimplantation of autologous cells (for reducing foreign body reactions), which are possibly genetically engineered to maximise GF production (*combined cell and gene therapy*).

2. The exploitation of *'pro-signalling molecules'* (Bonadio, 2000; Luo *et al.*, 2005; Kofron and Laurencin, 2006), whose task is to induce the production of the signalling molecules *in situ*, generally by forcing cells in and around the implanted material to (over-)express GFs; this may happen:
   - by modulating the efficiency of specific biosynthetic paths, but without any change to nucleic material (transcription factors); or
   - by the use of nucleic acids that encode them (*in vivo* gene therapy).

Transcription factors and nucleic acids for growth factor production

The expression of a transcription factor (proteins that bind to specific DNA sequences and modulate the efficiency and rate of transcription) may be confined to a specific cell type, controlling the expression of genes that characterise the differentiated state of these cells; this control is therefore essential, e.g. in embryo topobiology and in progenitor cell differentiation. In a logical analogy, the local delivery of transcription factors, or of drugs that

Table 7.1 Most commonly used growth factors in tissue engineering[a]

Growth factor (family)	Acronym	Effects	Applications
Bone morphogenetic proteins	BMPs[b]	Differentiation and migration of bone-forming cells	Osteoinduction (BMP-2, BMP-7 for promoting spine fusion and tibial fracture union; less, BMP-3) (Reddi, 1998)
Transforming growth factor beta	TGF-βs	Proliferation and differentiation of bone-forming cells; chemoattraction of fibroblasts; extracellular matrix synthesis	Osteoinduction (TGF-β1 and, less, TGF-β2 and TGF-β3) (Erlebacher *et al.*, 1998), cartilage regeneration (TGF-β1) (Jakob *et al.*, 2001), wound healing (TGF-β1) (Luhmann *et al.*, 2005)
Fibroblast growth factors	FGFs	Wide range cell proliferation	Osteoinduction (bFGF) (Marie, 2003), angiogenesis (aFGF and bFGF) (Folkman and Shing, 1992), nerve regeneration (aFGF and bFGF) (Eckenstein, 1994)
Insulin-like growth factors	IGFs	Proliferation of many cell types, primarily of foetal origin	Osteoinduction (IGF-I and, less, IGF-II) (Ohlsson *et al.*, 1998)
Platelet-derived growth factors	PDGFs	Proliferation and chemoattraction of smooth muscle cells; extracellular matrix synthesis and deposition	Osteoinduction (PDGF BB) (Heldin *et al.*, 1999), angiogenesis (Heldin *et al.*, 1999; Richardson *et al.*, 2001), wound healing (Heldin *et al.*, 1999)
Vascular endothelial growth factor	VEGF	Migration, proliferation and survival of endothelial cells	Vascular engineering (Folkman and Shing, 1992)
Epidermal growth factor	EGF	Proliferation of epithelial, mesenchymal and fibroblast cells	Wound healing (Bennet and Schulz, 1993), migration and differentiation of neural stem cells (Kuhn *et al.*, 1997)
Keratinocyte growth factor	KGF	Mitogenic for epithelial cells	Wound healing (Werner *et al.*, 1992)
Nerve growth factor	NGF	Axonal growth and neural cell survival	Neurite extension in central and peripheral nervous system (Thoenen, 1995)

[a] *GF production.* The growth factors used in tissue regeneration or therapy, even if they are mammalian or human proteins, are recombinantly expressed in bacteria, such as *Escherichia coli*; they are therefore identified with the prefix rh, as for the recombinant human bone morphogenetic proteins (rhBMPs), which have recently (10/05) FDA approval in clinical osteoinduction trials. In a number of cases truncated forms are produced: e.g. palifermin (commercially known as Kepivance) is a truncated form of the keratinocyte growth factor (KGF).

[b] *Bone morphogenetic proteins (BMPs).* It is noteworthy that, although not termed GFs (their discovery is antecedent to the use of the expression GF (Urist, 1965)), these proteins not only operate as growth factors, but, owing to structural similarity and analogous binding interactions, can also be considered members of TGF-β family.

Table 7.2 Some transcription factors and nucleic acids used in growth factor-based therapies

Agent[a]	Effects	Applications
PDGF (encoded molecule)	Gene delivery *in vitro* and *in vivo* from collagen matrix	Wound healing (Chandler *et al.*, 2000)
PDGF (encoded molecule)	PDGF plasmid DNA released from PLGA matrices, leading to enhanced vascularisation of tissue	Wound healing (Salvay and Shea, 2006)
BMP-2 (encoded molecule)	Bone production by adenoviral-mediated direct gene delivery	Bone regeneration (Musgrave *et al.*, 1999)
VEGF (encoded molecule)	Adenoviral-mediated gene transfer induced pericardial VEGF expression in dogs	Angiogenesis (Lazarous *et al.*, 1999)
Human parathyroid hormone (hPTH 1-34) (encoded molecule)	Predictable formation of normal new bone in canine models, obtained by gene delivery from collagen sponges	Bone regeneration (Bonadio, 2000)
Early growth response factor 1 (Egr-1)	Production of several genes encoding cytokines, adhesion molecules, members of the coagulation cascade, and growth factors such as aFGF, bFGF, TGF-β1, PDGF-A and -B, VEGF, and IGF-II	Wound healing, angiogenesis, bone formation (Braddock *et al.*, 2001)
fos-related antigen-1 (Fra-1)	Fra-I belongs to the activator protein-1 (AP-1) family of transcription factors, which are involved in osteoblast maturation *in vitro* and *in vivo*	Bone regeneration (Jochum *et al.*, 2000)

[a] *Agent:* we here report the encoded molecule for plasmid DNAs or the name of molecule for transcription factors.

modulate their action, could allow the local control of gene expression in differentiating cells within a developing tissue. At a hierarchically superior level, nucleic acids, too, can be used for forcing the (over)expression of specific molecules (e.g. GFs) at the site of application.

In the strategies based on the exploitation of cells (A2, B1 and B2), the effects are prolonged not only in time but also in space. This is specifically true when cells in the tissues surrounding the implant are involved (B1 and B2): in a simplistic view, cells act as catalysts, providing a multiplicative effect. Indeed GF-only therapy is often predicted to be largely ineffective in the treatment of large defects, which could theoretically be treated by cell-based therapies (Nussenbaum and Krebsbach, 2006). On the other hand, the use of non-autologous and/or genetically engineered cells poses a number of problems (immunogenic and regulatory) that have substantially hindered the clinical applications of this approach.

Hereafter, we will focus mostly on the direct delivery of signalling molecules (GFs). Gene and cell therapies are complex fields that require a separate and long description, we will here present only a brief overview (Table 7.2) of some systems among the most common ones, while the reader is advised to refer to the literature (Bonadio, 2000; Luo *et al.*, 2005; Sohier *et al.*, 2006) for a more complete review of the field.

A note on the delivery of complex actives (e.g. GFs) vs. more 'classical' drugs

The vast majority of the aforementioned signalling or pro-signalling (macro)molecules are readily water soluble. Their delivery could therefore be performed at the site of tissue engineering, e.g. by simple parenteral route (intravenous or topical injection of their solutions), of by using macro-, micro- or nano-sized carriers, but always without the stringent need of a solubilising structure. On the other hand, the delivery of a GF aims at solving the problem of its sustained availability, therefore must improve its stability and/or modify its time and space distribution.

Therefore, the choice to load/solubilise a proteinaceous active, such as a GF, in a carrier is mainly related to the wish to modulate its release. Most low molecular weight drugs, on the other hand, are poorly soluble in water, therefore in their case the task of a carrier is also to increase the overall amount of drug administered.

7.3 Signalling molecules in solution (parenteral administration)

An appropriate choice of site and of modality of administration of active molecules could allow modulation of the rate at which they reach systemic circulation, that is how long they could remain in a peripheral tissue, or how quickly they can reach a remote location. Topical release (e.g. subcutaneous or intratumour injection) could therefore allow for longer permanence at a target site, while systemic release (e.g. intravenous injection) is better suited for affecting not easily accessible or very dispersed targets. In fact, parenteral (→ systemic) administration has not been frequently employed for promoting local effects in peripheral tissues, while a number of examples can be found for systemic applications.

7.3.1 Examples of systemic administration

FGF-2 and VEGF have been the object of extensive studies for promoting therapeutic angiogenesis after intravenous administration (e.g. to patients soon after thromboembolic stroke) but the number of adverse side effects has made their clinical trials unsuccessful (Saltzman and Olbricht, 2002). Nerve growth factor (NGF) has been tested in different clinical applications, including systemic administration in the brain and subcutaneous administration to reduce a sensory neuropathy associated with human immunodeficiency virus (HIV) (McArthur *et al.*, 2000), although no consistent clinical results were obtained for FDA approval. Topical application of NGF led to healing of corneal neurotrophic ulcers, although this local therapy has not yet been through extensive trials (Lambiase *et al.*, 1998).

7.3.2 Performance

Several mechanisms can be identified to explain the general failure for the direct therapeutic applications of soluble factors in solution form. Firstly, growth factors typically have a short half-life once they are introduced into the body and are rapidly eliminated. Although very small quantities (picograms to nanograms) are necessary to generate a cellular response, growth factors are rapidly degraded once secreted. The biological half-lives of platelet-derived growth factor (PDGF), basic fibroblast growth factor (bFGF or FGF-2), and VEGF, for example, are limited to a range of 2–50 min when intravenously injected (Chen and Mooney, 2003). Secondly, a systemic delivery of growth factors can lead to uncontrolled distribution of these molecules throughout the body with unpredictable concentrations in peripheral organs, thus generating undesirable systemic effects and toxicity (Yancopoulos *et al.*, 2000).

7.4 Signalling molecules physically entrapped in a matrix

7.4.1 What do we want to happen out of the matrix?

Controlling spatial localisation (gradients)

A key issue in the delivery of soluble factors from an implanted matrix is the *penetration depth*, i.e. how far GFs can diffuse while retaining a measurable activity. It should be also noted that in this region the GFs will always exhibit a *concentration profile*, being more dilute the further they are from the implant. Since cell response is generally sensitive to concentration gradients, their presence may indeed be advantageous for stimulating directional or spatial changes during tissue morphogenesis, e.g. directed cell recruitment and migration.

Using a homogeneously loaded matrix, the extent and the depth of the concentration gradient are determined by the balance between free diffusion and clearance (= molecules removed by degradation, metabolism, binding or partitioning into capillaries and then removal through blood circulation). For example, a more diffusible and stable molecule will therefore cover a larger region of tissue, but with a less steep concentration gradient.

In the absence of vascularisation, diffusion takes place in and through interstitial liquids and for most agents, including proteins and small drugs, this penetration depth is confined within ~1 mm from the delivery source (Saltzman and Olbricht, 2002). In many cases, this limited penetration may be beneficial: fewer adverse effects and confinement of the GF-induced processes to specific anatomic regions. In other cases, however, it could be desirable to deliver an agent over larger regions, e.g. for a larger recruitment of cells. The most effective, and probably the only, way to obtain deeper penetration in peripheral tissues would be to engineer the signalling molecule structure, e.g. decreasing the likelihood of enzymatic degradation by chemical derivatisation (succinylation, acetylation), by conjugation to fusion proteins or PEG, or by complexation; on the other hand, these actions may affect the signalling activity of the molecule too.

A completely different and often undesired situation arises when these factors may reach a far-reaching liquid circulation (blood, lymph), as could easily happen, e.g. in vascular engineering. When a GF reaches blood, dilution and/or systemic effects are difficult to avoid and the problems already seen in parenteral administration arise again; lymph, due to its much slower circulation, should pose much less serious problems. In order to keep the GF delivery effects localised to a specific tissue, it is therefore often beneficial to avoid a massive transfer to blood circulation, a delicate issue in the engineering of strongly vascularised tissues (e.g. blood, liver, muscles and, clearly, blood vessels themselves).

Controlling delivery kinetics

Tissue morphogenesis, similar to what happens, for example in embryonic development (Lutolf and Hubbell, 2005), is very sensitive not only to the concentration of a morphogen and its spatial modulation, but also to its time behaviour. In a general case, tissues need to be exposed to appreciable concentrations of these factors for days or weeks in order to obtain an effective regeneration. This is probably the best-known problem in any field of drug release: most encapsulating matrices (and, specifically, the vast majority of those employed in tissue engineering) provide first-order release kinetics with a more or less pronounced decrease of the release rate with time, while zero-order kinetics (rate independent of time) would be desirable for most applications. The details of the release process, however, matter and they strongly depend on the composition and morphology of the matrix.

7.4.2 What do we want to happen in the matrix?

The essence of the matrix lies in providing protection to actives as long as they are encapsulated, and in releasing them in a sustained fashion. The release of any encapsulated active is controlled by its diffusion properties in the matrix, which may be affected by controlled process matrix degradation (Boontheekul and Mooney, 2003); both diffusion and degradation are essentially controlled by matrix morphology and composition.

What is the matrix? (morphology and composition)

As a general requirement, the matrix has to allow easy diffusion of dispersed molecules: for example, nutrients and oxygen must quickly reach cells that are either entrapped in or migrating to and through the matrix, and signalling molecules, too, may be required to move quickly, once released. The simplest and also most common way to achieve quick, long-range diffusion of substrates that, in the most general case, are hydrosoluble is through the presence of a water-based phase; more specifically, of a water continuous phase throughout the tissue engineering matrix. This approach is indeed followed in virtually any tissue engineering application, but it is exploited through two main different classes of morphologically different materials:

- *Microscopically homogeneous matrices*, which are composed of a water-swollen materials without phase boundaries in the >1 μm range.
- *Microscopically heterogeneous matrices*, where at least two phases coexist, one being water-insoluble, and have characteristic lengths >1 μm. The long-range transport properties of materials phase-separated at a sub-micrometre level (nano-phase separation, as it happens, e.g. in amphiphilic copolymers) are similar to those of truly homogeneous matrices, therefore they fall in the former case.

A list of the most commonly used materials of both classes is provided in Table 7.3.

Homogeneous matrices

In essence, we can define these materials as hydrogels, i.e. water-swollen elastic materials based on physically or chemically crosslinked networks of natural or

Table 7.3 Most common polymers used as matrices for the release of growth factors

Polymer	Matrix morphology	Signalling molecule	Applications
Poly(lactic acid) (PLLA & PLDLA), poly(glycolic acid) (PGA), poly(lactide-co-glycolide) (PLGA), poly(caprolactone)	Scaffolds, fibres, microparticles	BMPs, TGFβ, IGF-I, PDGF-BB, VEGF, NGF	Bone repair (Anderson and Langone, 1999; Schmidmaier *et al.*, 2001), cartilage formation (Lohmann *et al.*, 2000), soft tissue engineering (Eiselt *et al.*, 1998), nerve regeneration (Camarata *et al.*, 1992)
Poly(ethylene-co-vinyl acetate) (EVAc)	Hydrogels	aFGF	Angiogenesis for treatment of myocardial ischaemia (Lopez *et al.*, 1998)
Poly(vinyl alcohol) (PVA)	PVA-thread in platinum coil	bFGF	Aneurysm treatment (Matsumoto *et al.*, 2003)
Poly(ethylene glycol) (PEG)	Hydrogels	BMP-2, TGF-β, VEGF	Bone regeneration (Lutolf *et al.*, 2003), extracellular matrix stimulation (Jakob *et al.*, 2001), angiogenesis (Lutolf *et al.*, 2003)
Fibrin	Hydrogels	bFGF,VEGF, β-NGF	Wound healing (Sakiyama-Elbert and Hubbell, 2000b), angiogenesis (Zisch *et al.*, 2001), peripheral nerve regeneration (Sakiyama-Elbert and Hubbell, 2000a)
Collagen	Hydrogels, sponges	BMP, TGF-β, bFGF, PDGF	Bone formation (Lee *et al.*, 2001), cartilage regeneration, wound healing (Seal *et al.*, 2001, Friess, 1998)
Hyaluronic acid	Hydrogels	bFGF, VEGF, ang-1, KGF	Angiogenesis (Peattie *et al.*, 2004)
Alginate	Hydrogels	BMP-2, bFGF, VEGF	Osteoinduction, angiogenesis (Drury and Mooney, 2003)
Chitosan	Hydrogels	BMP,TGF-β, PDGF-BB	Bone and cartilage formation (Chenite *et al.*, 2000), periodontal bone regeneration (Luo *et al.*, 2005)

synthetic polymers. Among the natural polymers, a number of them can be degraded by mammalian cells through enzymatic mechanisms, e.g. hyaluronic acid, fibrin, collagen, gelatin. Alginate and chitosan (less frequently used) are of natural origin, but not intrinsically degradable by mammalian cells, although they can be destroyed by phagocytic cells or, as for calcium alginate gels, dissolved by calcium extraction.

In terms of synthetic polymers, the golden standard is offered by poly(ethylene glycol) (PEG, often referred to as poly(ethylene oxide) or PEO), mostly due to its protein-repellent character that makes it suitable for avoiding foreign body reactions (Stolnik *et al.*, 1995; Sofia *et al.*, 1998). PEG is not degradable per se, but its hydrogels have been made degradable through the introduction, e.g. of either hydrolysable ester units (Han and Hubbell, 1997), or protease-sensitive peptide groups (Lutolf *et al.*, 2003). Much more rarely, copolymers of ethylene and vinyl acetate (EVAc), poly(vinyl alcohol) and poly(hydroxylethyl methacrylate) have also been used.

How do GFs leave the matrix? (Release)

When large, hydrophilic molecules, such as GFs, are physically embedded in a hydrogel, they find themselves in a mesh of crosslinked polymer chains or fibres, whose connectivity will ultimately determine what kind of release kinetics could be achieved (Fig. 7.2):

- If the average mesh size of this network is much larger than the molecular dimension of the GF, this will diffuse similarly to a solution (diffusion coefficient inversely proportional to size in a linear fashion: Stokes–Einstein equation), determining a fairly fast (Whitaker *et al.*, 2001), first-order release of the GF; in principle, this is not the most desirable release profile.
- If mesh size and molecular dimensions are comparable, the release kinetics is considerably slowed down; one can see it as a reversible entrapment of the GF, which does not substantially change the diffuse nature of its motions but in principle allows for a more sustained, and more attractive, delivery.
- If the GF is larger than the mesh size of the network and the crosslinks are substantially irreversible, at least in the time frame considered (if they are reversible, the mesh size fluctuates dynamically and one falls back to the second case), the molecule is entrapped irreversibly. This case may look unattractive, but, on the other hand, introduces a very interesting possibility, namely to release a GF only when the network is degraded. As a result, GF could be effectively entrapped and protected, in order to be released through a kinetics that can be based solely on the properties of the matrix (if it is hydrolytically degradable), or on the ingress of cells and on their activity (if it is enzymatically degradable). In summary, the entrapment of large molecules in small mesh size, degradable networks opens the way to a more sophisticated and possibly responsive release kinetics.

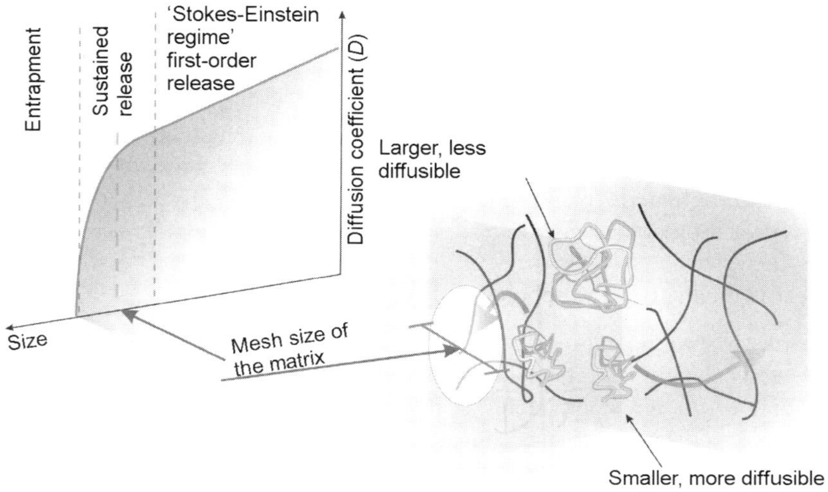

7.2 Molecules smaller than the mesh size of a hydrogel network present a linear relationship between hydrodynamic radius (size) and diffusion coefficient (Stokes–Einstein equation). Above that size, the molecules start being entrapped in the mesh and their diffusion becomes more and more problematic.

How do GFs enter the matrix? (Loading)

If it is true that interesting perspectives are offered by the possibility to reversibly immobilise GFs in a matrix where diffusion is problematic (see above), it is also clear that, correspondingly, loading is a problem, too, in such a material. The apparent contradiction is solved by forming the matrix *around* the large molecules: precursors of the matrix may be co-dissolved with GFs and *then* crosslinked to yield a hydrogel. Besides allowing loading of molecules too large to diffuse, this approach (also termed *in situ* gelation) has another important advantage: it also allows hydrogels to be produced from liquids and therefore matrices can be produced from injected precursors (minimally invasive implantation of a tissue engineering construct) and through a conformal process (perfect shape-matching with an internal cavity).

The *in situ* gelation can be performed through a variety of techniques, e.g. enzymatic reactions as for fibrin, chemical or photochemical as for PEG or hyaluronic acid gels.

Summary – homogeneous matrices

In the absence of strong interactions with the matrix, the release of complex, large and hydrosoluble actives (GFs) depends on the mesh size of the polymer network.

Heterogeneous matrices

The structure of these materials can be seen as a scaled-up version of that of homogeneous matrices: diffusion takes place in a water environment, and it is fast because the mesh size of the material is orders of magnitude larger (>1 μm) than the size of the diffusing species. In this case, the 'network' is therefore composed of material phase separated from water and this phase separation can be revealed by optical microscope or even naked eye observation.

What is the matrix? (Morphology and composition)

In order to ensure long-range diffusion, the water phase must be continuous; on the other hand, the water-insoluble phase (the 'network') can be either continuous, therefore producing scaffolds, or dispersed, typically the case of microspheres that are in mechanical contact (thus forming a 'network') but do not coalesce. In any case, the two phases must have a large interface, i.e. they must be well interpenetrated, in order to ensure an easy penetration of cells in every part of the material and also a capillary diffusion of nutrients and oxygen.

Despite the morphological difference, which profoundly affects mechanical properties (materials composed of microspheres can be robust only in compression, but flow under shear), scaffolds and microspheres share similar compositions.

Degradable constructs are mostly composed of hydrolytically erodible polyesters or of (derivatised) natural polymers. Poly(lactic acid) (PLLA and PLDLA), poly(glycolic acid) (PGA), their poly(lactide-*co*-glycolide) copolymers (PLGA) and poly(caprolactone), which are described as degradable because they depolymerise and eventually dissolve through hydrolysis (Whitaker *et al.*, 2001), have found widespread application in this field.

These materials are easily processable (for example, both solution casting and melt extrusion or moulding are possible) and have been extensively applied in surgery and in controlled drug release (Uhrich *et al.*, 1999): they therefore combine (1) approved clinical use, (2) physicochemical properties suitable for loading and release, and (3) erosion, which could result in a natural replacement by a growing tissue and this apparently ideal behaviour has boosted their use also as tissue engineering scaffolds (Anderson and Langone, 1999; Babensee *et al.*, 2000; Whitaker *et al.*, 2001). Furthermore, the rate of degradation of these polyesters can be easily modulated, possibly allowing erosion to be synchronised with appropriate developmental stages in tissue growth. Erosion, however, cannot always be predicted: besides easily controllable factors, such as degree of crystallinity (amorphous regions allow faster water diffusion and therefore faster degradation), molecular weight and copolymer ratios, stress factors at the site of implantation contribute too and may cause some variability in the performance (Whitaker *et al.*, 2001).

A more critical examination shows that polyesters suffer from a number of problems, which, however, have not discouraged their application in tissue engineering: firstly, these materials degrade through a bulk hydrolysis mechanism (as opposed to surface degradation), which at the same time results in a first order release kinetics for any encapsulated materials. Secondly, polyester degradation generates acidic species, which potentially create local inflammation in tissues and initiate enzyme hydrolysis (Whitaker *et al.*, 2001). Despite these drawbacks, FDA approval and an established history of use ensure that these synthetic polymers will still represent the leading materials in tissue engineering for years.

Non-degradable constructs are mostly based on ceramic materials, and are essentially used in bone tissue engineering where the inorganic matrix is then to be integrated in the calcified structure of the bone. However, purely ceramic, non-degradable matrices have not been used alone for the release of growth factors, while they have been employed as reinforcing or mineralisation-inducing elements in combination with degradable polymers (Murphy *et al.*, 2000; Laurencin *et al.*, 2001).

We here will concentrate on the release from polyester degradable constructs, neglecting the ceramic (non-degradable)–polymer (degradable) composites. In the latter systems, although the inorganic components are essential in determining the biomechanical interactions with encapsulated cells and surrounding tissues, they do not have a direct role in the release performance.

How do GFs leave the matrix? (Release) But, above all, how are GFs loaded and dispersed in the matrix?

It is apparent that if GFs are dissolved in the water phase, they are to diffuse out of the matrix fairly quickly, providing a poorly sustained release. A more prolonged delivery could, on the contrary, only be achieved by dispersing the GFs in the water-insoluble phase. There, two mechanisms are possible: release due to diffusion of GFs through, for example, the polyester matrix or, if diffusion is not possible, release after erosion of the matrix. In order to discriminate between the two possibilities, it is first necessary to understand which state GFs are in when loaded in a polymer matrix.

It is noteworthy to recall that GFs (and other proteinaceous signalling molecules, as well as nucleic acids) are large molecules with a hydrophilic exterior, at least when they are not denatured; their dissolution in a water-insoluble matrix must therefore take into account the following two thermodynamics-related points:

- An entropy-related factor: an increase molar weight decreases the entropic contribution to the free energy of dissolution (ΔS_{mix}, at constant volume fraction, decreases linearly with the molecular weight). The diffusion coefficient decreases too: according to the Stokes–Einstein equation, D decreases linearly with coil size, which is related to the molecular weight through a power law with exponent variable between 0.5 and 0.7 depending on the polymer conformation.

- An enthalpy-related factor: the Hildebrand parameter for water has a value of $\delta_{water} = 48\,MPa^{0.5}$; a number of proteins (but not all, and experimental conditions matter) can be solubilised without denaturation also by methanol, ethanol, dimethyl sulphoxide or dimethyl formamide, whose Hildebrand parameters have, respectively, values of 29.7, 26.2, 36.4, 24.7 $MPa^{0.5}$ (Barton, 1991). According to the classical solubility theory, a medium that would solubilise the GFs in a molecular and non-denatured form should therefore show a Hildebrand parameter at least >25. On the contrary, PLDLA (Siemann, 1992) and PLGA, as much as most common polar organic polymers, show Hildebrand parameter values in the range 20–25 $MPa^{0.5}$, which do not make them ideal solvents for complex macromolecular and hydrophilic actives, such as GFs.

Therefore, both for entropic and enthalpic reasons it is very unlikely for proteins to be molecularly dissolved and maintain their original conformation (exposing hydrophilic groups) in a fairly apolar, polymeric matrix such as that of degradable polyesters. Proteins are in contrast likely to form aggregates (with a non-negligible possibility of denaturation or degradation) and, since aggregates will present even lower diffusion coefficients than the isolated proteins, release could happen only upon erosion of the material (Fig. 7.3).

The methods employed for protein loading in scaffolds or microspheres can also worsen the situation, with organic solvents, heat and presence of interfaces (proteins can act as emulsifiers) causing even more denaturation (van de Weert *et al.*, 2000) . A number of strategies for minimising these drawbacks have been proposed and most of them (double emulsions, salt leaching, gas foaming) hinge around keeping the proteins in a protected environment, dispersed but not dissolved in the polymer matrix: the most simple example of protective environment for a protein is its hydration layer and indeed mild drying techniques (e.g. freeze drying with the use of cryoprotectants) are used for minimising its loss.

Summary – heterogeneous matrices

For protection purposes most processing techniques increase the already thermodynamics-driven tendency to the formation of protein depots, and this further decreases the possibility for GFs to diffuse through the matrix. GFs loaded in a degradable matrix (scaffold or microsphere) are therefore released only when this is degraded.

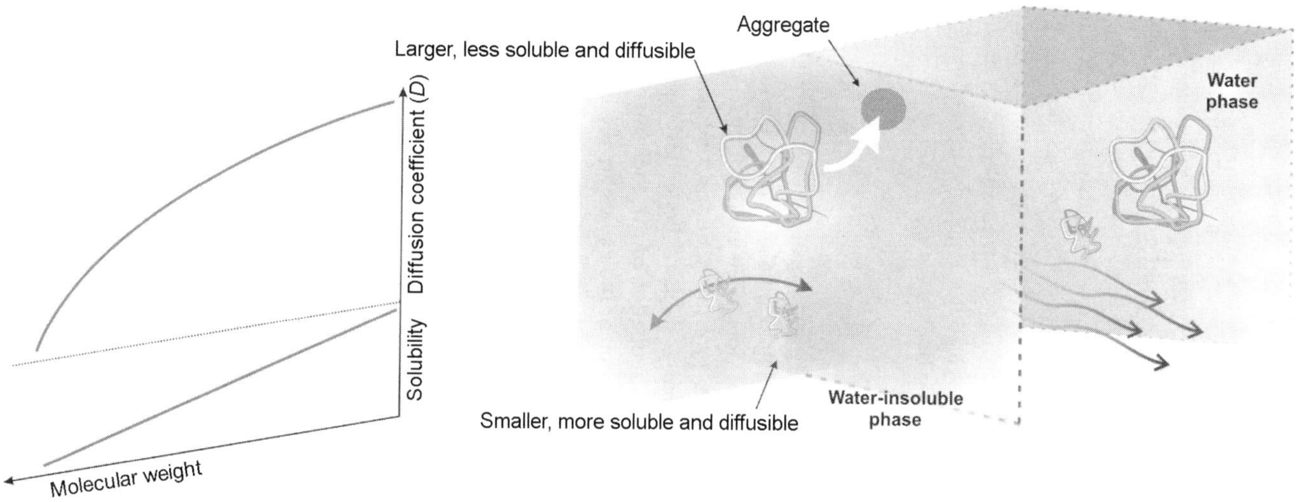

7.3 When dispersed in a non-hydrophilic (= non-water-swollen) matrix, a high molecular weight active would experience a dramatic drop in mobility, due mainly to aggregation (low solubility in the matrix) but also to reduction in diffusion coefficient with molar mass. In contrast, when in the surrounding fluids, its molecules will quickly diffuse away. The release from a hydrophobic matrix will therefore happen significantly upon matrix disruption/degradation.

7.4.3 Examples

Wound healing

Wound healing and skin regeneration have been among the first targets of tissue engineering to employ the release of growth factors. It is well understood that EGF and TGF-α induce proliferation and differentiation of epithelial cells, while PDGF stimulates the production of various components of the extracellular matrix, such as collagen and fibronectin. PDGF also stimulates macrophages to secrete other growth factors essential to the overall healing process. PVA sponges have been used to deliver these factors while maintaining their activity (Buckley *et al.*, 1985). Regranex, a commercial product based on PDGF encapsulated in a carboxymethylcellulose gel, is topically applied for the treatment of diabetic foot ulcers (Buckley *et al.*, 1985).

Bone regeneration

Bone regeneration is another deeply investigated area for growth factor delivery, with the primary goal of accelerating healing and enhancing new bone formation. Among the large number of factors identified in bone formation and remodelling, the TGF-β superfamily occupies probably the most relevant position: all these proteins are involved in the early stage of bone development after trauma, regulating proliferation and differentiation of mesenchymal precursor cells.

Within this family, the most popular molecules are the BMPs, which play an essential role in embryonic development and tissue repair (Elima, 1993). BMP-2 and BMP-7 are among the first human recombinant growth factors available for clinical use. They have been used to treat patients with a number of bone diseases, since they act as differentiation factors for both bone and cartilage precursor cells. A number of matrices have been used: collagen (which is also the principal non-mineral component of bone), PLGA, processed either as porous scaffold or in capsular form, and injectable PEG-based hydrogels (Chen and Mooney, 2003). Most of these systems, however, are not fully osteo-conductive, in that they do not provide good cell anchorage to the matrix and do not reinforce the mechanical stability during cell-operated remodelling. For this purpose, inorganic materials such as hydroxyapatite and calcium phosphate are quite often integrated in polymeric mineralised scaffold for growth factor delivery (Chen and Mooney, 2003).

Among the non-BMP growth factors, TGF-β1 is probably the most used, e.g. for modulating bone marrow stromal osteoblast functions (Lu *et al.*, 2001).

Angiogenesis

Angiogenesis is a critical requirement in the development of most tissues (but not e.g. for cartilage), as well as in critical vascular situations (e.g. the aftermath of a stroke). This multistep process involves endothelial cell activation and migration, leading to a generation of neovessels, which are eventually surrounded by smooth muscle cells and pericytes in different layers (Hirschi *et al.*, 2002). The complex succession of events is regulated through the cooperation of different growth factors; for example VEGF, ang-1 and ang-2 specifically activate endothelial cells; bFGF and VEGF then control their proliferation and migration; PDGF-BB and TGF-β recruit muscle cell and pericytes and promote ECM deposition.

So far, alginate hydrogels, PLGA microspheres and scaffolds have been used to deliver bFGF and VEGF, obtaining some *in vivo* positive results (Ennett and Mooney, 2002). A still open challenge is, however, that current strategies are apparently able to generate only small new blood vessels lacking pericytes, a situation that leads to long-term loss of functionality and regression (Chen and Mooney, 2003).

7.5 Signalling molecules released from a bound state

In most cases the release of growth factors from solid matrices suffers from two main problems:

- The release profile, which is often not ideal: in a first-order release (hydrogels) the GF is released mostly immediately after implantation, when inflammatory reactions are maximal, and it is hard to obtain a sustained action. In hydrolytically degradable materials the kinetics is more appropriate, but largely predetermined (not responsive to local conditions), and stability in the matrix is an issue.
- Stability out of the matrix: the GFs are released 'naked'.

Alternative approaches have been developed to overcome at least one of these problems, or possibly both. In the essence, they are all inspired by how natural extracellular matrices (EMCs) work: Nature has already developed an efficient and sophisticated method to store and deliver morphogens, in order to modulate signaling events in tissue development at the right tissue area and at the right time. In fact, ECMs are generally able to locally bind, accumulate and finally (responsively) release soluble factors, thus limiting their diffusion, increasing their concentration and providing a local storage (as a consequence of binding) when their production ends. Finally, even if this can happen in other occasions, GFs are mostly released when the matrix is degraded (through enzymatic action, e.g. by migrating cells) and this is an ideal bio-responsive trigger for GF release.

GFs may interact with ECMs by binding proteoglycans, and specifically their glycosoaminoclycan (GAG) components: the most common case is offered by GFs binding to heparan sulphate or chondroitin sulphate (heparin-binding growth factors), a phenomenon recognised for more than 20 years (Burgess and Maciag, 1989). This phenomenon is particularly well described for members of the fibroblast and endothelial growth factor families; it is important to note that the binding action is not only functional to storage of GFs in a matrix: heparan sulphate, for example, facilitates the GFs docking on their receptors (Pellegrini, 2001) and at the same time is thought to decrease their protease-mediated degradability. Another possibility is that GFs (e.g. of the TGF-β family) bind other proteic components, such as collagen and fibronectin.

Modern approaches to GF release utilise these growth factor–matrix interactions; biomaterials are therefore designed to mimic the natural ECM way of binding morphogens (Lutolf and Hubbell, 2005), for example introducing heparin-like motifs in their structures or adding heparin-binding groups, in order to use heparan sulphate as a kind of sandwich between matrix and GF. Although these mostly electrostatic interactions are weaker than covalent bonds, they are sufficiently strong to control the overall delivery mechanism, and to prolong the sustained release by more than 100 times the standard GF release profiles (Hubbell, 2006). When the growth factor is not strongly or not all heparin-binding, it can be engineered to interact directly with the matrix polymers either through affinity interactions or through covalent (but enzymatically degradable) bond.

During tissue regeneration, migrating cells activate plasmin and matrix metalloproteinases, which are enzymes that allow cells to penetrate the extracellular matrix, thus liberating the growth factors from their bound state. As an example, plasmin controls fibrin (the matrix) and heparin degradation, thus accelerating the release of the soluble factors immobilised through affinity interactions (Sakiyama-Elbert and Hubbell, 2000a; Sakiyama-Elbert *et al.*, 2001; Zisch *et al.*, 2001). Similar effects are also recorded when the growth factors are covalently bound to the matrix (Ehrbar *et al.*, 2005).

7.5.1 Example

A number of growth factors can be engineered to directly bind to fibrin. In angiogenesis, for instance, VEGF is used to stimulate existing blood vessels to branch and form new vasculature, but when VEGF concentration is uncontrolled, new blood vessels have shown to become highly permeable, e.g. very similar to a leaky tumour vasculature. In order to improve the mechanism of blood vessel formation, a fibrin-binding variant of VEGF has been produced, which also contains a plasmin sensitive linker. This variant of VEGF can remain bound within a fibrin gel until liberated by active plasmin, whereas standard VEGF simply mixed with fibrin diffuses out within few hours. *In vivo* studies have shown an increased quantity of new vasculature generated, with an improved morphology of the blood vessels, much less tumour-like than those induced by free diffusing VEGF (Ehrbar *et al.*, 2005).

7.6 References

Anderson, J. M. and Langone, J. J. (1999) Issues and perspectives on the biocompatibility and immunotoxicity evaluation of implanted controlled release systems. *Journal of Controlled Release,* **57**, 107–113.

Babensee, J. E., Mcintire, L. V. and Mikos, A. G. (2000) Growth factor delivery for tissue engineering. *Pharmaceutical Research,* **17**, 497–504.

Barton, A. (1991) *Handbook of Solubility Parameters and Other Cohesion Parameters, Second Edition,* Boca Raton, CRC.

Bennett, N. T. and Schultz, G. S. (1993) Growth-factors and wound-healing – biochemical-properties of growth-factors and their receptors. *American Journal of Surgery,* **165**, 728–737.

Bonadio, J. (2000) Tissue engineering via local gene delivery: update and future prospects for enhancing the technology. *Advanced Drug Delivery Reviews,* **44**, 185–194.

Boontheekul, T. and Mooney, D. J. (2003) Protein-based signaling systems in tissue engineering. *Current Opinion in Biotechnology,* **14**, 559–565.

Braddock, M., Houston, P., Campbell, C. and Ashcroft, P. (2001) Born again bone: tissue engineering for bone repair. *News in Physiological Sciences,* **16**, 208–213.

Buckley, A., Davidson, J. M., Kamerath, C. D., Wolt, T. B. and Woodward, S. C. (1985) Sustained-release of epidermal growth-factor accelerates wound repair. *Proceedings of the National Academy of Sciences of the United States of America,* **82**, 7340–7344.

Burgess, W. H. and Maciag, T. (1989) The heparin-binding (fibroblast) growth-factor family of proteins. *Annual Review of Biochemistry,* **58**, 575–606.

Camarata, P. J., Suryanarayanan, R., Turner, D. A., Parker, R. G. and Ebner, T. J. (1992) Sustained-release of nerve growth-factor from biodegradable polymer microspheres. *Neurosurgery,* **30**, 313–319.

Chandler, L. A., Doukas, J., Gonzalez, A. M., Hoganson, D. K., Gu, D. L., Ma, C. L., Nesbit, M., Crombleholme, T. M., Herlyn, M., Sosnowski, B. A. and Pierce, G. F. (2000) FGF2-targeted adenovirus encoding platelet-derived growth factor-B enhances *de novo* tissue formation. *Molecular Therapy,* **2**, 153–160.

Chen, R. R. and Mooney, D. J. (2003) Polymeric growth factor delivery strategies for tissue engineering. *Pharmaceutical Research,* **20**, 1103–1112.

Chenite, A., Chaput, C., Wang, D., Combes, C., Buschmann, M. D., Hoemann, C. D., Leroux, J. C., Atkinson, B. L., Binette, F. and Selmani, A. (2000) Novel injectable neutral solutions of chitosan form biodegradable gels in situ. *Biomaterials,* **21**, 2155–2161.

Drury, J. L. and Mooney, D. J. (2003) Hydrogels for tissue engineering: scaffold design variables and applications. *Biomaterials,* **24**, 4337–4351.

Eckenstein, F. P. (1994) Fibroblast growth-factors in the nervous-system. *Journal of Neurobiology,* **25**, 1467–1480.

Ehrbar, M., Metters, A., Zammaretti, P., Hubbell, J. A. and Zisch, A. H. (2005) Endothelial cell proliferation and progenitor maturation by fibrin-bound VEGF variants with differential susceptibilities to local cellular activity. *Journal of Controlled Release,* **101**, 93–109.

Eiselt, P., Kim, B. S., Chacko, B., Isenberg, B., Peters, M. C., Greene, K. G., Roland, W. D., Loebsack, A. B., Burg, K. J. L., Culberson, C., Halberstadt, C. R., Holder, W. D. and Mooney, D. J. (1998) Development of technologies aiding large-tissue engineering. *Biotechnology Progress,* **14**, 134–140.

Elima, K. (1993) Osteoinductive proteins. *Annals of Medicine,* **25**, 395–402.

Ennett, A. B. and Mooney, D. J. (2002) Tissue engineering strategies for *in vivo* neovascularisation. *Expert Opinion on Biological Therapy,* **2**, 805–818.

Erlebacher, A., Filvaroff, E. H., Ye, J. Q. and Derynck, R. (1998) Osteoblastic responses to TGF-beta during bone remodeling. *Molecular Biology of the Cell,* **9**, 1903–1918.

Folkman, J. and Shing, Y. (1992) Angiogenesis. *Journal of Biological Chemistry,* **267**, 10931–10934.

Friess, W. (1998) Collagen – biomaterial for drug delivery. *European Journal of Pharmaceutics and Biopharmaceutics,* **45**, 113–136.

Giannoudis, P. V. and Tzioupis, C. (2005) Clinical applications of BMP-7 – the UK perspective. *Injury – International Journal of the Care of the Injured,* **36**, 47–50.

Han, D. K. and Hubbell, J. A. (1997) Synthesis of polymer network scaffolds from L-lactide and poly(ethylene glycol) and their interaction with cells. *Macromolecules,* **30**, 6077–6083.

Heldin, C. H. and Westermark, B. (1999) Mechanism of action and *in vivo* role of platelet-derived growth factor. *Physiological Reviews,* **79**, 1283–1316.

Hirschi, K. K., Skalak, T. C., Peirce, S. M. and Little, C. D. (2002) Vascular assembly in natural and engineered tissues (in Reparative Medicine: Growing Tissues and Organs), *Annals of the New York Academy of Sciences*, **961**, 223–242..

Hubbell, J. A. (2006) Matrix-bound growth factors in tissue repair. *Swiss Medical Weekly,* **136**, 387–391.

Jakob, M., Demarteau, O., Schafer, D., Hintermann, B., Dick, W., Heberer, M. and Martin, I. (2001) Specific growth factors during the expansion and redifferentiation of adult human articular chondrocytes enhance chondrogenesis and cartilaginous tissue formation *in vitro*. *Journal of Cellular Biochemistry,* **81**, 368–377.

Jochum, W., David, J. P., Elliott, C., Wutz, A., Plenk, H., Matsuo, K. and Wagner, E. F. (2000) Increased bone formation and osteosclerosis in mice overexpressing the transcription factor Fra-l. *Nature Medicine,* **6**, 980–984.

Kofron, M. D. and Laurencin, C. T. (2006) Bone tissue engineering by gene delivery. *Advanced Drug Delivery Reviews,* **58**, 555–576.

Kuhn, H. G., Winkler, J., Kempermann, G., Thal, L. J. and Gage, F. H. (1997) Epidermal growth factor and fibroblast growth factor-2 have different effects on neural progenitors in the adult rat brain. *Journal of Neuroscience,* **17**, 5820–5829.

Lambiase, A., Rama, P., Bonini, S., Caprioglio, G. and Aloe, L. (1998) Topical treatment with nerve growth factor for corneal neurotrophic ulcers. *New England Journal of Medicine,* **338**, 1174–1180.

Laurencin, C. T., Attawia, M. A., Lu, L. Q., Borden, M. D., Lu, H. H., Gorum, W. J. and Lieberman, J. R. (2001) Poly(lactide-co-glycolide)/hydroxyapatite delivery of BMP-2-producing cells: a regional gene therapy approach to bone regeneration. *Biomaterials,* **22**, 1271–1277.

Lazarous, D. F., Shou, M., Stiber, J. A., Hodge, E., Thirumurti, V., Goncalves, L. and Unger, E. F. (1999) Adenoviral-mediated gene transfer induces sustained pericardial VEGF expression in dogs: effect on myocardial angiogenesis. *Cardiovascular Research,* **44**, 294–302.

Lee, C. H., Singla, A. and Lee, Y. (2001) Biomedical applications of collagen. *International Journal of Pharmaceutics,* **221**, 1–22.

Lohmann, C. H., Schwartz, Z., Niederauer, G. G., Carnes, D. L., Dean, D. D. and Boyan, B. B. (2000) Pretreatment with platelet derived growth factor-BB modulates the ability of costochondral resting zone chondrocytes incorporated into PLA/PGA scaffolds to form new cartilage *in vivo*. *Biomaterials,* **21**, 49–61.

Lopez, J. J., Edelman, E. R., Stamler, A., Hibberd, M. G., Prasad, P., Thomas, K. A., Disalvo, J., Caputo, R. P., Carrozza, J. P., Douglas, P. S., Sellke, F. W. and Simons, M. (1998) Angiogenic potential of perivascularly delivered aFGF in a porcine model of chronic myocardial ischemia. *American Journal of Physiology – Heart and Circulatory Physiology,* **43**, H930–H936.

Lu, L. C., Yaszemski, M. J. and Mikos, A. G. (2001) TGF-beta 1 release from biodegradable polymer microparticles: its effects on marrow stromal osteoblast function. *Journal of Bone and Joint Surgery – American Volume,* **83A**, S82–S91.

Luhmann, S. J., Bridwell, K. H., Cheng, I., Imamura, T., Lenke, L. G. and Schootman, M. (2005) Use of bone morphogenetic protein-2 for adult spinal deformity. *Spine,* **30**, S110–S117.

Luo, J., Sun, M. H., Kang, Q., Peng, Y., Jiang, W., Luu, H. H., Luo, Q., Park, J. Y., Li, Y., Haydon, R. C. and He, T. C. (2005) Gene therapy for bone regeneration. *Current Gene Therapy,* **5**, 167–179.

Lutolf, M. P. and Hubbell, J. A. (2005) Synthetic biomaterials as instructive extracellular microenvironments for morphogenesis in tissue engineering. *Nature Biotechnology,* **23**, 47–55.

Lutolf, M. P., Raeber, G. P., Zisch, A. H., Tirelli, N. and Hubbell, J. A. (2003) Cell-responsive synthetic hydrogels. *Advanced Materials,* **15**, 888–892.

Marie, P. J. (2003) Fibroblast growth factor signaling controlling osteoblast differentiation. *Gene,* **316**, 23–32.

Matsumoto, H., Terada, T., Tsuura, M., Itakura, T. and Ogawa, A. (2003) Basic fibroblast growth factor released from a platinum coil with a polyvinyl alcohol core enhances cellular proliferation and vascular wall thickness: an *in vitro* and *in vivo* study. *Neurosurgery,* **53**, 402–407.

McArthur, J. C., Yiannoutsos, C., Simpson, D. M., Adornato, B. T., Singer, E. J., Hollander, H., Marra, C., Rubin, M., Cohen, B. A., Tucker, T., Navia, B. A., Schifitto, G., Katzenstein, D., Rask, C., Zaborski, L., Smith, M. E., Shriver, S., Millar, L. and Clifford, D. B. (2000) A phase II trial of nerve growth factor for sensory neuropathy associated with HIV infection. *Neurology,* **54**, 1080–1088.

Murphy, W. L., Peters, M. C., Kohn, D. H. and Mooney, D. J. (2000) Sustained release of vascular endothelial growth factor from mineralized poly(lactide-co-glycolide) scaffolds for tissue engineering. *Biomaterials,* **21**, 2521–2527.

Musgrave, D. S., Bosch, P., Ghivizzani, S., Robbins, P. D., Evans, C. H. and Huard, J. (1999) Adenovirus-mediated direct gene therapy with bone morphogenetic protein-2 produces bone. *Bone,* **24**, 541–547.

Nussenbaum, B. and Krebsbach, P. H. (2006) The role of gene therapy for craniofacial and dental tissue engineering. *Advanced Drug Delivery Reviews,* **58**, 577–591.

Ohlsson, C., Bengtsson, B. A., Isaksson, O. G. P., Andreassen, T. T. and Slootweg, M. C. (1998) Growth hormone and bone. *Endocrine Reviews,* **19**, 55–79.

Peattie, R. A., Nayate, A. P., Firpo, M. A., Shelby, J., Fisher, R. J. and Prestwich, G. D. (2004) Stimulation of *in vivo* angiogenesis by cytokine-loaded hyaluronic acid hydrogel implants. *Biomaterials,* **25**, 2789–2798.

Pellegrini, L. (2001) Role of heparan sulfate in fibroblast growth factor signalling: a structural view. *Current Opinion in Structural Biology,* **11**, 629–634.

Reddi, A. H. (1998) Role of morphogenetic proteins in skeletal tissue engineering and regeneration. *Nature Biotechnology,* **16**, 247–252.

Richardson, T. P., Peters, M. C., Ennett, A. B. and Mooney, D. J. (2001) Polymeric system for dual growth factor delivery. *Nature Biotechnology,* **19**, 1029–1034.

Sakiyama-Elbert, S. E. and Hubbell, J. A. (2000a) Controlled release of nerve growth factor from a heparin-containing fibrin-based cell ingrowth matrix. *Journal of Controlled Release,* **69**, 149–158.

Sakiyama-Elbert, S. E. and Hubbell, J. A. (2000b) Development of fibrin derivatives for controlled release of heparin-binding growth factors. *Journal of Controlled Release,* **65**, 389–402.

Sakiyama-Elbert, S. E., Panitch, A. and Hubbell, J. A. (2001) Development of growth factor fusion proteins for cell-triggered drug delivery. *Faseb Journal,* **15**, 1300–1302.

Saltzman, W. M. and Olbricht, W. L. (2002) Building drug delivery into tissue engineering. *Nature Reviews Drug Discovery,* **1**, 177–186.

Salvay, D. M. and Shea, L. D. (2006) Inductive tissue engineering with protein and DNA-releasing scaffolds. *Molecular Biosystems,* **2**, 36–48.

Schmidmaier, G., Wildemann, B., Bail, H., Lucke, M., Fuchs, T., Stemberger, A., Flyvbjerg, A., Haas, N. P. and Raschke, M. (2001) Local application of growth factors (insulin-like growth factor-1 and transforming growth factor-beta 1) from a biodegradable poly(D,L-lactide) coating of osteosynthetic implants accelerates fracture healing in rats. *Bone,* **28**, 341–350.

Seal, B. L., Otero, T. C. and Panitch, A. (2001) Polymeric biomaterials for tissue and organ regeneration. *Materials Science & Engineering R – Reports,* **34,** 147–230.

Siemann, U. (1992) The solubility parameter of poly(DL-lactic acid). *European Polymer Journal,* **28**, 293–297.

Sofia, S. J., Premnath, V. and Merrill, E. W. (1998) Poly(ethylene oxide) grafted to silicon surfaces: grafting density and protein adsorption. *Macromolecules,* **31**, 5059–5070.

Sohier, J., Vlugt, T. J. H., Cabrol, N., Van Blitterswijk, C., De Groot, K. and Bezemer, J. M. (2006) Dual release of proteins from porous polymeric scaffolds. *Journal of Controlled Release,* **111**, 95–106.

Stolnik, S., Illum, L. and Davis, S. S. (1995) Long circulating microparticulate drug carriers. *Advanced Drug Delivery Reviews,* **16**, 195–214.

Thoenen, H. (1995) Neurotrophins and neuronal plasticity. *Science,* **270**, 593–598.

Uhrich, K. E., Cannizzaro, S. M., Langer, R. S. and Shakesheff, K. M. (1999) Polymeric systems for controlled drug release. *Chemical Reviews,* **99**, 3181–3198.

Urist, M. (1965) Bone: formation by autoinduction. *Science,* **150**, 893–899.

van de Weert, M., Hennink, W. E. and Jiskoot, W. (2000) Protein instability in poly(lactic-co-glycolic acid) microparticles. *Pharmaceutical Research,* **17**, 1159–1167.

Werner, S., Peters, K. G., Longaker, M. T., Fullerpace, F., Banda, M. J. and Williams, L. T. (1992) Large induction of keratinocyte growth-factor expression in the dermis during wound-healing. *Proceedings of the National Academy of Sciences of the United States of America,* **89**, 6896–6900.

Whitaker, M. J., Quirk, R. A., Howdle, S. M. and Shakesheff, K. M. (2001) Growth factor release from tissue engineering scaffolds. *Journal of Pharmacy and Pharmacology,* **53**, 1427–1437.

Yancopoulos, G. D., Davis, S., Gale, N. W., Rudge, J. S., Wiegand, S. J. and Holash, J. (2000) Vascular-specific growth factors and blood vessel formation. *Nature,* **407**, 242–248.

Zisch, A. H., Schenk, U., Schense, J. C., Sakiyama-Elbert, S. E. and Hubbell, J. A. (2001) Covalently conjugated VEGF-fibrin matrices for endothelialization. *Journal of Controlled Release,* **72**, 101–113.

8
Carrier systems and biosensors for biomedical applications

F DAVIS and S P J HIGSON, Cranfield University, UK

8.1 Introduction

This chapter addresses both carrier systems and biosensors which are often applied directly to tissues, either as skin patches, implanted or ingested by a variety of routes. It follows that there is a common theme between these applications and many of those discussed elsewhere within this book. Any device, scaffold or implant within the body must usually display extreme biocompatibility if it is not to cause harm to the patient. The techniques of tailoring surfaces to ensure no adverse reactions are a common theme running throughout this work on tissue engineering.

The first section of this chapter will describe the use of carrier systems in biomedical applications. Initially a discussion is provided as to why carrier systems are required and this is followed by descriptions of different classes of material of natural and synthetic origin. This first section will continue with a description of how nanotechnology is becoming utilised in carrier systems and will close with a discussion of current and future applications of these materials.

The second section of this chapter will be devoted to biosensors, beginning with a history and descriptions of basic sensor formats. Glucose biosensors will be described in more detail due to their dominance of the biosensor market. The construction of first, second and third generation biosensors and their principle of operation will be described. This will be followed by a description of the work progressing to the latest developments for implantable glucose sensors and their potential for continuous glucose monitoring. The chapter will close with a description of future needs including warfarin monitoring and implantable organs.

8.2 Carrier systems

8.2.1 Introduction

In many instances a simple one-off administration of a drug will suffice, e.g. taking an analgesic in response to a headache. However this simple approach is often insufficient since often the condition may require continuous medication over a period of time, with many conditions requiring long-term drug therapy. Often a drug may have an optimum concentration within the body; too low and no benefit is derived, too high and unwanted side effects or toxicity can occur, endangering the health of the patient. However, when a drug is ingested or injected, the level within the patient's blood tends to rise up towards a plateau and then after some time falls again to zero unless more drug is given. In many instances a more efficient application method would be one that gives a stable level of treatment. Other problems can occur with drug stability, for example for a drug to be taken orally it must not be degraded by digestive processes. At the time of writing approximately 15% of the current world pharmaceutical market consists of products which utilise a carrier system and the US market for drug delivery systems is expected to reach $82 billion by 2007 (Roco and Bainbridge, 2002).

To minimise drug degradation and optimise drug delivery, a wide variety of drug carrier and delivery systems are under investigation and several reviews have been published on this subject (Langer, 1995, 1998; Kaparissides *et al.*, 2006).

A number of methods can be utilised to apply various carrier systems:

- Ingestion – the drug can be incorporated into a carrier system so that the composite material can be swallowed. This method is the most popular owing to its simplicity and convenience. The carrier system must be capable of protecting the drug against the highly acidic medium of the stomach and/or enzymatic degradation throughout the digestive system. Sometimes the digestive process itself can be utilised to cause release of the drug. There are, however, a number of problems with this technique, since it can lead to irritation of the bowel and also in many cases the drug must still penetrate the stomach/intestine wall.
- Inhalation – the drug can be dispersed within an aerosol, usually via a nebuliser and in this way can be directly inhaled. Especially suitable for treating respiratory diseases, this technique is widely used for the treatment of asthma. However delivery via this method can still be adversely affected by the barrier between air and blood within the lung.
- Transdermal – the drug is incorporated within a patch, similar to the nicotine patches used to relieve 'cravings' of people attempting to stop smoking. This method avoids the problems of degradation of drugs by digestive process and can be used to provide local delivery, e.g. to a wound or skin conditions. Patches often need to be applied only once every several days and the method is non-invasive and painless. It also has the advantage that unlike oral and inhalation routes, patches can be easily and safely applied to unconscious patients. The skin barrier does lead to slow penetration rates and therefore in many instances only relatively low dosage levels can be attained. Other problems may include a lack of dosage flexability.

- Injection – this technique has the disadvantage of being invasive and can be painful. Drug carrier systems can be utilised within this method, for example to prevent degradation of the drug within the bloodstream. As an alternative, a drug/carrier composite can be surgically implanted close to an affected site.

8.2.2 Classes of materials

Hydrophilic polymers

The pharmaceutical industry has shown great interest in the development of controlled release systems based on hydrophilic polymers. Hydrogels represent a common class of polymers used for drug delivery. In essence a hydrogel is a material based on polymers such as poly(vinyl alcohol) or poly(acrylic acid) which would normally be soluble in water. However, either during or after the polymer synthesis, a degree of crosslinking converts the linear polymer chains into a polymer network. This process renders the polymers insoluble; however, the high presence of hydrophilic groups within the network gives the structure a high affinity for water. Although the network does not dissolve, it is capable of adsorbing water with consequent swelling of the polymer matrix.

The nature of the polymer and degree of crosslinking affect the swelling behaviour. A network with few crosslinks and a large number of hydrophilic groups will adsorb large amounts of water with a high degree of swelling. Less hydrophilic monomers, incorporation of hydrophobic co-monomers or a high degree of crosslinking all act to reduce water adsorption, usually leading to a firmer, more rigid gel. Since they include a high water content, hydrogels often show high degrees of biocompatibility. The polymer network itself can be either bioinert or be a biodegradable polymer network. Natural polymers can also be used, with, for example, hydrogels based on chitosan, alginase or collagen having been utilised. The application of a wide variety of hydrophilic biodegradable polymers to delivery of proteins has been extensively reviewed (Gombotz and Pettit, 1995). Typical polymers include hydrogels based on poly(vinyl alcohol), poly(vinylpyrrolidinone) or cellulose and other natural polymers such as alginase or collagen. Frequently the drugs, especially if they are biologically derived (such as proteins) can be quite unstable. Besides enabling the controlled release of the active material, these hydrogels often act as a stabilising medium for these unstable agents. For example, a drug could be incorporated into a hydrogel either as it is synthesised or post-synthesis. This can then be applied to the patient via any of the techniques described above. Once *in vivo*, the drug is released by a number of means. These might include simple diffusion or alternatively the polymer may be eroded or dissolved, for example by digestion. In an ideal situation this process will occur at a constant rate leading to continual release of controlled amounts of the drug. If the polymer is

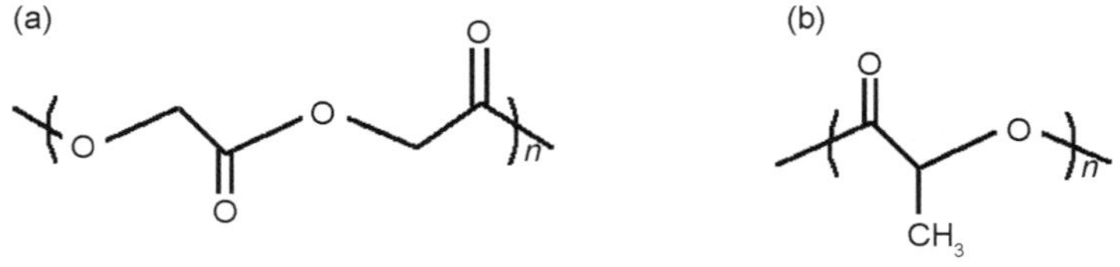

8.1 Structures of (a) poly(glycolide) and (b) poly(lactide).

utilised as a transdermal patch or is surgically implanted close to an affected site, the drug can be delivered where it is most needed.

A similar method utilises polymers which are degraded rather than just swollen. Examples include polymers which are based on poly(lactide) or poly(glycolide) (Fig. 8.1), and which are slowly hydrolysed *in vivo* to release an active agent, e.g. leuprolide acetate (Ogawa *et al.*, 1988).

A polymer must address several criteria before it is suitable for use as a carrier system. In the case of polymers that are effective *in vivo*, they must be both biocompatible and must biodegrade within a reasonable period of time. Irrespective of the manner of application, the polymer itself and any degradation products must be non-toxic and must not create any allergic or inflammatory response. The method of release is often dependent on the nature of the drug itself; low molecular weight drugs are capable of diffusing out of the polymeric matrix and if water soluble, will be rapidly released. However, larger molecules such as proteins will not diffuse as readily and often remain within the hydrogel matrix until the polymer itself is either degraded or enzymatic digestion releases them. Release rates depend on several factors including the quantity/dosing of drugs within the composite, the rate of degradation of the polymer, the water content (if it is a hydrogel-type material), and the presence and degree of crosslinking.

A property that makes hydrogels exceedingly suitable for drug encapsulation is that many of the materials used for these systems can be synthesised so as to be responsive to their environment. A change of pH, for example, can lead to protonation or deprotonation of active groups within the polymer. This can change its affinity for water, giving us a material whose swelling is pH responsive, thereby affecting many of their other physical properties such as permeability. This gives us the opportunity to design drug release agents which are selectively triggered by certain conditions, e.g. their responsiveness to pH means they can be designed to release active agents within a selected part of the digestive tract. Similar smart materials can be designed that respond to changes in physiological conditions and therefore only release the drug at times when it is needed. Hydrogels can be designed that respond to other stimuli as well as pH, such as ionic strength, temperature and electric field. They can even be designed to respond to the presence or absence of specific analytes.

As mentioned earlier, polymer-based delivery systems allow application of active agents in many ways including ingestion, transdermal patches, suppositories, ocular and subcutaneous methods. A few examples are given here: nitroglycerine is often used in the treatment of angina; however, volatilisation of the active component can lead to loss of tablet activity (Markovich *et al.*, 1997). This problem can be mitigated by the use of a acrylic-based hydrogel as a host for the nitroglycerine and incorporation of this composite in the construction of a transdermal patch.

There has been wide research into utilising these materials in the treatment of cancer for, for example, the delivery of ara-C for leukaemia. This drug gives rise to a number of side effects but these are mitigated when a constant infusion of the drug is introduced subcutaneously. As an alternative approach, crosslinked polyhydroxyethyl acrylate can be utilised as a host for this material. Discs of this composite display a steady controllable release of the active material (Teijon *et al.*, 1997). A similar composite based upon polycaprolactone/polyethylene glycol has been used as a matrix for the anticonvulsant drug clonazepam. Stable constant release properties were displayed for over 45 days (Cho *et al.*, 1999).

Similar polymers are suitable for localised delivery of pharmaceuticals. In the treatment of brain cancer, one approach that has been successfully used is surgery to remove as much of the tumour as possible, followed by placing in the surgical site, small wafers based on polyanhydrides. These contain the anticancer drug carmustine which is slowly released over a 1 month period to kill any remaining tumour cells (Brem *et al.*, 1995). Alternatively thermosensitive hydrogels based on block copolymers of poly(ethylene oxide) and poly(lactide) have been made which are liquid at 45 °C and can be injected directly to the required site. Upon cooling to body temperature, the gel immediately sets, trapping any pharmaceutical compounds in the solution, so allowing them to be released slowly (Jeong *et al.*, 1997).

The materials mentioned so far are usually based on simple polymeric systems; however, more complex reactive systems can be designed. As mentioned earlier, hydrogels can be synthesised which respond to environmental conditions. An example of this is a 'smart' porous membrane made from a polymethacrylic acid/polyethylene glycol copolymer. This is used as a host for insulin but also contains encapsulated glucose oxidase. Glucose oxidase specifically catalyses the oxidation of glucose to gluconic acid, which causes a subsequent pH drop. This leads to shrinkage of the membrane and the controlled release of insulin in response to hyperglycaemia (Gander *et al.*, 2001). Apart from drug delivery, recent work has focused on the use of hydrophilic polymers in gene therapy and has shown that encapsulation of the DNA inside a hydrophilic polymer can increase the transfection efficiency (Ottenbrite, 1999).

8.2.3 Natural polymers

Besides synthetic polymers, natural polymers have also been widely studied as drug carrier agents, especially as for many of them their biocompatibility and non-toxicity are well known since they form major components of our diets. Proteins are of great interest due to their wide variety of structures and their easy availability from sources such as, for example, egg white, soybean and whey (Chen *et al.*, 2006). Hydrogels can be easily generated from a wide variety of proteins and possess the desirable qualities of their synthetic analogues. Usually proteins can be gelated by methods such as heating, which unfolds the polypeptide chains within the protein structure. These then tend to aggregate forming a three-dimensional structure crosslinked by hydrogen bonding and/or hydrophobic effects (Clark *et al.*, 2001). Other methods for inducing gelation include crosslinking with calcium ions (Maltais *et al.*, 2005). A wide variety of hydrogel morphologies and microstructures can be generated by variation of the preparation technique (Chen *et al.*, 2006). Applications of these gels include controlled release of tocopherol (Chen *et al.*, 2006) from a lactoglobulin-based emulsion under simulated gastric conditions. Other examples include the incorporation of drug compounds within albumin (Sokolowski and Royer, 1984; Tomlinson and Burgen, 1985) or corn protein (Liu *et al.*, 2005) and a recent review has been published on this topic (Chen *et al.*, 2006).

8.2.4 Micelles, vesicles and liposomes

Micelles are formed by a wide range of amphiphilic surfactant molecules in aqueous solution. A typical amphiphile contains a long alkyl chain and a polar headgroup. In solution the interactions between the alkyl chains and water are extremely unfavourable and this drives aggregation of the alkyl chains together to minimise this interaction. A structure then forms as shown schematically in Fig. 8.2, where a spherical aggregate spontaneously assembles, with the polar headgroups on the outside of the sphere and a hydrophobic interior, stabilised by van der Waals interactions between the chains.

What makes micelles so useful as carrier agents is that hydrophobic species can be incorporated within the core of the micelle. This enables the transport of hydrophobic active agents at concentrations much higher than would be possible for a simple aqueous solution of these compounds. The hydrophobic environment of the core moreover protects the guest against hydrolysis and enzymatic degradation.

As an alternative to the classical long chain surfactants shown in Fig. 8.2, amphiphilic block copolymers are also very adept at forming micellar structures. For example, a block copolymer containing a hydrophobic block such as polystyrene and a hydrophilic block such as polyethylene oxide, can dissolve in water to give a micellar structure with a polystyrene core surrounded by a polyethylene oxide corona, with the whole structure being typically 5–50 nm in diameter. Block copolymers are especially suitable for this purpose since they are available in a wide variety of molecular weights. Since a wide variety of

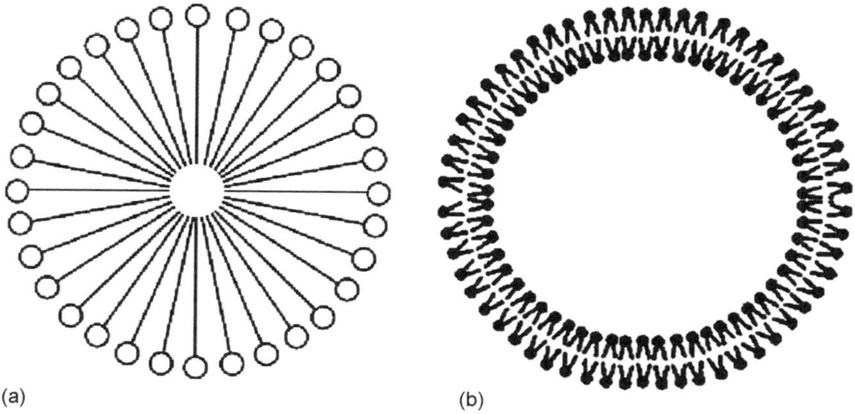

(a) (b)

8.2 Structures of (a) a micelle and (b) a vesicle.

monomers and the composition ratios can easily be selected, this allows fine control of the size, composition and morphology of the micelles. Incorporation of many functional groups is easily attained; for example, crosslinking groups can be utilised to stabilise the micelle structure and the surface properties, selectivity and biocompatibility of the micelles can all be tailored.

In a similar manner, a variety of compounds can be utilised to form vesicles. These are somewhat more complex structures than micelles (Fig. 8.2) in which a spherical membrane is formed containing an aqueous core. Liposomes are types of vesicles which are widely used within drug transport and consist of spherical phospholipid bilayers with an aqueous core and are the most commonly used of these carrier systems (Langer, 1998; Kaparissides *et al.*, 2006). They are capable of transporting either hydrophilic agents within the core or hydrophobic agents which are incorporated into the membrane. The membrane itself can be tailored to prevent access of enzymes to the core, thereby preventing enzymatic degradation, but do, however, allow the diffusion out of the active agent. For example, channel-type proteins may be incorporated within the membrane and retain their activity, thereby allowing transport of ions or other small solutes such as drugs through the membrane.

The use of liposomes can lead to a large increase in drug carrying capacity compared with the use of polymeric systems; however, disadvantages can be encountered such as shorter shelf-life and rapid destruction of the vesicle within the body. The biocompatibility of these systems, however, can be improved by modifying their surface properties, such as by the attachment of polyethylene glycol units (Park *et al.*, 1997; Lasic *et al.*, 1999; Moghimi and Szebeni, 2003). Also, antibodies can be attached to the surfaces of these systems, allowing them to specifically target sites that require treatment. Tumours, for example, can be targeted by substituting liposomes with antibodies against the *HER2* proto-oncogene, found in breast and other cancers (Kirpotin *et al.*, 1997; Park *et al.*,

1998). Similarly amylopectin can be attached to the surface of liposomes to enable targeting of lung tissue (Vyas *et al.*, 2004).

8.2.5 Nanotechnology

In recent times, nanotechnology has become an intensely studied field of research. The application of nanosized materials is expected to make significant advances in biomedical applications such as drug delivery and gene therapy (Moghimi *et al.*, 2001; Moghimi and Szebeni, 2003; Sahoo and Labhasetwar, 2003).

Earlier in this chapter the use of biodegradable polymers as drug carriers was discussed. Nanoparticles of these types of materials with sizes in the range 10–1000 nm can be synthesised and incorporate drug molecules by way of entrapment or binding. Similarly, nanocapsules can be synthesised in which a polymer membrane forms a vesicle-like structure with the drug molecules confined within a central cavity. Because of their small size, these systems are capable of passing through small capillaries and be taken up by cells, allowing efficient drug accumulation at a target size (Desai *et al.*, 1997). In these materials size and surface properties determine their distribution in the body (Sahoo and Labhasetwar, 2003). Localised application of these nanoparticles can be achieved by their tendency to accumulate in tumours due to enhanced permeation effects; this has been extensively reviewed elsewhere (Maeda, 2001). Alternatively the nanoparticles can be delivered locally to the site of interest such as within a specific artery after balloon angioplasty (Guzman *et al.*, 1996) to deliver long-term release (14 days) of dexamethasone.

Dendrimers are branched polymers grown from a central core with very precisely controlled degrees of polymerisation. Figure 8.3 shows a schematic of a fourth generation dendrimer. This makes their size, composition and molecular structure controllable to an exact degree and is an alternative form of polymer nanoparticle to those made by classical emulsion-type polymerisations. There has been great interest in utilising dendrimers as drug carrier agents because not only can the bulk of the dendrimer be synthesised exactly but a wide variety of surface groups can be utilised, enabling further surface functionalisation. Dendrimers, for example, can be made with hydrophilic surfaces and hydrophobic interiors or vice versa. Recent reviews on this subject (Svenson and Tomalia, 2005; Gupte *et al.*, 2006) extensively detail many of the most pertinent advances in this field so only a few highlights will be given here.

Dendrimers have been utilised to carry a variety of small molecule pharmaceuticals. Commercial poly(amidoamine) (PAMAM) dendrimers have been used to encapsulate the anti-cancer drug cisplatin, giving conjugates that exhibited slower release, higher accumulation in solid tumours, and lower toxicity compared with the free drug (Malik *et al.*, 1999). PAMAM dendrimers which had been functionalised with poly(ethylene glycol) chains had their encapsulation behaviour for the anti-cancer drugs adriamycin and methotrexate

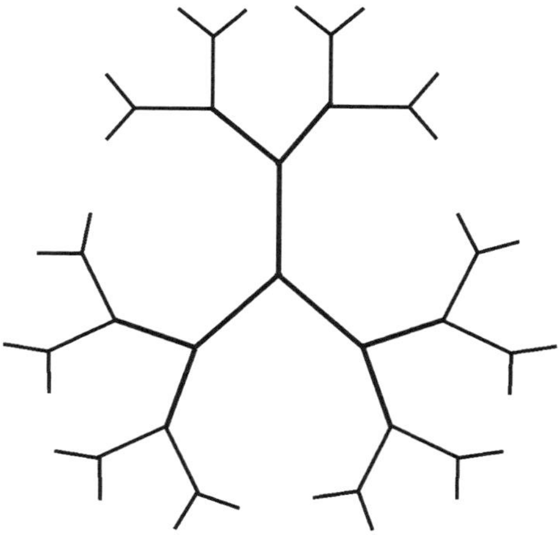

8.3 Schematic structure of a fourth generation dendrimer.

studied. Up to 6.5 adriamycin molecules or 26 methotrexate molecules per dendrimer could be incorporated for one of the materials studied. The drug release from this dendrimer was slow at low ionic strength but fast in isotonic solution (Kojima *et al.*, 2000). Similar dendrimers encapsulated the anticancer drug 5-fluorouracil, showing reasonable drug loading, and reduced release rate and haemolytic toxicity (Bhadra *et al.*, 2003). Up to 78 molecules of the anti-inflammatory drug ibuprofen were complexed by PAMAM dendrimers through electrostatic interactions between the dendrimer amines and the carboxyl group of the drug. The drug was successfully transported into lung epithelial carcinoma cells by the dendrimers (Kohle *et al.*, 2003). Other recent studies have also indicated that low-generation PAMAM dendrimers cross cell membranes (El-Sayed *et al.*, 2003).

Ceramic nanoparticles have also been studied owing to their inherent advantages of ease of synthesis in a wide variety of sizes, shapes and porosities by methods similar to sol–gel processes. They are available in very small (<30 nm) sizes making them capable of crossing cell membranes, possess surfaces which are easily chemically modified and do not display swelling processes with changes in pH (Sahoo and Labhasetwar, 2003). For example, silica nanoparticles can be constructed with an anti-cancer drug entrapped within their cores (Roy *et al.*, 2003) which are stable in aqueous systems. Tumour cells take up these nanoparticles, which upon irradiation generate singlet oxygen, which significantly damages the tumour cells.

8.3 Commercial systems

There are a wide variety of commercial drug carrier agents available. Since a complete review would form an article itself, only a few are discussed.

As previously mentioned, polyanhydrides are widely used as slow release agents due to the fact they biodegrade reproducibly and with no toxic by-products. Commercial products include materials such as Decapeptyl SR®, manufactured by Ipsen Ltd, which contains the active ingredient triptorelin acetate encapsulated in poly(lactide-co-glycolide). This is utilised in the treatment of prostate cancer. Similar products include Lupron Depot® (TAP Pharmaceuticals Inc., active ingredient leuprolide acetate) and Sandostatin LAR® (Novartis, active ingredient octreotide acetate).

Nitroglycerine is a problematic drug due to its loss of tablet activity, often by volatilisation of the active component (Markovich *et al.*, 1997). This can be avoided by incorporation of the nitroglycerine into an acrylic-based hydrogel which is then incorporated into a transdermal patch, as exemplified by products such as Deponit® (Schwarz Pharma), Minitran® (3M Pharma) and Nitrodisc® (G.D. Searle Company).

8.4 Biosensors

8.4.1 History and format of biosensors

The purpose of this chapter is to introduce the concept of using biological molecules as the selective recognition elements within biosensors. Most sensors consist of three principal components, as described below and shown in Fig. 8.4:

- The first of these includes a receptor species, usually biological in origin, such as an enzyme, antibody or DNA strand capable of recognising the analyte of interest with a high degree of selectivity; this is usually concurrent with a binding event between receptor and analyte.
- The second component that must be present is a transducer, enabling the translation of the binding event into a measurable physical change; possible events include the generation of electrons, protons, an electrochemically active chemical species such as hydrogen peroxide or simple physical changes such as a change in conductivity, optical absorbance or fluorescence.
- Thirdly there must be inclusion of a method of measuring the change detected at the transducer and converting this into useful information.

There are several advantages associated with using biological molecules as the active recognition entity within a sensor. Usually they display unsurpassed selectivities; for example glucose oxidase will interact with glucose and no other sugar, and in this way will act as a highly selective receptor. In the case of glucose oxidase, the electrochemically inactive substrate glucose is oxidised to form gluconolactone along with the concurrent generation of the electroactive species hydrogen peroxide. Enzymes also generally display rapid turnover rates and this is often essential (a) to avoid saturation and (b) to allow sufficient generation of the active species in order to be detectable.

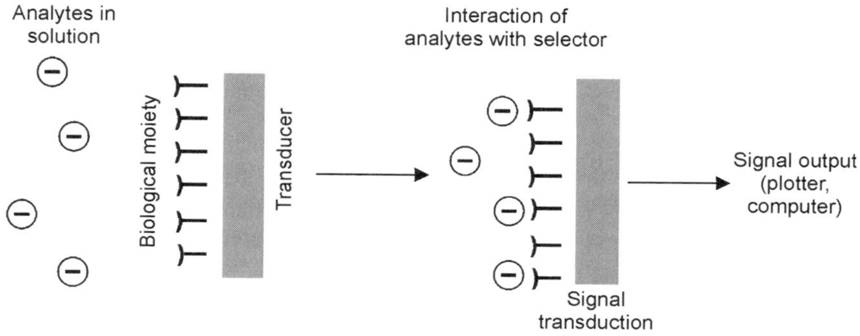

8.4 Schematic of a biosensor.

Antibodies bind solely to their antigens and achieve specificity via a complex series of multiple non-covalent bonds. Since the principle of immunoassay was first published by Yalow and Berson in 1959, there has been an exponential growth in both the range of analytes to which the technique has been successfully applied and the number of novel assay designs. Development of enzyme-labelled immunoanalytical techniques, e.g. enzyme-linked immunosorbent assay (ELISA) has provided analytical tests without the safety risks associated with radiolabelling-based techniques.

The rapid measurement of analytes of clinical significance, e.g. towards various disease markers, would permit earlier intervention, which in a medical setting is frequently of utmost importance. There has been much research on the development of direct immunosensors that do not rely on the use of a detectable label. Such a system will lead to simpler assay formats and ideally lower times of detection. A reusable and rapid detection system would, moreover, allow for continuous real-time measurement, so helping to maintain optimal homeostatic conditions.

Unfortunately there are also some disadvantages related to the construction and use of biosensors. Often the biological species can either be extremely expensive or difficult to isolate in sufficient purity. Immobilisation of these species can lead to loss of activity and the presence of various chemical species in the test solution can also cause loss of activity, e.g. enzymes can be easily poisoned by heavy metals. In biological samples such as blood or saliva, there can also be solutes that are electrochemically active and interfere with determinations of the target species. Again in physiological fluids such as blood, various species may be present which bind to the surface so causing fouling and loss of sensor response.

Antibody–antigen binding is based on multiple non-covalent interactions, and therefore there are no newly formed molecules, protons or electrons that are easily detectable, which has limited the development of direct antibody affinity-type sensors. Also affinity binding constants typically range from 10^5 to

10^{11} mol l^{-1} meaning that the antibody–antigen binding event is often irreversible. As a consequence of this, many contemporary immunosensors are of use only for 'single-shot' analyses and must be disposable in nature.

8.4.2 Glucose biosensors

A series of extensive reviews on biosensors and their history have been published elsewhere (Hall, 1990; Eggins, 1996; Wang, 2001) and therefore only a brief history will be given here. Easily the most intensively researched area has been towards the development of glucose biosensors (Wang, 2001; Newman *et al.*, 2004). The reason for this is the prevalence of diabetes, which has become a worldwide public health problem. Diabetes represents an increasing epidemic with, at the time of writing, 170 million sufferers worldwide (World Health Organization, www.who.org), and this is estimated to reach 300 million by 2045 (Newman *et al.*, 2004). Diabetes is related to a number of factors such as obesity and heart disease all of which make this disease one of the leading causes of death and disability in the world. The world market for biosensors is approximately \$5bn with, at the time of writing, approximately 85% of the world commercial market for biosensors being for blood glucose monitoring (Newman *et al.*, 2004).

These factors have led to the development of a number of inexpensive disposable electrochemical biosensors for glucose, incorporating glucose oxidase (GOD) bound immobilised at various electrodes. They are generally amperometric sensors, with electrodes polarised at a set potential; the oxidation or reduction of a chosen electroactive species at the surface will then lead to generation of a detectable current. The principal classes of glucose and other types of biosensors are described below.

First generation biosensors

The first electrochemical glucose biosensor was based on an oxygen electrode (Clark and Lyons, 1962). A film of immobilised GOD was laid down upon the oxygen electrode, which was overlaid with a semipermeable dialysis membrane. Upon exposure to glucose, the enzymatically catalysed oxidation reaction occurs, causing a localised consumption of oxygen.

$$\text{glucose} + O_2 \xrightarrow{\text{Glucose oxidase}} \text{gluconolactone} + H_2O_2 \tag{8.1}$$

This then leads to a drop in the current generated at the oxygen electrode. This device was subject to fluctuations caused by variable oxygen levels; however, further work (Updike and Hicks, 1967) utilised two oxygen electrodes, one of which was coated with glucose oxidase. Measurement of differential current between the two electrodes served to cancel out these effects. The first commercial glucose analyser was the Model 23 YSI analyser, launched by the

Yellow Spring Instrument Company in 1975 and based on the Clark electrode. This device was capable of measuring the glucose level in 25 ml of whole blood.

The Clark oxygen electrode to monitor the depletion of oxygen caused by the oxidation of glucose is an example of a first generation biosensor, where the reaction causes depletion of an electrochemically active compound (oxygen) which can then be measured. An alternative technique is based on the amperometric monitoring of the product of the enzyme-catalysed reaction, H_2O_2:

$$H_2O_2 \xrightarrow{+650\ \text{mV vs Ag/AgCl}} 2H^+ + O_2 + 2e^- \tag{8.2}$$

The results from these types of sensors can be affected by fluctuations in the ambient oxygen concentration or by the presence of electroactive species such as ascorbate, that are capable of being oxidised at +650 mV – giving rise to an erroneous result. Concentration of interferents at the electrode surface can be minimised by application of a permselective coating to the sensor, thereby reducing interference from electroactive species. Polymeric materials have led the way with materials such as the fluorinated ionomer Nafion (Turner and Sherwood, 1994) and cellulose acetate (Maines *et al.*, 1996) being two of the most commonly used. A beneficial side effect is that these materials can also confer a degree of biocompatibility. An alternative approach has been to electropolymerise suitable monomers to form protective coatings. 1,2-Diaminobenzene (Myler *et al.*, 1997), for example, when deposited at the bioelectrode surface serves to both stabilise the electrode due to its inherent high biocompatibility while also imparting selective exclusion of interferents such as ascorbate.

The robustness of these types of sensors and, after suitable cleaning procedures, capability for multiple analyses makes them suitable for use within hospitals. However, diabetes as a condition requires regular monitoring of blood glucose levels, so hospital analysis is impractical for a normal lifestyle. The obvious solution has been the development of inexpensive home detection methods where the physiological sample, usually blood, can be analysed by the patient. The problems of cleaning the sensor are moreover negated by using disposable sensor strips.

Second generation biosensors

As shown earlier, the reaction of GOD with glucose gives rise to the formation of gluconolactone and the reduced form of the enzyme, which is then reoxidised by oxygen. Direct transfer of electrons from the electrode would circumnavigate this reaction; however, the active site of GOD is encased in a protein sheath which inhibits this transfer. To facilitate this transfer, a suitable chemical species can be utilised to 'shuttle' electrons back and forth between active site and

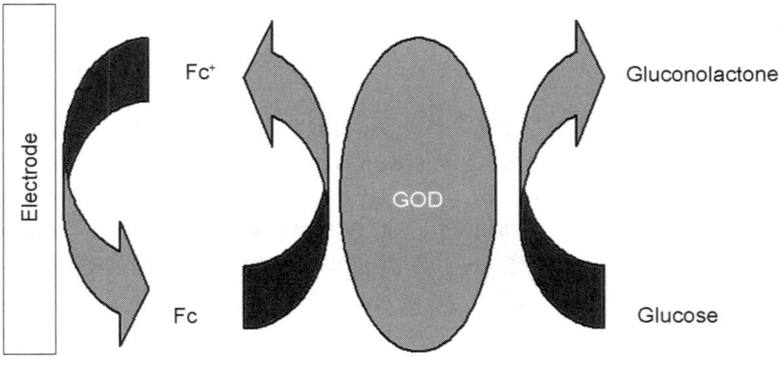

Fc = ferrocene derivative

8.5 The oxidation of glucose at an electrode, mediated by a ferrocene derivative.

electrode. This moiety, known as a mediator, must react readily with the enzyme to avoid competition by ambient oxygen and in both its reduced and oxidised forms, be stable and preferably require as low an over-potential to be oxidised as is feasible. This method sidesteps problems associated with detecting species such as oxygen or peroxide and also lowers the potential required for measurement of the enzyme catalysed reaction, thereby reducing the inference by redox active species present within the sample to be studied. A typical reaction scheme (Fig. 8.5) where a ferrocene compound acts as mediator is shown (Cass *et al.*, 1984). The first home glucose testing kit, the Exactech® glucose biosensor, is based on this chemistry. The actual pen-sized device is produced by Medisense® and utilises a disposable strip upon which a single drop of blood is placed. A wide range of glucose sensors (Wang, 2001) based on this method have since become commercially available for home glucose testing. Further work has concentrated on reducing blood volumes and produced devices such as the Pelikan® device which only require microlitre blood volumes (Newman *et al.*, 2004).

Third generation sensors

Second generation biosensors are still somewhat limited in that they require use of a suitable mediator. Attempts have been made to circumvent this by developing methods to electronically directly connect or 'wire' the enzyme to the electrode, thereby allowing simple electron transfer from the enzyme to the electrode without the requirement for a mediator. Although these types of devices have not as yet been developed commercially, they provide a possible alternative to mediated electron transfer.

Typical approaches involve the use of a polymeric coating which both immobilises the GOD and allows electron transfer to the electrode. One example is the use of polyvinyl pyridine where the synthetic polymer has been modified with a large number of osmium-based electron transfer relays (Degani and Heller, 1987; Ohara *et al.*, 1994). The polymer is co-immobilised at an electrode with glucose oxidase. When the enzyme reacts with glucose and is converted to the reduced form, the polymer allows facile transfer of electrons to convert it back to the oxidised form. Similar materials based on polyvinyl imidazole have also been utilised (Mano *et al.*, 2005). GOD and other enzymes have also been immobilised on various electrodes by a variety of chemical and physical methods (Davis and Higson, 2005).

Conducting polymers are especially suitable for the immobilisation of enzymes at electrode surfaces and have been reviewed in greater detail elsewhere (Barisci *et al.*, 1996; Gerard *et al.*, 2002). A variety of monomers such as pyrrole or aniline can be electropolymerised on an electrode surface and under correct conditions form stable conductive films. If during this electropolymerisation process, enzymes are present in the solution, they can be entrapped within the film during the deposition process, alternatively they can be adsorbed onto, or be chemically grafted to, the film following deposition (Barisci *et al.*, 1996; Gerard *et al.*, 2002). The close association between the conductive polymer and the enzyme facilitates rapid electron transfer between the enzyme and an electrode surface.

One of the first methods involved the simple entrapment of enzymes such as glucose oxidase within polyaniline films (Cooper and Hall, 1992). Our group has taken this process further, utilising both non-conductive and conductive polymers to fabricate arrays of conductive microelectrodes with entrapped biological molecules such as GOD (Barton *et al.*, 2004).

8.5 Continuous monitoring

One problem with the current commercial biosensors is the invasive procedure, requiring frequent withdrawal of blood for testing which can be both tedious and painful. Devices which could be implanted within the human body would negate this aspect of glucose testing. Implantable sensors suitable for *in vivo* glucose monitoring require the device to be extremely small and show long-term stability (with minimal drift, thereby removing the need for frequent calibration), display no oxygen dependency and also show high biocompatibility. The problem of biocompatibility has been the most elusive of these targets and as yet *in vivo* glucose sensors have only limited lifetimes. Effects on sensor performance include fouling by protein deposition on the surface or formation of fibrous tissue around the sensor (D'Orazio, 2003) leading to loss of sensor performance. Rejection by the immune system or thrombus formation if used intravascularly can also degrade device performance and risk harm to the user.

Subcutaneously applied sensors have been developed (Bindra *et al.*, 1991; Henry, 1998) which could monitor glucose concentrations and also be changed by the user. However, the effects of biofouling mean their use was limited to periods of 1–2 weeks. Again, fouling of the sensor and its effect on sensor performance is a major consideration. Attempts have been made to improve this using polymeric coatings. For example, a needle-based electrochemical probe coated with Nafion has been developed (Moussy *et al.*, 1993) which is just 0.5 mm in diameter and can be inserted subcutaneously through an 18-gauge needle. Again *in vivo* measurements could be carried out over periods of 2 weeks.

A different approach has been to implant a microdialysis fibre into subcutaneous tissue and an iso-osmotic electrolyte solution pumped through the fibre. Glucose diffuses into the fibre and the electrolyte. If flow rates are kept constant the glucose concentration in the outflow from the fibre can be directly related to the glucose concentration in the interstitial fluid. Rapid changes in blood glucose concentration can be determined although there is a time delay, typically about 8 minutes (Jansson *et al.*, 1988) between changes in blood and interstitial fluid concentrations. The use of these probes combined with glucose biosensors to monitor the outflow have achieved 4–7 days of continuous use for glucose monitoring in humans (Myerhoff *et al.*, 1992; Hashiguchi *et al.*, 1994).

A subcutaneously implantable device, the CGMS® (Continuous Glucose Monitoring System) has been commercialised by Minimed (www.minimed.com). The glucose biosensor probe is inserted just beneath the skin, usually in the abdomen, and can be used to monitor glucose for up to 72 h, with a reading every 5 min. Traditional blood sampling of glucose is used to calibrate the device. Other data such as meal times and exercise periods can also be recorded using the device and all data then downloaded to a computer.

Non-invasive glucose monitoring systems are an active area of research. Techniques such as reverse iontophoresis, where a low electric current 'pulls' interstitial fluid through the skin and onto a glucose biosensor and negates the need for blood withdrawal. This technique has been utilised in the GlucoWatch® G2 Biographer (Cygnus), a commercially available biosensor device for continuous glucose monitoring. The electric current pulls the interstitial fluid through the skin and collects it into two gel discs. These are in contact with an electrochemical glucose biosensor, which determines the glucose level. Again the device is calibrated using traditional blood testing. The device records glucose levels every 10 minutes for up to 13 h and can also sound an alert should the level stray outside predetermined levels. After 13 h, the active part of the sensor (i.e. gel pads and biosensor) can be replaced and the device reused. Some 95% of GlucoWatch® G2™ Biographer readings are clinically accurate or acceptable (Newman *et al.*, 2004). However there are some problems, according to the company brochure the Glucowatch may be affected by excessive perspiration and can cause minor skin irritation. The Glucowatch is now supplied by Animas Technologies.

8.6 Future trends

This is a brief section just to highlight several potential future applications which, although not in the current marketplace, are the subjects of intensive research efforts.

8.6.1 Point of care tests

Although the biosensor market is dominated by glucose testing applications, there is great interest in developing simple tests for a wide variety of medical problems which can be easily applied by the patient and detect physical problems which are beginning to occur while still in an early stage, thereby enabling the patient to seek early medical help. An example of this is the Multisense® system developed by Oxford Biosensors (www.oxfordbiosensors.com) and soon to be commercialised, which detects early markers of coronary heart disease. Heart disease is a major killer in the Western world and the risk of developing this condition is known to be related to factors such as diet, smoking, weight and high blood cholesterol.

The Multisense® system is based on a hand-held device with disposable electrochemical test strips. Each strip contains several sensors which individually sense for cholesterol (high-density and low-density lipoprotein) and triglycerides. Testing utilises a single drop of blood in the same manner as a glucose sensor. This will enable fast and simple screening, diagnosis and monitoring of patients who are at risk of heart disease.

8.6.2 Artificial pancreas

For many diabetes type 1 sufferers, life is a constant round of glucose testing and insulin injections. Externally worn insulin pumps have been developed, such as those by Medtronic (www.minimed.com). These have the advantage that they remove the need for injections by introducing insulin subcutaneously though a cannula. When coupled together with a glucose biosensor, there is then the potential for the device to inject insulin 'on demand'. This was the first device of its type to receive FDA approval (Newman *et al.*, 2004). At present, current commercial devices usually can only be used for a few days at a time without maintenance and injection, and sensing sites need to be changed regularly. An implanted device that automatically responds to changes in glucose levels would in effect act as an artificial pancreas and improve quality of life for many millions of people.

8.6.3 Warfarin monitors

Warfarin is routinely prescribed as an anti-coagulation agent for patients with an increased tendency for thrombosis or as prophylaxis in those individuals who have already formed a blood clot (thrombus) which required treatment. However warfarin when given in too high a dose inhibits natural clotting, leaving the patient subject to uncontrolled bleeding, both externally and gastrointestinal bleeding. At present, patients placed on warfarin require their blood to be regularly tested to ensure the clotting ability of their blood falls within required ranges (Poller *et al.*, 2003). This requires withdrawal of blood and testing within a laboratory environment. Since warfarin is one of the world's most prescribed drugs, a simple home test which could monitor warfarin levels or alternatively monitor blood clotting ability would have a huge global market.

8.7 Conclusions

Drug delivery spans a wide variety of fields such as medicine, chemistry, biology and materials science. As fields such as genomics, proteomics and immunology progress, we will see these approaches being utilised to target drugs to specific sites with greater specificity. Further investigation into the transport of drugs across obstacles such as the blood–brain or air–blood barriers will enable development of simplified delivery systems for these compounds.

There are a wide number of potential techniques that would prove highly beneficial to medical practitioners and are currently being widely researched, some of which are listed below:

- Biodegradable nanoparticles and methods for producing them.
- More complex controlled release characteristics, for example in response to changing physiological conditions.
- Delivery of large quantities of drugs in controlled and highly localised manner, for example to destroy a tumour without harming surrounding healthy tissue.
- Biodegradable coatings for implants, for example, for the release of agents that minimise rejection and/or promote healthy cell growth.
- Functionalising the delivery system to enable selective targeting of sites via immunochemical response.

There are many future opportunities in this field, with one of the most exciting being the application of nanotechnology to drug delivery. The potential advantages associated with nanoparticles such as easy tailoring of properties such as size, often easy transition across cell membranes, and potential site specificity ensures these approaches will be intensively studied. Problems to be overcome include the fact that these materials can be cytotoxic and often display poor stability in biochemical environments (Kaparissides *et al.*, 2006). However, these are challenges to be overcome and it can be seen that with the advances taking place in chemistry, materials, biology and other sciences that sooner or later they are likely to be achieved.

As can be seen biosensors are a rapidly expanding field of research. Currently the market is dominated by glucose sensing. Many other applications such as cholesterol monitoring, monitors for various drug treatments and many other analytes are being intensively researched. It is likely that some of these applications will soon be commercialised.

8.8 References

Barisci J N, Conn C, Wallace G G (1996) 'Conducting polymer sensors', *Trends Polym. Sci.*, **4**, 301–311.

Barton A C, Collyer S D, Davis F, Gornall D D, Law K A, Lawrence E C D, Mills D W, Myler S, Pritchard J A, Thompson M, Higson S P J (2004) 'Sonochemically fabricated microelectrode arrays for biosensors offering widespread applicability. Part I', *Biosens. Bioelec.*, **20**, 328–337.

Bhadra D, Bhadra S, Jain S, Jain N K (2003) 'A PEGylated dendritic nanoparticulate carrier of fluorouracil', *Int. J. Pharm.*, **257**, 111–124.

Bindra D, Zhang Y, Wilson G (1991) 'Design and *in vitro* studies of a needle type glucose sensor for subcutaneous monitoring', *Anal. Chem.*, **63**, 1692–1696.

Brem H, Piantadosi S, Burger P C, Walker M, Selker R, Vick N A, Black K, Sisti M, Brem S, Mohr G, Muller P, Morawetz R, Schold S C (1995) 'Placebo-controlled trial of safety and efficacy of intraoperative controlled delivery by biodegradable polymers of chemotherapy for recurrent gliomas', *Lancet*, **345**, 1008–1012.

Cass A E G, Davis G, Francis G D, Hill H A O, Aston W J, Higgins I J, Plotkin E V, Scott L D L, Turner A P F (1984) 'Ferrocene-mediated enzyme electrode for amperometric determination of glucose', *Anal. Chem.*, **56**, 667–671.

Chen L, Remondetto G L, Subirade M (2006) 'Food-based protein materials as nutraceutical delivery systems', *Food. Sci. Tech.*, **17**, 272–283.

Cho C S, Han S Y, Ha J H, Kim S H, Lim D Y (1999) 'Clonazepam release from bioerodible hydrogels based on semi-interpenetrating polymer networks composed of poly(epsilon-caprolactone) and poly(ethylene glycol) macromer', *Int. J. Pharm.*, **181**, 235–242.

Clark A H, Kavanagh G M, Ross-Murphy S B (2001) 'Globular protein gelation – theory and experiment', *Food Hydrocolloids*, **15**, 383–400.

Clark L, Lyons C (1992) 'Electrode systems for continuous monitoring in cardiovascular surgery', *Ann. NY Acad. Sci.*, **102**, 29–45.

Cooper J C, Hall E A H (1992) 'Electrochemical response of an enzyme-loaded polyaniline film', *Biosens. Bioelec.*, **7**, 473–485.

Davis F, Higson S P J (2005) 'Structured thin films as components in biosensors', *Biosens. Bioelec.*, **21**, 1–20.

Degani Y, Heller A (1987) 'Direct electrical communication between chemically modified enzymes and metal-electrodes. 1. Electron-transfer from glucose-oxidase to metal-electrodes via electron relays, bound covalently to the enzyme', *J. Phys. Chem.*, **91**, 1285–1289.

Desai M P, Labhasetwar V, Walter E, Levy R J, Amidon G L (1997) 'The mechanism of uptake of biodegradable microparticles in Caco-2 cells is size dependent', *Pharm. Res.*, **14**, 1568–1573.

D'Orazio P (2003) 'Biosensors in clinical chemistry', *Clin. Chim. Acta.*, **334**, 41–69.

Eggins B R (1996) *Biosensors*, Chichester, Wiley.

El-Sayed M, Rhodes C A, Ginski M, Ghandehari H (2003) 'Transport mechanism(s) of poly(amidoamine) dendrimers across Caco-2 cell monolayers', *Int. J. Pharm.*, **265**, 151–157.

Gander B, Meinel L, Walter E, Merkle H P (2001) 'Polymers as a platform for drug delivery: reviewing our current portfolio on poly(lactide-co-glycolide) (PLGA) microspheres', *Chimia*, **55**, 212–217.

Gerard M, Chaubey A, Malhotra B D (2002) 'Application of conducting polymers to biosensors', *Biosens. Bioelec.*, **17**, 345–359.

Gombotz W R, Pettit D (1995) 'Biodegradable polymers for protein and peptide drug-delivery', *Bioconjugate Chem.*, **6**, 332–351.

Gupta U, Agashe H B, Asthana A, Jain N K (2006) 'Dendrimers: novel polymeric nanoarchitectures for solubility enhancement', *Biomacromolecules*, **7**, 649–658.

Guzman L A, Labhasetwar V, Song C X, Jang Y S, Lincoff A M, Levy R, Topol E J (1996) 'Local intraluminal infusion of biodegradable polymeric nanoparticles – a novel approach for prolonged drug delivery after balloon angioplasty', *Circulation*, **94**, 1441–1448.

Hall E A C (1990) *Biosensors*, Maidenhead, Open University Press.

Hashiguchi Y, Sakakida M, Nishida K, Uemura T, Kajiwara K, Shichiri M (1994) 'Development of a miniaturized glucose monitoring system by combining a needle-type glucose sensor with microdialysis sampling method', *Diabetes Care*, **17**, 387–396.

Henry C (1998) 'Getting under the skin', *Anal. Chem.*, **70**, 594A–598A.

Jansson PA, Fowelin J, Smith U, Lonnroth R (1988) 'Characterization by microdialysis of intercellular glucose level in subcutaneous tissue in humans', *Am. J. Physiol.*, **255**, E218–E220.

Jeong B, Bae Y H, Lee D S, Kim S W (1997) 'Biodegradable block copolymers as injectable drug-delivery systems', *Nature*, **388**, 860–862.

Kaparissides C, Alexandridou S, Kotti K, Chaitidou S (2006) 'Recent advances in novel drug delivery systems', *J. Nanotech. Online*, 10.2240/azojono0111.

Kirpotin D, Park J W, Hong K, Zalipsky S, Li W L, Carter P, Benz C C, Papahadjopoulos D (1997) 'Sterically modified anti-HER2 immunoliposomes: design and targeting to human breast cancer cells *in vitro*', *Biochemistry*, **36**, 66–75.

Kojima C, Kono K. Maruyama K, Takagishi T (2000) 'Synthesis of polyamidoamine dendrimers having poly(ethylene glycol) grafts and their ability to encapsulate anticancer drugs', *Bioconjug. Chem.*, **11**, 910–917.

Kolhe P, Misra E, Kannan R M, Kannan S, Lieh-Lai M (2003) 'Drug complexation, *in vitro* release and cellular entry of dendrimers and hyperbranched polymers', *Int. J. Pharm.*, **259**, 143–160.

Langer R (1995) 'Biomaterials and biomedical engineering', *Chem. Eng. Sci.*, **50**(24) 4109–4121.

Langer R (1998) 'Drug delivery and targeting', *Nature*, **392**[supp], 5–10.

Lasic D D, Vallner J J, Working P K (1999) 'Sterically stabilized liposomes in cancer therapy and gene delivery', *Curr. Opin. Mol. Ther.*, **1**, 177–185.

Liu X, Sun Q, Wang H, Zhang L, Wang J (2005) 'Microspheres of corn protein zein for an ivermectin drug delivery system', *Biomaterials*, **26**, 109–115.

Maeda H (2001) 'The enhanced permeability and retention (EPR) effect in tumor vasculature: the key role of tumor-selective macromolecular drug targeting', *Adv. Enz. Reg.*, **41**, 189–207.

Maines A, Ashworth D, Vadgama P (1996) 'Diffusion restricting outer membranes for greatly extended linearity measurements with glucose oxidase enzyme electrodes', *Anal. Chim. Acta.*, **333**, 223–231.

Malik N, Evagorou E G, Duncan R (1999) 'Dendrimer–platinate: a novel approach to cancer chemotherapy', *Anticancer Drugs*, **10**, 767–776.
Maltais A, Remondetto G E, Gonsalves R, Subirade M (2005) 'Formation of soy protein isolate cold-set gels: protein and salt effects', *J. Food. Sci.*, **70**, 67–73.
Mano N, Mao F, Heller A (2005) 'On the parameters affecting the characteristics of the "wired" glucose oxidase anode', *J. Electroanal. Chem.*, **574**, 347–357.
Markovich R J, Taylor A K, Rosen J (1997) 'Drug migration from the adhesive matrix to the polymer film laminate facestock in a transdermal nitroglycerin system', *J. Pharm. Biomed. Anal.*, **16**(4), 651–660.
Moghimi S M, Szebeni J (2003) 'Stealth liposomes and long circulating nanoparticles: critical issues in pharmacokinetics, opsonization and protein-binding properties', *Prog. Lipid Res.*, **42**, 463–478.
Moghimi S M, Hunter A C, Murray J C (2001) 'Long-circulating and target-specific nanoparticles: theory to practice', *Pharmol. Rev.*, **53**, 283–318.
Moussy F, Harrison D J, O'Brien D W, Rajotte R V (1993) 'Performance of subcutaneously implanted needle-type glucose sensors employing a novel trilayer coating', *Anal. Chem.*, **65**, 2072–2077.
Myerhoff C, Bischof F, Sternberg F, Zier H, Pfeiffer E F (1992) 'On line continuous monitoring of subcutaneous tissue glucose in men by combining portable glucose sensor with microdialysis', *Diabetologia*, **35**, 1087–1092.
Myler S, Eaton S, Higson S P J (1997) 'Poly(o-phenylenediamine) ultra-thin polymer-film composite membranes for enzyme electrodes', *Anal. Chim. Acta.*, **357**, 55–61.
Newman, J D, Tigwell L J, Turner A P F, Warner P J (2004) 'Biosensors: a clearer view', in *Biosensors 2004 – The 8th World Congress on Biosensors*, Elsevier, New York.
Ogawa Y, Yamamoto M, Okada H, Yashiki T, Shimamoto T (1988) 'A new technique to efficiently entrap leuprolide acetate into microcapsules of polylactic acid or copoly(lactic glycolic) acid', *Chem. Pharm. Bull.*, **36**, 1095–1103.
Ohara T, Rajagopalan R, Heller A (1994) 'Wired enzyme electrodes for amperometric determination of glucose or lactate in the presence of interfering substances', *Anal. Chem.*, **66**, 2451–2457.
Ottenbrite, R M (1999) *Frontiers in Biomedical Polymer Applications*, Vol. 2, Lancaster, Technomic.
Park J W, Hong K, Kirpotin D, Papahadjopoulos D, Benz C C (1997) 'Immunoliposomes for cancer treatment', *Adv. Pharmacol.*, **40**, 399–435.
Park JW, Kirpotin D, Hong K, Colbern G, Shalaby R, Shao Y, Meyer O, Nielsen U, Marks J, Benz CC, Papahadjopoulos D (1998) 'Anti-HER2 immunoliposomes for targeted drug delivery', *Med. Chem. Res.*, **8**, 383–391.
Poller L, Keown M, Chauhan N, van den Besselaar A M H P, Tripodi A, Shiach C, Jespersen J (2003) 'Reliability of international normalised ratios from two point of care test systems: comparison with conventional methods', *Brit. Med. J.*, **327**, 30–32A.
Roco M C, Bainbridge W S (eds) (2002) *Converging Technologies for Improving Human Performance*, National Science Foundation Report, Dordrecht, Kluwer Academic.
Roy I, Ohulchanskyy T Y, Pudavar H E, Bergey E J, Oseroff A R, Morgan J, Dougherty T J, Prasad P N (2003) 'Ceramic-based nanoparticles entrapping water-insoluble photosensitizing anticancer drugs: a novel drug-carrier system for photodynamic therapy', *J. Am. Chem. Soc.*, **125**, 7860–7865.
Sahoo S K, Labhasetwar V (2003) 'Nanotech approaches to drug delivery and imaging', *Drug Discovery Today*, **8**, 1112–1120.
Sokoloski T D, Royer G P (1984) 'Drug entrapment within native albumin beads', in

Davis S S, Illum L, McVie J G, Tomlinson E (eds), *Microspheres and Drug Therapy, Pharmaceutical, Immunological and Medical Aspects*, Amsterdam, Elsevier, 295–307.

Svenson S, Tomalia D A (2005) 'Commentary – dendrimers in biomedical applications – reflections on the field', *Adv. Drug. Delivery. Rev.*, **57**, 2106–2129.

Teijon J M, Trigo R M, Garcia O, Blanco M D (1997) 'Cytarabine trapping in poly (2-hydroxyethyl methacrylate) hydrogels: drug delivery studies', *Biomaterials*, **18**, 383–388.

Tomlinson E, Burger J J (1985) 'Incorporation of water soluble drugs in albumin microspheres', in Widder J, Green R (eds), *Methods in Enzymology*, New York, Academic Press, 27–43.

Turner R B F, Sherwood C S (1994) *ACS. Symp. Ser.*, **556**, 211–221.

Updike S, Hicks G (1967) 'Electrode systems for continuous monitoring in cardiovascular surgery', *Nature*, **214**, 986.

Vyas S P, Kannan M E, Jain S, Mishra V, Singh P (2004) 'Design of liposomal aerosols for improved delivery of rifampicin to alveolar macrophages', *Int. J. Pharmaceut.*, **269**, 37–49.

Wang J (2001) 'Glucose biosensors: 40 years of advances and challenges', *Electroanalysis*, **13**, 983–988.

Yalow R S, Berson S A (1959) 'Assay of plasma insulin in human subjects by immunological methods', *Nature*, **184**, 1648–1649.

9
Characterisation using X-ray photoelectron spectroscopy (XPS) and secondary ion mass spectrometry (SIMS)

A J URQUHART and M R ALEXANDER,
University of Nottingham, UK

9.1 Introduction

Surface chemistry, topography and mechanical properties all influence biological response to materials (Curtis and Riehle, 2001; Vitte *et al.*, 2004; Even-Ram *et al.*, 2006). This chapter will introduce the techniques of X-ray photoelectron spectroscopy (XPS) and secondary ion mass spectrometry (SIMS), which provide complementary information on the chemistry of surfaces. Examples are chosen from the literature on biomaterials production and characterisation, specifically polymers, to highlight the issues involved in data acquisition and to illustrate some important concepts of data analysis.

It is necessary to first specify that the surface is different from the bulk. For many materials, the chemistry at the surface is very different, e.g. distinct oxide films are formed under ambient conditions on many metals such as stainless steel. The surface chemistry of pure single phase polymers, analysed in the vacuum of SIMS or XPS spectrometers, is similar to the bulk structure. Large differences have, however, been observed between the bulk and surface of multicomponent polymer formulations due to the mobility of polymer chains combined with the thermodynamic driving force to minimise surface energy. Such multicomponent systems include polymer blends, block copolymers, polymers containing unreacted monomer, degradation or depolymerisation products (e.g. oligomeric methoxy silanes) and additives (e.g. plasticisers, antioxidants, fillers, processing and release agents, etc.). The concentration of these components is low when measured as a proportion of the bulk material volume, and therefore they are often neglected in the specification of materials. However, if these components segregate to the surface they can dominate the surface chemistry and properties. Lastly, adventitious* surface contamination is found on almost all surfaces, most commonly airborne volatile organic compounds (VOCs) which is often referred to as hydrocarbon contamination.

* *def:* not inherent but added extrinsically.

The practical implication of the difference between surface and bulk compositions in laboratory and commercially formulated polymers (and ceramics, glasses and metals), is that surface analysis is required to characterise the actual surface. Surface analysis is also necessary when surface modification is carried out to confirm the desired chemistry has been achieved, although all too often this is neglected. It has been said that this omission is equivalent to cooking a meal without tasting it during the preparation, an approach that is rarely likely to yield the desired results.

9.1.1 A brief history of biomaterial surface analysis

To understand the current development of biomaterial surface studies, it is worth considering the emergence of the predominant techniques and their relatively recent application to biomaterials. XPS was developed by Kai Siegbahn in the late 1960s for which he was awarded the Nobel Prize for Physics (Fahlman *et al.*, 1966). In its early days, XPS was commonly known as electron spectroscopy for chemical analysis (ESCA) before other electron spectroscopic techniques were developed, after which the more specific acronym XPS became the norm. In the late 1970s the first paper containing XPS of a biomaterial was published, by Buddy Ratner at the University of Washington in Seattle, who reported on the surface characterisation of radiation grafted hydrogels (Ratner *et al.*, 1978). Much work had been carried out on the subject of biomaterial surfaces using techniques such as wettability measurements prior to the development of surface spectroscopy (Andrade, 1985). The application of XPS to biopolymer surfaces came shortly after the first publication of polymer surface analysis from the group of Dave Clark in Durham (UK) and the first of an ongoing series on surface analysis of polymer surfaces by Dave Briggs (Clark *et al.*, 1972; Briggs *et al.*, 1976, 2003). The applications of SIMS instruments to the analysis of polymers started in the early 1980s, with the technique developing to the present where it is not only used to analyse synthetic surfaces, but is also useful in elucidating problems as complex as the orientation of proteins on surfaces and the distribution of molecular species in brain matter (Gardella and Hercules, 1980; Briggs and Wootton, 1982; Johansson, 2006; Michel and Castner, 2006). Worldwide, many groups have taken up the study of surfaces with XPS and SIMS to meet the global need for characterisation in the development of polymer, metal, glass and ceramic biomaterial surfaces (Wilson *et al.*, 1994; Brunette *et al.*, 2001; McArthur, 2006).

The oldest form of surface analysis may be the measurement of water contact angle (WCA), with practical application undoubtedly predating the first traceable scientific reports (Ablett, 1923). Placing a drop of water onto a surface and observing the angle with which it intersects the surface is sufficient to have a numerical measurement of the surface wettability. With an analysis depth of ca. 1 nm, the WCA is still a very commonly used characteristic of a surface, most probably because of its ease of measurement and historical prominence. Although it may be a useful and quick measurement of the interaction of water or another liquid with a surface, it is important to realise that explicit information on the chemistry causing the wettability to change is not provided by this property, i.e. many different chemical changes could bring about the same WCA. If the surface chemistry is well characterised then subtle changes in chemistry can be inferred from the contact angle, e.g. self-assembled monolayer packing density (Bain and Whitesides, 1988; Whitesides and Laibinis, 1990; Pertays *et al.*, 2004; Brewer *et al.*, 2004; Foster *et al.*, 2006) However, the complementary combination of elemental and functional quantification of XPS combined with molecular specificity of SIMS is required to fully characterise the surface chemistry of organic materials. Unexpected contaminant species are also often identified when chemical analysis is carried out, which the WCA cannot differentiate from small deviations from predicted changes. Surface compositional data thus allows the behaviour of surfaces (such as wettability, protein adsorption and cell response) to be understood when combined with information on the surface topography, provided by atomic force microscopy (AFM) and scanning electron microscopy (SEM).

9.2 X-ray photoelectron spectroscopy (XPS)

XPS is arguably the most popular technique within the domain of surface science and has been successfully applied to a wide range of scientific disciplines from heterogeneous catalysis to biomaterials. The technique provides quantitative elemental analysis of a solid surface for all known elements apart from hydrogen and helium, as well as information on chemical bonding and oxidation states. It has a typical sampling depth of ca. 10 nm. The analysis depth is dictated by attenuation of the photoelectrons emitted from the solid which results in an exponential decay of the number reaching the surface with increasing depth. The technique is carried out under ultra-high vacuum conditions (UHV), usually in the region of 10^{-9}–10^{-10} torr.

The performance and capabilities of XPS spectrometers have developed significantly since the first commercial instruments in the early 1970s. However, the essential elements remain the same; these are the X-ray source, the sample manipulator, the electron transfer optics, the energy analyser and the detector. Details of the design and capabilities of the various commercial systems that are available can be found on manufacturers' websites listed in Table 9.1 (see Section 9.5).

Electron emission due to X-rays is commonly referred to as photoemission and the ejected electrons as photoelectrons. Photoelectrons come from both the core and valence orbitals of atoms. Core orbitals are the inner quantum shells which do not participate in chemical bonding and have discrete energy values, while valence orbitals are in the partially filled outer quantum shells involved in chemical bonding and have broad energy values referred to as the valence band. The highest level of the valence band is called the Fermi level.

Since core orbitals have discrete energy values and are less affected by their surrounding chemical environments, they have characteristic binding energies – the energy required to move an electron from an orbital to the Fermi level – that identifies the atom of origin. Thus, while a C 1s core level has a binding energy (BE) of ~285 eV, an O 1s core level has a BE of ~530 eV. This may be seen in the survey of the full BE range acquired from poly (D,L-lactide) (PLA) presented in Fig. 9.1a.

The BE of a core level photoelectron is determined by equation 9.1:

$$BE = h\nu - KE - \phi \tag{9.1}$$

where $h\nu$ is the energy of the photon which causes photoemission, KE is the kinetic energy of the photoelectron and ϕ is the work function. The photoemission process is graphically illustrated in Fig. 9.2. Looking closely at this equation we can see that the photon energy ($h\nu$) is equal to the total energy of the system (following the principles of energy conservation) but we must not forget that some energy is required to overcome the energy barrier caused by the attraction between the electron and the nucleus ($BE + \phi$). Using the above equation, BE can be determined provided that ϕ and the photon energy ($h\nu$) of the X-rays are known, while KE is determined experimentally. The work function (ϕ) is simply the minimum energy barrier that an electron has to overcome before photoemission can occur. In practical applications, the sample is in electrical contact with the spectrometer, which means that their work functions can be taken to be the same (this is due to their Fermi levels being aligned) and can be generally ignored (Attard and Barnes, 2003).

Even though photons can penetrate to depths measured in micrometres into the sample, causing photoemission from various depths, only electrons leaving the solid and still preserving their initial kinetic energy on photoemission contribute to the core level peaks observed in the spectrum. Electron photoemission from atoms below the top layer of the sample will have a certain probability of undergoing inelastic scattering before leaving the sample. Electrons that suffer inelastic collisions contribute to the background of the spectrum and to satellite structures.

The number of photoelectrons that reach a surface without suffering energy loss through collisions is described by the Beer–Lambert law (equation 9.2):

$$I_S = I_O e^{-d/\lambda} \tag{9.2}$$

where I_O is the intensity of electrons at a sample depth of d from the surface, I_S is the intensity of the electrons at the surface and λ is the inelastic mean free path (IMFP) of an electron in a solid. The IMFP (λ) represents the average distance travelled by an electron through the sample before it suffers inelastic scattering and is usually in the order of 1–3.5 nm. The sampling depth for XPS is

(a)

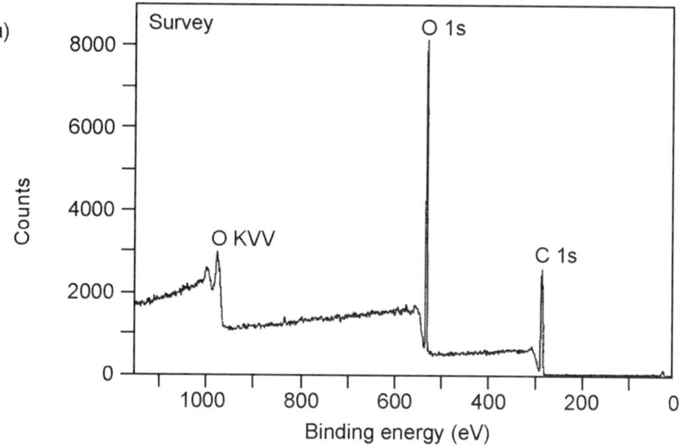

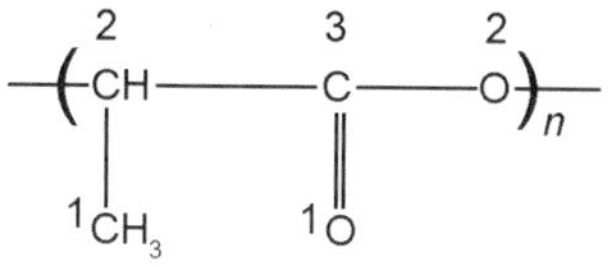

Poly(D,L-lactide) (PL)

	C 1s			O 1s	
	1	2	3	1	2
BE (eV)	285.00	286.98	289.06	532.25	533.66
FWHM (eV)	1.19	1.07	0.97	1.23	1.31
Area (%)	36	33	31	50	50
A	0.15	0.13	0.11	0.13	0.11
m	0.86	0.80	0.94	0.80	0.85

A = asymmetry parameter (O = symmetric, 1 = asymmetric);
m = mixing ration (O = pure gaussian, 1 = pure Lorentzian).

(b)

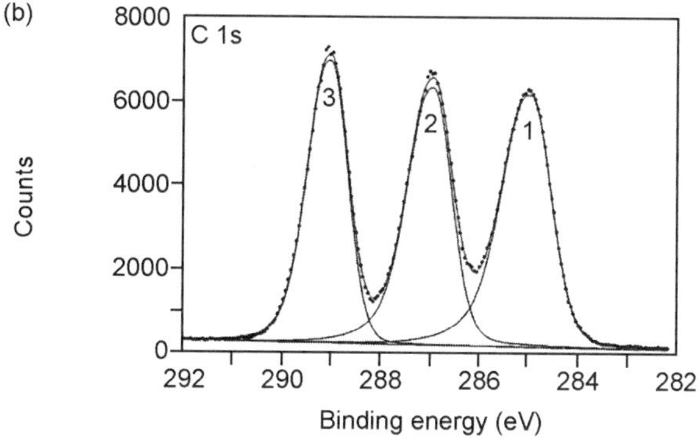

(c)

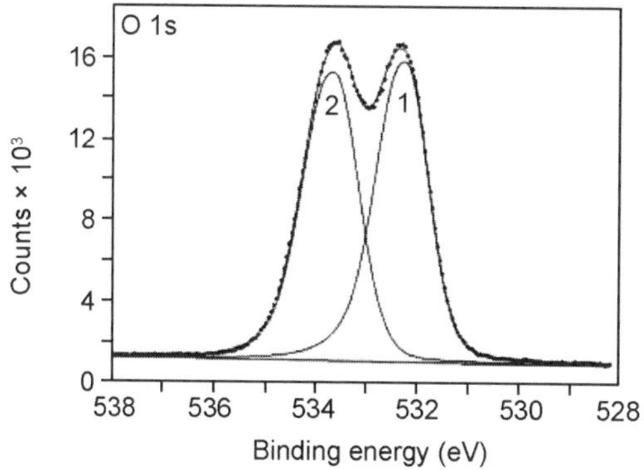

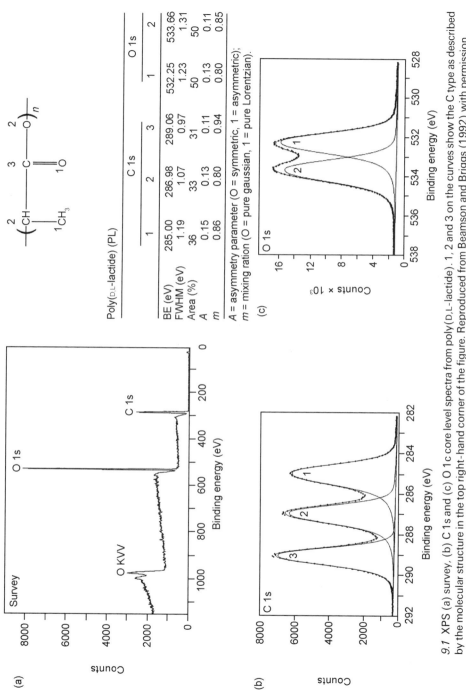

9.1 XPS (a) survey, (b) C 1s and (c) O 1c core level spectra from poly(D,L-lactide). 1, 2 and 3 on the curves show the C type as described by the molecular structure in the top right-hand corner of the figure. Reproduced from Beamson and Briggs (1992) with permission.

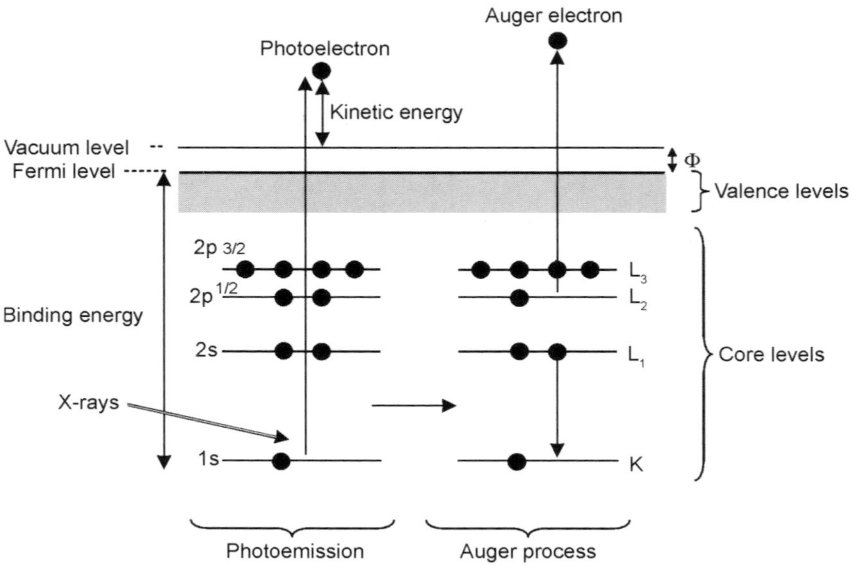

9.2 Schematic energy level diagram illustrating X-ray stimulated photoemission and resultant Auger emission.

usually defined as the depth where 95% of all electrons are scattered (corresponding to 3λ) and for most systems this gives a sampling depth of 3–10 nm, although as we will show later this is dependent on the take-off angle of the photoelectrons. When combined with appropriate values of λ, accurate overlayer thickness measurements may be made using this relationship (Petrovykh *et al.*, 2004).

9.2.1 Elemental composition

XPS is a quantitative technique since the number of photoelectrons from a core level is proportional to the amount of that element in a surface. To account for the different emission probabilities from different orbitals and elements, relative sensitivity factors (RSF) are introduced. Using appropriate RSF values, often provided by the instrument supplier, quantification of the elemental composition can be achieved once the intensity of electrons has been measured by insertion of an appropriate background. When more than one core level is detected for each element, the most intense peak will often be chosen to maximise the precision of the measurement. This may be carried out using the *survey scan*, a term used to designate a scan that includes the full range of binding energy, which may include XPS data, i.e. 0–1400 eV.

For insulating materials, photoemission results in sample charging which increases the apparent binding energy values of core levels. When non-monochromated X-ray sources are used, the secondary electrons emitted from the aluminium window on the end of the X-ray gun normally go some way to balancing the charge at the sample surface. However, this window is not used for monochromated sources and it is therefore necessary to balance the surface charge state by flooding the surface with low-energy electrons via an electron flood gun, commonly referred to as *charge neutralisation*. In practice this normally results in overcompensation, apparent from the appearance of the C 1s core level at binding energies 1 or 2 eV lower than 285.0 eV. In both cases a *charge correction* procedure is required to position the core levels at the binding energy of an electrically earthed surface. By convention, for polymeric surfaces, the binding energy of aliphatic hydrocarbon (285.0 eV) is used.

9.2.2 Chemical state information

Not only can XPS determine the elemental composition of a surface, it can also provide information on the chemical functionalities and oxidation states of elements. This is often called the chemical shift effect, which results because the binding energy of an electron from a specific element and energy level varies with the chemical environment surrounding its parent atom. For example, if in a particular chemical environment an atom is electron deficient, perhaps due to the loss of core level electrons to the valence level via bonding with a more electronegative atom, then the energy of its electronic levels is lowered due to the higher effective nuclear charge. This results in an increase in binding energy for photoejected electrons from that atom. This may be illustrated using the C 1s core level acquired from PLA, presented in Fig. 9.1b. A methyl group, $-CH_3-$, and an ester group, –C(=O)O–, it can be seen that the C in the methyl group exists in a non-polar environment while the C in carbonyl exists in a polar environment and is itself electron deficient. Thus the C 1s core level from an ester functionality will have a higher binding energy than the C 1s core level of methyl carbon.

In order to quantify the functional composition of surfaces, curve fitting of the core levels is undertaken. This process involves using a number of model peaks to represent the contribution of the separate functionalities and comparing the sum of their intensities with the experimental data. There are a number of parameters that are important when fitting XP spectra:

- For most core levels from insulators, electrons that have undergone energy losses contribute to the background of the spectrum at energies outside the peak range. It is therefore common to use a linear background for peaks deriving from insulating materials. Other mathematical operators which can describe a background, such as Shirley or Tougaard, are more appropriate for peaks with a large increase in background intensity within the range of the core level, e.g. as often seen in the Fe 2p core level.

- The shape of a peak is usually described as being Gaussian or Lorentzian or a combination of the two. XPS peaks are usually between the two extremes and described in terms of a Gaussian : Lorentzian ratio (expressed as a ratio or percentage) with peaks typically having values between 20 and 50%. It is sometimes necessary to introduce high binding energy asymmetry into peak shapes to account for processes such as vibrational fine structure (e.g. resolved in poly(ethylene)) (Beamson and Briggs, 1992).
- The full width at half maximum (FWHM) is a commonly used measure of peak width. Within one core level, the FWHM of peaks from the same phase are generally maintained to be about equal.

The goodness of fit may be assessed visually or quantitatively using a single parameter such as χ^2 or a plot of the residual versus binding energy to highlight areas of poor fit. Normally the addition of more peaks improves the fit iterated to by the software, but it does not necessarily improve the accuracy of the surface quantification. It is therefore good practice to have a specific reason for peak introductions, e.g. the different chemical environments in the structure of the PLA in Fig. 9.1. Assignment of core levels components is aided by reference to spectra from standard polymers (Beamson and Briggs, 1992).

Curve fitting is one of the most daunting procedures for those new to XPS and we therefore detail the issues for fitting data using the example of PLA. There are multiple solutions to the BE, intensity and width of model peaks in most XPS core levels, necessitating the use of care and supporting evidence to justify the inclusion of model peaks. The PLA structure is presented in Fig. 9.1 where the three carbon environments have been identified on the basis of the known structure of the material and then introduced at binding energies (or shifts) of related model compounds.

Environment ^{1}C is the lowest binding energy assigned to 285.0 eV. This is determined by convention and used to charge correct the core level spectra. The quality of the chosen peak shape to low BE may be used to determine its suitability. The core level envelope provides clear evidence for another two component peaks at a shifts of 2 and 4 eV from ^{1}C, labelled ^{2}C and ^{3}C respectively. The latter may be used to determine the suitability of the chosen peak shape to high BE. In this case it has been judged necessary to include a degree of high BE asymmetry. In the same way that FWHM are maintained at similar values, the asymmetry is also applied to all peaks to the same degree. By comparison with other model compounds these component peaks may be assigned to C–O–C(=O) and C(=O)O–C environments respectively. The latter assignment is made with reference to the known polymer structure. Without such information, based on the spectral data alone, an alternative carboxylic acid assignment is possible; C(=O)OH. When this structural evidence is not available, comparison with other data, i.e. other core levels, derivatisation, other techniques, e.g. Fourier transform infrared (FTIR), is necessary to provide a definite assignment. An alternative approach is derivatisation, which is reaction of the functionality of interest with a compound specific to that group which also provides a functional and/or elemental marker (Briggs and Seah, 1990; Chilkoti and Ratner, 1991; Fally *et al.*, 1995; Alexander *et al.*, 1996).

Oxygen environments may be assigned similarly from the O 1s core level. For the O 1s spectrum from PLA two environments are clearly indicated by the data (Fig. 9.1c). Again the peak shape is readily determined from this data. The components ^{1}O and ^{2}O are assigned to O=$\underline{\text{C}}$ and C(=$\underline{\text{O}}$)–O–C respectively. Correlation between the stoichiometry determined from the different core levels should be checked, although the precision of this approach relies on the accuracy of the RSFs. Excess carbon is often attributable to hydrocarbon adsorption, in preparation or during analysis. Silicon is often picked up during polymer or sample production in the form of silicones from sealing grease or release agents.

When spectral features are less clearly resolved than in the example of PLA, as a result of a greater range of chemical environments, inter-peak constraints may be used to fit broad core level envelopes (Alexander *et al.*, 1999). Typically the shifts are fixed relative to the C–C components and the FWHM are constrained to be approximately equal. The relative peak intensities may also be constrained based on physical requirements, e.g. in an ester, the intensity of the $\underline{\text{C}}$(=O)OC and C(=O)O$\underline{\text{C}}$ environments must be equal.

Fitting may be carried out on as-acquired data but is normally displayed charge corrected using the BE of the hydrocarbon peak. When curve fitting of core level data is presented it is good practice to state the charge compensation method, binding energy of the synthetic components, their peak width (FWHM), the peak shape and any constraints that have been imposed on the fit and the resultant functional composition.

9.2.3 Depth information from XPS analysis

Altering the angle of photoelectrons that are collected, usually by altering the angle of the sample with regards to the electron detector, alters the surface sensitivity of the analysis. This angle is called the take-off angle, and while the standard convention is to define it relative to the sample normal (θ), it is often defined relative to the sample surface (φ) (Fig. 9.3). Using Fig. 9.3 as an example of a tilted sample with a photoelectron take-off angle of θ it can be seen that, with a sampling depth of 3λ, the amount of actual sample analysed d corresponds to:

$$d = 3\lambda\cos\theta \tag{9.3}$$

When $\theta = 90°$ then d is at its maximum. Therefore by altering θ, the depth of sample analysed is modified. Qualitatively, this effect may be used to determine which species are located at the surface, e.g. elemental or functional components

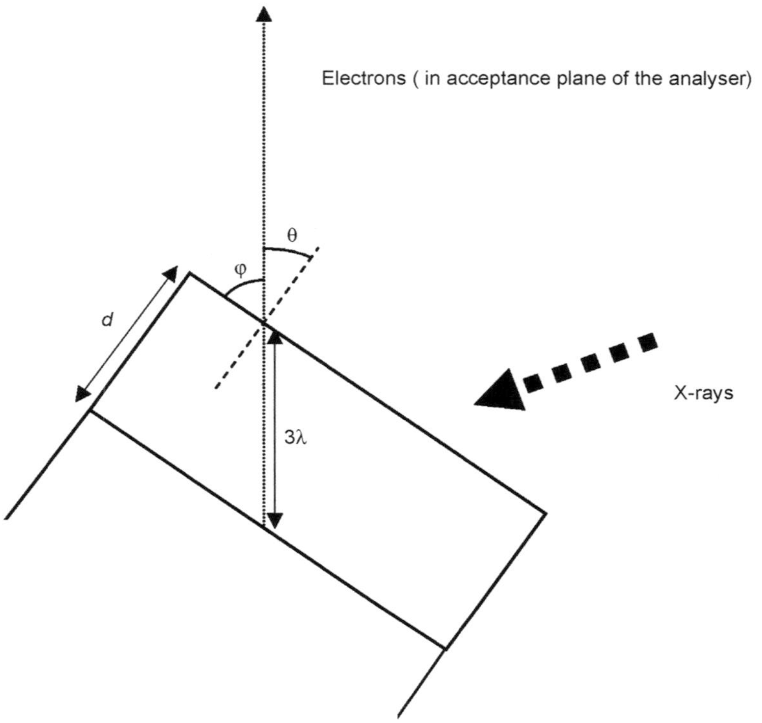

9.3 Schematic showing take-off angle effect in XPS.

(Beamson and Alexander, 2004). There are a number of algorithms that aim to *reconstruct* a depth profile from angle resolved X-ray photoelectron spectroscopy (AR-XPS) data, reviewed by Cumpson (1995), some of which can provide quantitative information on overlayer thickness and other surface structures.

The shape of the background on the high BE side of a core level holds information on the position of the core level atoms. If the species is buried below other material, then more photoelectrons will be inelastically scattered than if it is at the surface, and therefore there will be a rising background comprising the photoelectrons that have lost energy on their way to the surface, and no longer appear in the photoelectron peak. In many instances such a qualitative observation is satisfactory to place a species in the near surface region. Quantitative analysis of the background shapes have been made possible by detailed measurements on electron losses during passage through matter and are employed by Tougaard in software called QUASES (Tougaard, 1997, 1998). This approach enables not only quantitative overlayer measurements to be made, but also surface structures to be determined from a single survey spectrum.

9.2.4 Imaging XPS

Imaging XPS is now a standard capability on many new instruments. This is achieved either by focusing the emitted photoelectrons onto a position-sensitive detector, or rastering a focused X-ray beam. Thus far, the former approach has provided the images with the best lateral resolution, approaching 1 μm on samples with high photoemission cross-section such as gold.

Recent work has shown that principal component analysis (PCA) can be usefully applied to the problem of noise reduction in imaging XPS (Walton and Fairley, 2005). Noise in XPS images is a problem since full quantitative images cannot reliably be produced from intensity counting at single binding energies. Instead, acquisition of intensity maps at small binding energy steps over the core level of interest is necessary such that spectra can be constructed from the images enabling quantitative maps to be produced (Walton and Fairley, 2004). A problem with doing this for organic materials is that damage can become significant at the long acquisition times required to produce intensity maps of sufficiently high signal to noise. Thus, Walton and Fairley have developed a PCA-based approach to reduce the noise. This is anticipated to be of great use in the imaging of biomaterials, yet at the time of going to press no examples of multispectral imaging of polymer structures had been published.

9.2.5 Other spectral features

Numerous features appear in XP spectra, ranging from peak doublets to the background signal, and although the full scope of possible features is beyond this chapter, a number of common ones will be explained.

Doublets

Certain core levels appear as doublets, while others will appear as single component peaks. This is due to the interaction of the spin momentum (s) and the orbital angular momentum (l) of the ejected electron. This interaction, commonly referred to as spin–orbit coupling, gives rise to two final levels that differ in energy and degeneracy, e.g. the Cl $2p_{3/2}$ which is twice the intensity of the Cl $2p_{1/2}$ (Briggs and Seah, 1990).

Auger peaks

An Auger feature arises due to the de-excitation process that can occur after photoejection when an electron from a higher-energy level decays to fill the core hole vacancy (Fig. 9.2). This release of energy can result in either the emission of a photon (i.e. X-ray fluorescence) or the energy can be transferred to an electron residing in a similar energy state. This transfer of energy results in the ejection of an electron called an Auger electron. Auger peaks have their own notation depending on the core hole, the level of the electron that fills the core hole and the Auger electron level. Thus in Fig. 9.2 the core hole resides in the 1s level (K notation), while the 'fill' electron and Auger electron come from the 2s and 2p levels (L_1 and L_2 notation). This means the Auger signal would be a KL_1L_2 or more usually KLL. These electrons form the basis of Auger electron spectroscopy where electrons, rather than X-ray irradiation, is used to stimulate emission. This technique is principally applied to conducting samples since electron beam-induced charging of insulators prevents ready spectral acquisition by Auger spectroscopy.

Satellites

The appearance of high BE satellites in XP spectra arises from certain well-defined energy loses of outgoing photoelectrons. These processes include shake-up, shake-off and plasmon loss processes, which are phenomena related to the loss of a specific amount of energy by photoelectrons. Two key processes result in satellite peaks on the high binding energy side of the core level signal: (1) the shake-up process occurs when a valence electron is excited into a higher energy level by the removal of energy from the primary ejected photoelectron; (2) plasmon losses arise from the collective oscillation of electrons in the valence band caused by the outgoing photoelectron which suffers a discrete energy loss. In the case of the shake-off process the excited valence electron is completely ejected from the atom (unlike shake-up), leading to a doubly ionised system.

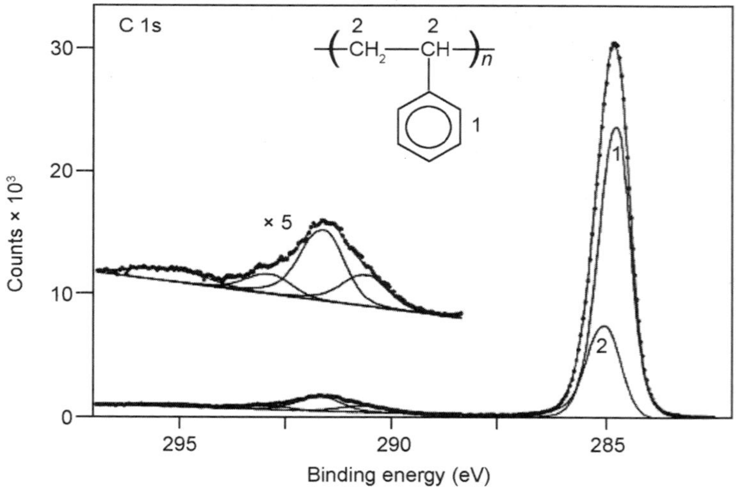

9.4 PS C 1s core level spectrum. The insert shows the π to π^* satellites for PS. 1 and 2 on the curves are described by the numbered molecular structure in the figure. Adapted from Beamson and Briggs (1992) with permission.

This causes broad features within the background of the core photoelectron signal. In organic solids the most prevalent satellites are shake-up/off processes involving π to π^* transitions associated with aromatic ring structures as can be seen at a shift of about 7 eV in the C 1s core level acquired from poly(styrene) (Fig. 9.4).

Charging

Ineffectual surface charge compensation during data acquisition can manifest as a high BE broadening of core levels reflecting the range of electrical charge states within the XPS analysis depth. It can be identified by comparison of all core levels from the same sample since it will occur to the same degree in all. Some workers have cleverly used this phenomenon to determine the electrical properties of surfaces (Cohen, 2004), although for most purposes it is an annoyance necessitating reacquisition of the data.

9.3 Static secondary ion mass spectrometry (SIMS) (Briggs and Seah, 1996; Vickerman and Briggs, 2001)

SIMS is a technique that is applicable to the analysis of most solids. Historically, as a technique it was first developed to obtain depth profiles through semiconductors and metals using a process of high primary ion dose bombardment. This approach is still widely applied to such materials, including the characterisation of semiconductor device structures. The term *static* SIMS refers to the regime utilising a low primary ion dose for surface analysis (e.g. $<10^{12}$ ions cm^{-2}). Under these conditions it is also possible to obtain molecular ions from organic surfaces. The technique has a depth analysis of ~1 nm and in common with XPS is carried out under UHV conditions. Time-of-flight secondary ion mass spectrometry (ToF-SIMS) instruments are now the most common type used to acquire static SIMS data. Unlike XPS, SIMS it is not a readily quantifiable technique as a result of the wide range of charged species that may be formed and their associated large range of ionisation probabilities.

9.3.1 Fundamentals

It is useful to understand how secondary ions are generated from a sample. A surface subjected to bombardment by ions, referred to as primary ions (with energies of 10–30 keV), causes atoms in the sample to be set into motion via direct collision with primary ions or indirectly through knock-on collisions (commonly referred to as the collision cascade). Primary ion collision results in extensive fragmentation at the atomic level of the material due to the high kinetic energies of these ions at the point of primary ion impact (Fig. 9.5). Thus,

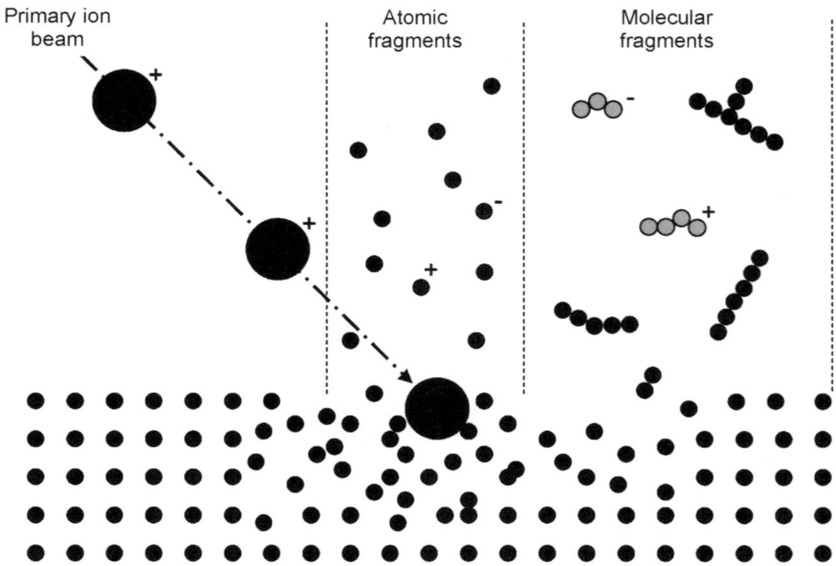

9.5 Schematic illustration of primary ion impact in SIMS.

only small and atomic species are emitted from this area of the surface. As the resulting transfer of kinetic energy radiates out from the impact site, energy dissipation occurs, which in turn produces molecular fragments of varying size, depending on the distance from the impact site, rather than atomising the surface. These fragments consist of neutral atoms and molecules, cations, anions and radicals – out of this menagerie only a small number are charged fragments (typically of the orders of 10^{-6}–10^{-1}). Charged secondary ions can be collected separately and mass analysed to produce positive and negative ion mass spectra.

Experience tells us that much useful molecular information may be obtained by SIMS, yet inherent in this process is surface damage due to primary ion bombardment. In the analysis of polymers, molecular fragments are generally more useful than atomics in surface diagnosis. In order to minimise such modification of the initial surface and promote desorption of large molecule fragments, the primary ion dose is minimised to ensure that there is a low probability of a given region being struck by a primary ion more than once. This is referred to as the *static limit* which is defined as below 10^{13} ions cm^{-2} or sometime 10^{12} ions cm^{-2} (Brown and Vickerman, 1986). The ability to acquire useful spectra within the static SIMS limit depends on the primary ion being used and the sample being analysed. Measuring the intensities of charged fragments from a sample versus increasing primary ion dose provides information on the interrelationship of these parameters.

Materials that are insulators accumulate charge on the surface during analysis due to charge transfer from primary ions and the emission of secondary electrons and ions. If this surface charge is not dissipated it reduces or eliminates the secondary ion signal, a problem that is particularly notable when negative secondary ions are extracted from the surface. To combat this effect an electron gun is used, which floods the surface with low-energy electrons in between the pulses of primary ions.

The technique of ToF mass analysis relies upon the principle that ions with the same kinetic energies have different velocities depending on their mass, thus altering their flight time between two points. If ions are accelerated by an electrostatic field to a common kinetic energy and travel to a detector, then the lighter ions will reach the detector before the heavier ones do. By measuring the differences in flight times for secondary ions, i.e. from a primary ion pulse to detection, ion masses can be determined. Figure 9.6 shows the positive ion spectrum for PLA (Ogaki *et al.*, 2007). A number of peaks are labelled as examples of ion fragments that are produced on ion bombardment, they include the monomeric unit ($M = C_3H_4O_2$), a number of deoxygenated fragments ($nM - O$, C_2H_3O and C_2H_3) and various dimeric units as highlighted by ($2M - CO_2$). Typical primary ion pulse rates are within the region of 10 kHz. Using this type of system mass spectra can be acquired with a resolution (commonly measured as $\Delta m/m$ in parts per million) sufficient to separate peaks of nominally the same unit mass, e.g. SiH^+, CHO^+ and $C_2H_5^+$ in Fig. 9.6b.

A common request is to 'depth profile' into a sample to obtain an analysis of the bulk or the distribution of a species in the near surface. This is readily possible in ceramic and metallic systems, but until recently the damage accumulation in organic systems precluded this approach. The emergence of a new generation of ion sources based on polyatomic primary ions such as buckminsterfullerene (C_{60}) suggest that depth profiling of organic systems may be a viable analytical approach (Fletcher *et al.*, 2006). While application to biomaterial systems is in its infancy at the time to going to press, it is anticipated that the wider availability of polyatomic sources for depth profiling will result in much progress in this area (Shard *et al.*, 2007).

9.3.2 Interpretation of SIMS spectra

Unlike conventional mass spectrometry, SIMS of simple organic compounds routinely yields numerous fragments which often do not include a molecular ion. Identification of SIMS spectra is therefore often a considerable challenge to those new to the field. Polymers normally yield fragments indicative of the repeat unit, but for reasons of surface entanglement, bulk samples rarely yield more than a few repeat units. The situation is normally more simple for metallic and oxide surfaces since identification of atomic and small ions is relatively easy using mass assignment accurate to three decimal places and isotopic pattern comparison. Spectra are generally interpreted with reference to the compound analysed, when known, and for identification of unknown species using a 'stare and compare'

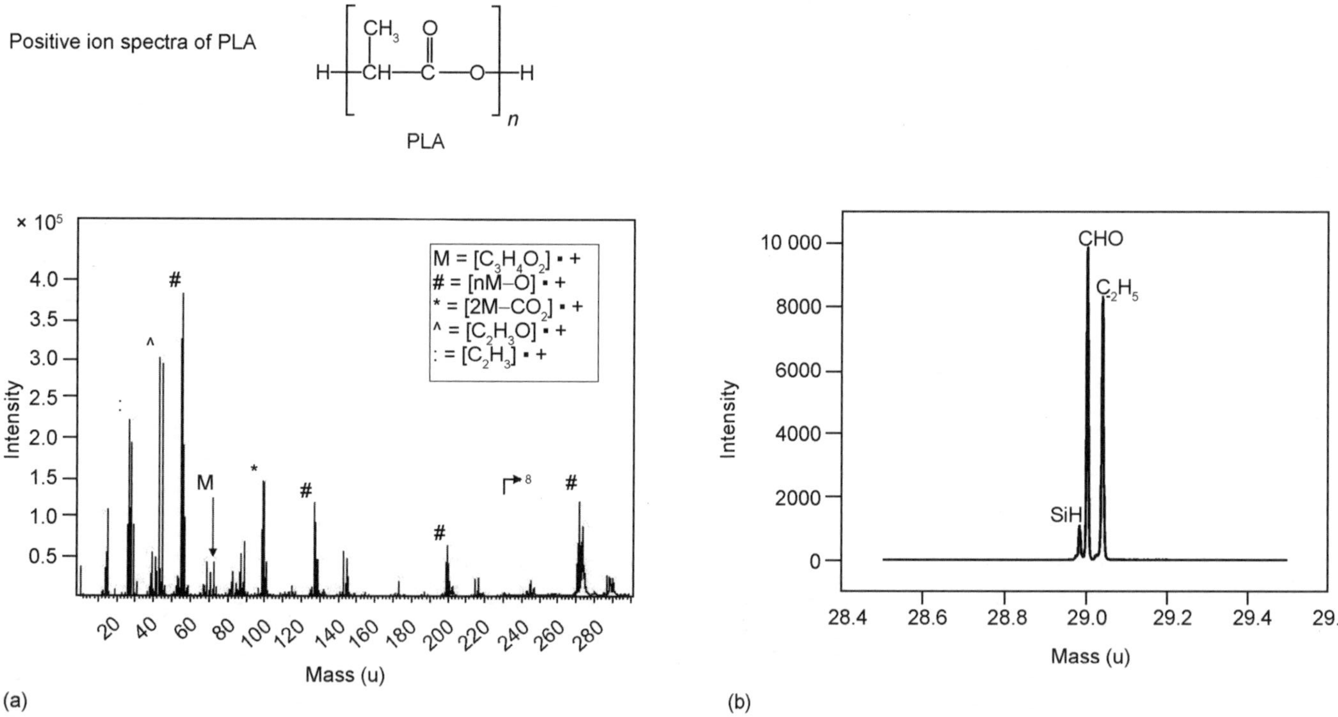

9.6 (a) ToF-SIMS spectrum – from LBSA polymer library reproduced with permission from Ogaki *et al.* (2006). (b) S-SIMS accurate mass assignment. Here three species with $m/z = 29$ can be resolved owing to accurate elemental masses.

approach to matching spectral fingerprints with a range of spectral databases has traditionally been used (Briggs *et al.*, 1989; Vickerman *et al.*, 2000).

With the notable exception of copolymers of lactic acid and glycolic acid (Shard *et al.*, 1996), secondary ion yields are generally not readily related to the concentration of a materials within the sample. One reason for this is the matrix effect which describes the relationship between ion formation and the local environment.

Recent years have seen the rise of multivariate data analysis (MVDA) within a wide range of subjects, from genomics to infrared spectroscopy. MVDA have been successfully applied to ToF-SIMS data sets in the form of PCA with the aim of extracting the important data from the large and complex spectra (Wagner *et al.*, 2004). Briefly, PCA is a statistical method that can be viewed as identifying the directions of greatest variance in multivariant space in the data. These directions (referred to as principal components) can be used as a new set of axes on which to consider how samples and variables (in ToF-SIMS these are ion fragments) relate to each other. In the case of different polymer samples, any correlation or grouping seen within a PC plot indicates commonality of the surface chemistry represented by the PC in question. Large deviations between PC scores indicate dissimilarity in surface chemistry. The identified chemistry can be identified by the PC loadings (variables) plot, which can be engineered to resemble a SIMS spectrum. This approach can be useful in removing interpretation bias and dealing with the large amount of data produced by SIMS. The methods of data preparation prior to PCA have a significant influence of the results, and should thus be chosen with care and attention to the intended use of the output.

Information on the collision cascade process, specifically the relationship between energy and probability of fragmentation, has been used to develop an approach called gentle SIMS or G-SIMS (Gilmore and Seah, 2003). Acquisition at different primary ion energies, using different primary ion sources and/or acceleration voltages, is used to provide a number of spectra with different degrees of fragmentation. Extrapolation from these spectra using an appropriate theoretical model is used to produce a spectrum representative of the unfragmented surface compounds. The utility in identifying the structurally characteristic components in the spectra of polymer biodegradable polymers has been investigated by Ogaki *et al.* (2006).

9.3.3 Imaging SIMS

One of the most useful aspects of ToF-SIMS is the ability to image the spatial distribution of a material, in other words providing a 'mass spectra map' of the surface. This is achieved by rastering the focusing the primary ion beam across the surface. In modern ToF instruments a mass spectrum is obtained at each point which can be converted into a pixel. After data acquisition, specific masses can be selected from the raw data and converted into an image. The resolution of

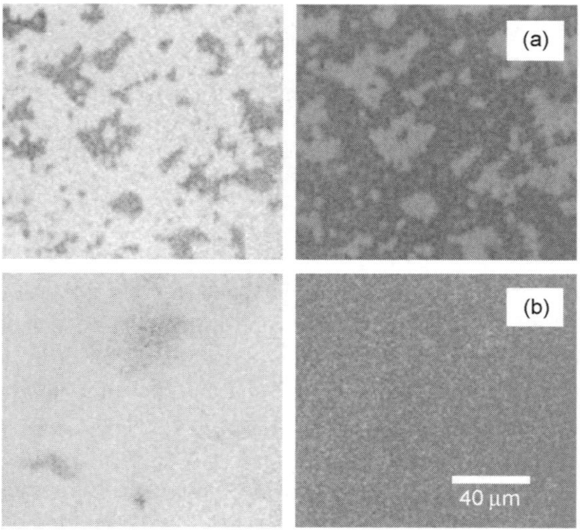

9.7 ToF-SIMS images of PLL-entrapped in PLA left PLA map (O^- plus $[M+H]^-$ and $[M+OH]^-$) and right PLL (CN^- and Br^-) at (a) high solvent concentration and (b) reduced solvent concentration. Reproduced from Quirk *et al.* (2001).

the image is ultimately determined by the amount of ions generated compared with the damaged caused to the sample (Tyler, 2001). The development of new primary ion sources continues to increase the available resolution with the recent introduction of bismuth ions producing a significant improvement on the initial argon and gallium ions used on the first static SIMS instruments.

To illustrate the capabilities of imaging SIMS we can consider the case where it has been used to readily reveal micrometre-scale poly-L-lysine (PLL)-rich domains physically entrapped in the PLA surface (Fig. 9.7). The presence of PLL areas was revealed to vary with solvent entrapment processing conditions using the summed intensity of ions exclusive to the PLA; O^- plus $[M\pm H]^-$ and $[M\pm OH]^-$ and PLL; CN^- and Br^- (Quirk *et al.*, 2001). More recently PEG and PLA blends have recently been imaged using SIMS with the contrast of PEG-enriched domains in a PLA matrix enhanced using deuterated PEG (Tsourapas *et al.*, 2006). This technique of incorporating isotopically modified compounds is not restricted to this system. A further interesting example of imaging SIMS is the physical entrapment of proteins within pre-formed alginate fibres which has been illustrated with bovine serum albumin (BSA) (Hou *et al.*, 2005).

9.3.4 Complementarity of SIMS with XPS

SIMS and XPS provide complementary information and as such are often employed together, i.e. qualitative molecular information is provided by SIMS and elemental and functional quantitative information is provided by XPS. Although the surface sensitivity of SIMS is greater than XPS, providing this is borne in mind in the interpretation, the two often provide thorough material surface analysis. An example of this is the structural information determined on the plasma polymerised acrylic acid, a material now commercially applied as part of a cell delivery system for chronic wounds (Haddow *et al.*, 2006). XPS was used to identify the elemental and functional composition, while the relation of the carboxylic acid group retention to the presence of oligomeric material was identified through the detection of molecular ions up to five repeat units of the monomer in the SIMS spectra (Alexander and Duc, 1998).

9.4 Specific sample preparation and acquisition procedures

Valid data may be generated from many samples simply by introducing them into the XPS or SIMS spectrometer and acquiring spectra. Specific preparation and acquisition considerations are required for certain classes of material and sample format which are covered in this section.

9.4.1 Cryogenics

Polymer surfaces that are above or near their glass transition temperature at ambient or physiological conditions are mobile. In some cases, rearrangement between an aqueous and the UHV analysis environment of the spectrometer is significant. Thus, if the analysis is to be representative of biological conditions, this can mean that the surface state in the aqueous environment needs to be 'frozen' by cooling to below the glass transition temperature. Substances that are volatile near room temperature may also be analysed using this technique. Surface-active components of biological tissue samples can be prevented from segregating to the surface using cooling, e.g. surface segregation of cholesterol, has been shown to mask other molecular species in SIMS of brain sections (Johansson, 2006).

Surface segregation has been well illustrated for poly(hydroxyethyl methacrylate (HEMA)-isoprene) copolymer with transmission electron microscopy (TEM) images using staining, dramatically visualising the inferred changes based on XPS and water contact angle data (Senshu *et al.*, 1999). Hydrophilic blocks have been found to orient away from the surface in the vacuum environment of XPS analysis, therefore providing an incorrect analysis of the surface in solution. Cryoscopic sample handling can be used to slow or stop these surface rearrangements, thereby enabling analysis in ultra high vacuum (UHV) of the surface that was present in an aqueous environment.

This phenomenon may also occur when species are grafted to homopolymers, e.g. immobilisation of hydroxyethyl methacrylate (a hydrogel in polymer form)

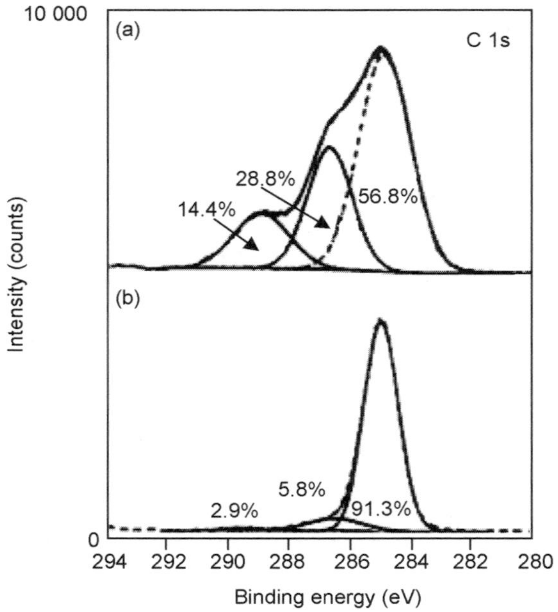

9.8 High-resolution C1s XPS spectra from HEMA radiation grafted onto a silicone elastomer after (a) hydration, freeze drying and analysis at <120 °C and (b) raising the sample temperature to 25 °C and vacuum-drying *in situ*. Reproduced with permission from Lewis and Ratner (1993).

by surface grafting to a silicone elastomer. The successful development of a cryoscopic sample handling procedure allowed Ratner *et al.* (1978) to capture the HEMA structure at the surface as seen in Fig. 9.8a, that was not observed when the surface had been exposed to room temperature in a dry purged environment (Fig. 9.8b). Thus C–O, and ester, C(=O)O, environments from the HEMA were seen as 28.8 and 14.4% of the carbon respectively in the cooled samples, whereas, when analysed at room temperature the functionalities representative of the silicone elastomer dominated the C 1s core level, i.e. C–Si (Fig. 9.8b). Similar reorientations were characterised at the surface of a functionalised poly(urethane) by cryo-XPS (Magnani *et al.*, 1995). If the surface rearrangements are sufficiently slow, cooling is not necessary (Senshu *et al.*, 1999). Hydrogels have been analysed in the hydrated state by cooling the sample prior to and during analysis by SIMS (Sosnika *et al.*, 2006).

A spectrometer suitable for cryscopic analysis would be equipped with cooling in both the entry and analysis chambers such that the sample can be kept cool between the atmosphere and the main chamber. A dry glove box on the entry chamber to facilitate sample mounting can be useful in preventing condensation from the ambient laboratory atmosphere, although some groups manage adequately with an inert gas purged entry lock. Samples are entered into a nitrogen purged entry chamber prior to cooling to well below both 0 °C to solidify any water and the glass transition temperature of the polymer to immobilise the chains. The purged environment prevents condensation of ambient water vapour on the cooled surfaces of the sample holder and heat transfer elements of the spectrometer. Sublimation of the ice by raising the temperature to a predefined point or in stages in combination with analysis are used to define the point at which the surface is free of ice. Ice sublimes at significant rates above 130 K (Fraser *et al.*, 2001). The increase in pressure caused by water vapour must be within the operational requirement of the system, and therefore a minimum of unnecessary water (e.g. condensed to the holder or within a thick polymer section) and sufficiently slow heating are requirements. To reduce water introduction to the vacuum chamber, the thickness of the polymer should be minimised, e.g. by preparing a film by spin casting. Reduction of the amount of water condensed on the sample holder can be achieved by first cooling the cold finger in the entry lock to encourage condensation on this prior to contact with the sample.

Contamination is an issue when cooling samples since volatile species will be encouraged to condense on cool surfaces, within and during entry to the spectrometer. Thus, protocols have been developed using variable angle XPS to detect surface contaminants from the bulk material (Lewis and Ratner, 1993). These need to be developed for each system, considering hydration, mounting in hydrated state, freezing to appropriate sublimation temperature in dry atmosphere, evacuation until sublimation is complete, transfer to analysis chamber, analysis at cryoscopic temperature, exit and increase of temperature to allow any rearrangement to occur and reanalysis at reduced temperature.

9.4.2 Porous bodies, powders and fibres

In contrast to studies of model polymers where flat films are often spun cast or solution coated onto substrates, tissue engineering scaffolds often have significant topography. XPS spectra can still readily be acquired from such surfaces providing care is taken when positioning the sample. The need for greater care in sample position optimisation originates from the possibility of shadowing of charge neutralising electrons and X-ray illumination of different areas of sample at different heights. If the sample position is optimised using only the peak intensity to maximise counts, as is common practice for flat samples, distorted peak shapes can result due to lateral differential charging across the analysis area. Monitoring of the peak shape during position optimisation by lateral as well as vertical sample movement is sufficient to eliminate such effects. The C 1s and O 1s core level from the top surface of a super critical fluid processed poly(D,L-lactic acid) scaffold is presented in Fig. 9.9a and b (Barry *et al.*, 2005). Well-resolved components representative of the $\underline{C}H_3$–, $\underline{C}$(=O)OC and $\underline{C}$–O–C(=O) environments are observed in the C 1s, and

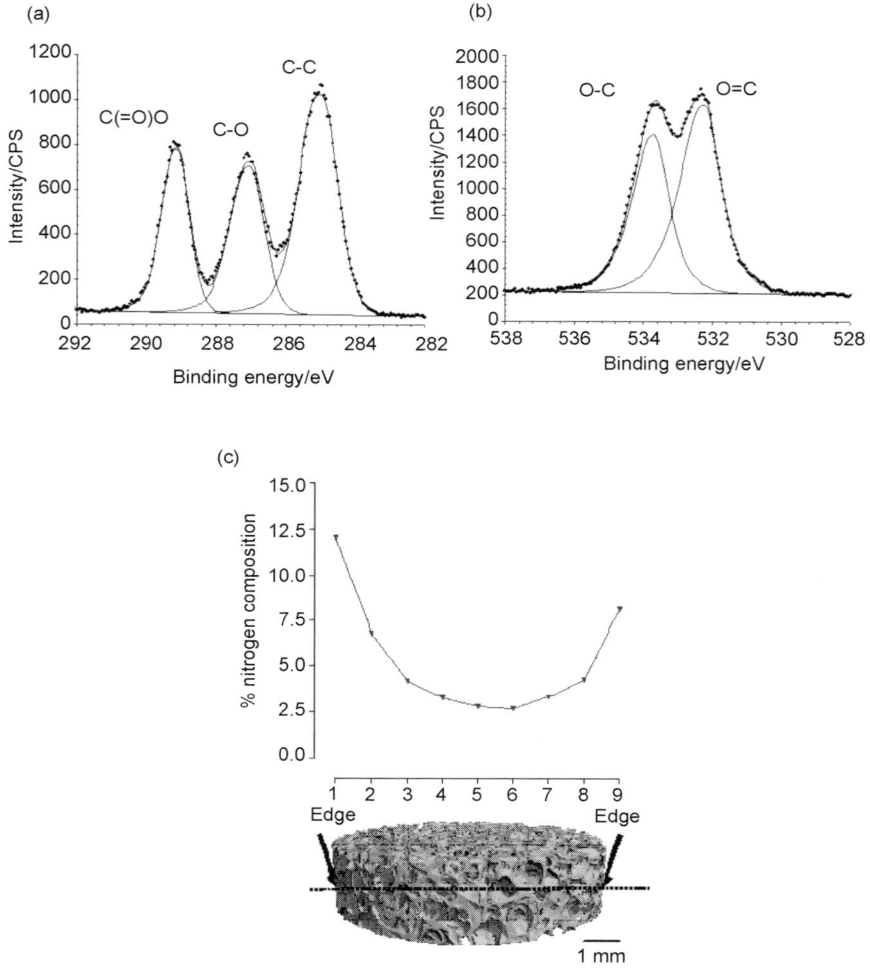

9.9 XPS (a) C1s and (b) O1s core level from PDLLA charge corrected to CC at 285.0 eV. (c) Elemental composition cross the surface formed by mechanical sectioning after ppAAm coating. Reproduced with permission from Barry *et al.* (2006).

the from the perspective of the oxygen at the surface in the O 1s; C–O–C and O=C. These are comparable to the spectra acquired from a flat PLA sample presented in Fig. 9.1.

Chemical modification of the internal surfaces of porous objects is a critical issue in production of porous scaffolds. Plasma-polymerised allylamine (ppAAm) coating has been shown to provide a surface chemistry conducive to cell adhesion and the retention of function equivalent to that of collagen I for some primary hepatocytes and 3T3 fibroblasts (Barry *et al.*, 2006; Dehili *et al.*, 2006). This is deposited by a vapour phase technique that is has recently been applied to coating the interior of porous objects. Sectioning and point analysis across the objects thickness is an ideal method to gain a measure of treatment penetration. In Fig. 9.9c nitrogen has been used as an elemental marker of a ppAAm coating. It is apparent that the ppAAm is depleted towards the centre of the body as a result of depletion of depositing species during diffusion through the pores. It is important to note that while the nitrogen composition of the surface is given, this corresponds to the surface analysed by XPS (ca. 10 nm) which comprises both the ppAAm and the underlying PLA.

Particulates may be analysed by XPS and SIMS, although the sample mounting required to avoid powder loss in the spectrometer and substrate contribution to the spectrum requires a trial and error approach for each powder type. Pressing or sprinkling onto insulating and conductive double-sided tape, indium liquefied by heating, deposition onto a silicon wafer from liquid suspension and pressing into a solid body using IR pellet presses have all been successful in certain cases. Issues to avoid are diffusion of silicone release agents from the tape surface and incomplete coverage and poor adhesion leading to particle loss through gravity or electrical charge imparted by the analysis technique. Such samples are prone to poor charge neutralisation (see 'Charging' in Section 9.2.5), although it is possible to obtain good quality spectra if care is taken in sample mounting and alignment. It has been shown that curve fitting of XPS C 1s core levels can be used to differentiate the presence of polyvinyl alcohol (PVA), poly(ethylene glycol) (PEG), and PLA on microparticle surfaces (Shakesheff *et al.*, 1997).

9.5 Conclusions

Although other chemical analysis techniques are beginning to find application in the analysis of surfaces, the complementary pairing of XPS and SIMS remain the workhorses in chemical characterisation of polymeric surfaces (Alexander, 2006). Instrumental developments are allowing previously unthought-of depth profiling experiments to be undertaken in polymeric systems (Shard *et al.*, 2007). This is an area that will surely continue to expand along with application of statistical methods of data handling to improve data quality, especially in XPS imaging (Walton and Fairley, 2004).

For those in search of more in-depth information, there are a number of text books describing XPS in far more detail than is presented here (Briggs and Seah, 1990; Briggs and Grant, 2003; Wolstenholme, 2003). There are also a number of textbooks dedicated to SIMS that the reader may wish to consult for a more complete description of these techniques (Briggs and Seah, 1992; Tyler, 2001). Table 9.1 gives a list of a few conferences, UK meetings and web resources. Review articles are constantly appearing for those wishing to find more opinion

Table 9.1 Organisations, conferences and Internet resources

Description	Web address
XPS manufacturer: Kratos Analytical, UK	http://www.kratosanalytical.net/surface/
XPS manufacturer: Specs, USA	http://www.specs.com/
XPS manufacturer: Thermo VG, UK	http://www.thermo.com/
XPS and SIMS instrument manufacturer: Physical Electronics, USA.	http://www.phi.com/
SIMS instrument manufacturer: ION-TOF, Germany	http://www.ion-tof.com/
List of UK, European and Worldwide surface sites including academic and commercial equipment and service providers	http://www.chem.qmul.ac.uk/surfaces/
Popular surface science mailing list for general discussion of all aspects of surface science	http://www.uwo.ca/ssw/mailinglist.html
Surface science of biologically important interfaces SSBII UK network with annual meetings	http://www.ssbii.org.uk
UK Surface Analysis Forum, biannual UK meetings	http://www.uksaf.org/
UK Society for Biomaterials site, holds annual meetings	http://www.uksb.org.uk/
AVS meeting, annual US meeting with well-attended bio- sessions and plenary events.	http://www.avs.org/
European Conference on Applications of Surface and Interface Analysis, biannual conference	http://www.ecasia.org/
Biosurf, biannual conference published in the online journal *European Cells and Materials*	http://www.ecmjournal.org/

and comment on application of surface analysis in bioengineering (Ratner, 1995; McArthur, 2006).

9.6 Future trends

Surface analysis techniques have not been combined with high throughput approaches until recently. The work by Urquhart *et al.* has taken advantage of the automated sample movement and acquisition features, now a standard option on most spectrometers, to automate XPS and SIMS analysis from microarrayed combinatorial polymer libraries (Urquhart *et al.*, 2007). The first such analysis deals with a set of 576 polymers on which embryonic stem cell differentiation has previously been assessed by cell culture (Anderson *et al.*, 2004). Thus, the role of surface chemistry can be rapidly assessed in cell behaviour for a wide range of materials. This marks a significant advance since previously it was possible to infer surface only from bulk chemistry of such arrays, an assumption that the initial work on these systems has revealed is rarely valid. Gradients are also under investigation as high-throughput platforms for monitoring interactions with surfaces (Whittle *et al.*, 2003; Parry *et al.*, 2006). It is anticipated that arrays and gradients will be a growth area in the assessment and development of biomaterials beyond the current relatively limited suite available to the tissue engineer.

The acquisition of increasingly more data drives the introduction of statistical methods to the analysis of the large data sets. This is an area predicted to gain greater importance as indicated through application of statistical approaches to XPS and SIMS data interpretation to aid utilisation of all aspects of the data (Michel and Castner, 2006).

9.7 Acknowledgement

Richard France (Regentec) is thanked for his contributions in checking aspects of this manuscript.

9.8 References

Ablett, R. (1923) An investigation of the angle of contact between paraffin wax and water. *Philosophical Magazine*, **46**, 244–256.

Alexander, M. R. (2006) Editorial: bio surface and interface analysis. *Surface and Interface Analysis*, **28**, 1379.

Alexander, M. R. and Duc, T. M. (1998) The chemistry of deposits formed from acrylic acid plasmas. *Journal of Materials Chemistry*, **8**, 937–943.

Alexander, M. R., Wright, P. V. and Ratner, B. D. (1996) Trifluoroethanol derivatization of carboxylic acid-containing polymers for quantitative XPS analysis. *Surface and Interface Analysis*, **24**, 217–220.

Alexander, M. R., Short, R. D., Jones, F. R., Michaeli, W. and Blomfield, C. J. (1999) A study of HMDSO/O-2 plasma deposits using a high-sensitivity and -energy resolution XPS instrument: curve fitting of the Si 2p core level. *Applied Surface Science*, **137**, 179–183.

Anderson, D., Levenberg, S. and Langer, R. (2004) Nanotiter-scale synethesis of arrayed biomaterials and application to human embryonic stem cells. *Nature Biotechnology Letters*, **22**, 863–866.

Andrade, J. D. (1985) *Surface and Interfacial Aspects of Biomedical Polymers*, New York, Plenum Press.

Attard, G. and Barnes, C. (2003) *Surfaces*, Oxford, Oxford University Press.

Bain, C. D. and Whitesides, G. M. (1988) Depth sensitivity of wetting – monolayers of omega-mercapto ethers on gold. *Journal of the American Chemical Society*, **110**, 5897–5898.

Barry, J. J. A., Silva, M., Shakesheff, K. M., Howdle, S. M. and Alexander, M. R. (2005) Using plasma deposits to promote cell population of the porous interior of three-dimensional poly(D,L-lactic acid) tissue-engineering scaffolds. *Advanced Functional Materials*, **15**, 1134–1140.

Barry, J. J., Howard, D., Shakesheff, K. M., Howdle, S. M. and Alexander, M. R. (2006) Cover picture: using a core–sheath distribution of surface chemistry through 3D tissue engineering scaffolds to control cell ingress. *Advanced Materials*, **18**, 1406–1410.

Beamson, G. and Alexander, M. R. (2004) Angle-resolved XPS of fluorinated and semi-fluorinated side-chain polymers. *Surface and Interface Analysis*, **36**, 323.

Beamson, G. and Briggs, D. (1992) *High Resolution XPS of Organic Polymers – The Scienta ESCA300 Database*, Chichester, Wiley; now avalable from http://www.surfacespectra.com/.

Brewer, N. J., Foster, T. T., Leggett, G. J., Alexander, M. R. and MCALPINE, E. (2004) Comparative investigations of the packing and ambient stability of self-assembled monolayers of alkanethiols on gold and silver by friction force microscopy. *Journal of Physical Chemistry B*, **108**, 4723–4728.

Briggs, D. and Grant, J. T. (2003) *Surface Analysis by Auger and X-Ray Photoelectron Spectroscopy*, Chichester, IM Publications.

Briggs, D. and Seah, M. P. (1990) *Practical Surface Analysis: Auger and X-ray Photoelectron Spectroscopy*, Chichester, Wiley.

Briggs, D. and Seah, M. (1992) *Practical Surface Analysis: Ion and Neutral Spectroscopy*, New York, John Wiley & Sons.

Briggs, D. and Seah, M. P. (1996) *Practical Surface Analysis: Ion and Neutral Spectroscopy*, Chichester, John Wiley & Sons.

Briggs, D. and Wootton, A. B. (1982) Analysis of polymer surfaces by SIMS. 1. An investigation of practical problems. *Surface and Interface Analysis*, **4**, 109–115.

Briggs, D., Brewis, D. M. and Konieczo, M. B. (1976) X-ray photoelectron-spectroscopy studies of polymer surfaces. 1. Chromic acid etching of polyolefins. *Journal of Materials Science*, **11**, 1270–1277.

Briggs, D., Brown, A. and Vickerman, J. C. (1989) *Handbook of Static Secondary Ion Mass Spectrometry (SIMS)*, Chichester, Wiley.

Briggs, D., Brewis, D. M., Dahm, R. H. and Fletcher, I. W. (2003) Analysis of the surface chemistry of oxidized polyethylene: comparison of XPS and ToF-SIMS. *Surface and Interface Analysis*, **35**, 156–167.

Brown, A. and Vickerman, J. C. (1986) A comparison of positive and negative-ion static SIMS spectra of polymer surfaces. *Surface and Interface Analysis*, **8**, 75–81.

Brunette, D. M., Tengvall, P., Textor, M. and Thomsen, P. (2001) *Titanium in Medicine*, Berlin/Heidelberg/New York, Springer-Verlag.

Chilkoti, A. and Ratner, B. D. (1991) An X-ray photoelectron spectroscopic investigation of the selectivity of hydroxyl derivatization reactions. *Surface and Interface Analysis*, **17**, 567–574.

Clark, D. T., Musgrave, W. K., Kilcast, D. and Feast, W. J. (1972) Applications of ESCA to polymer chemistry – studies of some nitroso rubbers. *Journal of Polymer Science Part A – 1 – Polymer Chemistry*, **10**, 1637.

Cohen, H. (2004) Chemically resolved electrical measurements using X-ray photoelectron spectroscopy. *Applied Physics Letters*, **85**, 1271–1273.

Cumpson, P. J. (1995) Angle-resolved XPS and AES – depth-resolution limits and a general comparison of properties of depth-profile reconstruction methods. *Journal of Electron Spectroscopy and Related Phenomena*, **73**, 25–52.

Curtis, A. and Riehle, M. (2001) Tissue engineering: the biophysical background. *Physics in Medicine and Biology*, **46**, R47–R65.

Dehili, C., Lee, P., Shakesheff, K. and Alexander, M. (2006) Comparison of primary rat hepatocyte attachment to collagen and plasma polymerised allylamine on glass. *Plasmas Processes and Polymers*, **3**, 474–484.

Even-Ram, S., Artym, V. and Yamada, K. M. (2006) Matrix control of stem cell fate. *Cell*, **126**, 645–647.

Fahlman, A., Hagstrom, S., Hamrin, K., Nordberg, R., Nordling, C. and Siegbahn, K. (1966) An apparatus for ESCA method. *Arkiv for Fysik*, **31**, 479–485.

Fally, F., Doneux, C., Riga, J. and Verbist, J. J. (1995) Quantification of the functional-groups present at the surface of plasma polymers deposited from propylamine, allylamine, and propargylamine. *Journal of Applied Polymer Science*, **56**, 597–614.

Fletcher, J., Lockery, N. and Vickerman, J. C. (2006) C60, buckminsterfullerene: its impact on biological ToF-SIMS analysis. *Surface and Interface Analysis*, **38**, 1393–1400.

Foster, T. T., Alexander, M. R., Leggett, G. J. and McAlpine, E. (2006) Friction force microscopy of self-assembled monolayers on aluminium surfaces – influence of alkyl chain length. *Langmuir*, **22**, 9254–9259.

Fraser, H. J., Collings, M. P., McCoustra, M. R. S. and Williams, D. A. (2001) Thermal desorption of water ice in the interstellar medium. *Monthly Notices of the Royal Astronomical Society*, **327**, 1165.

Gardella, J. A. and Hercules, D. M. (1980) Static secondary ion mass spectrometry, *Analytical Chemistry*, **22**, 226–232.

Gilmore, I. S. and Seah, M. P. (2003) G-SIMS of biodegradable homo-polyesters. *Applied Surface Science*, **252**, 6797–6800.

Haddow, D. B., MacNeil, S. and Short, R. D. (2006) A cell therapy for chronic wounds based upon a plasma polymer delivery surface. *Plasma Processes and Polymers*, **3**, 419–430.

Hou, Q. P., Rutten, F. J. M., Smith, E. F., Briggs, D., Davies, M. C., Buttery, L. D. K., Freeman, R. and Shakesheff, K. M. (2005) Surface characterization of pre-formed alginate fibres incorporated with a protein by a novel entrapment process. *Surface and Interface Analysis*, **37**, 1077–1081.

Johansson, B. (2006) ToF-SIMS imaging of lipids in cell membranes. *Surface and Interface Analysis*, **38**, 1401–1412.

Lewis, K. B. and Ratner, B. D. (1993) Observation of surface rearrangement of polymers using ESCA. *Journal of Colloid and Interface Science*, **159**, 77–85.

Magnani, A., Barbucci, R., Lewis, K. B., Leachscampavia, D. and Ratner, B. D. (1995) Surface-properties and restructuring of a cross-linked polyurethane-poly(amido-amine) network. *Journal of Materials Chemistry*, **5**, 1321–1330.

McArthur, S. L. (2006) XPS in bioengineering. *Surface and Interface Analysis*, **38**, 1381–1385.

Michel, R. and Castner, D. G. (2006) Advances in ToF SIMS analysis of protein films. *Surface and Interface Analysis*, **38**, 1386–1392.

Ogaki, R., Green, F., Li, S., Vert, M., Alexander, M. R., Gilmore, I. S. and Davies, M. C. (2006) G-SIMS of biodegradable homo-polyesters. *Applied Surface Science*, **252**, 6797–6800.

Ogaki, R., Green, F. M., Gilmore, I. S., Shard, A. G., Alexander, M. R. and Davies, M. C. (2007 in press) Study of the end group contribution to ToF-SIMS and G-SIMS spectra of poly(lactic acid) using deuterium labelling. *Surface and Interface Analysis*.

Parry, K., Shard, A., Short, R. D., White, R., Whittle, J. and Wright, A. (2006) ARXPS characterisation of plasma polymerised surface chemical gradients. *Surface and Interface Analysis*, **38**, 1497–1504.

Pertays, K., Thompson, G. E. and Alexander, M. R. (2004) Self assembly of stearic acid on aluminium: the importance of oxide surface chemistry. *Surf. Interface Anal.*, **36**, 1361–1366.

Petrovykh, D. Y., Kimura-Suda, H., Tarlov, M. J. and Whitman, L. J. (2004) Quantitative characterization of DNA films by X-ray photoelectron spectroscopy. *Langmuir*, **20**, 429–440.

Quirk, R. A., Briggs, D., Davies, M. C., Tendler, S. J. B. and Shakesheff, K. M. (2001) Characterization of the spatial distributions of entrapped polymers following the surface engineering of poly(lactic acid). *Surface and Interface Analysis*, **31**, 46–50.

Ratner, B., Weathersby, P., Hoffman, A. and Kelly Mascharpen, L. (1978) Radiation-grafted hydrogels for biomaterial applications as studied by ESCA technique. *Journal of Polymer Science*, **22**, 643–664.

Ratner, B. D. (1995) Advances in the analysis of surfaces of biomedical interest. *Surface and Interface Analysis*, **23**, 521–528.

Senshu, K., Yamashita, S., Mori, H., Ito, M., Hirao, A. and Nakahama, S. (1999) Time-resolved surface rearrangements of poly(2-hydroxyethyl methacrylate-block-isoprene) in response to environmental changes. *Langmuir*, **15**, 1754–1762.

Shakesheff, K. M., Evora, C., Soriano, I. and Langer, R. (1997) The adsorption of poly(vinyl alcohol) to biodegradable microparticles studied by X-ray photoelectron spectroscopy (XPS). *Journal of Colloid and Interface Science*, **185**, 538–547.

Shard, A. G., Volland, C., Davies, M. C. and Kissel, T. (1996) Information on the monomer sequence of poly(lactic acid) and random copolymers of lactic acid and glycolic acid by examination of static secondary ion mass spectrometry ion intensities. *Macromolecules*, **29**, 748–754.

Shard, A. G., Brewer, P. J., Green, F. M. and Gilmore, I. S. (2007) Measurement of sputtering yields and damage in C60 SIMS depth profiling of model organic materials. *Surface and Interface Analysis*, **38**, 294–298.

Sosnika, A., Sodhia, R. N. S., Brodersend, P. M. and Seftona, M. V. (2006) Surface study of collagen/poloxamine hydrogels by a 'deep freezing' ToF-SIMS approach. *Biomaterials*, **27**, 2340–2348.

Tougaard, S. (1997) Universality classes of inelastic electron scattering cross-sections. *Surface and Interface Analysis*, **25**, 137–154.

Tougaard, S. (1998) Accuracy of the non-destructive surface nano-structure quantification technique based on analysis of the XPS or AES peakshape. *Surface Interface Analysis*, **26**, 249–269.

Tsourapas, G., Rutten, F. J. M., Briggs, D., Davies, M. C. and Shakesheff, K. M. (2006) Surface spectroscopic imaging of PEG-PLA tissue engineering constructs with ToF-SIMS. *Applied Surface Science*, **252**, 6693–6696.

Tyler, B. (2001) ToF-SIMS image analysis. In Vickerman, J. C. and Briggs, D. (eds) *ToF-SIMS: Surface Analysis by Mass Spectrometry.* http://www.surfacespectra.com/ and IM.

Urquhart, A., Taylor, M., Anderson, D., Langer, R., Alexander, M. and Davies, M. C. (2007 in press) High through-put surface characterisation of arrayed biomaterials. *Advanced Materials*.

Vickerman, J. C., Briggs, D. and Henderson, A. (2000) *The Static SIMS Library*, Surface

Science, surfacespectra.com.
Vickerman, J. C. and Briggs, D. (eds) (2001) *ToF-SIMS: Surface Analysis by Mass Spectroscopy*. http://www.surfacespectra.com/and IM.
Vitte, J., Benoliel, A. M., Pierres, A. and Bongrand, P. (2004) Is there a predictable relationship between surface physical-chemical properties and cell behaviour at the interface? *European Cells and Materials*, **7**, 52–63.
Wagner, M. S., Graham, D. J., Ratner, B. D. and Castner, D. G. (2004) Maximizing information obtained from secondary ion mass spectra of organic thin films using multivariate analysis. *Surface Science*, **570**, 78–97.
Walton, J. and Fairley, N. (2004) Quantitative surface chemical-state microscopy by x-ray photoelectron spectroscopy. *Surface and Interface Analysis*, **36**, 89–91.
Walton, J. and Fairley, N. (2005) Noise reduction in X-ray photoelectron spectromicroscopy by a singular value decomposition sorting procedure. *Journal of Electron Spectroscopy and Related Phenomena*, **148**, 29–40.
Whitesides, G. M. and Laibinis, P. E. (1990) Wet chemical approaches to the characterization of organic-surfaces – self-assembled monolayers, wetting, and the physical organic-chemistry of the solid liquid interface. *Langmuir*, **6**, 87–96.
Whittle, J. D., Barton, D., Alexander, M. R. and Short, R. D. (2003) A method for the deposition of controllable chemical gradients. *Chemical Communications*, **14**, 1766–1767.
Wilson, J., Hench, L. L. and Greenspan, D. (1994) *Bioceramics*, Oxford, Pergamon Press.
Wolstenholme, J. F. W. J. (2003) *An Introduction to Surface Analysis by XPS and AES*, Chichester, John Wiley and Sons Ltd.

10
Characterisation using environmental scanning electron microscopy (ESEM)

A M DONALD, University of Cambridge, UK

10.1 Introduction

Environmental scanning electron microscopy (ESEM), or its more general variant of low vacuum scanning electron microscopy (LVSEM), is a technique which has been around for more than 15 years. However, in many ways it is still in its infancy in this field of tissue engineering as in many others across both the physical and biological sciences. Despite the many advantages of the technique, it has been slow to make its mark in university departments and in industry. In part this is because of the strength of conventional scanning electron microscopy (CSEM) which has so many applications and advantages (see Table 10.1 and Goldstein *et al.*, 2002). In part it would appear to have been due to failing marketing strategies within the companies that manufacture the different varieties of the instrument, initially implying the instrument was much simpler to use than many researchers found it, and subsequently failing to appreciate fully its strengths for hydrated samples, including those of biological origin. However, it also has to be recognised there are limitations of the instrument, as will be described later in this chapter, and some of these – including the fundamental challenge of beam damage – may ultimately be insuperable for imaging of live cells or following dynamic processes.

10.2 The instrument: a comparison with CSEM

So what is it that makes ESEM so different and advantageous compared with conventional scanning electron microscopy? The basic design is not so different. There is an electron source, which may (as in CSEM) be a simple tungsten filament, a LaB_6 filament or a field emission gun (FEG), each with its own advantages of brightness, durability and cost. This gun operates under the standard conditions of high vacuum, although exactly how high a vacuum is required depends on the type of gun used. The electron beam travels down the column towards the sample chamber. However, whereas in a CSEM this is a simple path, with lenses to focus the beam, but the entire trajectory being under

Table 10.1 A comparison of the operating conditions of CSEM and ESEM

	CSEM	ESEM
Gun	Tungsten, LaB_6 or field emission	Tungsten, LaB_6 or field emission
Sample environment	High vacuum	Gas (typically water), pressures up to $\sim$2000 Pa
SE detector	Everhardt–Thornley	Gaseous secondary electron detector
BSE detector	Yes	Not yet very good at discriminating
X-ray detection	Yes	Yes, but rather poor spatial resolution
Imaging insulators?	Need conductive coating, unless working at very low voltages	Yes
Imaging hydrated samples?	Only after dehydration	Yes
Imaging cells?	Yes, but after fixing, staining, etc.	Yes in principle without fixation, but optimisation still to be carried out

the same high vacuum as the gun, this is not the case in the ESEM (on which I will primarily focus in this chapter, rather than on the other LVSEM variants. These are simply less extreme variations, albeit with different detector types). Instead the column is divided up into a series of pressure zones, as shown in Fig. 10.1, each separated from its neighbour by a pressure-limiting aperture (PLA). The point about these is that they allow a pressure differential to develop between neighbouring zones, so that the pressure increases steadily down the column. This permits the last zone, the sample chamber itself, to contain a significant pressure of gas, up to around 10–15 torr (~2000 Pa). The maximum upper limit is set by a number of considerations including the desired resolution, the type of detector used and the accelerating voltage, and will be discussed further below.

It might be thought that the presence of the gas degrades the image quality owing to scattering of the incident electron beam by the gas molecules before the electrons ever reach the sample. In practice, although the beam is necessarily broadened, this may not prove too much of a problem, if the working distance (the distance from the bottom of the last aperture to the sample) is kept small, and the pressure in the chamber not too large, such that most electrons are not scattered; those that are, scattered only a few times. In this regime the shape of the incident beam is still strongly peaked above a rather flat background, as shown in Fig. 10.2. There will be some small signal arising from the further

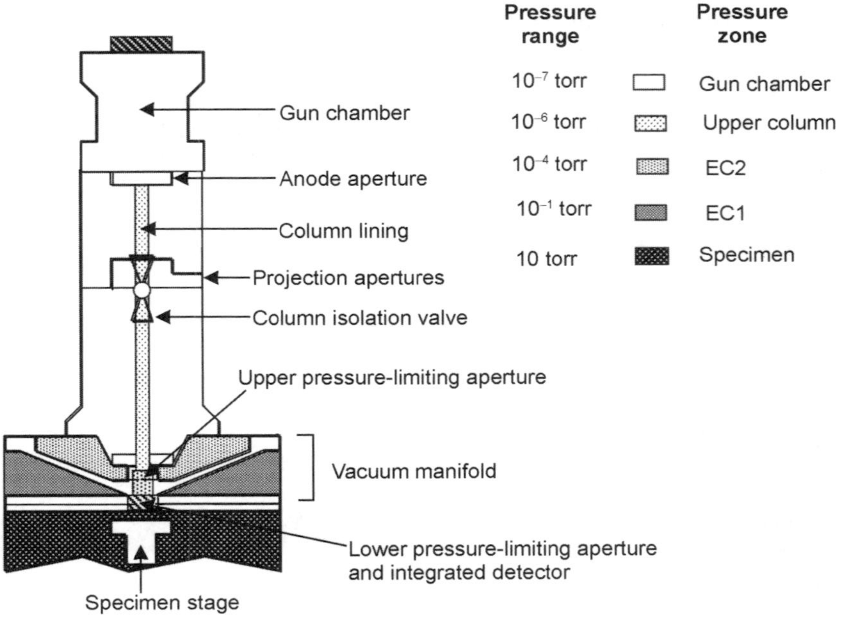

10.1 Schematic drawing of the ESEM column, showing the different pressure zones. 1 torr ~133 Pa.

reaches of this flat background, but the majority of the signal will be initiated by the sharp peak at the centre. The net effect is to introduce what is essentially a background DC signal on the signal of interest. In practice, resolution for samples of interest to the reader of this book is more likely to be limited by beam

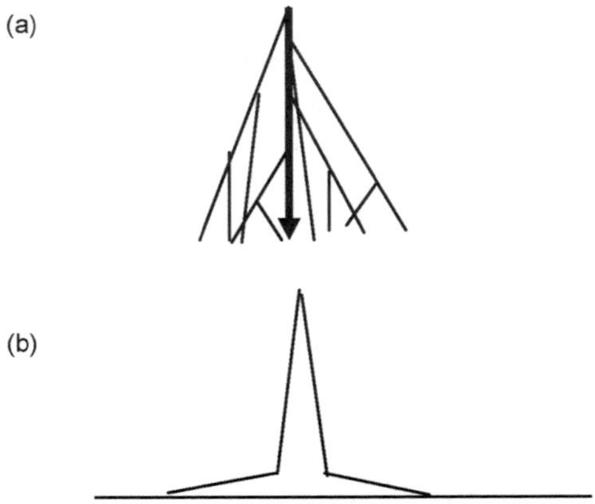

10.2 In the so-called oligo-scattering regime electrons in the incident probe undergo few, if any, collisions: (a) schematic of the scattering undergone by the incident beam (heavy arrowed line) in this regime; (b) the resultant probe has a sharp intensity peak superimposed on a low background 'skirt'.

damage than by the inherent resolution of the technique. Quoted instrumental resolution may be <10 nm, but for hydrated samples this is unlikely to be realised. The type of experiments for which the existence of the broad background 'skirt' is likely to matter is when X-ray microanalysis is attempted (Mansfield, 2000).

It is the presence of the gas around the sample that makes the key difference with CSEM, and which confers ESEM's many advantages. Frequently – but not necessarily – this gas is water vapour. If it is, in principle it permits a hydrated sample to be maintained in its hydrated state. When achievable, this confers a huge benefit on samples of biological origin which do not therefore need to undergo complex sample preparation routes. However, in order to maintain the state of hydration, two conditions have to be met. First the pressure in the sample chamber has to be high enough to maintain the level of hydration. At its crudest this implies that saturated vapour pressure must be maintained, but it has been demonstrated (Stokes *et al.*, 2003) that this is in many cases an over-estimate. For instance, for cells, owing to the presence of organelles and solutes within the cell as well as the cell membrane itself, it is possible to maintain a cell in a perfectly stable state against loss of water from the inside of the cell, at a lower vapour pressure around the cell than that corresponding to 100% RH (relative humidity). The second condition is more subtle: the pumpdown of the chamber prior to imaging has to be carried out so that the sample is not dehydrated during this process. A protocol to achieve this was described some years ago (Cameron and Donald, 1994), which can be appropriately modified for different instruments and sample types.

If one considers the simplest criterion of maintaining saturated vapour pressure (SVP) around the sample to prevent dehydration, then it is clear that the sample temperature is also important. Figure 10.3 shows the form of the SVP curve for water as a function of temperature. Whereas at ambient temperatures SVP is significantly higher than the useful working pressure of the chamber, if the temperature is dropped to just above freezing (say to 2–3 °C), a very modest sample pressure will suffice. Thus, when working with hydrated samples it is usually necessary to work with a cooling stage (frequently this is a Peltier stage) and carry out imaging at significantly lower than room temperature. For many samples of non-biological origin this is perfectly adequate (see, for example, the discussion in Donald *et al.*, 2000), for biological samples such as cells it obviously poses a potential limitation.

There is a second advantage conferred by the presence of the gas, one that again is particularly beneficial for samples of interest to this readership, and that is its effect on the build-up of charge. For insulators in general in the CSEM it is necessary to coat the surface with some conductive coating prior to imaging, in order to prevent sample charging and degradation of image quality (although modern low-voltage instruments can to some extent overcome this problem). In the ESEM this is not usually required: the gas molecules in the sample chamber

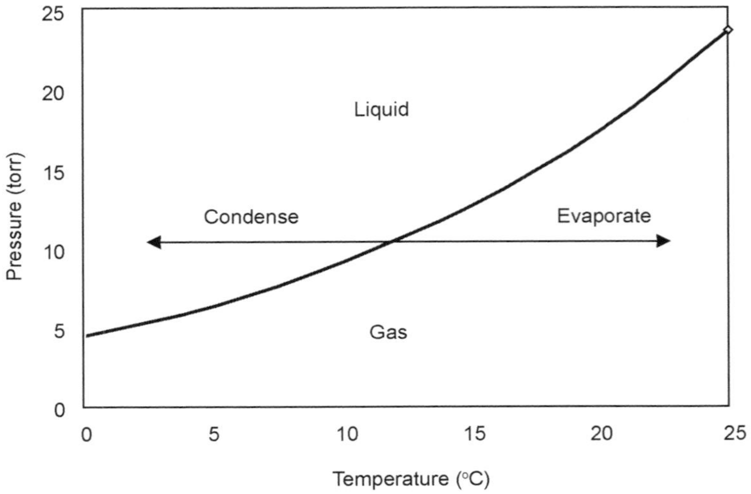

10.3 The saturated vapour pressure (SVP) curve for water vapour, indicating the regimes in which evaporation or condensation will occur.

are ionised by electrons emitted by the sample (secondary electrons predominantly, since their collision cross-section with the gas molecules is higher than the more energetic backscattered electrons). The corresponding cascade of electron–gas molecule ionisations is shown schematically in Fig. 10.4. Each ionising collision produces a positively charged ion together with an additional so-called daughter electron, leading to an amplification in the number of electrons travelling towards the detector. This amplification can be very significant. However, in the context of charging, what matters is the generation of a substantial number of positive gaseous ions. These tend to drift towards the negatively charged sample – negatively charged because of the implantation of incident electrons – and as they accumulate at the surface they tend to compensate for the negative charge build-up, and so reduce the effect of charging. The processes involved are complex, and the interested reader should look at the review (Thiel and Toth, 2005) for more details. In general, however, the effect means that insulators do not need a conductive coating. Furthermore, if a new surface is opened up, for instance during a dynamic experiment to study fracture, charging is still not an issue. In general this means ESEM is particularly powerful for dynamic experiments of insulators. Dynamic experiments will be discussed in more detail later.

So there are two fundamental advantages of the ESEM over CSEM: the sample can be imaged in a hydrated state, and insulators do not need to be coated. Let us now look at the signal that is detected. In CSEM, the detector works by having a positive bias to which the electrons are attracted. Conventionally there are two detectors for detecting the low-energy secondary electrons (SEs), usually defined as those with energies <50 eV, and the high-energy

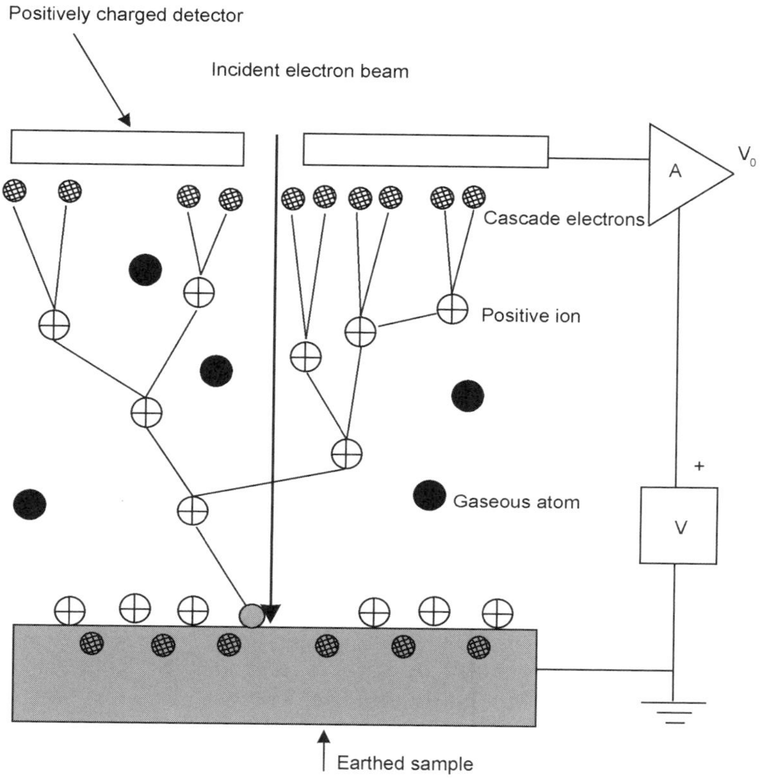

10.4 Schematic of the cascade amplification process. Electrons emitted from the sample collide with gaseous atoms and can ionise them, producing additional daughter electrons which contribute to the signal detected at the positively charged detector.

backscattered electrons (BSEs). These latter are electrons deflected through close to 180° on hitting the sample, whereas the low-energy SEs are electrons emitted from the sample due to inelastic processes (ionisation, etc.). Because the SEs possess low energy, they can only escape from close to the sample surface, and therefore are particularly useful at providing topographic information. In contrast, BSEs can come from much deeper within the sample, and the number generated depends on atomic number. Thus BSEs typically can provide information on chemical composition. In CSEM the two detectors are set up to distinguish the different signals, and therefore topographic and chemical information can be obtained separately. However, these detectors do not work in the low vacuum conditions of an ESEM, and existing detectors for the ESEM are much less successful at distinguishing the different signals. An additional problem arises because of the nature of the cascade amplification itself. This process amplifies all signals, thus amplifying noise such as that generated by

BSEs hitting other parts of the chamber and causing additional electrons to be generated which contain no useful information. These electrons are known as SE2s, and tend simply to degrade image quality. A discussion of signal composition can be found in Fletcher *et al.* (1999). In practice, good BSE detectors are still not available for the ESEM, and the SE detector will not as completely discriminate between topographic and compositional contrast.

10.3 Static experiments

10.3.1 Insulators

The first class of samples, relevant to readers of this book, for which ESEM offers clear advantages are insulators such as ceramics and polymers. As indicated above, the presence of the gas in the chamber means that charging is no longer an issue, as in CSEM, so that these samples – which often form the fabric of tissue scaffolds and prosthetic devices, as well as bone itself – can be studied without a conducting coating obscuring surface detail. For samples such as these the advantage of being able to obtain high-quality images with essentially zero sample preparation can be very helpful. Recent relevant examples include prosthetic meshes (Gil *et al.*, 2003, Arbos *et al.*, 2006), characterisation of pharmaceutical formulations such as drug delivery vehicles and actives (Elvira *et al.*, 2004; Möschwitzer and Müller, 2006; Fu *et al.*, 2006) and bone cement (Vazquez *et al.*, 2005). An additional advantage of the lack of coating is that contrast, often obscured by the presence of a coating, becomes visible. This was initially observed for a rather different type of sample, that of water–oil emulsions (Stokes *et al.*, 1998). The origin of the contrast lies in the differences in electronic properties between the two components. Similar contrast effects have also been observed in ceramics such as gibbsite (Griffin, 2000). However, care must be taken in carrying out such imaging to ensure the operating conditions of the microscope are correct to enable this rather low level of contrast to be observed. Figure 10.5 shows how variations in growth conditions, and hence local concentrations of atoms, and variations in local potentials can give rise to strong contrast when the correct instrumental parameters are chosen (Craven *et al.*, 2002).

10.3.2 Hydrated samples

The second class of samples that is likely to prove of interest to the reader are those in which biological samples are imaged. Ideally one wants to be able to observe these alive, but there are many challenges here. A useful review of ESEM for biological applications, with particular reference to the biomedical arena, can be found in Muscariello *et al.* (2005). Without concern about keeping the cells viable, useful information can be obtained on issues such as confluence,

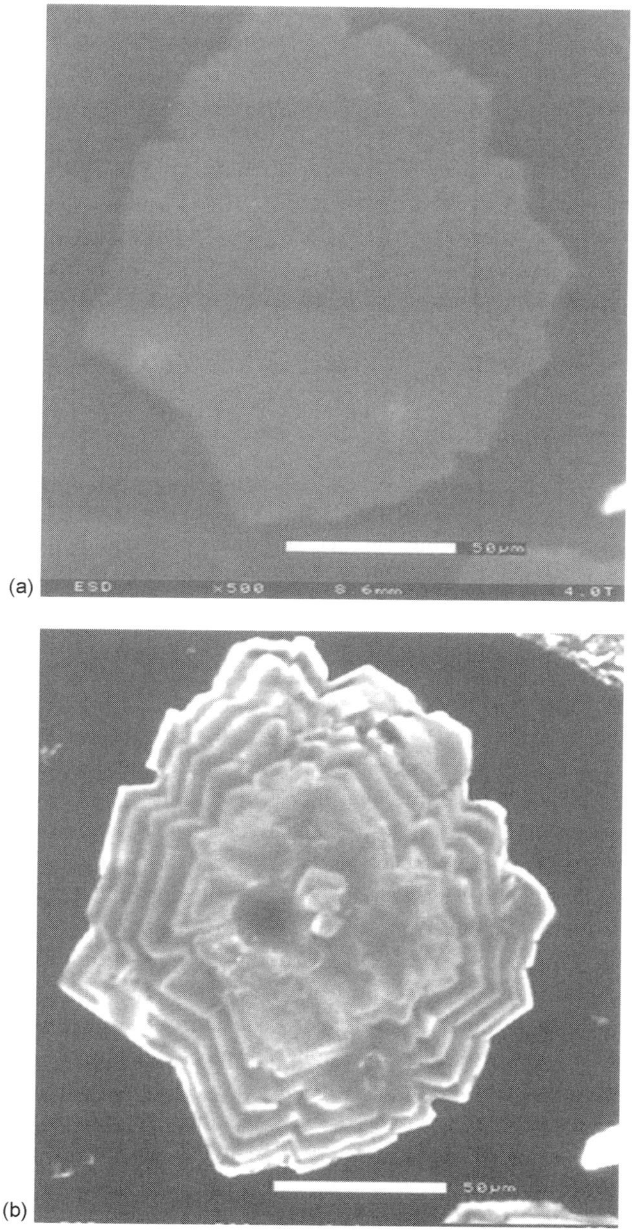

10.5 ESEM image of polished gibbsite ($Al(OH)_3$): (a) beam energy 20 keV, gas pressure = 4 torr, working distance 8.8 mm, scan speed 2.1 frames s^{-1}, with 8 frame integration; (b) as (a) but now with a grid of copper wires 1 mm above the sample surface. This grid attracts positive ions, preventing them from scavenging the electron signal. From Craven *et al.* (2002).

and the relative abilities of different substrates to permit proliferation (Wiedmann-al-Ahmad *et al.*, 2005). In order to observe cells alive, it is necessary not just to avoid many of the standard sample preparation stages used for conventional SEM (fixing, staining, critical point drying, etc.), but also to keep beam damage to a minimum and maintain the environment around the sample in a state that does not put undue stress on the cell. To the author's knowledge, no one has yet conclusively demonstrated this can be done.

Although there are a growing number of papers making comparisons between ESEM and CSEM, and demonstrating the existence of artefacts which the CSEM sample preparation route may have introduced (Callow *et al.*, 2003; Habold *et al.*, 2003; McKinlay *et al.*, 2004; Castillo *et al.*, 2005; Muscariello *et al.*, 2005), that in itself is not sufficient. Keeping the cell hydrated is a necessary requirement, but as has been shown (Tai and Tang, 2001; Stokes *et al.*, 2003; Muscariello *et al.*, 2005) this can be achieved without requiring 100% relative humidity in the chamber, because the cell contents do not correspond to pure water. The vapour pressure of the cell is reduced due to the presence of salts and other solutes, even without taking into account the protective membrane which will additionally slow moisture loss. It is clearly an advantage to be able to work at the lowest possible vapour pressure to keep the image quality as high as possible. Nevertheless, it is still necessary to work at sub-ambient temperatures in general, at least when using standard detectors on the instrument, and for many organisms/cells this low temperature will in itself pose a stress and may cause cells to become non-viable or die. An additional limitation is the absence of carbon dioxide in the chamber when water vapour is used. This is in itself likely to affect the state of a sample. In principle gas mixing can be achieved, but it has not yet been directed towards the introduction of physiologically relevant carbon dioxide levels to remove this source of stress for the cell.

Thus we see that using the ESEM for examination of biological samples is still a technique in its infancy. As an example of the issues that may arise when a direct comparison is made between standard CSEM imaging and ESEM, Fig. 10.6 shows an example of an algal cell from *Enteromorpha*, which is of interest because of its propensity to stick to surfaces causing biofouling. The adhesive pad, when imaged following dehydration, appears to show a quite coarse fibrillar appearance which is completely absent in the ESEM images of a sample hydrated *in situ.* By correctly imaging the nature of the pad, it becomes possible to build a better understanding of its hydrophobicity, and hence mechanism of adhesion (Callow *et al.*, 2005).

10.4 Dynamic experiments

The ESEM offers a wide range of opportunities for dynamic experiments, some of a kind inconceivable in a conventional SEM. Again, the field is young and relatively few groups are attempting to do this. The kinds of experiments

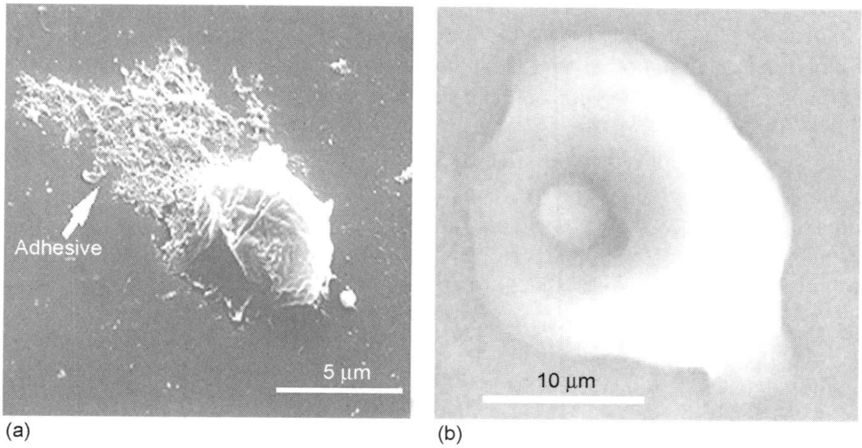

10.6 Images of *Enteromorpha*: (a) conventional SEM of fixed and dried spore, the adhesive pad looks fibrillar; (b) hydrated spore imaged in the ESEM, showing that the adhesive pad is not actually fibrillar, but appears to be structureless at this level of resolution. From Callow *et al.* (2003).

possible can be divided up into several types:

- *In situ* mechanical testing, where it is the absence of a coating on an insulator that is of interest, and the ability to image a hydrated sample.
- Experiments where the state of hydration is altered by changing either the temperature of the stage or the pressure in the chamber; in this case either dehydration or hydration can be carried out.
- Reactions within the chamber, for instance due to introducing water or some other gas.

Each of these types will be discussed in turn.

10.4.1 *In situ* mechanical testing

For insulators, mechanical testing in a CSEM is limited by the fact that, even if the sample is initially coated with a conductor, as fresh surface opens up, this new surface is uncoated and will potentially itself charge, causing imaging problems. This is improved by working at low voltages, but this poses other problems. As indicated above, in the ESEM this complication is completely removed, making it an ideal tool for the study of failure in real time. Since stress–strain (or at least load–extension) curves can be recorded simultaneously, much new insight is potentially available. However, a fundamental limitation remains that only the surface can be imaged. Owing to the plane stress conditions at the surface for a bulk insulator, this may mean that failure occurs invisibly beneath the surface, although not necessarily so. However, when dealing with cellular materials – such as many tissue scaffolds or plant tissue – this is much less of a restriction.

Studies have been carried out on both onion epidermis (Donald *et al.*, 2003) and phloem parenchyma of carrot (Thiel and Donald, 1998), as well as breadcrumbs (Stokes and Donald, 2000), where the ability to correlate stages in failure with the stress–strain curve has proved very helpful. In all these cases, the ability to maintain the sample in its hydrated state (none of the above samples is a dry insulator) has proved crucial. As an example of the utility of this approach in the area of wound care, there is the work on carboxymethyl cellulose fibres, on which dynamic tensile tests were performed *in situ* under conditions of changing relative humidity (Wei *et al.*, 2003; Wei and Wang, 2005).

However, it is no mean feat to carry out such tests, since the mechanical stage is less able to keep the sample at a fixed temperature and hence keep the sample hydrated, than a standard Peltier stage. If good thermal contact is maintained with the sample, there is a real danger of interfering with the mechanical response; if good thermal contact is not maintained, some loss of moisture may occur. A good example of this was shown during studies of gelatin gels. The top surface of the gels lost water during testing, creating a locally higher modulus material in a thin layer at the surface. In turn this impacted on the mechanical response, and the surface of the sample exhibited a rather beautiful instability (Rizzieri *et al.*, 2006), reflecting the complex behaviour of a bilayer structure. Nevertheless, this approach of *in situ* mechanical deformation could be used to study the failure of tissue scaffolds under different conditions and via different sample preparation routes.

10.4.2 Dynamic hydration and dehydration

Changing the sample environment opens up the way to carry out a wide range of dynamic experiments. In order to achieve this, careful control needs to be maintained over the sample stage and chamber environment. As with the experiments described above, it is important that there is good thermal contact between the sample and the stage, so that the conditions thought to be applied by the operator actually do pertain to the sample surface (or with a known correction). The crucial issue is to appreciate the state of the sample relative to the SVP curve shown in Fig. 10.3. If initially the sample sits right on the line, then by dropping the stage temperature/raising the chamber pressure condensation of the water will occur. Conversely, raising the stage temperature/dropping the chamber pressure leads to evaporation. In practice it is often easier to raise the temperature to provide good control. Some degree of imprecision in the known conditions (e.g. the actual temperature at the top of the sample) may be acceptable because in reality one can tell whether evaporation or condensation is occurring by observing the sample.

Figure 10.7 shows a particular example of drying an aqueous dispersion of a lacquer containing silica particles (Royall and Donald, 2002). In the early stages, when the water content is high, little can be seen of the silica particles which are

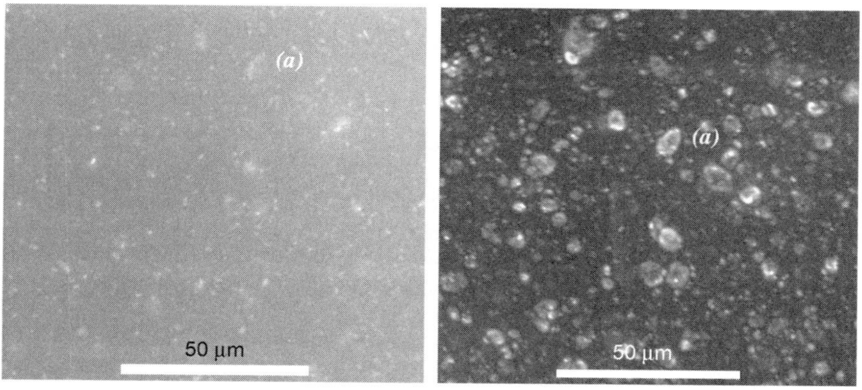

10.7 Lacquer containing precipitated silica undergoing drying in the ESEM. As drying proceeds, silica particles are revealed by their strong contrast, and the overall signal intensity drops due to the loss of water. The area marked *(a)* shows how a single particle emerges during the drying process. Left-hand image 3 mins after drying initiated in the ESEM; right-hand image 59 min after drying initiated.

largely still submerged. As drying proceeds, they become increasingly visible. Silica has a significantly higher backscattered electron yield than the bulk of the polymeric matrix, and hence the particles are easy to see as the covering water layer evaporates and film formation proceeds. The overall signal intensity of the matrix drops, however, because water has a rather high secondary electron yield (Meredith and Donald, 1996) and the water content is obviously falling.

Figure 10.7 therefore shows a typical progression of a drying film, with the drying being carried out *in situ* in the ESEM chamber. The literature contains other examples of similar processes (Keddie *et al.*, 1995; Donald *et al.*, 2000; He and Donald, 1996). This is an approach that is also beginning to be used for drug formulations. For instance Mohammed *et al.* (2004) used ESEM to explore the changes in liposome morphology during dehydration, and demonstrated that the addition of ibuprofen improved the stability of the egg-phosphatidylcholine/cholestorol liposomes. Meredith *et al.* (1996) coupled dehydration with cryo-ESEM to mimic freeze drying conditions. In the context of work utilising dehydration of cells, the work of Addadi's group is particularly interesting (Cohen *et al.*, 2003). They studied the pericellular coat around chondrocytes, although the coat was only visible after staining with uranyl acetate. As the chamber vapour pressure was reduced, the cells shrunk to around 80% of their fully hydrated value, and the gel corresponding to the hyaluronan coating disappeared.

One can also imagine (re)hydrating a sample (or indeed moving through hydration/dehydration cycles). This could, for instance, correspond to dissolution of a powder (e.g. some drug formulation), the rehydration of a dried gel or desiccated structure (Lavoie *et al.*, 1995) or determination of local contact angles on a heterogeneous substrate. This is an approach that has been particularly picked up by the cement and clay communities, although many of these hydration processes are actually equivalent to reactions occurring, discussed further below, but other research areas have also utilised the approach (e.g. Roman-Gutierrez *et al.*, 2002). There are many potential applications within the field of tissue engineering, particularly with respect to exploring the effect of rehydration under controlled conditions of pre-prepared gels and powders, but the field has yet really to respond to the opportunities.

As an example of utilising the ESEM for hydration experiments, Fig. 10.8 shows a comparison of contact angle measurements on a cellulose fibre as seen

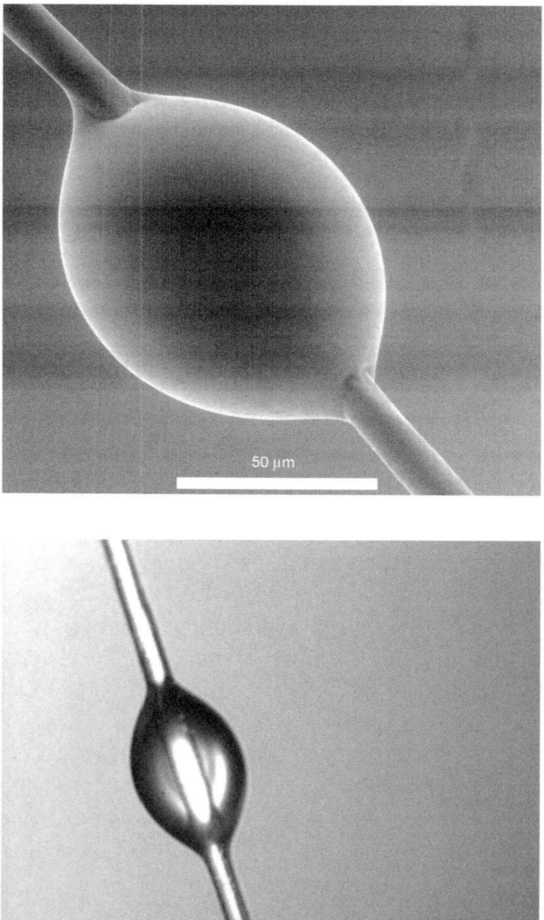

10.8 Micrographs showing water droplets condensed onto a cellulosic fibre at high magnification in the ESEM (upper image) and at low magnification in the optical microscope (lower image).

in the ESEM and a standard optical microscopy image. For a cylindrical fibre, the water drop adopts a so-called 'unduloidal' shape, partially wrapping around the fibre. The scattering from the droplet edges makes accurate determination with an optical microscope difficult to accomplish. In contrast, the precise shape of the contact of the water with the fibre is easy to determine in the ESEM image. In principle this means that for systems with more than one component one can measure local contact angles too. In practice things can be slightly more complicated. The fibre is actually a rather easy geometry to study. For a planar surface, one will typically view the droplet from above, in which case the contact angle cannot be easily determined. Attempts have been made to get around this problem, both by analysing how one would expect contrast to vary across a spherical cap with different contact angles, and by altering the sample orientation so that the drops can be seen 'edge on' (Stelmashenko *et al.*, 2001). For cellular scaffolds this issue may be less significant, since the individual struts in the scaffold will resemble the case of the fibre described above.

10.4.3 Reactions in the chamber

Finally, closely akin to the above class of experiments, one can consider ones where a reaction is allowed to proceed in the chamber during imaging, or to use the dynamic changes available as a measure of processes. The simplest case to consider is that where the water vapour pressure is altered to allow not simply hydration to occur, but some change in state. The growth and morphology of nitric acid trihydrate grown in the ESEM at cryo temperatures under nitrogen gas were studied by Grothe *et al.* (2006). The growth of polymeric films at an air–water interface was followed in real time by Miller and Cooper (2002). Many of the instances of studies on cement and its components cited above are actually following a complex sequence of reactions in real time in the ESEM. For instance, the different stages of calcium trisilicate hydration correspond to a chain of reactions, interspersed with periods of dormancy (Meredith *et al.*, 1995). Hung *et al.* (2005) use hydration to explore the effect of changes in processing on how hygroscopic oleic acid droplets are by looking at changes in structures as the relative humidity in the chamber is altered, and *in situ* wetting was used by de la Parra (1993) to explore heterogeneity of microporous polymer membranes. Superdisintegration of pharmaceutical formulations upon addition of water vapour was explored by Thibert and Hancock (1996). In principle, the reaction need not be prompted by water itself, but there are complications in using other liquids/gases owing to the problems of introducing the reactant in a controllable way. Micromanipulators offer the possibility of introducing other fluids (i.e not from the gas phase) but have the danger of pushing the sample out of the field of view, since it is very difficult to introduce sufficiently small volumes of fluid. Using gases other than water vapour has been successfully demonstrated, e.g oxygen at high temperatures has been introduced to the chamber to explore changes in silver catalysts (Uwins *et al.*, 1997) and high-temperature studies of metals under methanol vapour and hydrogen-carbon monoxide were carried out by Schmid *et al.* (2001). Thus the ESEM chamber can, in some senses, be thought of as a reaction chamber in which many different types of processes could be followed. There is clearly scope for many more studies of this type.

10.5 Dual beam instruments – an emerging technique

So far this chapter has focused on imaging in the ESEM. New instrumental improvements are constantly occurring, however, and along with changing accessories such as better detectors and new stages, a significant development in the past few years has been the arrival of so-called dual beam instruments. These are familiar within the CSEM community, but now an ESEM variant is also available. These instruments have two beams, the electron source as in a standard ESEM, but also an ion beam consisting of a focused beam of high-energy gallium ions (the usual abbreviation for this beam is FIB, standing for focused ion beam). These ions can mill through a sample to reveal sub-surface structures in a controlled way, and at a position chosen by imaging with the electron beam. Conventional dual beam (FIB-SEM) instruments have been available for a number of years, and have found particular application within the semiconductor industry. A useful introduction can be found in Krueger (1999). The ESEM variant of this instrument offers many of the advantages of the two techniques.

The two beams in the dual beam instrument have separate columns inclined at around 50° to each other, as shown in Fig. 10.9. The specimen is tilted to an angle of 52° such that its surface is perpendicular to the ion beam and tilted with respect to the electron beam. In this way, the ion beam can be used to sputter or mill atoms from the specimen while the electron beam provides a means of monitoring progress, all without having to translate the specimen. A cross-section is made by allowing the ion beam to dwell for varying lengths of time, ultimately creating a flat vertical face deep into the specimen for viewing with the electron beam. This has the effect of revealing heterogeneities buried within the sample, such as an internal interface, and moves on from conventional (E)SEM which can only probe the surface of a sample. Hence, it opens up the way for investigating the full three-dimensional structure of appropriate samples, at a resolution beyond that obtainable by conventional optical microscopy, such as confocal approaches. This can be accomplished by alternately imaging and milling a little deeper into the sample to build up a series of slices which can then be reconstructed with appropriate software.

Whereas for semiconductors FIB-SEM works fine, for insulators the problems of imaging after milling remain, because of the charging issue. Thus there is a niche for FIB-ESEM. Although the gallium source requires a high

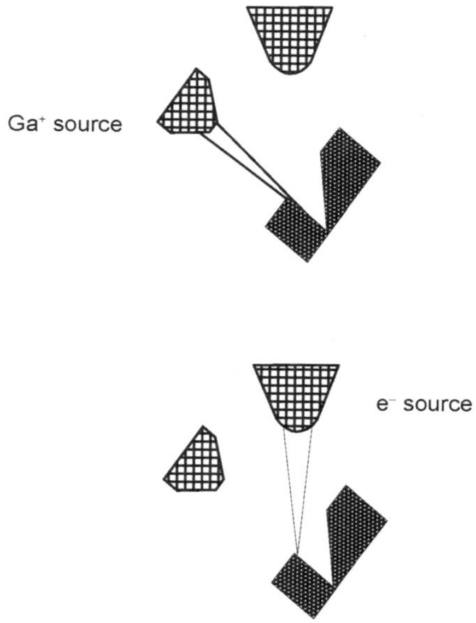

10.9 Schematic operation of the dual beam FIB/ESEM. The ion beam and electron beam are inclined at 52°. The upper image shows the gallium ion beam in operation, and a section of the sample is milled away; this process must be performed in high vacuum. The lower image shows the path of the electron beam for imaging into the exposed, milled surface; imaging can occur under standard ESEM conditions.

vacuum, so the capability of imaging wet samples cannot be utilised, for polymers or ceramics, the ESEM mode of imaging after milling remains an attractive advantage. Very limited work has been so far published on this new generation of instruments (Stokes *et al.*, 2006), but the potential is there for significant advances in the study of complex scaffolds, etc. to be made.

10.6 Potential and limitations

Throughout this brief overview attempts have been made to attempt to stress the pros and cons of this emerging technique. Although the basic instrument has been around for quite a few years now, its take-up has been quite slow. In some instances people have used the equipment without appreciating what it can really do. As an example of this one could note researchers who use ESEM on cells after fixing, drying and gold coating (Janssen *et al.*, 2006), so that they are actually using it simply as a conventional SEM. Clearly the scope for significantly extending the range of samples that can be studied without the standard complex preparation routes, and the accompanying danger of the introduction of artefacts, is substantial. For dry insulators the situation is clearest: fine surface detail may be observable due to the absence of gold coating, and contrast due to structural heterogeneity may be visible as the electrons detected originate from the sample itself and not from the covering coating. However, even for samples such as these, there are various issues to be considered: beam damage, sample environment and detector limitations.

10.6.1 Beam damage

Beam damage of organic material has long been known to be an issue for most material types. Grubb, many years ago, referred to a typical dose in a transmission electron microscope as being equivalent to 'exploding a 10 MT H-bomb about 30 yards away' (Grubb, 1974). The dose in an SEM is less, but still very significant. For organic materials there is usually a balance between breaking bonds and causing crosslinking, but either of these processes can significantly affect the structures observed and lead to misinterpretation. Again the careful work of Grubb indicates the sorts of issues that may arise for polymers. A recent review pertinent to conventional SEM can be found in Egerton *et al.* (2004). The reason that beam damage does not appear to feature more in the literature of ESEM is probably that typically high-resolution work has not been carried out. The resolution of the ESEM is in general not as good as for conventional SEM, for a variety of reasons discussed by Stokes (2003), and because the extension of the range of experiments and sample types that can be studied by ESEM is able to open up new avenues even without pushing the resolution, most published work has been carried out at comparatively low magnifications when the electron dose is moderate. For instance, for the ESEM images of cells published in Stokes *et al.* (2003), it was stated that beam damage was not found to be significant under the conditions used.

Nevertheless, it will be present and has been considered. In particular, when water vapour is present in the chamber, this will itself contribute to the damage process; radiolysis of the water molecules by the electron beam may occur, causing the production of free radicals, which can then contribute directly to the beam damage of the sample. This has been explored theoretically by Royall *et al.* (2001) and experimentally by Kitching and Donald (1998). Experimental work examining the consequences of the damage can be found in Forsberg and Lepoutre (1994), Kifetew and Sandberg (2000), Royall and Donald (2002) and Turkulin *et al.* (2005). Thus, beam damage is an issue to be borne in mind when studying fragile samples in the ESEM, but is often not a limiting factor in imaging many sample types.

10.6.2 Sample environment

As has been discussed above, it is possible to control the sample temperature via a Peltier stage (or indeed a hot stage, although this is less relevant to readers of this book), and thereby the state of the sample relative to the saturated vapour pressure curve. However, this can only be accomplished within certain bounds. Firstly, there is the question of accuracy that can be achieved. For thick, thermally insulating samples, poor heat conduction means that the stage temperature may differ slightly from the temperature of the sample surface. In certain situations this may matter, and needs to be factored in. The sample chamber pressure also has bounds. In practice this requires most imaging of hydrated samples to be carried out at sub-ambient temperatures, which may lie well away from physiologically relevant conditions. For some situations this may be a crucial limitation, but not for all. Nevertheless it must be borne in mind when designing experiments.

10.6.3 Detector design

New detectors are constantly being developed. Nevertheless it must be recognised that the standard detectors supplied with an ESEM have limitations in the pressure range at which they can operate (thereby affecting the temperature at which samples can be imaged, as described above), the field of view accessible at different magnifications, and the discrimination of signals achievable. One new type of detector being developed is that of a STEM (scanning transmission electron microscopy detector) which sits below the sample and permits a transmitted signal to be acquired simultaneously with the standard SE signal (Williams *et al.*, 2005; Doehne and Baken, 2006). Further detector developments are to be anticipated which may expand the experimental conditions under which useful information can be extracted from the ESEM.

10.7 Conclusions

It is hoped that this chapter has highlighted how useful the ESEM approach may be for the materials of interest – insulating and often hydrated – for tissue engineering. The burgeoning interest in this general area can be seen from the Proceedings of Scanning 2006, published in *Scanning* **28**(2), in which a session was dedicated to the applications of environmental and low-vaccuum SEM in the pharmaceutical industry. Whether the possibility of additionally imaging living cells is feasible remains an open question, but there can be no doubt that examination of interfaces between biological material and synthetic substrate is much easier in the ESEM than following the complex sample preparation routes required in a conventional SEM. It is to be hoped that the technique will be adopted far more widely by this community in the future.

10.8 References

Arbos, M. A., Ferrando, J. M., Quiles, M. T., Vidal, J., Lopez-Cano, M., Gil, J., Manero, J. M., Pena, J., Huguet, P., Schwartz-Riera, S., Reventos, J. and Armengol, M.

(2006), 'Improved surgical mesh integration into the rat abdominal wall with arginine administration', *Biomaterials*, **27**, 756–68.

Callow, J. A., Osborne, M. P., Callow, M. E., Baker, F. and Donald, A. M. (2003), 'Use of ESEM to image the spore adhesive of the marine alga *enteromorpha* in its natural hydrated state.' *Coll and Surf B*, **27**, 315–21.

Callow, J. A., Callow M E, Ista L K, Lopez G and Chaudhury, M. K. (2005), 'The influence of surface energy on the wetting behaviour of the spore adhesive of the marine alga *Ulva linza* (synonym *Enteromorpha linza*)', *Journal of the Roy Soc Interface*, **2**, 319–25.

Cameron, R. E. and Donald, A. M. (1994), 'Minimising sample evaporation in the ESEM', *J Micros*, **173**, 227–37.

Castillo, U., Myers, S., Browne, L., Strobel, G., Hess, W. M., Hanks, J. and Reay, D. (2005), 'SEM of some endophytic streptomycetes in snakevine – *Kennedia nigricans*', *Scanning*, **27**, 305–11.

Cohen, M., Klein, E., Geiger, B. and Addadi, L. (2003), 'Organization and adhesive properties of the hyaluronan pericellular coat of chondrocytes and epithelial cells', *Biophys J*, **85**, 1996–2005.

Craven, J. P., Baker, F. S., Thiel, B. L. and Donald, A. M. (2002), 'Consequences of positive ions upon imaging in low vacuum scanning electron microscopy', *J Micros*, **205**, 96–105.

Doehne, E. and Baken, E. (2006), 'An environmental STEM detector for ESEM: new applications for humidity control at high resolution', *Scanning*, **28**, 103–4.

Donald, A., Baker, F., Smith, A. and Waldron, K. (2003), 'Fracture of plant tissues and walls as visualized by environmental scanning electron microscopy', *Ann Bot*, **92**, 73–7.

Donald, A. M., He, C., Royall, C. P., Sferrazza, M., Stelmashenko, N. A. and Thiel, B. L. (2000), 'Application of ESEM to colloidal aggregation and film formation', *Coll Surf A*, **174**, 37–53.

Egerton, R. F., Li, P. and Malac, M. (2004), 'Radiation damage in the TEM and SEM', *Micron*, **35**, 399–409.

Elvira, C., Fanovich, A., Fernandez, M., Fraile, J., San Roman, J. and Domingo, C. (2004), 'Evaluation of drug-delivery characteristics of microspheres of PMMA-PCL-cholesterol obtained by super-critical CO_2 impregnation and by dissolution-evaporation techniques', *J Cont Rel*, **99**, 231–40.

Fletcher, A. L., Thiel, B. L. and Donald, A. M. (1999), 'Signal components in the ESEM', *J Micros*, **196**, 26–34.

Forsberg, P. and Lepoutre, P. (1994), 'Environmental scanning electron-microscope examination of paper in high-moisture environment – surface structural-changes and electron-beam damage', *Scanning Micros*, **8**, 31–34.

Fu, G., Li, H., Yu, H., Liu, L., Yuan, Z. and He, B. (2006), 'Synthesis and lipoprotein sorption properties of porous chitosan beads grafted with poly(acrylic acid)', *React Funct Poly*, **66**, 239–46.

Gil, F. J., Manero, J. M., Planell, J. A., Vidal, J., Ferrando, J. M., Armengol, M., Quiles, M. T., Schwartz, S. and Arbos, M. A. (2003), 'Stress relaxation tests in polypropylene monofilament meshes used in the repair of abdominal walls', *J Mat Sci Mater Med*, **14**, 811–15.

Goldstein, J. I., Newbury, D. E., Joy, D. C., Lyman, C. E., Echlin, P., Lifshin, E., Sawyer, L. and Michael, J. R. (2002), *Scanning Electron Microscopy and X-ray Microanalysis*, New York, Springer.

Griffin, B. (2000), 'Charge contrast imaging of material growth and defects in environmental scanning electron microscopy – linking electron emission and cathodoluminescence', *Scanning*, **22**, 234–42.

Grothe, H., Tizek , H., Waller, D. and Stokes, D. J. (2006), 'The crystallisation kinetics and morphology of nitric acid trihydrate', *Phys Chem Chem Phys*, **8**, 2232–9.

Grubb, D. R. (1974), 'Radiation damage and electron microscopy of organic polymers', *J Mat Sci*, **9**, 1715.

Habold, C., Dunel-Erb, S., Chevalier, C., Laurent, P., Le Maho, Y. and Lignot, J.-H. (2003), 'Observations of the intestinal mucosa using ESEM; comparison with CSEM', *Micron*, **34**, 373–9.

He, C. and Donald, A. M. (1996), 'Morphology of core–shell polymer latices during drying', *Langmuir*, **12**, 6250–56.

Hung, H.-M., Katrib, Y. and Martin, S. T. (2005), 'Products and mechanisms of the reaction of oleic acid with ozone and nitrate radical', *J Phys Chem A*, **109**, 4517–30.

Janssen, F. W., Oostra, J., Van Oorschot, A. and Van Blitterswijk, C. A. (2006), 'A perfusion bioreactor system capable of producing clinically relevant volumes of tissue-engineered bone: *in vivo* bone formation showing proof of concept', *Biomaterials*, **27**, 315–23.

Keddie, J. L., Meredith, P., Jones, R. A. L. and Donald, A. M. (1995), 'Kinetics of film formation in acrylic latices studied with multiple-angle-of-incidence ellipsometry and environmental SEM', *Macromolecules*, **28**, 2673–82.

Kifetew, G. and Sandberg, D. (2000), 'Material damage due to electron beam during testing in the environmental scanning electron microscope (ESEM)', *Wood Fiber Sci*, **32**, 44–51.

Kitching, S. and Donald, A. M. (1998), 'Beam damage of polypropylene in the environmental scanning electron microscope: an FTIR study', *J Micros*, **190**, 357–65.

Krueger, R. (1999), 'Dual column (FIB-SEM) wafer applications', *Micron*, **30**, 221–6.

Lavoie, D. M., Little, B. J., Ray, R. I., Bennett, R. H., Lambert, M. W., Asper, V. and Baerwald, R. J. (1995), 'Environmental scanning electron microscopy of marine aggregates', *J Micros*, **178**, 101–6.

Mansfield, J. F. (2000), 'X-ray analysis in the ESEM: a challenge or a contradiction', *Mikrochim Acta*, **132**, 137–43.

McKinlay, K. J., Allison, F. J., Scotchford, C. A., Grant, D. M., Oliver, J. M., King, J. R., Wood, J. V. and Brown, P. D. (2004), 'Comparison of ESEM with high vacuum SEM as applied to the assessment of cell morphology', *J Biomed Mater Res A*, **69A**, 359–66.

Meredith, P. and Donald, A. M. (1996), 'Study of "wet" polymer latex systems in environmental scanning electron microscopy: some imaging considerations', *J Micros*, **181**, 23–35.

Meredith, P., Donald, A. M. and Luke, K. (1995), 'Pre-induction and induction hydration of tricalcium silicate: an environmental scanning electron microscopy study', *J Mat Sci*, **39**, 1921–30.

Meredith, P., Donald, A. M. and Payne, R. S. (1996), 'Freeze drying: *in situ* observations using cryoESEM and DSC', *J Pharm Sci*, **85**, 631–7.

Miller, A. F. and Cooper, S. J. (2002), '*In situ* imaging of Langmuir films of nylon 6-6 polymer using ESEM', *Langmuir*, **18**, 1310–17.

Mohammed, A. R., Weston, N., Coombes, A. G. A., Fitzgerald, M. and Perrie, Y. (2004), 'Liposome formation of poorly water soluble drugs: optimisation of drug loading and ESEM analysis of stability', *Int J Pharm*, **285**, 23–34.

Möschwitzer, J. and Muller, R. H. (2006), 'Spray coated pellets as carrier system for

muchoadhesive drug nanocrystals', *Eur J Pharm Biopharm*, **62**, 282–7.
Muscariello, L., Rosso, F., Marino, G., Giordano, A., Barbarasi, M., Cafiero, G. and Barbarisi, A. (2005), 'A critical overview of ESEM applications in the biological field', *J Cell Phys*, **205**, 328–34.
de la Parra, R. E. (1993), 'A method to detect variations in the wetting properties of microporous polymer membranes', *Micros Res Technique*, **25**, 362–73.
Rizzieri, R., Mahadevan, L., Vaziri, A. and Donald, A. M. (2006), 'Superficial wrinkles in stretched, drying gelatin films', *Langmuir*, **22**, 3622–6.
Roman-Gutierrez, A. D., Guilbert, S. and Cuq, B. (2002), 'Description of microstructural changes in wheat flour and flour components during hydration by using environmental scanning electron microscopy', *Lebensm.-Wiss. u.-Technol*, **35**, 730–40.
Royall, C. P. and Donald, A. M. (2002), 'Optimisation of environmental SEM for observation of drying of matt water-based lacquers', *Scanning*, **24**, 305–13.
Royall, C. P., Thiel, B. L. and Donald, A. M. (2001), 'Radiation damage of water in ESEM', *J Micros*, **204**, 185–95.
Schmid, B., Aas, N. and Odegard, G. (2001), 'High-temperature oxidation of nickel and chromium studies with an *in-situ* environmental scanning electron microscope', *Scanning*, **23**, 255–66.
Stelmashenko, N. A., Craven, J., Donald, A. M., Thiel, B. L. and Terentjev, E. M. (2001), 'Topographic contrast of partially wetting water droplets in environmental scanning microscopy', *J Micros*, **204**, 172–83.
Stokes, D. J. (2003), 'Recent advances in electron imaging, image interpretation and applications: environmental scanning electron microscopy', *Phil Trans Roy Soc A*, **361**, 2771–87.
Stokes, D. J. and Donald, A. M. (2000), '*In situ* mechanical testing of hydrated breadcrumbs in the ESEM', *J Mat Sci*, **35**, 599–607.
Stokes, D. J., Thiel, B. L. and Donald, A. M. (1998), 'Direct observation of water-oil emulsion systems in the liquid stated by environmental SEM', *Langmuir*, **14**, 4402–8.
Stokes, D. J., Rea, S. M., Best, S. and Bonfield, W. (2003), 'Electron microscopy of mammalian cells in the absence of fixing, freezing, dehydration or specimen coating', *Scanning*, **25**, 181–4.
Stokes, D. J., Morrissey, F. and Lich, B. H. (2006), 'A new approach to studying biological and soft materials using focused ion beam scanning electron microscopy (FIB SEM)', *J Phys: Conference Series*, **26**, 50–53.
Tai, S. and Tang, X. (2001), 'Manipulating biological samples for environmental scanning electron microscopy observation', *Scanning*, **23**, 267–72.
Thibert, R. and Hancock, B. C. (1996), 'Direct visualization of superdisintegrant hydration using environmental scanning electron microscopy', *J Pharmaceutical Sci*, **85**, 1255–8.
Thiel, B. L. and Donald, A. M. (1998), '*In situ* mechanical testing of fully hydrated carrots (*Daucus carota*) in the ESEM', *Ann Botany*, **82**, 727–33.
Thiel, B. L. and Toth, M. (2005), 'Secondary electron contrast in low-vacuum/environmental scanning electron microscopy of dielectrics', *J Appl Phys*, **97**, 051101.
Turkulin, H., Holzer, L., Richter, K. and Sell, J. (2005), 'Application of the ESEM technique in wood research: Part I. Optimization of imaging parameters and working conditions', *Wood Fiber Sci*, **37**, 552–64.
Uwins, P. J. R., Millar, G. J. and Nelson, M. L. (1997), 'Dynamic imaging of structural

changes in silver catalysts by environmental scanning electron microscopy', *Micros Res Technique*, **36**, 382–9.

Vazquez, B., Ginebra, M. P., Gil, X., Planell, J. A. and San Roman, J. (2005), 'Acrylic bone cements modified with beta-TCP particles encapsulated with PEG', *Biomaterials*, **26**, 4309–16.

Wei, Q. F. and Wang, X. Q. (2005), 'Dynamic characterisation of carboxymethyl cellulosic nonwoven material in the ESEM', *Mat Char*, **55**, 148–52.

Wei, Q. F., Mather, R. R., Fotheringham, A. F. and Yang, R. D. (2003), 'Dynamic wetting of fibres observed in an ESEM', *Text Res J*, **73**, 557–61.

Wiedmann-al-Ahmad, M., Gutwald, R., Gellrich, N.-C., Hübner, U. and Schmelzeisen, R. (2005), 'Search for ideal biomaterials to cultivate human osteoblast-like cells for reconstructive surgery', *J Mat Sci: Mat Med*, **16**, 57–66.

Williams, S. J., Donald, A. M., Thiel, B. L. and Morrison, D. E. (2005), 'Imaging of semi-conducting polymer blend systems using ESEM and ESTEM', *Scanning*, **27**, 190–8.

11 Characterisation using imaging techniques

S I ANDERSON, University of Nottingham, UK

11.1 Introduction

The field of tissue engineering delivers a distinct set of samples to the microscopist. These comprise both biomaterial samples (polymers, metals, ceramics) in a two- or three-dimensional structure, with or without the addition of biological material (cells and extracellular matrix components) around and within the material scaffold. This composite structure is challenging as, although microscopy is used as a tool to examine both biological and material samples independently, the typical protocols utilised are generally different. For example the chemicals used to process tissue for histology or electron microscopy may dissolve some polymers or the force required to section engineered scaffolds may shred the delicate cell–matrix layers attached to their surfaces. It is therefore necessary to adapt techniques and alter experimental design to address the unique challenge these samples pose. This chapter is principally concerned with a practical approach to microscopy of biomaterial samples and tissue engineering constructs. A key issue, crucial to the efficiency and success of a cell experiment in tissue engineering is experimental design and the general considerations here will be described so that the appropriate technique can be chosen on the basis of what is optimal for the samples and also what is available in the imaging laboratory. Other chapters in this book are concerned with histology and both transmission and scanning electron microscopy so they will only be briefly dealt with here. A core topic will be the use of confocal laser scanning microscopy (CLSM) to image cells in 3D scaffolds and the different options available within this technique. This section will also contain a guide to live cell imaging using vital dyes and a section on combining CLSM with other techniques to maximise the amount of information gained from a given experiment.

11.2 General considerations and experimental design

In the most common experiments where cultured cells are to be imaged microscopically, cells are seeded on glass coverslips or Thermanox plastic coverslips (Nunc) in a well plate format. This allows easy monitoring of the cells' growth, allowing the experimenter to visually confirm the state of health and the stage of confluence/maturity of the cells prior to fixation, staining and viewing. Biomaterial surfaces do not conform to this ideal. They are often opaque and can have rough or even irregular surfaces, they may be different sizes from each other and they may be porous. These factors all cause problems both for cell seeding and in monitoring the cell's progress.

It is therefore essential to design an experiment that will compensate for these factors. Extra samples may be required per experiment because it is not possible to monitor cells cultured on opaque surfaces (for level of confluence/maturity, etc.) as for a straightforward cell culture. This makes it difficult to visually establish the timescale involved compared with control cultures, so additional samples allow for the establishment of appropriate experimental time-points. Owing to the non-standard nature of many of the samples more replicates may be required to reduce the effect of different sizes and potentially seeding densities. It will not, with opaque samples, be possible to evaluate initial cell growth on the sample visually, so a 24 h time point is advisable to check that cells are growing as expected over all the surface and to confirm that any subsequent differences in staining/labelling are not due to original seeding problems.

It is important to consider the format of the experiment when designing the individual surfaces for microscopy studies. If the materials are cast so that they fit into a standard well plate format, for example, the subsequent seeding of the biomaterial surface will be much easier and the experimental data more reliable. Materials can be cast in syringes and cut to a workable thickness using a blade/microtome so that they almost fill the space. Many polymer and ceramic samples float in tissue culture medium and a suitable means of weighing them down is necessary such as sterile plastic/rubber 'O' rings. The samples should be of a similar size, thickness and surface area where possible (unless this is what is under investigation) and should fit neatly into the culture space.

11.2.1 Cell seeding

If material samples are ill fitting in the culture space, many of the cells may adhere to the base of the tissue culture plate in preference to the scaffold/surface. It is important to ensure that cells adhere to the material rather that the surrounding tissue culture plastic, which is optimised for their attachment. In formats where there is space around the sample this should be monitored by phase contrast microscopy in the tissue culture lab to ensure that a large amount of cells have not adhered to the base of the well in preference to the sample. Alternatively, a thin layer of agarose gel can be used to fill in the space around the sample, as cells do not easily grow on this substrate. To avoid colonisation on any remaining, uncovered base of the well cells can be seeded in a small volume of medium on the sample surface for 90 min, to give the cells time to attach to the surface, before carefully adding the remaining amount of medium. It is important to ensure that materials are of a regular thickness so that all cell colonies are exposed to an equal volume of culture medium. If enough space does not remain in the well, after addition of the samples, to add a sufficient volume of medium, cells will suffer nutrient deficiency as the colony expands. The seeding density may need to be adjusted to account for increases in surface area due to grooves/roughness, etc. In any case it is important to seed some cells onto Thermanox/glass as a positive control to monitor the status of the cells and to act as an indication of when cell confluence or differentiation may occur. It may be beneficial to run two sets of identical control samples to ensure that there is no significant difference between them as the experiment progresses.

For 3D scaffolds, consisting of a porous or fibrous network, cell seeding poses different problems as cells may attach and grow preferentially on the outer surface, forming a crust around the scaffold and preventing cell growth into the scaffold. This in turn can prevent adequate nutrient diffusion into the scaffold for those cells that manage to colonise the internal area. This area has received particular emphasis in recent years (Wu *et al.*, 1999; Almarza and Athanasiou, 2004; Chen *et al.*, 2004; Griffon *et al.*, 2005; Hu and Athanasiou, 2006) and new developments in scaffold design, such as the introduction of a macroporous channel in the centre of a scaffold to facilitate cell ingrowth (Rose *et al.*, 2004) and in the use of bioreactors have helped to alleviate the problem and must be given due consideration when planning an experiment using 3D scaffolds.

It may be that samples are very expensive to produce or that the yield is low. In this case it is important to design the experiment so that a definite result is obtained. This is often best achieved by limiting the number of variables (e.g. time-points) rather than using fewer replicates per variable and ending up with poor results. Imaging will be much more reliable and it will be easier to take representative and accurate data as a result of good experimental design.

11.3 CLSM

Imaging cell growth and matrix development in a tissue-engineering environment often utilises fluorescence techniques due to the opaque nature of ceramic and polymer samples. Fluorescently labelled cells can be imaged on an opaque surface where conventional transmitted light microscopy would not be suitable. Fluorescence microscopy is useful for many tissue-engineering applications but has limitations. For example, where many cell layers are present the resolution of the image will suffer due to the contribution of out-of-focus information to the image ('out-of-focus blur'). Where scaffold or roughened surfaces are used, cells growing in deeper grooves or pores will not be easily imaged by higher magnification lenses as they will be out of range for the working distance of the lens. Additionally many ceramics and polymers are auto-fluorescent, making conventional fluorescence microscopy difficult. In these cases, and where 3D information is required, confocal microscopy has become a routine technique for cellular imaging.

11.3.1 Principle

Confocal laser scanning microscopy is an advanced method of studying fluorescently labelled samples. The principle of CLSM is relatively straightforward. Excitation of the fluorescent label is achieved by means of a laser specific to the excitation maxima of the dye. This has the advantage of maximising the signal to background ratio. In addition, the inclusion of confocally aligned pinholes in the optical path means that only information for the plane of focus of the microscope is collected (Fig. 11.1). Information from above and below the plane of focus is excluded, which eliminates the 'out of focus blur' seen in fluorescence microscopes and provides an increased resolution over conventional fluorescence microscopy. The image is collected by point scanning the area and collected using a photomultiplier tube so the image is easily quantifiable. By altering the plane of focus stepwise, a series of images, all in focus, can be collected throughout the depth of the tissue. This is called optical sectioning and allows images to be collected at several depths within the tissue (z series). By collecting optical sections at the correct intervals (see later) it is possible to create a 3D reconstruction of the entire image series. These systems have a powerful optical zoom, allowing much better resolution and magnification over the conventional light microscope. Simultaneously scanning multiple fluorescent labels at high magnification allows confirmation of the presence and localisation of the stained molecule to a particular organelle/membrane, etc. or to co-localise two labelled molecules. For detailed reading on CLSM see Pawley (2006) and Robinson (2001). For the tissue engineer, the extent of colonisation of a surface, the viability of the cells, the production of extracellular matrix and the presence of phenotypic markers can be examined and compared using CLSM. An additional advantage for the biomaterials scientist/tissue engineer is that the problem of auto-fluorescence, which troubles many polymers/ceramics, can be almost eliminated. On a flat auto-fluorescent material a z series is collected of the cell layers only and the auto-fluorescent surface is excluded from the z scan. In a 3D scaffold, if the intensity of staining is optimal, it can be used as an advantage, allowing the structure of the material to be visualised, yet not strong enough to overpower the staining.

It is important to note that, because CLSM is fluorescence based, only what is stained can be seen. This means the experimental outcomes need careful consideration so the number of dyes required to visualise all the elements needed is not beyond what is practical. For tissue engineers the depth of focus of the lenses and the non-transparency of scaffolds means that 3D data from only a small depth of the surface of the scaffold can be collected and cells within deeper pores will fall beyond the limits of the microscope. The depth of information that can be

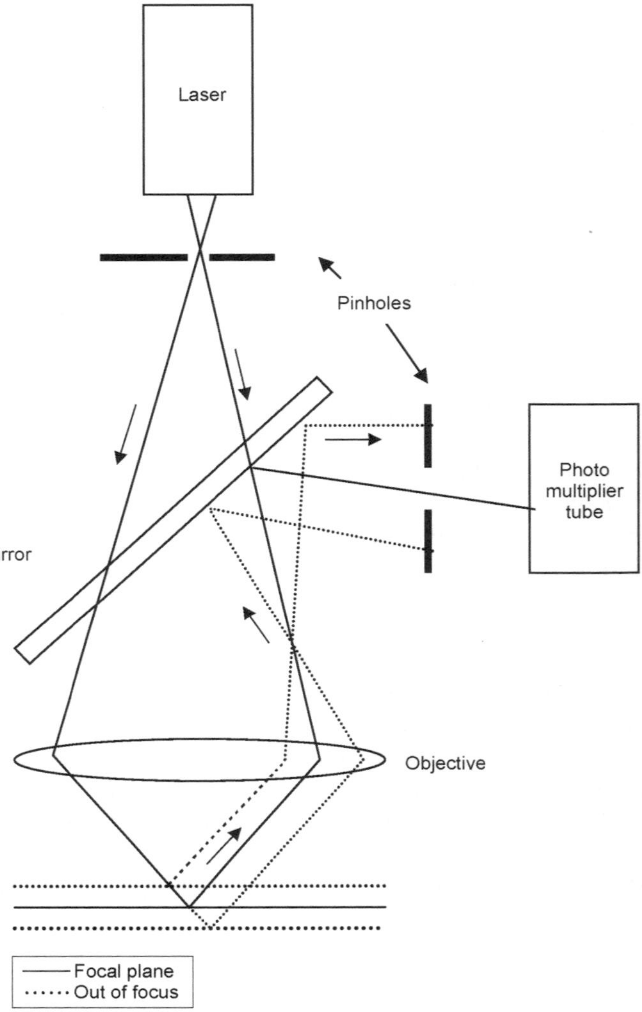

11.1 The optical pathway in a CLSM showing how information from only the plane of focus is collected. Information from above and below the plane of focus is excluded from the final image.

collected depends not only on optical properties but also on the transparency of the sample (e.g. 3D information can be collected from sub-surface pores in some gel/transparent polymer structures) and the density of the network (a fibrous polymer network will allow information to be collected from deeper than a dense porous ceramic simply because of the interconnectivity of the structure). It is, however, a very useful tool in tissue engineering and provided the experimental design and approach are tailored to utilise the strengths of the system, can be used to give hitherto inaccessible information about the status of cells on, or in, materials. CLSM is the method of choice to study what cells are doing – 'are they alive or dead?', 'are they producing the right growth factors or matrix?', 'are they maturing?', etc. There are a number of practical considerations, which must be taken into account when planning a CLSM experiment, using tissue-engineered constructs and these will be dealt with in the following section.

11.3.2 Setting up an experiment

The way in which an experiment is planned and executed will depend on several factors such as the type of equipment available, the type and shape of material under investigation or the availability of fluorescent labels. Each experiment will need a positive control (this is used to validate the staining technique and is usually a frozen tissue section where the antigen of interest is definitely present) and a negative control (to ensure absence of non-specific staining by the secondary antibody, usually the same tissue or cell sample where the primary antibody is replaced by buffer) as well as a control to illustrate how the cells would perform on a straight tissue culture comparison. It is not possible to recover cells growing in a well plate for fluorescence labelling without the use of cell scraping or chemical means such as trypsin so control samples are most often grown on glass or tissue culture plastic coverslips. Thermanox plastic coverslips (Nunc) are available in a range of diameters suitable for the well plate format. Care needs to be taken if glass is used as a control sample as many cell types do not behave in the same way on glass as tissue culture plastic, which is an optimised surface for cell growth and often outperforms the biomaterial surface under investigation! For ease of use negative controls for the labelling procedure (i.e. where the primary antibody is excluded from the labelling procedure) can be carried out on cells growing on the flat surfaces of glass or Thermanox. A frozen section of an appropriate tissue is most often used as a positive control so that the efficacy of the label and the staining method can be checked in the natural *in vivo* situation (e.g. skin dermis for collagen 1). Thermanox/glass coverslips are usually treated in the same way as a tissue section for subsequent labelling and placed cell side up on a slide under a coverslip using anti-fade mounting medium. Photobleaching results from reaction of the excited-dye molecule with oxygen and the anti-fade component of a mountant mops up reactive oxygen species. The refractive index of the mountant should be as close as possible to that of biological tissue (1.5). Commonly used anti-fade mountants include Vectashield (which contains *p*-phenylenediamine) and DABCO (1,4-diazabicyclo[2.2.2]octane; refractive index 1.4). DABCO can be made in the laboratory by adding 2% DABCO to PBS and adding 1 part of this to 9 parts glycerol. For optimal image quality the thickness of coverslip used must be matched to the lens. The required coverslip thickness is engraved on each lens (e.g. 0.17 = 0.17 μm), which corresponds to a no. 11/2 coverslip. It is inadvisable to invert the glass/Thermanox coverslip onto the slide as the underside of the Thermanox coverslip is not clean, having been incubated in tissue culture medium and undergone staining procedures. In addition Thermanox is auto-fluorescent and much thicker than a glass coverslip and therefore does not have appropriate optical properties. Use of small round coverslips also carries the risk that the nail varnish often used to seal the sample obscures much of the available viewing area. Hard setting mountants are available, such as Vectashield Hard Set (www.vectorlabs.com), and may be more appropriate for 2D biomaterial surfaces. Note that shrinkage occurs, with the potential to cause tissue damage, though, in practical terms, this may not be any more than is caused by fixation and staining of the sample.

11.3.3 Upright versus inverted microscopy

Initially the upright microscope may appear to be the best choice. Most biomaterials are opaque and it may be assumed that cells growing on the surface will need to be imaged from above. The samples are mounted on a glass slide with the cells uppermost and a coverslip added with anti-fade mounting medium used to minimise photobleaching. The samples are adhered to the slides by means of a double-sided sticky tab (carbon tabs used to mount SEM samples are ideal, from Agar Scientific/TAAB). It is not possible to seal these samples so they may not store well and it is advisable to image them straight after making them unless a hard setting mounting medium is used. Placing the slides in a suitable carrier and wrapping loosely in tin foil allows short-term storage at 4 °C.

As the samples are imaged from above, the uneven nature of many surfaces can cause artefacts. In an effort to bring deeper objects into focus, to visualise the bottom of pores or to push at a fibrous network, it is possible to exert pressure on the coverslip and hence compress the sample. In this case false information about the depth, size and shape of stained cells and surfaces will be recorded (see Fig. 11.2). In a system designed to give 3D information about a sample, awareness of the potential of these artefacts is essential. Additionally, it is possible to inadvertently drag the coverslip across the surface of the sample, while locating a suitable area for imaging, and this will disturb the cells. Often the best way to prevent these potential pitfalls is to use an inverted system.

Looking at samples mounted in the conventional manner in an inverted system can be difficult: the sample must be inverted, risking losing the coverslip and distorting the cells by putting pressure on them from above and below. This can best be avoided by placing the sample/scaffold in a glass-bottomed culture dish, where a coverslip is inserted into a hole in the base of the Petri dish. These can be made in the laboratory by modifying small Petri dishes or purchased from a range of suppliers (e.g. Iwaki). The base of the dish is removed using a heated cork borer and a glass coverslip, of the appropriate thickness, glued onto the base from beneath using surgical glue. The dishes can be sterilised using UV light. Using

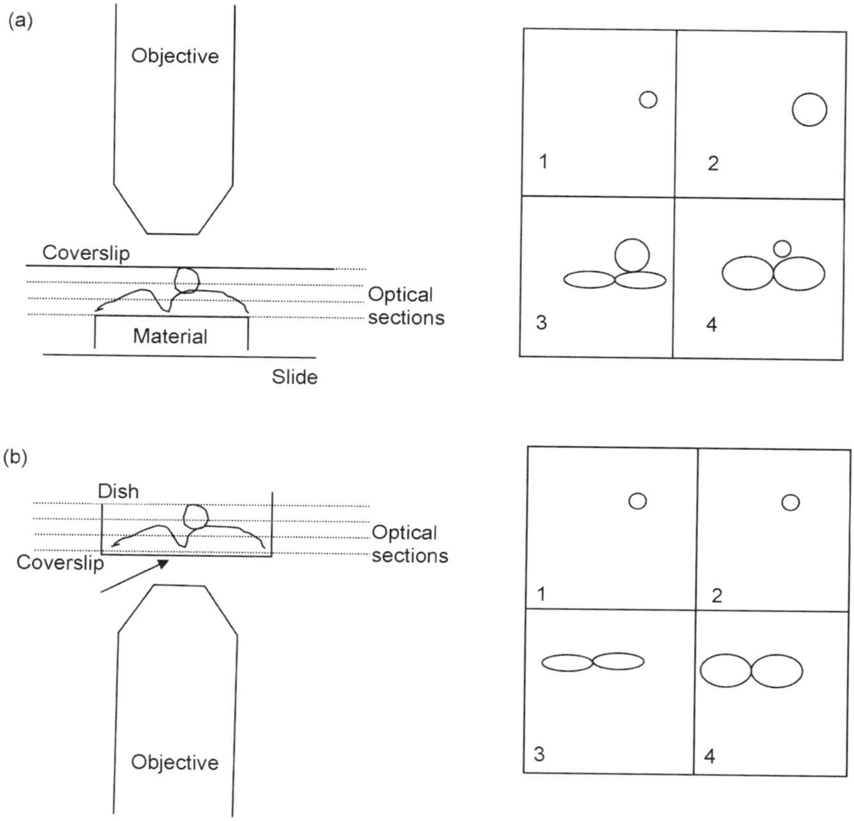

11.2 Depth shape and size artefacts in CLSM due to pressure exerted on the coverslip during imaging. In (a) optical sections from the line drawing on the left are displayed as a gallery of slices on the right. The circular cell is represented in four optical slices, instead of two, meaning that it is being recorded as larger than it really is and the pressure exerted on it has caused it to be represented as oval rather than circular. Analysis of the gallery would also suggest the cell is part of the basal layer of the cells, rather than part of an upper layer. In (b) the use of an inverted system with a glass bottomed petri dish means that the cells are accurately represented.

glass-based dishes has the advantage that the lens cannot exert pressure from the coverslip onto the sample so depth information will be accurate. In addition, live cell imaging is also possible as the cells can remain in tissue culture medium.

11.3.4 Flatness of field and surface roughness of sample

In an upright system the upper and lower surface of any material needs to be as parallel as possible, otherwise depth information will be inaccurate and cell in-growth into a scaffold may be reported when in fact it is simply a sloped surface

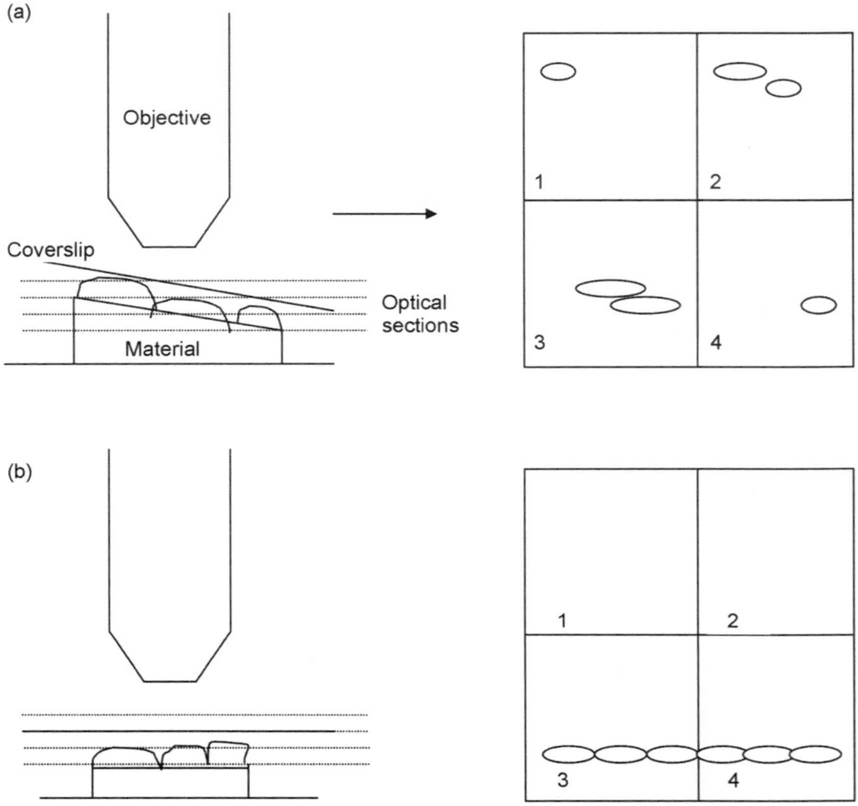

11.3 The effect of slope on data acquired in CLSM z series. Optical sections from the line drawing on the left are displayed as a gallery of slices on the right. Cells from a single cell layer appear in four optical slices due to slope (a) when they should only appear in one or two slices in a correctly presented surface as shown in (b).

that has been presented (Fig. 11.3). The same applies to dishes being placed in an inverted system – if they are not absolutely level the depth information will be inaccurate. Corrections for this, i.e. linear and parabolic flattening tools, are often present in the CLSM software but in the author's experience can overcompensate and so it is best to avoid having to use them where possible. In an inverted system, as long as the culture dish is level, the sample will be presented correctly. However, samples that are porous/roughened will present the microscopist with a new challenge. In the same way as depression slides, or indeed slides with excess mounting medium, cannot be focused except at very low magnifications (due to the concave surface exceeding the working distance of higher magnification objectives), material samples with a surface that is excessively rough, concave or convex will present the same problems. In this case only a certain proportion of the surface of the sample will be within the range of the working distance of the lens. If the sample is too uneven only a small proportion of the entire surface may be brought into focus, reducing the area of the sample that can be examined. This limits the operator in two ways. Firstly, it is not possible to survey the whole sample in order to ensure what is recorded is representative of the surface as a whole and secondly, the element of choice is removed so that the collected image is dictated by what can be focused rather than by what is representative or valid. To minimise this, attention to detail during and after casting (e.g. trimming of excess material from the edges of the sample, ensuring a sample is not excessively concave or convex) will ensure that the roughness of the sample is part of its intrinsic properties rather than bad casting. If the scaffold is produced in such a way that the surface is as level as possible, much more of it will be in contact with the coverslip and optical sectioning will allow deeper areas of the sample to be collected; however, the cells growing in the bases of very deep pores or crevices in the material will not contribute to the final image.

The CLSM can also be used to measure the surface roughness of a sample and this is a useful tool. Optical sections are acquired in reflectance mode (see below) and the Ra value is calculated (Tomovich and Peng, 2005). Samples of standard roughness ($Ra = 3$ and $Ra = 0.5$) have been used to validate the use of CLSM for roughness measurements. The magnification used is critical and one should ideally see 10–20 repeats of the roughness feature (grooves, pits, etc.) in a scanned area, and ensure that the surface is entirely level. Using the standard samples the author has shown that a 16× lens is most accurate for the $Ra = 3$ sample and a 63× lens is most accurate for imaging the $Ra = 0.5$ sample.

11.3.5 Opacity and shape of sample

The opacity of a solid sample presents no problems for imaging. The cells are all on the surface and as long as this is level (as outlined above) information can be collected from many cell layers and cells growing in shallow grooves, pitted or roughened surfaces. In a porous or fibrous scaffold structure the depth of penetration into the sample depends on the opacity of the sample. If it is transparent, the laser can penetrate into pores and spaces below the surface and provide information on cell colonisation, viability, extracellular matrix production and presence of phenotypic markers. Similarly if the sample is a 3D cell culture, such as a spheroid (Thomas *et al.*, 2005, 2006) then it is possible, with good staining, to image depths of up to 300 μm. Often in these cases the penetration of the stain is the limiting factor rather than the size of the sample. Using vital dyes incorporated into living cells (such as DiI or Di8-Annepps for membranes, or 'live/dead' assay for cells, etc.), or matrix (such as tetracycline for bone mineral) or green fluorescent protein (GFP) constructs transfected into cells can overcome the depth of staining issues.

In an opaque 3D open network the depth of penetration will be limited below the surfaces. Occasionally a large surface pore will allow access to a deeper structure but generally these samples need to be cut in order to view cell colonisation/matrix deposition/maturation in the interior. Initially image series can be collected from the upper and lower surfaces as well as the sides by turning the sample in the glass-based dish so that the surface of interest is downwards. Then the sample can be cut into a series of slices (with a sharp scalpel blade) to show cell in growth and activity in several areas. These, in turn, can be imaged on both sides and by using a systematic approach quantifiable information on the colonisation of the construct can be achieved. This use of the 'live/dead' assay is useful to assess cell viability in the centre of the construct. Combining this technique with others allows the colonisation of the scaffold to be mapped, using e.g. microCT or SEM and related to the cell behaviour as recorded from specific areas using CLSM.

11.3.6 Fluorescent labels

Traditionally the means of fluorescently labelling cells has been to primarily label the antigen of interest with an antibody raised in a different species. The cells are then incubated in a species-specific secondary antibody conjugated to a fluorescent probe. The fluorescence is excited using a laser matched to the maximum excitation wavelength of the dye. The choice of label is dependent on the individual experiment, but many cellular organelles, membranes and cyto-skeletal elements are now available, already conjugated to a range of different fluorochromes. These are summarised in the molecular probes handbook (http://probes.invitrogen.com) with detailed technical information, protocols and references for fluorescence labelling and a valuable resource to help in the selection of appropriate probes. In this way labelling of a cell with three or four dyes, to identify the structure and localisation of many cellular components, can be readily achieved. For a typical example in the biomaterials field see Recknor *et al.* (2006).

The increasing range of lasers matched to the increased number of fluorescent proteins available have resulted in a huge increase in the number of wavelength dyes available so that the microscopist is no longer dependent on the traditional blue/green/red fluorescence (for example cyan fluorescent protein (CFP) and yellow fluorescent protein (YFP)). This has meant that, apart from the most common GFP, other fluorescent proteins now span the entire visible spectrum, greatly increasing the opportunities for multiple labelling.

11.3.7 Reflectance microscopy

Reflectance is a way of confocal imaging that uses reflected light instead of fluorescence. The laser interrogates the sample in the usual way and light reflected from the sample is collected. In this way unlabelled specimens can be imaged. This is an extremely useful tool for the tissue engineer and depends on the nature of the sample. Highly reflective materials such as metals are shown in detail (Fig. 11.4a). Even if the material is not highly reflective (as in the case of some polymers) they still may be imaged if there is a significant fluctuation of refractive index at certain boundaries, for example in a polymer fibre network structure (Fig. 11.4b). Collagen fibres have also been imaged using this method (Paddock, 2002). Generation of simple images of cells growing in a scaffold is relatively straightforward. For example, one can stain cells with 'cell tracker' or the 'live/dead' stain or a nuclear label nuclei and use reflectance to image the scaffold for an informative result without any difficulty or expense. The DAB (diaminobenzidine) reaction, often used in histology can be detected using reflectance microscopy, further enlarging the number of labels that can be imaged simultaneously (Robinson and Batten, 1989).

11.3.8 Number of optical sections, 3D reconstruction and localisation

For CLSM imaging of tissue engineered and biomaterial samples, the casting of the samples, seeding of the cells and presentation of the samples to the microscope are key to getting a good set of data. The actual process of imaging the samples is routine and straightforward. However, it is useful to review some essential information on the selection of the correct number of optical sections. If the number of sections is selected accurately then a 3D reconstruction and animation can be prepared, either using the confocal software itself or one of the other available programs such as Imaris (Bitplane), Volocity (Improvision) or Metamorph (Molecular Devices). This can provide a useful means of interpreting data and make it easier to present such data in a meaningful way. In addition, having gathered the data in an *XY* plane, it is often useful to view it in an *XZ* and *YZ* direction. Figure 11.5 shows a z series of a blood vessel where the interpretation is easier using with *XZ* and *YZ* views. The interaction of these vessels with neuronal cell processes was easily monitored using this view. The resolution of these views will entirely depend on having collected enough depth information in the z series. Additionally, it is crucial to collect the correct number of optical sections, especially if very small molecules are being imaged (e.g. synaptic vesicles) or if imaging labels that are likely to be in a single plane in the cell (e.g. focal adhesions). If there is not a sufficient degree of over-sampling then information will be missed from the series. It is not necessary, however, to exhaustively gather the correct number of sections for applications where 3D reconstructions are not required. In a sample where a large depth of larger objects (e.g. cells/nuclei) is to be scanned and if these results are to be displayed as a gallery of images from the top to the bottom of several cell layers, for example, much time will be saved by spacing the z slices by several micrometres.

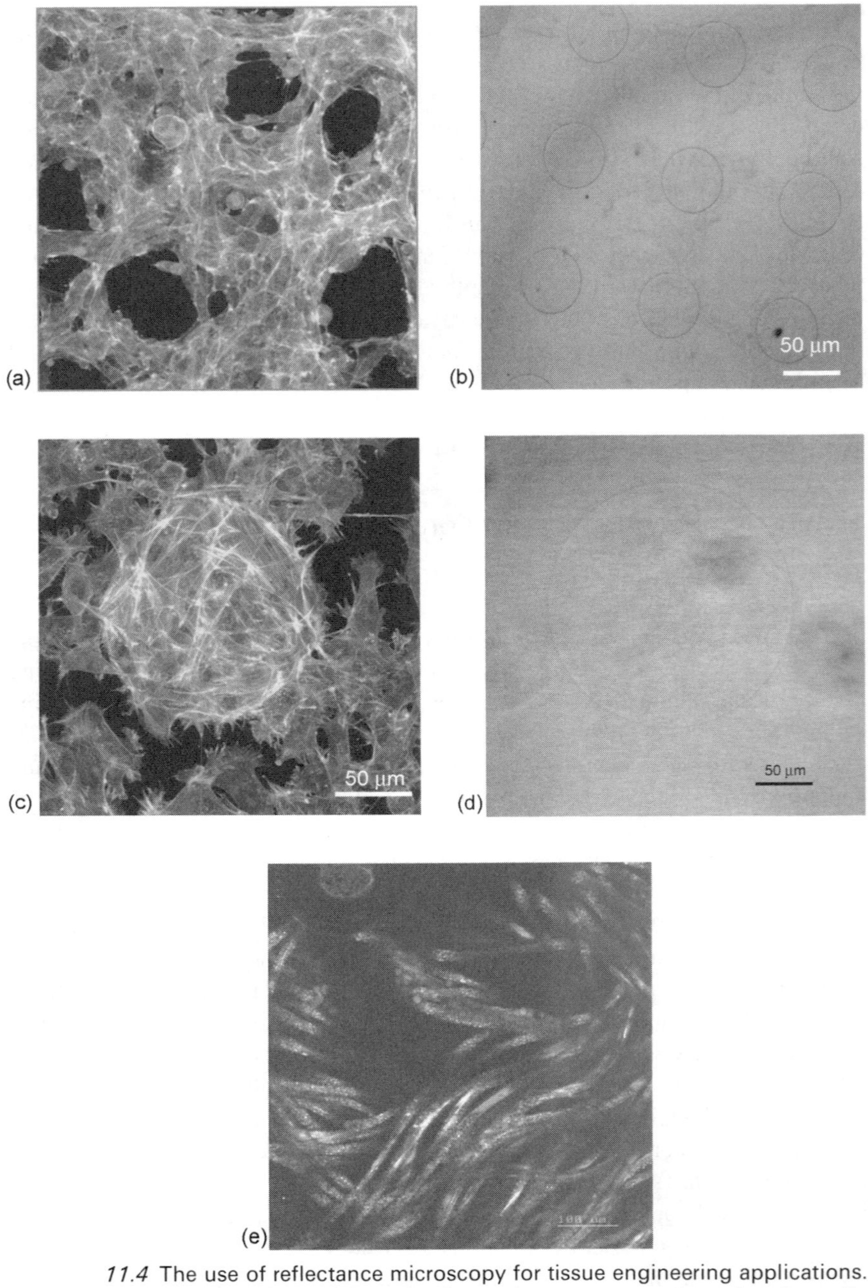

11.4 The use of reflectance microscopy for tissue engineering applications. CLSM images of osteoblasts (a, c) growing on patterned metal surfaces (b, d). The underlying pattern, in this case dots of 50 (b) and 150 (d) μm diameter, can be visualised using reflectance microscopy. In (e), reflectance has been used to visualise a fibrous polymer network.

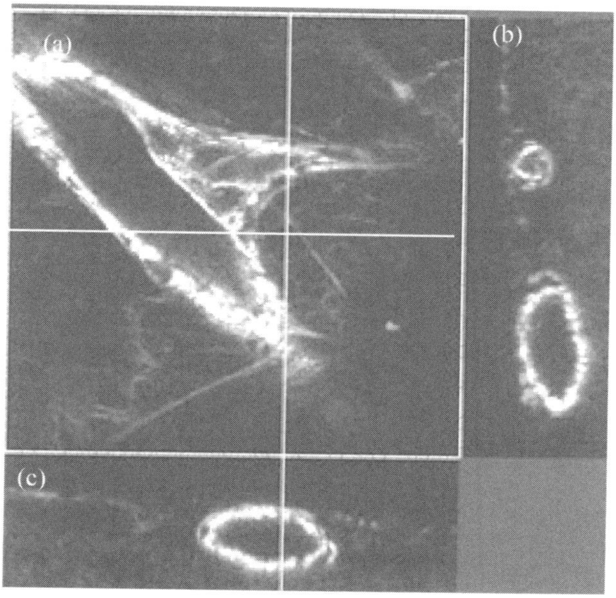

11.5 Visualisation of 3D objects in the CLSM using *XY* and *XZ* view. An optical slice of fluorescently labelled blood vessels is shown in (a). Three-dimensional information from the image stack is represented by *YZ* (b) and *XZ* (c) views.

The correct number of optical sections (z step size) is calculated according to the Nyquist sampling theory, which describes the sampling frequency required to represent the true identity of the sample. As a rule of thumb the z step between sections should be half the objective resolution, which is calculated by:

$$\lambda/2(\text{NA})$$

where λ = wavelength of the laser and NA = the numerical aperture of the lens which is engraved onto the lens casing. So, using the 488 nm laser line, for a 63× objective with a numerical aperture of 1.4 the resolution is 174 nm and the z step should be 87 nm. In practice the minimum step size is often 100 nm. For a 10× objective with a NA of 0.4, the resolution is 610 nm and the z step should be 305 nm.

11.3.9 Live cell imaging

For the tissue engineer some useful information can be gleaned using relatively simple live cell imaging and in this way CLSM provides an entirely different but complementary set of information from SEM. Cell viability studies are perhaps the most straightforward of these. The basis of the propidium iodide stain is that it can only enter cells with damaged membranes and therefore it will label only dead (or fixed) cells (red). PI can be combined with calcein, which is actively taken up by metabolising cells, and in this way the living cells are also labelled (green), giving a quantitative analysis of viability over the whole culture. For tissue engineering this is especially valuable as it gives viability information that is specific to the location within a sample, e.g. in a large tissue-engineered sample are the cells in the middle doing worse than those near the surface? Over time are the sells seeded into a scaffold alive and thriving or just slowly dying off? Live cell imaging can also be carried out on 3D culture systems such as hepatic spheroids (Thomas *et al.*, 2006).

Cell tracker (Invitrogen) is another useful simple method of fluorescently labelling live cells and has the advantage that it persists in daughter cells after mitosis, so that the introduction of a few labelled cells in a culture can be tracked. In this way the progress of labelled cells into a scaffold could be monitored.

The structure of cellular components can also be visualised in live cells using a range of vital markers for subcellular components, a selection of which is shown in Table 11.1 (see http://probes.invitrogen.com for extensive listings and protocols) but the distinct advantage of live cell imaging is the ability to also monitor cellular function.

The use of ratiometric dyes to measure ionic flux (such as calcium indicators (e.g. Fura-2) or pH indicators (e.g. SNARF)) facilitate more complex investigations on cellular function. These probes bind available ions in a ratiometric manner so that the wavelength or fluorescent intensity is increased or decreased in a quantifiable manner. Lohr (2003) describes a method for using Fura Red for CLSM imaging of calcium signalling in neurons. Other cellular functions, such as measuring oxidation can also be performed, e.g. hydrogen peroxide activity can be measured using dichlorofluorescein diacetate (Carter *et al.*, 1994).

Table 11.1 Table of vital dyes commonly used for live cell imaging using CLSM and fluorescence applications

Organelle	Vital dye
Nucleus	Hoechst 33342
	SYTO 13
Mitochondria	Rhodamine 123
	Mitotracker
Endoplasmic reticulum	ER Tracker
	$DiOC_6$
Golgi apparatus	BODIPY FL C5 Ceramide
	C6-NBD-Ceramide
Lysosomes	Lysotracker
	Lysosensor (for pH)
Plasma membrane	Di-I
	Wheat germ agglutinin
Intracellular membrane	Cell trace BODIPY TR methyl ester
Cytoskeletal actin	NBD Phalloidin
Microtubules	Oregon Green paclitaxel

GFP is a product of the naturally chemiluminescent jellyfish *Aequoria Victoria* and is used primarily as a marker for specific molecular sequences. GFP can be retrovirally expressed in specific colonies of cells and these are studied under various conditions (such as cell division, cell death, or in mutant versus wild-type strains of the same cell type), often using time lapse imaging (Jessel *et al.*, 2006), but can also be fixed and subsequently labelled with a number of other fluorescent markers. In the biomaterials field GFP has been used to study adult hippocampal progenitor cell cultures on micropatterned polymer surfaces (Recknor *et al.*, 2006). GFP can also be used with other vital dyes in order to multiple label live cells. Many other fluorescent proteins have been developed in recent years, allowing the possibility of multiple labelling of specific moieties in live cells or of monitoring the interaction of two cell types.

Dynamic information from live cell applications can be gained by using 'fluorescence resonance energy transfer' (FRET), 'fluorescence recovery after photobleaching' (FRAP) and 'total internal reflectance microscopy' (TIRF). A full review of these techniques is beyond the scope of this text but a basic description will be given here. FRET allows detection of the energy transfer between two molecules which are 10–100 Å apart. One molecule acts as the donor and one the acceptor and each is labelled with fluorescent tags where the emission spectrum of the donor must significantly overlap the excitation spectrum of the acceptor. By exciting the donor molecule and detecting the fluorescence emission from the acceptor, it can be confirmed that FRET is occurring. This gives information on the proximity and communication between the two molecules. FRAP is a means of measuring transport and diffusion gradients within a cell. The cell is labelled with a fluorescent tag and a small region of interest is scanned with a laser on maximum power until the fluorescence in the region of interest has been completely bleached out. A time-lapse series is then recorded and the rate of recovery of the fluorescence is measured. Perhaps the most useful technique currently for the tissue engineer is TIRF. This requires a separate set of optics but gives high-resolution imaging to facilitate the study of interfaces such as cell–substrate or cell material interactions (Burmeister *et al.*, 1998).

11.4 Combining techniques

The nature of tissue-engineered samples, which are a composite of biological and material moieties often in a complex arrangement, means that no one technique is likely to provide all the information required from an experiment. Often experiments are run in parallel for biochemical assay to provide quantitative information on the function of a cell population. Combining imaging techniques provides a way of extracting further information from the samples, which may be in limited supply or expensive to produce. Additionally, imaging has the added advantage of investigating cells *in situ*, i.e. they are not removed from the scaffold and remain intact. In this way the nuances of a cell culture can be identified. For example, in a typical osteoblast culture cells form nodules as they mature and this is the site of mineral formation. It is therefore possible, using imaging techniques, to characterise the distinction between nodules versus non-nodular areas within a single culture.

Simply combining fluorescence and reflection techniques within CLSM provides additional useful information on the location of cells within a scaffold as previously shown. Another useful combination is live and fixed cell imaging. Having imaged cells growing in a scaffold using a vital dye it is then possible to fix the cells and stain for conventional dyes. In our experience, this has proved to be a very useful tool and has we have answered questions arising during live cell imaging by fixing and labelling while the dish was still in the microscope, allowing visualisation of the same cell in both live and fixed states. Figure 11.6 shows cultured human osteoblasts labelled with the vital membrane dipole dye Di-8-Anepps (Invitrogen). This was used to assess the potential interaction of a probe with osteoblasts by serial additions of the probe to the culture dish and time-lapse confocal imaging of intensity fluctuations in the dye. At a certain concentration there was an influx of the dye into the cell. By carefully removing the medium, washing and adding fixative to the dish and then labelling the nuclei with propidium iodide, it was confirmed that the dye had relocated to the cytoplasm and not the nucleus. It is also possible to label receptors or focal adhesions in this way, which gives an opportunity to further study 'interesting' areas found during live cell experiments.

Scanning electron microscopy has been dealt with at length in another chapter but it is worth reiterating a few facts here. Firstly, the resolving power of electron microscopy with the added bonus of a large depth of field makes SEM a powerful technique for looking at cells on materials. It allows the user to look at cell colonisation deep in surface features in a 3D porous scaffold and can give information on sub-surface cell in growth in a fibrous network (Chua *et al.*, 2005) while maintaining focus. Additionally, energy dispersive X-ray micro-analysis (EDX) or elemental mapping can be used to assist in the identification of elemental composition of composites (e.g. particle size and distribution of a second phase within a polymer). Also, backscattered electron imaging gives great contrast to, for example cells on metal, as the atomic number for the major elements present are far apart in the periodic table. Harris *et al.* (1999) compared methods of imaging cellular cytoskeletal elements using backscattered electrons. These factors make SEM an invaluable tool for imaging cells on materials. It can be difficult in the case of some materials to distinguish cells from the material, especially if the cell is spread out to the extent that is very thin, and when this is the case CLSM is a good option, as it also allows the quantification of cell viability. Fluorescently labelled cells have excellent contrast against unstained materials. This works best when the material itself is auto-fluorescent (in the

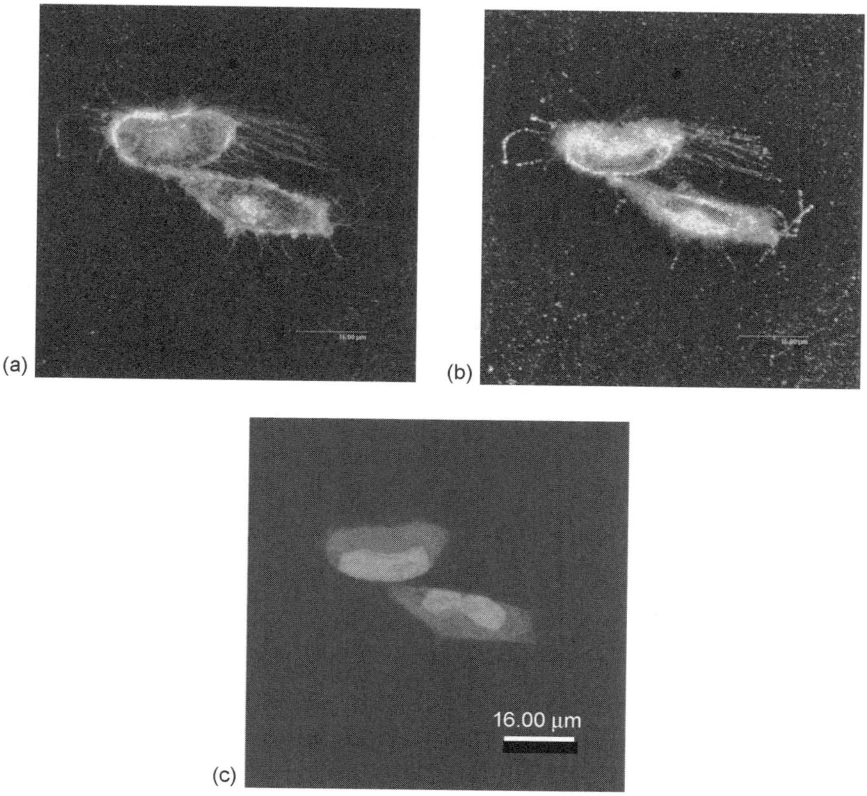

11.6 Combining live and fixed cell imaging. Live human osteoblasts, stained with Di-8-Annepps, were imaged before (a) and after (b) intervention showing the altered localisation of the fluorochrome. By fixing the samples in the dish and staining with a nuclear label (propidium iodide) it was confirmed that the Di-8-Annepps was confined to the cell cytoplasm (c) and not the stained nucleus (*).

case of hydroxyapatite and some polymers), to facilitate viewing or reflective, to facilitate reflectance imaging. The combination of SEM and CLSM augments the amount of information that can be gathered from a single experiment. By labelling specific receptors or molecules, CLSM gives the additional information on cell viability, cell behaviour, expression of phenotypic markers, cell maturation or reaction to changes in topography or surface chemistry. The combination of these techniques allows the characterisation of the material surface, the distribution of cells and the characterisation of cell phenotype and behaviour in a single experiment (Recknor *et al.*, 2006).

Given the limitation of both CLSM and SEM in imaging cells deep within the scaffold, it is perhaps unsurprising that an increasing number of groups are using microCT to assess cell in growth within a 3D porous network (Jones *et al.*, 2004;

Gauthier *et al.*, 2005; Knackstedt *et al.*, 2006). The resolution achievable with microCT (2 μm) is sufficient to give a good indication of colonisation of a scaffold and data processing allows the scaffold and cells to be distinguished. This is becoming an invaluable technique for tissue engineers and consideration should be given to combining this technique with both SEM and CLSM.

11.5 Future trends

Combining techniques has proved to be very useful in the imaging of tissue engineered samples as has already been discussed. A potential future trend for the field also lies in the combination of two techniques, i.e. CLSM and TEM.

TEM is a very powerful technique as it allows the highest achievable resolution in microscopy, achieving magnifications typically up to 300 000× and resolving fine ultrastructural detail. Electromagnetic coils assume the function of lenses, focusing the beam and magnifying the section. The electron beam is propelled through a high vacuum by an accelerating voltage of 60–100 kV and passes through a thin (70–90 nm) section projecting the final image on a phosphorescent screen. The main limiting factors for TEM in tissue engineering are whether the scaffold (especially polymers) is stable throughout a processing schedule that uses alcohols, acetone and propylene oxide and whether the material component can be cut without damage to the knife or biological content. Optimal tissue processing is needed to prevent excessive shrinkage of the tissue from the pores of the scaffold. TEM allows identification of cells, organelles and membranes and can also distinguish cell types and extracellular matrix components (gags, collagen) and also types of intercellular communication (tight/gap junctions/desmosomes). Additionally it is possible to immunolabel targets of interest using tiny gold particles instead of coloured or fluorescent tags, and double labelling is achieved by using gold probes of different sizes. The challenge to the tissue engineer is to develop a processing schedule, which does not destabilise the material while retaining antigenicity for subsequent immunolabelling. One option of dealing with this problem is use of the new generation of nanoprobes (1.4 nm), which allows the labelling of fresh/fixed tissue prior to embedding. This overcomes the problem of the routine TEM processing schedule being incompatible with antigen retention. It is necessary to use silver enhancement techniques (e.g. HQ Silver, Nanoprobes) to build these probes into a visible size by nucleation of silver grains uniformly around the gold particle. A second option is the use of cryo techniques. In this case the scaffold–cell composite would be frozen and sectioned using a cryo-ultramicrotome to produce ultra-thin sections of polymer and cell for subsequent immunolabelling. This, although complex technically, removes the need to subject either the cells or the material to chemical intervention and opens up the possibility to conduct immunocytochemical investigations of intracellular, cell–matrix, cell–cell and cell–material interactions at the ultrastructural level.

Combining TEM and CLSM has been made easier by the emergence of fluoronanogold probes (Nanoprobes http://www.Nanoprobes.com via UK agent Universal Biologicals http://www.universalbiologicals.ltd.uk). It is now possible to label an antigen of interest with a label containing both a fluorescent tag and a 1.4 nm gold probe. This allows a cell to be examined firstly by CLSM and subsequently by TEM to zone in on areas of interest and ultrastructurally localise the target to within a cell type/organelle. The small size of the gold probe allows for good tissue and cell penetration and the presence of a fluorescent tag allows both confirmation of the presence of the target of interest and its broad localisation within the larger structure (Robinson and Vandre, 1997; Robinson *et al.*, 2001; Takizawa and Robinson, 2003).

Immunocytochemical techniques are also possible with SEM. Therefore, not only is it possible to augment the knowledge gained from SEM studies using CLSM, it is also possible to further enhance information gained using CLSM by using immuno-SEM to localise immunostaining of a particular antigen to a distinct subcellular component or membrane. This can be achieved for cell surface and extracellular matrix components and also for internal cellular components using cellular 'unroofing' techniques (Heuser, 2000; Mishra *et al.*, 2004). Filmon *et al.* (2002) has combined backscattered SEM imaging with CLSM and immuno-TEM to study the interaction between osteoblastic cells and polymer hydroxyapatite composite materials. The future for imaging cells in a 3D scaffold offers huge potential both in terms of advances in individual techniques and in combining these techniques to get the maximum amount of information from a given experiment.

11.6 References

Almarza AJ, Athanasiou KA (2004), 'Seeding techniques and scaffolding choice for tissue engineering of the temporomandibular joint disk tissue engineering', *Tissue Engineering*, **10**(11–12), 1787–1795.

Burmeister JS, Olivier LA, Reichert WM, Truskey GA (1998), 'Application of total internal reflection fluorescence microscopy to study cell adhesion to biomaterials', *Biomaterials*, **19**(4–5), 307–325.

Carter WO, Narayanan P, Robinson JP (1994), 'Intracellular superoxide and hydrogen peroxide anion detection in endothelial cells', *Journal of Leukocyte Biology*, **55**, 253–258.

Chen HC, Lee HP, Sung ML, Liao CJ, Hu YC (2004), 'A novel rotating-shaft bioreactor for two-phase cultivation of tissue-engineered cartilage', *Biotechnology Progress*, **20**(6), 1802–1809.

Chua K-N, Lima W-S, Zhanga P, Lua H, Wend J, Ramakrishna S, Leonga KW, Maoa H-Q (2005), 'Stable immobilization of rat hepatocyte spheroids on galactosylated nanofiber scaffold', *Biomaterials*, **26**, 2537–2547.

Filmon R, Basle MF, Atmani H, Chappard D (2002), 'Adherence of osteoblast-like cells on calcospherites developed on a biomaterial combining poly(2-hydroxyethyl) methacrylate and alkaline phosphatase', *Bone*, **30**(1), 152–158.

Gauthier O, Muller R, von Stechow D, Lamy B, Weiss P, Bouler JM, Aguado E, Daculsi G (2005), '*In vivo* bone regeneration with injectable calcium phosphate biomaterial: a three-dimensional micro-computed tomographic, biomechanical and SEM study', *Biomaterials*, **26**(27), 5444–5453.

Griffon DJ, Sedighi MR, Sendemir-Urkmez A, Stewart AA, Jamison R (2005), 'Evaluation of vacuum and dynamic cell seeding of polyglycolic acid chitosan scaffolds for cartilage engineering', *American Journal of Veterinary Research*, **66**, 599–605.

Harris LG, ap Gwynn I, Richards RG (1999), 'Contrast optimisation for backscattered electron imaging of resin embedded cells', *Scanning Microscopy*, **13**(1), 71–81.

Heuser J (2000), 'The production of "cell cortices" for light and electron microscopy', *Traffic*, **1**, 545–552.

Hu JC, Athanasiou KA (2006), 'A self-assembling process in articular cartilage tissue engineering', *Tissue Engineering*, **12**(4), 969–979.

Jessel N, Oulad-Abdelghani M, Meyer F, Lavalle P, Haikel Y, Schaaf P, Voegel J-C (2006), 'Multiple and time-scheduled *in situ* DNA delivery mediated by cyclodextrin embedded in a polyelectrolyte multilayer', *Proceedings of the National Academy of Sciences of the USA*, **103**, 8618–8621.

Jones A, Milthorpe B, Sakellariou A, Limaye A, Arns C, Sheppard AP, Sok RM, Senden T, Knackstedt M (2004) 'Analysis of 3D bone ingrowth into polymer scaffolds via micro-computed tomography', *Biomaterials*, **25**, 4947–4954.

Knackstedt M, Arns C, Senden TJ, Gross K (2006) 'Structure and properties of clinical coralline implants measured via 3D imaging and analysis', *Biomaterials*, **27**(13), 2776–2786.

Lohr C (2003), 'Monitoring neuronal calcium signalling using a new method for ratiometric confocal calcium imaging', *Cell Calcium*, **34**(3), 295–303.

Mishra SK, Hawryluk MJ, Brett TJ, Keyel PA, Dupin AL, Jha A, Heuser JE, Fremont DH, Traub LM (2004), 'Dual engagement regulation of protein interactions with the AP-2 adaptor α appendage', *Journal of Biological Chemistry*, **279**(44), 46191–46203.

Paddock S (2002), 'Confocal reflection microscopy', *BioTechniques*, **32**(2), 274–278.

Pawley, J. (ed.) (2006), *Handbook of Biological Confocal Microscopy*, 3rd edn, Springer-Verlag, New York.

Recknor JB, Donald S. Sakaguchib DS, Surya K, Mallapragadaa SK (2006), 'Directed growth and selective differentiation of neural progenitor cells on micropatterned polymer substrates', *Biomaterials*, **27**, 4098–4108.

Robinson JM, Batten BE (1989), 'Detection of diaminobenzidine reactions using scanning laser confocal reflectance microscopy', *Journal of Histochemistry and Cytochemistry*, **37**(12), 761–765.

Robinson JM, Vandre DD (1997), 'Efficient immunocytochemical labeling of leukocyte microtubules with FluoroNanogold: an important tool for correlative microscopy', *Journal of Histochemistry and Cytochemistry*, **45**(5), 631–642.

Robinson JM, Takizawa T, Pombo A, Cook PR (2001), 'Correlative fluorescence and electron microscopy on ultrathin cryosections: bridging the resolution gap', *Journal of Histochemistry and Cytochemistry*, **49**(7), 803–808.

Robinson JP (2001), 'Principles of confocal microscopy', in *Methods in Cell Biology*, Vol. 63, 89–105, Academic Press, San Diego.

Rose FR, Cyster LA, Grant DM, Scotchford CA, Howdle SM, Shakesheff KM (2004), '*In vitro* assessment of cell penetration into porous hydroxyapatite scaffolds with a central aligned channel', *Biomaterials*, **25**, 5507–5514.

Takizawa T, Robinson JM (2003), 'Correlative microscopy of ultrathin cryosections is a powerful tool for placental research', *Placenta*, **24**(5), 557–565.

Thomas RJ, Bhandari R, Barrett DA, Bennett AJ, Fry JR, Powe D, Thomson BJ, Shakesheff KM (2005), 'The effect of three-dimensional co-culture of hepatocytes and hepatic stellate cells on key hepatocyte functions *in vitro*', *Cells Tissues Organs*, **181**, 67–79.

Thomas RJ, Bennett A, Thomson B, Shakesheff KM (2006), 'Hepatic stellate cells on poly(DL-lactic acid) surfaces control the formation of 3D hepatocyte co-culture aggregates *in vitro*', *European Cells and Materials*, **11**, 16–26.

Tomovich SJ, Peng Z (2005), 'Optimised reflection imaging for surface roughness analysis using confocal laser scanning microscopy and height encoded image processing', *Journal of Physics: Conference Series*, **13**, 426–429.

Wu F, Dunkelman N, Peterson A, Davisson T, De La Torre R, Jain D (1999), 'Bioreactor development for tissue-engineered cartilage', *Annals of the New York Academy of Sciences*, **875**, 405–411.

12
Characterisation using Raman micro-spectroscopy

I NOTINGHER, University of Nottingham, UK

12.1 Introduction

The concept of tissue engineering is to make living cell-based components for failing human tissues and organs that can be supplied on demand for therapeutic purposes. The engineered tissues or organs are built by growing cells on scaffolds which support cell growth in three dimensions and also enhance and stimulate cell migration, proliferation and differentiation in order to obtain specific tissue constructs that mimic the complex structures and physiological behaviour of natural tissues. During this process, non-invasive observation of cells is very important for monitoring their viability, phenotype and interaction with the scaffold material. Additionally, monitoring the properties of the scaffold materials enables the evaluation of bioactivity, biodegradation, surface reactions, etc. Understanding and controlling the parameters on both the biological and material sides is critical in producing reliable cell–scaffold systems.

Recent technological developments have led to a wide range of analytical techniques for studying cells *in vitro*. Many biochemical methods are available to detect biochemical changes in cells that can be related to intracellular processes, phenotype changes, disease or interaction with drugs and toxins. It has now become possible to measure the expression of thousands of genes simultaneously by using high-throughput microarrays. Polymerase chain reaction, immunofluorescence and immunochemistry are common biochemical methods used routinely for characterisation of cells. These methods can provide specific and detailed biochemical information about the cells, and represent useful tools for cell biologists. However, they all share severe limitations when it comes to studying individual living cells or parts of a living cell without inducing cell damage. These methods cannot be applied *in situ* since they are invasive, require cell fixation, staining or lysis. They are also limited with regard to the minimum number of cells that can be analysed, as they all routinely require large numbers of cells. Additionally, they need long sample preparation times, and are expensive in terms of labour and chemicals. Thus, it is not possible to use these techniques to monitor the development of engineered tissues over long periods of time and then use the tissue for a medical application, because the monitoring techniques require the destruction of the tissue.

Optical techniques using light to probe live cells may provide unique advantages as they may not require invasive procedures and can achieve the high-spatial resolution needed for studying single cells. Conventional optical microscopy allows cell biologists to observe living cells but the information obtained (e.g. morphology) is limited and difficult to quantify. In some cases, microscopic imaging based on intrinsic fluorescence of certain cellular components can be successfully used to study cells non-invasively. However, fluorescence experiments have limited application when applied to single living cell studies because the number of fluorescent cellular components is small. The broad bands present in fluorescence spectra limit the amount of information that can be obtained. Techniques based on transfection of cells with plasmids designed to ligate proteins and peptides to fluorescent labels, such as green fluorescent protein, have become very popular in the past two decades. However, genetic manipulation of cells is expensive and labour intensive as selection protocols need to be developed for particular genes. Additionally, genetic modification may also interfere with normal behaviour of cells. Another severe limitation common to all fluorescence imaging methods is the difficulty in quantifying the results due to intensity variations of the fluorescence emission caused by photobleaching as well as variations in staining protocols.

Optical techniques based on vibrational spectroscopy are well suited for non-invasive biochemical analysis of samples at a micrometre scale. In particular, Raman micro-spectroscopy has a clear advantage over infrared spectroscopy due to reduced background signals from the culture medium and intracellular water. Additionally, well-established optical instrumentation developed to work in the visible range of the electromagnetic range, such as optical microscopes, can be attached to Raman spectrometers in order to obtain the micro-scale spatial resolution required to study individual cells. In Raman spectroscopy, both excitation and detection are performed optically without need of cell staining or labelling. The information obtained from Raman spectra represents a fingerprint of molecular vibrations, which have high chemical specificity.

During the past decades, Raman micro-spectroscopy has been extensively used in biology and medicine, from studies of isolated biopolymers to complex tissues (Tu, 1982; Manoharan *et al.*, 1996; Hanlon *et al.*, 2000; Gremlich and Yan, 2001). The potential of Raman micro-spectroscopy arises from its ability to detect biochemical changes in cells at a molecular level, and therefore, can be used for diagnosis, or as a tool for developing therapies, as well as testing and evaluation of drugs.

Apart from studying cellular behaviour, the optimisation of scaffold materials used for supporting, directing and promoting the growth of the tissue *in vitro* is of equal importance. *In vitro* studies are often used to understand the factors that determine the properties of biomaterials in order to improve their properties. There are a large number of techniques that can be used to characterise the surface reactions of bioactive samples immersed in simulated body fluids, including X-ray diffraction, inductively coupled plasma, optical microscopy and scanning electron microscopy (Hench and Wilson, 1993). These techniques can provide information regarding material structure, dissolution kinetics and imaging of material surface with high spatial resolution. Raman spectroscopy has become a common technique for studying chemical composition and surface properties of solid samples. Its main advantages are reduced cost, rapidity, quantitative results and non-destructive non-contact sampling of either large or small areas, as well as enabling *in situ* measurements, especially for samples immersed in aqueous solutions.

Owing to its ability to study both live cells and properties of biomaterials, Raman spectroscopy may be further developed for non-invasive monitoring of the overall development of engineered tissues and follow in real-time the interactions between cells and scaffold materials. It is well known that biomaterials have an active role in the development of tissues *in vitro* as they are designed to influence cell adhesion, differentiation and proliferation, and modulate gene expression (Hench and Polak, 2002). Considering the high chemical specificity of Raman spectroscopy, it may be possible to sample individual cells and the biomaterial for hours, days or weeks, in order to monitor spectral changes that can be correlated with changes in the cell phenotype, cell growth as well as biodegradability of the scaffold, release of bio- and chemical stimuli, etc.

This chapter provides an introduction to Raman spectroscopy and applications to both live cells and tissue engineering scaffolds. Section 12.2 represents a brief description of Raman scattering and the instrumentation required for spectroscopic measurements. The main advantageous features of this technique are emphasised, in particular through comparison with other techniques used for studying cells and tissue engineering scaffolds. Applications to studying various cellular processed in live cells, including cell proliferation, cell death, differentiation and bone nodule mineralisation, are described in Section 12.3. This section describes the detection of various biochemical changes within the cells during these cellular processes without damaging the cells. The last section deals with applications to studying properties of biomaterials used for building tissue engineering scaffolds. The ability of Raman spectroscopy to detect the formation of hydroxyapatite layer at the surface of bioactive sol–gel glasses is shown. Additionally, a brief description of using Raman spectroscopy to monitor the degradation of polymeric scaffolds is also presented. A comprehensive list of references is included for readers who want to learn more about various applications of this technique.

12.2 Principles of Raman spectroscopy

The Raman scattering effect was discovered in 1928 and is based on inelastic scattering (Raman scattering) of photons following their interaction with vibrating molecules of the sample (Raman and Krishnan, 1928). The inelastic interactions lead to a frequency shift of the incident photons as they transfer/receive energy to/from the sample molecules. Therefore, the energy loss/gain of the scattered photons corresponds to the vibrational energy levels of the molecules (Fig. 12.1). Since the vibrational energy spectrum depends on the physical and chemical properties of the sample (type of atoms, bond strength, bond angles, symmetry, etc.), a Raman spectrum represents a physiochemical fingerprint of the sample. For more detailed description of the physics of the Raman effect see Ferraro *et al.* (2003), Lewis and Edwards (2001) and Long (1977).

The intensity of the Raman scattering is proportional with the fourth power of the frequency of the incident laser photons, but it is still typically 15 orders of magnitude lower than fluorescence emission of dye molecules. Certain molecules present in cells, such as carotenoids and haem proteins, can produce enhanced Raman signals when specific laser wavelengths are used for excitation (Spiro, 1985; Myers and Mathies, 1987). If the laser wavelength corresponds to a strong electronic absorption band of a molecule, the intensity of some Raman-active vibrations of the molecule can be enhanced by a factor of 10^2–10^4. Higher enhancements of Raman spectra (typically 10^7 or even higher) can be achieved when molecules are adsorbed on rough metallic substrates or added to metal colloids (surface enhanced Raman spectra, SERS) (Moskovits, 1985; Otto *et al.*,

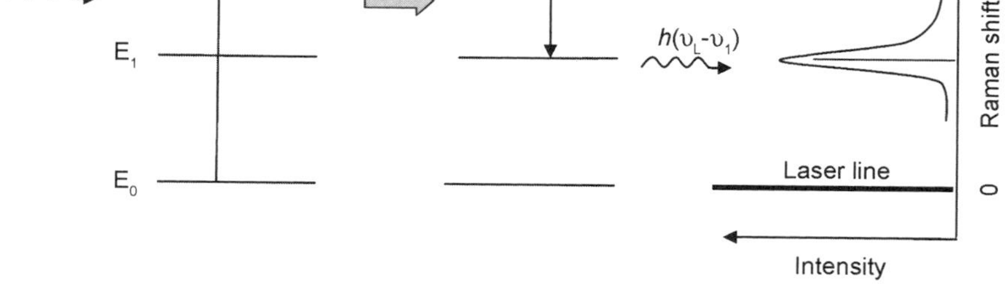

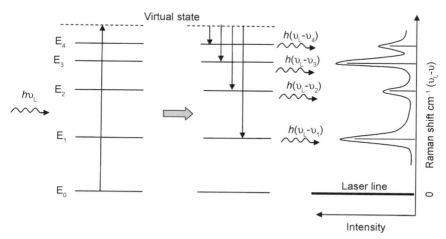

12.1 Schematic description of the Raman scattering (Stokes). Raman spectrum represents a fingerprint of the vibrational energy level of molecules (frequency shifts $\nu_L - \nu$ of incident photons, ν_L is the laser frequency and ν represents a molecular vibrational frequency).

1992). The most popular metals for SERS when excitation is realised in the visible or near-infrared regions are silver and gold. Recent studies proved that SERS is able to study even single molecules, suggesting Raman scattering enhancements as large as 10^{14}–10^{15} (Nie and Emory, 1997; Kneipp *et al.*, 2000). Although these enhancing mechanisms produce stronger Raman signals, their applications in tissue engineering has been yet rather limited compared with non-resonant Raman spectroscopy.

In modern Raman spectrometers (Fig. 12.2), a laser in the visible or near-infrared regions is used as the monochromatic excitation source. The frequency spectrum (the shift from the laser frequency) of the scattered photons is usually analysed using a dispersion spectrometer. Spectrometers based on Fourier transformation can also be used, but they usually have lower sensitivity due to the longer wavelength (low-frequency) lasers used. In the dispersive spectrometers, a diffraction grating is used to disperse the Raman scattered beam into multiple beams corresponding to specific frequencies which are subsequently focused on an array of detectors, such as a high-sensitivity charge coupled device (CCD). A notch or edge filter is used in all modern spectrometers to reject the elastically scattered photons (Rayleigh photons), which have the same

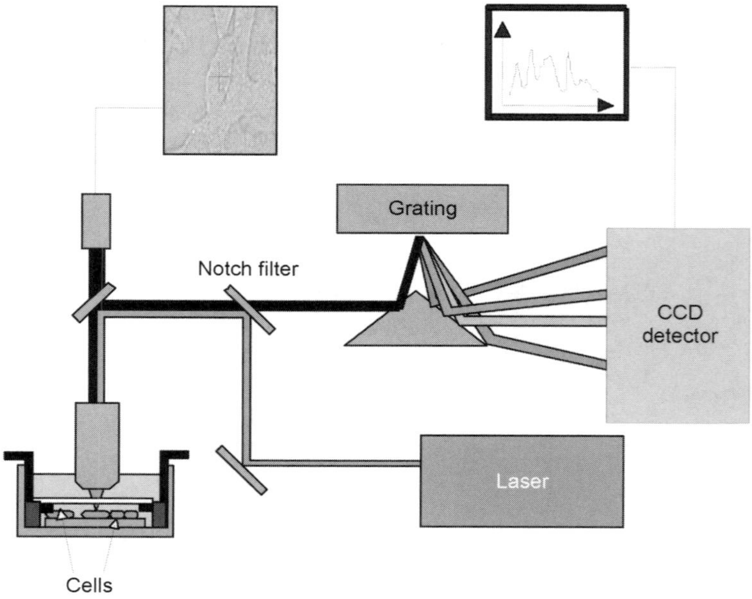

12.2 Schematic diagram of a Raman micro-spectrometer (upright microscope). Cells or growing tissue can be enclosed in a sterile temperature-controlled chamber to ensure cell viability during measurements.

energy as the laser photons and would produce an intense background covering the Raman photons.

To achieve the high spatial resolution required for studying living cells, a Raman spectrometer is coupled to an optical microscope. Thus, spatial resolution limited by optical diffraction to approximately half wavelength of the excitation laser (~250 nm if a 514 nm laser is used for excitation) can be achieved in the horizontal plane. The axial resolution can be improved by using a pinhole of approximately 100 μm (confocal set-up) to reject the out-of-focus photons. Certain Raman micro-spectrometers can achieve improved spatial resolution by using the actual detector pixels or the aperture of the collection optical fibre as a confocal pinhole. Confocal Raman instruments have reported sampling volumes as small as 1.4 fL from inside single cells (Uzunbajakava *et al.*, 2003b). Both upright and inverted optical microscopes have become common in Raman micro-spectrometers and are used both for focusing the laser on a cell/biomaterial and at the same time collect the Raman scattered photons (see Fig. 12.2).

The recent developments in laser technology and charged coupled detectors (CCD) during the last decades have had a high impact on Raman micro-spectroscopy. Raman spectroscopy was previously known to be a weak signal technique that required long integration time. Additionally, the fluorescence background commonly observed when exciting biological molecules with visible lasers limited Raman spectroscopy to biomedical applications. The use of modern near-infrared lasers, cooled silicon CCD, holographic filters and high-throughput dispersive spectrographs makes it possible to obtain high-quality Raman spectra of cells within tens of seconds (Puppels *et al.*, 1990).

Using near-infrared lasers (700–850 nm) significantly reduces the fluorescence background while still fitting in the high quantum efficiency spectral region of silicon CCDs. It was also shown that near-infrared lasers induce less photodamage compared with UV or visible lasers (Puppels *et al.*, 1991; Neuman *et al.*, 1999; Notingher *et al.*, 2002). This finding allowed use of higher laser power for the excitation of Raman photons, increasing the signal strength and subsequently reducing the measurement time.

Another important feature of this technique is the ability to build Raman spectral maps in two and three dimensions. Such maps can be achieved by representing the intensity of a certain spectral peak (Uzunbajakava *et al.*, 2003a,b), score of a principal component (Krafft *et al.*, 2003) or weight obtained in a least square fitting analysis (Shafer-Peltier *et al.*, 2002) for each individual location in the three-dimensional region where Raman spectra were acquired. Alternatively, wide-field images can be obtained by using narrow-band interference filters to obtain a global image for a specific Raman peak over the entire field of view (Puppels *et al.*, 1993).

12.3 Characterisation of living cells

12.3.1 Peak assignment of Raman spectra of biomolecules and cells

Cellular biopolymers (nucleic acids, proteins, lipids and carbohydrates) have been studied by Raman spectroscopy for more than three decades (Tu, 1982). These studies have led to the assignment of most Raman peaks to various molecular vibrations, as summarised in Table 12.1.

Table 12.1 Assignment of the most important peaks in Raman spectra of biopolymers and cells

Peak (cm^{-1})	Assignment			
	DNA/RNA	Proteins	Lipids	Carbohydrates
1736			C=O ester	
1680–1655		Amide I	C=C str	
1617		C=C Tyr, Trp		
1607		C=C Phe, Tyr		
1578	G, A			
1480–1420	G, A, CH def	CH def	CH def	CH def
1342	A, G	CH def		CH def
1320	G	CH def		
1301			CH_2 twist	
1284–1220	T, A	Amide III	=CH bend	
1209		C–C_6H_5 str Phe, Trp		
1176		C–H bend Tyr		
1158		C–C/C–N str		
1128		C–N str		C–O str
1095–1060	PO_2^- str		Chain C–C str	C–O, C–C str
1033		C–H in-plane Phe		
1005		Sym. Ring br Phe		
980		C–C BK str β-sheet	=CH bend	
937		C–C BK str α-helix		C–O–C glycos
877			C–C–N^+ sym str	C–O–C ring
854		Ring br Tyr		
828	O–P–O asym str	Ring br Tyr		
811	O–P–O str RNA			
788	O–P–O str DNA			
782	U, C, T ring br			
760		Ring br Trp		
729	A			
717			$CN^+(CH_3)_3$ str	
667	T, G			
645		C–C twist Tyr		
621		C–C twist Phe		

def = deformation; str = stretching; br = breathing; bend = bending; sym = symmetric; asym = asymmetric

The vibrations of the sugar–phosphate backbone of nucleic acids produce strong peaks in the Raman spectra of cells. These vibrations can provide useful information regarding the secondary structure of DNA, and can also be used to discriminate between RNA and DNA. For example, the Raman peak corresponding to the phosphodiester bond is found at 788 cm^{-1} for DNA and at 813 cm^{-1} for RNA. The vibration of the phosphodioxy group produces a relatively strong Raman peak at 1095 cm^{-1}, but which is rather insensitive to conformational changes of the nucleic acids. Additionally, Raman peaks corresponding to nucleotides can also be identified at 782 cm^{-1} (thymine, cytosine and uracil) and 1578 cm^{-1} (guanine and adenine).

The Amide I and Amide III vibrations produce the most intensive peaks in the Raman spectra of proteins. The peaks occur in the ranges 1660–1670 cm^{-1} (Amide I) and 1200–1300 cm^{-1} (Amide III), and the exact positions can be used for determining the secondary structure of the proteins. Certain amino acids produce strong Raman peaks, such as phenylalanine (1005 cm^{-1}), tyrosine (854 cm^{-1}) and tryptophan (760 cm^{-1}). The large number of C–H bonds in proteins also leads to a strong Raman band around 1449 cm^{-1}.

Lipids are characterised by intense Raman peaks at 1449 and 1301 cm^{-1} corresponding to C–H vibrations. Peak assigned to the stretching vibrations of unsaturated C=C bonds may also appear around 1660 cm^{-1}. Additional Raman peaks corresponding to head groups of phospholipids can also be found, such as the 719 cm^{-1} peak corresponding to the C–C–N^{+} symmetric stretching in phosphatidylcholine, a major constituent of cellular membranes.

C–O–C vibrations of the glycosidic bonds and ring vibrations in carbohydrates produce specific Raman peaks in the 800–1100 cm^{-1} range. However, in cells, the intensities of the Raman peaks corresponding to the vibration of carbohydrates are usually smaller compared with the other biomolecules.

12.3.2 Cell proliferation and death

Detection of cell death is important in tissue engineering for detecting potential toxic effects of biomaterials. Apoptosis, i.e. programmed death, involves biochemical changes within cells including protein distribution and DNA condensation and fragmentation. Raman micro-spectroscopy can be used to compare the spectra of healthy live cells and dying cells in order to determine reliable spectral markers for cell viability (Notingher *et al.*, 2003). A decrease by 80% and 66% in dying cells of the Raman peaks at 788 cm^{-1} and 782 cm^{-1} respectively, corresponding to DNA vibrations reflected the breakdown of DNA phosphodiester bonds and bases. Nucleotide condensation within nuclear fragments in apoptotic cells was detected by confocal Raman micro-spectroscopy (Uzunbajakava *et al.*, 2003a). The intensity of the 788 cm^{-1} peak was used to build 2D spectral images of nuclear fragments. The higher intensity of the 788 cm^{-1} in nuclear fragments compared with healthy cells indicated the

DNA condensation during apoptosis. Similar methods were also used to image the reorganisation and distribution of proteins in apoptotic cells.

Cell proliferation is an important factor in developing scaffold materials as it has recently been found that certain types of bioactive glasses stimulate cell proliferation by activating specific genes. Therefore, developing Raman spectral markers which may be used for monitoring cell growth and cell cycle are important in assessing the biomaterials. Studies comparing Raman spectra of tumorogenic and non-tumorogenic rat embryo fibroblasts were measured in the non-proliferating and the exponential phases of growth (Short *et al.*, 2005). The results showed that for both tumorogenic and non-tumorogenic exponential cells the ratio between the concentrations of lipids and RNA decreased while the ratio between amounts of protein and lipids increased compared to non-proliferating cells that have plateaued.

12.3.3 Differentiation

Cell differentiation is an important factor in tissue engineering. It is important for cells growing on scaffolds to have and maintain the appropriate phenotype during the growth of the tissue *in vitro*. It has recently been recognised that stem cells have great potential in tissue engineering, as given the right environment and stimuli, they can differentiate into the required mature cell type. Cell differentiation involves many intracellular biochemical and biophysical changes, including production of specific proteins which allow them to perform their functions. The high chemical specificity of Raman micro-spectroscopy enables monitoring of the biochemical changes undergoing in cells during their differentiation-specific phenotype.

Raman micro-spectroscopy was used to identify spectral markers for monitoring the differentiation status of embryonic stem (ES) as function of differentiation time (Fig. 12.3) (Notingher *et al.*, 2004). The main biochemical changes observed were related to higher RNA concentration in undifferentiated ES cells than in differentiated cells. The massive down-regulation is consistent with the hypothesis that ES cells 'keep their options open' by maintaining many genes at intermediate levels and selecting only a few for continuous expression that are needed for differentiation to a specific phenotype. The rest will be down-regulated after commitment to a cell fate for which they are not needed.

12.3.4 Bone nodule formation and mineralisation *in vitro*

Raman micro-spectroscopy can be used to study mineralisation of bone and bone nodules produced in cultures *in vitro*. The mineral environment in bone tissue and the CaP species can be determined by measuring the vibration frequency and intensity of the symmetric stretching vibration (ν_1) of the PO_4^{3-} group (Timlin *et al.*, 2000; Tarnowski *et al.*, 2002). The frequency of ν_1 vibration changes with

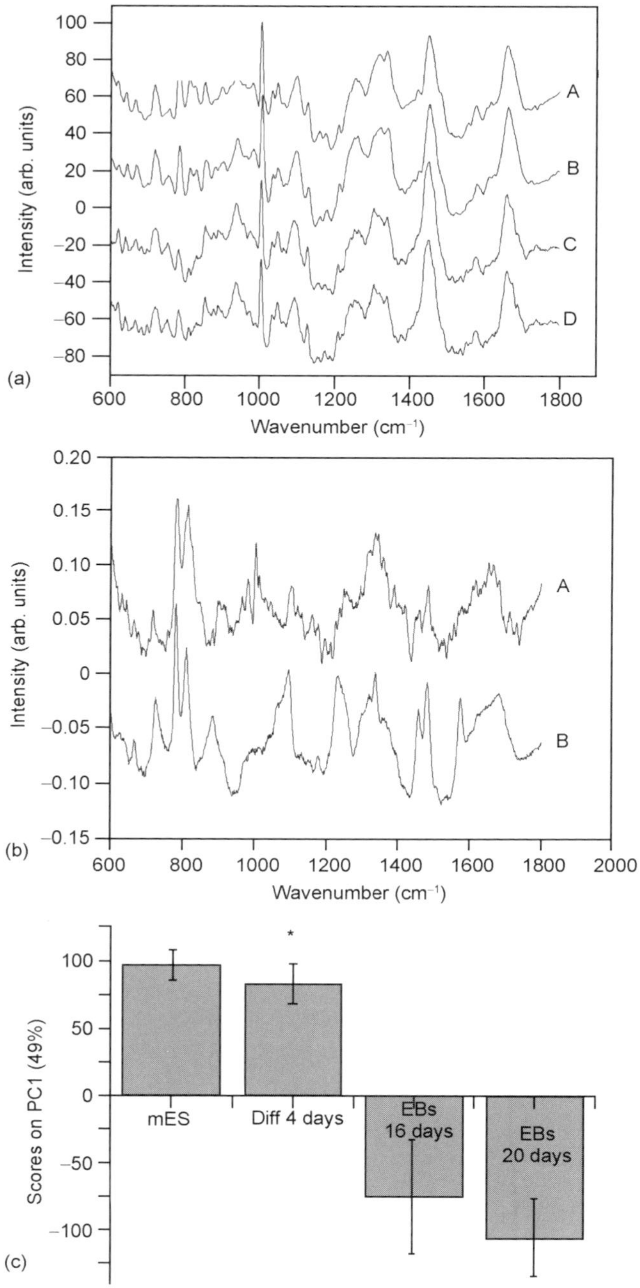

12.3 Monitoring differentiation of mouse ES cells: (a) Raman spectra of (A) undifferentiated, (B) 4 days, (C) 16 days, (D) 20 days. (b) PCA analysis: comparison between PC1 Loading (A) and Raman spectrum of reference RNA (B). (c) PCA analysis: scores of PC1.

ionic incorporation and crystallinity of the apatite: 955–959 cm^{-1} in B-type apatite (carbonate substituted phosphate in apatite lattice), 962–964 cm^{-1} in nonsubstituted apatite, and 945–950 cm^{-1} in disordered HA lattice (probably A-type carbonate substitution, i.e. carbonate for hydroxide or amorphous calcium phosphate) (Timlin *et al.*, 2000; Tarnowski *et al.*, 2002). These peaks were also used to understand ionic incorporation and crystallinity of the apatite at different stages of development of the mineralisation process in mouse calvaria (Tarnowski *et al.*, 2002). The same spectroscopic markers have been used to analyse fatigue-related microdamage in bovine bone (Timlin *et al.*, 2000). In regions of undamaged tissue, the phosphate $\nu1$ band is found at 957 cm^{-1}, as expected for the carbonated hydroxyapatic bone mineral. However, in regions of visible microdamage, an additional phosphate $\nu1$ band is observed at 963 cm^{-1} and interpreted as a more stoichiometric, less carbonated mineral species. Similar methods were used to determine the apatic species of CaP deposited *in vitro* by bone cells exposed to growth factors (Kale *et al.*, 2000). The osteoblasts cultured on planar surfaces exposed to TGF-beta1 produced a large number of micro-spicules which contained phosphate and carbonate substituted apatite similar to that found in mature bone. The effect of surface roughness on the formation and mineralisation of bone nodules by human primary osteoblast can be investigated using Raman spectroscopy (Gough *et al.*, 2004). Raman spectroscopic measurement showed higher concentration of CaP in the bone nodules formed by osteoblasts on rough bioactive glass (45S5 Bioglass®) compared with similar but smooth substrates (Fig. 12.4) (Gough *et al.*, 2004).

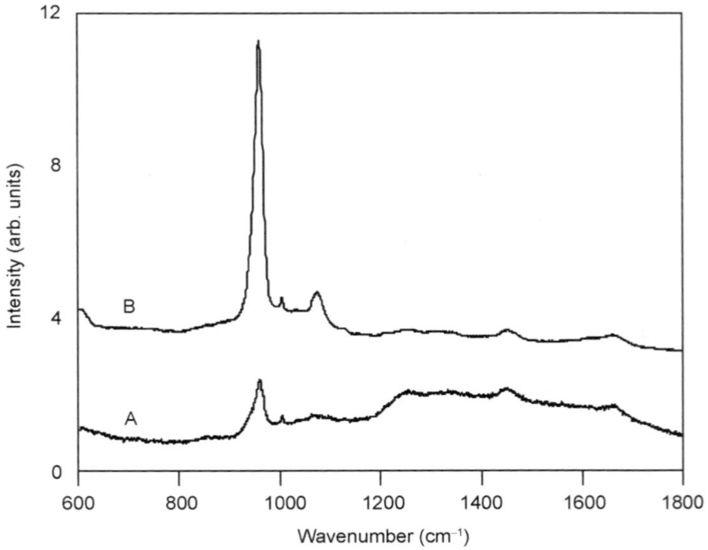

12.4 Raman spectra of bone nodules on 45S5 Bioglass® formed by osteoblasts fed with β-glycerophosphate and dexamethasone: (A) smooth surface, (B) rough surface.

12.4 Characterisation of tissue engineering scaffolds

12.4.1 Study of bioactivity

Tissue engineering scaffolds manufactured from bioactive materials offer many advantages over bio-inert materials, including improved cell attachment and stimulation of proliferation and differentiation. Bioactive materials also provide a more stable and stronger bond with living tissue compared with bioinert implants due to reactions occurring at the material surface exposed to body fluids. For bioactive materials, surface reactions lead to formation of a biologically active hydroxycarbonate apatite layer to which living tissue can strongly adhere. For bone tissue engineering, osteoblasts attach and proliferate on this apatite layer, which is biologically equivalent to the apatite present in bones (Rehman *et al.*, 1995). Further, collagen produced by bone cells bind to the apatite layer, leading to bone regeneration (Hench and West, 1996). Bioactive materials also stimulate a biochemical response from the living tissue in order to obtain a strong bond ('biological fixation') between the material and the tissue, therefore facilitating the regeneration of tissue instead of replacing it (Jones and Hench, 2001).

The bioactive properties of a material are usually determined by the rate of the formation of the hydroxyapatite layer at its surface. The time required for the formation of this layer depends on many factors, such as composition, structure, geometry, solution composition and ratio of surface area to solution volume. Raman spectroscopy can be used to determine the bioactive properties of porous bioactive glass-foams produced by sol–gel technique by measuring the strong 960 cm^{-1} peak characteristic to the PO_4 vibration of apatite, which does not overlap to other Raman peaks originating from other vibrations of the glass.

Glasses based on SiO_2, Na_2O, CaO and P_2O_5 elicit a quick biochemical response when placed in physiological fluids leading to the formation of an apatite layer (Hench and West, 1996; Jones and Hench, 2001). Sol–gel processing of bioactive gel-glasses allows the control of bioactive properties by tailoring the composition and microstructure (Hench and West, 1990). The nanometre scale porosity in sol–gel glasses offers better control of the dissolution rate due to a higher surface area-to-volume ratio (more than 50 times) compared with the melt-derived glasses. The sol–gel process provides a very attractive technique to fabricate highly porous, foam-like scaffolds that reproduce the 3D structure of trabecular bone (Sepulveda *et al.*, 2002).

Figure 12.5(a) presents the Raman spectra of a 58S glass foam sample (60SiO_2, 36CaO, 4P_2O_5 mol%) as a function of immersion time in simulated body fluid (SBF). These foams have interconnected pore windows with diameter up to 100 μm and macropores up to 500 μm. The peak at 960 cm^{-1} increases with increasing immersion time in SBF, which indicates that the amount of apatite formed on the sample increases with time. The main changes occur between 1 h and 8 h, after which a saturation level is achieved. The amount of

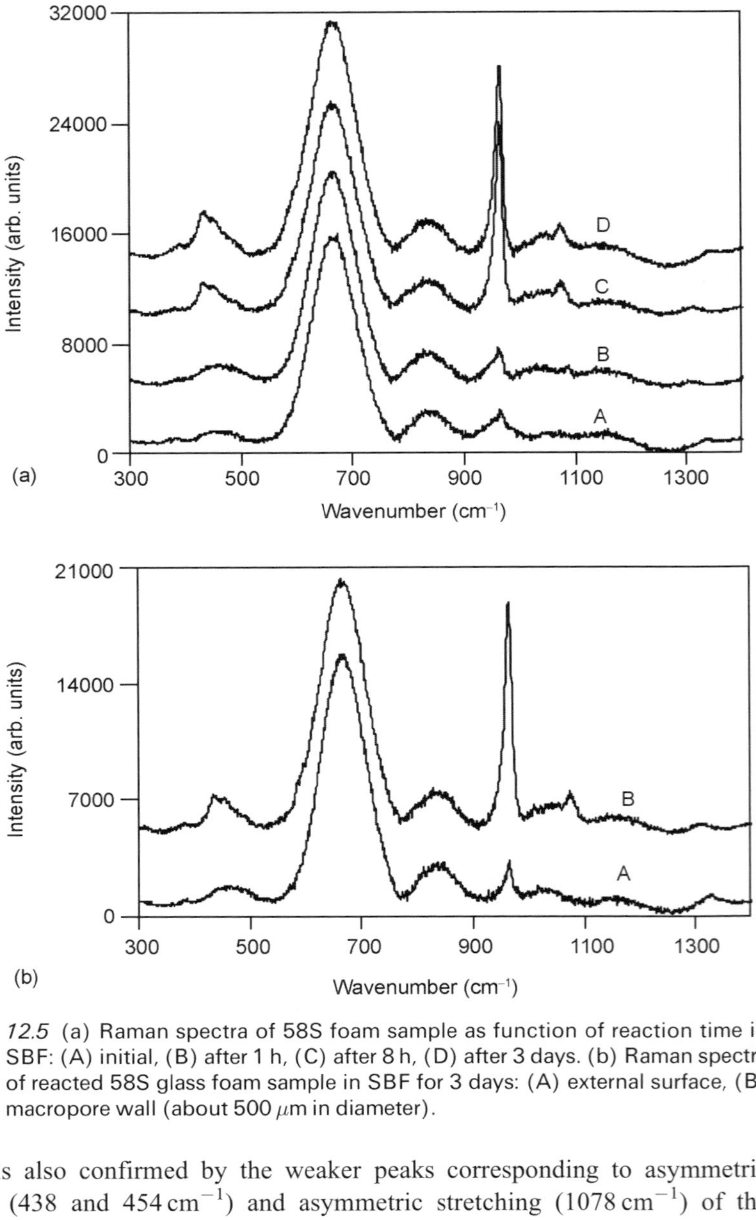

12.5 (a) Raman spectra of 58S foam sample as function of reaction time in SBF: (A) initial, (B) after 1 h, (C) after 8 h, (D) after 3 days. (b) Raman spectra of reacted 58S glass foam sample in SBF for 3 days: (A) external surface, (B) macropore wall (about 500 μm in diameter).

apatite is also confirmed by the weaker peaks corresponding to asymmetric bending (438 and 454 cm^{-1}) and asymmetric stretching (1078 cm^{-1}) of the PO_4^{3-} groups.

The high spatial resolution of the Raman micro-spectrometer makes it possible to characterise the 58S reactivity inside the macropores. Figure 12.5(b) compares the Raman spectra measured from the external surface of the sample and from the internal surface of a macropore of a foam immersed in SBF for 3 days. The spectra in Fig. 12.5B indicate that the apatite does not develop uniformly. The external surface, being more exposed to the solution, reacts much faster, while the pore walls react more slowly, leading to a lower amount of hydroxyapatite being formed.

12.4.2 Degradation of polymeric foams

Scaffolds made of biodegradable polymers, such as poly(lactic acid) (PLA), poly(glycolic acid) (PGA) and their copolymers poly(lactic acid-co-glycolic acid) (PLGA), have been extensively investigated because of their attractive characteristics (Rezwan *et al.*, 2006): (i) they are biocompatible, (ii) 3D foams with interconnected macropores can be produced to facilitate cell ingrowth and flow of nutrients and metabolic by-products, (iii) it is possible to alter the degradation rates by tailoring the composition and morphology and (iv) they can be easily fabricated into complex structures. It has been widely accepted that the biodegradation of PLA, PGA and PLGA *in vivo* is a non-enzymatic process and occurs by hydrolytic degradation. Therefore, *in vitro* degradation studies are relevant for estimating the degradation of these scaffolds *in vivo*, in order to optimise the composition and manufacturing procedure for a required degradation rate.

Raman spectroscopy can be used to monitor the degradation process of polymeric foams. Raman spectra measured for pure PLGA scaffolds obtained by compression moulding method and measured after various degradation times are shown in Fig. 12.6. All spectra show peaks characteristic to both its components, PLA (Cassanas *et al.*, 1993; Kister *et al.*, 1998a) and PGA (Cassanas *et al.*, 1993; Kister *et al.*, 1997; Epple and Herzberg, 1998). Table 12.2 summarises the peak assignment of the most important Raman peaks based on previous studies (Cassanas *et al.*, 1993; Epple and Herzberg, 1998; Kister *et al.*, 1992, 1997, 1998a,b; Taddei *et al.*, 2001).

The degradation of the PLGA foams is a hydrolytic reaction leading to the scission of the macromolecular chains into smaller molecules, especially monomers. The degradation process can be identified in the measured Raman spectra by the decrease in the magnitude of the peaks corresponding to stretching vibrations of ester groups and repeat units (Cassanas *et al.*, 1993). The C–COO stretching peak of PLA (872 cm^{-1}) has the highest sensitivity to hydrolytic degradation and has been used as a measure of polymer chain length (Cassanas *et al.*, 1993; Bertoluzza *et al.*, 1997; Taddei *et al.*, 2001, 2002). In lactic acid monomers, the frequency of this vibration is lower and occurs around 830 cm^{-1} (Cassanas *et al.*, 1993). The Raman spectra of PLGA copolymers also show peaks at 892 cm^{-1} from C–COO vibrations in the amorphous GA units (Cassanas *et al.*, 1993; Kister *et al.*, 1998b). Raman peaks corresponding to vibrations of the ester groups, such as the O–C–O and C=O stretching, were also shown to be sensitive to the length of polymer chains (Cassanas *et al.*, 1993). An increase in

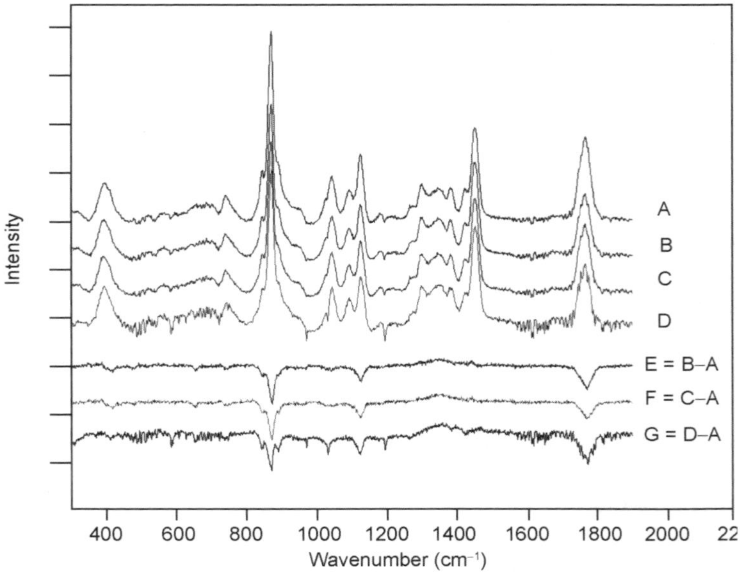

12.6 Raman spectra of PLGA foams at 0 weeks (A), 4 weeks (B), 8 weeks (C) and 16 weeks (D) of degradation in PBS, difference spectra: (E)–(G) (spectra shifted vertically for clarity).

the O–C–O peak at 1128 cm^{-1} was observed during the polymerisation of the PDLLA while the frequency of the C=O vibration was observed at ~1725 cm^{-1} in the COOH groups of the lactic acids and increased to ~1766 cm^{-1} in amorphous polymers (Cassanas *et al.*, 1993; Kister *et al.*, 1998a). This band also split in several components in crystalline PLA samples (Kister *et al.*, 1998a).

Table 12.2 Peak assignment of the Raman spectra of PLGA foams

Peak (cm^{-1})	Assignment	
	PLA	PGA
1766	C=O stretch	C=O strech
1453	CH_3 asymmetric bend	CH_2 bend
1424		CH_2 bend
1384	CH_3 symmetric bend	
1301	CH bend	
1127	CH_3 bend	
1093	O–C–O stretch	O–C–O stretch
1044	C–CH_3 stretch	
892		C–COO
872	C–COO stretch	
846		CH_2 rocking
396	C=O	

The different Raman spectra in Fig. 12.6 show a significant decrease in the peaks corresponding to C–COO and to the ester vibrations C=O and O–C–O during the soaking in phosphate-buffered saline (PBS), indicating hydrolytic degradation of PLGA foams with time. The C=O bands at 1766 and 400 cm^{-1} remain unstructured compared with crystalline PLA (Kister *et al.*, 1998a), suggesting that no crystalline domains formed. The Raman spectra indicate that the degradation occurs more rapidly as early as 4 weeks (~80%) with no further chain breaks at 8 or 16 weeks (~80%).

12.5 Conclusions and future trends

Raman micro-spectroscopy is a powerful technique for analysing both live cells and scaffold materials used in tissue engineering. This technique is non-invasive and does not require labelling, staining or other forms of sample preparation. A Raman spectrum is produced following the interaction of the exciting laser photons with vibrating molecules in the sample. Consequently, the inner molecular composition of the sample is probed, providing a high chemical specificity. Owing to high transparency and low Raman signal of physiological solutions (buffers, culture media, etc.), this technique can be used *in situ* for studying cells and scaffolds maintained in physiological conditions during measurements. Additionally, the coupling of the lasers and spectrographs to optical microscopes, measurements with diffraction-limited spatial resolution can be performed for producing chemical images of cells and scaffolds.

From the large range of analytical techniques used by tissue engineering scientists, Raman spectroscopy is the only method that allows monitoring of cellular processes or biomaterial surface reactions over extended periods of time in a non-invasive way and without removing the sample from the correct environment. This would enable, for example, the phenotype of the cells during growth, development of extracellular matrix, and onset of angiogenesis to be assessed, which are all required for achieving stable tissues. Raman spectroscopy may be also useful for detecting changes in cells related to differentiation, phenotype and viability following attachment on scaffolds. Establishing and maintaining a mature phenotype is a key requirement for a tissue to be implanted from an *in vitro* environment to *in vivo*.

Raman spectroscopy can be used in assessing the bioactivity of various biomaterials by monitoring the formation of the hydroxycarbonate apatite. The degradation of biodegradable polymer foams can be detected by measuring the Raman peaks corresponding to vibrations of the chemical bonds between the monomers.

Raman spectroscopy has great potential in tissue engineering and it may provide in the future an integrated solution for monitoring the cell–scaffold constructs from the culturing moment until the implantation surgery.

12.6 References

Bertoluzza, A., Fagnano, C., Mietti, N., Tinti, A., Giannini, S., Giardino, R. and Cacciari, G. L. (1997): In P. Carmona, R. Navarro, and A. Hernanz *Spectroscopy of Biological Molecules: Modern Trends*, Dordrecht, Kluwer Academic Publishers.

Cassanas, G., Kister, G., Fabregue, E., Morssli, M. and Bardet, L. (1993): Raman spectra of glycolic acid, L-lactic acid and D,L-lactic acid oligomers. *Spectrochim Acta* **49**, 271–279.

Epple, M. and Herzberg, O. (1998): Porous polyglycolide. *J Biomed Mater Res B* **43**, 83–88.

Ferraro, J. R., Nakamoto, K. and Brown, C. W. (2003): *Introductory Raman Spectroscopy*, San Diego, Academic Press.

Gough, J. E., Notingher, I. and Hench, L. L. (2004): Osteoblast attachment and mineralized nodule formation on rough and smooth 45S5 bioactive glass monoliths. *J Biomed Mater Res A* **68**, 640–650.

Gremlich, H. U. and Yan, B. (2001): *Infrared and Raman Spectroscopy of Biological Materials*, New York, Marcel Dekker.

Hanlon, E. B., Manoharan, R., Koo, T. W., Shafer, K. E., Motz, J. T., Fitzmaurice, M., Kramer, J. R., Itzkan, I., Dasari, R. R. and Feld, M. S. (2000): Prospects for *in vivo* Raman spectroscopy. *Phys Med Biol* **45**, R1–59.

Hench, L. L. and Polak, J. L. (2002): Third-generation biomedical materials. *Science* **295**, 1014–1017.

Hench, L. L. and West, J. K. (1990): The sol–gel process. *Chem. Rev.* **90**, 33–72.

Hench, L. L. and West, J. K. (1996): Biological applications of bioactive glasses. *Life Chem Rep* **13**, 187–241.

Hench, L. L. and Wilson, J. (1993): *Introduction to Bioceramics*, Singapore, Singapore World Scientific.

Jones, J. R. and Hench, L. L. (2001): Biomedical materials for new millennium: perspective on the future. *Mater Sci Technol* **17**, 891–900.

Kale, S., Biermann, S., Edwards, C., Tarnowski, C., Morris, M. and Long, M. W. (2000): Three-dimensional cellular development is essential for *ex vivo* formation of human bone. *Nature Biotechnol* **18**, 954–958.

Kister, G., Cassanas, G., Fabregue, E. and Bardet, L. (1992): Vibrational analysis of ring-opening polymerizations of glycolide, L-lactide and D,L-lactide. *European Polym J* **28**, 1273–1277.

Kister, G., Cassanas, G. and Vert, M. (1997): Morphology of poly(glycolic acid) by IR and Raman spectroscopies. *Spectrochim Acta Part A: Molec Biomolec Spectrosc* **53**, 1399.

Kister, G., Cassanas, G. and Vert, M. (1998a): Effects of morphology, conformation and configuration on the IR and Raman spectra of various poly(lactic acid)s. *Polymer* **39**, 267.

Kister, G., Cassanas, G. and Vert, M. (1998b): Structure and morphology of solid lactide-glycolide copolymers from 13C N.M.R., infra-red and Raman spectroscopy. *Polymer* **39**, 3335–3340.

Kneipp, K., Kneipp, H., Corio, P., Brown, S. D., Shafer, K., Motz, J., Perelman, L. T., Hanlon, E. B., Marucci, A., Dresselhaus, G. and Dresselhaus, M. S. (2000): Surface-enhanced and normal Stokes and anti-Stokes Raman spectroscopy of single-walled carbon nanotubes. *Phys Rev Lett* **84**, 3470–3473.

Krafft, C., Knetschke, T., Siegner, A., Funk, R. H. W. and Salzer, R. (2003): Mapping of single cells by near infrared Raman microspectroscopy. *Vibrational Spectroscopy* **32**, 75–83.

Lewis, I. and Edwards, H. (2001): *Handbook of Raman Spectroscopy*, New York, Marcel Dekker.

Long, D. A. (1977): *Raman Spectroscopy*. New York, McGraw-Hill.

Manoharan, R., Wang, Y. and Feld, M. S. (1996): Histochemical analysis of biological tissues using Raman spectroscopy. *Spectrochim Acta Part A Molec Biomolec Spectros* **52**, 215–249.

Moskovits, M. (1985): Surface-enhanced spectroscopy. *Rev Modern Physics* **57**, 783–828.

Myers, A. B. and Mathies, R. A. (1987): Resonance Raman spectra of polyenes and aromatics. In T. G. Spiro (Ed.): *Biological Applications of Raman Spectroscopy*, New York, John Wiley and Sons.

Neuman, K. C., Chadd, E. H., Liou, G. F., Bergman, K. and Block, S. M. (1999): Characterization of photodamage to *Escherichia coli* in optical traps. *Biophys J* **77**, 2856–2863.

Nie, S. and Emory, S. R. (1997): Probing single molecules and single nanoparticles by surface-enhanced Raman spectroscopy. *Science* **275**, 1102–1106.

Notingher, I., Verrier, S., Romanska, H., Bishop, A. E., Polak, J. M. and Hench, L. L. (2002): In situ characterisation of living cells by Raman spectroscopy. *Spectroscopy Int J* **16**, 43–51.

Notingher, I., Verrier, S., Haque, S., Polak, J. M. and Hench, L. L. (2003): Spectroscopic study of human lung epithelial cells (A549) in culture: living cells versus dead cells. *Biopolymers* **72**, 230–240.

Notingher, I., Bisson, I., Bishop, A. E., Randle, W. L., Polak, J. M. P. and Hench, L. L. (2004): *In situ* spectral monitoring of mRNA translation in embryonic stem cells during differentiation *in vitro*. *Anal Chem* **76**, 3185–3193.

Otto, A., Mrozek, I., Grabhorn, H. and Akemann, W. (1992): Surface-enhanced Raman scattering. *J Phys: Condensed Matter* **4**, 1143–1212.

Puppels, G. J., de Mul, F. F., Otto, C., Greve, J., Robert-Nicoud, M., Arndt-Jovin, D. J. and Jovin, T. M. (1990): Studying single living cells and chromosomes by confocal Raman microspectroscopy. *Nature* **347**, 301–303.

Puppels, G. J., Olminkhof, J. H., Segers-Nolten, G. M., Otto, C., de Mul, F. F. and Greve, J. (1991): Laser irradiation and Raman spectroscopy of single living cells and chromosomes: sample degradation occurs with 514.5 nm but not with 660 nm laser light. *Exp Cell Res* **195**, 361–367.

Puppels, G. J., Grond, M. and Greve, J. (1993): Direct imaging Raman microspectroscopy based on tunnable wavelength excitation and narrow band emission detection. *Appl Spectrosc* **47**, 1256–1267.

Raman, C. V. and Krishnan, K. S. (1928): A new type of secondary radiation. *Nature* **121**, 501–502.

Rehman, I., Smith, R., Hench, L. L. and Bonfield, W. (1995): Structural evaluation of human and sheep bone and comparisons with hydroxyapatite by FT-Raman spectroscopy. *J Biomed Mater Res* **29**, 1287–1294.

Rezwan, K., Chen, Q. Z., Blaker, J. J. and Boccaccini, A. R. (2006): Biodegradable and bioactive porous polymer/inorganic composite scaffolds for bone tissue engineering. *Biomaterials* **27**, 3413–3431.

Sepulveda, P., Jones, J. R. and Hench, L. L. (2002): Bioactive sol–gel foams for tissue repair. *J Biomed Mater Res* **59**, 340–348.

Shafer-Peltier, K. E., Haka, A. S., Motz, J. T., Fitzmaurice, M., Dasari, R. R. and Feld, M. S. (2002): Model-based biological Raman spectral imaging. *J Cell Biochem Suppl* **39**, 125–137.

Short, K. W., Carpenter, S., Freyer, J. P. and Mourant, J. R. (2005): Raman spectroscopy detects biochemical changes due to proliferation in mammalian cell cultures. *Biophys J* **88**, 4274–4288.

Spiro, T. G. (1985): Resonance Raman spectroscopy as a probe of heme protein structure and dynamics. *Adv Protein Chem* **37**, 111–159.

Taddei, P., Tinti, A. and Fini, G. (2001): Vibrational spectroscopy of polymeric biomaterials. *J Raman Spectrosc* **32**, 619–629.

Taddei, P., Monti, P. and Simoni, R. (2002): Vibrational and thermal study on the in vitro and *in vivo* degradation of a poly(lactid acid)-based bioabsorbable periodontal membrane. *J Mater Sci Mater Med* **13**, 469–475.

Tarnowski, C. P., Ignelzi, M. A., Jr and Morris, M. D. (2002): Mineralization of developing mouse calvaria as revealed by Raman microspectroscopy. *J Bone Miner Res* **17**, 1118–1126.

Timlin, J. A., Carden, A., Morris, M. D., Rajachar, R. M. and Kohn, D. H. (2000): Raman spectroscopic imaging markers for fatigue-related microdamage in bovine bone. *Anal Chem* **72**, 2229–2236.

Tu, A. T. (1982): *Raman Spectroscopy in Biology: Principles and Applications*. New York, John Wiley and Sons.

Uzunbajakava, N., Lenferink, A., Kraan, Y., Volokhina, E., Vrensen, G., Greve, J. and Otto, C. (2003a): Nonresonant confocal Raman imaging of DNA and protein distribution in apoptotic cells. *Biophys J* **84**, 3968–3981.

Uzunbajakava, N., Lenferink, A., Kraan, Y., Willekens, B., Vrensen, G., Greve, J. and Otto, C. (2003b): Nonresonant Raman imaging of protein distribution in single human cells. *Biopolymers* **72**, 1–9.

Part II

Tissue and organ generation

13
Engineering of tissues and organs

A ATALA, Wake Forest University, USA

13.1 Introduction

Tissue engineering, stem cells, and cloning are three areas of technology encompassed by the field of regenerative medicine. Tissue engineering, a major component of regenerative medicine, follows the principles of cell transplantation, materials science, and engineering to develop biological substitutes that can restore and maintain normal function. Tissue engineering strategies generally fall into two categories: the use of acellular matrices, depend on the body's natural ability to regenerate for proper orientation and direction of new tissue growth, and the use of matrices with cells. Acellular tissue matrices can be prepared by manufacturing artificial scaffolds or by removing cellular components from tissues by mechanical and chemical manipulation to produce collagen-rich matrices (Dahms *et al.*, 1998; Yoo *et al.*, 1998; Chen *et al.*, 1999). These matrices slowly degrade on implantation and are generally replaced by the extracellular matrix (ECM) proteins secreted by the in-growing cells. Cells can also be used for therapy via injection either with carriers, such as hydrogels, or alone.

A small piece of donor tissue is dissociated into individual cells when cells are used for tissue engineering. These cells are either implanted directly into the host or are expanded in culture, attached to a support matrix, and then reimplanted into the host after expansion. The source of donor tissue can be heterologous (such as bovine), allogeneic (same species, different individual), or autologous (from the host). Ideally, both structural and functional tissue replacement will occur with minimal complications. The preferred cells to use are autologous cells, where a biopsy of tissue is obtained; the cells are dissociated and expanded in culture; and the expanded cells are implanted into the same host (Atala *et al.*, 1994, 1999; Yoo and Atala, 1997; Fauza *et al.*, 1998; Machluf and Atala, 1998; Yoo *et al.*, 1998, 1999; Oberpenning *et al.*, 1999; Atala and Lanza, 2001; Godbey and Atala, 2002). Although it can cause an inflammatory response, the use of autologous cells avoids rejection, and the deleterious side effects of immunosuppressive medications can be avoided.

Most current strategies for tissue engineering depend upon a sample of autologous cells from the diseased organ of the host. In certain situations, stem cells are envisioned as being an alternative source of cells from which the desired tissue can be derived. For example, in many patients with extensive end-stage organ failure, a tissue biopsy may not yield enough normal cells for expansion and transplantation. In other instances, primary autologous human cells cannot be expanded from a particular organ, such as the pancreas. Stem cells can be derived from discarded human embryos (human embryonic stem cells), from fetal-related tissue (amniotic fluid or placenta), or from adult sources (bone marrow, fat, skin). Therapeutic cloning has also played a role in the development of the field of regenerative medicine.

This chapter describes cells and materials used in tissue engineering, including regeneration of specific structures in the body. Therapies at the cellular, tissue, and organ levels are described, with their specific challenges and applications. Definitions of the various types of stem cell research and therapies are explained, including the latest findings on alternate sources of stem cells.

13.2 Native cells

The difficulty of growing specific cell types in large quantities is one of the limitations of applying cell-based regenerative medicine techniques to organ replacement. Even when some organs, such as the liver, have a high regenerative capacity *in vivo*, cell growth and expansion *in vitro* may be complex. By studying the privileged sites for committed precursor cells in specific organs, as well as exploring the conditions that promote differentiation, one may be able to overcome the obstacles that limit cell expansion *in vitro*. For example, urothelial cells could be grown in the laboratory setting in the past, but only with limited expansion. Several protocols were developed over the past two decades that identified the undifferentiated cells in bladder tissue and kept them undifferentiated during their growth phase (Cilento *et al.*, 1994; Liebert *et al.*, 1997; Puthenveettil *et al.*, 1999). These methods of cell culture make it possible to expand a urothelial strain from a single specimen that initially covered a surface area of 1 cm^2 to one covering a surface area of 4202 m^2 (the equivalent of one football field) within 8 weeks (Cilento *et al.*, 1994). These studies indicated that it should be possible to collect autologous bladder cells from human patients, expand them in culture, and return them to the donor in sufficient quantities for reconstructive purposes (Cilento *et al.*, 1994; Liebert *et al.*, 1997; Solomon *et al.*, 1998; Nguyen *et al.*, 1999; Rackley *et al.*, 1999). Within the past decade, major advances have been made in the expansion of a variety of primary human cells, with specific techniques that make the use of autologous cells feasible for clinical application.

13.3 Biomaterials

For cell-based tissue engineering, the expanded cells are seeded onto a scaffold synthesized with the appropriate biomaterial. In tissue engineering, biomaterials replicate the biologic and mechanical function of the native extracellular matrix (ECM) found in tissues in the body by serving as an artificial ECM. Biomaterials provide a three-dimensional space for the cells to form new tissues with appropriate structure and function and also can allow for the delivery of cells and appropriate bioactive factors (such as cell adhesion peptides and growth factors), to desired sites in the body (Kim and Mooney, 1998). As the majority of mammalian cell types are anchorage-dependent and will die if no cell-adhesion substrate is available, biomaterials provide a cell-adhesion substrate that can deliver cells to specific sites in the body with high loading efficiency. Biomaterials can also provide mechanical support against *in vivo* forces to maintain the predefined 3D structure during tissue development.

To support the replacement of normal tissue without inflammation, the ideal biomaterial should be biodegradable and bioresorbable. Incompatible materials are destined for an inflammatory or foreign-body response that eventually leads to rejection and/or necrosis. Degradation products, if produced, should be removed from the body via metabolic pathways at an adequate rate that keeps the concentration of these degradation products in the tissues at a tolerable level (Bergsma *et al.*, 1995). The biomaterial should also provide an environment in which appropriate regulation of cell behavior (adhesion, proliferation, migration, and differentiation) can occur so that functional tissue can form. Cell behavior in the newly formed tissue has been shown to be regulated by multiple interactions of the cells with their microenvironment, including interactions with cell-adhesion ligands (Hynes, 1992) and with soluble growth factors (Deuel, 1997). Since biomaterials provide temporary mechanical support while the cells undergo spatial tissue reorganization, the properly chosen biomaterial should allow the engineered tissue to maintain sufficient mechanical integrity to support itself in early development. In late development, the biomaterial begins degradation and thus does not hinder further tissue growth (Kim and Mooney, 1998).

Generally, three classes of biomaterials have been utilized for engineering tissues: naturally derived materials (e.g. collagen and alginate), acellular tissue matrices (e.g. bladder submucosa and small intestinal submucosa), and synthetic polymers (e.g. polyglycolic acid (PGA), polylactic acid (PLA), and poly(lactic-co-glycolic acid) (PLGA)). These classes of biomaterials have been tested in respect to their biocompatibility (Pariente *et al.*, 2002). Naturally derived materials and acellular tissue matrices have the potential advantage of biological recognition. However, synthetic polymers can be produced reproducibly on a large scale with controlled properties of their strength, degradation rate, and microstructure.

13.3.1 Naturally derived materials

Collagen is the most abundant and ubiquitous structural protein in the body, and may be readily purified from both animal and human tissues with enzyme treatment and salt/acid extraction (Li, 1995). Collagen implants degrade through a sequential attack by lysosomal enzymes. The *in vivo* resorption rate can be regulated by controlling the density of the implant and the extent of intermolecular crosslinking. The lower the density, the greater the interstitial space and the larger the pores for cell infiltration, leading to a higher rate of implant degradation. Collagen contains cell adhesion domain sequences (e.g. RGD, arginine, glycine, aspartate) that may assist to retain the phenotype and activity of many types of cells, including fibroblasts (Silver and Pins, 1992) and chondrocytes (Sams and Nixon, 1995).

Alginate, a polysaccharide isolated from seaweed, has been used as an injectable cell delivery vehicle (Smidsrod and Skjak-Braek, 1990) and a cell immobilization matrix (Lim and Sun, 1980) owing to its gentle gelling properties in the presence of divalent ions such as calcium. Alginate is a family of copolymers of D-mannuronate and L-guluronate. The physical and mechanical properties of alginate gel are strongly correlated with the proportion and length of polyguluronate block in the alginate chains (Smidsrod and Skjak-Braek, 1990). Alginate is relatively biocompatible and it is approved by the Food and Drug Administration (FDA) for human use as wound dressing material.

13.3.2 Acellular tissue matrices

Acellular tissue matrices are collagen-rich matrices prepared by removing cellular components from tissues. The matrices are often prepared by mechanical and chemical manipulation of a segment of tissue (Yoo *et al.*, 1998; Chen *et al.*, 1999). The matrices slowly degrade upon implantation and are replaced and remodeled by ECM proteins synthesized and secreted by transplanted or in growing cells.

13.3.3 Synthetic polymers

Polyesters of naturally occurring α-hydroxy acids, including PGA, PLA, and PLGA, are widely used in tissue engineering. These polymers have gained FDA approval for human use in a variety of applications, including sutures. The ester bonds in these polymers are hydrolytically labile, and these polymers degrade by non-enzymatic hydrolysis. The degradation products of PGA, PLA, and PLGA are non-toxic natural metabolites and are eventually eliminated from the body in the form of carbon dioxide and water (Gilding, 1981). The degradation rate of these polymers can be tailored from several weeks to several years by altering crystallinity, initial molecular weight, and the copolymer ratio of lactic to glycolic acid. Since these polymers are thermoplastics, they can be easily formed into a 3D scaffold with a desired microstructure, gross shape, and dimension by various techniques, including molding, extrusion (Freed *et al.*, 1994), solvent casting (Mikos *et al.*, 1994), phase separation techniques, and gas foaming techniques (Harris *et al.*, 1998). Many applications in tissue engineering often require a scaffold with high porosity and ratio of surface area to volume. Other biodegradable synthetic polymers, including poly(anhydrides) and poly(ortho-esters), can also be used to fabricate scaffolds for tissue engineering with controlled properties (Peppas and Langer, 1994).

13.4 Alternate cell sources: stem cells and nuclear transfer

Human embryonic stem cells exhibit two remarkable properties: the ability to proliferate in an undifferentiated but pluripotent state (self-renewal), and the ability to differentiate into many specialized cell types (Brivanlou *et al.*, 2003). They can be isolated from the inner cell mass of the embryo during the blastocyst stage (5 days post-fertilization) and are usually grown on feeder layers consisting of mouse embryonic fibroblasts or human feeder cells (Richards *et al.*, 2002). More recent reports have shown that these cells can be grown without the use of a feeder layer (Amit *et al.*, 2003), and this avoid the exposure of these human cells to mouse viruses and proteins. The cells have demonstrated longevity in culture by maintaining their undifferentiated state for at least 80 passages when grown using current published protocols (Thomson *et al.*, 1998).

Human embryonic stem cells have been shown to differentiate into cells from all three embryonic germ layers *in vitro*. Skin and neurons have been formed, indicating ectodermal differentiation (Reubinoff *et al.*, 2001; Schuldiner *et al.*, 2001). Blood, cardiac cells, cartilage, endothelial, and muscle cells have been formed, indicating mesodermal differentiation (Kaufman *et al.*, 2001; Kehat *et al.*, 2001; Levenberg *et al.*, 2002). And pancreatic cells have been formed, indicating endodermal differentiation (Assady *et al.*, 2001). In addition, as further evidence of their pluripotency, embryonic stem cells can form embryoid bodies, which are cell aggregations that contain all three embryonic germ layers while in culture and can form teratomas *in vivo* (Itskovitz-Eldor *et al.*, 2002).

While there has been tremendous interest in the field of nuclear cloning since the birth of Dolly in 1997, the first successful nuclear transfer was reported over 50 years ago by Briggs and King (1952). Cloned frogs, which were the first vertebrates derived from nuclear transfer, were subsequently reported by Gurdon in 1962, but the nuclei were derived from non-adult sources. Tremendous advances in nuclear cloning technology have been reported since 2000, indicating the relative immaturity of the field. Dolly was not the first cloned mammal to be produced from adult cells: live lambs were produced in 1996 using nuclear transfer and differentiated epithelial cells derived from embryonic discs (Campbell *et al.*, 1996). The significance of Dolly was that she was the first mammal to be derived from an adult somatic cell using nuclear transfer (Wilmut *et al.*, 1997). Since then, animals from several species have been grown using nuclear transfer technology.

Two types of nuclear cloning, reproductive cloning and therapeutic cloning, have been described, and a better understanding of the differences between the two types may help to alleviate some of the controversy that surrounds these technologies (Vogelstein *et al.*, 2002). Banned in most countries for human applications, reproductive cloning is used to generate an embryo that has the identical genetic material as its cell source. This embryo can then be implanted into the uterus of a female to give rise to an infant that is a clone of the donor. On the other hand, therapeutic cloning is used to generate early stage embryos that are explanted in culture to produce embryonic stem cell lines whose genetic material is identical to that of its source. These autologous stem cells have the potential to become almost any type of cell in the adult body, and thus could be useful in tissue and organ replacement applications (Hochedlinger and Jaenisch, 2003). Therefore, therapeutic cloning, which has also been called somatic cell nuclear transfer, may provide an alternative source of transplantable cells. Figure 13.1 shows the strategy of combining therapeutic cloning with tissue engineering to develop tissues and organs. With current allogeneic tissue transplantation protocols, rejection is a frequent complication because of immunologic incompatibility, and immunosuppressive drugs are usually administered to treat and hopefully prevent host-versus-graft disease (Hochedlinger and Jaenisch, 2003). The use of transplantable tissue and organs derived from therapeutic cloning may potentially lead to the avoidance of immune responses that typically are associated with transplantation of non-autologous tissues (Lanza *et al.*, 1999).

While promising, somatic cell nuclear transfer technology has certain limitations that require further improvements before therapeutic cloning can be applied widely in replacement therapy. Currently, the efficiency of the overall cloning process is low. The majority of embryos derived from animal cloning do not survive after implantation (Solter, 2000). In practical terms, multiple nuclear transfers must be performed in order to produce one live offspring for animal cloning applications. The potential for cloned embryos to grow into live offspring is between 0.5 and 18% for sheep, pigs, and mice (Tsunoda and Kato, 2002). However, greater success (80%) has been reported in cattle (Kato *et al.*, 1998), which may be in part due to the availability of advanced bovine supporting technologies, such as *in vitro* embryo production and embryo transfer, which have been developed for this species for agricultural purposes. To improve cloning efficiency, further improvements are required in the multiple complex steps of nuclear transfer, such as enucleation and

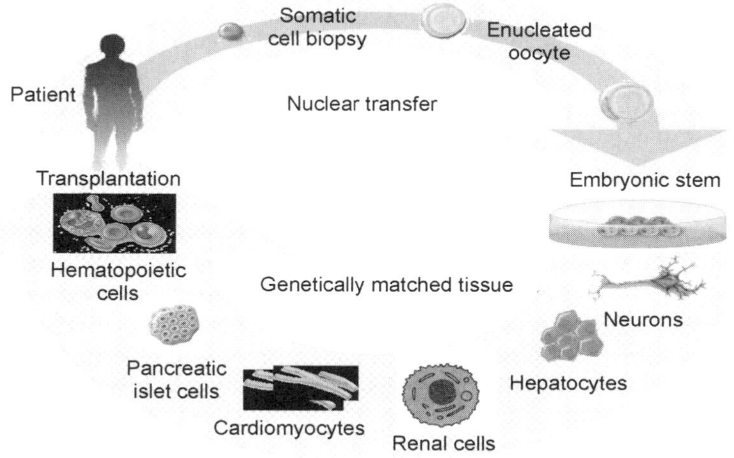

13.1 Strategy for therapeutic cloning and tissue engineering. Reactivation of key embryonic genes at the blastocyst stage is usually not present in embryos cloned from somatic cells, while embryos cloned from embryos consistently express early embryonic genes. Proper epigenetic reprogramming to an embryonic state may help to improve the cloning efficiency and reduce the incidence of abnormal cloned cells.

reconstruction, activation of oocytes, and cell cycle synchronization between donor cells and recipient oocytes (Dinnyes *et al.*, 2002).

Furthermore, common abnormalities have been found in newborn clones if they survive to birth, including enlarged size with an enlarged placenta (large-offspring syndrome) (Young *et al.*, 1998), respiratory distress, defects of the kidney, liver, heart, and brain (Cibelli *et al.*, 2002), obesity (Tamashiro *et al.*, 2002), and premature death (Ogonuki *et al.*, 2002). These may be related to the epigenetics of the cloned cells, which involve reversible modifications of the DNA or chromatin while the actual DNA (genetic) sequences remain intact. Faulty epigenetic reprogramming in clones, where the DNA methylation patterns, histone modifications, and the overall chromatin structure of the somatic nuclei are not being reprogrammed to an embryonic pattern of expression, may explain the above abnormalities (Hochedlinger and Jaenisch, 2003). Reactivation of key embryonic genes at the blast stage is usually not present in embryos cloned from somatic cells, while embryos cloned from embryos consistently express early embryonic genes (Bortvin *et al.*, 2003). Proper epigenetic reprogramming to an embryonic state may help to improve the cloning efficiency and reduce the incidence of abnormal cloned cells.

An alternate source of stem cells is the amniotic fluid and placenta. Amniotic fluid and the placenta are known to contain multiple partially differentiated cell types derived from the developing fetus. We isolated stem cell populations from these sources, called amniotic fluid and placental stem cells (AFPSC) that express embryonic and adult stem cell markers. The undifferentiated stem cells expand extensively without feeders and double every 36 hours. Unlike human embryonic stem cells, the AFPSC do not form tumors *in vivo*. Lines maintained for over 250 population doublings retained long telomeres and a normal karyotype. Amniotic fluid-derived stem (AFS) cells are broadly multipotent. Clonal human lines verified by retroviral marking can be induced to differentiate into cell types representing each embryonic germ layer, including cells of adipogenic, osteogenic, myogenic, endothelial, neuronal, and hepatic lineages. In this respect, they meet a commonly accepted criterion for pluripotent stem cells, without implying that they can generate every adult tissue. Examples of differentiated cells derived from AFS cells that display specialized functions include neuronal lineage cells secreting the neurotransmitter L-glutamate or expressing G-protein-gated inwardly rectifying potassium (GIRK) channels, hepatic lineage cells producing urea, osteogenic lineage cells forming tissue-engineered bone and myogenic lineage cells forming myocardial cells (Fig. 13.2). These types of stem cells can be obtained either from amniocentesis or chorionic villous sampling in the developing fetus (AFS cells), or from the placenta at the time of birth (placental stem cells). The cells could be preserved for self-use, and used without rejection, or they could be banked. A bank of 100 000 specimens could potentially supply 99% of the US population with a perfect genetic match for transplantation. Such a bank may be easier to create than with other cell sources, since there are approximately 4.5 million births per year in the US (De Coppi *et al.*, 2007).

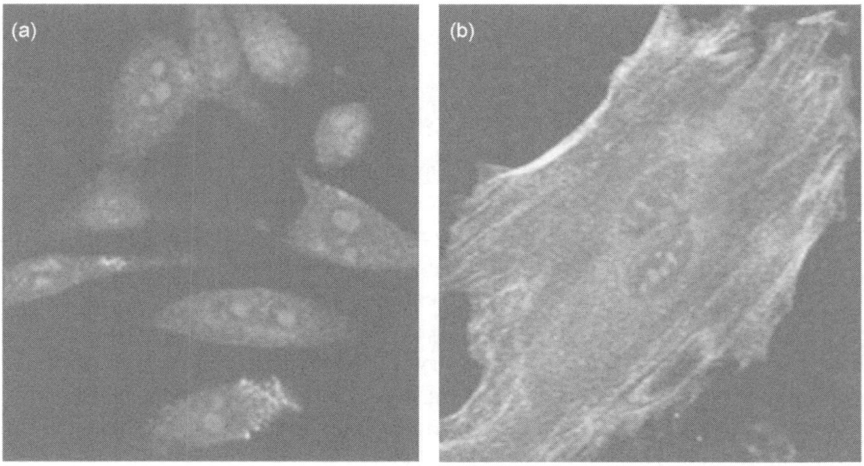

13.2 Amniotic stem cell differentiation: (a) undifferentiated mouse amniotic fluid stem (AFS) cells; (b) myogenic differentiated mouse AFS cell at 72 h after differentiation. Note the increase in light colored material in the cytoplasm; this represents FITC staining for tropomyosin, a cardiac marker.

13.5 Tissue engineering of specific structures

Investigators around the world, including our laboratory, have been working towards the development of several cell types and tissues and organs for clinical application.

13.5.1 Urethra

Various biomaterials without cells, such as PGA and acellular collagen-based matrices from small intestine and bladder, have been used experimentally in animal models for the regeneration of urethral tissue (Atala *et al.*, 1992; Chen *et al.*, 1999; Sievert *et al.*, 2000). Some of these biomaterials, such as acellular collagen matrices derived from bladder submucosa, have also been seeded with autologous cells for urethral reconstruction. Our laboratory has been able to replace tubularized urethral segments with cell-seeded collagen matrices.

Acellular collagen matrices derived from bladder submucosa by our laboratory have been used experimentally and clinically. In animal studies, segments of the urethra were resected and replaced with acellular matrix grafts in an onlay fashion. Histological examination showed complete epithelialization and progressive vessel and muscle infiltration, and the animals were able to void through the neo-urethras (Chen *et al.*, 1999). These results were confirmed in a clinical study of patients with hypospadias and urethral stricture disease (Yoo *et al.*, 1999). Decellularized cadaveric bladder submucosa was used as an onlay matrix for urethral repair in patients with stricture disease and hypospadias (see Fig. 13.3). Patent, functional neo-urethras were noted in these patients with up to a 7-year follow-up. The use of an off-the-shelf matrix appears to be beneficial for patients with abnormal urethral conditions and obviates the need for obtaining autologous grafts, thus decreasing operative time and eliminating donor site morbidity.

Unfortunately, the above techniques are not applicable for tubularized urethral repairs. The collagen matrices are able to replace urethral segments only when used in an onlay fashion. However, if a tubularized repair is needed, the collagen matrices should be seeded with autologous cells to avoid the risk of stricture formation and poor tissue development (De Filippo *et al.*, 2002). Therefore, tubularized collagen matrices seeded with autologous cells can be used successfully for total penile urethra replacement.

13.5.2 Bladder

Problems are encountered with the current method of bladder replacement or repair. Gastrointestinal segments are commonly used for this purpose, but they are designed to absorb specific solutes, whereas bladder tissue is designed for the excretion of solutes. Numerous investigators have attempted alternative materials and tissues for bladder replacement or repair due to the problems encountered with the use of gastrointestinal segments.

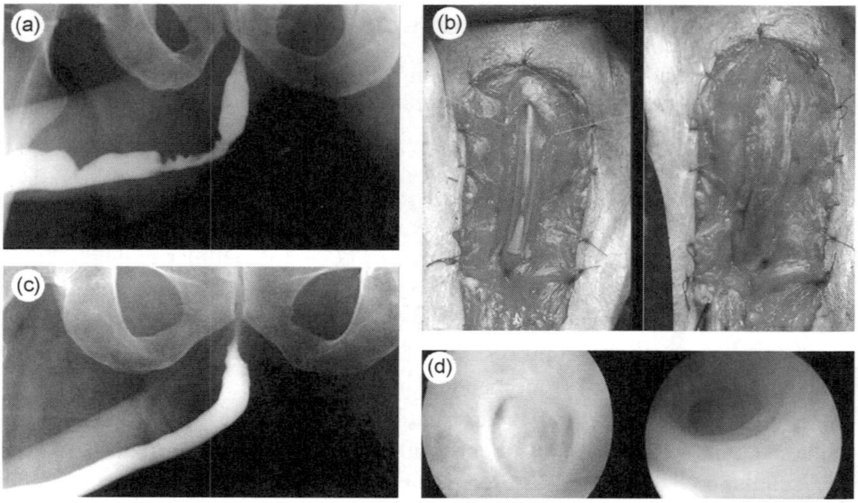

13.3 Tissue engineering of the urethra using a collagen matrix. (a) Representative case of a patient with a bulbar stricture. (b) In a urethral repair, strictured tissue is excised, preserving the urethral plate on the left side, and matrix is anastomosed to the urethral plate in an onlay fashion on the right. (c) Urethrogram 6 months after repair. (d) Cystoscopic view of urethra before surgery on the left side, and 4 months after repair on the right side.

The ability to use donor tissue efficiently and to provide the right conditions for long-term survival, differentiation, and growth determine the success of cell transplantation strategies for bladder reconstruction. These principles were applied in the creation of tissue engineered bladders in an animal model that required a subtotal cystectomy with subsequent replacement with a tissue engineered organ in beagle dogs (Oberpenning *et al.*, 1999). Urothelial and muscle cells were separately expanded from an autologous bladder biopsy and seeded onto a bladder-shaped biodegradable polymer scaffold. The results from this study showed that it is possible to tissue engineer bladders that are anatomically and functionally normal (see Fig. 13.4). Clinical trials for the application of this technology are currently being conducted.

A clinical experience involving engineered bladder tissue for cystoplasty reconstruction was conducted starting in 1999. A small pilot study of seven patients was reported, using a collagen scaffold seeded with cells either with or without omentum coverage, or a combined PGA–collagen scaffold seeded with cells and omental coverage (see Fig. 13.5). The patients reconstructed with the engineered bladder tissue created with the PGA–collagen cell-seeded scaffolds showed increased compliance, decreased end-filling pressures, increased capacities and longer dry periods (Atala *et al.*, 2006). Although the experience is promising in terms of showing that engineered tissues can be implanted safely, it is just a start in terms of accomplishing the goal of

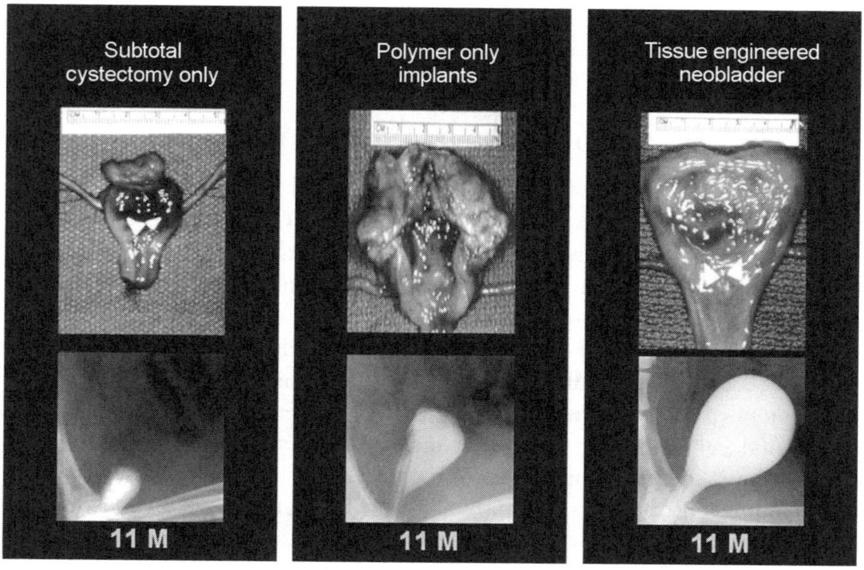

13.4 Comparison of tissue engineered neo-bladders in dogs. Gross specimens and cystograms at 11 months of the cystectomy-only, non-seeded controls, and cell-seeded tissue engineered bladder replacements. The cell-seeded tissue engineered bladder replacements achieved an average bladder capacity of 95% of the original precystectomy volume, and the compliance showed almost no difference from preoperative values. The others showed considerable loss of capacity and compliance.

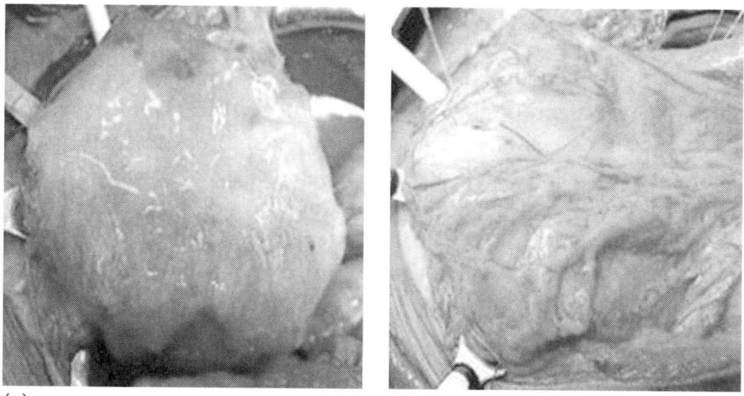

(a) (b)

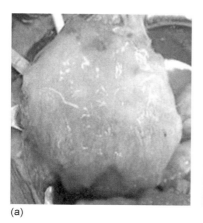

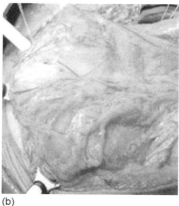

13.5 Construction of engineered bladder: (a) engineered bladder anastamosed to native bladder with running 4-0 polyglycolic sutures; (b) implant covered with fibrin glue and omentum.

engineering fully functional bladders. Further experimental and clinical work is being conducted.

13.5.3 Male and female reproductive organs

Reconstructive surgery is required for a wide variety of pathologic penile conditions, such as penile carcinoma, trauma, and severe erectile dysfunction, and congenital conditions such as ambiguous genitalia, hypospadias, and epispadias. One of the major limitations of phallic reconstructive surgery is the scarcity of sufficient autologous tissue.

The major components of the phallus are corporal smooth muscle and endothelial cells. The creation of autologous functional and structural corporal tissue *de novo* would be beneficial. Autologous cavernosal smooth muscle and endothelial cells were harvested, expanded, and seeded on acellular collagen matrices and implanted in a rabbit model (Kwon *et al.*, 2002). Histologic examination confirmed the appropriate organization of penile tissue phenotypes. Structural and functional studies, including cavernosography, cavernosometry, and mating studies, demonstrated that it is possible to engineer autologous functional penile tissue. Our laboratory is currently working on increasing the size of the engineered constructs.

Several pathologic conditions, including congenital malformations and malignancy of the vagina, can adversely affect normal development or anatomy. Vaginal reconstruction has traditionally been challenging due to the paucity of available native tissue. The feasibility of engineering vaginal tissue *in vivo* was investigated. Vaginal epithelial and smooth muscle cells of female rabbits were harvested, grown, and expanded in culture. These cells were seeded onto biodegradable polymer scaffolds, and the cell-seeded constructs were then implanted into nude mice for up to 6 weeks. Immunocytochemical, histological, and western blot analyses confirmed the presence of vaginal tissue phenotypes. Electrical field stimulation studies in the tissue-engineered constructs showed similar functional properties to those of normal vaginal tissue. When these constructs were used for autologous total vaginal replacement, patent vaginal structures were noted in the tissue engineered specimens, while the non-cell-seeded structures were noted to be stenotic (De Filippo *et al.*, 2003).

Congenital malformations of the uterus may have profound clinical implications. Patients with cloacal exstrophy and intersex disorders may not have sufficient uterine tissue present for future reproduction. We investigated the possibility of engineering functional uterine tissue using autologous cells (Wang *et al.*, 2003). Autologous rabbit uterine smooth muscle and epithelial cells were harvested, then grown and expanded in culture. These cells were seeded onto preconfigured uterine-shaped biodegradable polymer scaffolds, which were then used for subtotal uterine tissue replacement in the corresponding autologous animals. Upon retrieval 6 months after implantation, histological, immunocytochemical, and western blot analyses confirmed the presence of normal uterine tissue components. Biomechanical analyses and organ bath studies showed that the functional characteristics of these tissues were similar to those of normal uterine tissue. Breeding studies using these engineered uteri are being performed.

13.5.4 Kidney

We applied the principles of both tissue engineering and therapeutic cloning in an effort to produce genetically identical renal tissue in a large animal model, the cow (*Bos taurus*) (Lanza *et al.*, 2002). Bovine skin fibroblasts from adult Holstein steers were obtained by ear notch, and single donor cells were isolated and microinjected into the perivitelline space of donor enucleated oocytes (nuclear transfer). The resulting blastocysts were implanted into progestin-synchronized recipients to allow for further *in vivo* growth. After 12 weeks, cloned renal cells were harvested, expanded *in vitro*, then seeded onto biodegradable scaffolds. The constructs, which consisted of the cells and the scaffolds, were then implanted into the subcutaneous space of the same steer from which the cells were cloned to allow for tissue growth.

The kidney is a complex organ with multiple cell types and a complex functional anatomy that renders it one of the most difficult to reconstruct (Auchincloss and Bonventre, 2002). Previous efforts in tissue engineering of the kidney have been directed toward the development of extracorporeal renal support systems made of biological and synthetic components (Aebischer *et al.*, 1987; Lanza *et al.*, 1996; Humes *et al.*, 1999; Amiel *et al.*, 2000; Joki *et al.*, 2001), and *ex vivo* renal replacement devices are known to be life-sustaining. However, there would be obvious benefits for patients with end-stage kidney disease if these devices could be implanted long term without the need for an extracorporeal perfusion circuit or immunosuppressive drugs.

Cloned renal cells were seeded on scaffolds consisting of three collagen-coated cylindrical polycarbonate membranes (see Fig. 13.6(a)). The ends of the three scaffolds were connected to catheters that terminated into a collecting reservoir. This created a renal neo-organ with a mechanism for collecting the excreted urinary fluid (see Fig. 13.6(b)). These scaffolds with the collecting devices were transplanted subcutaneously into the same steer from which the genetic material originated, and then retrieved 12 weeks after implantation.

Chemical analysis of the collected urine-like fluid, including urea nitrogen and creatinine levels, electrolyte levels, specific gravity, and glucose concentration, revealed that the implanted renal cells possessed filtration, reabsorption, and secretory capabilities. Histological examination of the retrieved implants revealed extensive vascularization and self-organization of the cells into glomeruli and tubule-like structures. A clear continuity between the glomeruli, the tubules, and the polycarbonate membrane was noted, which allowed the

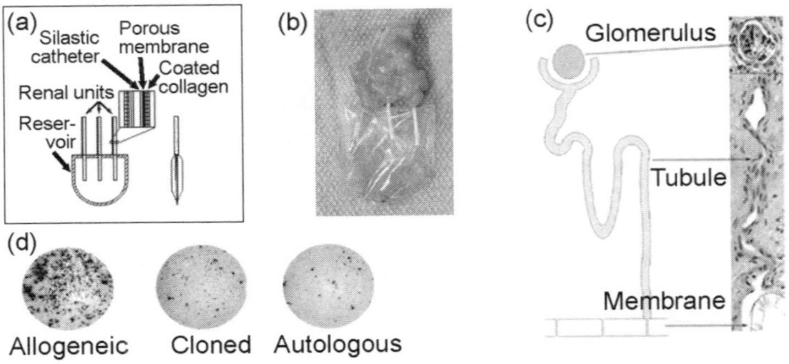

13.6 Combining therapeutic cloning and tissue engineering to produce kidney tissue. (a) Illustration of the tissue-engineered renal unit. (b) Renal unit seeded with cloned cells, three months after implantation, showing the accumulation of urinelike fluid. (c) Clear unidirectional continuity between the mature glomeruli, their tubules, and the polycarbonate membrane. (d) Elispot analyses of the frequencies of T cells that secrete IFN-gamma after primary and secondary stimulation with allogeneic renal cells, cloned renal cells, or nuclear donor fibroblasts.

passage of urine into the collecting reservoir (see Fig. 13.6(c)). Immunohistochemical analysis with renal-specific antibodies revealed the presence of renal proteins. Reverse transcriptase polymerase chain reaction (RT-PCR) analysis confirmed the transcription of renal specific RNA in the cloned specimens, and western blot analysis confirmed the presence of elevated renal-specific protein levels.

Since previous studies have shown that bovine clones harbor the oocyte mitochondrial DNA (Steinborn *et al.*, 2000), the donor egg's mitochondrial DNA (mtDNA) was thought to be a potential source of immunologic incompatibility. Differences in mtDNA-encoded proteins expressed by cloned cells could stimulate a T-cell response specific for mt-DNA-encoded minor histocompatibility antigens when the cloned cells are implanted back into the original nuclear donor (Fischer *et al.*, 1991). We used nucleotide sequencing of the mtDNA genomes of the clone and fibroblast nuclear donor to identify potential antigens in the muscle constructs. Only two amino acid substitutions were noted to distinguish the clone and the nuclear donor and, as a result, a maximum of two minor histocompatibility antigens could be defined. Given the lack of knowledge regarding peptide-binding motifs for bovine MHC class I molecules, there is no reliable method to predict the impact of these amino acid substitutions on bovine histocompatibility.

Oocyte-derived mtDNA was also thought to be a potential source of immunologic incompatibility in the cloned renal cells. Maternally transmitted minor histocompatibility antigens in mice have been shown to stimulate both skin allograft rejection *in vivo* and cytotoxic T lymphocytes expansion *in vitro* (Fischer *et al.*, 1991). These antigens could prevent the use of the cloned constructs in patients with chronic rejection of major histocompatibility matched human renal transplants (Yard *et al.*, 1993). We tested for a possible T-cell response to the cloned renal devices using delayed-type hypersensitivity testing *in vivo* and Elispot analysis of interferon-gamma secreting T cells *in vitro*. Both analyses revealed that the cloned renal cells showed no evidence of a T-cell response, suggesting that rejection will not necessarily occur in the presence of oocyte-derived mtDNA (see Fig. 13.6D). This finding may represent a step forward in overcoming the histocompatibility problem of stem cell therapy (Auchincloss and Bonventre, 2002).

These studies demonstrated that cells derived from nuclear transfer can be successfully harvested, expanded in culture, and transplanted *in vivo* with the use of biodegradable scaffolds on which the single suspended cells can organize into tissue structures that are genetically identical to that of the host. This was the first demonstration of the use of therapeutic cloning for regeneration of tissues *in vivo*.

13.5.5 Blood vessels

Xenogenic or synthetic materials have been used as replacement blood vessels for complex cardiovascular lesions. However, these materials typically lack growth potential and may place the recipient at risk for complications such as stenosis, thromboembolization or infection (Matsumura *et al.*, 2003).

Tissue-engineered vascular grafts have been constructed using autologous cells and biodegradable scaffolds and have been applied in dog and lamb models (Shinoka *et al.*, 1995, 1997; Watanabe *et al.*, 2001). The key advantage of using these autografts is that they degrade *in vivo*, and thus allow the new tissue to form without the long-term presence of foreign material (Matsumura *et al.*, 2003).

Movement of these techniques from the laboratory to the clinical setting has begun, with autologous vascular cells harvested, expanded, and seeded onto a biodegradable scaffold (Shin'oka *et al.*, 2001). The resultant autologous construct was used to replace a stenosed pulmonary artery that had been previously repaired. Seven months after implantation, no evidence of graft occlusion or aneurysmal changes was noted in the recipient.

13.5.6 Articular cartilage and trachea

Full-thickness articular cartilage lesions have limited healing capacity and thus represent a difficult management issue for the clinicians who treat adult patients with damaged articular cartilage (Hunter, 1995; O'Driscoll, 1998). Large defects can be associated with mechanical instability and may lead to degenerative joint disease if left untreated (Buckwalter and Mankin, 1998). Chondrocytes have been expanded and cultured onto biodegradable scaffolds to create engineered cartilage for use in large osteochondral defects in rabbits (Schaefer *et al.*, 2002). When sutured to a subchondral support, the engineered cartilage was able to withstand physiologic loading and underwent orderly remodeling of the large osteochondral defects in adult rabbits, providing a biomechanically functional template that was able to undergo orderly remodeling when subjected to quantitative structural and functional analyses.

Few treatment options are currently available for patients who suffer from severe congenital tracheal pathology, such as stenosis, agenesis, and atresia, due to the limited availability of autologous transplantable tissue in the neonatal period. Tissue engineering in the fetal period may be a viable alternative for the surgical treatment of these prenatally diagnosed congenital anomalies, as cells could be harvested and grown into transplantable tissue in parallel with the remainder of gestation. Chondrocytes from both elastic and hyaline cartilage specimens have been harvested from fetal lambs, expanded *in vitro*, then dynamically seeded onto biodegradable scaffolds. The constructs were then implanted as replacement tracheal tissue in fetal lambs. The resultant tissue-engineered cartilage was noted to undergo engraftment and epithelialization, while maintaining its structural support and patency. Furthermore, if native tracheal tissue is unavailable, engineered cartilage may be derived from bone marrow-derived mesenchymal progenitor cells as well (Fuchs *et al.*, 2003).

13.6 Cellular therapies

Several therapies on the cellular level have been investigated for direct treatment of medical conditions or use as enhancers of implanted engineered tissue. A variety of delivery systems have been tested, including injection, gene delivery, implantation, encapsulation, or a combination of systems. Applications of the following therapies are described: bulking agents, injectable muscle cells, endocrine replacement modalities, angiogenic agents, and anti-angiogenic agents.

13.6.1 Bulking agents

Injectable bulking agents can be endoscopically used in the treatment of both urinary incontinence and vesico-ureteral reflux. The advantages in treating urinary incontinence and vesico-ureteral reflux with this minimally invasive approach include the simplicity of this quick outpatient procedure and the low morbidity associated with it. Investigators are seeking alternative implant materials that would be safe for human use.

The ideal substance for the endoscopic treatment of reflux and incontinence should be injectable, non-antigenic, non-migratory, volume stable, and safe for human use. Toward this goal long-term studies were conducted to determine the effect of injectable chondrocytes *in vivo* (Atala *et al.*, 1993). It was initially determined that alginate, a liquid solution of glucuronic and mannuronic acid, embedded with chondrocytes, could serve as a synthetic substrate for the injectable delivery and maintenance of cartilage architecture *in vivo*. Alginate undergoes hydrolytic biodegradation, and its degradation time can be varied depending on the concentration of each of the polysaccharides. The use of autologous cartilage for the treatment of vesico-ureteral reflux in humans would satisfy all the requirements for an ideal injectable substance.

Chondrocytes derived from an ear biopsy can be readily grown and expanded in culture. Neocartilage formation can be achieved *in vitro* and *in vivo* using chondrocytes cultured on synthetic biodegradable polymers. In these experiments, the cartilage matrix replaced the alginate as the polysaccharide polymer underwent biodegradation. This system was adapted for the treatment of vesico-ureteral reflux in a porcine model (Atala *et al.*, 1994). These studies showed that chondrocytes can be easily harvested and combined with alginate *in vitro*; the suspension can be injected cystoscopically; and the elastic cartilage tissue formed is able to correct vesico-ureteral reflux without any evidence of obstruction.

Two multicenter clinical trials were conducted using this engineered chondrocyte technology. Patients with vesico-ureteral reflux were treated at ten centers throughout the US. The patients had a similar success rate as with other injectable substances in terms of cure (see Fig. 13.7). Chondrocyte formation was not noted in patients who had treatment failure. It is supposed that the patients who were cured have a biocompatible region of engineered autologous tissue present, rather than a foreign material (Diamond and Caldamone, 1999). Patients with urinary incontinence were also treated endoscopically with injected chondrocytes at three different medical centers. Phase 1 trials showed

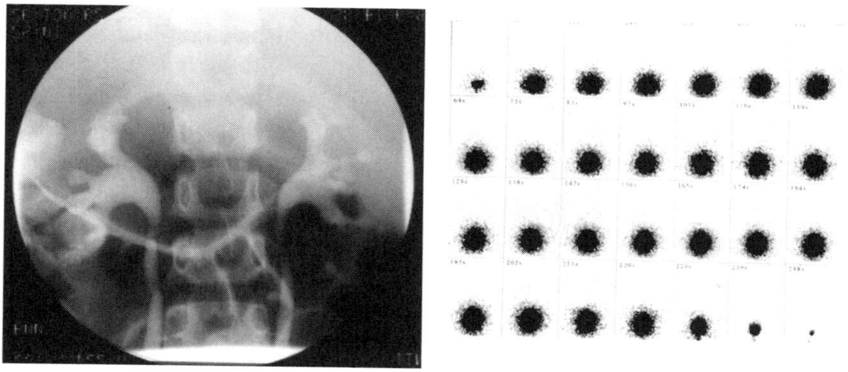

(a) (b)

13.7 Autologous chondrocytes for the treatment of vesico-ureteral reflux: (a) preoperative voiding cysto-urethrogram of a patient with bilateral reflux; (b) postoperative radionuclide cystogram of the same patient 6 months after injection of autologous chondrocytes.

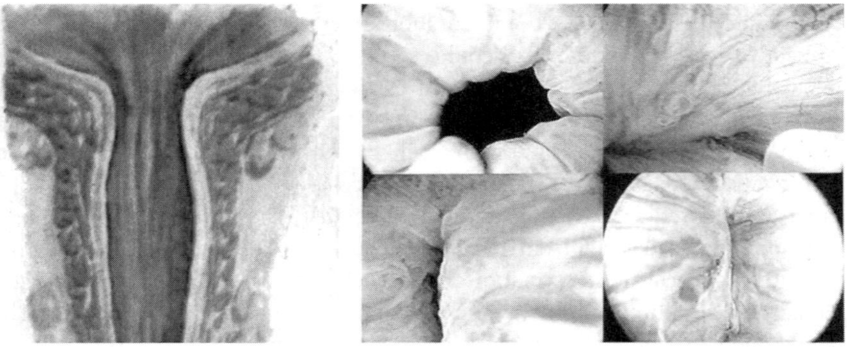

13.8 Chondrocytes are harvested and combined with alginate *in vitro*, and the suspension is injected cystoscopically as a bulking agent to treat urinary incontinence in adults.

an approximate success rate of 80% at follow-up 3 and 12 months post-operatively (Bent *et al.*, 2001). Therefore, human application of cell-based tissue engineering technology for urologic applications occurred with the injection of chondrocytes for the correction of vesico-ureteral reflux in children and for urinary incontinence in adults (see Fig. 13.8) (Diamond and Caldamone, 1999; Bent *et al.*, 2001).

13.6.2 Injectable muscle cells

The potential use of injectable cultured myoblasts for the treatment of stress urinary incontinence has been investigated (Chancellor *et al.*, 2000). Labeled myoblasts were directly injected into the proximal urethra and lateral bladder walls of nude mice with a micro-syringe in an open surgical procedure. Tissue harvested up to 35 days post-injection contained the labeled myoblasts, as well as evidence of differentiation of the labeled myoblasts into regenerative myofibers. The authors reported that a significant portion of the injected myoblast population persisted *in vivo*. Similar techniques of sphincteric-derived muscle cells have been used for the treatment of urinary incontinence in a pig model (Strasser *et al.*, 2003). The fact that myoblasts can be labeled and survive after injection and begin the process of myogenic differentiation further supports the feasibility of using cultured cells of muscular origin as an injectable bioimplant.

The use of injectable muscle precursor cells has also been investigated for use in the treatment of urinary incontinence due to irreversible urethral sphincter injury or maldevelopment. Muscle precursor cells are the quiescent satellite cells found in each myofiber that proliferate to form myoblasts and eventually myotubes and new muscle tissue. Intrinsic muscle precursor cells have previously been shown to play an active role in the regeneration of injured striated urethral sphincter. In a subsequent study, autologous muscle precursor cells were injected into a rat model of urethral sphincter injury, and both replacement of mature myotubes as well as restoration of functional motor units were noted in the regenerating sphincteric muscle tissue (Yiou *et al.*, 2003). This is the first demonstration of the replacement of both sphincter muscle tissue and its innervation by the injection of muscle precursor cells. As a result, muscle precursor cells may be a minimally invasive solution for urinary incontinence in patients with irreversible urinary sphincter muscle insufficiency.

13.6.3 Endocrine replacement

Patients with testicular dysfunction and hypogonadal disorders are dependent on androgen replacement therapy to restore and maintain physiological levels of serum testosterone and its metabolites, dihydrotestosterone and estradiol (Machluf *et al.*, 2003). Currently available androgen replacement modalities, such as testosterone tablets and capsules, depot injections, and skin patches may be associated with fluctuating serum levels and complications such as fluid and nitrogen retention, erythropoiesis, hypertension, and bone density changes (Santen and Swerdloff, 1990). Since Leydig cells of the testes are the major source of testosterone in men, implantation of heterologous Leydig cells or gonadal tissue fragments has previously been proposed as a method for chronic testosterone replacement (Tai *et al.*, 1989). But these approaches were limited by the failure of the tissues and cells to produce testosterone.

Encapsulation of cells in biocompatible and semipermeable polymeric membranes has been an effective method to protect against a host immune response as well as to maintain viability of the cells while allowing the secretion of desired therapeutic agents (De Vos *et al.*, 1997; Tai and Sun, 1993). Alginate poly-L-lysine-encapsulated Leydig cell microspheres were used as a novel method for testosterone delivery *in vivo* (Machluf *et al.*, 2003). Elevated stable serum testosterone levels were noted in castrated adult rats over the course of the study, suggesting that microencapsulated Leydig cells may be a potential therapeutic modality for testosterone supplementation.

13.6.4 Angiogenic agents

The engineering of large organs will require a vascular network of arteries, veins, and capillaries to deliver nutrients to each cell. One possible method of vascularization is through the use of gene delivery of angiogenic agents such as vascular endothelial growth factor (VEGF) with the implantation of vascular endothelial cells (EC) in order to enhance neovascularization of engineered tissues. Skeletal myoblasts from adult mice were cultured and transfected with an adenovirus encoding VEGF and combined with human vascular endothelial cells (Nomi *et al.*, 2002). The mixtures of cells were injected subcutaneously in nude mice, and the engineered tissues were retrieved up to 8 weeks after implantation. The transfected cells were noted to form muscle with neovascularization by histology and immunohistochemical probing with maintenance of their muscle volume, while engineered muscle of non-transfected cells had a significantly smaller mass of cells with loss of muscle volume over time, less neovascularization, and no surviving endothelial cells. These results indicate that a combination of VEGF and endothelial cells may be useful for inducing neovascularization and volume preservation in engineered tissue.

13.6.5 Anti-angiogenic agents

The delivery of anti-angiogenic agents may help to slow tumor growth for a variety of neoplasms. Encapsulated hamster kidney cells transfected with the angiogenesis inhibitor endostatin were used for local delivery on human glioma cell line xenografts (Joki *et al.*, 2001). The release of biologically active endostatin led to significant inhibition of endothelial cell proliferation and substantial reduction in tumor weight. Continuous local delivery of endostatin via encapsulated endostatin-secreting cells may be an effective therapeutic option for a variety of tumor types.

13.7 Conclusions and future trends

As regenerative medicine incorporates the fields of tissue engineering, cell biology, nuclear transfer, and materials science, personnel who have mastered the techniques of cell harvest, culture, expansion, transplantation, as well as polymer design are essential for the successful application of these technologies. Various tissues are at different stages of development, with some already being used clinically. Recent progress suggests that engineered tissues may have an expanded clinical applicability in the future and may represent a viable therapeutic option for those who require tissue replacement or repair. The future outlook is positive with the recent discovery of an alternate source of stem cells in the amniotic fluid and placenta derived from the developing fetus. These cells meet the accepted criterion for pluripotent stem cells without the implication that they can generate every tissue in the adult. Some advantages are that they do not form tumors *in vivo*; they are broadly multipotent; and they can be routinely harvested from amniocentesis sampling during pregnancy or obtained from the placenta at birth. In the future, banking these stem cells may provide a convenient source for autologous therapy and for matching recipients with histocompatible donors (De Coppi *et al.*, 2007). Regenerative medicine efforts are currently underway experimentally for virtually every type of tissue and organ within the human body and hold great promise for the future of medicine.

13.7 References

Aebischer P, Ip T K, Panol G, Galletti P M. (1987), 'The bioartificial kidney: progress towards an ultrafiltration device with renal epithelial cells processing', *Life Support Systems*, **5** (2), 159–168.

Amiel G E, Yoo J J, Atala A (2000), 'Renal therapy using tissue-engineered constructs and gene delivery', *World J Urol*, **18** (1), 71–79.

Amit M, Margulets V, Segev H, *et al.* (2003), 'Human feeder layers for human embryonic stem cells', *Biol Reprod*, **68** (6), 2150–2156.

Assady S, Maor G, Amit M, *et al.* (2001), 'Insulin production by human embryonic stem cells', *Diabetes*, **50** (8),1691–1697.

Atala A, Lanza R P (2001). Preface. In Atala A, Lanza RP, eds, *Methods of Tissue Engineering*, San Diego, Academic Press.

Atala A, Vacanti J P, Peters C A, *et al.* (1992), 'Formation of urothelial structures *in vivo* from dissociated cells attached to biodegradable polymer scaffolds *in vitro*', *J Urol*, **148** (2), 658–662.

Atala A, Cima L G, Kim W, *et al.* (1993), 'Injectable alginate seeded with chondrocytes as a potential treatment for vesicoureteral reflux', *J Urol*, **150** (2), 745–747.

Atala A, Kim W, Paige K T, *et al.* (1994), 'Endoscopic treatment of vesicoureteral reflux with a chondrocyte-alginate suspension', *J Urol*, **152** (2), 641–643.

Atala A, Guzman L, Retik A B (1999), 'A novel inert collagen matrix for hypospadias repair', *J Urol*, **162** (3),1148–1151.

Atala A, Bauer S B, Soker S, *et al.* (2006), 'Tissue-engineered autologous bladders for patients needing cystoplasty', *The Lancet*, **367** (9518), 1241–1246.

Auchincloss H, Bonventre J V (2002), 'Transplanting cloned cells into therapeutic promise' [comment], *Nature Biotechnol*, **20** (7), 665–666.

Bent A, Tutrone R, McLennan M, *et al.* (2001),'Treatment of intrinsic sphincter deficiency using autologous ear chondrocytes as a bulking agent', *Neurourol Urodynam*, **20** (2),157–165.

Bergsma J E, Rozema F R, Bos R R, *et al.* (1995), '*In vivo* degradation and biocompatibility study of *in vitro* pre-degraded aspolymerized polyactide particles' [comment], *Biomaterials*, **16** (4), 267–274.

Bortvin A, Eggan K, Skaletsky H, *et al.* (2003), 'Incomplete reactivation of Oct4-related genes in mouse embryos cloned from somatic nuclei', *Development*, **130** (8),1673–1680.

Briggs R, King T J (1952), 'Transplantation of living nuclei from blastula cells into enucleated frogs' eggs', *Proc Natl Acad Sci USA*, **38** (5), 455.

Brivanlou A H, Gage F H, Jaenisch R, Jessell T, *et al.* (2003), 'Stem cells: setting standards for human embryonic stem cells' [comment], *Science*, **300** (5621), 913–916.

Buckwalter J A, Mankin H J (1998), 'Articular cartilage repair and transplantation', *Arthritis Rheum*, **41** (8), 1331.

Campbell K H, McWhir J, Ritchie W A, Wilmut I (1996), 'Sheep cloned by nuclear transfer from a cultured cell line' [comment], *Nature*, **380** (6569), 64–66.

Chancellor M B, Yokoyama T, Tirney S, *et al.* (2000), 'Preliminary results of myoblast injection into the urethra and bladder wall: a possible method for the treatment of stress urinary incontinence and impaired detrusor contractility', *Neurourol Urodynam*, **19** (3), 279–287.

Chen F, Yoo J J, Atala A (1999), 'Acellular collagen matrix as a possible "off the shelf" biomaterial for urethral repair', *Urology*, **54** (3), 407–410.

Cibelli J B, Campbell K H, Seidel G E, *et al.* (2002), 'The health profile of cloned animals', *Nature Biotechnol*, **20** (1), 13–14.

Cilento B G, Freeman M R, Schneck F X, *et al.* (1994), 'Phenotypic and cytogenetic characterization of human bladder urothelia expanded *in vitro*', *J Urol*, **152** (2), 665–670.

Dahms S E, Piechota H J, Dahiya R, *et al.* (1998), 'Composition and biomechanical properties of the bladder acellular matrix graft: comparative analysis in rat, pig and human', *Br J Urol*, **82** (3), 411–419.

De Coppi P, Bartsch Jr G, Siddiqui M M, *et al.* (2007), 'Isolation of amniotic stem cell lines with potential for therapy', *Nature Biotechnol*, **25** (1), 100–106.

De Filippo R E, Yoo J J, Atala A (2002), 'Urethral replacement using cell seeded tubularized collagen matrices', *J Urol*, **168** (4), 1789–1793.

De Filippo R E, Yoo J J, Atala A (2003), 'Engineering of vaginal tissue for total reconstruction', *J Urol*, **169A** (3), 1057.

Deuel T F (1997), 'Growth factors', in Lanza R P, Langer R, Chick W L, *Principles of Tissue Engineering*, New York, Academic Press, 133–149.

De Vos P, De Haan B, Van Schilfgaarde R (1997), 'Effect of the alginate composition on the biocompatibility of alginate–polylysine microcapsules', *Biomaterials*, **18** (3), 273–278.

Diamond D A, Caldamone A A (1999), 'Endoscopic correction of vesicoureteral reflux in children using autologous chondrocytes: preliminary results', *J Urol*, **162** (3), 1185–1188.

Dinnyes A, De Sousa P, King T, *et al.* (2002), 'Somatic cell nuclear transfer: recent progress and challenges', *Cloning Stem Cells*, **4** (1), 81–90.

Fauza D O, Fishman S J, Mehegan K, *et al.* (1998), 'Videofetoscopically assisted fetal tissue engineering: skin replacement', *J Pediatr Surg*, **33** (2), 357–361.

Fischer Lindahl K, Hermel E, Loveland B E, *et al.* (1991), 'Maternally transmitted antigen of mice: a model transplantation antigen', *Annu Rev Immunol*, **9**, 351–372.

Freed L E, Vunjak-Novakovic G, Biron R J, *et al.* (1994), 'Biodegradable polymer scaffolds for tissue engineering', *Biotechnology NY*, **12** (7), 689–693.

Fuchs J R, Hannouche D, Terada S, *et al.* (2003), 'Fetal tracheal augmentation with cartilage engineered from bone marrow-derived mesenchymal progenitor cells', *J Pediatr Surg*, **38** (6), 984–987.

Gilding D K. (1981), 'Biodegradable polymers', in Williams D F, *Biocompatibility of Clinical Implant Materials*, Boca Raton, CRC Press, 209–232.

Godbey W T, Atala A (2002), '*In vitro* systems for tissue engineering', *Ann NY Acad Sci*, **961**,10–26.

Gurdon J B (1962), 'The developmental capacity of nuclei taken from intestinal epithelium cells of feeding tadpoles', *J Embryology and Exp Morph*, **10**, 622–640.

Harris L D, Kim B S, Mooney D J (1998), 'Open pore biodegradable matrices formed with gas foaming', *J Biomed Mater Res*, **42** (3), 396–402.

Hochedlinger K, Jaenisch R (2003), 'Nuclear transplantation, embryonic stem cells, and the potential for cell therapy' [comment], *N Engl J Med*, **349** (3), 275–286.

Humes H D, Buffington D A, MacKay S M, *et al.* (1999), 'Replacement of renal function in uremic animals with a tissue-engineered kidney' [comment], *Nature Biotechnol*, **17** (5), 451–455.

Hunter W (1995), 'Of the structure and disease of articulating cartilages', *Clin Orthop*, **317**, 3–6.

Hynes R O (1992). 'Integrins: versatility, modulation, and signaling in cell adhesion', *Cell*, **69** (1), 11–25.

Itskovitz-Eldor J, Schuldiner M, Karsenti D, *et al.*(2002), 'Differentiation of human embryonic stem cells into embryoid bodies compromising the three embryonic germ layers', *Mol Med*, **6** (2), 88–95.

Joki T, Machluf M, Atala A, *et al.* (2001), 'Continuous release of endostatin from microencapsulated engineered cells for tumor therapy' [comment], *Nature Biotechnol*, **19** (1), 35–39.

Kato Y, Tani T, Sotomaru Y, *et al.* (1998), 'Eight calves cloned from somatic cells of a single adult' [comment], *Science*, **282** (5396), 2095–2098.

Kaufman D S, Hanson E T, Lewis R L, *et al.* (2001), 'Hematopoietic colony-forming cells derived from human embryonic stem cells', *Proc Natl Acad Sci USA*, **98** (19), 10716–10721.

Kehat I, Kenyagin-Karsenti D, Snir M, *et al.* (2001), 'Human embryonic stem cells can differentiate into myocytes with structural and functional properties of cardiomyocytes' [comment], *J Clin Invest*, **108** (3), 407–414.

Kim B S, Mooney D J (1998), 'Development of biocompatible synthetic extracellular matrices for tissue engineering', *Trends Biotechnol*, **16** (5), 224–230.

Kwon T G, Yoo J J, Atala A (2002), 'Autologous penile corpora cavernosa replacement using tissue engineering techniques', *J Urol*, **168** (4),1754–1758.

Lanza R P, Hayes J L, Chick W L (1996), 'Encapsulated cell technology', *Nature Biotechnol*, **14** (9), 1107–1111.

Lanza R P, Cibelli J B, West M D (1999), 'Prospects for the use of nuclear transfer in human transplantation', *Nature Biotechnol*, **17** (12), 1171–1174.

Lanza R P, Chung H Y, Yoo J J, *et al.* (2002), 'Generation of histocompatible tissues using nuclear transplantation' [comment], *Nature Biotechnol*, **20** (7), 689–699.

Levenberg S, Golub J S, Amit M, *et al.* (2002), 'Endothelial cells derived from human embryonic stem cells', *Proc Natl Acad Sci USA*, **99** (7), 4391–4396.

Li S T (1995), 'Biologic biomaterials: tissue-derived biomaterials (collagen)', in Bronzino J D, *The Biomedical Engineering Handbook*, Boca Raton, CRC Press, 627–647.

Liebert M, Hubbel A, Chung M, *et al.* (1997), 'Expression of mal is associated with urothelial differentiation *in vitro*: identification by differential display reverse-transcriptase polymerase chain reaction', *Differentiation*, **61** (3), 177–185.

Lim F, Sun A M (1980), 'Microencapsulated islets as bioartificial endocrine pancreas', *Science*, **210** (4472), 908–910.

Machluf M, Atala A (1998), 'Emerging concepts for tissue and organ transplantation', *Graft*, **1** (1), 31–37.

Machluf M, Orsola A, Boorjian S, *et al.* (2003), 'Microencapsulation of Leydig cells: a system for testosterone supplementation', *Endocrinology*, **144** (11), 4975–4979.

Matsumura G, Miyagawa-Tomita S, Shin'oka T, *et al.* (2003), 'First evidence that bone marrow cells contribute to the construction of tissue-engineered vascular autografts *in vivo*', *Circulation*, **108** (14),1729–1734.

Mikos A G, Thorsen A J, Czerwonka L A, *et al.* (1994), 'Preparation and characterization of poly(L-lactic acid) foams', *Polymer*, **35**, 1068–1077.

Nguyen H T, Park J M, Peters C A, *et al.* (1999), 'Cell-specific activation of the HB-EGF and ErbB1 genes by stretch in primary human bladder cells', *In Vitro Cell Dev Biol* Anim, **35** (7), 371–375.

Nomi M, Atala A, Coppi P D, *et al.* (2002), 'Principles of neovascularization for tissue engineering', *Mol Aspects Med*, **23** (6), 463–483.

Oberpenning F, Meng J, Yoo J J, *et al.* (1999), '*De novo* reconstitution of a functional mammalian urinary bladder by tissue engineering' [comment], *Nature Biotechnol*,

17 (2), 149–155.
O'Driscoll S W (1998), 'The healing and regeneration of articular cartilage', *J Bone Joint Surg Am*, **80** (12), 1795–1812.
Ogonuki N, Inoue K, Yamamoto Y, *et al.* (2002), 'Early death of mice cloned from somatic cells', *Nat Genet*, **30** (3), 253–254.
Pariente J L, Kim B S, Atala A (2002), '*In vitro* biocompatibility evaluation of naturally derived and synthetic biomaterials using normal human bladder smooth muscle cells', *J Urol*, **167** (4), 1867–1871.
Peppas N A, Langer R (1994), 'New challenges in biomaterials' [comment], *Science*, **263** (5154), 1715–1720.
Puthenveettil J A, Burger M S, Reznikoff CA (1999), 'Replicative senescence in human uroepithelial cells', *Adv Exp Med Biol*, **462**, 83–91.
Rackley R R, Bandyopadhyay S K, Fazeli-Matin S, *et al.* (1999), 'Immunoregulatory potential of urothelium: characterization of NF-kappaB signal transduction', *J Urol*, **162** (5), 1812–1816.
Reubinoff B E, Itsykson P, Turetsky T, *et al.* (2001), 'Neural progenitors from human embryonic stem cells' [comment], *Nature Biotechnol*, **19** (12),1134–1140.
Richards M, Fong C Y, Chan W K, *et al.* (2002), 'Human feeders support prolonged undifferentiated growth of human inner cell masses and embryonic stem cells', [comment], *Nature Biotechnol*, **20** (9), 933–936.
Sams A E, Nixon A J (1995), 'Chondrocyte-laden collagen scaffolds for resurfacing extensive articular cartilage defects', *Osteoarthritis Cartilage*, **3** (1), 47–59.
Santen R J, Swerdloff R S (1990), 'Clinical aspects of androgen therapy', presented at the Workshop Conference on Androgen Therapy: Biologic and Clinical Consequences, Penn State Workshop Proceedings, Philadelphia, PA.
Schaefer D, Martin I, Jundt G, *et al.* (2002), 'Tissue-engineered composites for the repair of large osteochondral defects', *Arthritis Rheum*, **46** (9), 2524–2534.
Schuldiner M, Eiges R, Eden A, *et al.* (2001), 'Induced neuronal differentiation of human embryonic stem cells', *Brain Res*, **913** (2), 201–205.
Shin'oka T, Imai Y, Ikada Y (2001), 'Transplantation of a tissue-engineered pulmonary artery', *N Engl J Med*, **344** (7), 532–533.
Shinoka T, Breuer CK, Tanel R E, *et al.* (1995), 'Tissue engineering heart valves: valve leaflet replacement study in a lamb model', *Ann Thorac Surg*, **60** (6S), S513–S516.
Shinoka T, Shum-Tim D, Ma P X, *et al.* (1997), 'Tissue-engineered heart valve leaflets: does cell origin affect outcome?', *Circulation*, **96** (9S), II-102–107.
Sievert K D, Bakircioglu M E, Nunes L, *et al.* (2000), 'Homologous acellular matrix graft for urethral reconstruction in the rabbit: histological and functional evaluation', *J Urol*, **163** (6), 1958–1965.
Silver F H, Pins G (1992), 'Cell growth on collagen: a review of tissue engineering using scaffolds containing extracellular matrix', *J Long Term Eff Med Implants*, **2** (1), 67–81.
Smidsrod O, Skjak-Braek G (1990), 'Alginate as immobilization matrix for cells', *Trends Biotech*, **8** (3), 71–78.
Solomon L Z, Jennings A M, Sharpe P, *et al.* (1998), 'Effects of short-chain fatty acids on primary urothelial cells in culture: implications for intravesical use in enterocystoplasties' [comment], J *Lab Clin Med*, **132** (4), 279–283.
Solter D (2000), 'Mammalian cloning: advances and limitations', *Nat Rev Genet*, **1** (3), 199–207.
Steinborn R, Schinogl P, Zakhartchenko V, *et al.* (2000), 'Mitochondrial DNA heteroplasmy in cloned cattle produced by fetal and adult cell cloning', *Nat Genet*, **25** (3), 255–257.

Strasser H, Marksteiner R, Eva M, *et al.* (2003), 'Transurethral ultrasound guided injection of clonally cultured autologous myoblasts and fibroblasts: experimental results', presented at the Proceedings of the 2003 International Bladder Symposium, Arlington, VA.

Tai I T, Sun A M (1993), 'Microencapsulation of recombinant cells: a new delivery system for gene therapy', *FASEB J*, **7** (11), 1061–1069.

Tai J, Johnson H W, Tze W J (1989), 'Successful transplantation of Leydig cells in castrated inbred rats', *Transplantation*, **47** (6), 1087–1089.

Tamashiro K L, Wakayama T, Akutsu H, *et al.* (2002), 'Cloned mice have an obese phenotype not transmitted to their offspring' [comment], *Nature Medicine*, **8** (3), 262–267.

Thomson J A, Itskovitz-Eldor J, Shapiro S S, *et al.* (1998), 'Embryonic stem cell lines derived from human blastocysts' [comment], *Science*, **282** (5391), 1145–1147. Erratum in *Science* (1998) **282**, 1827.

Tsunoda Y, Kato Y (2002), 'Recent progress and problems in animal cloning', *Differentiation*, **69** (4–5), 158–161.

Vogelstein B, Alberts B, Shine K (2002), 'Genetics: please don't call it cloning!' [comment], *Science*, **295** (5558), 1237.

Wang T, Koh CJ, Yoo J J, *et al.* (2003), 'Creation of an engineered uterus for surgical reconstruction' [abstract], presented at the Proceedings of the American Academy of Pediatrics Section on Urology, New Orleans, LA.

Watanabe M, Shin'oka T, Tohyama S, *et al.* (2001), 'Tissue engineered vascular autograft: inferior vena cava replacement in a dog model', *Tissue Eng*, **7** (4), 429–439.

Wilmut I, Schnieke A E, McWhir J, *et al.* (1997), 'Viable offspring derived from fetal and adult mammalian cells' [comment], *Nature*, **385** (6619), 810–813. Erratum in *Nature* (1997) **386**, 200.

Yard BA, Kooymans-Couthino M, Reterink T, *et al.* (1993), 'Analysis of T cell lines from rejecting renal allografts', *Kidney Int*, **39**, S133–S138.

Yiou R, Yoo J J, Atala A. (2003), 'Restoration of functional motor units in a rat model of sphincter injury by muscle precursor cell autografts', *Transplantation*, **76** (7), 1053–1060.

Yoo J J, Atala A (1997), 'A novel gene delivery system using urothelial tissue engineered neo-organs', *J Urol*, **158** (3), 1060–1070.

Yoo J J, Meng J, Oberpenning F, *et al.*(1998), 'Bladder augmentation using allogenic bladder submucosa seeded with cells', *Urology*, **51** (2), 221–225.

Yoo J J, Park H J, Lee I, *et al.* (1999), 'Autologous engineered cartilage rods for penile reconstruction', *J Urol*, **162** (3), 1119–1121.

Young L E, Sinclair K D, Wilmut I (1998), 'Large offspring syndrome in cattle and sheep', *Rev Reprod*, **3** (3), 155–163.

14
Bone regeneration and repair using tissue engineering

P WOŹNIAK, Medical University of Warsaw, Poland and
A J EL HAJ, Keele University Medical School, UK

14.1 Introduction

Bone is a metabolically active, highly vascularised tissue with a unique ability to regenerate without creation of a scar (Sommerfeldt and Rubin, 2001). The functions of bone encompass the body's physiological, structural and biological storage demands. As such, the remarkable plasticity of this organ's ability to remodel and respond to external and internal cues is a key feature. Repair strategies for bone which rely heavily on metal and non-degradable implant materials have existed for centuries. Although able to replace the structural functions, these implants do not address the other important functions of the skeleton in maintaining ion homeostasis, storing biological factors and cues within the matrices or remodelling to the external load bearing. In addition, the lifespan of the repair is short owing to loosening of the implant and a lack of integration with the living tissue, which leads ultimately to failure. With this in mind, regenerative medicine seeks to meet all the biological functions within a repair strategy replacing metal implants with donor cells and degradable materials. If it were possible to create a biological substitute that would go on to remodel and respond in a normal way to the surrounding environmental cues, the need for revision operations would be drastically reduced.

In this chapter, we set out to describe the characteristics of bone and its ability to remodel and restructure to internal hormonal cues and external mechanical environments. We outline some of the new strategies for tissue engineering bone; sources of donor cells and new generation 'smart materials'. Finally, we identify the challenges and the scope for improvement in the field.

14.2 Principles of bone biology

14.2.1 Bone functions

Bone is a type of hard, very dense, endoskeletal connective tissue derived from mesoderm (Baron, 2003). Together with cartilage tissue, bones create a skeletal system. There are three general functions of bone tissue: (1) support of body structures and the muscles' locomotor activities; (2) protection of delicate critical organs, such as brain, spine and heart; and (3) accumulation of ions, mainly calcium and phosphate, which is essential for the maintenance of serum homeostasis (Baron, 2003; Shea and Miller, 2005). Bone tissue is composed of cells and the extracellular matrix (ECM). However, in contrast to other connective tissues the ECM of bone has a unique ability to become calcified, which is crucial for the mechanical properties of the tissue (Einhorn, 1996).

14.2.2 Structure of bone

Macroscopic features

By scanning the human skeleton, it is possible to distinguish two types of bones: long bones (femur, tibia, etc.) and flat bones (mandible, skull bones and scapula). The obvious anatomical difference between both types of bones results in divergences in their function. While long bones are generally involved in locomotion, flat bones usually serve a protective role (Sommerfeldt and Rubin, 2001; Baron, 2003).

Examination of long bones shows that they are tubular in structure. The central section (shaft) of the long bone is called the diaphysis, the extremities are termed the epiphyses. Between the terminus ends and the middle shaft there is a developmental zone, the metaphysis. Additionally, in the case of growing long bone, there is a layer of cartilage separating the epiphase and the diaphase. This so-called growth plate is responsible for the longitudinal growth of bones. The ends of the bones that form the joints are covered with hyaline cartilage, the 'articular cartilage' (Baron, 2003; Shea and Miller, 2005).

The adult skeleton consists of two types of bone tissue: cortical (or compact) bone and cancellous (or trabecular) bone. The ratio between cortical and trabecular bone vary at different locations in the skeleton. The overall ratio between compact and trabecular bone is 80% : 20% respectively; however, there are bones with predominantly cancellous bone.

Human cortical bone is dense with a porosity of approximately 10%. It is usually found in the external part of long bones and flat bones. In contrast, trabecular bone is 50–90% porous, making its mechanical properties, including compressive strength, around 10 times lower than that of cortical bone. Cancellous bone is organised into sponge-like structure made of bars and rods of various sizes called trabeculae. It is mainly present in the metaphysis of long bones. Quite often, to improve the mechanical properties of the whole structure, spongy bone is covered by a layer of compact bone (Baron, 2003; Sommerfeldt and Rubin, 2001).

Microscopic composition

The composition of bone depends on the anatomical location, age and general health condition. Basically, bone mineral represents about 50–70% of adult bone, the organic matrix about 20–40%, water about 5–10% and lipids about 1–5% (Shea and Miller, 2005).

Bone matrix and mineral

The most abundant component of bone organic matrix is type I collagen fibres (90% of total protein content) (Baron, 2003; Shea and Miller, 2005). Collagen is synthesised and secreted by the osteoblasts and then deposited in the form of preferentially orientated layers called lamellae (Baron, 2003; Shea and Miller, 2005). The above organisation of fibres allows for a much higher amount of collagen to be accumulated per volume of tissue. In trabecular bone and periosteum, the lamellae are comparable to each other while in the Haversian system, they are concentric. In some metabolic diseases, fracture healing and during bone development, the collagen fibres lose their preferential orientation and become 'woven bone'. The ground substance of bone matrix consists of proteoglycans and glycoproteins. They create highly anionic complexes with an ability to bind and accumulate ions, and are thought to be crucial for the calcification of bone mineral (Baron, 2003; Shea and Miller, 2005).

Bone matrix is also composed of non-collagenous proteins. Their origin is both endo- and exogenous. The endogenous proteins synthesised by bone-forming cells are primarily growth factors and osteocalcin. It is assumed, that they may play important role in bone growth, metabolism and turnover. The exogenous proteins include plasma-derived albumin and alpha2-HS-glycoprotein that may be involved in the matrix mineralisation (Baron, 2003; Shea and Miller, 2005; Fernandez-Tresguerres-Hernandez-Gil *et al.*, 2006a).

The mineral component of bone represents about 70% bone tissue by weight. It is mainly formed by small hydroxyapatite crystals $Ca_{10}(PO_4)_6(OH)_2$ and contains many impurities such as carbonate and magnesium. These apatite crystals are positioned around the collagen fibres and in the ground substance. They are thought to give bone its rigidity and compressive strength (Shea and Miller, 2005; Fernandez-Tresguerres-Hernandez-Gil *et al.*, 2006a).

The skeletal cell population

Despite its rigidity, bone is a living tissue that is continuously remodelled through the lifetime. It is believed that four different types of cells are involved in the formation, resorption, and maintenance of bone: osteoblasts, osteoclasts, osteocytes and bone lining cells (Baron, 2003; Lian *et al.*, 2003; Mundy *et al.*, 2003; Shea and Miller, 2005).

The osteoblast lineage

Osteoblasts are the cells responsible for the synthesis of the bone matrix (osteoid) components and they also play an important role in bone mineralisation (Lian *et al.*, 2003; Mackie, 2003). Osteoblasts originate from local mesenchymal stem cells, which under the influence of growth factors, including fibroblast growth factors (FGFs), parathyroid hormone (PTH), bone morphogenetic proteins (BMPs)/TGFβ superfamily and the transcription factors such as Runx2 and Osterix, differentiate into preosteoblasts and then into mature osteoblasts (Ducy and Karsenty, 1998; Fernandez-Tresguerres-Hernandez-Gil *et al.*, 2006a). A differentiated osteoblast is quite a large cell (20–30 μm), in a form of a polyhedron, with a visible rough endoplasmic reticulum and Golgi apparatus. Mature osteoblasts are localised in cell clusters on the surface of the bone. Because of the time delay between formation and calcification of osteoid, osteoblasts are usually found lining the layer of unmineralised bone matrix that they are producing (Baron, 2003; Lian *et al.*, 2003; Fernandez-Tresguerres-Hernandez-Gil *et al.*, 2006a). Osteoblasts synthesise organic matrix, and then participate in its mineralisation due to high activity of alkaline phosphatase (ALP) (Mackie, 2003; Fernandez-Tresguerres-Hernandez-Gil *et al.*, 2006a), a key osteoblastic marker of differentiation. The first stage of osteogenic differentiation is expression of core binding protein a1 Cbfa1/Runx2. Collagen I and osteopontin (OPN) are expressed continuously, starting early in the osteo-progenitor cells. In general, ALP expression increases in levels up to the progressed mineralisation stage at which stage levels decline. Osteocalcin (OCN) and bone sialoprotein (BSP) expression are up-regulated in bone forming osteo-blasts, providing useful osteogenic markers in the final stages of differentiation (Lian *et al.*, 2003). The activity of osteoblasts is regulated by many endocrine mediators due to the presence of cell membrane receptors for prostaglandins, PTH, RANKL and 1,25$(OH)_2D_3$. They also have the ability to secrete factors participating in osteoclastogenesis and bone turnover such as osteoprotegrin, colony-stimulating factor 1 (CSF-1) and some cytokines (Udagawa *et al.*, 1999; Mackie, 2003). The fate of the osteoblast is either to be entrapped in the mineralised matrix and differentiate into an osteocyte or to stay on the bone surface as a flat lining cell (Lian *et al.*, 2003; Mackie, 2003).

Terminally differentiated osteoblasts are termed osteocytes, which have been entrapped within bone matrix before mineralisation (Baron, 2003; Knothe Tate *et al.*, 2004; Shea and Miller, 2005). These adundant cells have the ability to maintain direct contact with the bone surface (bone lining cells and osteoblasts) as well as with other osteocytes. It happens through the numerous cellular extensions of filopodial processes, which are created before and during the matrix synthesis (Lian *et al.*, 2003; Knothe Tate *et al.*, 2004; Shea and Miller, 2005) forming a functional syncytium of cells connected by gap junctions (Knothe Tate *et al.*, 2004). In mineralised bone, cells are localised inside small osteocytic lacunae, while their processes are found within channels (canaliculi) filled with extracellular bone fluid. The cell–cell connection is crucial for osteocyte maturation and activity (Lian *et al.*, 2003). It is thought, that the lacuno-canalicular network participates in sensing mechanical stimuli and osteocytes play an important mechanosensory function in bone (Klein-Nulend *et al.*, 2005). The flow of bone extracellular fluid in the canalicular system transfers the mechanical loads to the osteocytes (Knothe Tate *et al.*, 2004; Klein-Nulend *et al.*, 2005). Osteocytes, as a representative of the final stage of the osteoblastic linage, are not able to proliferate and self-renew (Knothe Tate *et al.*, 2004). They express a specific marker, CD44 (Jamal and Aubin, 1996), which is strongly expressed in osteocytes and negative in osteoblasts and bone lining cells (Lian *et al.*, 2003; Fernandez-Tresguerres-Hernandez-Gil *et al.*, 2006a).

Some cells of the osteoblastic linage, instead of becoming embedded in the matrix like osteocytes, differentiate into flattened cells, which cover the majority of the inactive/non-remodelling surface of bone. These specialised cells, called 'bone lining cells' form a very thin 'protective' layer along the bone surface. In contrast to osteoblasts, bone lining cells contain a very flat nuclei with few cytoplasmic organelles. They are able to proliferate and communicate with each other through the gap junctions, which may be essential for their functions (Mackie, 2003; Shea and Miller, 2005; Fernandez-Tresguerres-Hernandez-Gil *et al.*, 2006a). Bone lining cells are thought to be involved in the coordination of bone resorption and bone deposition, playing, possibly, a crucial role in modulating osteoclasts' activity (Everts *et al.*, 2002).

The osteoclast lineage

Osteoclasts are highly specialised cells attached to the bone surface which are thought to derive from a different lineage to osteoblasts and their derivatives. They originate from haematopoietic stem cells known as 'granulocyte-macrophage colony-forming units' (GM-CFU), precursors of macrophages and monocytes (Boyle *et al.*, 2003; Shea and Miller, 2005; Fernandez-Tresguerres-Hernandez-Gil *et al.*, 2006a). Osteoclasts are large (100 μm), multinucleated cells containing a number of mitochondria and vacuoles. One of the unique features of this cell type is the ability to form specialised membrane structures involved in the resorption process. Following the migration to a resorption site, the osteoclast's plasma membrane attaches to the bone matrix creating specific membrane domain – the sealing zone. It separates ('seals') the resorption compartment from its surrounding using integrin and actin filaments. Subsequently, another functional 'domain-ruffled border' is formed. It is a specialised, invaginated, deeply folded cell membrane involved in release of proteolytic enzymes and hydrogen ions into the resorption site (Baron, 2003; Shea and Miller, 2005). Osteoclasts synthesise specific enzymes, such as tartrate-resistant acid phosphatase (TRAP) as well as many others, including collagenases, metalloproteases, cathepsin K and glucuronidase. The resorption starts with the solubilisation of firstly, organic and than the mineral components of the bone matrix (Fernandez-Tresguerres-Hernandez-Gil *et al.*, 2006a).

The whole life cycle of osteoclasts, especially differentiation, resorption activity and apoptosis, seems to be regulated by a number of factors. The crucial systemic hormones include parathyroid hormone (PTH), PTH-related peptide, calcitonin, glucocorticoids, 1,25$(OH)_2D_3$, prolactin, and others. There is also an abundance of local hormones and cytokines, which may influence osteoclast functions and bone resorption, for instance: colony-stimulating factor (CFS or M-CSF), interleukin-1(IL-1), interleukin-6 (IL-6), tumor necrosis factor (TNF), prostaglandins, interferon-gamma and members of the transforming growth factor beta superfamily including bone morphogenetic proteins (BMPs) (Mundy *et al.*, 2003; Shea and Miller, 2005).

14.3 Basics of bone remodelling

Bone remodelling is a continuous process of bone resorption and formation carried out by different types of bone cells, which team up or work individually to renew the skeleton (Mundy *et al.*, 2003; Robling *et al.*, 2006). The process takes place through the whole life but is only in positive balance up to the age of 30. After this, the bone mass begins to decrease due to the deficit of osteoblast matrix formation relative to osteoclast matrix resorption (Parfitt, 1984a). The remodelling occurs in microscopic, discrete cavities of old bone, distributed throughout the skeleton and separated from each other geometrically as well as chronologically (Mundy *et al.*, 2003). These so-called 'bone remodelling units' or 'basic multicellular units' (BMU) (Frost, 1964) are found in both trabecular and compact bone. It is estimated that the turnover of each 'pocket' takes about 3–4 months, but differs among different bones of the skeleton and different locations within each bone (probably longer in cancellous bone) (Mundy *et al.*, 2003). BMUs are composed of cells derived from different sources: osteoclasts, osteoblasts and uncharacterised mononuclear cells (Robling *et al.*, 2006). The cells always participate in the same sequence of events: quiescence, activation, resorption, reversal, formation, mineralisation and return to quiescence (Parfitt, 1984b). The new bone tissue formed during the above stages is called a bone structural unit (BSU) (Frost, 1964) (Fig. 14.1).

It is thought that the most important physiological roles of bone remodelling are, on the one hand, adaptation of the skeleton to the variety of mechanical loads (environment) and, on the other hand, a maintenance of the skeleton's ability to take part in mineral homeostasis (Shea and Miller, 2005).

14.3.1 Remodelling phases

Bone remodelling is a very complex process, described as a cycle of five stages: quiescence, activation, resorption, reversal, formation and mineralisation before returning to the quiescence stage (Fig. 14.2) (Parfitt, 1984b; Fernandez-Tresguerres-Hernandez-Gil *et al.*, 2006b).

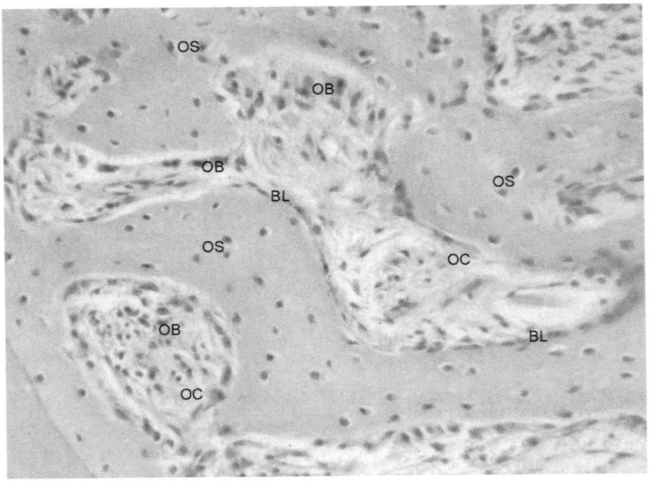

14.1 Histological cross-section of the bone showing new bone formation and cell types responsible for bone remodelling. BL, bone lining cells; OS, osteocytes; OB, ostoblasts; OC, osteoclasts.

Quiescence

It is well known that approximately 80% of the cancellous bone surface and about 95% of the intracortical bone surface is inactive (Parfitt, 1984b). As mentioned previously, remodelling takes place only in microscopic cavities (BMU), predominantly along the endosteal surface of bones. So far, the mechanisms of the induction of the remodelling remain unknown.

Activation

Many studies indicate the presence of non-mineralised, collagenous 'membrane' on the quiescence bone surfaces (Chow and Chambers, 1992). This thin ($<1\ \mu m$) layer separates mineralised bone and bone cells (mostly lining cells and osteoblasts) and it is widely accepted that removal of this collagenous membrane is required before osteoclast resorption take place (Bord *et al.*, 1996). The first step towards the resorption is 'activation' of bone surface, where bone lining cells are retracted and the non-mineralised collagenous layer is removed from endosteal surface. Following this, the mononuclear cells, originally derived from hematopoietic stem cells, travel to the activation zone in the circulation and pass through the vessel wall. When the precursor cells attach to the surface, they start phagocytosis and finally fuse to become functional, multinuclear osteoclasts (Parfitt, 1984b). It is believed that the key players in the activation stage are lining cells. Special capabilities, such as release of collagenase and signalling, involve them in many activities, for instance: degradation of a non-mineralised

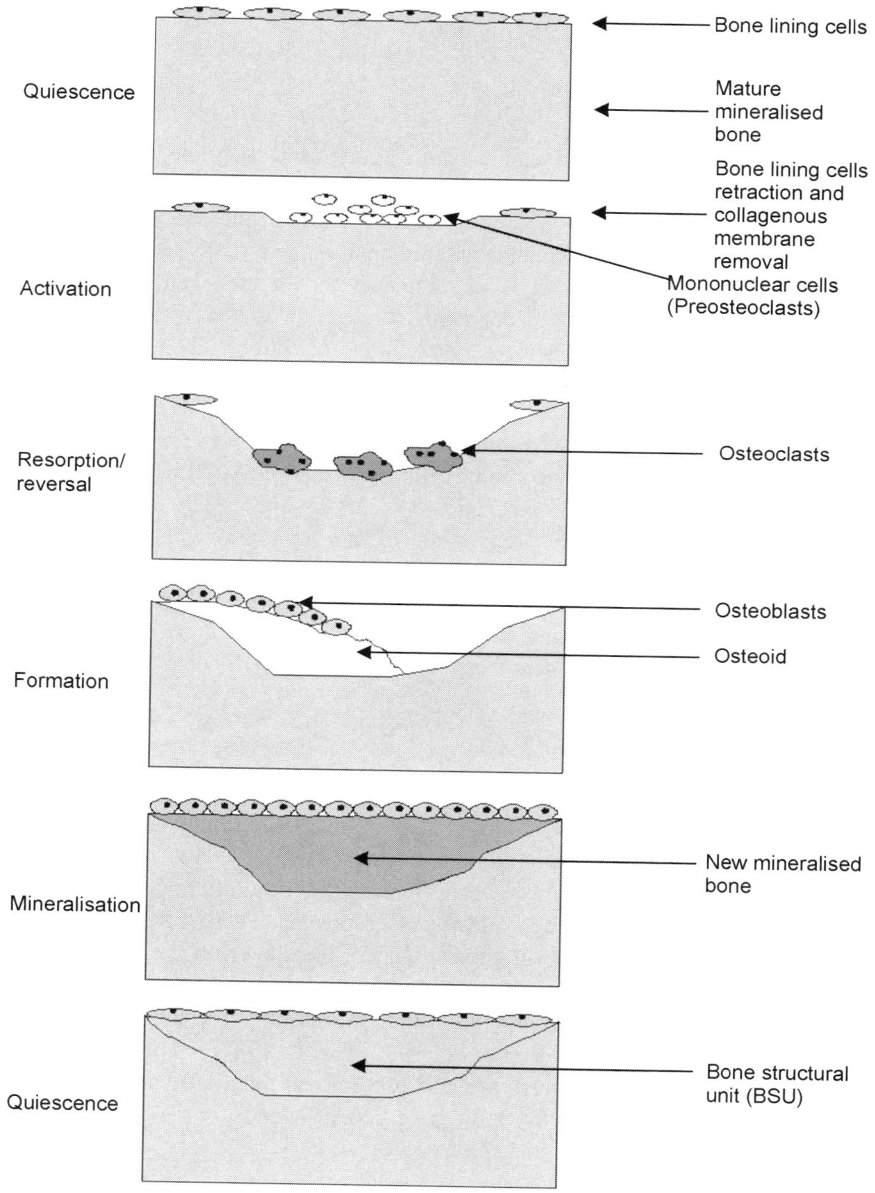

14.2 The sequence of events during bone remodelling (modified from Parfitt, 1984b).

layer from the endosteum, chemotaxis and stimulation of preosteoclasts proliferation, osteoclast formation and mediation of the initial response of resorbing cells to hormones (Parfitt, 1984b).

Resorption

The new mature osteoclast begins to migrate and degrade organic and mineralised bone matrix, resulting in the creation of a cavity either termed a Howship's lacuna or a cutting cone (Parfitt, 1984b). In the case of trabecular bone, the erosion of the bone matrix is rapid up to a depth of about two-thirds of the final cavity. The last part is decomposed by mononuclear cells, which potentially, might be non-fused precursor osteoclasts or macrophages (Parfitt, 1984b; Fernandez-Tresguerres-Hernandez-Gil *et al.*, 2006b). The termination of the resorption phase is thought to occur when the cavity reaches a mean depth of about 60–100 μm from the surface (Parfitt, 1984b); however, the detailed mechanisms of that process is still not fully understood. It was proposed that, as in the case of activation, bone lining cells may also play a crucial role at the stage of termination of the resorption (Parfitt, 1984b). The other process, potentially engaged into the regulation of the resorption phase could be osteoclast apoptosis (Mundy *et al.*, 2003). The resorption of bone may be modified by many different factors including: systemic hormones, local hormones/cytokines and many others (for review see Mundy *et al.*, 2003; Robling *et al.*, 2006; Fernandez-Tresguerres-Hernandez-Gil *et al.*, 2006b).

Reversal

The reversal phase is the period between the end of resorption and beginning of formation at a particular BMU. This is also the time when the coupling of bone creation to previous bone degradation occurs (Parfitt, 1984b; Mundy *et al.*, 2003; Gasser, 2006). During reversal phase the bone remodelling unit is occupied by mononuclear cells, mostly macrophages. These cells deposit and then mineralise a thin, dense granular collagen-free layer on the bone surface (Tran Van *et al.*, 1982). This so-called cement substance is supposed to be crucial for preparation of the surface for bone formation phase and binds old and new synthesised bone (Tran Van *et al.*, 1982; Parfitt, 1984b). The mechanisms controlling the coupling process are still not defined although various theories exist (Rodan and Martin, 1981; Smit *et al.*, 2002; Mundy *et al.*, 2003; Gasser, 2006).

Formation, mineralisation and quiescence

In this phase, differentiated osteoblasts fill the bone remodelling unit and commence to form new bone matrix (Parfitt, 1984b; Mundy *et al.*, 2003). Bone formation may be divided into two stages: osteoid synthesis and mineralisation. In the first step, osteoblasts make and deposit components of ECM, mainly collagen type I, organising the collagen fibres into the lamellar structure of bone (Parfitt, 1984b). The unmineralised matrix undergoes a poorly characterised process of maturation. At the beginning, due to a high density of osteoblasts, the osteoid is deposited very rapidly, at the rate of 2–3 μm per day (Parfitt, 1984b). After the unmineralised matrix reaches a width of 20 μm, the second stage of bone formation begins. Initially, the rate of mineral apposition is 1–2 μm per day (Parfitt, 1984b). Despite termination of the matrix synthesis, the mineralisation continues until the osteoid seam disappears and the cells remaining on the bone surface completely transform to lining cells (Parfitt, 1984b). This is the end of the remodelling loop. The bone surface has returned to its original state of quiescence except that the bone is younger (Parfitt, 1984b).

14.3.2 Regulation of bone remodelling

It is believed that bone turnover is regulated by numerous factors including: genetic, mechanical, vascular, nervous, nutritional, hormonal and local growth factors and cytokines (Mundy *et al.*, 2003; Fernandez-Tresguerres-Hernandez-Gil *et al.*, 2006b). At present it is widely believed that the coupling is mediated by the systemic hormones (PTH, $1{,}25(OH)_2D_3$), local humoral factors and possibly local strain levels (functional adaptation). The mononuclear cells mentioned above may be key players. They are probably responsible for the release of the growth factors accumulated inside the matrix including: TGF-β, BMPs, FGF, as well as the factors, platelet-derived growth factor (PDGF), and insulin-like growth factor I and II (IGF-I and II). These factors attract and activate preosteoblasts and osteoblasts, leading ultimately to new bone formation (Mundy *et al.*, 2003; Fernandez-Tresguerres-Hernandez-Gil *et al.*, 2006b; Gasser, 2006). An alternative to the above is a simultaneous stimulation of osteoblastic and osteoclastic cell lineages. In this case, transcriptional factors important for both osteoclastogenesis and bone formation, like Runx2/Cbfa1 may be engaged into the coupling process (Enomoto *et al.*, 2003).

All the above factors affect the quality and amount of the bone produced. Of particular interest is mechanical loading, which is thought to be a potent stimulus for bone cells, improving bone strength and inhibiting bone loss with age (Meyer *et al.*, 2006b; Robling *et al.*, 2006). The effects of loading on bone remodelling are characterised by a U-shaped curve – increase in remodelling rate is observed in disuse (insufficient loading) or overuse (overloading causing damage). The range of loading within which bone remodelling is minimised is often called the physiological range (Robling *et al.*, 2006).

It is generally accepted, that remodelling allows bone tissue to adapt its macroscopic organisation and mass to mechanical demands. As a result, functional adaptation guarantees maximal strength with minimal bone mass (Klein-Nulend *et al.*, 2005). This fundamental 'form follows function' concept was proposed for the first time in 1892, and it is widely referred to as Wolff's law (Wolff, 1892). Over a century later it is still not fully understood how the bone tissue senses the environment but the process is thought to be multi-factorial. There is growing support for the theory in which osteocytes and bone lining cells are the main mechano-sensors of loads within bone tissue, and that these cells propagate signalling cascades that ultimately influence the behaviour of neighbouring osteoblasts and osteoclasts that are engaged in bone resorption and formation (Klein-Nulend *et al.*, 2005).

It has been shown that mechanical loading at physiologically relevant magnitudes initiate bone formation and remodelling in both *in vivo* (Rubin and Lanyon, 1984) as well as *in vitro* studies (El Haj *et al.*, 1990). The last point indicates that mechano-transduction plays an essential role in bone tissue remodelling and repair (El Haj *et al.*, 2005). It seems to be clear that the creation of appropriate environment for proliferation and differentiation of cells into functional tissue is crucial for 'engineering' bone tissue *in vitro* (El Haj *et al.*, 2005).

The ultimate aim of a tissue engineering is to create a bone environment that includes the mixed cell types and can remodel throughout the life of the patient. Thus, the challenge is to re-engineer the environment outlined above. New strategies have been proposed and in the remainder of this chapter, we outline the key features of some of these approaches.

14.4 Skeletal tissue reconstruction – a tissue engineering approach

14.4.1 Clinical needs

Do we need tissue engineering?

Musculoskeletal trauma is thought to be one of the major health concerns in both USA and EU. The main reason for this increasing crisis is the ageing of the population combined with a rise in 'an active healthy lifestyle', which causes an increase in joint damage and sport-related injuries (Laurencin *et al.*, 1999). In the UK alone, there are about 600 000 bone and joint operations annually (DoH, 2005). A huge proportion involve bone grafting. It is estimated, that in USA and EU together, there are over 1 million bone graft procedures each year (Anon., 2002). The principle of reconstructive medicine is to replace the damage or defective tissue with a new functional and viable substitute (Laurencin *et al.*, 1999). In the case of bone reconstructive surgery, there are two main types of bone grafts used currently: autografts and allografts (Laurencin *et al.*, 1999; Finkemeier, 2002).

Current strategies – advantages and disadvantages

Autologous bone grafts have osteogenic (i.e. the capability to form bone tissue *de novo*), osteoconductive (i.e. the capacity of a material to guide bone forming tissue into a defect) and osteoinductive properties (i.e. the ability to induce bone formation by attracting and stimulating bone-forming cells of the recipient) (Finkemeier, 2002; Kneser *et al.*, 2006). This type of graft is still thought to be the 'gold standard' in bone reconstruction; however, there are some serious limitations to this method. The amount of the tissue which could be safely harvested is limited and the procedure requires 'opening' of the donor site, which often cause complications such as bleeding, haematoma and chronic pain (Arrington *et al.*, 1996).

The majority of the above problems might be avoided by using allogenic grafts. These so-called biostatic implants are stored in tissue banks. Despite using very strict procedures of harvesting, preservation and sterilisation there is still the possibility of transmitting disease and availability of the tissue is still 'not unlimited' (Laurencin *et al.*, 1999; McCann *et al.*, 2004; Lewandowska-Szumiel, 2005).

As an alternative to the above types of graft, synthetic bone graft substitutes, such as metals, ceramics, polymers and composites, have been introduced. The first generation of 'biomaterials' was born in 1960s and 1970s. They were designed to 'achieve a suitable combination of physical properties to match those of the replaced tissue with a minimal toxic response in the host' (Hench, 1980). These materials were 'nearly inert', and with time they were separated from the surrounding natural tissue by thin, fibrous capsule. Ultimately, and as a result of non-integration, the implants often loosen (Hench, 1980; Lewandowska-Szumiel, 2005). The second generation of artificial implantable materials, introduced in the mid 1980s, was bioactive and bioresorbable. A variety of bioactive glasses, ceramics, glass-ceramics, and composites have reached clinical use (Hench and Polak, 2002) and there has been a great improvement in bone graft substitute properties. Unfortunately, the detailed analyses of the implants' lifespan indicated, that about half of the prostheses would fail within 10–20 years, and revision surgery would be required (Hench and Polak, 2002).

The third generation of biomaterials was designed to interact with cells and modify their reactions. It was accomplished by molecular modifications of 'old biomaterials', which became much more responsible to the 'changes in physiological loads and biochemical stimuli' (Hench and Polak, 2002).

The future of the implantable materials seems to be tightly connected with the term 'biomimetic'. All around the world, researchers from different fields such as biology, material sciences, chemistry, physics and biotechnology cooperate, are researching to create 'bio-inspired' materials which mimic nature (Sanchez *et al.*, 2005). This area of research of bone reconstructive medicine has been termed tissue engineering (TE). Tissue engineering aspires to generate a biocompatible product, which would have all the required properties of the 'gold standard' (allograft) and additionally, would be available 'off-the-shelf' (Lewandowska-Szumiel, 2005; Kneser *et al.*, 2006).

14.4.2 Tissue engineering

The concept

Over the past ten years, the tissue engineering field has been developing intensively. One of the first and the most acceptable definitions of tissue engineering was given by Langer and Vacanti in 1993 as 'an interdisciplinary field that applies the principles of engineering and the life sciences toward the development of biological substitutes that restore, maintain, or improve tissue function'. For a tissue engineering approach, the crucial issue is understanding the underlying processes of tissue formation and regeneration which will point us towards the creation of a new functional tissue (Kneser *et al.*, 2006). To achieve this goal, biologists, clinicians, chemists, engineers, material scientists and physicians integrate their knowledge, working together in 'tissue engineering' laboratories all around the world (Langer and Vacanti, 1993; Laurencin *et al.*, 1999; Salgado *et al.*, 2004).

Bone repair was proposed to be one of the first, major applications of tissue engineering (DoH, 2000). The general concept of bone tissue engineering is based on the formation of a 'tissue engineering construct' (TEC) to encourage the regeneration of the damaged tissue (Hutmacher and Garcia, 2005). Typically, TEC is composed of the scaffold (matrix), viable cells and biologically active agents (Langer and Vacanti, 1993; Salgado *et al.*, 2004; Hutmacher and Garcia, 2005). The scaffold acts as a temporary three-dimensional (3D) support for seeding, proliferation and differentiation of cells, while bioactive molecules (incorporated within the scaffold or supplied from 'outside') manage these processes. The cultivation of the cell–scaffold construct takes place in a precisely defined environment, provided and controlled within appropriate growth chambers termed 'bioreactors' (El Haj *et al.*, 2005). The crucial issue is to understand that the required properties of the construct depend on the intended clinical application (Kneser *et al.*, 2006). Despite the above, the generally acceptable properties of the 'ideal' bone TEC include osteogenicity, osteoconductivity, osteoinductivity, mechanical stability and 'ease of handling' (Kneser *et al.*, 2006).

Scaffold properties

In vivo, tissues cannot exist without an ECM which creates an optimal, 3D environment for cells. The main physiological functions of the ECM include storage of the nutrients, growth factors and cytokines as well as mechanical stabilisation for anchorage-dependent cells (Kneser *et al.*, 2006). In the context of bone tissue engineering, the scaffold should possess the following properties (Hutmacher, 2000; Salgado *et al.*, 2004):

- biocompatibility,
- bioresorbability/biodegradability,
- open/interconnected porosity,
- suitable topography and surface chemistry, and
- appropriate mechanical properties.

To fulfil the above requirements, several different types of the materials have been proposed (Laurencin *et al.*, 1999; Vats *et al.*, 2003). Based on the origin, the scaffold materials may be divided into two main groups: (1) naturally derived materials such as collagen, glycosaminoglycans (GAGs), starch, chitosan and alginates; and (2) synthetic ones, including metals, ceramics, bioactive glasses and polymers (Table 14.1) (Hutmacher, 2000; Vats *et al.*, 2003; Lewandowska-Szumiel, 2005; Yang and El Haj, 2006). In addition, the surface properties of the scaffold will influence cell adhesion and activity. Recent research has demonstrated how the properties on the surface of the scaffold can also affect the ways in which cells adhere and perceive load signals and ultimately their ability to remodel to their environment (Yang *et al.*, 2004; Yang and El Haj, 2006). Polymer surfaces such as PLLA are hydrophilic and do not promote cell attachment. Therefore addition of synthetic peptide sequences from the ECM such as RGD (arginine, glycine, aspartate) or complete

Table 14.1 Types of biomaterials used for preparation of scaffolds for bone tissue engineering

Name	Degradation time	3D architecture
Naturally derived materials		
Collagen	$<$ A few months	Fibrous, sponge, hydrogel
Starch	A few months	Porous
Chitosan	$<$ A few months	Sponge, fibres
Alginates	Weeks to a few months	Hydrogel, sponge
Hyaluronic acid (HA)	$<$ A few months	Hydrogel
Polyhydroxyalkanotes (PHA)	Months	Porous, hydrogen
Synthetic materials		
Polyurethanes (PU)	$>$ 1 year	Porous
Poly(α-hydroxy acids) (i.e. PLLA, PGLA)	1 month to a few years	Porous
Poly(ϵ-caprolactone) (PCL)	$>$ 1 year	Sponge, fibres
Poly(propylene fumarates) (PPF)	Weeks to a few months	Hydrogel
Titanium	Non-degradable	Mesh
Calcium phosphate	Varied	Porous

extracellular matrix coatings such as fibronectin, laminin or collagen can improve cell attachment (Yang and El Haj, 2006). Furthermore, surface topography of the materials can play a role in affecting cell alignment, activity and matrix production (Yang *et al.*, 2004).

Cell type and source

Which cell source is suitable for bone tissue engineering applications? There is no simple answer to this question. In each case, there are advantages and disadvantages (Table 14.2) (Heath, 2000; Salgado *et al.*, 2004; Lewandowska-Szumiel, 2005). Generally, it is believed that cells derived from 'the ideal source' should be easily expandable, non-immunogenic and express proteins similar to the regenerated tissue or be capable of differentiating down this pathway *in vitro* or *in vivo* (Heath, 2000; Salgado *et al.*, 2004). The source of cells will also vary according to the complexity of the constructs. Tissues such as bone may require multiple cell types for tissue engineered constructs and therefore the choice of cell donor or donors will have to reflect this. It is also possible that the cell source may vary, depending on the each individual clinical case (Lewandowska-Szumiel, 2005; Kneser *et al.*, 2006).

The first, natural, choice seems to be isolation of autologous bone cells from bone tissue during biopsies. The second, alternative, sources are allogenic and xenogenic cells. The last, and probably the best alternative, is to use stem and/or progenitor cells (Heath, 2000; Salgado *et al.*, 2004; Caplan, 2005; Lewandowska-Szumiel, 2005). In the last case, it is possible to distinguish several locations, from which stem cells could be isolated including embryos (embryonic stem cells, ESCs), peripheral blood (peripheral blood mesenchemal stem cells, PBMSCs), cord blood, adult bone marrow (bone marrow mesenchemal stem cells, BMMSCs), adipose tissue and many others (Heath, 2000; Kowalczyk *et al.*, 2004; Salgado *et al.*, 2004; Lewandowska-Szumiel, 2005).

Growth factors

Growth factors/cytokines are highly specialised molecules, secreted by the majority of cell types. They work as 'molecular messengers' controlling almost all possible cell functions (Deuel and Zhang, 2000). Binding of signalling molecules such as growth factors to its receptor initiates a cascade of intracellular signal transduction that may lead to the changes in cell proliferation, differentiation, adhesion, migration, and expression of the key genes (Deuel and Zhang, 2000; Salgado *et al.*, 2004; Kneser *et al.*, 2006). As mentioned previously, local growth factors are crucial for bone tissue formation and remodelling (Compston, 2001; Fernandez-Tresguerres-Hernandez-Gil *et al.*, 2006b). Therefore, it seems to be clear that this kind of molecule has the potential to play 'a leading role' in bone tissue engineering (Salgado *et al.*,

Table 14.2 Potential human cell sources for bone tissue engineering

Cell source	Advantages	Disadvantages
Autologous cells (derived from small samples of mature tissue) – tissue-specific stem cells	Non-immunogenic after transplantation Source of the optimal growth factors	Time-consuming procedure Limited cell number available Small proliferative potential Variability in source-tissue quality
Autologous adult bone marrow mesenchymal stem cells (BMMSCs)	Multipotency Satisfactory proliferative potential	Low initial cell number (1 cell per 100 000 mononucleated cells)
Peripheral blood mesenchymal stem cells (PBMSCs)	Simple cell harvesting method Cells maintain their homing potential Multipotency	Low initial cell number (around 20 times lower than BMMSCs)
Stem cells isolated from cord blood	Relatively simple cell harvesting method	Concerns about the quantity of available stem cell in cord blood samples
Human embryonic stem cells (ESCs)	Nearly unlimited self-renewal capability 'Unlimited' differentiation potential via precursor cells (pluripotency)	Tumorogenicity induction Ethical and social limitation

2004). Among the vast number of local bone growth factors, the most promising, from the bone TE point of view, are the factors outlined earlier, namely BMPs, TGFβ, FGFs, IGF I/II and PDGF (Lieberman *et al.*, 2002; Salgado *et al.*, 2004).

The use of growth factors in bone TE is closely connected with scaffold properties. The 3D structure of the growth frame could be used as a storage and delivery vehicle for BMPs or FGFs. The design and creation of 'biologically' active scaffolds may be especially used for the direct induction of stem cell differentiation *in vitro* and, perhaps even more importantly, *in vivo* at the implantation site (Salgado *et al.*, 2004; Lewandowska-Szumiel, 2005; Seto *et al.*, 2006). To induce such a specific biological response, an appropriate carrier for a growth factor must be selected (Lieberman *et al.*, 2002). Many different material scaffolds have been used for immobilisation and delivery of 'osteogenic' growth factors to the 'place of interest' including hydroxyapatite, collagen, tricalcium phosphate and titanium (Hu *et al.*, 2003; Sachse *et al.*, 2005; Saito *et al.*, 2006). The challenge lies in giving an appropriate and effective signal in the correct concentration at the correct time. It may also be useful to spatially orient different cues within a scaffold in order to deliver local signals to generate a profile of required cell types from one donor source, for example endothelial and bone cells for a vascularised bone tissue construct. Unfortunately, despite a large number of animal experiments and some clinical trials, an effective system for growth factor delivery, which maintains the biological activity of the agents, has yet to be clearly defined (Lieberman *et al.*, 2002; Lewandowska-Szumiel, 2005).

Bioreactors and the control of the environment

One of the main challenges in bone tissue engineering approach is to create an *in vitro* environment that provides both the biochemical and mechanical signals that will allow and enable the control of new tissue development and maintenance prior to implantation (Freed and Vunjak-Novakovic, 2000). To achieve this goal and create functional 3D TEC, bioreactors have been developed. Depending on the design, it is possible to control different aspects of the '*in vitro* environment' including temperature, pH, concentration of gases, supply of nutrients and removal of the wastes. Bioreactors also may control the mechanical environment surrounding the cell–scaffold construct, allowing biomechanical stimuli to be applied in the form of shear stress, compression and tension, etc. (Martin *et al.*, 2004; El Haj *et al.*, 2005).

The invention of different types of bioreactor systems for bone tissue engineering has also supplied us with new tools for the next generation of tissue culture, and the transition from monolayer to 3D constructs. Static culture, which is still one of the most commonly used systems for culturing cell/tissues *in vitro* is very simple, and consistent reproducibility is a key feature. However, it has some obvious limitations, e.g. low efficiency of cell seeding, no effective nutrient supply, a limitation in the size of the construct due to mass transfer issues and a lack of mechanical stimulation (Martin *et al.*, 2004).

Surprisingly, the cell seeding technique remains one of the most important factors affecting 3D cell culture *in vitro* (Vunjak-Novakovic *et al.*, 1998; Wendt *et al.*, 2003). The basic seeding requirements include: (1) high yield (reducing donor cell loss); (2) high kinetic rate (limitation of the time in suspension for anchorage-dependent cells); and (3) high and uniform cell distribution (to ensure uniformity of tissue regeneration) (Vunjak-Novakovic *et al.*, 1998). To fulfil the above needs, different types of bioreactors, for example spinner flasks, rotating wall vessels and flow perfusion bioreactors, have been developed (Vunjak-Novakovic *et al.*, 1998; Wendt *et al.*, 2003). Using these systems it is possible to obtain better cell distribution throughout the 3D structure of the scaffold (Vunjak-Novakovic *et al.*, 1998; Woźniak *et al.*, 2005) and many of these systems are now commercially available for use.

It is well known that another critical issue in 3D cultivation is the supply of oxygen and soluble nutrients (Martin *et al.*, 2004; El Haj *et al.*, 2005). To improve the mass transfer during the culture period, different fluid-circulation regimes may be applied to the engineered constructs. Media flow has been introduced using bioreactors similar to those used during the seeding step, for instance: spinner flasks, rotating wall vessels and perfusion systems (Rucci *et al.*, 2002; Sikavitsas *et al.*, 2002; Gomes *et al.*, 2006).

The spinner flask represents one of the simplest designs, and has been in use for many decades in the recombinant protein industry. Despite this, it has been shown that the improved mixing of media accelerates proliferation and differentiation of mesenchymal stem cells (MSCs). However, a problem with spinner flasks is that homogenous tissue development is reduced because of heterogeneous shear forces distributed over the cell–scaffold construct (Sikavitsas *et al.*, 2002; Wendt *et al.*, 2003).

Another type of bioreactor for use in bone tissue engineering is based on the dynamic laminar flow of media. As an alternative to the spinner flask, the rotating wall vessel (RWV) bioreactor allows for an efficient reduction in mass transfer limitations, as well as lower levels of shear stress (Martin *et al.*, 2004). This system was successfully used for 3D bone tissue formation *in vitro* and to study the effects of microgravity on bone cells (Rucci *et al.*, 2002). However, in another study the RWV bioreactor was found to be less efficient than a stirrer flask, which was evaluated on the basis of lower proliferation rate and reduced level of differentiation of human MSCs (Sikavitsas *et al.*, 2002).

Flow perfusion bioreactors were originally developed to maintain the function of highly metabolised cells (e.g. hepatocytes) (Martin *et al.*, 2004). The idea was introduced to the bone tissue engineering and is based on the perfusion of culture media directly through the pores of the cell-seeded scaffold. Research has demonstrated that is possible to achieve homogeneous mass transfer characteristics within the centre and periphery of the scaffold (Wendt *et al.*, 2003). It was demonstrated in several studies that direct perfusion bioreactors stimulate growth, differentiation and mineralised matrix deposition by bone cells on 3D scaffolds (Bancroft *et al.*, 2002). One of the most important factors involved is the flow rate of the culture media (Bancroft *et al.*, 2002; Martin *et al.*, 2004). Optimisation of the flow dynamics is necessary to obtain the required TEC properties. Computational fluid dynamic (CFD) models, which have been used to calculate the optimal flow rate, demonstrated that the architecture of the scaffold is crucial for the proper prediction of shear stresses and, thus, matrices with a homogeneous pores distribution would enable more precise control over the shear stresses (Martin *et al.*, 2004). One challenge for these systems that rely on perfusion through the pore is the changing dynamic of the pores with time. As the pores fill with ECM they restrict perfusion through the scaffolds, thus effecting mass transfer. This highlights the balance between time to develop a mature construct and the appropriate time for implantation, which has yet to be established.

Bioreactors and mechanical conditioning

In vivo, many experimental studies have shown that static loading regimes do not stimulate bone formation, while cyclic loading can increase it significantly (Duncan and Turner, 1995). Thus, dynamic loading is thought to be the most effective cue for bone formation (Klein-Nulend *et al.*, 2005). In addition, applying appropriate loading regimes to a developing construct will influence where and how the matrix is laid down to provide mechanical strength to the construct. This 'fit for function' is argued to be particularly appropriate with bone tissue. In consequence, an appropriate mechanical conditioning system is thought to be vital for engineering functional bone tissue constructs *in vitro* (El Haj *et al.*, 2005). In the case of bone cells, the most frequently used stress profiles include hydrostatic compression (Roelofsen *et al.*, 1995), strain (Lewandowska-Szumiel *et al.*, 2007), stretch (Walboomers *et al.*, 2004) and fluid flow (Klein-Nulend *et al.*, 1998). It is well accepted that these types of loads influence proliferation, orientation/deformation, gene expression and other activities of cells *in vitro*, and thus the above factors might be potentially used for stimulation of the formation of bone tissue constructs *in vitro* (Duncan and Turner, 1995; Meyer *et al.*, 2006a).

An increasing number of mechanical bioreactors are being developed worldwide. The difficulties lie in scale-up for use clinically and the lack of commercialisation of many products being developed. New systems developed by Bose Enduratec are being marketed worldwide with some effect. In addition, new systems for application of mechanical forces to cells in a closed, bioreactor environment are still being developed (Dobson *et al.*, 2002; Janssen *et al.*, 2006; Meyer *et al.*, 2006a). For instance, in our group, we have developed a new technology based on the use of magnetic particles and magnetic field (Magnetic

14.3 A prototype of the magnetic force bioreactor (Magnecell Ltd) used for tissue culture polystyrene systems.

Force Bioreactor, MFB) (Fig. 14.3) (Cartmell *et al.*, 2002; Dobson *et al.*, 2002). This novel technology employs nano and micro-magnetic particles, coated with various proteins, which specifically attach to the cell membrane mechano-receptors. In this way, cells cultured in monolayer, as well as cells seeded on the 3D scaffolds, can be exposed to an oscillating magnetic field, which results in the translational and rotational movement of the particles (Cartmell *et al.*, 2002, 2005). Eventually, the application of a magnetic field leads to the induction of a stretch and torque directly on the surface of the cells. In this way, a mechanical force in the range of piconewtons is subjected to the cells. In the case of long-term 2D studies, this type of conditioning resulted in the up-regulation of characteristic osteoblastic gene expression and stimulation of matrix mineralisation (Cartmell *et al.*, 2002). In another study, the magnetic particles technique was used to apply mechanical stimuli to the human bone cells seeded

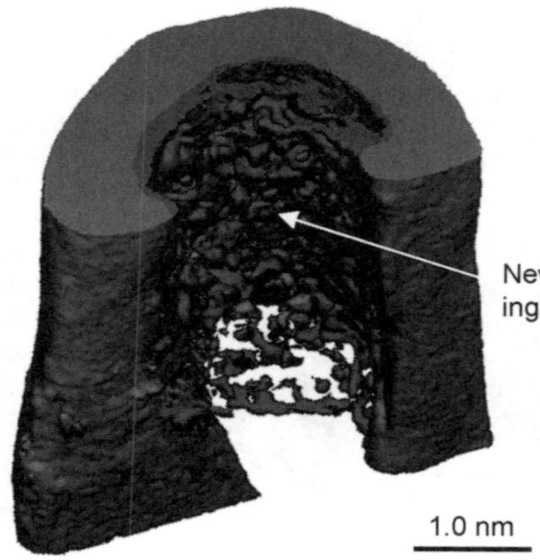

14.4 Rat tibia with a repaired defect. μCT image of a new bone growth after 3 weeks following implantation of dihydropyridine release PLLA scaffold seeded with MSCs.

on 3D porous PLLA scaffolds. It was reported that in case of experimental group (particles attached to the cells + magnetic field loading), up-regulation of bone-related gene expression – osteocalcin, osteopontin and collagen type I – occurred (Cartmell *et al.*, 2003). This new technology could be a valuable tool for mechanical stimulation of bone cells *in vitro* and might be used in the preparation of tissue engineering constructs. One of the proposed advantages of magnetic particles-based mechanical stimulation of the cell-seeded scaffolds is opportunity to utilise even mechanically 'weak' scaffolds. Furthermore, alteration of the type (magnetic properties), the size or the number of particles attached to each cell allows applying different magnitudes of force throughout the tissue-engineered construct (El Haj *et al.*, 2005).

Other approaches to enhancing mechanical signals have been developed, including release strategies, biomaterial approaches and autologous sources of cells. Recent research has demonstrated a technique for a novel scaffold releasing dihydropyridine agonists, which act with mechanical load on the membranes of cells to attenuate the load signal on calcium (Fig. 14.4) (Wood *et al.*, 2006a,b). These scaffolds can be used in bioreactor systems outlined above to increase and optimise tissue growth *in vitro* or delivered directly *in vivo*. This release approach demonstrates how tissue engineering strategies can be designed which attempt to control key mechano-responsive target receptors involved in growth and differentiation.

14.5 Conclusions

The ultimate goal of bone tissue engineering is a creation of the bone graft substitute. The essential requirements of such a tissue engineered construct include adequate biomechanical properties. Although 'the challenge of imitating nature' has been taken, development of tissues and organs available 'off-the-shelf' is still ahead. It is believed, that the future of the bone tissue engineering is: (1) optimisation of the scaffold materials, their coatings and the processing techniques; (2) utilisation of mesenchymal stem cells as the 'biological' component of the engineered construct; (3) incorporation of bioactive factors for stimulation and control of tissue growth; and, finally, (4) detailed characterisation of the biomechanical conditioning of constructs in bioreactors.

In summary, it must to be underlined, that, in spite of being 'overpromised and underdelivered' (Nerem, 2006), and, despite that, the industry is still in its fledgling state (Lysaght and Hazlehurst, 2004), tissue engineering has enormous potential to overcome the current strategies for bone regeneration.

14.6 Acknowledgements

This work was supported by Grant MEST-CT-2004-8104 and the Foundation for Polish Science (Domestic Grants for Young Scientists – 2006).

14.7 References

Anon. (2002) In *Clinical Reports*, PJB Publications Limited.

Arrington, E. D., Smith, W. J., Chambers, H. G., Bucknell, A. L. and Davino, N. A. (1996) *Clin Orthop Relat Res*, **329**, 300–9.

Bancroft, G. N., Sikavitsas, V. I., van den Dolder, J., Sheffield, T. L., Ambrose, C. G., Jansen, J. A. and Mikos, A. G. (2002) *Proc Natl Acad Sci USA*, **99**, 12600–5.

Baron, R. (2003) In *Primer on the Metabolic Bone Diseases and Disorders of Mineral Metabolism* (Ed, Favus, M. J.), American Society for Bone and Mineral Research, Washington, DC, pp. 1–9.

Bord, S., Horner, A., Hembry, R. M., Reynolds, J. J. and Compston, J. E. (1996) *Bone*, **19**, 35–40.

Boyle, W. J., Simonet, W. S. and Lacey, D. L. (2003) *Nature*, **423**, 337–42.

Caplan, A. I. (2005) *Tissue Eng*, **11**, 1198–211.

Cartmell, S. H., Dobson, J., Verschueren, S. B. and El Haj, A. J. (2002) *IEEE Trans Nanobiosci*, **1**, 92–7.

Cartmell, S. H., Magnay, J. L., Dobson, J. and El Haj, A. J. (2003) *European Cells Mater*, **6**, 7.

Cartmell, S. H., Keramane, A., Kirkham, G. R., Verschueren, S. B., Magnay, J. L., El Haj, A. J. and Dobson, J. (2005) *J Phys: Conference Series*, **17**, 77–80.

Chow, J. and Chambers, T. J. (1992) *Calcif Tissue Int*, **50**, 118–22.

Compston, J. E. (2001) *Physiol Rev*, **81**, 419–47.

Deuel, T. F. and Zhang, N. (2000) In *Principles of Tissue Engineering* (Eds, Lanza, R. P., Langer, R. and Vacanti, J. P.), Academic Press, London, pp. 129–41.

Dobson, J., Keramane, A. and El Haj, A. J. (2002) *European Cells Mater*, **4**, 42–4.

DoH (2000) *Science*, **289**, 1498–500.

DoH (2005) In *Government Statistical Service*, Department of Health, London.

Ducy, P. and Karsenty, G. (1998) *Curr Opin Cell Biol*, **10**, 614–19.

Duncan, R. L. and Turner, C. H. (1995) *Calcif Tissue Int*, **57**, 344–58.
Einhorn, T. A. (1996) In *Principles of Bone Biology* (Eds, Bilezikian, J. P., Raisz, L. G. and Rodan, G. A.), Academic Press, San Diego, pp. 25–37.
El Haj, A. J., Minter, S. L., Rawlinson, S. C., Suswillo, R. and Lanyon, L. E. (1990) *J Bone Miner Res*, **5**, 923–32.
El Haj, A. J., Wood, M. A., Thomas, P. and Yang, Y. (2005) *Pathol Biol (Paris)*, **53**, 581–9.
Enomoto, H., Shiojiri, S., Hoshi, K., Furuichi, T., Fukuyama, R., Yoshida, C. A., Kanatani, N., Nakamura, R., Mizuno, A., Zanma, A., Yano, K., Yasuda, H., Higashio, K., Takada, K. and Komori, T. (2003) *J Biol Chem*, **278**, 23971–7.
Everts, V., Delaisse, J. M., Korper, W., Jansen, D. C., Tigchelaar-Gutter, W., Saftig, P. and Beertsen, W. (2002) *J Bone Miner Res*, **17**, 77–90.
Fernandez-Tresguerres-Hernandez-Gil, I., Alobera-Gracia, M. A., del-Canto-Pingarron, M. and Blanco-Jerez, L. (2006a) *Med Oral Patol Oral Cir Bucal*, **11**, E47–51.
Fernandez-Tresguerres-Hernandez-Gil, I., Alobera-Gracia, M. A., del-Canto-Pingarron, M. and Blanco-Jerez, L. (2006b) *Med Oral Patol Oral Cir Bucal*, **11**, E151–7.
Finkemeier, C. G. (2002) *J Bone Joint Surg Am*, **84-A**, 454–64.
Freed, L. E. and Vunjak-Novakovic, G. (2000) In *Principles of Tissue Engineering* (Eds, Lanza, R. P., Langer, R. and Vacanti, J. P.), Academic Press, London, pp. 143–56.
Frost, H. M. (1964) In *Bone Biodynamics*, Little & Brown, Boston.
Gasser, J. A. (2006) *J Musculoskelet Neuronal Interact*, **6**, 128–33.
Gomes, M. E., Holtorf, H. L., Reis, R. L. and Mikos, A. G. (2006) *Tissue Eng*, **12**, 801–9.
Heath, C. A. (2000) *Trends Biotechnol*, **18**, 17–19.
Hench, L. L. (1980) *Science*, **208**, 826–31.
Hench, L. L. and Polak, J. M. (2002) *Science*, **295**, 1014–17.
Hu, Y., Zhang, C., Zhang, S., Xiong, Z. and Xu, J. (2003) *J Biomed Mater Res A*, **67**, 591–8.
Hutmacher, D. W. (2000) *Biomaterials*, **21**, 2529–43.
Hutmacher, D. W. and Garcia, A. J. (2005) *Gene*, **347**, 1–10.
Jamal, H. H. and Aubin, J. E. (1996) *Exp Cell Res*, **223**, 467–77.
Janssen, F. W., Oostra, J., Oorschot, A. and van Blitterswijk, C. A. (2006) *Biomaterials*, **27**, 315–23.
Klein-Nulend, J., Helfrich, M. H., Sterck, J. G., MacPherson, H., Joldersma, M., Ralston, S. H., Semeins, C. M. and Burger, E. H. (1998) *Biochem Biophys Res Commun*, **250**, 108–14.
Klein-Nulend, J., Bacabac, R. G. and Mullender, M. G. (2005) *Pathol Biol (Paris)*, **53**, 576–80.
Kneser, U., Schaefer, D. J., Polykandriotis, E. and Horch, R. E. (2006) *J Cell Mol Med*, **10**, 7–19.
Knothe Tate, M. L., Adamson, J. R., Tami, A. E. and Bauer, T. W. (2004) *Int J Biochem Cell Biol*, **36**, 1–8.
Kowalczyk, P., Olkowski, R. M., Sienkiewicz-Latka, E., Lisik, W., Kosieradzki, M., Sinski, M., Wierzbicki, Z., Przybylski, J. and Lewandowska-Szumiel, M. (2004) *Ann Transplant*, **9**, 61–3.
Langer, R. and Vacanti, J. P. (1993) *Science*, **260**, 920–6.
Laurencin, C. T., Ambrosio, A. M., Borden, M. D. and Cooper, J. A., Jr. (1999) *Annu Rev Biomed Eng*, **1**, 19–46.
Lewandowska-Szumiel, M. (2005) In *Biomaterials in the Orthopaedic Practice*, Vol. 5 (Eds, Lekszycki, T. and Maldyk, P.), IPPT PAN, CoE ABIOMED, Warsaw, pp. 9–30.
Lewandowska-Szumiel, M., Sikorski, K., Szummer, A., Lewandowski, Z. and Marczynski, W. (2007) *J Biomech*, **40**, 554–60.

Lian, J. B., Stein, G. S. and Aubin, J. E. (2003) In *Primer on the Metabolic Bone Diseases and Disorders of Mineral Metabolism* (Ed, Favus, M. J.), American Society for Bone and Mineral Research, Washington, DC, pp. 13–28.

Lieberman, J. R., Daluiski, A. and Einhorn, T. A. (2002) *J Bone Joint Surg Am*, **84-A**, 1032–44.

Lysaght, M. J. and Hazlehurst, A. L. (2004) *Tissue Eng*, **10**, 309–20.

Mackie, E. J. (2003) *Int J Biochem Cell Biol*, **35**, 1301–5.

Martin, I., Wendt, D. and Heberer, M. (2004) *Trends Biotechnol*, **22**, 80–6.

McCann, S., Byrne, J. L., Rovira, M., Shaw, P., Ribaud, P., Sica, S., Volin, L., Olavarria, E., Mackinnon, S., Trabasso, P., VanLint, M. T., Ljungman, P., Ward, K., Browne, P., Gratwohl, A., Widmer, A. F. and Cordonnier, C. (2004) *Bone Marrow Transplant*, **33**, 519–29.

Meyer, U., Buchter, A., Nazer, N. and Wiesmann, H. P. (2006a) *Br J Oral Maxillofac Surg*, **44**, 134–40.

Meyer, U., Kruse-Losler, B. and Wiesmann, H. P. (2006b) *Br J Oral Maxillofac Surg*, **44**, 289–95.

Mundy, G. R., Chen, D. and Oyajobi, B. O. (2003) In *Primer on the Metabolic Bone Diseases and Disorders of Mineral Metabolism* (Ed, Favus, M. J.), American Society for Bone and Mineral Research, Washington, DC, pp. 46–58.

Nerem, R. M. (2006) *Tissue Eng*, **12**, 1143–50.

Parfitt, A. M. (1984a) *Calcif Tissue Int*, **36 Suppl 1**, S123–8.

Parfitt, A. M. (1984b) *Calcif Tissue Int*, **36 Suppl 1**, S37–45.

Robling, A. G., Castillo, A. B. and Turner, C. H. (2006) *Annu Rev Biomed Eng*, **8**, 6.1–6.44.

Rodan, G. A. and Martin, T. J. (1981) *Calcif Tissue Int*, **33**, 349–51.

Roelofsen, J., Klein-Nulend, J. and Burger, E. H. (1995) *J Biomech*, **28**, 1493–503.

Rubin, C. T. and Lanyon, L. E. (1984) *J Bone Joint Surg Am*, **66**, 397–402.

Rucci, N., Migliaccio, S., Zani, B. M., Taranta, A. and Teti, A. (2002) *J Cell Biochem*, **85**, 167–79.

Sachse, A., Wagner, A., Keller, M., Wagner, O., Wetzel, W. D., Layher, F., Venbrocks, R. A., Hortschansky, P., Pietraszczyk, M., Wiederanders, B., Hempel, H. J., Bossert, J., Horn, J., Schmuck, K. and Mollenhauer, J. (2005) *Bone*, **37**, 699–710.

Saito, A., Suzuki, Y., Kitamura, M., Ogata, S., Yoshihara, Y., Masuda, S., Ohtsuki, C. and Tanihara, M. (2006) *J Biomed Mater Res A*, **77**, 700–6.

Salgado, A. J., Coutinho, O. P. and Reis, R. L. (2004) *Macromol Biosci*, **4**, 743–65.

Sanchez, C., Arribart, H. and Guille, M. M. (2005) *Nat Mater*, **4**, 277–88.

Seto, I., Marukawa, E. and Asahina, I. (2006) *Plast Reconstr Surg*, **117**, 902–8.

Shea, J. E. and Miller, S. C. (2005) *Adv Drug Deliv Rev*, **57**, 945–57.

Sikavitsas, V. I., Bancroft, G. N. and Mikos, A. G. (2002) *J Biomed Mater Res*, **62**, 136–48.

Smit, T. H., Burger, E. H. and Huyghe, J. M. (2002) *J Bone Miner Res*, **17**, 2021–9.

Sommerfeldt, D. W. and Rubin, C. T. (2001) *Eur Spine J*, **10 Suppl 2**, S86–95.

Tran Van, P., Vignery, A. and Baron, R. (1982) *Cell Tissue Res*, **225**, 283–92.

Udagawa, N., Takahashi, N., Jimi, E., Matsuzaki, K., Tsurukai, T., Itoh, K., Nakagawa, N., Yasuda, H., Goto, M., Tsuda, E., Higashio, K., Gillespie, M. T., Martin, T. J. and Suda, T. (1999) *Bone*, **25**, 517–23.

Vats, A., Tolley, N. S., Polak, J. M. and Gough, J. E. (2003) *Clin Otolaryngol Allied Sci*, **28**, 165–72.

Vunjak-Novakovic, G., Obradovic, B., Martin, I., Bursac, P. M., Langer, R. and Freed, L. E. (1998) *Biotechnol Prog*, **14**, 193–202.

Walboomers, X. F., Habraken, W. J., Feddes, B., Winter, L. C., Bumgardner, J. D. and Jansen, J. A. (2004) *J Biomed Mater Res A*, **69**, 131–9.

Wendt, D., Marsano, A., Jakob, M., Heberer, M. and Martin, I. (2003) *Biotechnol Bioeng*, **84**, 205–14.

Wolff, J. (1892) *Das Gesetz der Transformation des Knochens*, Hirschwald, Berlin.

Wood, M. A., Hughes, S., Yang, Y. and El Haj, A. J. (2006a) *J Control Release*, **112**, 96–102.

Wood, M. A., Yang, Y., Thomas, P. B. and Haj, A. J. (2006b) *Tissue Eng*, **12**(9), 2489–97.

Woźniak, P., Chroscicka, A., Olkowski, R. and Lewandowska-Szumiel, M. (2005) *Eng Biomater*, **47–53**, 190–2.

Yang, Y. and El Haj, A. J. (2006) *Expert Opin Biol Ther*, **6**, 485–98.

Yang, Y., Magnay, J., Cooling, L., Cooper, J. J. and El Haj, A. J. (2004) *Med Biol Eng Comput*, **42**, 22–9.

15

Bone tissue engineering and biomineralization

L DI SILVIO, Kings College London, UK

15.1 Introduction

Bone loss as a result of injury or disease results in poor quality of life at significant socio-economic cost. Although a variety of biomaterials are available for clinical application, autologous bone grafts are still regarded as the gold standard for bone reconstruction. Commonly used autograft procedures in orthopaedics include spinal fusion and the treatment of non-union fractures, where bone is usually taken from the patient's iliac crest and used to stabilize the defect site. Autologous bone grafts have proved highly successful, due to the immunological compatibility, and direct transfer of osteogenic cells and osteo-inductive cues. However, the main limiting factor is short supply, but collection is painful and there are risks of donor site morbidity and infection with the patient having two operations. Furthermore, it is highly dependent on the quality and quantity of patient bone that can be harvested. Allograft, tissue transplanted from a different donor host, provides an abundant supply of both cortical and trabecular bone, but can cause an immune response, and potentially transmit viral diseases. In addition, its processing removes all cells and bioactive molecules compromising its incorporation into existing host tissue. The natural skeletal repair process of fracture healing utilizes bone-forming cells, a cartilaginous scaffold upon which the new woven bone forms, and bioactive molecules to direct the repair sequence. An ideal augmentation would be '*de novo*' tissue which has the same function, structure and mechanical integrity equal as the original tissue that was lost. Bone morphogenesis is a sequential cascade with three key phases: chemotaxis and mitosis of mesenchymal cells, differentiation of the mesenchymal cells initially into cartilage and replacement of the cartilage by bone. Hence a successful strategy for bone tissue engineering requires all of these factors.

The current generation of synthetic bone substitutes is helping to overcome the problems associated with availability and donor site morbidity. In recent years tissue engineering strategies have emerged with the potential to develop tissue replacement parts. As a result of advances in the field of biomaterials, novel developments include scaffolds that mimic the natural tissue in question, stem cell technologies and growth-stimulating factors, all of which have created a unique mode to fabricate tissues *ex vivo* and then transplant them back into the patient. A successful strategy to develop tissue replacement is achievable only by adopting a multidisciplinary approach that combines the principles of engineering and the life sciences towards the development of biological, biofunctional tissue replacement substitutes that can either fully restore or improve tissue function. Recent advances in cellular and molecular biology have created the opportunity to isolate pluripotent cells from embryonic tissue, adult bone marrow and umbilical cord blood. Methods are being developed and refined to generate sufficient numbers of cells able to be differentiated and retain their phenotype and perform the required biological functions, in the case of bone, the production of extracellular matrix, secretion of cytokines and other signalling molecules to enable mineralization to take place. Another important requirement of tissue engineered systems is the scaffold, which has to be able to sustain and promote the growth of the relevant cells and provide a temporary template for tissue growth and incorporation with the host tissue. This area is rapidly progressing with many novel biomaterials being developed to mimic natural tissues and induce biomineralization. In addition, these scaffold materials are being developed to have dual functions: to act as carriers for cells and also to deliver growth-stimulating factors locally to the appropriate site. Concepts of tissue engineering with particular reference to scaffolds for biomineralization are discussed in this chapter.

15.2 Tissue engineering

There is an ever-increasing need for tissue and organ substitutes as a result of trauma, age-related diseases, degenerative conditions and end-stage organ failure.[1] Currently available methods include mechanical devices or artificial prostheses, which do not repair the tissue or organ function and do not integrate with the host tissue. In the ageing population worldwide, the incidence of bone-related illness is more prevalent than ever before and bone transplantation is a commonly required surgical procedure. In the UK there are approximately 150 000 fractures (wrist, vertebral and hip) per year due to osteoporosis, with high rates of morbidity and mortality associated with them. In the United States the total national healthcare cost for these patients exceeds $120 billion per year. Hence, there is an increasing demand for developing bone augmentation strategies.[2]

The requirement of new bone for these conditions has led to the need for procedures to generate bone for skeletal use. However, one of the major challenges confronted by orthopaedic surgeons is the repair and restoration of large defects, for example the resection of malignant bone neoplasms and trauma.[3] Bone transplantation has evolved considerably over the past 50 years, with successful and widespread use of osteogenic and osteoconductive materials.[4] Currently, the mainstay of skeletal reconstruction is bone grafting, using either autograft or allograft. Although still considered the 'gold standard' for bone restoration, autograft and allograft have associated problems.[5] These include limited supply, the potential to transmit infectious diseases, the requirement of second site surgery and inferior osteoinductivity with limited ability to incorporate with host bone. Alternatives to autografts and allograft preparations have included calcium-phosphates, bioactive glass, polymers and many other composite materials.[6–8]

Over the years, many materials have been described for application in bone repair (Table 15.1). Organic and inorganic synthetic polymers have been used in a wide variety of biomedical applications. A desirable feature is the synchronization of the rate of degradation of a scaffold with that of replacement by the natural host tissue. The polymers can be biodegradable, for example polylactic acid and polyglycolic acid.[9] Biodegradable synthetic polymers offer many advantages over other materials as scaffolds for tissue engineering. The key advantages include the ability to tailor mechanical properties and degradation kinetics to suit various applications. Synthetic polymers are also attractive because they can be fabricated into various shapes with desired pore morphologic features conducive to tissue in-growth. Furthermore, polymers can be designed with chemical functional groups that can induce tissue in-growth (Fig. 15.1). These polymers are broken down in the body hydrolytically to produce lactic acid and glycolic acid, respectively.[10] Other biodegradable polymers currently being studied for potential tissue engineering applications include polycaprolactone, polyanhydrides and polyphosphazenes.[11,12] Polymethyl-

Table 15.1 Common biomaterials in use

Polymer	Ceramics	Composite/natural
Polylactic acid	Bioglass®	Poly(D,L-lactide-co-glycolide) + bioactive glass
Polyglycolic acid	Sintered hydroxyapatite	Extracellular matrix (ECM)
Polycaprolactone	Glass-ceramic A-W	Hyaluronan – linear glycosaminoglycan (GAG)
Polyanhydrides	Hydroxyapatite (HA) – calcium phosphate-based ceramic	Demineralized bone matrix (DBM)
Polyphosphazenes	Collagraft – commercial graft = HA + tricalcium phosphate ceramic + fibrillar collagen	
Polymethylmethacrylate (PMMA)	Bioactive glass	
Polytetrafluoroethylene (PTFE)	Sol–gel-derived bioactive glass	

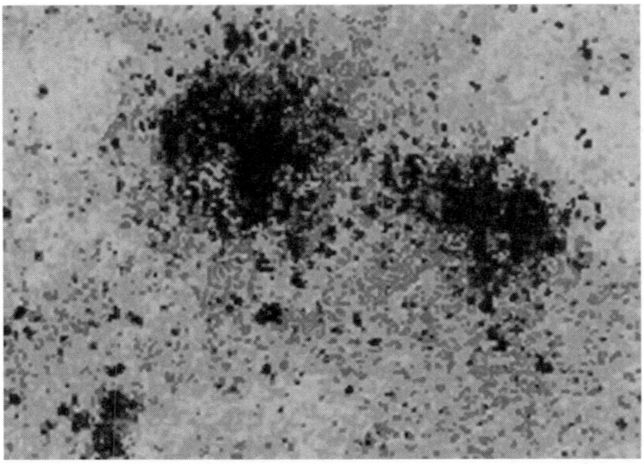

15.1 Human mesenchymal stem cells cultured in PEG gel for 14 days in osteogenic supplemented medium, showing mineralization nodules following staining with Von Kossa stain (black areas).

methacrylate (PMMA) is another example of a synthetic polymer; however, it is a non-biodegradable polymer which has been used for over 30 years in orthopaedic surgery to fix prosthetic implants and incorporation of hydroxyapatite has resulted in its increasing biological activity.[13,14] PMMA has also been widely used in dentistry. Other polymers such as polytetrafluoroethylene (PTFE) have also been used for augmentation and guided bone regeneration. This acts to encourage new bone to grow and also prevents the in-growth of fibrous scar tissue into the grafted site; thus new bone formation occurs.[15,16]

Ceramics have also been widely used in orthopaedic and dental applications.[17] Some ceramics, such as Bioglass®, sintered hydroxyapatite and glass-ceramic A-W, spontaneously form a bone-like apatite layer on their surface in the living body, and bond to bone through the apatite layer. These materials are called bioactive ceramics, and are clinically important for use as bone-repairing materials.[18] Hydroxyapatite (HA) is a calcium-phosphate-based ceramic and has been used for over 20 years. HA is biocompatible, and stimulates osseoconduction,[19,20] by recruiting osteoprogenitor cells and causing them to differentiate into osteoblast-like bone-forming cells, it is resorbed and replaced by bone at a slow rate[21] (Fig. 15.2).

Properties such as sintering temperature, configuration and pore size all affect the performance of HA.[22] Low strength and poor mechanical properties limit clinical application to low or non-load-bearing applications as loading can lead to fatigue fractures.[23] The necessity for unique material property combinations with respect to biological and physiochemical functionality has resulted in a large number of composite biomaterials being developed. Composites can result in substitutes with properties between each of the respective materials; for example, bovine collagen has been manufactured with HA. Collagraft, a

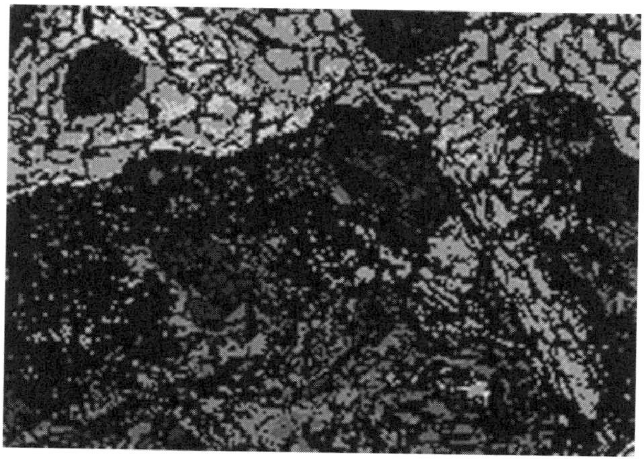

15.2 Materials such as hydroxyapatite (HA) provide stable porous structures to serve as pre-designed three-dimensional scaffolds. Porous HA scaffold, seeded with human osteoblasts, was completely covered in 48 h.

commercial graft, consists of a mixture of porous beads composed of hydroxyapatite, tricalcium phosphate ceramic and fibrillar collagen. When mixed with autogenous bone marrow, it serves as an effective bone graft.[24] It is generally accepted that the combination of collagen and calcium-based ceramics provides a bone-like matrix that supports the adhesion, migration, growth and differentiation of bone-forming cells.

Bioactive glasses are another class of interesting materials as they elicit a specific biological response at the interface of the material, which results in the formation of a bond between tissues and the material.[25] Three-dimensional scaffolds have been developed with suitable properties for bone tissue engineering; for example, sol–gel-derived bioactive glasses have been foamed to produce interconnected pore morphologies similar to trabecular bone.[26] Bioactive glasses have been shown to bind to soft tissue and bone. Composite scaffolds comprising macroporous biodegradable poly(D,L-lactide-co-glycolide) and bioactive glass, have demonstrated angiogenic properties and possibly have potential for tissue engineering for enhancement of vascularization.[27]

The extracellular matrix (ECM) of bone is a composite material made up of an organic phase reinforced by an inorganic phase. ECM proteins are commonly exploited as natural bone graft materials. Collagens make up 90% of the organic matrix, also known as osteoid. The remaining fraction consists of non-collagenous proteins. Various collagen-based products are currently under development. Work by Sachlos *et al.*, described a collagen scaffold reinforced using biomimetic composite nano-sized carbonate-substituted with hydroxyapatite crystals. The scaffold produced had microchannel features which could favourably assist the mass transport of essential nutrients of metabolites through the inner sections of the scaffolds, thus mimicking similar features to bone.[28]

Another promising matrix material for bone tissue engineering is Hyaluronan (HyA). HyA is a linear glycosaminoglycan (GAG) and has been experimentally determined to induce chondrogenesis and angiogenesis during remodelling and is being studied both individually and in combination with collagen as a matrix for bone repair. HyA use exhibits unique properties that bestow many possible functions; its future in tissue engineering is well described in the review by Allison and Grande-Allen.[29]

Demineralized bone matrix (DBM) is produced by acid extraction of bone, which involves the removal of the mineral phase, but retention of growth factors such as the bone morphogenetic proteins (BMPs), non-collagenous proteins and collagen type I. Since the initial study by Urist in 1965 the osteoinductive capacity of DBM has been well documented.[30] Thus, there has been increasing orthopaedic interest in DBM because of its therapeutic potential in the treatment of musculoskeletal injury.[31] Russell and Block evaluated 21 studies that used DBM in the treatment of a variety of clinical and experimental orthopaedic situations that would have normally have warranted treatment with standard bone graft.[32] They concluded that over 80% of the authors reported favourable results with the use of DBM. However, in some of these studies, it was argued that new bone formation was generated by osteoconduction rather than osteoinduction. This infers that bone regeneration occurred by growth of existing host bone on the DBM granules, which acted as a scaffold, rather than by *de novo* differentiation of bone. The scepticism regarding the osteoinductive properties of DBM still exists in the science community. Nonetheless DBM has been shown to be a valid carrier for bone marrow cells and bone morphogenetic proteins (Fig. 15.3). The variables influencing clinical outcome from DBM therapy include non-standardized procurement and preparation techniques, donor age and gender, reagents and quality control, and sterilization (e.g. gamma irradiate, ethylene oxide sterilize).[33] Clinical reports on combinations of DBM and autograft favour this composition over DBM alone and underscore DBM as an autograft expander.[34] Owing to the strength limitations of collagen-based products such as these, defect dimensions are limited and must be complemented by skeletal fixation. Osteoinductive properties are imparted to these matrices when combined with bone marrow aspirates, thus greatly enhancing the repair process. Combination of these materials with growth-stimulating factors has also been developed for clinical application to restore and repair of bone tissue.[35,36]

Skeletal reconstruction techniques play an integral role in the management of disorders involving local bone loss. Tissue engineering (TE) is a rapidly evolving discipline that seeks to repair, replace or regenerate tissues or organs by translating fundamental knowledge in physics, chemistry and biology into practical and effective materials, or devices and clinical strategies.[37,38]

Among the many tissues in the body, bone has the highest potential for regeneration, thus making it a suitable tissue for engineering. Many different approaches have been described in the literature for bone tissue engineering, but

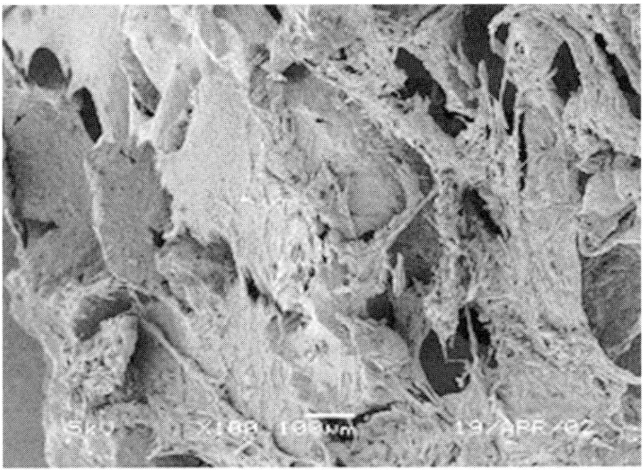

15.3 The osteoinductive properties of DBM can be further enhanced by addition of bone morphogenetic proteins (OP-1). When seeded with MSCs and OP-1, DBM provides an osteoconductive and osteoinductive graft material resulting in *de novo* bone formation.

all essentially involve at least one or more key elements. These include harvested cells, recombinant signalling molecules and three-dimensional (3D) matrices. For example, cells can be seeded into a highly porous biodegradable matrix or scaffold which has a similar architecture to that of the tissue being replaced. The coordinated function of cells is regulated by extracellular signals derived from soluble factors such as growth factors, and insoluble molecules of the ECM.[39]

These signalling molecules can be incorporated into the scaffolds in order to enhance cell growth, differentiation and cell migration. The cell-seeded scaffold can then be cultured to allow cell growth and finally be implanted into the defect to induce and direct the growth of new bone. The objective is that after the cells have attached to the scaffold, they proliferate and differentiate into cells exhibiting the bone-specific phenotype, and organize into normal, healthy bone as the scaffold degrades. Ideally, this type of scaffold will promote repair and in some cases regeneration of damaged or diseased tissues.[40]

Modification of cell phenotypes through genetic alteration is also a potential future strategy, to restore deficient osseous form and function. The alteration of cells through genetic modification requires a gene-delivery vector. Viral vectors are most commonly used because the life cycle of viruses involves gene transfer to cells. This transfection process results in a high level of efficiency in the expression of the protein by the transfected cells; since several types of viral vectors have been reported in the literature with the potential to be used in osseous tissue engineering, it will not be discussed in this chapter.[41] Cell-based therapies are emerging as an alternative therapeutic approach to many diseases. Tissue engineering strategies work on the concept that those cells capable of initiating and sustaining the regeneration process are *'switched on'* either through biological signals such as growth factors (GFs) or by genes, so that they generate new functional tissue of the required type.[42] Stem cells that reside in the bone marrow and periosteum have the potential to differentiate towards the osteoblastic lineage, through the up-regulation of biological signalling cues such as bone morphogenetic proteins present in extracellular matrix or by biomechanical cues. In the case of a fracture, these signalling cues direct the stem cells to progress towards the osteogenic lineage. However, the most important element to the success of TE is the ability to generate sufficient cell numbers that are able to maintain the appropriate phenotype and perform the required biological function of the tissue in question. Stem cells can be defined as undifferentiated cells that have the capacity to proliferate and self-renew and to differentiate to one or more types of specialized cells.[43] Stem cells can be isolated from embryos, foetuses or adult tissue. Both have advantages and drawbacks; embryonic stem cells have a high malleability, infinite lifespan and unlimited supply. However, the drawbacks include the potential for undesired development (teratomas), high ethical burden and uncertain legal status. In contrast adult stem cells are easier to manage and have less moral ambiguity; however, they have limited developmental potential and can lose their ability to proliferate/differentiate after a time in culture.[44] The main challenge, however, remains the ability of the cell-seeded scaffold to vascularize and integrate with the host tissue (Fig. 15.3). For this to be successful cells need to interact with neighbouring cells to produce growth factors and cytokines in order to produce extracellular matrix in an organized manner.

The coordinated function of bone cells is regulated by the integration of extracellular signals derived from soluble factors, such as growth factors, and insoluble molecules of the ECM which have the capability to control growth, differentiation and cell migration.[45] These proteins stimulate proliferation and/or differentiation, and the most potent osteoinductive proteins are bone morphogenetic proteins (BMPs).[46] BMPS were originally discovered following Urist's landmark study in 1965 which showed the osteoinductive capacity of demineralized bone matrix.[30] Since then, a great deal of research has been carried out using BMPs. Thus there is a plethora of evidence-based data regarding their properties. Most research concerning BMPs has focused on BMP-7, otherwise known as osteogenic protein–1 (OP-1)/recombinant human OP-1 (rhOP-1). The osteoinductive capacity of rhOP-1 has been successfully demonstrated in numerous preclinical trials.[47] These studies have combined rhOP-1 with a collagen carrier and demonstrated the induction of new bone formation in heterotopic sites as well as in the repair of critical sized skeletal defects. BMP-2 and BMP-7 have been used clinically for fractures and non-unions, but only in selected patients.[48] The localized use of osteoinductive growth factors *in vivo*, however, requires a controlled delivery systems and optimized pharmacokinetics that mimic physiological release.[49]

15.3 Scaffolds and biomineralization

Central to the formation of new bone tissue is the role of the scaffold, and a large number of scaffolds from different biomaterials are available for clinical use. Scaffold materials for making matrices for bone tissue engineering include several classes of biomaterials: synthetic polymers, ceramics, natural materials and composites. The primary function of the scaffold is that of tissue conduction, therefore it must allow cell attachment, migration onto or within the scaffold, cell proliferation and differentiation. It should provide an environment in which cells can maintain their phenotype and synthesize required proteins, in some cases acting as a signalling cue to the cells. Furthermore, its characteristics should include high porosity, high surface area, structural strength, specific 3D shape and ideally biodegradability. Porous 3D scaffolds promote new tissue formation by providing a surface that promotes the attachment and subsequent migration of cells that is necessary for proliferation and desired differentiation.[50,51] Critical variables in the scaffold design and function include the bulk material from which it is made, the architecture, surface chemistry, mechanical properties and also the local environment, which is influenced by the degradation products of the scaffold. Synthetic biomaterials continue to play an important role in the implementation of many new regenerative therapies and exhibit several attractive features for applications in controlled drug delivery, tissue repair and tissue engineering. Synthetic biomaterial guidance provided by biomaterials may facilitate restoration of structure and function of damaged or diseased tissues. Biomaterial scaffolds can be used in cell-based therapies, for example as carriers for cells expanded *ex vivo*, and in acellular therapies, such as those where the scaffold acts locally to induce growth and differentiation of cells *in situ*.[52]

Natural bone is a highly complex biomineralized system with intricate hierarchical structure. It is assembled by the orderly deposition of apatite minerals within a type I collagen matrix. The ECM microenvironment, which surrounds the cells and comprises the molecular signals, is composed of fibrillar and non-fibrillar components. The major fibrillar proteins are collagen and elastin. These proteins are important because they are responsible for tissue strength and resilience and also because of their dynamic role in promoting cell growth and differentiation. Scaffolds are being designed to act as artificial matrices that temporarily take over the role of the ECM in bone. They function as support structures to the host tissue, and as an adhesion site for the recruitment of bone cells and also act as delivery systems for the controlled release of biologically active molecules.[53] The temporal role of the biomaterial scaffold is critical: it must be designed to degrade into biocompatible products throughout the bone healing process, eventually resulting in repair or regenerated bone tissue.

Although naturally derived biomaterials have been used effectively in many clinical applications, there is a need for custom-made matrices or scaffolds for tissue-specific and site-specific application. A new generation of biomaterials is emerging however, that aims to mimic the appropriate tissue architecture, ECM and normal tissue regeneration. In the case of TE, a highly porous synthetic or natural ECM or scaffold is required to accommodate mammalian cells, direct their growth and the regeneration of the tissue in three dimensions. While reconstruction of small defects seems feasible using tissue engineering principles, reconstruction of larger defects remains a challenge. Vascularization concepts become a concern and the combination of TE approaches with flap prefabrication techniques may result in the application of bone substitutes grown *in vivo*, where the patient is acting as the bioreactor, with the advantage of minimal donor site morbidity, compared to conventional vascularized bone grafts.[54]

HA-based materials have been described for bone reconstruction, with the requirement that they should be porous to accommodate the ingrowth of blood vessels, and to guide the deposition of bone in a natural manner mimicking the Haversian structure of normal compact bone. HA is known to be osteoconductive, that is, it allows bone deposition in the presence of differentiated osteoblasts present at the site of implantation.[55] It does this by adsorbing osteogenic growth factors from the local milieu, thus providing favourable conditions for bone formation. Its rigidity provides some mechanical integrity and also a template onto which bone can be deposited. HA however, can also be rendered osteoinductive by the incorporation of growth factors such as bone morphogenetic proteins. That is, it has the potential to induce *de novo* differentiation of competent osteogenic cells from non-committed stem cells.

Biomineralization is a highly regulated process by which living organisms produce minerals, and is an essential requirement for normal skeletal development of the mechanical properties of hard tissues such as bone and teeth. In the case of vertebrate skeletons, the mineral phases include calcium carbonates and calcium phosphates. On a volume basis the mineral constitutes approximately 50% of bone, with the remaining ECM, comprising a hydrated mixture of collagen and non-collagenous matrix proteins. These proteins are associated with the mineral phase of bone, and are mostly acidic proteins rich in glutamic acid, aspartic acid, and phosphorylated serin/threonin residues, with a high capacity for binding calcium ions and HA crystal surfaces. The non-collagenous proteins have been implicated in the regulation of mineral deposition during osteogenesis and dentinogenesis.

Biomineralization occurs only in highly regulated biological environments in the presence of organic macromolecules or proteins, which controls numerous cell functions and also determine the rate of formation and the orientation of the inorganic crystals.[56] *In vivo* biomineralization is known to occur via the interaction of immobilized, negatively charged functional groups with calcium and phosphate ions. Acidic ECM proteins that are attached to the collagen scaffold are believed to play an important role in mineralization by serving as binding sites for calcium.[57]

A new paradigm in the assessment of factors that control biomineralization is that the variety of ions that exist in the mineralizing milieu play a significant role in regulating the precipitation process and influencing the nature of the apatite formed. Studies in this area have confirmed the ability of cells to alter the mineral formation process *in vivo*, in response to external factors. This control is accomplished through a feedback mechanism that allows the alteration of the composition of the ECM.[58] Consequently, to mimic biomineralization, we first require a method that mimics the actual organization of the proteins in the ECM. Attempts have been made to identify and classify proteins that promote or inhibit biomineralization. For example, Linde *et al.* showed that small quantities of different non-collagenous proteins immobilized by covalent binding to a substrate promoted mineral formation.[57] In contrast, work by Hunter and Goldberg on the nucleation of HA by bone sialoprotein revealed a promoter effect of unbound polyanionic proteins on mineral formation.[59] Opposing effects that are both inhibitory and promoter, have also been reported on apaptite formation of dentine and bone sialoprotein.[60] However, despite numerous studies designed to elucidate the mechanisms that determine where, when and how mineral crystals form, the precise mechanisms of biomineralization remain uncertain, in particular the essential components involved in the initiation of mineralization and in the regulation of crystal growth.

A scaffold in the context of bone tissue engineering is the extracellular matrix (ECM). The 3D geometry of tissue is determined both by cell–matrix and also cell–cell interactions. The ECM provides a unique microenvironmental niche for bone morphogenesis and the presence or absence of tissue-organizing cues in the ECM is a determinant of the tissue regeneration capacity. In the case of bone tissue engineering, the scaffolds should closely mimic the *in vivo* environment for bone growth.[61] Most natural biomaterials possess a composite microstructure. That is, they contain two or more chemically and structurally defined components, each of which play a specific role and in combination generate the physical and biological properties characteristic of the particular matrix or tissue.[58] An important consideration in the case of hybrid or composite material is the balance between the desirable mechanical properties of one component with the biological compatibility and physiological relevance of the other component.[60] The aim of biomimetic materials is to mimic the natural way of producing minerals such as apatites and the native ECM of bone, and include collagens, proteoglycans, glycosaminoglycans and hyaluronan, and also adhesive proteins fibronectin and laminin both of which are critical for the attachment of cells to the ECM.[61]

Biomimetic porous nano-fibrous poly(L-lactic acid) (PLLA) scaffolds have been developed to mimic a morphological function of collagen fibrils and to provide a more favourable microenvironment for cells. This study showed that biomineralization was enhanced substantially on the nano-fibrous scaffolds compared to solid-walled scaffolds, and this was confirmed by von Kossa staining, measurement of calcium content.[62] Collagen scaffolds have also been described. They have numerous advantages over other materials for cell-seeded scaffolds for tissue engineering. A major benefit is that collagen is found in abundance in tissue and is highly organized as a structure. Furthermore, collagen has both osteoconductive and osteoinductive properties, and has good adhesive properties as well as high tensile strength. Collagen scaffold systems designed to mimic the *in vivo* environment have been described with the potential to incorporate stem and other cells and provide the essential biological cues for cell migration, proliferation and differentiation.[63]

Calcium phosphate (CaP)-based biomaterials have found many applications for bone substitution and repair. The bioactivity of CaP and other similar biomaterials has been linked to their ability to nucleate carbonate apaptite crystals similar to bone minerals. Nanocrystalline calcium phosphates have also been described for use as scaffolds for bone tissue engineering. These materials show excellent *in vivo* biocompatibility, cell proliferation and resorption. Phase composition and microstructure have been shown to be very important, with nanocrystalline calcium phosphate releasing Ca^{2+} to the medium due to the higher solubility of the β-TCP component, thus stimulating cell response.[64]

15.4 Conclusions and future trends

Developing mineralized tissue scaffolds that mimic the microstructure of bone and associated biomechanical properties can only be achieved by a greater understanding of the interactions and factors controlling mineralization at a molecular level. For example, understanding the role of mineral-associated proteins such as osteopontin and bone sialoproteins will influence the choice of material used for bone tissue engineering. The cell biology underlying biomineralization is still, however, not well understood. In order to fully understand how mineralization really occurs at this level, there is a need to study the formation of biomineralized structures using approaches of more sophisticated cell, molecular and developmental biology. In addition, the precise relationship of matrix molecules to nucleation, growth and patterning of the various minerals, require in-depth understanding of protein structure–function relationships. Much of the current research in biomineralization is currently directed towards identifying and mimicking organized biomineralization in order to produce tailor-made inorganic materials. There is little doubt that the development of 3D, biomimetic cell–seeded scaffolds, incorporating biological signalling cues, will bring us closer to developing tissues of practical size, predetermined shapes and biofunctionality that will define the field of regenerative medicine.

15.5 References

1. Langer R and Vicente J P (1993), 'Tissue engineering', *Science* **260**, 920–5.
2. Vicar A R (2002), 'The role of the osteoconductive scaffold in synthetic bone graft', *Orthopaedics* **25**(5), 571–80.
3. Hollinger J O, Winn S and Bandai J (2000), 'Options for tissue engineering to address challenges of the aging skeleton', *Tissue Eng* **6**(4): 341–50.
4. NorAm R M (2006), 'Tissue engineering: the hope, the hype and the future', *Tissue Eng* **12**, 1143–50.
5. Lower G L, Maxwell K M, Kara sick D, Block J E and Russo R (1995), 'Comparison of autograft and composite grafts of demineralized bone matrix and autologous bone to posterior lateral fusions: an interim report', *Innovation Technol Biol Medicine* **16**(S1).
6. Burg K J, Porter S and Elam JF (2000), 'Biomaterial developments for bone tissue engineering', *Biomaterials*, **21**(23), 2347–59.
7. Hench L L and Paschall H A, (1973), 'Direct chemical bond of bioactive glass-ceramic materials to bone and muscle', *J Biomed Mater Res* **7**, 25–42.
8. Spector M (2006), 'Biomaterials-based tissue engineering and regenerative medicine solutions to musculoskeletal problems', *Swiss Med Wkly* **136**, 293–301.
9. Holy C E, Fialkov J A, Davies J E and Shoichet M S (2003), 'Use of biomimetic strategy to engineer bone', *J Biomed Mater Res.* **65A**(4), 447–53.
10. Gunatillake P A and Adhikari R (2003), 'Biodegradable synthetic polymers for tissue engineering', *Eur Cells Mater* **5**, 1–16.
11. Bass E, Kuiper J H, Wood M A, Yang Y and El Haj A J (2006), 'Micro-mechanical analysis of PLLA scaffolds for bone tissue engineering', *Eur Cells Mater* **11**(3), 34.
12. Cohn D and Salomon A H (2005), 'Designing biodegradable multiblock PCL/PLA thermoplastic elastomers', *Biomaterials* **26**, 2297–305.
13. Dalby M J, Di Silvio L, Harper L and Bonfield W (2002), 'Increasing the hydroxyapatite incorporation into poly(methylmethacrylate) cement increases osteoblast adhesion and response', *Biomaterials* **23**, 569–76.
14. Opara T N, Dalby M J, Harper E J, Di Silvio L and Bonfield W (2003) 'The effect of varying percentage hydroxyapatite in poly(ethylmethacrylate) bone cement on human osteoblast-like cells', *J Mater Sci: Mater Med* **14**(3), 277–82.
15. Santhosh Kumar T R and Krishnam L K (2002), 'A stable matrix for generation of tissue engineered non-thrombogenic vascular grafts', *Tissue Eng* **8**(5), 763–70.
16. Dahklin C, Anderrsson L and Linde A (1991), 'Bone augmentation at fenestrated implants by an osteopromotive membrane technique. A controlled clinical study', *Clin Oral Implants Res.* **2**(4), 159–65.
17. Jarcho M (1981), 'Calcium phosphate ceramics as hard tissue prostheses', *Clin Ortho Rel Res* **157**, 259–66.
18. Hench L L (1991), 'Bioceramics from concept to clinic', *J Am Ceramic Soc* **1**(74), 1487–510.
19. Di Silvio L, Dalby M and Bonfield W (1998), '*In vitro* response of osteoblasts to hydroxyapatite reinforced polyethylene composites', *J Mater Sci: Mater Med* **9**, 845–8.
20. Schnettler R, Alt V, Dingeldein E, Pfefferle H J, Kilian O, Meyer C, Heiss C and Wenisch S (2003), 'Bone ingrowth in βFGF-coated hydroxyapatite ceramic implants', *Biomaterials* **24**, 4603–8.
21. Kizuki T, Ichinose M O S, Nakamura S, Hashimoto K, Toda Y, Yokogawa Y and Yamashita K (2006), 'Specific response of osteoblast-like cells on hydroxyapatite

layer containing serum protein', *J Mater Sci: Mater Med* **17**(9), 859–67.
22. Gauthier O, Bouler J.-M, Aguado E, Pilet P and Daculsi G (1998), 'Macroporous biphasic calcium phosphate ceramics: influence of macropore diameter and macroporosity percentage on bone ingrowth', *Biomaterials* **19**, 133–9.
23. Geesink R G, de Groot K and Klein CP (1988), 'Bonding of bone to apatite coated implants', *J Bone Joint Surgery* **70**(1), 17–22.
24. Cornell C N, Lane J M, Chapman M, Merkow R, Seligson D, Henry S, Gustilo R and Vincent K (1991), 'Multicenter trial of Collagraft as bone graft substitute', *J Orthop Trauma* **5**(1), 1–8.
25. Hench L L, Splinter R J, Allen W C and Greenlee J TK (1972), 'Bonding mechanisms at the interface of ceramic prosthetic materials', *J Biomed Mater Res* **2**(1), 117–41.
26. Jones J R, Ehrenfried L M and Hench L L (2006), 'Optimising bioactive glass scaffolds for bone tissue engineering biomaterials', *Biomaterials* **27**(7), 964–73.
27. Day R M, Maquet V, Boccaccini A R, Jerome R and Forbes A (2005), '*In vitro* and *in vivo* analysis of macroporous biodegradable poly(D,L-lactide-co-glycolide) scaffolds containing bioactive glass', *J Biomed Mater Res Part A* **75A**(4), 778–87.
28. Sachlos E, Gotora D and Czernuszka J T (2006), 'Collagen scaffolds reinforced with biomimetic composite nano-sized carbonate-substituted hydroxyapatite crystals and shaped by rapid prototyping to contain internal microchannels', *Tissue Eng* **12**(9), 2479–87.
29. Allison D D and Grande-Allen K J (2006), 'Review: Hyaluronan: a powerful tissue engineering tool', *Tissue Eng* **12**(8), 1–10.
30. Urist M R (1965), 'Bone: formation by autoinduction', *Science* **159**, 893–9.
31. Bostrom M P G and Seigerman D A (2005), 'The clinical use of allografts, demineralized bone matrices, synthetic bone graft substitutes and osteoinductive growth factors: a survey study', *HSS Journal*, **1**, 9–18.
32. Russell J L and Block J E (1999), 'Clinical utility of demineralized bone matrix for osseous defects, arthrodesis, and reconstruction: impact of processing techniques and study methodology', *Orthopaedics* **22**, 524–31.
33. Yasko A W, Lane J M and Fellinger E J (1992), 'The healing of segmental bone defects induced by recombinant human bone morphogenetic protein (rhBMP-2)', *J Bone Joint Surg* **74A**, 659–70.
34. Maddox E, Zhan M, Mundy G R, Drohan W N and Burgess W H (2000), 'Optimizing human demineralised bone matrix for clinical application', *Tissue Eng* **6**(4), 441–8.
35. Cook S D and Reuger D C (1996), 'Osteogenic protein 1: biology and applications', *Clin Orthop* **324**, 29–38.
36. Geesink R G T, Hoefnagels N H M and Bulstra S K (1999), 'Osteogenic activity of OP-1 bone morphogenetic protein (BMP 7) in a human fibular defect', *J Bone Joint Surg (Br)*, **1-B**, 710–18.
37. Di Silvio L, Gurav N and Tsiridis E (2004), 'Tissue Engineering, Bone', *Encyclopedia of Biomaterials and Biomedical Engineering*, Marcel Dekker, New York, 1500–07.
38. Reddi A H (2000), 'Morphogenesis and tissue engineering of bone and cartilage: Inductive signals, stem cells and biomimetic biomaterials', *Tissue Eng* **6**, 351–9.
39. Juliano R L and Huskill S (1993), 'Signal transduction from the extracellular matrix', *J Cell Biol* **120**, 577–85.
40. Tsiridis E, Bhalla A, Ali Z, Gurav N, Heliotis M, Deb S and Di Silvio L (2006) 'Enhancing the osteoinductive properties of hydroxyapatite by the addition of human

mesenchymal stem cells, and recombinant human osteogenic protein-1 (BMP-7) *in vitro*', *Injury, In. J Care Injured* **37S**, S25–S32.

41. Fang J, Zhu Y Y, Smiley E, Bonadio J, Rouleau J P, Goldstein S A, McCauley L K, Davidson B L and Roessler B J (1996), 'Simulation of new bone formation by direct transfer of osteogenic plasmid genes', *Proc Natl Acad Sci USA* **93**(12), 5753–8.
42. Polak J M and Bishop A (2006), 'Stem cells and tissue engineering: past, present and future', *Ann NY Acad Sci* **1068**, 352–66.
43. Jaiswal N, Haynesworth S E, Caplan A I and Bruder S P (1997), 'Osteogenic differentiation of purified, culture-expanded human mesenchymal stem cells *in vitro*', *J Cell Biochem* **64**(2), 295–312.
44. McLaren A (2001), 'Ethical and social considerations of stem cell research', *Nature* **414**(6859), 129–31.
45. Caplan A I and Bruder S P (2001), 'Mesenchymal stem cells: building blocks for molecular medicine in the 21st century', *Trends Mol Med* **7**(6), 259–64.
46. Ripamonti U, Crooks J and Rueger D C (2001), 'Induction of bone formation by recombinant osteogenic protein-1 and sintered hydroxyapatite in adult primates', *Plast Reconstruc Surg* **107**(4), 977–88.
47. Blokhuis T J, den Boer F C, Bramer J A, Jenner J M, Bakker F C, Patka P and Haarman H J (2001), 'Biomechanical and histological aspects of fracture healing, stimulated with osteogenic protein-1', *Biomaterials* **22**(7), 725–30.
48. Kain M S and Einhorn T A (2005), 'Recombinant human bone morphogenetic proteins in the treatment of fractures', *Foot Ankle Clin* **4**, 639–50.
49. Hollinger J O, Uludag H and Winn S R (1998), 'Sustained release emphasizing recombinant human bone morphogenetic protein – 2', *Adv Drug Deliv Rev* **31**, 303–18.
50. Annaz B, Hing K A, Kayser M, Buckland T and Di Silvio L (2004), 'Porosity variation in hydroxyapatite and osteoblast morphology: a scanning electron microscopy study', *J Microsc* **215**, 100–10.
51. Herath H M T U, Di Silvio L and Evans J R G (2005), 'Porous hydroxyapatite ceramics for tissue engineering', *J Appl Biomater Biomech* **43**(3), 1–7.
52. Lutolf M P and Hubbell J A (2005), 'Synthetic biomaterials as instructive extracellular microenvironments for morphogenensis in tissue engineering', *Nature Biotechnol* **23**(1), 47–55.
53. Green D W, Mann S and Oreffo R O C (2006), 'Mineralized polysaccharide capsules as biomimetic microenvironments for cell, gene and growth factor delivery in tissue engineering', *Royal Soc Chem* **2**, 732–7.
54. Heliotis M, Lavery K M, Ripamonti U, Tsiridis E and Di Silvio L (2006), 'Transformation of a prefabricated hydroxyapatite/osteogenic protein-1 implant into a vascularised pedicled bone flap in the human chest', *Int J Oral Maxillofac Surg* **35**(3), 265–9.
55. Yang S, Leong K F, Du Z and Chua C K (2002), 'The design of scaffolds for use in tissue engineering. Part II. Rapid prototyping techniques', *Tissue Eng* **8**, 1–11.
56. He G, Ramachandran A, Dahl T, George S, Schultz D, Cooksen D, Veis A and George A (2005), 'Phosphorylation of phosphophoryn is crucial for its function as a mediator of biomineralization', *J Biol Chem* **280**(39), 33109–14.
57. Linde A, Lussi A and Crenshaw M A (1998), 'Mineral induction by immobilized polyanionicproteins', *Calcif Tissue Int* **44**, 286–95.
58. Boskey A L (1996), 'Matrix proteins and mineralization: an overview', *Connect Tissue Res* **35**, 357–63.
59. Hunter G K and Goldberg H A (1993), 'Nucleation of hydroxyapaptite by bone

sialoprotein', *Proc Natl Acad Sci USA* **90**, 8562–5.

60. Mann S (1988), 'Molecular recognition in biomineralization', *Nature* **261**, 1286–92.
61. Chaikof E L, Matthew H, Kohn J, Mikos A G, Prestwich G D and Yip C M (2002), 'Biomaterials and scaffolds in reparative medicine', *Ann NY Acad Sci* **961**, 96–105.
62. Daamean W F, van Moerkerk H ThB, Hafmans T, Buttafoco L, Poot A A, Veerkamp J H and van Kuppevelt T H (2003), 'Preparation and evaluation of molecularly-defined collagen-elastin-glycosaminoglycan scaffold for tissue engineering', *Biomaterials* **24**, 4001–9.
63. Gurav N, Bailey G, Sambrook R, Tsiridis E and Di Silvio l (2006), 'A novel *ex-vivo* culture system for studying bone repair', *Injury, Int J Care Injured* **37S**, S10–S17.
64. Song J, Saiz E and Bertozzi C R (2003), 'A new approach to mineralization of biocompatible hydrogel scaffolds: an efficient process toward 3-dimensional bone like composites', *J Am Chem Soc* **125**, 1236–43.

16
Cardiac tissue engineering

Q Z CHEN, S E HARDING, N N ALI, H JAWAD and A R BOCCACCINI, Imperial College London, UK

16.1 Introduction

Heart disease remains the leading cause of death and disability in the industrialised nations. It afflicts 1.8 million Britons and 25 million people worldwide, with approximately 120 000 new cases diagnosed each year in the UK. Heart failure, which is characterised by a dilated or hypertrophied heart, is a syndrome of breathlessness, oedema and fatigue resulting from damaged or defective myocardium. The enlargement in ventricular volume leads to progressive structural and functional changes in ventricles (called ventricular remodelling), representing a predisposing factor towards the end stage of heart failure. The most frequent initiating cause of heart failure is myocardial infarction, also known as heart attack, which is the single most common cause of death in economically developed countries, including the US and Western Europe. Myocardial infarction typically results in fibrous (collagen) scar formation and permanently impaired cardiac function because, after a massive cell loss due to ischaemia, the myocardial tissue lacks significant intrinsic regenerative capability. Eventually, heart transplantation is the ultimate treatment option to end-stage heart failure. Owing to the lack of organ donors and complications associated with immune suppressive treatments, scientists and surgeons constantly look for new strategies to repair the injured heart.

Around the mid-1990s studies veered to an intriguing strategy: cell therapy (Koh *et al.*, 1993). A number of studies carried out so far (e.g. Li *et al.*, 1996; Scorsin *et al.*, 1997; Taylor *et al.*, 1998; Reinecke *et al.*, 1999) have indicated that cell implantation (Fig. 16.1a) in models of myocardial infarction can improve contractile (mostly diastolic) function. An alternative approach to deliver isolated cells into the heart is to use a synthetic biodegradable construct (Fig. 16.1b), which is manipulated *in vitro* with cells and implanted later *in vivo*. So far, many studies have been published using different cells and different synthetic materials (Cima *et al.*, 1991; Zammaretti and Jaconi, 2004; Leor *et al.*, 2005). In this chapter, we present a review on the achievements of myocardial tissue engineering, focusing on construct (scaffold)-based strategies. It must be

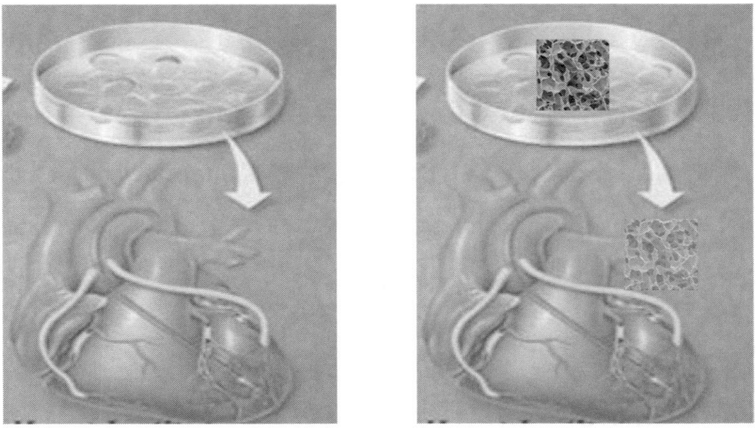

(a) (b)

16.1 Two strategies in myocardial tissue engineering: (a) direct transplantation of isolated cells, and (b) implantation of an *in vitro* engineered tissue construct (adapted from: http://www.umm.edu/news/releases/myoblast.htm).

mentioned that a number of excellent reviews focusing on different aspects of cardiac tissue engineering, including cell-based therapies for myocardial regeneration, have been published recently (Eschenhagen, 2005; Leor *et al.*, 2005; Gerecht-Nir *et al.*, 2006; Zimmermann *et al.*, 2006).

16.2 Cell sources

The selection of cell sources for myocardial tissue engineering should be based on the studies of related cell-based approaches. A variety of cell models have been under intensive investigation with the hope of improving myocardial function. They can be categorised into three groups: (1) somatic muscle cells, such as fetal or neonatal cardiomyocytes and skeletal myoblasts; (2) myocardium-regenerating stem cells, such as embryonic stem cells and (possibly) bone marrow-derived mesenchymal stem cells; and (3) angiogenesis-stimulating cells, including fibroblasts and endothelial progenitor cells. Owing to their relevance, the first two cell types are briefly reviewed in the following sections.

16.2.1 Somatic muscle cells

Fetal or neonatal cardiomyocytes

Early cell transplantation studies focused on using fetal or neonatal rodent (rat or mouse) cardiomyocytes, as these cells have the inherent electrophysiological, structural and contractile properties of cardiomyocytes and still retain some proliferative capacity. In their pioneering study, Soonpaa *et al.* (1994)

established the principles of cardiac cell implantation in the heart. They demonstrated that fetal or neonatal cardiomyocytes could be transplanted and integrated within both the healthy and dystrophic myocardium of mice and dogs, and that the surviving donor cells were aligned with recipient cells and formed cell-to-cell contacts. Similar results have later been reported by other research groups (Li *et al.*, 1996; Reinecke *et al.*, 1999; Murry *et al.*, 2002). All results showed that early-stage cardiomyocytes (fetal and neonatal) were better candidates than more mature cardiac cells due to their superior *in vivo* survival capability (Reinecke *et al.*, 1999). Several mechanisms have been proposed for improved heart function following cardiac myocyte transplantation (Kessler and Byrne, 1999; Etzion *et al.*, 2001a):

- direct contribution of the transplanted myocytes to contractility;
- attenuation of infarct expansion by virtue of the elastic properties of cardiomyocytes; and
- angiogenesis induced by growth factors secreted from the cells resulting in improved collateral flow.

Skeletal myoblasts

Theoretically, skeletal muscle cells may be superior to cardiomyocytes for infarct repair, because skeletal myoblasts have almost all the properties of the ideal donor cell type except their non-cardiac origin: (1) autologous sources, which obviate the need for immune suppression; (2) rapid expandability in an undifferentiated state *in vitro*; and (3) capability to withstand ischemia (Caspi and Gepstein, 2006; Etzion *et al.*, 2001b).

Although it was originally hoped that skeletal myoblasts would adopt a cardiac phenotype, it is now clear that in the heart cardiac phenotype myoblasts remain committed to form only mature skeletal muscle cells that possess completely different electromechanical properties from those of heart cells. However, studies in small and large animal models of infarction demonstrated beneficial effects of grafting of these cells on ventricular performance (Scorsin *et al.*, 2000). Clinical trials with these cells were initiated some years ago but have now been discontinued, largely because of the pro-arrhythmic effect of skeletal myoblast transplantation.

16.2.2 Stem cell-derived myocytes

In principle, stem cells are the optimal cell source for tissue regeneration, including myocardium. Firstly, they are capable of self-replication throughout life such that an unlimited number of stem cells of similar properties can be produced via expansion *in vitro*. Secondly, the stem cells are clonogenic, and thus each cell can form a colony in which all the cells are derived from this single cell and have identical genetic constitution. Thirdly, they are able to

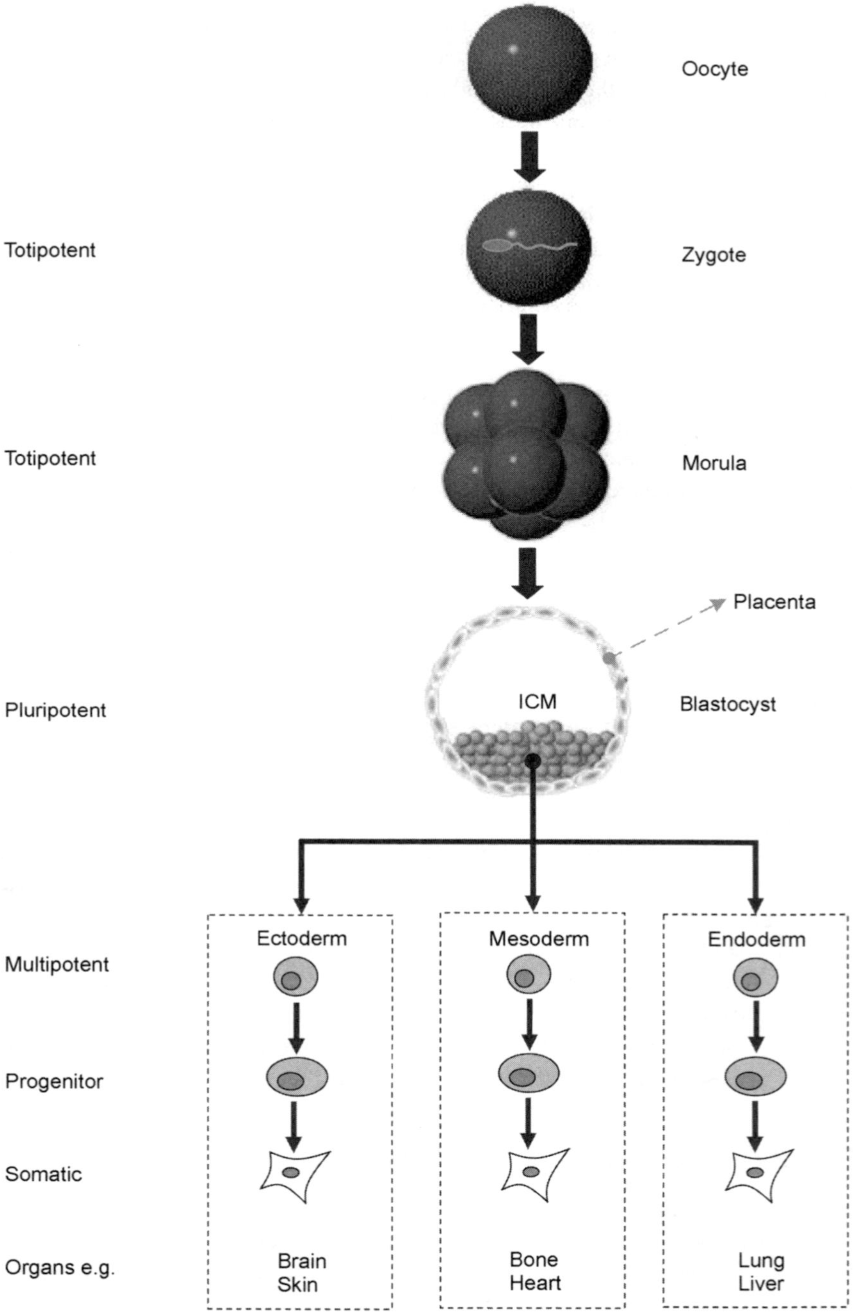

16.2 Types of stem cells in terms of potency.

differentiate into one or more specific cell types. Hence, after expansion stem cells can be directed to differentiate into cardiomyogenic lineage. For these reasons, stem cell-based therapies for cardiac muscle regeneration have been under intensive research since 1995. Figure 16.2 shows types of stem cells in terms of potency. In this section, we briefly describe the applications of three types of stem cells in myocardial tissue engineering: bone marrow-derived stem cells (multipotent somatic or cord blood stem cells), native cardiac progenitor cells (multipotent somatic stem cells) and embryonic stem cells (pluripotent).

Bone marrow-derived stem cells

Bone marrow stem cells are the most primitive cells in the marrow. These cells can be classified into: (1) bone marrow-derived mesenchymal stem cells (MSC) and (2) haematopoietic stem cells (HSC) or called haematopoietic progenitor cells (HPC).

Bone marrow-derived mesenchymal stem cells

Bone marrow-derived mesenchymal stem cells are a subset of bone marrow stromal cells. This potential multipotent stem cell is derived from the non-haematopoietic, stromal compartment of the bone marrow which can grow into non-marrow cells, such as bone, cartilage, adipose, endothelial, and myogenic cells. A number of studies suggested that bone marrow-derived MSCs could differentiate into cardiomyocytes both *in vitro* and *in vivo* (Orlic *et al.*, 2001; Pittenger and Martin, 2004; Amado *et al.*, 2005). One possible advantage of mesenchymal stem cells is their ability to be either autotransplanted or allotransplanted, as some reports suggested that they may be relatively privileged in terms of immune compatibility (Le Blanc, 2003).

Haematopoietic stem cell

It has been hypothesised that bone marrow stem cells might be able to differentiate into myocardium *in vivo*. Studies in the animal models of ischaemia and clinical trials suggested that delivery of haematopoietic stem cells and circulating endothelial progenitor cells (they originate from bone marrow stem cells) may result in improvement in the ventricular function in ischaemic heart disease patients (Caspi and Gepstein, 2006). However, Balsam *et al.* (2004) and Murry *et al.* (2004) demonstrated that the haematopoietic stem cells continued to differentiate along the haematopoietic lineage, suggesting the functional improvement observed may not be related to transdifferentiation into the cardiac lineage, but rather from indirect mechanisms.

Clinical trials have been carried out with autologous bone marrow-derived cells. These are often unfractionated, and therefore contain an unquantified amount of the HSCs and MSCs. At present, the results of three medium size clinical trials (100–200 patients) show a variable and modest improvement in cardiac function (Assmus *et al.*, 2006; Lunde *et al.*, 2006; Schachinger *et al.*, 2006).

Native cardiac progenitor cells

It had long been accepted that the adult mammalian heart, a terminally differentiated organ, had no self-renewal potential. This notion about the adult heart, however, has been challenged by accumulated evidence that myocardium itself contains a resident progenitor cell population capable of giving rise to new cardiomyocytes (Beltrami *et al.*, 2003; Oh *et al.*, 2003; Messina *et al.*, 2004; Pfister *et al.*, 2005; Laugwitz *et al.*, 2005). The existence of these progenitor cells will no doubt open new opportunities for myocardial repair, though many issues still need to be addressed.

Embryonic stem cells

Embryonic stem cells are thought to have much greater translational potential than other stem cells because of their several advantages over other stem cells, since they are: (1) pluripotent, which means they have a broader multilineage expressing profile, (2) robust, which means they have the long-term proliferation ability with a normal karyotype and (3) genetically manipulable (Thomson *et al.*, 1998).

The first study using embryonic stem cells as a source for cell transplantation into the myocardium was reported by Klug *et al.* in 1996. By using genetically selected mouse embryonic stem cell-derived cardiomyocytes, it was shown that the differentiated cells developed myofibrils and gap junctions between adjacent cells and performed synchronous contractile activity *in vitro*. This study proved the feasibility of guiding an unlimited number of embryonic stem cells into cardiomyogenic cell linage and to utilise them for myocardial regeneration. Later studies (Menard *et al.*, 2005), utilising the infarcted rat heart model, demonstrated that transplantation of differentiated mouse embryonic stem cell-derived cardiomyocytes can result in short- and long-term improvement of myocardial performance.

The vast biomedical potential of human ESCs has stirred enthusiasm in the field of tissue engineering. An overview of the utilisation of human ESCs in cardiac tissue engineering is given in Table 16.1. These studies demonstrated that the human ESC-derived cardiomyocytes displayed structural and functional properties of early-stage cardiomyocytes (Snir *et al.*, 2003; Laflamme *et al.*, 2005).

Table 16.1 Human ES cells in myocardial tissue engineering

Method of hES differentiation	State before transplantation	Major result	Group and reference
In vitro; via EBs in suspension	NA	ES cells differentiated into cardiomyocytes, even after long-term culture. Upon differentiation, beating cells were observed after 1 week, increased in numbers with time, and retained contractility for >70 days. The beating cells expressed markers of cardiomyocytes.	Xu *et al.* (2002), Laflamme *et al.* (2005)
In vitro; via EBs in suspension	NA	ES cells showed consistence in phenotype with early-stage cardiomyocytes, and expression of several cardiac-specific genes and transcription factors.	Kehat *et al.* (2001)
In vitro; via EBs in suspension	NA	ES cells showed a progressive ultrastructural development from an irregular myofibrillar distribution to an organised sarcomeric pattern at late stages.	Snir *et al.* (2003)
In vitro; via EBs in suspension	NA	ESC-derived myocytes at mid-stage development demonstrated the stable presences of functional receptors and signalling pathways, and the presence of cardiac-specific action potentials and ionic currents.	Satin *et al.* (2004)
In vitro; via EBs in suspension	NA	ES cells differentiated into cardiomyocytes. Upon differentiation, beating cells were observed after 9 days, and retained contractility for longer than 6 months.	Harding *et al.* (2007)

Table 16.1 Continued

Method of hES differentiation	State before transplantation	Major result	Group and reference
In vitro; via EBs in suspension	*In vitro*: co-culture of differentiated rat cardiomyocyte and hESCM	Tight electrophysiological coupling between the engrafted human ESCms and rat cardiomyocytes was observed.	Kehat *et al.* (2004)
	In vivo transplantation of hESCM into swine	The transplanted hES cell-derived cardiomyocytes paced the hearts of swine.	
In vitro (co-culture of undifferentiated hESC with mouse endoderm-like cells)	Differentiated	ESCs differentiated to beating muscle. Sarcomeric marker proteins, chronotropic responses, and ion channel expression and function were typical of cardiomyocytes. Electrophysiology demonstrated that most cells resembled human fetal ventricular cells.	Mummery *et al.* (2003)
In vitro via Ebs in suspension	*In vitro* co-culture of hESCM with rat myocytes Differentiated hESCM transplanted into guinea pig	Electrically active, hESC-derived cardiomyocytes are capable of actively pacing quiescent, recipient, ventricular cardiomyocytes in vitro and ventricular myocardium *in vivo*	Xue *et al.* (2005)

NA = Not applicable

16.3 Construct-based strategies in myocardial tissue engineering

In tissue engineering strategies based on the use of scaffolds, the regenerative ability of the host body should be increased through a designed construct that is populated with isolated cells and signalling molecules, aiming at regenerating functional tissue as an alternative to conventional organ transplantation and tissue reconstruction. In the following paragraphs, we review and discuss the design criteria and potential biomaterials used in myocardial tissue engineering approaches involving the use of scaffolds.

16.3.1 Design criteria of myocardial tissue engineering constructs

One of the major challenges in tissue engineering is the design and fabrication of tissue-like materials to provide a scaffold or template for cells. An ideal scaffold should mimic extracellular matrix of the tissue that is to be engineered, which means that it must meet several stringent criteria. The specific requirements for myocardial tissue engineering scaffolds are as follows:

- *Ability to deliver cells*. The material should not only be biocompatible, but also foster cell attachment, survival, differentiation and proliferation.
- *Biodegradability*. The composition of the material, combined with the porous structure of the scaffold, should lead to biodegradation *in vivo* at a rate that matches the tissue regeneration rate. In other words, a synthetic scaffold should remain in the body for as short a period as possible; at the same time it must maintain its viability long enough for the cells to make their own matrix.
- *Mechanical properties*. The principle in the mechanical design of the scaffold is that the scaffold should not interrupt the normal beating process of cardiomyocytes, while providing mechanical support for the cells to attach and to secrete their own matrix.
- *Porous structure*. The scaffold should have an interconnected porous structure with porosity $> 90\%$ and diameters between 300–500 μm for cell penetration, tissue ingrowth and vascularisation, and nutrient and waste transportation.
- *Commercialisation*. The synthesis of the material and fabrication of the scaffold should be cost-effective, being suitable for commercialisation.

Among these criteria, biocompatibility and cell-supporting and fostering ability are of highest importance for tissue engineering. So far, numerous biocompatible and biodegradable materials, including polymers, ceramics and composites, have been developed to support and foster cells, as described in other chapters in this book. Based on these available biomaterials, the foremost criteria to be considered are the mechanical properties and the porous structure of scaffolds, which are discussed in the following two sections.

Mechanical design

Ideally, a construct material should display mechanical and functional properties of native myocardium, such as coherent contractions, low diastolic tension and syncytial propagation of action potentials. Since the extracellular matrix (ECM) of myocytes is collagen, it is reasonable to hypothesise that cardiomyocytes would be able to beat adequately in a scaffolding material with mechanical properties similar (if impossible to be the same) to those of myocardial collagen. While natural collagens are an obvious option, their variable physical properties, including mechanical properties, with different sources of the protein matrices have hampered their application. Concerns have also arisen regarding immunogenic problems associated with the introduction of foreign collagen. As such, it is not surprising that much attention has been paid to synthetic polymers which have reproducible properties and are considered highly reliable materials for tissue engineering.

The myocardium collagen matrix mainly consists of type I and III collagens, which form a structural continuum. Synthesised by cardiac fibroblasts, type I and III collagens have different physical properties. Type I collagen mainly provides rigidity, whereas type III collagen contributes to elasticity. The two types of collagens together support and tether myocytes to maintain their alignment, whereas their tensile strength and resilience resist the deformation, maintain the shape and thickness of the construct, prevent the rupture and contribute to the passive and active stiffness of the myocardium. The ratio of collagen types (type I/type III) within a healthy heart is typically about 0.5 (Pauschinger *et al.*, 1999) and the stiffness of the collagen matrix of heart muscle is of several tens to hundreds kPa.

Design of porous structures

Tissue regeneration can be induced to take place *in vitro*. However, in the absence of true vascularisation, *in vitro* tissue engineering approaches face the problem of critical thickness: mass transportation into tissue is difficult beyond a thin peripheral layer of a tissue construct even with artificial means to supply engineered tissue constructs with nutrients and oxygen (<100 μm without the help of a bioreactor, and <200 μm with the support of a bioreactor) (Radisic and Vunjak-Novakovic, 2005). Diffusion barriers that are present *in vitro* are most likely to become more deleterious *in vivo* owing to lack of vascularisation. Once the engineered tissue construct is placed in the body, vascularisation becomes a key issue for further remodelling in the *in vivo* environment. It was thought that a pore size in the range of 50–100 μm was sufficient to allow the vascularisation of a scaffold following transplantation (Shachar and Cohen, 2003). Recently, Radisic and Vunjak-Novakovic (2005) suggested a larger (100 μm) pore size for cardiac tissue constructs. There are other authors who have suggested that pore sizes larger than 300 μm are necessary for long-term survival, i.e. vascularisation (Temenoff *et al.*, 2000). It is suggested that pore size in the range of 300–500 μm will not obstruct either vascularisation or mass transportation of nutrients and waste products, while the scaffold can maintain adequate mechanical integrity during *in vitro* culture and *in vivo* transplantation.

In summary, a potentially advantageous material for myocardial tissue engineering scaffold should be a soft elastomer with stiffness of the order of several tens to hundreds kPa, and the scaffold should be made to exhibit pore sizes in the rage of 300–500 μm.

16.3.2 Potential scaffolding biomaterials

So far, a number of polymeric biomaterials have been developed or are under development for myocardial tissue engineering. An overview of biomaterials applied in myocardial tissue engineering is given in Table 16.2. In this section, the studies conducted on each of these polymers are discussed.

Collagen gel matrix

In 1997, Eschenhagen *et al.* (1997) reported, for the first time, an artificial heart tissue, which was termed engineered heart tissue (EHT) (Fig. 16.3). In this work, embryonic chick cardiomyocytes were mixed with collagen solution and allowed to gel between two Velcro-coated glass tubes. By culturing the cardiomyocytes in the collagen matrix, they produced a spontaneously and coherently contracting 3D heart tissue *in vitro*. Immunohistochemistry and electron

Table 16.2 Overview of biomaterials used in myocardium tissue engineering

Biomaterial	Physical state	Reference
Natural		
Collagen	Liquid/gel	Zimmermann *et al.* (2000)
Collagen mesh (or sponge)	Solid	Kofidis *et al.* (2003)
Gelatine mesh	Solid	Li *et al.* (1999)
Alginate mesh	Solid	Leor *et al.* (2000)
Synthetic		
PGA and copolymer with PLA	Solid	Radisic and Vunjak-Novakovic (2005)
PLLA	Solid	Radisic and Vunjak-Novakovic (2005)
PCL and copolymer with PLA	Solid	Shin *et al.* (2004)
PGS, PGA	Solid	Vunjak-Novakovic *et al.* (2006)

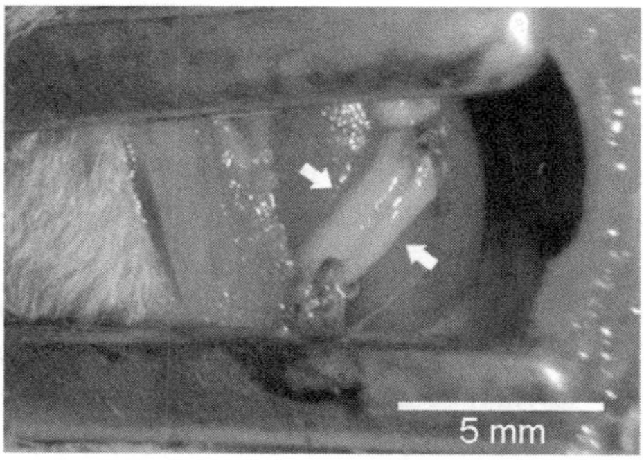

16.3 Engineered heart tissue for regeneration of diseased hearts. (Reprinted from: Zimmermann WH, Melnychenko I, Eschenhagen T., *Biomaterials*, **25**(9) 1639–47 (2004), with permission from Elsevier.)

microscopy revealed a highly organised myocardium-like structure exhibiting typical cross-striation, sarcomeric myofilaments, intercalated discs, desmosomes and tight junctions.

More recently, large (thickness 1–4 mm and diameter 15 mm), mechanically supportive EHTs were produced with neonatal rat heart cells, and were implanted on hearts with myocardial infarction in immune-suppressed rats (Zimmermann *et al.*, 2006). When evaluated 28 days later, EHTs showed several beneficial effects:

- developing electrical coupling to the native myocardium without evidence of arrhythmia induction;
- preventing further dilation;
- inducing systolic wall thickening of infarcted myocardial segments; and
- faster fractional area shortening of infarcted hearts, compared with controls (sham operation and non-contractile constructs).

In summary, the research has confirmed that EHTs have many structural, functional and physiological characteristics of cardiac tissue, and that EHTs can be implanted to both healthy and infarcted hearts and can survive *in vivo* in both situations. Although this is important proof-of-concept work, there is no realistic possibility of human or rat neonatal cardiomyocytes coming to clinical application. The potential for expansion of this work to the construction of large implants with embryonic stem cell-derived cardiomyocytes remains to be established.

Collagen fibrous mesh (or collagen sponge)

The application of collagen gel matrix is limited by insufficient mechanical strength. This has led researchers to look for new approaches based on solid scaffolds. Kofidis *et al.* (2003) seeded neonatal rat cardiomyocytes *in vitro* into a

3D solid collagen mesh and called the product artificial myocardial tissue (AMT). The artificial myocardial tissue was shown to possess structural, mechanical, physiological and biological characteristics similar to native cardiac tissue (Kofidis *et al.*, 2003). More recently, Kofidis *et al.* (2005) utilised undifferentiated embryonic stem cells as the substrate of artificial myocardial tissue. The bioartificial mixtures were implanted in the infarcted area of the rat hearts. Studies revealed that embryonic stem cells formed stable intramyocardial grafts that were incorporated into the surrounding area without distorting myocardial geometry, thereby preventing ventricular wall thinning. In contrast to collagen gel, solid collagen mesh is too stiff to match host heart muscle. This drawback is in fact shared by all existing solid scaffolds discussed below.

Gelatine mesh

Gelatine mesh is the second type of solid scaffold applied in cardiac muscle engineering. Li *et al.* (1999) seeded fetal rat ventricular muscle (not isolated cells) into a gelatine foam to form grafts. The grafts were cultured *in vitro* for 7 days, forming a beating cardiac graft. The grafts implanted into the subcutaneous tissue contracted regularly and spontaneously. When implanted onto myocardial scar tissue, the cells within the grafts survived and formed junctions with the recipient heart cells.

Alginate mesh

In addition to protein-based materials, there is intensive activity in the area of natural polysaccharides. Alginate, a negatively charged polysaccharide from seaweed that forms hydrogels in the presence of calcium ions, was initially developed for drug delivery and it is now under development for tissue engineering scaffolds. The group of Cohen (Zmora *et al.*, 2002) produced an alginate sponge using a freeze-drying technique, with porosity being 90% and pore size 50–150 μm. Moreover, Leor *et al.* (2000) seeded fetal rat myocardial cells into this sponge to form an engineered heart construct. After 4-day culture *in vitro*, the engineered constructs were implanted into the rat hearts with myocardium infarct. Hearts were harvested 9 weeks after implantation. A large number of blood vessels were found in the grafting area, indicating intensive neovascularisation. The specimens showed almost complete disappearance of the scaffold and good integration into the host. In contrast to the control animals which developed significant left ventricular dilatation accompanied by progressive deterioration in left ventricular contractility, the graft-treated rats showed attenuation of left ventricular dilatation and unchanged contractility in left ventricle.

Poly(glycolic acid) (PGA) and its copolymer with poly(lactic acid) (PLA)

Synthetic biopolymers are thought to have a future in tissue engineering due to not only their excellent processing characteristics, which can ensure the off-the-shelf availability, but also their advantage of being biocompatible and biodegradable. The biodegradable synthetic polymers most often utilised for 3D scaffolds in tissue engineering are the poly(α-hydroxy acids), including poly(lactic acid) (PLA) and poly(glycolic acid) (PGA), as well as poly(lactic-co-glycolide) (PLGA) copolymers. The first published work on a synthetic polymer-based scaffold designed for cardiac muscle engineering was by Freed and Vunjak-Novakovic in 1997. In this pioneering work, it was demonstrated that cultivation of primary neonatal rat cardiomyocytes on highly porous (porosity being 97%) PGA scaffolds in bioreactors could result in contractile 3D cardiac-like tissues, which consisted of cardiomyocytes with cardiac-specific structural and electrophysiological properties, contracting spontaneously and synchronously.

In the subsequent studies, the group invested great efforts to overcome the limitation on the thickness of engineered tissue through improving cell-seeding and tissue culturing conditions (bioreactor). A maximal construct thickness of 1–2 mm has been reported (Carrier *et al.*, 2002).

Poly(ϵ-caprolactone) (PCL) and its copolymer with PLA

PCL is another important member of the aliphatic polyester family. It has been used to effectively entrap antibiotic drugs, and a construct made with PCL has been considered as a drug delivery system. The degradation of PCL and its copolymers involves similar mechanisms to PLA, proceeding in two stages: random hydrolytic ester cleavage and weight loss through the diffusion of oligomeric species from the bulk.

More recently, Vacanti's group (Shin *et al.*, 2004) has demonstrated the formation of contractile cardiac grafts *in vitro* using a nanofibrous PCL mesh. The nanofibrous mesh, which was produced by the electrospin technique, had an extracellular matrix-like topography. The average fibre diameter of the scaffold was about 250 nm, well below the size of an individual cardiomyocyte. After neonatal rat cardiomyocytes were seeded in the nanofibrous mesh, the construct was cultured, while being suspended across a wire ring that acted as a passive load to contracting cardiomyocytes. The cardiomyocytes started beating after 3 days and were cultured *in vitro* for 14 days. The cardiomyocytes attached well on the PCL meshes and expressed cardiac-specific proteins such as alpha-myosin heavy chain, connexin43 and cardiac troponin I. This work indicated that using this technique, cardiac grafts can be matured *in vitro* to obtain sufficient function prior to implantation.

Elastomers: poly(glycerol-sebacate) (PGS) and PET-DLA

To engineer the tissue of heart, which beats cyclically and constantly throughout life, the biomaterial should be as soft and elastic as heart muscle. These mechanical characteristics are impossible with polyester-based thermosetting polymers, such as PGA, PLA, PCL and their copolymers, because they undergo plastic deformation and they are prone to failure when exposed to long-term cyclic strains. The limitation of their use in engineering elastomeric and flexible tissues has made biomaterial scientists turn to elastomers for cardiac tissue engineering.

PGS was recently developed for the field of soft tissue engineering (Wang *et al.*, 2002). This polymer is a biodegradable, biocompatible and inexpensive elastomer, and has already shown potential in nerve (Sundback *et al.*, 2005) and vascular tissue engineering (Fidkowski *et al.*, 2005). PGS also has superior mechanical properties, being capable of sustaining and recovering from deformation due to its intrinsic elasticity, and is thus suited to work in a mechanically dynamic environment, such as heart. So far, no systematic *in vitro* and *in vivo* studies have been reported, regarding applications of PGS in cardiac tissue engineering.

Another elastomer being investigated for myocardial tissue engineering is a multiblock poly(aliphatic/aromatic-ester) containing phthalic acid sequences such as poly(ethylene terephthalate) (PET) and a dimmer fatty acid (DFA) (El Fray and Boccaccini, 2005). An initial *in vitro* assessment by the present authors using mouse ES cell-derived cardiomyocytes has shown that this material can support healthy, functional heart muscle cells (Fig. 16.4).

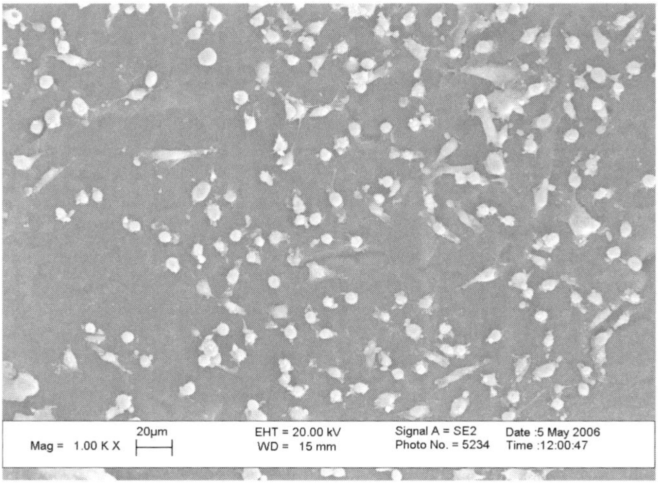

16.4 SEM image showing differentiating mouse ES cells seeded on PET-DLA substrate (authors' own results, PET-DLA material supplied by Prof. M. El Fray, Szczecin University of Technology, Poland).

16.4 Conclusions and future trends

Heart muscle engineering aims to regenerate functional myocardium to repair diseased and injured heart. Huge efforts have been invested in the development of cell sources for myocardial regeneration, including fetal cardiomyocytes, skeletal myoblasts, bone marrow stem cells, endothelial progenitors, native cardiac progenitor cells and embryonic stem cells. A number of biocompatible polymeric materials have also been investigated for cardiac regeneration strategies involving artificial constructs (patches). No matter what types of cells and biomaterials they are built from, the utilisation of artificial matrices to engineer myocardium has not been fully successful yet, with no examples of human application. The limitations are not only set by cell-related issues (such as scale-up in a rather short period, efficiency of cell seeding or cell survival rate and immune rejection), but also caused by the properties of the engineered tissue construct.

First of all, engineered heart muscle must develop systolic (contractive) force with appropriate compliance; at the same time it must withstand diastolic (expansive) load. The material used to build the construct has no ability to beat without cells. The contractile movement of the engineered construct is completely driven by the seeded myocardial cells that inherently have a beating ability. One can envisage that the transfer of mechanical signals from cells to the scaffold would be jeopardised if the scaffold material is too stiff. Most of the above reviewed biomaterials, including collagen fibres, are much stronger than myocardium (Table 16.3 and Fig. 16.5). This explains why solid engineered constructs lack a contractile function. On the other hand, collagen gels are too weak to sustain the required mechanical loads.

Another equally critical issue is size limitation, which is still a barrier in the field. Perfusion *in vitro* may improve the nutrient supply to augment the size of

Table 16.3 Properties of potential biomaterials for cardiac muscle engineering

Polymer	Elastomer (E) or thermoplastic (T)	Young's modulus (or stiffness)	Tensile strength	Degradation (month)
PGA	T	7–10 GPa	70 MPa	2–12
PLLA or PDLLA	T	1–4 GPa	30–80 MPa	2–12
PGS	E	0.282 MPa	0.5 MPa	Degradable
Collagen fibre (tendon/cartilage/ligament/bone)	E	2–46 MPa	1–7 MPa	Degradable
Collagen gel (calf skin)	E	0.002–0.022 MPa	1–9 kPa	Degradable
		Maximum stiffness		
Myocardium of rat	E	0.14 MPa	30–70 kPa	NA
Myocardium of human	E	0.2–0.5 MPa	3–15 kPa	NA

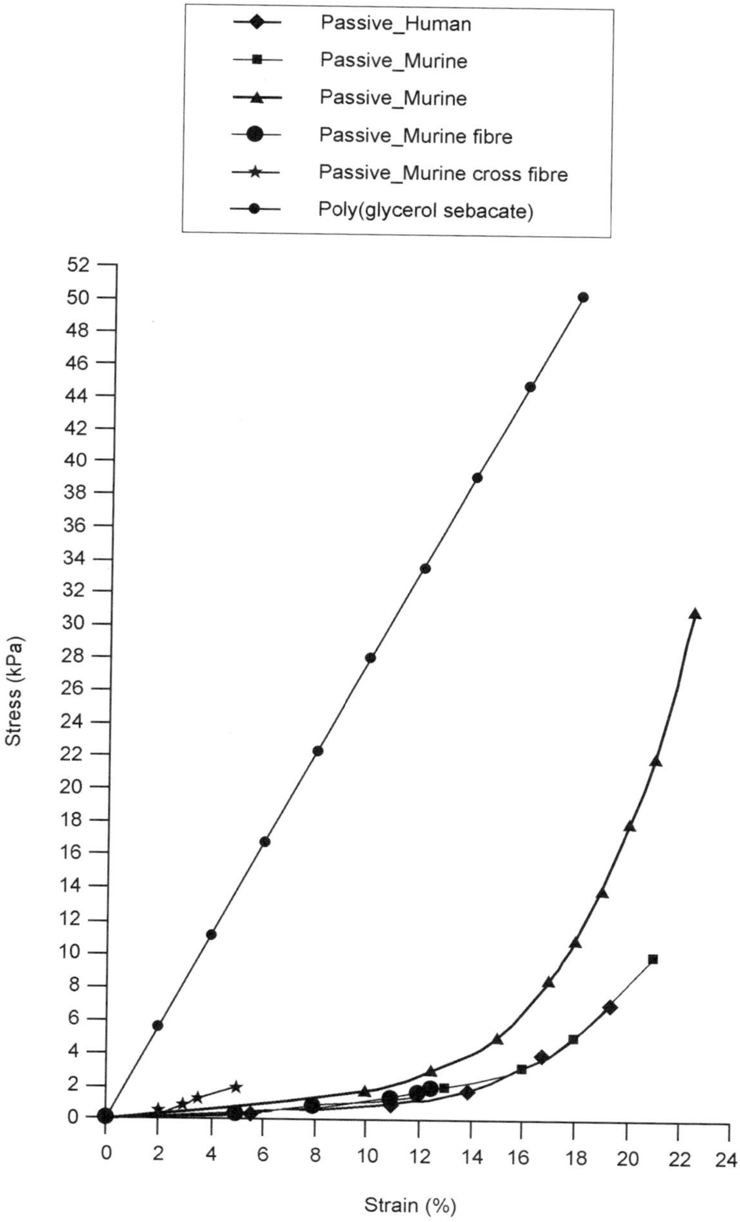

16.5 Typical stress–strain curves of synthetic polymer (PGS) and heart muscles from different species.

the engineered constructs. However, once they are implanted *in vivo*, the portion of thick tissue will need to be vascularised so that the cells within it will receive the necessary nutrients and waste products will be removed. Hence, pre-vascularisation is highlighted as a crucial step in the development of an engineered tissue construct. The use of novel materials to enhance cell growth and properties, to encourage vascularisation and to support the damaged heart, will be needed to bring this new technology into clinical application.

16.5 Acknowledgement

We acknowledge financial support from BBSRC/EPSRC (Grant number BB/D011027/1).

16.6 References and further reading

Amado L C, Saliaris A P, Schuleri K H, St John M, Xie J S, Cattaneo S, Durand D J, Fitton T, Kuang J Q, Stewart G, Lehrke S, Baumgartner W W, Martin B J, Heldman A W and Hare J M (2005), 'Cardiac repair with intramyocardial injection of allogeneic mesenchymal stem cells after myocardial infarction', *Proc Natl Acad Sci USA*, **102** (32), 11474–11479.

Assmus B, Honold J, Schachinger V, Britten M B, Fischer-Rasokat U, Lehmann R, Teupe C, Pistorius K, Martin H, Abolmaali N D, Tonn T, Dimmeler S and Zeiher A M (2006), 'Transcoronary transplantation of progenitor cells after myocardial infarction', *N Engl J Med*, **355** (12), 1222–1232.

Balsam L B, Wagers A J, Christensen J L, Kofidis T, Weissman I L and Robbins R C (2004), 'Haematopoietic stem cells adopt mature haematopoietic fates in ischaemic myocardium', *Nature*, **428** (6983), 668–673.

Beltrami A P, Barlucchi L, Torella D, Baker M, Limana F, Chimenti S, Kasahara H, Rota M, Musso E, Urbanek K, Leri A, Kajstura J, Nadal-Ginard B and Anversa P (2003), 'Adult cardiac stem cells are multipotent and support myocardial regeneration', *Cell*, **114** (6), 763–776.

Carrier R L, Rupnick M, Langer R, Schoen F J, Freed L E and Vunjak-Novakovic G (2002), 'Perfusion improves tissue architecture of engineered cardiac muscle', *Tiss Eng*, **8** (2), 175–188.

Caspi O and Gepstein L (2006), 'Stem cells for myocardial repair', *Eur Heart J Suppl*, **8** (Supplement E), E43–E54.

Chen M K and Beierle E A (2004), 'Animal models for intestinal tissue engineering', *Biomaterials*, **25** (9), 1675–1681.

Cima L G, Vacanti J P, Vacanti C, Ingber D, Mooney D and Langer R (1991), 'Tissue engineering by cell transplantation using degradable polymer substrates', *J Biomech Eng – Trans ASME*, **113** (2), 143–151.

El Fray M and Boccaccini A R (2005), 'Novel hybrid PET/DFA-TiO2 nanocomposites by in situ polycondensation', *Mater Lett*, **59** (18), 2300–2304.

Eschenhagen T (2005), 'Engineering myocardial tissue', *Circ Res*, **97** (12), 1220–1231.

Eschenhagen T, Fink C, Remmers U, Scholz H, Wattchow J, Weil J, Zimmerman W, Dohmen H H, Schafer H, Bishopric N, Wakatsuki T and Elson E L (1997), 'Three-dimensional reconstitution of embryonic cardiomyocytes in a collagen matrix: a new heart muscle model system', *Faseb J*, **11** (8), 683–694.

Etzion S, Battler A, Barbash I M, Cagnano E, Zarin P, Granot Y, Kedes L H, Kloner R A and Leor J (2001a), 'Influence of embryonic cardiomyocyte transplantation on the progression of heart failure in a rat model of extensive myocardial infarction', *J Mol Cellul Cardio*, **33** (7), 1321–1330.

Etzion S, Kedes L H, Kloner R A and Leor J (2001b), 'Myocardial regeneration: present and future trends', *Am J Cardiovasc Drugs*, **1** (4), 233–244.

Fidkowski C, Kaazempur-Mofrad M R, Borenstein J, Vacanti J P, Langer R and Wang Y D (2005), 'Endothelialized microvasculature based on a biodegradable elastomer', *Tiss Eng*, **11** (1–2), 302–309.

Freed L E and VunjakNovakovic G (1997), 'Microgravity tissue engineering', *In Vitro Cell Develop Biol – Animal*, **33** (5), 381–385.

Gerecht-Nir S, Radisic M, Park H, Cannizzaro C, Boublik J, Langer R and Vunjak-Novakovic G (2006), 'Biophysical regulation during cardiac development and application to tissue engineering', *Int J Develop Biol*, **50** (2–3), 233–243.

Harding S E, Gorelik J, Ali N N and Brito-Martins M (2007), 'The human embryonic stem cell-derived cardiomyocyte as a pharmacological model', *Pharmacol Therapeut*, **113**, 341–353.

Kehat I, Kenyagin-Karsenti D, Snir M, Segev H, Amit M, Gepstein A, Livne E, Binah O, Itskovitz-Eldor J and Gepstein L (2001), 'Human embryonic stem cells can differentiate into myocytes with structural and functional properties of cardiomyocytes', *J Clin Invest*, **108** (3), 407–414.

Kehat I, Khimovich L, Caspi O, Gepstein A, Shofti R, Arbel G, Huber I, Satin J, Itskovitz-Eldor J and Gepstein L (2004), 'Electromechanical integration of cardiomyocytes derived from human embryonic stem cells', *Nat Biotechnol*, **22** (10), 1282–1289.

Kessler P D and Byrne B J (1999), 'Myoblast cell grafting into heart muscle: cellular biology and potential applications', *Annu Rev Physiol*, **61**, 219–242.

Klug M G, Soonpaa M H, Koh G Y and Field L J (1996), 'Genetically selected cardiomyocytes from differentiating embryonic stem cells form stable intracardiac grafts', *J Clin Invest*, **98** (1), 216–224.

Kofidis T, Lenz A, Boublik J, Akhyari P, Wachsmann B, Stahl K M, Haverich A and Leyh R G (2003), 'Bioartificial grafts for transmural myocardial restoration: a new cardiovascular tissue culture concept', *Eur J Cardio-Thorac Surg*, **24** (6), 906–911.

Kofidis T, de Bruin J L, Hoyt G, Ho Y, Tanaka M, Yamane T, Lebl D R, Swijnenburg R J, Chang C P, Quertermous T and Robbins R C (2005), 'Myocardial restoration with embryonic stem cell bioartificial tissue transplantation', *J Heart Lung Transplant*, **24** (6), 737–744.

Koh G Y, Klug M G, Soonpaa M H and Field L J (1993), 'Differentiation and long-term survival of C2c12 myoblast grafts in heart', *J Clin Invest*, **92** (3), 1548–1554.

Laflamme M A, Gold J, Xu C H, Hassanipour M, Rosler E, Police S, Muskheli V and Murry C E (2005), 'Formation of human myocardium in the rat heart from human embryonic stem cells', *Am J Pathol*, **167** (3), 663–671.

Laugwitz K L, Moretti A, Lam J, Gruber P, Chen Y H, Woodard S, Lin L Z, Cai C L, Lu M M, Reth M, Platoshyn O, Yuan J X J, Evans S and Chien K R (2005), 'Postnatal isl1+cardioblasts enter fully differentiated cardiomyocyte lineages', *Nature*, **433** (7026), 647–653.

Le Blanc K (2003), 'Immunomodulatory effects of fetal and adult mesenchymal stem cells', *Cytotherapy*, **5** (6), 485–489.

Leor J, Aboulafia-Etzion S, Dar A, Shapiro L, Barbash I M, Battler A, Granot Y and

Cohen S (2000), 'Bioengineered cardiac grafts – a new approach to repair the infarcted myocardium?', *Circulation*, **102** (19), 56–61.

Leor J, Amsalem Y and Cohen S (2005), 'Cells, scaffolds, and molecules for myocardial tissue engineering', *Pharmacol Therapeut*, **105** (2), 151–163.

Li R K, Mickle D A G, Weisel R D, Zhang J and Mohabeer M K (1996), '*In vivo* survival and function of transplanted rat cardiomyocytes', *Circ Res*, **78** (2), 283–288.

Li R K, Jia Z Q, Weisel R D, Mickle D A G, Choi A and Yau T M (1999), 'Survival and function of bioengineered cardiac grafts', *Circulation*, **100** (19), 63–69.

Lunde K, Solheim S, Aakhus S, Arnesen H, Abdelnoor M, Egeland T, Endresen K, Ilebekk A, Mangschau A, Fjeld J G, Smith H J, Taraldsrud E, Grogaard H K, Bjornerheim R, Brekke M, Muller C, Hopp E, Ragnarsson A, Brinchmann J E and Forfang K (2006), 'Intracoronary injection of mononuclear bone marrow cells in acute myocardial infarction', *N Engl J Med*, **355** (12), 1199–1209.

Menard C, Hagege A A, Agbulut O, Barro M, Morichetti M C, Brasselet C, Bel A, Messas E, Bissery A, Bruneval P, Desnos M, Puceat M and Menasche P (2005), 'Transplantation of cardiac-committed mouse embryonic stem cells to infarcted sheep myocardium: a preclinical study', *Lancet*, **366** (9490), 1005–1012.

Messina E, De Angelis L, Frati G, Morrone S, Chimenti S, Fiordaliso F, Salio M, Battaglia M, Latronico M V G, Coletta M, Vivarelli E, Frati L, Cossu G and Giacomello A (2004), 'Isolation and expansion of adult cardiac stem cells from human and murine heart', *Circ Res*, **95** (9), 911–921.

Mummery C, Ward-van Oostwaard D, Doevendans P, Spijker R, van den Brink S, Hassink R, van der Heyden M, Opthof T, Pera M, de la Riviere A B, Passier R and Tertoolen L (2003), 'Differentiation of human embryonic stem cells to cardiomyocytes – role of coculture with visceral endoderm-like cells', *Circulation*, **107** (21), 2733–2740.

Murry C E, Whitney M L, Laflamme M A, Reinecke H and Field L J (2002), 'Cellular therapies for myocardial infarct repair', *Cold Spring Harb Symp Quant Biol*, **67**, 519–526.

Murry C E, Soonpaa M H, Reinecke H, Nakajima H, Nakajima H O, Rubart M, Pasumarthi K B S, Virag J I, Bartelmez S H, Poppa V, Bradford G, Dowell J D, Williams D A and Field L J (2004), 'Haematopoietic stem cells do not transdifferentiate into cardiac myocytes in myocardial infarcts', *Nature*, **428** (6983), 664–668.

Oh H, Bradfute S B, Gallardo T D, Nakamura T, Gaussin V, Mishina Y, Pocius J, Michael L H, Behringer R R, Garry D J, Entman M L and Schneider M D (2003), 'Cardiac progenitor cells from adult myocardium: homing, differentiation, and fusion after infarction', *Proc Natl Acad Sci USA*, **100** (21), 12313–12318.

Orlic D, Kajstura J, Chimenti S, Bodine D M, Leri A and Anversa P (2001), 'Transplanted adult bone marrow cells repair myocardial infarcts in mice', in *Hematopoietic Stem Cells 2000 Basic and Clinical Sciences*, Vol. 938, New York Acad. Sci., New York, pp. 221–230.

Pauschinger M, Knopf D, Petschauer S, Doerner A, Poller W, Schwimmbeck P L, Kuhl U and Schultheiss H P (1999), 'Dilated cardiomyopathy is associated with significant changes in collagen type I/III ratio', *Circulation*, **99** (21), 2750–2756.

Pfister O, Mouquet F, Jain M, Summer R, Helmes M, Fine A, Colucci W S and Liao R (2005), 'CD31(−) but not CD31(+) cardiac side population cells exhibit functional cardiomyogenic differentiation', *Circ Res*, **97** (1), 52–61.

Pittenger M F and Martin B J (2004), 'Mesenchymal stem cells and their potential as cardiac therapeutics', *Circ Res*, **95** (1), 9–20.

Radisic M and Vunjak-Novakovic G (2005), 'Cardiac tissue engineering', *J Serbian Chem Soc*, **70** (3), 541–556.

Reinecke H, Zhang M, Bartosek T and Murry C E (1999), 'Survival, integration, and differentiation of cardiomyocyte grafts – a study in normal and injured rat hearts', *Circulation*, **100** (2), 193–202.

Satin J, Kehat L, Caspi O, Huber I, Arbel G, Itzhaki I, Magyar J, Schroder E A, Perlman I and Gepstein L (2004), 'Mechanism of spontaneous excitability in human embryonic stem cell derived cardiomyocytes', *J Physiol – London*, **559** (2), 479–496.

Schachinger V, Erbs S, Elsasser A, Haberbosch W, Hambrecht R, Holschermann H, Yu J T, Corti R, Mathey D G, Hamm C W, Suselbeck T, Assmus B, Tonn T, Dimmeler S and Zeiher A M (2006), 'Intracoronary bone marrow-derived progenitor cells in acute myocardial infarction', *N Engl J Med*, **355** (12), 1210–1221.

Scorsin M, Hagege A A, Marotte F, Mirochnik N, Copin H, Barnoux M, Sabri A, Samuel J L, Rappaport L and Menasche P (1997), 'Does transplantation of cardiomyocytes improve function of infarcted myocardium?', *Circulation*, **96** (9), 188–193.

Scorsin M, Hagege A, Vilquin J T, Fiszman M, Marotte F, Samuel J L, Rappaport L, Schwartz K and Menasche P (2000), 'Comparison of the effects of fetal cardiomyocyte and skeletal myoblast transplantation on postinfarction left ventricular function', *J Thorac Cardiovasc Surg*, **119** (6), 1169–1175.

Shachar M and Cohen S (2003), 'Cardiac tissue engineering, *ex-vivo*: design principles in biomaterials and bioreactors', *Heart Failure Rev*, **8** (3), 271-276.

Shin M, Ishii O, Sueda T and Vacanti J P (2004), 'Contractile cardiac grafts using a novel nanofibrous mesh', *Biomaterials*, **25** (17), 3717–3723.

Snir M, Kehat I, Gepstein A, Coleman R, Itskovitz-Eldor J, Livne E and Gepstein L (2003), 'Assessment of the ultrastructural and proliferative properties of human embryonic stem cell-derived cardiomyocytes', *Am J Physiol – Heart Circul Physiol*, **285** (6), H2355–H2363.

Soonpaa M H, Koh G Y, Klug M G and Field L J (1994), 'Formation of nascent intercalated disks between grafted fetal cardiomyocytes and host myocardium', *Science*, **264** (5155), 98–101.

Sundback C A, Shyu J Y, Wang Y D, Faquin W C, Langer R S, Vacanti J P and Hadlock T A (2005), 'Biocompatibility analysis of poly(glycerol sebacate) as a nerve guide material', *Biomaterials*, **26** (27), 5454–5464.

Taylor D A, Atkins B Z, Hungspreugs P, Jones T R, Reedy M C, Hutchinson K A, Glower D D and Kraus W E (1998), 'Regenerating functional myocardium: Improved performance after skeletal myoblast transplantation', *Nat Med*, **4** (8), 929–33.

Temenoff J S, Lu L and Mikos A G (2000), 'Bone tissue engineering using synthetic biodegradable polymer scaffolds', In *Bone Tissue Engineering* (Ed. Davies, J. E.) EM Squared, Toronto, pp. 455–462.

Thomson J A, Itskovitz-Eldor J, Shapiro S S, Waknitz M A, Swiergiel J J, Marshall V S and Jones J M (1998), 'Embryonic stem cell lines derived from human blastocysts', *Science*, **282** (5391), 1145–1147.

Vunjak-Novakovic G, Radisic M and Obradovic B (2006), 'Cardiac tissue engineering: effects of bioreactor flow environment on tissue constructs', *J Chem Tech Biotech*, **81** (4), 485–490.

Wang Y D, Ameer G A, Sheppard B J and Langer R (2002), 'A tough biodegradable elastomer', *Nat Biotechnol*, **20** (6), 602–606.

Xu C H, Police S, Rao N and Carpenter M K (2002), 'Characterization and enrichment of cardiomyocytes derived from human embryonic stem cells', *Circ Res*, **91** (6), 501–508.

Xue T, Cho H C, Akar F G, Tsang S Y, Jones S P, Marban E, Tomaselli G F and Li R A (2005), 'Functional integration of electrically active cardiac derivatives from genetically engineered human embryonic stem cells with quiescent recipient ventricular cardiomyocytes – insights into the development of cell-based pacemakers', *Circulation*, **111** (1), 11–20.

Zammaretti P and Jaconi M (2004), 'Cardiac tissue engineering: regeneration of the wounded heart', *Curr Opin Biotechnol*, **15** (5), 430–434.

Zimmermann W H, Fink C, Kralisch D, Remmers U, Weil J and Eschenhagen T (2000), 'Three-dimensional engineered heart tissue from neonatal rat cardiac myocytes', *Biotechnol Bioeng*, **68** (1), 106–114.

Zimmermann W H, Didie M, Doker S, Melnychenko I, Naito H, Rogge C, Tiburcy M and Eschenhagen T (2006), 'Heart muscle engineering: an update on cardiac muscle replacement therapy', *Cardiovasc Res*, **71** (3), 419–429.

Zmora S, Glicklis R and Cohen S (2002), 'Tailoring the pore architecture in 3-D alginate scaffolds by controlling the freezing regime during fabrication', *Biomaterials*, **23** (20), 4087–4094.

17
Intervertebral disc tissue engineering

J HOYLAND and T FREEMONT, University of Manchester, UK

17.1 Introduction

This chapter will discuss the current and future uses of biomaterials in the management of the common disorder, degeneration of the intervertebral disc (IVD). This is clinically important as it is responsible for much of chronic low back pain. The chapter will describe:

- the impact of degeneration of the IVD on modern society;
- the normal anatomy, function and cell biology of the intervertebral disc;
- the pathobiology of IVD degeneration;
- treatment of IVD degeneration;
- use of biomaterials to reconstruct the various components of the degenerate intervertebral disc;
- tissue regeneration and the IVD.

17.2 The impact of disorders of the intervertebral disc (IVD) on modern society

Approximately 70% of the population will experience low back pain (LBP) during their lives (MacFarlane *et al.*, 1999). Back pain is a cause of considerable morbidity, accounting for about 15% of all sickness leave in the UK (Jayson, 1987) and costs the UK NHS and social services about £11 billion pounds a year. Despite being such an important public health issue, little is known of its pathogenesis. The consensus view is that much chronic LBP is related to disorders of the IVD (discogenic back pain). Furthermore, clinical, interventional and imaging studies on volunteers and patients (Kelgren, 1977; Luoma *et al.*, 2000) have produced evidence directly implicating mechanical or traumatic disorders of the IVD in up to 40% of chronic LBP patients. In almost every case when IVD tissue has been examined from patients with LBP it has shown the features typical of a disorder known as 'degeneration'. It is implicit that treatment strategies for back pain should involve reversing the pathology of degeneration.

17.3 The normal anatomy, function and cell biology of the IVD

Intervertebral discs are joints between the major components of the bones of the spine, the vertebral bodies. Each consists of three major components (Fig. 17.1):

- Nucleus pulposus (NP). The space between the end plates of adjacent vertebrae is filled by the nucleus pulposus, consisting of chondrocytes within a matrix of type II collagen and proteoglycan, mainly aggrecan (Feng *et al.*, 2006). Surprisingly little is known of the structure of the NP or the biology of its cells. The type II collagen fibres are not believed to give the same level of order to the structure or the same degree of mechanical stability to the matrix as in articular cartilage. The proteoglycans are hydrophilic, causing the NP to swell. The swelling pressure of the NP proteoglycans is constrained by the end plates above and below and the annulus fibrosus (see below) around the periphery. On H&E sections the NP appears homogeneous and pale lilac-blue, consistent with its complement of proteoglycans. In polarised light it exhibits little birefringence.
- Annulus fibrosus (AF). This comprises dense sheets of highly orientated collagen fibres (mainly type I but also types II and III) in which are cells with the morphology and phenotype of fibroblasts. Our experience is that when cultured in alginate they can be transformed into chondrocytes. Functionally the annulus fibrosus is a very strong ligament binding together the outer rims of adjacent vertebrae. On H&E sections it exhibits fairly uniform eosinophilia and in polarised light its constituent alternating bands of highly orientated collagen fibres are clearly seen.

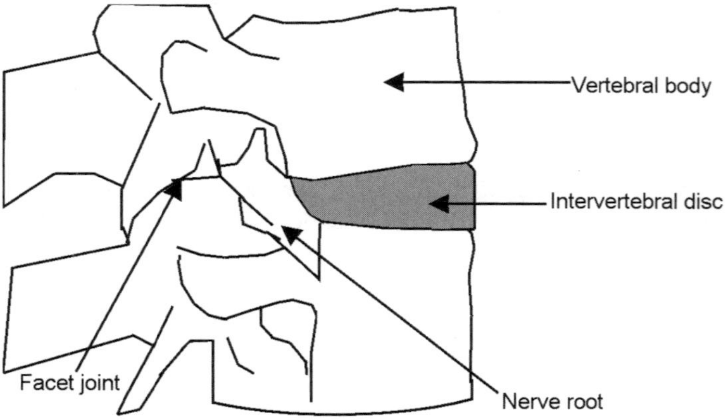

17.1 Drawing of two adjacent vertebrae and the intervening IVD. The diagram gives an idea of the complexity of the structures around the IVD and shows how the correct height of the IVD acts as a spacer giving a precise separation to the IVD.

- End plates (EP). These are plates of bone and cartilage that mark the interface between the vertebral bodies of the spine and the IVD. They are important as they interface with the NP (and therefore NP replacements and any mismatch between the modulus of the two structures will lead to implant, bone or disc failure).

With the exception of the outer third of the annulus fibrosus, the normal adult IVD is avascular and aneural, there is a relatively clear demarcation between the AF and the NP, and the end plate forms an intact layer of cartilage overlying cortical bone.

The IVD allows movement between bones, in this case adjacent vertebral bodies. In its normal fully hydrated state each IVD permits small degrees of flexion, extension, and lateral bending and twisting, but is resistant to compressive loads. When summated, the small movements at each IVD permit great mobility of the entire spine.

17.4 The pathobiology of IVD degeneration

It is believed that many of the disorders of the IVD are mechanical in origin. Humans are the only vertebrates to constitutively stand erect on their back legs. As a consequence the spine is subjected to a range, rate and degree of loading that is not experienced by other vertebrates and that, arguably, it is poorly designed to sustain. Whether or not this is the cause, in a high proportion of individuals, there is a pattern of morphological and histological changes, which increases with age, that together constitute discal degeneration.

In degeneration, the normal structure of the NP is disturbed, with changes in the proportion and types of proteoglycans and collagens and reduction in the total number of viable cells. In addition there is breakdown of the matrix with formation of permeative 'slit-like' spaces (Figs 17.2 and 17.3). There is often also disruption of the collagen fibre arrays in the annulus fibrosus, bone sclerosis and traumatic damage to the end plate, and vessel and nerve ingrowth into the inner AF and NP. Current understanding of the biology of connective tissues would causally implicate alterations in the function of local cells in these events.

From the perspective of relating the pathobiology of back pain to key events in degeneration, the loss of hydrophilic proteoglycans from the nucleus pulposus and the ingrowth of nerves and vessels into the deeper tissues seem critical (Freemont *et al.*, 2001). The former causes a decrease in the swelling pressure of the NP resulting in a reduction in the distance between adjacent vertebral bodies ('loss of disc height'). Loss of disc height causes widespread changes in the structural relationships between bones, joints and soft tissues such as nerves, in the vicinity of the degenerate IVD, leading to impingement (Fig. 17.4), abnormal motion and altered loading, all of which can be painful. In addition

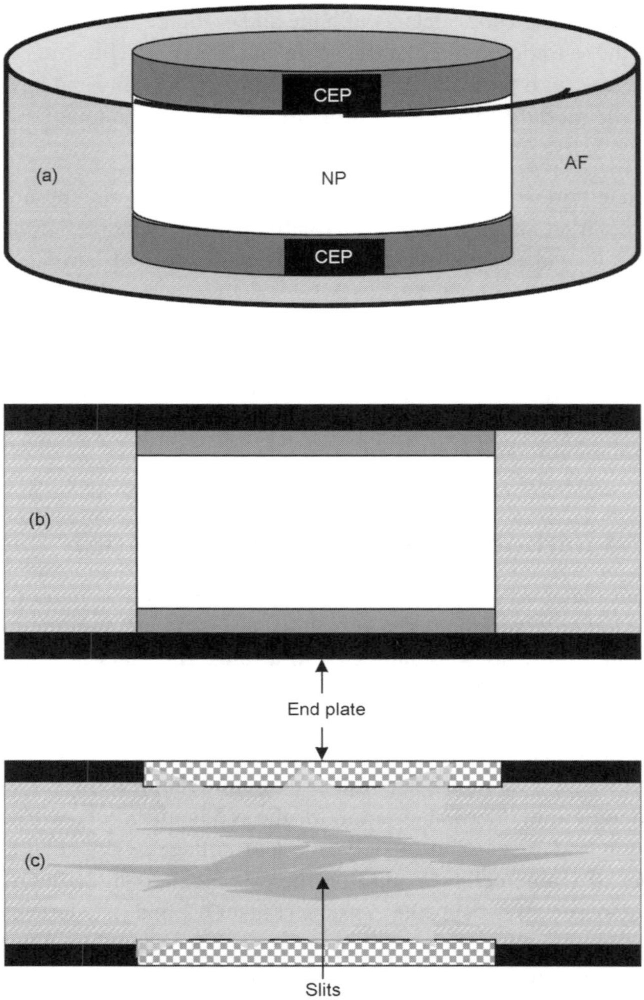

17.2 Diagrams showing the differences between the normal IVD (a and b) and the degenerate IVD (c). (a) 3D representation of an IVD without the bony end plate. (b) and (c) 2D representation of an IVD with the bony end plate (black), shading otherwise as in 3D image. (b) A normal IVD. (c) A degenerate IVD showing decreased disc height, damage to the cartilageous and bony parts of the EP. The NP comes to resemble the AF because it has lost proteoglycans and there are areas of tissue degradation (slits) in the NP and extending into the AF.

there is increased instability in disc structures, which in turn increases the risk of microtrauma at mechanically vulnerable areas of the IVD such as the EP and insertions of the AF into bone.

These observations have stimulated considerable interest in the processes leading to loss of disc height and in particular the molecular pathology of

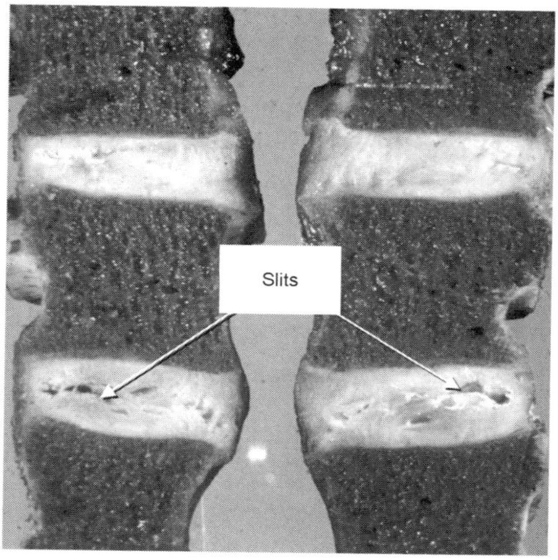

17.3 Transected human intervertebral discs. The upper image shows a normal, healthy disc and the lower one with the changes of early degeneration. The slits in the nucleus pulposus to potentially hold a bioengineered construct are arrowed.

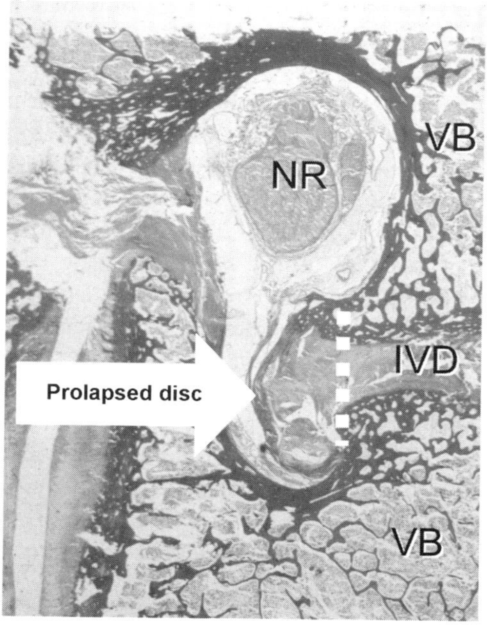

17.4 Histological section of two vertebral bodies (VB) and the back of an intervertebral disc (IVD). The AF of the degenerate IVD has prolapsed backwards towards the nerve root (NR). The original edge of the IVD is marked by a dotted line IVD prolapse causes pain by direct impingement on the nerve root and by the prolapsed disc producing pain stimulating molecules.

degeneration. This in turn has stimulated a re-evaluation of the management of IVD disease and low back pain.

Such is the nature of connective tissues that they undergo changes in structure in response to ageing and load (Handa *et al.*, 1997; Fujisawa *et al.*, 1999) In particular collagens and proteoglycans change with age in concert with alterations of the nutrition pathways to the tissue (Scott *et al.*, 1994). There is a real need to distinguish these changes from degeneration and, in terms of regeneration, to recognise what the end point of the regeneration process should be, as it may be different in patients of different ages or with different loading histories (e.g. differences in lifestyle or weight).

17.4.1 Altered matrix composition and integrity in degeneration

When compared with age- and sex-matched controls there are profound alterations in the cell biology of chondrocytes in degenerate IVD. Normal discal chondrocytes are characterised by expression of type II collagen and proteoglycans (LeMaitre *et al.*, 2004b) and the 'master chondro-regulatory gene' *SOX-9*. In discal degeneration chondrocyte synthesis of matrix molecules changes differentially with the degree of degeneration (Antoniou *et al.*, 1996), leading to an increase in synthesis of collagens I and III and decreased production of aggrecan. Exactly what effect these changes have on IVD function is unknown.

Furthermore the regulation of matrix turnover is deranged, affecting both synthesis and degradation (Freemont *et al.*, 2002a). In degeneration, there is a net increase in matrix-degrading enzyme activity over their natural inhibitors (TIMPs – tissue inhibitors of metalloproteinases), which leads to loss of discal matrix. Particular attention has been paid to the role of matrix metalloproteinases (MMPs) in these processes (Liu *et al.*, 1991; Kang *et al.*, 1996; Goupille *et al.*, 1998a; Le Maitre *et al.*, 2004a), making them a potential target for therapy designed to inhibit discal degeneration. More recently, following work on the mechanism of aggrecan degradation in articular cartilage, interest has grown in the possible role of aggrecanases in IVD degeneration (Roberts *et al.*, 2000; Le Maitre *et al.*, 2004a). Aggrecan has two cleavage sites, one acted upon by MMPs and the other by members of a group of enzymes called the ADAMTS family, after their hybrid function 'A Disintegrin And Metalloproteinase with ThromboSpondin motifs'. Current knowledge indicates that two ADAMTS, ADAMTS 4 and 5, are key aggrecanases and these too have become therapeutic targets (Malemud, 2006).

17.4.2 Cytokines and degeneration

The causes of these matrix changes are only just being uncovered. There is accumulating evidence that cytokines are critical regulators of matrix turnover.

Two cytokines, interleukin-1 (IL-1) and tumour necrosis factor alpha (TNFα) in particular have been the subject of some study. There is early evidence that the cytokine IL-1 is homeostatic, playing a key role in the normal turnover of matrix. In degeneration it increases in amount while there are no similar changes in its natural inhibitor (interleukin-1 receptor antagonist [IL-1Ra]). Unopposed IL-1 will induce increased production of matrix degrading enzymes and inhibit synthesis of matrix molecules (Goupille *et al.*, 1998b; Freemont *et al.*, 2002a; Le Maitre *et al.*, 2005) and adding IL-1Ra to degenerate tissue in culture will inhibit matrix degrading activity (Le Maitre *et al.*, 2006). TNFα has been implicated in disc herniation and sciatic pain (Olmarker and Rydevik, 2001; Cooper and Freemont, 2004), but its role in disc degeneration has still to be clarified.

The significance of these molecules is twofold. First, they are obvious targets for inhibiting the processes of disc degeneration. Second, and more important from the perspective of the use of materials in management of degeneration, these molecules will be present in the diseased tissues into which cell-seeded biomaterial regenerates may one day be placed for treating degeneration (this residual damaged tissue that surrounds the implanted regenerate is described as its 'niche'). They could adversely affect the function of the regenerate. Their inhibition might therefore be a key factor in ensuring that the regenerate 'takes' and new tissue is regenerated.

17.4.3 Reduced cell number

The reduction in chondrocytes that typifies IVD degeneration, has been ascribed to apoptosis (Gruber and Hanley, 1998; Anderson and Tannoury, 2005) a form of cell death in which there is a programmed sequence of events that requires energy and leads to elimination of cells from the pool within the tissue. Apoptosis is often induced by outside stimuli such as cytokines or load. Indeed there is some evidence of a 'dose-dependent' relationship between apoptosis and excessive load (lifestyle, body weight, age) (Lotz *et al.*, 1998).

17.4.4 Cellular senescence

A characteristic of connective tissues in which cell turnover is low is the development of cellular senescence in both mature and stem cell populations. This is an age-related phenomenon but also occurs prematurely in a tissue-specific manner in disorders such as degeneration of the IVD. There are two types of senescence, replicative senescence (RS) and stress-induced premature senescence (SIPS) (Carrington, 2005). The latter seems the more likely cause of senescence in degeneration of the IVD. The significance of senescence is that cells function abnormally and cannot undergo replication *in vivo* or *in vitro*. Thus it is not possible to replace disc cells lost by apoptosis, nor can cells extracted from the degenerate IVD be used to seed biomaterial scaffolds for tissue engineering and regenerative medicine purposes.

17.4.5 Nerve and blood vessel ingrowth

Although the normal adult IVD is avascular and aneural, nerves (Freemont *et al.*, 1997) and blood vessels (Kauppila, 1995) grow into diseased IVD. The nerves are nociceptive (i.e. pain nerves). Understanding how pain nerves grow into the IVD could hold a clue to the origins of back pain and identify targets for therapy. Two possibilities have been raised to explain nerve and blood vessel ingrowth:

- Local production of angiogenic and neurogenic molecules within degenerate IVD. Expression of the potent angiogenic factor, vascular endothelial growth factor (VEGF) (Tolonen *et al.*, 1997) has been demonstrated within the IVD as has the neurogenic cytokine NGF (Freemont *et al.*, 2002b).
- Decrease in a natural inhibitor of vessel or nerve growth. In this context it is interesting to note that normal human discal aggrecan is anti-angio and neurogenic and that the altered aggrecans that occur in degeneration lose some or all of this activity (Johnson *et al.*, 2002, 2005a).

17.4.6 The biology of degeneration and the causes of discogenic back pain

In the current state of our knowledge:

- the major processes of disc degeneration are loss of hydrophilic material from the nucleus pulposus driven by locally produced cytokines and mediated by alterations in cell function (mainly impaired matrix synthesis and breakdown);
- the causes of pain are ingrowth of nociceptive nerves into the usually aneural structures of the IVD, local production of TNFα inducing nerve root pain, and loss of disc height and instability leading to local traumatic damage to the disc.

This is relatively new knowledge and it is leading the way in which modern medicine regards the management of IVD degeneration and discogenic back pain.

17.5 Treatment of degeneration of the IVD

17.5.1 Current treatment

The degenerate IVD is a major therapeutic target in the management of chronic LBP (Singh *et al.*, 2005; Andersson *et al.*, 2006; Zhou and Abdi, 2006). But defining therapeutic end points in this clinical setting is a problem because the symptom complex of back pain with or without sciatica runs a fluctuating course. In addition treatments tend to be applied to large groups of patients with different symptom complexes. It is not known whether these patients have uniform pathologies, making analysis of the results of treatment open to criticism. Success is therefore expressed in terms of outcomes from cohort (large group) studies. In general terms the most successful treatments are either:

- symptomatic, such as pain management or muscle strengthening to obtain better spinal stability;
- minimally invasive procedures (i.e. ones that require the minimum necessary surgery) designed to remove small amounts of protruding disc tissue or bone, release local adhesions, or decompress the intervertebral disc;
- major surgery such as spinal fusion which stabilises the spine and increases disc height.

More recently there have been therapies designed to replace the diseased intervertebral disc using prostheses. These have applied a similar strategy to that for replacing joints such as the knee and hip. Currently there are several devices (de Kleuver *et al.*, 2003) in clinical use such as the PRODISC® (Spine Solutions, New York) and CHARITÉ™ Artificial Disc (DePuy Spine, Inc.). These devices are composite in design with a cobalt chromium alloy end plate and an ultra-high molecular weight polyethylene core. Both devices have been used for over 10 years. The published results have generally shown this approach to management to be reasonably successful at treating intractable back pain in cohorts of patients.

17.5.2 Potential treatments based on an understanding of the biology of IVD degeneration

Current therapies are incremental, starting with the least invasive (e.g. analgesia, muscle balancing exercises) and progressing to major surgery. But there is a big ‘gap’ between where the non-invasive technologies leave off and major surgery starts. There has been an upsurge in the use of minimally invasive spinal surgery which fills that gap. Not only has minimally invasive surgery generated new treatments (and a new language!) but it has also led to a remarkable degree of surgical sophistication and skill at micromanipulation of structures in and around the intervertebral disc with minimal trauma to the patient.

With increasing understanding of the biology of the degenerate IVD and the opportunities for using minimally invasive surgery to manipulate tissues inside the spine, there is a growing opinion that there might be better methods of managing disc degeneration that exploit this new knowledge and novel technologies. This view has gained momentum as the potential of regenerative medicine and tissue engineering becomes clearer, generating two exciting developments in therapeutics based on biomaterials. The more immediate is to use materials science to design implants that restore IVD function. The future vision is to develop regenerative medicine strategies effectively to re-grow functional intervertebral disc tissue.

Implant design is an emerging technology, with products already in clinical use and more being developed almost daily. The latter is the subject of major research but has yet to be tested in clinical (or even significant preclinical) applications. In terms of materials, in the former the physical properties of the materials are key. In the latter, however, although the physical properties are critical, they are equalled in importance by the material's ability to alter the biology of the tissue, by supporting and nurturing cells and/or delivering bioactive molecules to the niche or the regenerate.

17.6 The place of biomaterials in proposed strategies for managing IVD degeneration

Table 17.1 summarises the materials used in managing IVD disease.

17.6.1 Improving spinal fusion

An existing successful therapy for the unstable and foreshortened spinal segment is to replace the IVD with bone (spinal fusion). When applied with spinal distraction surgery to partially or completely restore the distance between adjacent vertebrae, this approach gives symptom control and stability. Unfortunately the stability is too great and throws abnormal load on adjacent IVD, leading to degeneration. Nevertheless the immediate symptom relief and the extended time it takes for onset of complications make this an acceptable technique.

17.6.2 Replacing the nucleus pulposus

The earliest uses of polymers for replacement of the NP were reported in the mid-1950s. In both, methylmethacrylate was injected into the degeneration-induced spaces in the NP of the IVD. It seems almost impossible, in this highly regulated environment of modern experimental medicine, that these and other similar treatments were used empirically.

Many initial treatments failed because of a lack of knowledge of biomechanics or the properties of biomaterials. For instance Fernstrom's treatments of humans with implanted steel ball bearings failed because of a modulus mismatch between the material and the bone/cartilage end plate of the vertebral body (Fernstrom, 1966). Only with the introduction of experiments such as those of Nachemson and Morris (1963), whose demonstration in cadaveric spines that silicon rubber implants used in spine surgery as spacers in the IVD could not

Table 17.1 A summary of the materials used in managing intervertebral disc disease

Therapeutic target in disc	Type of therapy	Type of material	Reference	Trade name
Whole disc	Total disc replacement	Cobalt chromium alloy end plate + ultra-high molecular weight polyethylene core	de Kleuver *et al.* (2003)	CHARITÉ™, PRODISC®
Nucleus pulposus	Metal	Steel ball bearings	Fernstrom (1966)	–
	Preformed polymers	Elastic memory coiling spiral made from polycarbonate urethane	Husson *et al.* (2003)	Newcleus
	In vivo curable polymers	Two-part curable polyurethane	http://www.discdyn.com/DASCOR.html	Dascor system
		Bovine albumin and glutaraldehyde	http://www.cryolife.com/pdf/Britspine.Poster.pdf	BioDisc
		Recombinant co-polymer made from amino acids derived from silk and elastin	Boyd and Carter (2006)	NuCore
	Hybrid	PNIPAAm-PEGDM (temperature-sensitive polymer)	Vernengo *et al.* (2006)	–
		Poly(EA/MAA/BDDA) (pH-sensitive microgel)	Saunders *et al.* (2007)	–
	Combination	Polyacrylamide and polyacrylonitrile pellet encased in a polyethylene jacket	Ray (2002)	Prosthetic disc nucleus
		Aquacryl polymer reinforced with a dacron mesh	Bertagnoli *et al.* (2005)	NeuDisc
Annulus fibrosus		PDLLA/45S5 Bioglass composite films	Wilda and Gough (2006)	–

withstand expected physiological loads, did need for a better understanding of the material and the environment start to make an impact on spinal research.

The current direction of therapeutic evolution is towards the use of implants. Currently a number are being evaluated. They are probably best considered as falling into three groups – preformed polymers, *in vivo* curable polymers, or combination implants made of more than one material (Ray, 2002; Thomas *et al.*, 2003).

Certain basic principles have influenced the choice of polymers. They should be:

- biocompatible (i.e. not cytotoxic or carcinogenic);
- resistant to loading, e.g. estimates suggest that a person in a sedentary occupation takes 1.5–2 million steps a year, so a practical implant might need to resist between 75 and 100 million episodes of loading from 300 kPa to 1 MPa, plus any additional loading from activities such as lifting (up to 2.5 MPa) jogging, etc. (Nachemson *et al.*, 1979);
- not undergo breakdown or wear (wear particles stimulate a massive 'foreign body-type' response which accelerates breakdown and can lead to painful inflammation and damage to adjacent tissues, such as nerves);
- relatively well fixed in position to prevent extrusion and enhance stability;
- of a similar modulus to the end plates.

In situ *curable polymers*

Currently, the most common injectable elastomers being used within the intervertebral disc space are silicone and polyurethane. These materials can be implanted through a minimally invasive approach. This had the advantage of causing the least damage to the annulus, which limits extrusion. The major advantage of these polymers is that as monomers they can spread through the irregular spaces left by degeneration or surgery. This maximises the amount of polymer within the disc and because of the complex shape of the spaces filled by polymer, improves stability. For this to be successful, the outer limits of the space (the AF and EP) have to be intact. As there are almost always small defects in the AF in degeneration it is an advantage if the curing process is fairly rapid (minutes rather than hours). Two recently introduced examples are the 'Dascor system' (Disc Dynamics, Inc., Eden Prairie, Minnesota, http://www.discdyn.com/DASCOR.html) consisting of a two-part curable polyurethane, and BioDisc (Cryolife, Kennesaw, GA) which consists of purified bovine albumin and glutaraldehyde which are combined in the disc in the ratio of 4:1 (http://www.cryolife.com/pdf/Britspine_Poster.pdf). The major problems with *in situ* polymerisation include heat generation during polymerisation which can be destructive of adjacent tissue and the presence of residual unpolymerised monomer, which can be cytotoxic.

A recent product currently undergoing clinical trials is the NuCore™ injectable nucleus being developed by Spine Wave (Shelton, CT, USA) avoids some of these problems. It employs a recombinant protein co-polymer consisting of amino acid blocks derived from silk and elastic structural proteins as an injectable material for replacement of the NP. The material forms a viscous gel following the addition of a crosslinker. It cures in 5 min and reaches optimal mechanical strength 30 min after addition of the crosslinker. To date this material appears promising as it: mimics the water content, protein content, pH and complex modulus of the natural NP; withstands loads experienced by the *in vivo* disc; and appears non-toxic (http://www.spinewave.com/products/nucore.html) (Boyd and Carter, 2006).

Preformed polymers

Most preformed NP polymers have the advantage of a predetermined size and shape, which provide consistent and predictable viscoelastic properties. The downsides are the requirement to match the size and shape of the implant to the cavity, and the need for a larger entry portal for the implant into the disc (although some implants are very cleverly fashioned to minimise the portal size). An interesting example of a preformed polymer is the Newcleus spiral implant. It utilises an elongated elastic memory-coiling spiral made of polycarbonate urethane, which can be inserted through a small incision in the annulus and then coils to fill the space left following removal of the NP. It has been tested up to 50 million loading cycles with 1200 N multidirectional loads without demonstrating significant wear or micro-cracks and has performed well in biocompatibility and early clinical trials (Husson *et al.*, 2003).

An interesting and potentially exciting development that combines the advantages of both the *in situ* curable and preformed polymers without their disadvantages, is to use a preformed polymer whose physical properties are different when in the body from those outside. Particularly interesting in this respect are temperature-sensitive polymers such as hydroxybutyl chitosan and PNIPAAm-PEGDM (a branched copolymer of poly(*N*-isopropylacrylamide) and poly(ethylene) glycol dimethacrylate; Vernengo *et al.*, 2006) and the pH-sensitive poly(EA/MAA/BDDA) (ethylacrylate, methacrylic acid and 1,4-butanediol diacrylate) microgel recently described by Saunders *et al.* (2007), which are liquid in the conditions prevailing outside the body, and as such can be easily introduced by injection, and form a hydrogel at body temperature/pH.

Combination implants

Most of these implants are composed of dehydrated hydrogels constrained by some sort of enclosing membrane to prevent excessive swelling. After implantation, water is imbibed by the hydrogel from the hydrated environment of the IVD (note that human tissues have a high proportion of water and although the degenerate IVD is said to be 'dehydrated' this is relative to the hyperhydration of the aggrecans within the normal IVD), the hydrogel swells and in the hydrated state replaces the function of the NP. Unfortunately, the evidence that all hydrogels swell consistently *in vivo* is sparse.

The following are examples of such hydrogels:

- The 'Prosthetic Disc Nucleus' (Raymedica, Inc., Bloomington, MN) which has been implanted in many hundreds of patients. It consists of a hydrogel pellet of polyacrylamide and polyacrylonitrile encased in a polyethylene jacket. It has been tested for biocompatibility, toxicity, carcinogenicity and genotoxicity, evaluated in human cadaveric models and undergone clinical trials with some success (Ray, 2002; Ray *et al.*, 2002).
- The 'NeuDisc' (Replication Medical Inc., New Brunswick, NJ) NP implant uses two grades of a modified poly-acrylonitrile Aquacryl polymer. The Aquacryl polymer is said to closely mimic key hydrogel properties of the NP. The polymer is reinforced with a dacron mesh for support and structure. The implant is inserted in a dehydrated state, which allows the implant to be folded and implanted in a minimally invasive manner. Once the component is inserted in the correct position and orientation, it can absorb 90% of its weight in water in an anisotropic fashion, allowing it to preferentially expand vertically (Bertagnoli *et al.*, 2005).

Repairing the AF

Frequently in late degeneration the effects of long-term microtrauma and local production of matrix degrading enzymes leads to weakness and perforation of the AF. The AF has a structure similar to ligament but with a lamellar arrangement with alternating rafts of fibres running in different directions. This is key to its strength and must guide technologies directed towards its repair. Although having an intact AF is key to the correct functioning of the IVD and the retention of NP implants *in situ*, there are very few strategies for its repair. Most commonly, suturing or diathermy are used to close defects or insertion portals. In has been suggested (Patent application, USPTO no: #20060015182) that a thin layer of biomaterial such as a sheet of highly compressed reconstituted collagen might be employed as a patch.

The real challenge in AF repair is to reproduce a matrix similar in quality to that of the normal AF and particularly one in which there is a fibre orientation similar to that of the natural AF. Although there are no reports of this having been achieved *in situ*, Gough has recently published (Wilda and Gough, 2006) data showing that bovine AF cells grown on PDLLA/45S5 (PDLLA–poly(D,L-lactide)) Bioglass composite films produce an AF-like matrix. Roberts' group from Keele, UK, has shown that producing a micro-grooved substrate topography was effective in aligning IVD cell growth. There is good evidence that this is a prerequisite for the production of an aligned collagen fibre matrix, although in their system the cells did not synthesise a matrix (Johnson *et al.*, 2005b).

17.7 Tissue regeneration and the IVD

The next generation of biomaterials-based IVD repair systems will be cell-seeded biomaterials used to regenerate IVD tissues (regenerates). At their most sophisticated these biomaterials will perform several roles:

- *Delivering biological molecules.* The key to regulating regeneration is to deliver one or more molecules at different times to a changing biological system. These will regulate both the abnormal biology of the niche and the differentiation and function of cells within the regenerate. Bioactive molecules can be linked to smart biomatrices through disease-specific enzyme-sensitive bonds or microspheres (Richardson *et al.*, 2001; Halling *et al.*, 2005). Biomaterials are of limited value in delivering agents into sites of early degeneration that from our knowledge of disease processes would respond to bioactive molecule therapy, because the still intact dense matrix does not lend itself to biomaterial-based delivery systems. However in theory the severe degenerate disc (which is the usual type found when patients come to surgery) would be an ideal environment for the use of these systems.
- *Supporting the differentiation of stem cells, directing differentiation and regulating matrix synthesis.* With current understanding of materials, regenerative medicine and spinal biomechanics, an ideal scenario for replacing the NP would be to introduce a cell-seeded biomaterial that immediately functions biomechanically as the normal human NP, but that supports the cells, allowing them to differentiate/behave as IVD cells and regenerate the NP as the original biomaterial progressively degrades. This 'holy grail' is still a long way off. However, the individual elements are being explored and progress made. The fact that cells within degenerate IVD are senescent prevents their use to seed biomaterials. Much interest has focused on the use of autologous adult mesenchymal stem cells (MSC) instead of differentiated cells to seed the regenerate. These cells need correct signals, usually from their environment, to differentiate into IVD cells. Chemical signals from the tissue can be enhanced by delivery of biomolecules from biomaterials, but the cells also require appropriate signals from their physical environment. Experiments with different biomaterials have shown that gels (e.g. alginate, chitosan) support both NP-derived cells and MSCs (Le Maitre *et al.*, 2004b; Risbud *et al.*, 2004). Indeed encapsulation of either cell type into gels induces a shape change ('rounding up') of the cell, which stimulates differentiation along chondroid lineages. There are still numerous problems to overcome, for instance: many of these gels do not have the load-bearing properties of the NP or the ability to direct collagen synthesis required to regenerate the AF.

Similarly, polymer scaffolds (e.g. PLLA) designed to have the appropriate physical properties for mimicking NP function, cannot, of themselves, sustain MSC survival and differentiation. However, strategies to improve survival have been described such as *SOX-9* transfection to enhance MSC survival on scaffolds (Richardson *et al.*, 2006). Bioabsorbable materials have been developed for use in the spine but most are designed as aids for spinal fusion (Vaccaro *et al.*, 2003). Most attention has focused on the alpha-polyesters or poly(alpha-hydroxy) acids (e.g. polylactides and polyglycolides). These have also been considered for IVD repair (Richardson *et al.*, 2006).

Regenerative medicine has yet to be applied to the IVD in humans. However, early experiments have been carried out in a rabbit model of IVD degeneration. In these experiments bone marrow-derived autologous MSCs were introduced into a nucleotomy (removal of the NP) site embedded in atelocollagen gel (Sakaia *et al.*, 2006). This gel is a liquid when cool and hardens at body temperature. The potential of the gel to restore the structure and function of the IVD was followed using X-rays, magnetic resonance imaging (MRI), histology, immunohistochemistry and matrix-associated gene expression. Data were compared between normal controls without operations, sham operated animals with induced disc degeneration only and the regenerate and MSC-transplanted animals. Animals were followed for a 24-week period.

Results showed that after 24 weeks the regenerate treated animals regained about 91% of their original disc height and MRI signal intensity (a measure of proteoglycan synthesis) of about 81%, compared with 67% and 60% respectively in the sham operated group. Examination of the tissue showed restoration of the NP and a clear demarcation between AF and NP, restoration of the proteoglycan content of the IVD and relatively normal cell function, which again was not seen in the sham operated group. These data indicate that transplantation of a regenerate consisting of MSCs in a suitable biomaterial can restore, at least in part, an effective IVD. These are excellent results, but translating them into the human may not be as easy as it first appears.

The animal model is not directly analogous to the human situation. It does not have such a biologically complex niche and unlike the human, the rabbit disc contains a stem cell population of notachordal cells, some of which will remain after nucleotomy. Nevertheless the data from this study are very encouraging. As newer materials are produced and more is known about specific aspects of the cell biology and pathology of the IVD regenerative medicine strategies will become more sophisticated until the concept becomes a therapeutic reality for use in humans.

17.8 Conclusions

Current treatments for low back pain targeted at the IVD fall into three categories:

- Symptomatic remedies (e.g. analgesics, bed rest, muscle-strengthening exercises).
- Minimally invasive surgery (e.g. microdiscectomy, foraminotomy).
- Major surgery (e.g. spinal fusion, disc replacement).

With the knowledge that many of the symptoms of chronic low back pain are caused by loss of disc height secondary to degeneration of the IVD and skills learned from minimally invasive surgery, new strategies are being developed to restore disc height using polymer implants seated in the centre of the IVD that replace the function of the NP. This technology is still developmental but offers exciting possibilities for the future. The challenges are to develop an implant that:

- can withstand the loads taken by the IVD in normal usage;
- will not deteriorate under the millions of loading cycles experienced over years of use;
- will restore disc height;
- is biocompatible;
- is seated stably;
- can be introduced without so damaging the AF that the risk of extrusion is increased.

Several implants are currently at different stages of preclinical and clinical investigations to determine whether they meet these challenges.

The NP implants differ from a complete IVD replacement in that positioning the NP replacement requires much less major surgery and less tissue damage than disc replacement, the biomechanics are much more similar to those of the native disc, allowing a greater range of movements particularly in the axial plane, and should the implant fail to relieve symptoms revision (primarily to intervertebral fusion) is much easier.

The future for these implants will be in refining the nature of the polymers, improving the surgery to allow accurate placement of the implant and repair of any damage to the AF, and new techniques (probably through improved imaging) to follow the performance and integrity of the implant.

17.9 Future trends

The next generation of implants will be different still. The target is to meet all the challenges listed in the summary by engineering the local production of new IVD tissue using the techniques of regenerative medicine. This will require new generations of polymers seeded with cells (probably stem cells). This is an ideal and still a long way off. However, there is no doubt that:

- within 5 years, cell-seeded biomaterial constructs will be in clinical use for replacing the NP in patients with an intact AF;

- within 10 years, there will be more focused regenerative medicine treatments. These will include:
  - strategies for repairing the annulus fibrosus,
  - regulating the biology of the niche (the environment in the diseased tissue into which the tissue engineering construct will be placed) to improve regenerate 'take' and function,
  - complex biodegradable, purpose-designed biomaterials to: deliver regulatory molecules to the construct; replace the function of the nucleus pulposus; facilitate differentiation of stem cells; support the function of mature cells.

What of the time beyond that? It is reassuring to know that a new generation is emerging who not only appreciate the importance of chronic back pain to modern society, but also have an unrestricted view of how modern biotechnology and biomaterials could combine to manage this condition. This is an extract from the winning essay in the 2004 Future City Competition. (http://www.futurecity.org/home_press_2004_essay.shtm). It comes from the pupils of Riverview Junior/Senior High School, Oakmont, Pennsylvania:

> Straightening Nana with Nanos: Medical Polymer Construction Using Nanotechnology
> Nanobots are carbon-based robots less than 100 nanometers in size that can be injected into the body. Since the late 20th century, scientists have envisioned using these miniature robots to manipulate and move individual atoms. Today, our nanotechnologists have made this vision a reality. They have engineered nanobots to act like a miniature chemical assembly line within the body. During *[annulus]* fibrosus repair, nanobots manipulate the molecules of injected ethylene monomer, placing them into the damaged regions of the fibrosus. The freshly applied monomer is then irradiated by a tiny gamma emitter embedded within the nanobot. This begins free radical polymerization, joining the monomers together to form high-density polyethylene. The damaged fibrosus is now 'patched' with a nonreactive biocompatible compound that is strong and flexible. Doctors can then inject silicone into the nucleus pulposus to return the deflated disc to its original volume and shape. The nanobots, having completed their task, are then processed and eliminated by the body's lymph system within several days. (Reproduced with kind permission of Carol D. Rieg, National Director, Future City Competition.)

Perhaps with imagination and polymers nothing is impossible!

17.10 Sources of further information and advice

The use of biomaterials to augment the function of the IVD is a relatively new area of science and there is still only a small amount of literature on the subject. There is a standard set of journals where this work tends to be published including *Spine* and *The European Spine Journal*. Work is also published in the

Rheumatology literature including *Arthritis and Rheumatism* and *Annals of the Rheumatic Diseases*.

The Internet is a marvellous source of information. Many of the evolving areas of materials development in the management of degeneration of the IVD are led by industry and either difficult to find or simply not available in the conventional literature, but can be found on commercial websites by using a search engine.

New advances in the management of IVD disease are first brought into the scientific arena as abstracts at major orthopaedic and spine research society meetings. Pre-eminent among these are the Orthopaedic Research Society (http://www.ors.org) and the International Society for the Study of the Lumbar Spine (http://www.issls.org), although there are many more.

Advances in biomaterials can be followed through societies such as the American Society of Chemical Engineers (AIChE, http://www.aiche.org) and the Society for Biomaterials (http://www.biomaterials.org).

17.11 References

Anderson D G and Tannoury C (2005), 'Molecular pathogenic factors in symptomatic disc degeneration', *Spine*, **5** (6 Suppl), 260S–66S.

Andersson G B, Mekhail N A and Block J E (2006), 'Treatment of intractable discogenic low back pain. A systematic review of spinal fusion and intradiscal electrothermal therapy (IDET)', *Pain Physician*, **9** (3), 237–48.

Antoniou J, Steffen T, Nelson F, Winterbottom N, Hollander A P, Poole R A, Aebi M and Alini M (1996), 'The human lumbar intervertebral disc: evidence for changes in the biosynthesis and denaturation of the extracellular matrix with growth, maturation, ageing, and degeneration', *J Clin Invest*, **98**, 996–1003.

Bertagnoli R, Sabatino C T, Edwards J T, Gontarz G A, Prewett A and Parsons J R (2005), 'Mechanical testing of a novel hydrogel nucleus replacement implant', *Spine J*, **5** (6), 672–81.

Boyd L M and Carter A (2006), 'Injectable biomaterials and vertebral endplate treatment for repair and regeneration of the IVD', *Eur Spine J*, **15** (suppl 15), S414–21.

Carrington J L (2005), 'Aging bone and cartilage: cross-cutting issues', *Biochem Biophys Res Commun*, **328** (3), 700–8.

Cooper R G and Freemont A J (2004), 'TNF-alpha blockade for herniated intervertebral disc-induced sciatica: a way forward at last?', *Rheumatology*, **43**, 119–21.

Feng H, Danfelter M, Stromqvist B and Heinegard D (2006), 'Extracellular matrix in disc degeneration', *J Bone Joint Surg Am*, **88** (Suppl 2), 25–9.

Fernstrom U (1966), 'Arthroplasty with intercorporal endoprosthesis in herniated disc and in painful disc', *Acta Chir Scand*, **357**, 154–9.

Freemont A J, Peacock T E, Goupille P, Hoyland J A, O'Brien J and Jayson M I (1997), 'Nerve ingrowth into diseased intervertebral disc in chronic back pain', *Lancet*, **350**, 78–181.

Freemont T J, LeMaitre C, Watkins A and Hoyland J A (2001), 'Degeneration of intervertebral discs: current understanding of cellular and molecular events, and implications for novel therapies', *Expert Rev Mol Med*, **2001**, 1–10.

Freemont A J, Watkins A, Le Maitre C, Jeziorska M and Hoyland J A (2002a), 'Current understanding of cellular and molecular events in intervertebral disc degeneration: implications for therapy', *J Pathol*, **196** (4), 74–9.

Freemont A J, Watkins A, Le Maitre C, Baird P, Jeziorska M, Knight M T, Ross E R, O'Brien J P and Hoyland J A (2002b), 'Nerve growth factor expression and innervation of the painful intervertebral disc', *J Pathol*, **197** (3), 286–92.

Fujisawa T, Hattori T, Takahashi K, Kuboki T, Yamashita A and Takigawa M (1999), 'Cyclic mechanical stress induces extracellular matrix degradation in cultured chondrocytes via gene expression of matrix metalloproteinases and interleukin-1', *J Biochem (Tokyo)*, **125**, 966–75.

Goupille P, Jayson M I, Valat J P and Freemont A J (1998a), 'Matrix metalloproteinases: the clue to intervertebral disc degeneration?' *Spine*, **23** (14), 1612–26.

Goupille P, Jayson M I, Valat J P and Freemont A J (1998b), 'The role of inflammation in disk herniation-associated radiculopathy', *Semin Arthritis Rheum*, **28** (1), 60–71.

Gruber H E and Hanley E N (1998), 'Analysis of aging and degeneration of the human intervertebral disc. Comparison of surgical specimens with normal controls', *Spine*, **23**, 751–7.

Halling P J, Ulijn R V and Flitsch S L (2005), 'Understanding enzyme action on immobilised substrates', *Curr Opin Biotechnol*, **16** (4), 385–92.

Handa T, Ishihara H, Ohshima H, Osada R, Tsuji H and Obata K (1997), 'Effects of hydrostatic pressure on matrix synthesis and matrix metalloproteinase production in the human lumbar intervertebral disc', *Spine*, **22**, 1085–91.

Husson J L, Korge A, Polard J L, Nydegger T, Kneubuhler S and Mayer H M (2003), 'A memory coiling spiral as nucleus pulposus prosthesis: concepts, specification, bench testing, and first clinical results', *J Spinal Disord Techniques*, **16**, 405–11.

Jayson M I (1987), *The Lumbar Spine and Back Pain*, 3rd edn, London, Churchill Livingstone.

Johnson W E, Caterson B, Eisenstein S M, Hynds D L, Snow D M and Roberts S (2002), 'Human intervertebral disc aggrecan inhibits nerve growth *in vitro*', *Arthritis Rheum*, **46** (10), 2658–64.

Johnson W E, Caterson B, Eisenstein S M and Roberts S (2005a), 'Human intervertebral disc aggrecan inhibits endothelial cell adhesion and cell migration *in vitro*', *Spine*, **30** (10), 1139–47.

Johnson W E B, Wootton A, El Haj A, Eisenstein S M, Curtis A S G and Roberts S (2005b), 'Topographical guidance of intervertebral disc cell growth *in vitro*: towards the development of tissue repair strategies for the annulus', *Europ Cells Mater*, **10** Suppl. 3, 44.

Kang J D, Georgescu H I, McIntyre-Larkin L, Stefanovic-Racic M, Donaldson W F and Evans C H (1996), 'Herniated lumbar intervertebral discs spontaneously produce matrix metalloproteinases, nitric oxide, interleukin-6, and prostaglandin E2', *Spine*, **21**, 271–7.

Kauppila L I (1995), 'Ingrowth of blood vessels in disc degeneration. Angiographic and histological studies of cadaveric spines', *J Bone Joint Surg Am*, **77**, 26–31.

Kelgren J H (1977), 'The anatomical source of back pain', *Rheumatol Rehab*, **16**, 3–12.

de Kleuver M, Oner F C and Jacobs W C (2003), 'Total disc replacement for chronic low back pain: background and a systematic review of the literature', *Eur Spine J*, **12** (2), 108–16.

Le Maitre C L, Freemont A J and Hoyland J A (2004a), 'Localization of degradative enzymes and their inhibitors in the degenerate human intervertebral disc', *J Pathol*, **204** (1), 47–54.

Le Maitre C L, Hoyland J A and Freemont A J (2004b), 'Studies of human intervertebral disc cell function in a constrained in vitro tissue culture system', *Spine*, **29** (11), 1187–95.

Le Maitre C L, Freemont A J and Hoyland JA (2005), 'The role of interleukin-1 in the pathogenesis of human intervertebral disc degeneration', *Arthritis Res Ther*, **7** (4), R732–45.

Le Maitre C L, Freemont A J and Hoyland J A (2006), 'A preliminary *in vitro* study into the use of IL-1Ra gene therapy for the inhibition of intervertebral disc degeneration', *Int J Exp Pathol*, **87** (1), 17–28.

Liu J, Roughley P J and Mort J S (1991), 'Identification of human intervertebral disc stromelysin and its involvement in matrix degradation', *J Orthop Res*, **9**, 568–75.

Lotz J C, Colliou O K, Chin J R, Duncan N A and Liebenberg E (1998), 'Compression-induced degeneration of the intervertebral disc, an *in vivo* mouse model and finite-element study', *Spine*, **23**, 2493–506.

Luoma K, Riihimaki H, Luukkonen R, Raininko R, Viikari-Juntura E and Lamminen A (2000), 'Low back pain in relation to lumbar disc degeneration', *Spine*, **25**, 487–92.

MacFarlane G J, Thomas E, Croft P R, Papageorgiou A C, Jayson M I and Silman A J (1999), 'Predictors of early improvement in low back pain amongst consulters to general practice: the influence of pre-morbid and episode-related factors', *Pain*, **80**, 113–19.

Malemud C J (2006), 'Matrix metalloproteinases (MMPs) in health and disease: an overview', *Front Biosci*, **11**, 1696–701.

Nachemson A and Morris J (1963), 'Lumbar discometry. Lumbar intradiscal pressure measurements *in vivo*', *Lancet*, **1**, 1140–2.

Nachemson A L, Schultz A B and Berkson M H (1979), 'Mechanical properties of human lumbar spine motion segments. Influence of age, sex, disc level, and degeneration', *Spine*, **4** (1), 1–8.

Olmarker K and Rydevik B (2001), 'Selective inhibition of tumor necrosis factor-alpha prevents nucleus pulposus-induced thrombus formation, intraneural edema, and reduction of nerve conduction velocity: possible implications for future pharmacologic treatment strategies of sciatica', *Spine*, **26**, 863–9.

Ray C D (2002), 'The PDN prosthetic disc-nucleus device', *Eur Spine J*, **11** (Suppl. 2), S137–42.

Ray C D, Sachs B L, Norton B K, Mikkelsen E S and Clausen N A (2002), 'Prosthetic disc nucleus implants; an update', in Gunzburg R and Szpalski M, *Lumbar disc herniation*, Philadelphia: Lippincott Williams & Wilkins, 222–33.

Richardson S M, Curran J M, Chen R, Vaughan-Thomas A, Hunt J A, Freemont A J and Hoyland J A (2006), 'The differentiation of bone marrow mesenchymal stem cells into chondrocyte-like cells on poly-L-lactic acid (PLLA) scaffolds', *Biomaterials*, **27** (22), 4069–78.

Richardson T P, Murphy W L and Mooney D J (2001), 'Polymeric delivery of proteins and plasmid DNA for tissue engineering and gene therapy', *Crit Rev Eukaryot Gene Expr*, **11** (1–3), 47–58.

Risbud M V, Albert T J, Guttapalli A, Vresilovic E J, Hillibrand A S, Vaccaro A R and Shapiro I M (2004), 'Differentiation of mesenchymal stem cells towards a nucleus pulposus-like phenotype *in vitro*: implications for cell-based transplantation therapy', *Spine*, **29** (23), 2627–32.

Roberts S, Caterson B, Menage J, Evans E H, Jaffray D C and Eisenstein S M (2000), 'Matrix metalloproteinases and aggrecanase: their role in disorders of the human intervertebral disc', *Spine*, **25** (23), 3005–13.

Sakaia D, Mochidaa J, Iwashinaa T, Hiyamaa K, Omia H, Imaia M, Nakaia T, Andob K and Hottab T (2006), 'Regenerative effects of transplanting mesenchymal stem cells embedded in atelocollagen to the degenerated intervertebral disc', *Biomaterials*, **27**, 335–45.

Saunders J M, Tong T, LeMaitre C L, Freemont T J and Saunders B R (2007), 'A study of pH-responsive microgel dispersions: from fluid-to-gel transitions to mechanical property restoration for loadbearing tissue', *Soft Matter*, **3**, 486–96.

Scott J E, Bosworth T R, Cribb A M and Taylor J R (1994), 'The chemical morphology of age-related changes in human intervertebral disc glycosaminoglycans from cervical, thoracic and lumbar nucleus pulposus and annulus fibrosus', *J Anat*, **184**, 73–82.

Singh K, Ledet E and Carl A (2005), 'Intradiscal therapy: a review of current treatment modalities', *Spine*, **30** (17 Suppl), S20–6.

Thomas J, Lowman A and Marcolongo M (2003), 'Novel associated hydrogels for nucleus pulposus replacement', *J Biomed Mater Res Part A*, **4**, 1 329–37.

Tolonen J, Gronblad M, Virri J, Seitsalo S, Rytomaa T and Karaharju E O (1997), 'Platelet-derived growth factor and vascular endothelial growth factor expression in disc herniation tissue: and immunohistochemical study', *Eur Spine J*, **6**, 63–9.

Vaccaro A R, Singh K, Haid R, Kitchel S, Wuisman P, Taylor W, Branch C and Garfin S (2003), 'The use of bioabsorbable implants in the spine', *Spine J*, **3** (3), 227–37.

Vernengo J, Fussell G, and Lowman A M (2006), 'Evaluation of novel injectable hydrogels', American Institute of Chemical Engineers 2006 Annual meeting, San Francisco, Injectable Biomaterials (08B07), Presentation 190c.

Wilda H and Gough J E (2006), '*In vitro* studies of annulus fibrosus disc cell attachment, differentiation and matrix production on PDLLA/45S5 Bioglass composite films', *Biomaterials*, **27** (30), 5220–9.

Zhou Y and Abdi S (2006), 'Diagnosis and minimally invasive treatment of lumbar discogenic pain – a review of the literature', *Clin J Pain*, **22** (5), 68–81.

18
Skin tissue engineering

S MACNEIL, University of Sheffield, UK

The laboratory expansion of skin cells and the clinical use of these initiated the field of tissue engineering in the early 1980s. Tissue-engineered skin is currently used to assist patients suffering from extensive skin loss or chronic non-healing ulcers. Some 25 years of work has led to considerable knowledge in the area, a range of tissue-engineered products which are benefiting patients, and also tissue-engineered models for *in vitro* use which are proving very valuable in furthering our understanding of normal and abnormal skin cell biology. However, there are many challenges yet to be met for tissue-engineered skin to achieve its full potential. This chapter first of all reviews the need for tissue-engineered skin, both clinical and *in vitro*, then summarises landmark events in developing tissue-engineered skin, before looking briefly at the remaining challenges.

18.1 Why do we need tissue-engineered skin?

The major function of skin is to act as a barrier to the outside world. Figure 18.1(a) and (b) show the structure of skin and (c)–(e) show examples of patients who have lost barrier function due to superficial (c) or extensive (d) burns or chronic non-healing ulcers (e). Skin consists of an upper epidermal barrier layer which is relatively thin (0.1–0.2 mm in depth), securely attached to an underlying dermis by a specialised basement membrane zone. The basement membrane zone is visible only at the electron microscope level. The dermis varies in thickness depending on the site of the body and is composed primarily of collagen I with dermal inclusions of hair shafts and sweat glands which are lined with epidermal keratinocytes. The dermis is well vascularised and also contains receptors for touch, temperature and pain. Keratinocytes in the epidermis rely solely on diffusion from the adjacent dermal capillary network. These cells progressively differentiate from cells in the basal layer on the basement membrane giving rise to daughter keratinocytes which are pushed upwards. These stratify, lose their nuclei and eventually become an integrated sheet of keratin which is finally shed. The ulcer illustrated in Fig. 18.1(e) shows

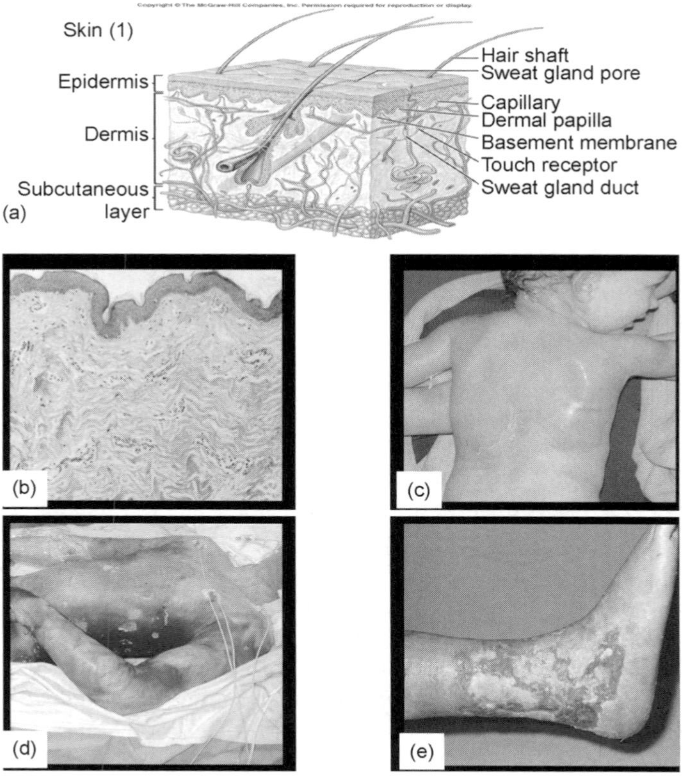

18.1 Skin structure and examples of wound healing problems. (a) A diagram of human skin (reprinted with permission from the McGraw-Hill Company). (b) A standard haematoxylin and eosin (H&E) stained section through normal human skin showing the convoluted epidermal layer attached to the much thicker spongy dermis. (c) The most common burn injury, that of superficial scalds. In this injury heat damage causes the epidermal layer to blister from the lower dermal layer. The majority of such scalds will heal well with epidermal keratinocytes from the dermal inclusions growing up to repopulate the epidermal layer. (d) Extensive full thickness burns where all epidermal and dermal layers have been permanently damaged. There are no residual cells for regenerating new skin and the barrier layer is lost, making the patient vulnerable to infection. (e) A chronic ulcer that has persisted for several decades.

that while there are islands of epidermis remaining the underlying wound bed is inflamed and such wound beds are often infected. Here it is not a lack of epithelial keratinocytes that delays healing but rather the condition of the underlying wound bed. Successful healing of chronic wounds often requires a combination of therapies to debride damaged and infected tissue, improve the vascularisation of the wound bed and combat local infection. Wounds may then heal spontaneously or, if not, healing can often be reinitiated with the addition of cultured cell products.

Any loss of full thickness skin more than 4 cm in diameter will not heal well without a graft taken from elsewhere on the body.[1] Where considerable amounts of skin are needed, the 'gold standard' approach is to take thin grafts from elsewhere on the body and use these to treat damaged areas. This is particularly challenging (Fig. 18.2). Surgeons have to avoid making the patient's condition acutely worse. Removing too much epidermal barrier from elsewhere on the body could result in very little remaining barrier function, leaving the patient vulnerable to bacterial entry and fluid loss. Fatalities (where patients have survived initial stabilisation) are usually due to this and the consequent onset of bacterial sepsis.

Laboratory expansion of sheets of cultured cells was developed to assist treatment of burn victims, giving surgeons a valuable extra option.[2–4] Figure 18.2 shows a diagram of two-stage wound reconstruction (a–d). Full thickness burn injuries require replacement of both epidermal and dermal skin elements. Figure 18.2(a)–(d) show that when full thickness skin is lost the most successful and rapid treatment is grafting with a split thickness skin graft (Fig. 18.2b) where this is available. In these cases the autologous skin graft will already have a secure epidermal–dermal junction and residual blood vessels and will

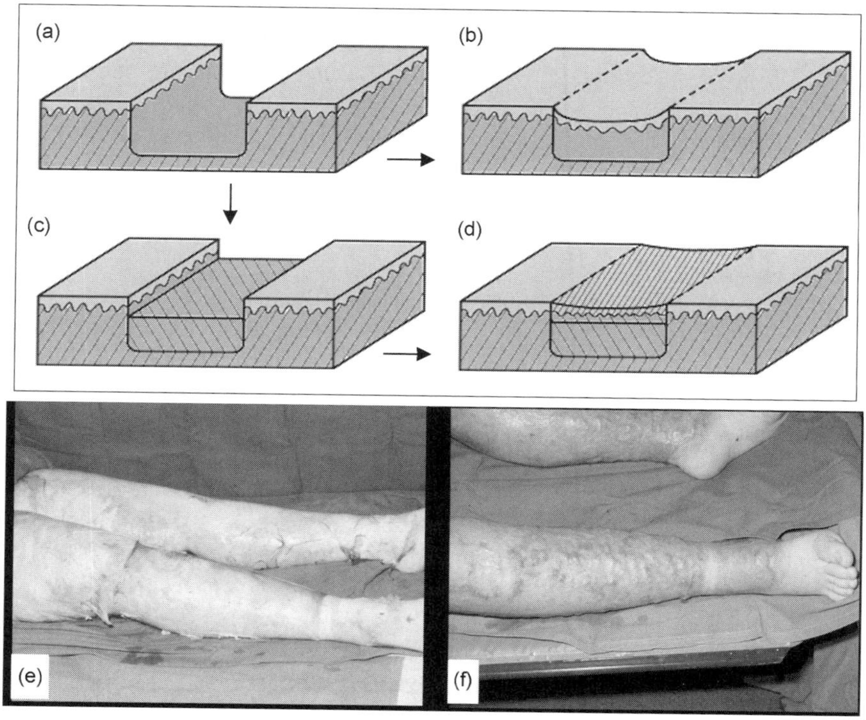

18.2 Two-stage reconstruction of full thickness burn injuries.

generally take well on clean wound beds. Where split thickness skin grafts are not available (usually because of the extent of burns injuries), then most commonly a two-stage approach to treating wounds is undertaken. Figure 18.2(c) illustrates stage 1 in which a dermal replacement material is initially used. This can be donor skin[50] or a product such as Integra®.[38] In each case the use of the dermal replacement material has two objectives – to provide some immediate barrier cover and to provide a dermal replacement material which will become well vascularised and provide support for an epithelial layer. Figure 18.2(d) illustrates stage 2: once the dermal replacement material has become well vascularised the patient's epithelium must be restored. This may be with a split thickness skin graft which in the case of major burns patients often requires recropping of healed donor sites and taking a very thin graft. Alternatively cultured autologous keratinocytes can be used, often delivered as cultured epithelial skin grafts (CEA)[44] or reconstructed tissue engineered skin such as Permaderm®.[57] Figure 18.2(e) and (f) show a patient with full thickness burns before (e) and after (f) treatment with donor skin and cultured cells. Although the majority of new cells have attached, this also illustrates that without a secure epidermal/dermal junction, blistering can occur.

It should be emphasised that in the treatment of patients with full thickness burns injuries, the gold standard remains the use of split thickness skin grafts. Tissue-engineered skin tends to be used in addition to these grafts, not instead of them. However, in economic terms extensive acute burns are not numerically significant in the Western world.[5,6] They have proven to be largely avoidable by education and lifestyle. However, they remain a major healthcare problem for developing countries. In these communities there are neither the economic resources nor the infrastructure to establish tissue engineering technologies, but a working knowledge of the landmarks in this field could make a significant impact today.

In a world where life expectancy and affluence have increased so dramatically, chronic wounds associated with ageing and with diabetes have become significantly more pressing than the treatment of burns.[7] Chronic wounds are very expensive, both economically to the healthcare system and in human terms for the patient.[8] The need to treat these provides the 'volume market' for tissue-engineered skin products. Repeated applications of skin cells, whether keratinocytes or fibroblasts, and whether these are autologous or allogeneic, can all offer some benefit to chronic non-healing wounds in prompting them to restart healing. Here cultured cells are being used as biological 'factories' to assist the body's own wound repair mechanisms.

Other areas of clinical potential are the growing number of applications in reconstructive surgery, scar revision, scar prevention, correction of pigmentation defects and the treatment of some blistering diseases and disorders (e.g. congenital nevi[9–11]).

18.2 Key events in the development of tissue-engineered skin

Development of tissue-engineered skin was possible only because there was already a mature understanding of keratinocyte culture and, relatively quickly, research using cultured cells alone became combined with cultured cells plus biomaterials to deliver materials that sought to replace the dermis or the whole of the epidermis and the dermis. As will be discussed shortly, tissue-engineered skin 'products' seek to replace the epidermal barrier layer, or the dermal layer or, less commonly, as this is more challenging, the epidermal and the dermal layer.

The key methodology which enabled the production of tissue-engineered skin was the development of a methodology for culturing adult keratinocytes.[12,13] Table 18.1 summarises current accepted knowledge. The predominant current culture methodology[12,13] uses murine fibroblasts and bovine serum. It predated knowledge of prion diseases and concerns that murine cells might contain viruses capable of transforming human cells. These issues represent significant

Table 18.1 Keratinocyte behaviour *in vitro* and *in vivo*

- Human keratinocytes can be cultured *in vitro* in a medium containing mitogens and a feeder layer of fibroblasts.[12,13] Alternative methodologies using low calcium media without a feeder layer have been developed and are often used experimentally but have not generally been taken to the clinic because of the lack of a defined media approved for clinical use, as discussed in Bullock *et al.*[14]
- Keratinocyte colony formation *in vitro* is predictable.[15]
- Keratinocytes exhibit the behaviour of a stem cell population *in vivo.* Transplantation of laboratory-expanded keratinocytes can result in an epidermis which continually renews throughout the patient's lifetime. *In vitro,* however, keratinocytes in physiological calcium will only go through a few passages before growth arresting.[16]
- Cells develop a predictable programme of differentiation even *in vitro*.
- Keratinocyte differentiation is dependent on a permissive level of extracellular/intracellular calcium (around 1 mM).[17]
- Keratinocyte differentiation is strongly stimulated by exposure to air and reduced by co-culture with fibroblasts.[18]
- Keratinocytes and fibroblasts are co-dependent – they influence each others' proliferation and production of growth factors and extracellular matrix proteins, e.g. keratinocytes in partnership with fibroblasts dictate the extracellular matrix of normal skin (e.g. Delvoye *et al.*[19,20])
- Both keratinocytes and fibroblasts are required for the formation, maintenance and repair of the basement membrane zone (e.g. Konig and Brucker-Tuderman,[21] Ralston *et al.*[22]).
- Keratinocytes are immunogenic – donor keratinocytes will not survive long term without immunosuppression.[23] However, donor keratinocytes can confer short-term healing benefit on patients and can be used in the treatment of chronic wounds.

risks for patients and a reliable alternative culture method free from animal products or cells and approved for clinical use is long overdue.

Meanwhile culture continues with feeder cells and bovine serum in physiological calcium (which mimics the natural environment of the keratinocyte), giving rapid expansion of cells which will not go beyond a few passages. To reduce risks for clinical use, murine feeder fibroblasts are subjected to lethal gamma radiation which limits their survival in the event of accidental transfer to the human wound bed. In the UK the risk of transmission of bovine spongiform encephalitis (BSE) is managed by sourcing the serum from Australasia where herds are free from BSE. However, should BSE spread worldwide, a serum-free methodology becomes a necessity.

Culture of keratinocytes for extended passages without serum or feeder layers is possible if one restricts media calcium to a level which does not permit keratinocyte differentiation. Early versions of serum-free low calcium media were unreliable but more promising media now exist, e.g. EpiLife.[24] Their compositions are proprietary, making transfer to the clinic difficult. However, establishing FDA master-files to register some such defined media for clinical use is underway.

An alternative approach to avoiding bovine serum and animal feeder cells while maintaining a physiological environment for keratinocyte expansion has been explored. Fibroblasts have the ability to promote keratinocyte proliferation in the absence of serum on a variety of substrates.[14,18,25,26] The fibroblast feeder layer reduces the degree of differentiation of the keratinocytes and does not need to be murine in origin. Substituting a well-established screened human fibroblast feeder cell line (e.g. MRC5 cells, used in human vaccine production for more than 30 years) may be an approach to avoiding the use of both animal feeder cells and bovine serum.[14]

Where human dermis is used this should be from screened donors via accredited skin banks. It is also possible to further reduce the risk of disease transmission to the patient by sterilising the dermis (without damaging either it or the basement membrane) using a combination of a gentle dehydrating agent (such as glycerol[27]) followed by a sterilising agent (e.g. ethylene oxide[28] or peracetic acid[29]).

By this stage it is clear that safety and risk management are key issues when culturing skin cells for clinical use. There is still a need for further improvements in keratinocyte culture at many levels although at the time of writing there are not to the best of the author's knowledge any reported cases of keratinocyte transformation. Patients have received cultured cells, and very often use of cultured cells has proved life-saving or had reduced the length of stay in hospital and achieved better clinical results for patients than would otherwise have been possible. Thus the issue of risk benefit in the use of cultured cells to treat patients should be considered throughout.

18.3 Do we need stem cells for tissue engineering of skin?

No-one doubts the totipotent stem cell at the start of life, but the concept of tissue- specific stem cells is currently under challenge. There are suspicions that progenitor cells circulate in the body not yet fully committed to any particular tissue. Also apparently tissue-specific stem cells may not after all be so committed and transdifferentiation may occur, e.g. Reynolds and Jahoda,[30] Gharzi *et al.*[31] and Fliniaux *et al.*[32]

There is evidence that there is a multipotent stem cell compartment in skin with cells located in niches distributed along the basement membrane and in the bulge region of the hair follicle.[33,34] Rapid and extensive expansion of adult keratinocytes is routine. If one pays attention to the culture conditions it is possible to maintain cells with a reasonable percentage of colony-forming cells. Patients who received cultured skin cells some 25 years ago as children are alive and well. There have been no reports of these transplanted skins suddenly breaking down. Thus at a pragmatic level autologous skin expanded in the laboratory and returned to the patient continues to renew suggesting that current culturing methods maintain sufficient cells with the ability to continually renew the skin post-transplantation.

18.4 Key steps in development of tissue-engineered skin for clinical use

Table 18.2 identifies key developments in the culture of skin cells for clinical use. It used to be believed that adult skin could not be cultured, but it has been shown that skin from 80-year-old patients grows as well in the laboratory as that from 20-year-olds. The breakthrough methodology published by Rheinwald and Green in 1975[12] remains today the 'gold standard' for reliable and rapid

Table 18.2 Landmark events in tissue engineering of human skin for clinical use

- 1975 Publication of a methodology for the rapid expansion of adult skin cells[12,13]
- 1979 Development of a method for transferring cultured cells from the laboratory to burns patients as sheets of cells – cultured epithelial autografts (CEA)[35]
- 1981 Clinical use of CEA in burns patients[2]
- 1981 Development of a synthetic dermal alternative – Integra[37,38]
- 1986 Reports on use of donor skin to provide a vascularised dermal alternative[36]
- 1995 Development of reconstructed skin using autologous keratinocytes and fibroblasts and bovine collagen for use in major burns patients[39–42]

expansion of keratinocytes. Keratinocytes are expanded on a lethally irradiated feeder layer of mouse fibroblasts in a media containing a range of mitogens (including cholera toxin and bovine foetal calf serum).

The next landmark in 1979 was the development of the technique of culturing cells into integrated sheets of cells some two to three cell layers thick – cultured epithelial autografts – which are then enzymically detached as sheets from the culture flask.[35] These were first used clinically in 1981.[2] Not long thereafter was the realisation that without a well-vascularised dermal wound bed keratinocytes alone would be of limited value for full thickness burns. Several approaches were developed – donor dermis derived from skin banks[36] and the development of dermal substitutes, of which the most useful to date has proven to be Integra.[37] This dermal substitute comprises bovine collagen and shark chondroitin sulphate with a silicone membrane attached to it which acts as a temporary barrier.[38] This was designed for the management of major burns. The material is grafted onto the wound bed and under this membrane vasculogenesis occurs. At the point that there is a well-vascularised wound bed (several weeks) the silicone barrier membrane is removed and replaced with a barrier of the patient's own cells, which can either be a split thickness skin graft or tissue engineered material. The use of a collagen dermal substitute has been very successfully developed by Steven Boyce and his colleagues.[39] The Boyce group used reconstructed tissue-engineered skin based on autologous keratinocytes, fibroblasts and bovine collagen sheets to treat extensive full thickness burns injuries with hard won success[40,41] as reviewed in Supp and Boyce.[42] Direct application of cultured cells to an Integra wound bed has been found to be problematic as cells fail to form a secure adhesion hence the need to develop a secure epidermal/dermal junction *in vitro*.[42]

It is now accepted practice[43] in the treatment of extensive full thickness burns to debride the burned tissue (often to the muscle fascia), provide a dermal equivalent[36,37] and, once this has become well vascularised, provide an epidermal layer (see Fig. 18.2).

Donor skin remains an excellent material for managing major burns and treating chronic wounds. It is currently an underused resource which could be especially valuable in the developing world. Skin banks can be established at much lower cost than purchasing dermal substitutes. With good banking practices and sterilisation[27–29] the risks of using donor skin can be reduced to that of using blood.

18.5 Challenges in converting research into products

It is worth pointing out that good research that results in new knowledge and therapies that may work clinically does not guarantee commercial success. This is perhaps self-evident but the field of tissue engineering has suffered from unrealistic commercial and clinical expectations. These then led to the criticism that the science had failed to deliver. A general pattern of development has been that innovative tissue-engineered products have often been developed within small companies who often manage to get the product as far as clinical proof of concept and then run out of money. At this point larger companies have sometimes stepped in to try to take the product through clinical trials and on to the market. This has been and continues to be particularly challenging because of a background of regulatory uncertainty with the same product being regulated differently in different countries throughout the world. All tissue-engineered 'products' for commercial uptake have been developed within a regulatory system that has evolved more slowly than the field it seeks to regulate.

Another major challenge is that if one is developing tissue-engineered skin for patients with extensive burns then permanent barrier function will be achieved only by using the patient's own epidermal cells, either from a skin graft or cultured autologous keratinocytes. All donor cells will be rejected by the immune system. The need to culture each patient's cells for that patient means that this approach is enormously expensive. This 'bespoke tailoring' approach is still relatively new and there are few economically viable models as yet for this type of activity. However, in the rapidly developing worlds of tissue engineering and regenerative medicine this is going to be a recurrent issue that needs to be tackled. When one is using allogeneic cells (for stimulating wound healing) then the issues of scale-up are less challenging.

The issues of how one designs the product, how one manufactures it and scales it up (in the case of allogeneic cells) or manages to produce many unique autologous products to the same recipe are ones which the fledgling tissue engineering industries are grappling with at present with varying degrees of success.

Accordingly, if one examines reviews of tissue-engineered products over the past 15 years, a consistent feature is the rate of change. Products seem to come and go, others change names and companies and relatively few appear to have survived intact throughout the last decade. Supp and Boyce[42] recently undertook a comprehensive review of the various skin substitutes. Table 18.3 lists some of these tissue-engineered 'products' categorised into whether one is seeking to achieve epithelial cover alone, replace the dermis using a dermal replacement material or replace both epidermis and dermis. This table does not claim to be exhaustive but rather is representative of the sorts of product that are available currently.

There are some clinical conditions where transferring laboratory-expanded keratinocytes or fibroblasts alone can benefit patients; however, the treatment of major full-thickness burns requires the replacement of both dermis and epidermis. If one requires an epidermis to take in a permanent sense then it must be autologous. If one does not require permanent take then allogeneic cells can be used to stimulate wound healing.

Irrespective of how keratinocytes are delivered to the wound bed, healing success depends 90% on the condition of the wound bed and only 10% on the condition of the keratinocytes. There are several ways of delivering cells to wound

Table 18.3 Some skin replacement materials available for clinical use in 2007

Epithelial cover	Cultured keratinocytes delivered as • an integrated sheet[2,44,45] or Epicel™[46] • subconfluent cells on a carrier, e.g. Myskin™[47] • Epidex[48] • subconfluent cells delivered in a spray, e.g. CellSpray®[49]
Dermal replacement materials	• Donor skin, e.g. Hermans[50] • Integra® DRM[38] • Alloderm®[51] • Dermagraft®[52] • Transcyte®[53] • Permacol®[54]
Epidermal/dermal replacement materials	• Autologous split thickness skin grafts • Donor skin, e.g. Hermans[50] • Apligraf®[55] • Orcel™[56] • Cincinnati skin substitute – Permaderm™[57] • Reconstructed human skin based on donor dermis[58]

beds – by growing them into sheets (cultured epithelial autografts, CEA)[2,46] in a cell spray[49] or on a carrier dressing.[47] The use of any one particular form of cell delivery will ultimately be determined by factors such as ease of use, cost, transportation and variables outside of the biological health of the cells.

In a recent audit of the use of CEA sheets over a 10-year period in the Burns Unit of the Northern General Hospital, Sheffield,[59] we found the overall take rate of the CEA sheets was 45%. However, a key point to note was that on the same patient take could be 80–100% in one area of the body and 0% on another area of the body. This more than anything emphasises that it is not the quality of the cells leaving the laboratory which determines take, but rather the underlying wound bed.

The issue of the condition of the wound bed is rarely emphasised sufficiently when one reads about tissue-engineered products. The best tissue-engineered product in the world will not achieve any clinical benefit on a heavily infected wound bed. Indeed some products using occlusive silicone membranes can actually encourage anaerobic bacterial growth if the wound bed is already infected. Also in many chronic wounds the question is not whether the wound bed is infected, but whether the wound bed is infected to a gross or minor degree. Another issue with the use of tissue-engineered products is that a feature limiting their clinical success is the fallacy that one can use the new tissue-engineered product instead of the previous clinical treatment regime. All new tissue-engineered products need to be used in conjunction with best clinical practices. A corollary to this is that if the use of a tissue-engineered product requires a totally new way of managing a patient then it is unlikely that there will be widespread uptake of new tissue-engineered products unless they bring unprecedented clinical benefit.

In Sheffield University, UK, where cultured cells have been delivered to patients since 1992, the preference is now for the use of a carrier dressing[60,61] to deliver cells to patients. This is a medical grade silicone disc coated with an extremely thin polymer containing acid functional groups using the technique of plasma polymerisation.[62–64] Figures 18.3(a) and 18.3(b) show the 'take' of these cells when placed over wide meshed autograft.[61] Repeated applications can promote wound healing in chronic wounds (Fig. 18.3(c) to 18.3(f)).[60,61] The carrier surface in Fig. 18.3(a)–(f) is a sterile medical grade silicone disc of 6 cm diameter coated with a plasma polymerised functional surface containing 20% carboxylic acid.[47,62] An initial biopsy of skin was used to expand keratinocytes in the laboratory. When sufficient cells were ready for clinical use (for full details see Zhu *et al.*[61] for (a), (b), (e) and (f), and Moustafa *et al.*[60] for (c) and (d)) cells were delivered to the wound bed. In (a) and (b) cultured keratinocytes were applied at 17 days post-burn injury over wide mesh split thickness skin graft on the anterior trunk. Figure 18.3(b) shows the anterior trunk 10 days post-split thickness skin grafting and autologous cell application when the carrier dressings were removed-clear transfer of cells can be seen. Figure 18.3(c) shows a non-healing diabetic foot ulcer that had resisted healing for 3 years, while Fig. 18.3(d) shows the appearance of the ulcer after eight applications of cells delivered once per week.[60] The effect of seven applications of cultured cells on the transfer dressing to a pretibial laceration on the right leg of an elderly patient is shown in Fig. 18.3(e) and (f). This wound had been non-healing for 6 weeks prior to cell applications. Figure 18.3(g) shows the appearance of reconstructed human skin based on de-epidermised acellular donor dermis to which the patient's laboratory expanded keratinocytes and fibroblasts were added and Fig. 18.3(h) shows the appearance of this 2 months after grafting for correction of earlier skin graft contraction in the axilla.[58] In common with other such strategies[65] a single application of cells rarely achieves healing in a wound which may have been stuck for a long period but repeated applications of healthy cells (using either autologous or allogeneic cells) can achieve healing.

In contrast, when delivering keratinocytes for permanent take in major burns patients, the stability of the attachment of the keratinocytes to the underlying dermis is the key issue. Sufficient research has been undertaken to show that it is very challenging for cultured keratinocytes to make a secure attachment (equivalent to early stage basement membrane formation) to a burns wound bed unless this has been well prepared to achieve a vascularised dermis or dermal equivalent prior to grafting. Even then, cultured cells can struggle to take on full thickness burns wounds and a better route is to make a reconstructed skin in the laboratory where attachment between keratinocytes and dermis can be secured.[39]

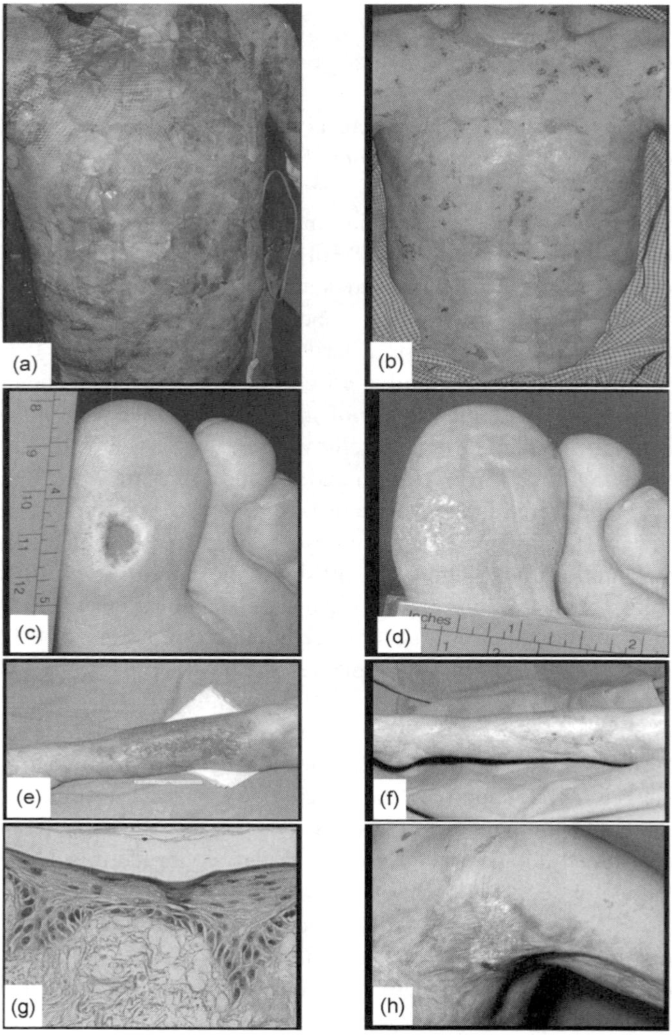

18.3 Clinical uses of tissue-engineered skin. In (a) to (f) laboratory-expanded autologous cultured keratinocytes were delivered to wound beds on a carrier surface.

Treatment of extensively burned patients is never cheap or easy. Even after the restoration of barrier function, aesthetic appearance can be far from ideal. Problems with pigmentation[66] and with contracture of grafts[58] can occur with tissue-engineered grafts as with conventional splitskin grafts. These difficulties, the relatively small number of major burns patients in the developed world and the challenging economics of treating these patients explain why so few commercial companies have developed products in this area. One notable exception is

Genzyme.[46] Most companies have focused on the chronic wounds market with off-the-shelf approaches using either cell-free materials as temporary wound dressings (e.g. Integra[38]) or the use of donor cells in some form of scaffold.

18.6 Clinical problems in the use of tissue-engineering skin

Failure to consider the maximum thickness of skin replacement material which can become easily vascularised (around 0.4 mm) can result in *delayed angiogenesis and loss of grafts*.[58] New blood vessels cannot penetrate quickly enough to feed the epidermal layer. Grafts must be placed on a well-vascularised wound bed. Reconstructed skin placed over fat or poorly vascularised wound beds will be lost. The Boyce group found that most reliable take of skin equivalents is obtained if the wound bed is pre-treated with a product such as Integra[38] in order to achieve good vascularisation before grafting. Although this is a two-stage process, no time is lost as the Integra can be used while the reconstructed skin is being grown in the laboratory.[39] In Sheffield, tissue-engineered oral mucosa[67] has been recently used to correct fibrosis of the urethra. In contrast to skin, the urethra has an excellent blood supply. Vasculogenesis of this reconstructed material was good in 5/5 patients (unpublished data).

Other problems are *contracture*[58] and *abnormal pigmentation*.[66] Clinicians and research scientists may consider that wound healing and restoration of barrier function with full mobility are great achievements and that contracture and abnormal pigmentation are lesser considerations. The patient, however, finds these issues important and more must be done to improve the quality of the final result. Fortunately these problems can be investigated *in vitro* using tissue engineered models of skin.

Not quite equivalent to the Teflon frying pan as a by-product of initial space exploration programmes, the development of 3D skin models for clinical use has led to physiologically relevant skin models which can be used for a wide range of applications. The versatility of these models again is only now becoming apparent and tissue-engineered skin models are already delivering value in skin biology research in dermatotoxicity studies, skin barrier penetration studies, wound healing, pigmentation, skin contraction, melanoma invasion, angiogenesis and other areas (Table 18.4, Fig. 18.4). Figure 18.4(a) and (b) shows the morphology of reconstructed human skin based on de-epidermised acellular dermis to which (a) keratinocytes and fibroblasts and (b) keratinocytes and fibroblasts and the human melanoma cell line (HBL) were added. Note the spontaneous pigmentation of this cell line in 3D culture which was not observed when these cells are grown in monolayer. Figure 18.4(c) and (d) shows histology of reconstructed human skin containing (c) keratinocytes and fibroblasts and (d) keratinocytes, fibroblasts and HBL cells. Reconstructed skin models were cultured for 14 days and were stained for H&E for general morphology in (c) and

Table 18.4 Some *in vitro* uses for tissue-engineered skin

- Replacement for some animal testing
- Studying cell–cell and cell–extracellular matrix interactions
- Dermatotoxicity and mucotoxicity
- Skin barrier penetration
- Wound healing
- Angiogenesis
- Pigmentation
- Melanoma invasion
- Skin contraction
- Inflammatory stress (e.g. UV radiation)

also with HMB45 using a red end point (as indicated by arrows) for detection of HBL cells in (d). Figure 18.4(e) shows the appearance of reconstructed skin containing keratinocytes and fibroblasts and melanocytes whereas Fig. 18.4(f) contains keratinocytes and melanocytes only. Melanocytes were obtained from a pale skin type donor 1/2. These reconstructed skin models were cultured for 10 days and in both (e) and (f) two 8 mm diameter discs were removed after 5 days and replaced with material lacking any additional melanocytes for comparison. Figure 18.4(g) shows the distribution of melanocytes added to reconstructed skin. Melanocytes were initially added as 1:20 ratio of keratinocytes. Melanocytes are identified by staining with S100. Figure 18.4(h) shows vascularisation of reconstructed human skin placed on the flank of a nude mouse at 7 days. This reconstructed skin contained keratinocytes and fibroblasts but no endothelial cells. Figure 18.4(i) shows a piece of de-epidermised acellular human dermis to which keratinocytes only have been added and cultured at an air–liquid interface for 10 days. The ability of keratinocytes to gather in the underlying dermis is clearly shown. Figure 18.4(j) and (k) shows the histology (j) and gross appearance (k) of tissue-engineered oral mucosa. A biopsy of buccal mucosa was taken to culture epithelial cells and fibroblasts and then added to de-epidermised acellular donor dermis of 10 by 2 cm. Figure 18.4(j) shows the histology of the biopsy of tissue engineered oral mucosa shown in (k) taken 2 days prior to clinical application for treatment of scarring of the urethra.[67]

The growth potential here is very significant, driven not only by more demanding clinical needs, but also by increased pressure to find *alternatives to animal testing*. In volume terms the biggest need is the testing of the many thousands of chemical additives to human skin products. Humanitarian reasons for reducing animal experimentation have now translated in Europe to directives from the European Union.[68]

3D *in vitro* models lack vasculature and an immune system and so are not a complete substitute for testing *in vivo*, but they are certainly much more appropriate for dermatotoxicity and mucotoxicity studies than examining the effects of xenobiotic agents in monolayer culture.

3D constructs of skin facilitate study of cell–cell and cell–extracellular matrix

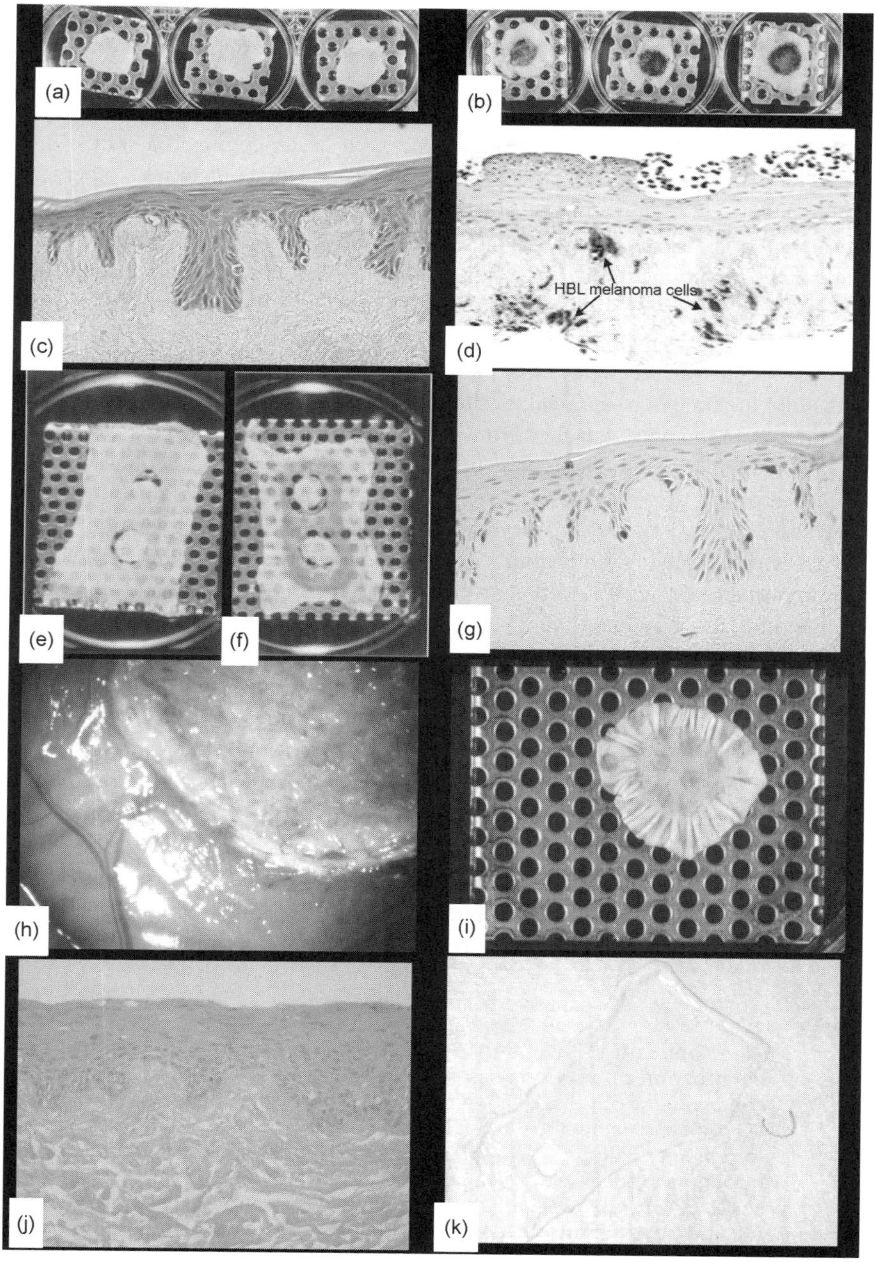

18.4 In vitro applications of tissue-engineered skin based on human dermis.

biology in response to threat: physical wounds, UV radiation, xenobiotic agents, bacterial contamination, etc. Although this research is in its infancy, already unexpected outcomes have been observed. The next decade promises to reveal major insights in several important areas, for example, new drug targets for pigmentation biology, assisting skin to cope with inflammatory stress, skin contraction, and the induction of angiogenesis in tissue-engineered skin.

Some uses of *in vitro* skin models have yielded expected outcomes – others are highly unexpected. The *influence of stromal cells on epithelial cell organisation* is overlooked in many studies, but in the absence of fibroblasts epithelial cells rarely achieve a physiological level of differentiation.[21] This is easily demonstrated by assessing the strength of keratinocyte attachment to the underlying dermal substrate. Reconstructed skin based on de-epidermised acellular sterilised human dermis reconstructed with both keratinocytes and fibroblasts successfully withstood the trauma of surgical meshing. In the absence of fibroblasts, however, the same keratinocytes detached from the underlying dermis.

In skin reconstruction *in vivo* the new *basement membrane* is synthesised by a combination of fibroblasts and keratinocytes. There is considerable published data showing that these cells communicate to produce soluble factors which influence each other's behaviour.[19–22,66] *Epidermal/mesenchymal interactions* are important. *In vitro*, even starting with a residual basement membrane architecture, this basement membrane was lost in the absence of fibroblasts, but retained when both keratinocytes and fibroblasts are present.[22] Although the need for a basement membrane equivalent remains a challenge, it appears that the native dermal collagen architecture per se is not required for cell–cell interactions. Fibroblast support of keratinocyte expansion and organisation can be demonstrated in a 3D electrospun polystyrene matrix which has none of the specialised collagen structure of the natural dermis.[26] Cell–cell interactions can clearly take place within an approximate 3D environment. This opens the possibility of using simplified synthetic scaffolds for dermal replacement *in vivo* and also for *in vitro* models of dermatoxicity (Fig. 18.5). A polystyrene scaffold with fibres of around 10 μm thickness with a packing fraction of 0.1 was

18.5 (opposite) Reconstruction of 3D tissue-engineered skin using an electrospun polystyrene scaffold. (a) shows the appearance of various combinations of cells, keratinocytes (K), fibroblasts (F) and endothelial cells (E) with and without foetal calf serum (FCS). All cells on scaffolds are stained for viability using an MTT-ESTA assay. (b) and (c) show cross-sections through scaffolds with all three cells present. (d) The upper surface of the scaffold. (e) The lower surface of the scaffold with all three cell types present. (f) and (g) Immunofluorescence micrographs of scaffolds containing keratinocytes, fibroblasts and endothelial cells under serum free conditions. (f) All cells identified by labelling with DAPI, whereas (g) specifically identifies keratinocytes by immunostaining with pancytokeratin. The scale bars shown are 100 μm. For full details see Sun *et al.*[26]

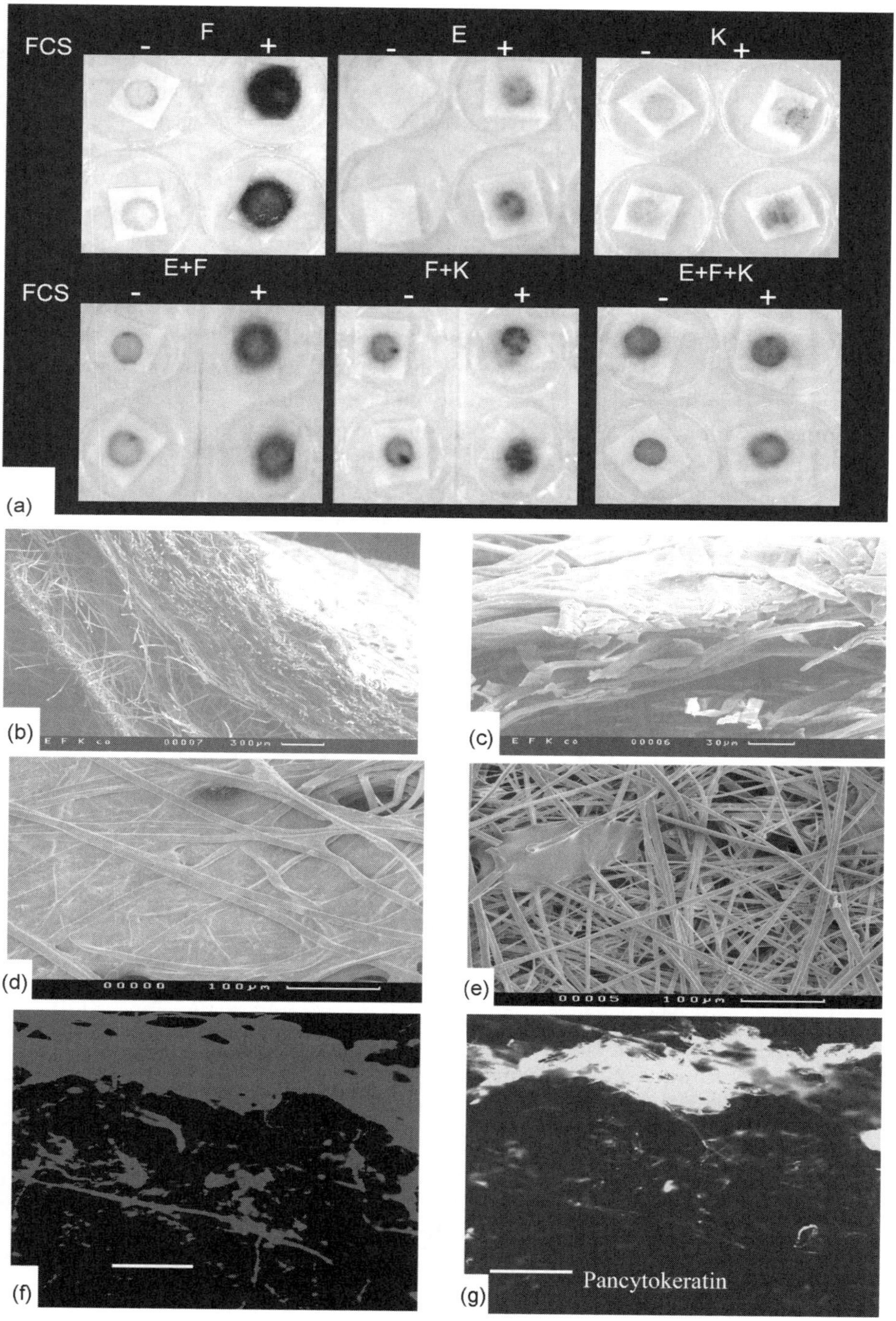
FCS
F
-
+
E
-
+
K
-
+
E+F
FCS
-
+
F+K
-
+
E+F+K
-
+
(a)
(b)
00007 300µm
(c)
00006 30µm
(d)
00000 100µm
(e)
00005 100µm
(f)
(g)
Pancytokeratin

produced by electrospinning. Figure 18.5(a) shows the viability of cells placed within this scaffold and cultured for 10 days. Single cultures and co-cultures of human keratinocytes (K), fibroblasts (F) and small vessel endothelial cells (E) with or without the presence of fetal calf serum are shown where + equals culture with fetal calf serum and – equals culture without serum (the greater the staining, the greater the number of viable cells). Figure 18.5(b)–(f) shows scanning electron micrographs of co-cultures of human keratinocytes, fibroblasts and endothelial cells in 3D electrospun polystyrene scaffolds without the presence of serum proteins.

Skin cells in 3D and in co-culture also cope more readily with potentially toxic agents than the same cells in 2D and in mono-culture.[69] *Product testing* in 2D culture will yield less favourable results. Product testing with keratinocytes alone in 3D in the absence of fibroblasts will also yield less favourable results compared with reconstructed skin containing both cell types.[69]

Skin barrier formation *in vitro* can be achieved with reconstructed skin. The key requirements are the presence of physiological calcium and culture at an air–liquid interface. Under these conditions an electrical resistance in cultured skin similar to that obtained in a split thickness skin graft[70] can be achieved. Supporting pragmatic evidence is that when cultured keratinocytes are transplanted to the patient the rapid recovery of barrier function is rarely a problem.

3D models can be used to examine some *wound healing responses*.[70] Reconstructed skin can be 'wounded' by the application of intense heat or cold or by using suction force to produce a blister. All of these can result in an epidermal thickness wound which can then be studied. Reconstructed skin can be used to examine *angiogenesis*. When small vessel endothelial cells are introduced into reconstructed human skin[58] hypoxia acts as a stimulus to endothelial cell entry.[71]

18.7 Unexpected results from using 3D skin models

18.7.1 Pigmentation

Excellent work on understanding the biology of both normal and vitiligo melanocytes in reconstructed skin has been published by the Taieb Laboratory.[72] The ability to do 'mix and match' experiments led to the finding that melanocytes in 3D culture spontaneously produce pigment reflecting their donor origin. This ability is not influenced by the keratinocyte and fibroblast environment. In Sheffield, the surprising finding was that the fibroblast appears to play a substantial role in determining whether spontaneous pigmentation occurs in reconstructed skin.[73,74] In the presence of fibroblasts, repopulating reconstructed skin with melanocytes taken from pale skin donors resulted in amelanotic skin. Omission of the fibroblasts led to spontaneous pigmentation in the same experiments[75] (see Fig. 18.4(e) and (f) – the centre of these reconstructed skins discs have deliberately not been seeded with melanocytes to provide contrast). The conclusion is that pigmentation develops as a result of an interaction between melanocytes and the other cells present – keratinocytes and fibroblasts (Fig. 18.4e) and keratinocytes only (Fig. 18.4f). Following from this was the finding that melanocytes cultured with deliberately stressed fibroblasts spontaneously pigment.[76] Speculation arises from these results. It is possible that there may be two layers of protection against UV and oxidative or inflammatory stress. The hypothesis proposed is that keratinocytes act as the primary barrier to cope with incoming stress while fibroblasts in the dermis respond secondarily. Only when the fibroblasts reach the limit of their ability to cope are melanocytes activated. Pigmented melanocytes are therefore under considerable control from keratinocytes and fibroblasts. The conclusion is that melanocyte behaviour is very different in the presence and absence of fibroblasts – a result which can be seen only with the appropriate combination of cells present in the 3D model. The role of fibroblasts in reducing pigmentation was not obvious before this.

18.7.2 Melanoma invasion

Using a reconstructed de-epidermised acellular sterilised dermis (DED) with keratinocytes, fibroblasts and a human melanoma cell line (HBL),[77] the melanoma cells were only able to invade the dermis when skin cells were added. For this melanoma line at least, *invasion was totally dependent on adjacent skin cells*. Extending this,[78] the behaviour of three different metastatic human melanoma cell lines (A375-SM, C8161 and HBL cells) interacting with keratinocytes and fibroblasts was studied, focusing on the activation of proteolytic enzymes. Again results were surprising. There was active communication between melanoma cells and adjacent skin cells and the invasive process was dictated by the melanoma cells. Unexpected was the finding that the melanoma cells showed little degradative enzyme activity but appeared instead to use the skin cell matrix – metalloproteinases (MMP2 and MMP9) for invasion. These enzymes are present but not activated in normal skin. In reconstructed skin (which as it is assembled is closer to a wound than mature skin) these enzymes were active and apparently provided an easier path for melanoma cell migration. Other studies in 2D[79,80] and 3D models[81] also suggested that inflammation accelerated melanoma migration plays a significant role in invasion. Both findings are consistent with the clinical observation that wound environments in which there is a high level of inflammation seem to provide more opportunities for melanoma migration and invasion. This is consistent with the hypothesis proposed by Balkwill and Mantovani[82] that inflammation plays a role in the progression of a range of solid tumours. This could help explain the phenomena of local recurrence in melanoma as discussed in Katerinaki *et al.*[83] For a proportion of patients melanoma recurs in the same site and one explanation is that melanoma cells may have been missed at the time of surgical excision.

However, another explanation is that circulating melanoma cells find an ideal site to attach to and invade the wound bed of the melanoma excision. Anti-inflammatories such as α-MSH[84] and aspirin[83] can retard TNF-stimulated melanoma cell invasion *in vitro* and our research in this area is now focused on developing topical anti-inflammatory approaches to retard melanoma migration and invasion.

18.7.3 Skin contraction

Populating human dermis with keratinocytes and fibroblasts leads to contraction of reconstructed skin *in vitro*[85–87] and also *in vivo*.[58] Analysis of the contraction showed that this is caused by keratinocytes and not fibroblasts.[85,86] Keratinocytes will contract human dermis in the absence of fibroblasts. Fibroblasts alone, and the addition of fibroblasts to keratinocytes, have no appreciable effect on dermal contraction induced by the keratinocytes, in marked contrast to the considerable body of literature dealing with the ability of cultured fibroblasts to contract dilute collagen gels.[88] Keratinocytes also contract these[89] but such gels differ significantly from a mature crosslinked human dermis. The finding that keratinocytes have a marked ability to contract relatively thick human dermis throws new light on the problem of skin graft contracture and this has led to the use of the tissue-engineered skin model to examine this.[85] Flowing from this are new pharmacological insights – for example, inhibitors of lysyl oxidase can block contraction to a large degree without having adverse effects on the skin morphology.[85] Research is underway in Sheffield to develop a topical approach to using these which potentially could translate to the clinic.

18.8 Future trends

There are still many significant challenges for the future. In terms of success post-grafting the two greatest challenges are persuading cultured cells to attach securely to often challenging wound beds, accelerating *basement membrane formation*, and new approaches to speed up *vascularisation*. Beyond this, there is the need to develop ways of blocking skin graft *contraction* and to achieving normal *pigmentation*. Also, to reduce risks for patients receiving cultured cells, a reliable and regulatory approved *xenobiotic-free* keratinocyte culture methodology is needed. Finally it is difficult to consider how to achieve sustained clinical success without considering the *commercial viability* of products, and alongside this a mature *regulatory* environment.

Those starting in this field will find that nearly 25 years of published research does provide a wealth of guidance well founded in cell biology and wound healing. There is a relatively mature understanding of what works well, what remains challenging and where to concentrate effort. Progress has been made, but there are still considerable challenges to be overcome. There are significant opportunities for clinical and non-clinical scientists to undertake both good quality basic and translational research in this very rewarding field.

18.9 References

1. Herndon, D.N. *et al.* A comparison of conservative versus early excision therapies in severely burned patients, *Ann Surg* **209**, 547–552 (1989).
2. O'Connor, N.E, Mulliken, J.B, Banks-Schlegel, S., Kehinde, 0. and Green, H. Grafting of burns with cultured epithelium prepared from autologous epidermal cells. *Lancet* **1**, 75–78 (1981).
3. Gallico, G.G, O'Connor, N.E., Compton, C.C., Kehinde, Q. and Green, H. Permanent coverage of large burn wounds with autologous cultured human epithelium. *New Eng J Med* **311**, 448–451 (1984).
4. Odessey, R. Addendum: multicenter experience with cultured epidermal autograft for treatment of burns. *J Burn Care Rehab* **13**, 174–180 (1992).
5. *Burn Incidence and Treatment in the United States: 1999 Fact Sheet*, The Burn Foundation, Philadelphia (PA) (1999).
6. Rose, J.K. and Herndon, D.N. Advances in the treatment of burn patients. *Burns* **23**, 721–725 (1997).
7. Falanga, V. Chronic wounds: pathophysical and experimental considerations. *J Invest Dermatol* **100**, 721–725 (1993).
8. Philips, T., Santon, B., Provan, A. and Lew, R. A study of the impact of leg ulcers on quality of life: financial, social, and psychologic implications. *J Am Acad Dermatol* **31**, 49–53 (1994).
9. Bittencourt, F.V. *et al.* Large congenital melanocytic nevi and the risk of development of malignant melanoma and neurocutaneous melanocytosis. *Pediatrics* **106**, 736–741 (2000).
10. Cooper, M.L., Spielvogel, R.L., Hansbrough, J.F., Boyce, S.T. and Frank, D.H. Reconstitution of the histological characteristics of a giant congenital nevomelanocytic nevus employing the athymic mouse and a cultured skin substitute. *J Invest Dermatol* **97**, 649–658 (1991).
11. Gallico, G.O. *et al.* Cultured epithelial auto grafts for giant congenital nevi. *Plast Reconstr Surg* **84**, 1–9 (1989).
12. Rheinwald, J.G. and Green, H. Serial cultivation of strains of human epidermal keratinocytes: the formation of keratinizing colonies from single cells. *Cell* **6**, 331–344 (1975).
13. Rheinwald, J.G. and Green, H. Epidermal growth factor and the multiplication of cultured human epidermal keratinocytes. *Nature* **265**, 421–424 (1977).
14. Bullock, A.J., Higham, M.C. and MacNeil, S. Use of human fibroblasts in the development of a xenobiotic-free cuture and delivery system for human keratinocytes. *Tissue Eng* **12**, 245–255 (2006).
15. Fuchs, E. Epidermal differentiation: the bare essentials. *J Cell Biol* **6**, 2807–2814 (1990).
16. *The Keratinocyte Handbook* (Supplement: Keratinocyte Methods), edited by I.M. Leigh, E.B. Lane and F.M. Watt, Cambridge University Press (1994).
17. Hennings, H. *et al.* Calcium regulation of growth and differentiation of mouse epidermal keratinocytes in culture. *Cell.* **19**, 245–254 (1980).
18. Sun, T. *et al.* Development in xenobiotic-free culture of human keratincytes for clinical use. *Wound Repair Regeneration* **12**, 626–634 (2004).

19. Delvoye, P. *et al.* Fibroblasts induce the assembly of the macromolecules of the basement membrane. *J Invest Dermatol* **90**, 276–282 (1990).
20. Demarchez, M., Hartmann, D.J., Regnier, M. and Asselineau, D. The role of fibroblasts in dermal vascularisation and remodelling of reconstructed human skin after transplantation onto the nude mouse. *Transplantation* **54**, 317–326 (1992).
21. Konig, A. and Brucker-Tuderman, L. Epithelial-mesenchymal interactions enhance expression of collagen VII *in vitro*. *J Invest Dermatol* **96**, 803–808 (1991).
22. Ralston, D.R. *et al.* The requirement for basement membrane antigens in the production of human epidermal/dermal composites *in vitro*. *Brit J Dermatol* **140**, 605–615 (1999).
23. Carver, N., Navsaria, H., Green, C.J. and Leigh, I.M. Acute rejection of cultured keratinocyte allografts in non immunosuppressed pigs. *Transplantation* **52**, 918–920 (1991).
24. EpiLife® media for keratinocytes from Cascade Biologics: www.cascadebio.com
25. Higham, M.C. *et al.* Development of a stable chemically defined surface for the culture of human keratinocytes under serum free conditions for clinical use. *Tissue Eng* **9**, 919–930 (2003).
26. Sun, T. *et al.* Self-organisation of skin cells in 3D-electrospun polystyrene scaffolds. *Tissue Eng* **11**, 1023–1033 (2005).
27. Marshall, L. *et al.* Effect of glycerol on intracellular virus survival: implications for the clinical use of glycerol preserved cadaver skin. *Burns* **21**, 356–361 (1995).
28. Chakrabarty, K.H. *et al.* Development of autologous human epidermal/dermal composites based on sterilised human allodermis for clinical use. *Brit J Dermatol* **141**, 811–823 (1999).
29. Huang, Q., Dawson, R., Pegg, D.E., Kearney, J.N. and MacNeil, S. Use of peracetic acid to sterilise human donor skin for production of acellular matrices for clinical use. *J. Wound Repair Regeneration* **12**, 276–287 (2004).
30. Reynolds, A.J. and Jahoda, C.A.B. Cultured dermal papilla cells induce follicle formation and hair growth by transdifferentiation of an adult epidermis. *Development* **115**, 587–593 (1992).
31. Gharzi, A., Reynolds, A.J .and Jahoda, C.A.B. Plasticity of hair follicle dermal cells in wound healing and induction. *Exp. Dermatol* **12**, 126–136 (2003).
32. Fliniaux, I., Viallet, J.P., Dhouailly, D. and Jahoda, C.A.B. Transformation of amnion epithelium into skin. *Differentiation* **72**, 558–565 (2004).
33. Watt, F.M. Out of Eden: stem cells and their niches. *Science* **287**, 1427–1403 (2000).
34. Braun, K.M. and Watt, F.M. Epidermal label-retaining cells: background and recent applications. *J Invest Dermatol Symp Proc* **9**, 196–201 (2004).
35. Green, H., Kehinde, O. and Thomas, I. Growth of cultured human epidermal cells into multiple epithelia suitable for grafting. *Proc Natl Acad Sci USA* **76**, 5665–5668 (1979).
36. Cuono, C.B., Langdon, R. and McGuire, J. Use of cultured epidermal auto grafts and dermal allografts as skin replacement after burn injury. *Lancet* **1** 1123–1124 (1986).
37. Burke, J., Yannas, I., Quinby, W.C., Bondoc, C.C .and Jung, W.K. Successful use of a physiologically acceptable artificial skin in the treatment of extensive burn injury. *Ann Surg* **194**, 413–427 (1981).
38. Integra® DRT (dermal replacement template) from Integra: www.integra-ls.com
39. Boyce, S.T. *et al.* The 1999 Clinical Research Award. Cultured skin substitutes combined with Integra to replace native skin autograft and allograft for closure of full-thickness burns. *J Burn Care Rehabil* **20**, 453–461 (1999).

40. Boyce, S.T. *et al.* Comparative assessment of cultured skin substitutes and native skin autograft for treatment of full-thickness burns, *Ann. Surg* **222** , 743–752 (1995).
41. Boyce, S.T. *et al.* Cultured skin substitutes reduce donor skin harvesting for closure of excised, full-thickness burns. *Ann. Surg* **235**, 269–279 (2002).
42. Supp, D.M. and Boyce, S.T. Engineered skin substitutes: practices and potentials. *Clinics Dermatology* **23**, 403–412 (2005).
43. Herndon, D.N. and Parks, D.H. Comparison of serial debridement and auto grafting and early massive excision with cadaver skin overlay in the treatment of large burns in children. *J Trauma* **26**, 149–152 (1986).
44. Compton, C.C. Current concepts in pediatric burn care: the biology of cultured epithelial autografts: an eight-year study in pediatric burn patients. *Eur J Pediatr Surg* **2**, 216–222 (1992).
45. Carsin, H.P. *et al.*, Cultured epithelial autografts in extensive burn coverage of severely traumatized patients: a five year single-center experience with 30 patients. *Burns* **26**, 379–387 (2000).
46. Epicel™ from Genzyme: www.genzymebiosurgery.com
47. Myskin® from Celltran Ltd: www.celltran.co.uk
48. *EpiDex™ from Modex Therapeutics: www.modex-t3r.com*
49. CellSpray® from C3: www.clinicalcellculture.com
50. Hermans, M.H. Clinical experience with glycerol-preserved donor skin treatment in partial thickness burns. *Burns Incl Therm Inj* **15**, 57–59 (1989).
51. AlloDerm® regenerative dermis from www.lifecell.com
52. Dermagraft® from Advanced BioHealing : www.AdvancedBioHealing.com
53. TransCyte® from Advanced BioHealing : www.AdvancedBioHealing.com
54. Permacol®, a natural matrix derived from decellularized porcine dermis from Tissue Science Laboratories: www.tissuescience.com.
55. Apligraf® from Organogenesis: www.organogenesis.com
56. Orcel™: www.ortecinternational.com
57. Cincinnati Skin Substitute – now referred to as Permaderm™ www.cambrex.com
58. Sahota, P.S. *et al.* Development of a reconstructed human skin model for Angiogenesis. *J Wound Repair Regeneration* **11**, 275–284 (2003).
59. Hernon, C.A., Dawson, R.A., Freedlander, E., Short, R., Haddow, D.B., Brotherston, M. and MacNeil, S. Clinical experience using cultured epithelial autografts leads to an alternative methodology for transfering skin cells from the laboratory to the patient. *Regenerative Medicine* **1**(6), 809–821 (2006).
60. Moustafa, M. *et al.* A new autologous keratinocyte dressing treatment for non-healing diabetic neuropathic foot ulcers. *Diabetic Medicine* **21**, 786–789 (2004).
61. Zhu, N. *et al.* Treatment of burns and chronic wounds using a new cell transfer dressing for delivery of autologous keratinocytes. *Europ J Plastic Surgery* **28**, 319–330 (2005).
62. France, R.M., Short, R.D., Dawson, R.A. and MacNeil, S. Attachment of human keratinocytes to plasma co-polymers of acrylic acid/octa-1,7-diene and allyl amine/octa-1,7-diene. *J Mater Chem* **8**, 37–42 (1998).
63. Haddow, D.B. *et al.* Comparison of proliferation and growth of human keratinocytes on plasma copolymers of acrylic acid/1,7-octadiene and self-assembled monolayers. *J Biomed Mater Res* **47**, 379–387 (1999).
64. Haddow, D.B. *et al.* Plasma polymerised surfaces for culture of human keratinocytes and transfer of cells to an *in vitro* wound bed model. *J Biomedical Mater Res*, **64** 80–87 (2003).

65. De Luca, M. *et al.* Multicentre experience in the treatment of burns with autologous and allogeneic cultured epithelium, fresh or preserved in a frozen state. *Burns* **15**, 303–309 (1989).
66. Medalie, D.A. *et al.* Differences in dermal analogs influence subsequent pigmentation, epidermal differentiation, basement membrane and rete ridge formation of transplanted composite skin grafts. *Transplantation* **64**, 454–465 (1997).
67. Bhargarva, S., Chappell, C., Bullock, A.J., Layton, C. and MacNeil, S. Tissue-engineered buccal mucosa for substitution urethroplasty. *Brit J Urology* **93**, 807–811 (2004).
68. Sauer, U.G., Spielmann, H. and Rusche, B. Fourth EU report on the statistics on the number of animals used for scientific purposes in 2002 – trends, problems, conclusions. *ALTEX*, **22**, 59–67 (2002). www.altex.ch/hauptseite_e.htm
69. Sun, T., Jackson, S., Haycock, J.W. and MacNeil, S. Culture of skin cells in 3D rather than 2D improves their ability to survive exposure to cytotoxic agents. *Biomaterials* **27**, 3459–3465 (2006).
70. Bullock, A.J., Barker, A.T., Coulton, L. and MacNeil, S. The effect of induced alternating electrical currents on re-epithelialisation of a novel wound healing model. *Bioelecromagnetics* **28**(1), 31–41 (2007).
71. Sahota, P.S., Burn, J.L., Brown, N.J. and MacNeil, S. Approaches to improve angiogenesis in tissue-engineered skin. *Wound Repair Regeneration*, **12**, 635–642 (2004).
72. Bessou, S. *et al. Ex vivo* study of skin phototypes. *J Invest Dermatol* **107**, 684-688 (1996).
73. Buffey, J.A. et al. Extracellular matrix derived from hair and skin fibroblasts stimulates human skin melanocyte tyrosinase activity. *Brit J Derm* **131**, 836–842 (1994).
74. Hedley, S., Gawkrodger, D.J., Weetman, A.P. and MacNeil, S. Investigation of the influence of extracellular matrix proteins on normal human melanocyte morphology and melanogenic activity. *Brit J Derm* **135**, 888–897 (1996).
75. Hedley, S.J. *et al.* Fibroblasts play a regulatory role in the control of pigmentation in reconstructed human skin from skin types I and II. *Pigment Cell Res* **15**, 49–56 (2002).
76. Balafa, C. *et al.* Dopa oxidase activity in the hair, skin and ocular melanocytes is increased in the presence of stressed fibroblasts. *Exp Dermatol* **14**, 363–372 (2005).
77. Eves, P. *et al.* Characterisation of an *in vitro* model of human melanoma invasion based on reconstructed human skin. *Brit J Dermatol* **142**, 210–222 (2003).
78. Eves, P. *et al.* Melanoma invasion in reconstructed human skin is influenced by skin cells – investigation of the role of proteolytic enzymes. *Clin Exp Metastasis* **20**, 685–700 (2003).
79. Zhu, N. *et al.* Melanoma cell attachment, invasion and integrin expression is upregulated by tumour necrosis factor-α and suppressed by α-melanocyte stimulating hormone. *J.Invest. Dermatol.* **119**, 1165–1171 (2002).
80. Canton, I. *et al.* TNF-α increases and α-MSH reduces uveal melanoma invasion through fibronectin. *J Invest Dermatol* **121**, 557–563 (2003).
81. Katerinaki, E., Evans, G.S., Lorigan, P.C and MacNeil, S. TNF-α increases human melanoma cell invasion and migration *in vitro*: the role of proteolytic enzymes. *Br J Cancer* **89**, 1123–1129 (2003).
82. Balkwill, F. and Mantovani, A. Inflammation and cancer: back to Virchow? *Lancet* **357**, 539–545 (2001).

83. Katerinaki, E., Haycock, J.W., Lalla, R., Carlson, K., Yang, Y., Hill, R.P., Lorigan, P.C. and MacNeil, S. Sodium salicylate inhibits TNF-α induced NF-κB activation, cell migration, invasion and ICAM-1 expression in human melanoma cells. *Melanoma Res* **16**(1), 11–22 (2006).
84. Eves, P. *et al.* Anti-inflammatory and anti-invasive effects of alpha-melanocyte stimulating hormone in human melanoma cells. *Brit J Cancer* **89**, 2004–2015 (2003).
85. Ralston, D.R. *et al.* Keratinocytes contract normal human dermal extracellular matrix and reduce soluble fibronectin production by fibroblasts in a skin composite model. *Brit J Plast Surg* **50**, 408–415 (1997).
86. Chakrabarty, K.H. *et al.* Keratinocyte-driven contraction of reconstructed human skin. *Wound Repair Regeneration* **9**, 95–106 (2001).
87. Harrison, C.A. *et al.* Use of an *in vitro* model of tissue-engineered human skin to investigate the mechanism of skin graft contraction. *Tissue Eng* **12**(11), 3119–3133.
88. Grinnell, F. Fibroblasts, myofibroblasts and wound contraction. *J Cell Biol* **124**, 401–404 (1994).
89. Souren, I.B.M., Ponec, M. and Van Wijk, R. Contraction of collagen by human fibroblasts and keratinocytes. *In Vitro Cell Dev Biol* **25**, 1039–1045 (1989).

19
Liver tissue engineering

K SHAKESHEFF, University of Nottingham, UK

19.1 Introduction

The liver is a fascinating tissue to attempt to recreate by engineering. Within the human body the liver can possess a remarkable ability to regenerate. For example, full functional recovery after 80% hepatectomy can occur in patients after removal of a tumour. In contrast, isolated liver cells are difficult to culture in the laboratory and display a rapid loss of function and viability. The gap between the ability of our bodies to engineer liver tissue and laboratory methods to replicate this engineering is wide.

The liver also offers tissue engineers an example of a structure in which complex architecture is essential to tissue function. The numerous cell types of the liver are constructed into 3D lobular structures in which blood and bile flood are directional and spatial patterning of gene expression are essential to the coordinated function of the lobule. Recreating complex spatial and flow relationships within an engineering liver model may prove essential.

The motivation to engineer liver tissue is strong because there are numerous clinical and commercial applications awaiting a robust method of regenerating or reconstructing a long-term functional liver system. Clinically, liver disease is a major cause of morbidity and mortality worldwide resulting in over 25 000 deaths per year in the US alone (Hammond *et al.*, 2006).

One potential attraction to liver tissue engineering is the numerous *in vitro* applications of a simple liver model. Within the pharmaceutical and chemicals industry there is a growing need for human models of the liver that can provide metabolic and toxicological information on new chemical entities and lead drug molecules.

As with other branches of tissue engineering the origins of liver tissue engineering can be traced back to seminal studies in the early 1990s. However, it is important to note that hepatocyte culture, including aspects of three-dimensional co-culture pre-date the advent of tissue engineering by many decades. A discussion of earlier developments in *in vitro* hepatocyte cell culture is beyond the scope of this chapter but the reader is encouraged to refer to earlier literature when considering new strategies to engineering liver (Yeoh *et al.*, 1979; Diegelmann *et al.*, 1983; Dunn *et al.*, 1989).

This chapter begins with a brief overview of the structure and function of the liver. Next I consider the clinical and commercial potential of engineered liver. The main sections on tissue engineering strategies for liver have been divided into the topics of multicellular aggregates, scaffolds, bioreactors, microtechnology and cell patterning and *in situ* regeneration. Across the chapter there are numerous examples of the use of scaffold materials. The important role of the extracellular matrix (ECM) in liver development and maintenance has led to the extensive use of ECM proteins, especially collagen. In addition, synthetic biodegradable polymers have been widely employed, Finally, the use of microfabrication and *ex vivo* application of engineered liver tissue has diversified the materials type used in this field to include material chosen for its applicability to fabrication route rather than biological response.

19.2 The structure of the liver lobule

The major cell type by number and volume within the human liver is the hepatocyte. Numerous hepatology textbooks provide comprehensive descriptions of the function of hepatocytes and morphology of the liver (see, for example, Zakim and Boyer, 1996). Here a brief overview will be provided.

On the organ length-scale, the liver is the largest human gland, weighing approximately 1.5 kg in healthy adults. The liver consists of four lobes. Each lobe is composed of numerous liver lobules. The lobules are at the length-scale at which the relationship between liver function and architecture are apparent. As shown in Fig. 19.1, the lobule consists of dual blood supply network (from the portal vein and hepatic artery) that flow from portal triads towards the central vein of each lobule. Hepatocytes and associated cells are arranged to form cords of tissue through which blood and bile flow.

The sinusoids form a rich vascular network lined by endothelial cells and Kupffer cells. The schematic in Fig. 19.2 illustrates the position of these cells in relation to each other. The perisinusoidal space between the sinusoidal wall and the hepatocytes is known as the Space of Disse. Blood plasma passes along the hepatic sinusoids and escapes into the Space of Disse via small pores or fenestrae within the endothelial lining of the sinusoids.

19.3 Clinical and commercial applications of engineered liver tissue

There are very powerful reasons to engineer liver tissue for both clinical and commercial applications. On the clinical side, toxic injury, infection, cancer or developmental problems cause acute and chronic liver diseases (Zakim and Boyer, 1996). In acute clinical situations, such as acetaminophen (paracetamol)

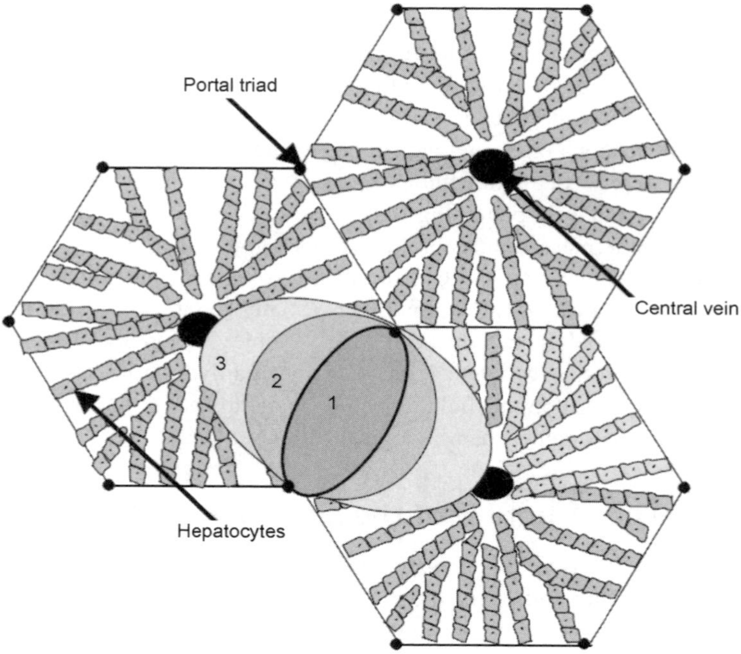

19.1 Schematic diagram of three classical liver lobules. The radiating pattern of hepatocyte plates is illustrated and the periportal zone (1), the centrolobular zone (2) and the perivenous zone (3) are designated. These zones indicate the relative position of the cells with respect to a gradient in oxygen concentration of blood flowing from the branches of the hepatic artery and portal vein.

overdose or major resection, an immediate bridging device is required to provide vital lost functions while the liver regenerates or possibly a liver transplant becomes available. Bioartificial liver devices have been developed for short-term use. These devices may be cell-free, contain primary animal cells, or immortalised human cells. Tissue engineering approaches to the design of these devices have been discussed by Selden and Hodgson (2004).

In chronic clinical situations where there is no prospect of regeneration the current treatment is a transplant. The lack of availability of livers for transplantation is an increasing problem, although the *in situ* split liver procedure offers a method to widen the pool of donors (Cintorino *et al.*, 2006). Therefore, alternative methods based on cell transplantation and tissue engineering are under intense investigation (Chan *et al.*, 2004; Kulig and Vacanti, 2004; Selden and Hodgson, 2004).

On the commercial side, the importance of the liver in chemical metabolism ensures that assays of liver interactions play a key role in drug development. Product safety represents a significant challenge in the success of potential new blockbuster drugs. It has been estimated that 75% of the cost of developing new

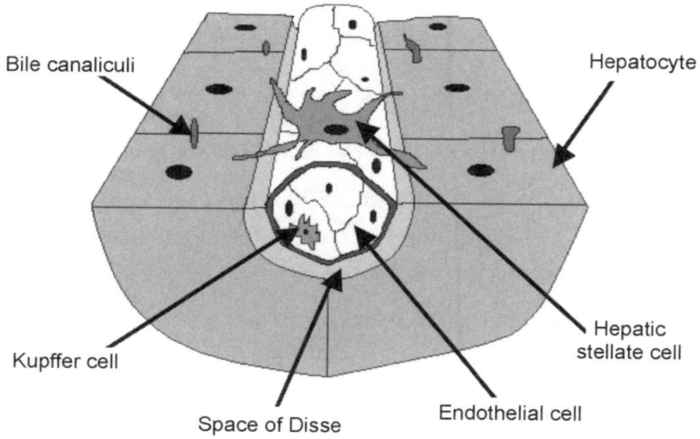

19.2 Schematic diagram of the hepatic sinusoid. Adapted from Bhatia *et al.* (1999).

drugs is due to the failure of compounds due to poor efficacy, safety or metabolic profiles. The most common reason for compound attrition is hepatotoxicity which is estimated to cost the pharmaceutical industry billions of pounds a year. Major challenges to the study of drug metabolism and hepatotoxicity are the de-differentiation of hepatocytes in culture, lack of supply of human cells and the lack of consideration of the role of non-parenchymal cells. Potentially, tissue-engineered models could address these challenges.

19.4 Approaches to liver tissue engineering

Most researchers working in tissue engineering integrate a number of strategies to try to replicate the complexity of liver tissue. As a result, most literature on liver tissue engineering does not fit into a single category. In the following sections the major themes of tissue engineering are introduced using examples from the literature. Most of the quoted papers could have been used to exemplify two or more of the themes.

19.4.1 Multicellular aggregates

The concept of forming spheroids of liver cells predates the advent of tissue engineering (Landry *et al.*, 1985). Spheroids are a simple 3D system in which cells self-assemble, or are mechanically forced, into small (typically less than 500 μm) aggregates. This culture method has been used extensively to develop models of cancer (Sutherland, 1980) and other tissues.

Initial studies with liver spheroids used homotypic systems containing purified hepatocytes only. Landry *et al.* (1985) showed that hepatocytes spontaneously re-aggregated and histotypic structures could be identified within aggregates. Long-term production of albumin and induction of tyrosine aminotransferase demonstrated maintenance of liver-specific functions.

In a tissue engineering context, a number of improvements to the homotypic aggregation have been developed. Powers *et al.* (1997) have explained the relationship between cell–substratum interactions and the spontaneity of spheroid formation. This paper showed that cells could be held in monolayers at higher concentrations of Matrigel coating (1 $\mu g/cm^2$) or encouraged to form spheroids at 0.01 $\mu g/cm^2$ Matrigel coverage. Abu-Absi *et al.* (2002) have demonstrated that homotypic aggregations formed in a stirred suspension culture can form functional bile canaliculi. They visualised the location of the bile canalicular enzyme dipeptidyl peptidase IV and found that it resided in tubular structures. Furthermore, bile acid secretion by polarised hepatocytes was shown. This study provides an example of ability to restore complex architectural features of the liver and hence restore numerous functions (Fig. 19.3).

Although the formation of bile canaliculi within hepatocyte aggregates is encouraging, these systems are still a major simplification of the true architecture

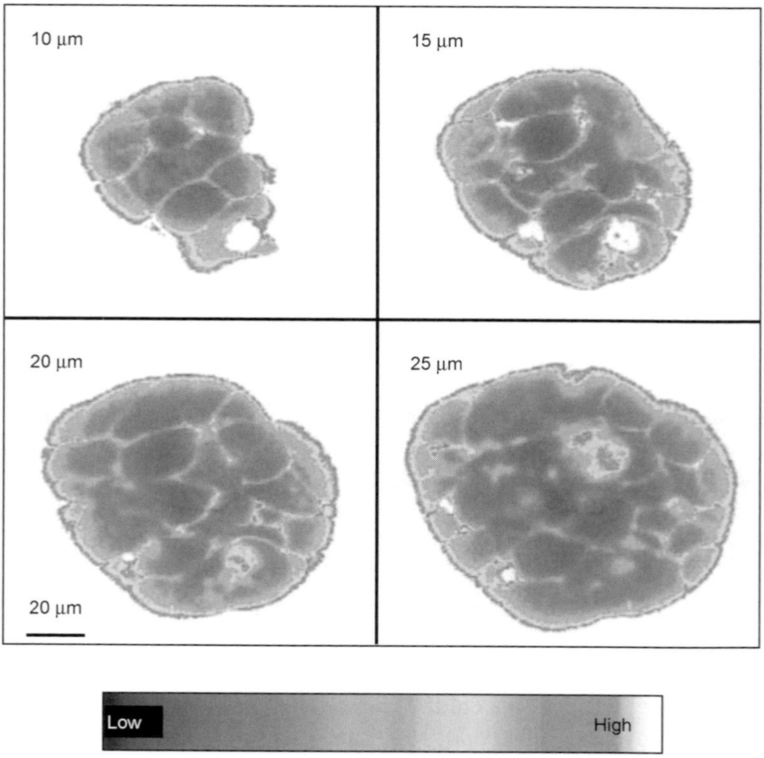

19.3 The location of bile canaliculi within hepatocyte spheroids is visualised by the diffusion of FITC-dextran and confocal microscopy. Reproduced with permission from Abu-Abii *et al.*, *Experimental Cell Research*, **274**, 56–67, 2002.

of the liver. Therefore, there is significant activity in the design of heterotypic aggregates in which other cell types of the liver are combined with hepatocytes. This field is the subject of an excellent article by Bhatia *et al.* (1999). Early studies on this subject used fibroblast or fibroblast cell lines. The use of this cell source simplifies the work because it eliminates the need to isolate and expand rare cell types from liver isolates. Khetani *et al.* (2004) have taken a functional genomics approach to elucidate the mechanism by which three murine fibroblast cell types interact with hepatocytes. Seventeen functionally characterised candidate genes involved in cell communication were identified and experimental validation for the role of N-cadeherin and decorin was performed. Chia *et al.* (2005) have demonstrated that a soluble factor TGFβ1 increased hepatocyte functionality in co-culture with NIH3T3 fibroblasts. Bhandari *et al.* (2001) used the co-culture of 3T3 fibroblasts and hepatocytes to prolong cytochrome P450 enzyme activity in models of liver drug metabolism. Takezawa *et al.* (1992) combined the concepts of co-culture and surface-enhanced spheroid formation by culturing layers of fibroblasts and hepatocytes on a thermally responsive polymer (poly-*N*-isopropyl acrylamide) and dissolving away the polymer by changing the culture temperature. The aggregates generated possessed complex histotypic architecture and enhanced functionality.

A further step in refining the co-culture concept is the replacement of generic fibroblast lines with non-parenchymal liver cells. A long-standing approach in this field is to mix hepatocytes with the crude mixture of non-hepatocyte cells isolated from the liver. From a tissue engineering perspective there are advantages in further refining this method to isolate individual cell types and expand those cells *in vitro* to scale-up the manufacture of the liver tissue. Sinusoidal endothelial cells are a logical target for these studies due to the intimate relationship between these cells and hepatocytes *in vivo*. For example, Lee *et al.* (2004) have shown enhanced and lengthened hepatocyte functionality if spheroids are coasted with an endothelial cell line. Further approaches to introduce endothelial cells will be discussed further in Section 19.4.4 (Griffith and Swartz, 2006). A second target cell for co-culture is the stellate cell. This cell type makes multiple physical connections with hepatocytes in the liver and produces extracellular matrix. Overactivity of the stellate cell underlies a number of clinical conditions of the liver including fibrosis. Thomas *et al.* demonstrated that stellate cells could contract to pull hepatocytes into spheroids (Thomas *et al.*, 2006) and that the hepatocytes displayed enhanced functionality including testosterone metabolism in these heterotypic aggregates (Thomas *et al.*, 2005).

19.4.2 Scaffolds

In Section 19.4.1, the 3D cell structure formed as a result of spontaneous interactions between cells and the resulting spheroids were dense balls of cells.

From an engineering perspective there are clear advantages in intervening in the process of spheroid formation and providing an improved architectural template for the cells. Porous scaffolds provide a method of defining the 3D space within which the tissue forms and influencing the internal porosity as tissues form. In addition, from a clinical perspective, a scaffold can be precisely located and secured within the body. *In vivo* applications of scaffolds are discussed in more detail in Section 19.4.5.

The design of scaffolds for liver tissue engineering is less advanced than scaffolds for orthopaedic and cardiovascular applications. Within the field of orthopaedics there is a long history of the use of scaffolds, in the form of demineralised bone matrix or ceramics. In contrast in the hepatology there is no history of clinical application of biomaterials in augmenting liver regeneration.

The groups led by Linda Cima-Griffith and Joseph Vacanti have perfomed the foundation work for the use of scaffolds in liver tissue engineering. Cima-Griffith *et al.* (1991) undertook the first comparison of hepatocyte behaviour on synthetic biodegradable polymers against Matrigel and collagen type I. They found that the synthetic materials could host long-term hepatocyte functioning. The poly(lactic acid-*co*-glycolic acid) (PLGA) materials used by Cima-Griffith were subsequently used by Mooney *et al.* (1996) to form a controlled release formulation for epidermal growth factor (EGF) to stimulate the engraftment of hepatocytes *in vivo*. EGF was encapsulated in microparticles of the polymer and then dispersed in a porous sponge made of a slower degrading polymer. Based on these studies and the success of this polymer class in other branches of tissue engineering, PLGA scaffolds have been widely used in applications where scaffold plus cells are implanted within the body and liver tissue regenerates *in situ* (Section 19.4.5).

Alginate scaffolds have also been widely investigated in liver tissue engineering. Glicklis *et al.* (2000) formed alginate sponges with pores ranging in diameter from 100 to 150 μm and seeded hepatocytes within the 3D matrix. The cells initially seeded onto the scaffolds and then detached to form spheroids that were embedded within the sponge. Long-term functionality studies revealed that the alginate system performed equivalently to conventional spheroid culture. Yang *et al.* (2001) improved on the alginate scaffold concept by including a galactosylated chitosan component with the alginate. The chitosan component improved the physical stability of the scaffold with an aqueous culture environment and the galactose groups provided anchorage points for hepatocytes. The composite scaffold outperformed the alginate-only scaffold in terms of cell-seeding efficiency and functionality. Confirmation of the beneficial environment created within alginate scaffolds has been confirmed by the studies of Dvir-Ginzberg *et al.* (2003) and Elkayam *et al.* (2006) in which enhanced drug metabolism by primary hepatocytes and cell lines was measured in high-density cultures.

An alternative method of using alginate materials to enhance liver cell function has been exemplified by Rahman *et al.* (2004) Here, the alginate material was used to fully encapsulate a hepatocyte cell line (HepG2). The cells maintained significantly higher expression of differentiated functions. An important advantage of this system was the protection offered against damage resulting from cryopreservation. Hence, this system can be stored for long periods within hospitals and thawed for use when required.

Enhancing hepatocyte adhesion to scaffolds has been achieved by including biological molecules within scaffolds. Cho *et al.* (2006) have discussed the use of galactose-carrying polymers that target hepatocytes via the asialoglycoprotein receptor. Quirk *et al.* (2003) used a surface engineering approach to selectively bind hepatocytes to poly(DL-lactic acid) (PLA) surfaces via terminal galactose groups. Lin *et al.* (2004) have adopted the approach of using liver extracellular matrix (biomatrix) as a scaffold for hepatocytes. This approach builds on the work of the Badylak group to utilise the fully complex of natural tissue extracellular matrix as a scaffold that restores numerous cell adhesion and soluble molecule signals. Lin *et al.* confirmed that the biomatrix could outperform collagen as a support for hepatocytes in terms of longevity of function of key cytochrome P450 enzymes.

19.4.3 Bioreactors

The liver lobule is a highly perfused tissue structure with a zonal structure and directional blood flow. The use of multicellular aggregates and/or scaffolds does not address the need to recreate perfusion conditions. Therefore, a bioreactor can serve at least four functions in liver tissue engineering; enhanced delivery of oxygen and nutrients, improved viability and functionality of hepatocytes if transportation is required, miniaturisation of engineered tissue for higher throughput assay and zonation of hepatocyte functioning.

A well-established application for bioreactors is to increase the availability of oxygen within high cell number cultures by circulating and replenishing an oxygenated cell culture medium. A major driving force for these studies has been the clinical goal of manufacturing bioartificial liver support devices that can maintain liver functions after massive liver failure. Within these systems, billions of hepatocytes must retain viability and certain functions for many days. Gerlach *et al.* (1990) demonstrated that direct provision of oxygen to cells via gas permeable hollow fibres that run through the 3D tissue space enhanced metabolic activity in hepatocytes. Bratch and Al-Rubeai (2001) confirmed that direct oxygenation maintained specific functions such as albumin production and cytochrome P450 activity.

The group led by Gerlach has built on its early work on oxygenation system to evolve a new bioreactor that promotes the reorganisation of hepatocytes and non-parenchymal cells into histotypic structures (Zeilinger *et al.*, 2004). This bioreactor system enabled cell metabolic activity to remain constant for at least 20 days. Both bile canaliculi and sinusoidal-like structures developed spontaneously. Of particular note was the detection of increased proliferative activity within the cell population. Proliferation combined with sustained liver-specific function offers the potential to increase tissue mass from a primary cell source and thereby, reduce the tissue requirement for each clinical procedure. This type of bioreactor also shows significant potential as an *in vitro* model for drug metabolism and toxicology studies (Zeilinger *et al.*, 2002). Sauer *et al.* (2005) demonstrated that this type of bioreactor could act as a transportable device to facilitate the movement of cells from a site of donation to the site of clinical need.

Gerlach's work elegantly demonstrates that the self-assembly of complex tissue structures can be promoted by perfusion of the cells with medium and direct access to oxygen. Another approach to recreate the complexity of liver architecture is force different cell types into to architectural arrangements that mimic the liver using microtechnology and fabrication. The groups of Griffiths, Bhatia and Vacanti have independently and collaboratively pioneered the use of microtechnology in this field and their work is highlighted in Section 19.4.4. One bioreactor paper of note from Bhatia's group has demonstrated that the zonation of hepatocyte function can be recreated *in vitro* by controlling oxygen gradients within co-cultures of hepatocytes and non-parenchymal cells (Allen *et al.*, 2005). Oxygen gradients were mathematically modelled and then experimentally created using a syringe pump to control the introduction of oxygenated medium at one end of flat bed bioreactor. By culturing hepatocytes within this bioreactor *in vivo*-like zonation of expression of two key drug metabolising enzymes was created. Furthermore, drug toxicity matched the zonation of the metabolising enzymes.

19.4.4 Microtechnology and cell patterning

As discussed previously in this chapter, one of the most intriguing aspects of liver tissue engineering is the need to replicate the architecture of the liver lobule to fully replicate function. Research on multicellular aggregates shows that liver cells have some ability to spontaneous organise *in vitro*; but spheroidal architectures do not allow the replication of complex fluid flow within tissue structures. Therefore, a number of leading groups in liver tissue engineering are now focusing on methods of building microscale architecture into their liver models.

The introduction of architecture can occur in one, two or three dimensions. In the previous section, the method of Allen *et al.* (2005) created one-dimensional architecture in which a monolayer of hepatocytes experienced a unidirectional flow. 2D patterning involves the use of fabrication to create zonation of surface chemistry (Bhatia *et al.*, 1998). Bhatia *et al.* demonstrated that islands of surfaces coated with collagen could attract hepatocytes to form isolated 2D discs. The area surrounding each hepatocyte island could be filled with fibroblasts to form a co-culture. The major advantage of this approach was that the ratio of hepatocytes to fibroblasts and the relative areas of both cell types and the length of interface between the two cell types could be precisely controlled by varying the micropattern. This allowed Bhatia *et al.* to quantify the effect of co-culture parameters on hepatocyte function.

An alternative method of micropatterning cells is to use field patterns within enhanced field-induced dielectrophoresis (DEP) traps. Ho *et al.* (2006) have demonstrated a heterogeneous liver cell on-chip system in which hepatocyte cell livers and endothelial cells are focussed into radial patterns that mimic the liver lobule structure. An example of their cell patterns is reproduced in Fig. 19.4.

Moving up in dimensionality a number of group have formed cultures that have 2.5D characteristics. Much of the work in this area has been pioneered by the group of Griffith at MIT. For example, Powers *et al.* (1997) have described a 3D perfused microarray bioreactor for the culture of heterotypic multicellular liver models. The microarray system provides thin transparent structure within which a silicon scaffold provides square ports that host the liver cells. This group has proved that preformed spheroids out-perform single cell suspensions as the format for seeding of the bioreactor. If hepatocytes and non-parenchymal cells were formed into spheroids before addition to the bioreactor then histotypic structures formed and albumin secretion better maintained that direct addition of the same cell types without the performing of spheroids.

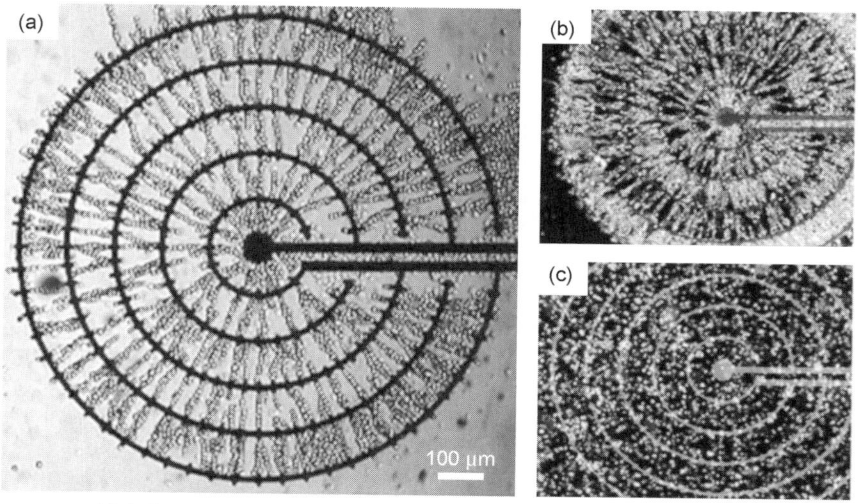

19.4 On-chip heterogeneous-cells patterning demonstration reproduced with permission from Ho *et al.* (2006): (a) formation of a radial pattern of HepG2 cells; (b) rapid heterogeneous-integration patterning of HepG2 cells and endothelial cells; (c) control study of location of HepG2 cells and endothelial cells in the absence of dielectrophoresis; in (a), (b) and (c), the medium flows from the right to the left.

Fukuda *et al.* (2006) combined microfabrication and microcontact printing to form 300 μm cavities in which spheroids formed with tight control of diameter. Within each cavity the central region was rendered cell adhesive using either collagen or peptide sequences. Long-term maintenance of key liver functions (albumin secretion and ammonia removal) was measured and liver-enriched transcriptional factors were maintained for extended culture times. This approach promises to create a high-throughput methodology of the formation and use of spheroids. Petronis *et al.* (2001) have used a ceramic template to control hepatocyte interactions within pit structures.

Finally, full 3D systems require the formation of a network that replicates the sinusoids of the liver with surrounding hepatocytes. Microfluidic structures offer the potential to expose small populations of cells to flowing medium with precise control of the location of multiple chemicals and the potential to recreate the relationships between fluid flow and liver architecture. Leclerc *et al.* (2003) have established a poly(dimethylsiloxane) microfluidic structure and proved that it functions as a closed loop bioreactor for cells. Recently, Vacanti's group (Khademhosseini *et al.*, 2006) have published a number of exciting papers on the potential to great intricate vascular patterns in silicon (Kaihara *et al.*, 2000) or polymer (Bettinger *et al.*, 2006). These patterns have been used *in vitro* to maintain hepatocyte functionality but the greatest promise of these devices is as implantable 3D organs in which a large mass of hepatocytes can be implanted with an in-built vascular network.

19.4.5 *In situ* regeneration

Many of the approaches described in the previous sections have targeted applications of liver tissue engineering in which the final culture is used *in vitro* as a drug screening device or *in situ* as an extracorporeal bioartificial liver. This final section of the chapter moves on to consider applications of liver tissue engineering in which the aim is restore functional liver tissue *in vivo*. This area of research has been introduced in an excellent article by Davis and Vacanti (1996).

A critical challenge in the use of hepatocyte cell-based therapies is the maintenance of viability of the cells and their successful engraftment to form a permanent tissue. Three major issues can be identified with the literature concerning engraftment: effect of engraftment site, requirement for a scaffold and benefits of co-delivery of agents that stimulate angiogenesis or hepatocyte proliferation.

Lee *et al.* (2003) provide a quantified comparison of different sites of implantation of hepatocytes within polymer scaffolds and with/without portal-venous shunt (PCS). They found that the subcutaneous space was the worst site for engraftment with the omentum proving the best site and intermediate results in the mesentery. PCS was effective in increasing engraftment in the omentum and mesentery only. Kneser *et al.* (1999) used a poly(vinyl alcohol) (PVA) scaffold as a host for transplanted hepatocytes in the small intestine mesentery of rats. Encouraging results were obtained if the transplantation was accompanied by a PCS. Initial cell number declined in all groups but recovered within the scaffolds.

Bruns *et al.* (2005) used a fibrin gel to inject hepatocytes directly into the liver and confirmed the formation of new tissue and a functional contribution from the new cells. Ohashi *et al.* have developed a method to form stable transplants of liver cells under the kidney capsule in the mice (Ohashi *et al.*, 2005b,c). They used a extracellular matrix gel to assist the initial engraftment (Ohashi *et al.*, 2005a). A thin layer of liver tissue was maintained in the kidney. Then, liver regeneration was stimulated by repeated partial hepatectomies were performed to stimulate regeneration of liver at the kidney site.

One approach to improve cell survival is to co-deliver growth factors within microparticles. Mooney *et al.* (1996) described early work in this area that generated a promising result that slow release of epidermal growth factor (EGF) improved hepatocyte engraftment. However, the authors of this paper noted that an additional surgical procesure, a portal-caval shunt (PCS), was required in addition to the EGF. This suggest that multiple factors in the portal blood supply are important in engraftment and hepatocyte survival. More recently, Mooney's group published a careful study on the effects of vascular endothelial growth factor (VEGF), EGF and hepatocyte growth factor (HGF) (Smith *et al.*, 2006). They found a short-term benefit of VEGF but no long-term benefit of EGF and HGF. The authors conclude that multiple signals are required to generate a useful delivery system.

Yokoyama *et al.* (2006) explored whether the subcutaneous cavity could provide an easily accessible site for engraftment of hepatocytes. They showed that a scaffold that released a powerful angiogenic factor could generate a highly vasularised area in which hepatocytes could be inserted.

An alternative to co-delivery of growth factors is the transfection of hepatocytes with genes for VEGF, HGF and tumor necrosis factor-alpha. Ajioka *et al.* (2001) have demonstrated that transfected cells simulated the formation of thick networks of blood vessels and albumin secretion was enhanced.

19.5 Conclusions

It is important to recognise the immense challenge posed to tissue engineers by the architectural and functional complexity of the liver. The scale of the challenge has stimulated groups to attempt some of the most complex and bold strategies to restore tissue architecture. At the other end of the spectrum there is the need for microscale liver for *in vitro* applications that could improve the quality of information available at early points in drug design and development.

A convergent theme across recent literature in this field is the need to balance self-assembly and engineering interventions. Self-assembly of liver structures in multicellular aggregates is a long-standing principle that drives improvements in functionality. However, it is clear that tissue engineers need to intervene to make self-assembly productive. *In vitro* the works of, for example, Bhatia, Griffith and Vacanti demonstrate that microarchitecture, imposed by scaffold design, can enhance self-assembly. This concept has been adopted *in vivo* with numerous studies to enhance blood vessel formation to, in turn, enhance hepatocyte engraftment.

19.6 Future trends

The current literature provides a strong scientific basis for complex co-cultures with internal flows that replicate blood flow. In the near future these systems will display zonal liver functions, be manufactured on small scales to allow high-throughout drug molecule screening, and will stimulate improvement in hepatocyte transplantation. However, major challenges still exist. For example, while simple functions of the liver may be maintained in current tissue-engineered liver systems, more complex functions are always lost. Undoubtedly, gene expression within hepatocytes is greatly disturbed by isolation from the tissue and we are yet to fully understand the molecular biology sufficiently to restore full homeostasis.

19.7 References

Abu-Absi, S. F., Friend, J. R., Hansen, L. K. and Hu, W. S. (2002) Structural polarity and functional bile canaliculi in rat hepatocyte spheroids. *Experimental Cell Research*, **274**, 56–67.

Ajioka, I., Nishio, R., Ikekita, M., Akaike, T., Sasaki, M., Enami, J. and Watanabe, Y. (2001) Establishment of heterotropic liver tissue mass with direct link to the host liver following implantation of hepatocytes transfected with vascular endothelial growth factor gene in mice. *Tissue Engineering*, **7**, 335–344.

Allen, J. W., Khetani, S. R. and Bhatia, S. N. (2005) *In vitro* zonation and toxicity in a hepatocyte bioreactor. *Toxicological Sciences*, **84**, 110–119.

Bettinger, C. J., Weinberg, E. J., Kulig, K. M., Vacanti, J. P., Wang, Y. D., Borenstein, J. T. and Langer, R. (2006) Three-dimensional microfluidic tissue-engineering scaffolds using a flexible biodegradable polymer. *Advanced Materials*, **18**, 165–169.

Bhandari, R. N. B., Riccalton, L. A., Lewis, A. L., Fry, J. R., Hammond, A. H., Tendler, S. J. B. and Shakesheff, K. M. (2001) Liver tissue engineering: a role for co-culture systems in modifying hepatocyte function and viability. *Tissue Engineering*, **7**, 345–357.

Bhatia, S. N., Balis, U. J., Yarmush, M. L. and Toner, M. (1998) Microfabrication of hepatocyte/fibroblast co-cultures: role of homotypic cell interactions. *Biotechnology Progress*, **14**, 378–387.

Bhatia, S. N., Balis, U. J., Yarmush, M. L. and Toner, M. (1999) Effect of cell–cell interactions in preservation of cellular phenotype: cocultivation of hepatocytes and nonparenchymal cells. *Faseb Journal*, **13**, 1883–1900.

Bratch, K. & Al-Rubeai, M. (2001) Culture of primary rat hepatocytes within a flat hollow fibre cassette for potential use as a component of a bioartificial liver support system. *Biotechnology Letters*, **23**, 137–141.

Bruns, H., Kneser, U., Holzhuter, S., Roth, B., Kluth, J., Kaufmann, P. M., Kluth, D. and Fiegel, H. C. (2005) Injectable liver: a novel approach using fibrin gel as a matrix for culture and intrahepatic transplantation of hepatocytes. *Tissue Engineering*, **11**, 1718–1726.

Chan, C., Berthiaume, F., Nath, B. D., Tilles, A. W., Toner, M. and Yarmush, M. L. (2004) Hepatic tissue engineering for adjunct and temporary liver support: critical technologies. *Liver Transplantation*, **10**, 1331–1342.

Chia, S. M., Lin, P. C. and Yu, H. (2005) TGF-beta 1 regulation in hepatocyte-NIH3T3 co-culture is important for the enhanced hepatocyte function in 3D microenvironment. *Biotechnology and Bioengineering*, **89**, 565–573.

Cho, C. S., Seo, S. J., Park, I. K., Kim, S. H., Kim, T. H., Hoshiba, T., Harada, I. and Akaike, T. (2006) Galactose-carrying polymers as extracellular matrices for liver tissue engineering. *Biomaterials*, **27**, 576–585.

Cima, L. G., Ingber, D. E., Vacanti, J. P. and Langer, R. (1991) Hepatocyte culture on biodegradable polymeric substrates. *Biotechnology and Bioengineering*, **38**, 145–158.

Cintorino, D., Spada, M., Gruttadauria, S., Riva, S., Luca, A., Volpes, R., Vizzini, G., Arcadipane, A., Henderson, K., Verzaro, R., Foglieni, C. S. and Gridelli, B. (2006) *In situ* split liver transplantation for adult and pediatric recipients: an answer to organ shortage. *Transplantation Proceedings*, **38**, 1096–1098.

Davis, M. W. and Vacanti, J. P. (1996) Toward development of an implantable tissue engineered liver. *Biomaterials*, **17**, 365–372.

Diegelmann, R. F., Guzelian, P. S., Gay, R. and Gay, S. (1983) Collagen formation by the hepatocyte in primary monolayer-culture and *in vivo*. *Science*, **219**, 1343–1345.

Dunn, J. C. Y., Yarmush, M. L., Koebe, H. G. and Tompkins, R. G. (1989) Hepatocyte dunction and extracellular-matrix geometry – long-term culture in a sandwich configuration. *Faseb Journal*, **3**, 174–177.

Dvir-Ginzberg, M., Gamlieli-Bonshtein, I., Agbaria, R. and Cohen, S. (2003) Liver tissue engineering within alginate scaffolds: effects of cell-seeding density on hepatocyte viability, morphology, and function. *Tissue Engineering*, **9**, 757–766.

Elkayam, T., Amitay-Shaprut, S., Dvir-Ginzberg, M., Harel, T. and Cohen, S. (2006) Enhancing the drug metabolism activities of C3A – a human hepatocyte cell line – by tissue engineering within alginate scaffolds. *Tissue Engineering*, **12**, 1357–1368.

Fukuda, J., Sakai, Y. and Nakazawa, K. (2006) Novel hepatocyte culture system developed using microfabrication and collagen/polyethylene glycol microcontact printing. *Biomaterials*, **27**, 1061–1070.

Gerlach, J., Kloppel, K., Stoll, P., Vienken, J. and Muller, C. (1990) Gas-supply across membranes in bioreactors for hepatocyte culture. *Artificial Organs*, **14**, 328–333.

Glicklis, R., Shapiro, L., Agbaria, R., Merchuk, J. C. and Cohen, S. (2000) Hepatocyte behavior within three-dimensional porous alginate scaffolds. *Biotechnology and Bioengineering*, **67**, 344–353.

Griffith, L. G. and Swartz, M. A. (2006) Capturing complex 3D tissue physiology *in vitro*. *Nature Reviews Molecular Cell Biology*, **7**, 211–224.

Hammond, J. S., Beckingham, I. J. and Shakesheff, K. M. (2006) Scaffolds for liver tissue engineering. *Expert Review of Medical Devices*, **3**, 21–27.

Ho, C. T., Lin, R. Z., Chang, W. Y., Chang, H. Y. and Liu, C. H. (2006) Rapid heterogeneous liver-cell on-chip patterning via the enhanced field-induced dielectrophoresis trap. *Lab on a Chip*, **6**, 724–734.

Kaihara, S., Borenstein, J., Koka, R., Lalan, S., Ochoa, E. R., Ravens, M., Pien, H., Cunningham, B. and Vacanti, J. P. (2000) Silicon micromachining to tissue engineer branched vascular channels for liver fabrication. *Tissue Engineering*, **6**, 105–117.

Khademhosseini, A., Langer, R., Borenstein, J. and Vacanti, J. P. (2006) Microscale technologies for tissue engineering and biology. *Proceedings of the National Academy of Sciences of the United States of America*, **103**, 2480–2487.

Khetani, S. R., Szulgit, G., Del Rio, J. A., Barlow, C. & Bhatia, S. N. (2004) Exploring interactions between rat hepatocytes and nonparenchymal cells using gene expression profiling. *Hepatology*, **40**, 545–554.

Kneser, U., Kaufmann, P. M., Fiegel, H. C., Pollok, J. M., Kluth, D., Herbst, H. and Rogiers, X. (1999) Long-term differentiated function of heterotopically transplanted hepatocytes on three-dimensional polymer matrices. *Journal of Biomedical Materials Research*, **47**, 494–503.

Kulig, K. M. and Vacanti, J. R. (2004) Hepatic tissue engineering. *Transplant Immunology*, **12**, 303–310.

Landry, J., Bernier, D., Ouellet, C., Goyette, R. and Marceau, N. (1985) Spheroidal aggregate culture of rat-liver cells – histotypic reorganization, biomatrix deposition, and maintenance of functional activities. *Journal of Cell Biology*, **101**, 914–923.

Leclerc, E., Sakai, Y. and Fujii, T. (2003) Cell culture in 3-dimensional microfluidic structure of PDMS (polydimethylsiloxane). *Biomedical Microdevices*, **5**(2), 109–114.

Lee, D. H., Yoon, H. H., Lee, J. H., Lee, K. W., Lee, S. K., Kim, S.K., Choi, J. E., Kim, Y. J. and Park, J. K. (2004) Enhanced liver-specific functions of endothelial cell-covered hepatocyte hetero-spheroids. *Biochemical Engineering Journal*, **20**(2–3), 181–187.

Lee, H. M., Cusick, R. A., Utsunomiya, H., Ma, P. X., Langer, R. and Vacanti, J. P. (2003) Effect of implantation site on hepatoctyes heterotopically transplanted on biodegradable polymer scaffolds. *Tissue Engineering*, **9**(6), 1227–1232.

Lin, P., Chan, W. C. A., Badylak, S. F. and Bhatia, S. N. (2004) Assessing porcine liver-derived biomatrix for hepatic tissue engineering. *Tissue Engineering*, **10**, 1046–1053.

Mooney, D. J., Kaufmann, P. M., Sano, K., Schwendeman, S. P., Majahod, K., Schloo, B., Vacanti, J. P. and Langer, R. (1996) Localized delivery of epidermal growth factor improves the survival of transplanted hepatocytes. *Biotechnology and Bioengineering*, **50**, 422–429.

Ohashi, K., Kay, M. A., Kuge, H., Yokoyama, T., Kanehiro, H., Hisanaga, M., Ko, S., Nagao, M., Sho, M. and Nakajima, Y. (2005a) Heterotopically transplanted hepatocyte survival depends on extracellular matrix components. *Transplantation Proceedings*, **37**, 4587–4588.

Ohashi, K., Kay, M. A., Yokoyama, T., Kuge, H., Kanehiro, H., Hisanaga, M., Ko, S. and Nakajima, Y. (2005b) Stability and repeat regeneration potential of the engineered liver tissues under the kidney capsule in mice. *Cell Transplantation*, **14**, 621–627.

Ohashi, K., Waugh, J. M., Dake, M. D., Yokoyama, T., Kuge, H., Nakajima, Y., Yamanouchi, M., Naka, H., Yoshioka, A. and Kay, M. A. (2005c) Liver tissue engineering at extrahepatic sites in mice as a potential new therapy for genetic liver diseases. *Hepatology*, **41**, 132–140.

Petronis, S., Eckert, K. L., Gold, J. and Wintermantel, E. (2001) Microstructuring ceramic scaffolds for hepatocyte cell culture. *Journal of Materials Science – Materials in Medicine*, **12**, 523–528.

Powers, M. J., Rodriguez, R. E. and Griffith, L. G. (1997) Cell-substratum adhesion strength as a determinant of hepatocyte aggregate morphology. *Biotechnology and Bioengineering*, **53**, 415–426.

Quirk, R. A., Kellam, B., Bhandari, R. N., Davies, M. C., Tendler, S. J. B. and Shakesheff, K. M. (2003) Cell-type-specific adhesion onto polymer surfaces from mixed cell populations. *Biotechnology and Bioengineering*, **81**, 625–628.

Rahman, T. M., Selden, C., Khalil, M., Diakonov, I. and Hodgson, H. J. F. (2004) Alginate-encapsulated human hepatoblastoma cells in an extracorporeal perfusion system improve some systemic parameters of liver failure in a xenogeneic model. *Artificial Organs*, **28**, 476-482.

Sauer, I. M., Schwartlander, R., van der Jagt, O., Steffen, I., Efimova, E., Pless, G., Kehr, D. C., Kardassis, D., Fruhauf, J. H., Gerlach, J. C. and Neuhaus, P. (2005) *In vitro* evaluation of the transportability of viable primary human liver cells originating from discarded donor organs in bioreactors. *Artificial Organs*, **29**, 144–151.

Selden, C. and Hodgson, H. (2004) Cellular therapies for liver replacement. *Transplant Immunology*, **12**, 273–288.

Smith, M. K., Riddle, K. W. and Mooney, D. J. (2006) Delivery of hepatotrophic factors fails to enhance longer-term survival of subcutaneously transplanted hepatocytes. *Tissue Engineering*, **12**, 235–244.

Sutherland, R. M. (1980) Multicellular spheroid system as a tumor-model for studies of radiation sensitizers. *Pharmacology & Therapeutics*, **8**, 105–123.

Takezawa, T., Yamazaki, M., Mori, Y., Yonaha, T. and Yoshizato, K. (1992) Morphological and immuno-cytochemical characterization of a hetero-spheroid composed of fibroblasts and hepatocytes. *Journal of Cell Science*, **101**, 495–501.

Thomas, R. J., Bhandari, R., Barrett, D. A., Bennett, A. J., Fry, J. R., Powe, D., Thomson, B. J. & Shakesheff, K. M. (2005) The effect of three-dimensional co-culture of hepatocytes and hepatic stellate cells on key hepatocyte functions *in vitro*. *Cells Tissues Organs*, **181**, 67–79.

Thomas, R. J., Bennett, A. J., Thomson, B. J. and Shakesheff, K. M. (2006) Hepatic stellate cells on poly(DL-lactic acid) surfaces control the formation of 3D hepatocyte co-culture aggregates *in vitro*. *European Cells & Materials*, **11**, 16–26.

Yang, J., Chung, T. W., Nagaoka, M., Goto, M., Cho, C. S. and Akaike, T. (2001) Hepatocyte-specific porous polymer-scaffolds of alginate/galactosylated chitosan sponge for liver-tissue engineering. *Biotechnology Letters*, **23**, 1385–1389.

Yeoh, G. C. T., Bennett, F. A. and Oliver, I. T. (1979) Hepatocyte differentiation in culture – appearance of tyrosine aminotransferase. *Biochemical Journal*, **180**, 153–160.

Yokoyama, T., Ohashi, K., Kuge, H., Kanehiro, H., Iwata, H., Yamato, M. and Nakajima, Y. (2006) *In vivo* engineering of metabolically active hepatic tissues in a neovascularized subcutaneous cavity. *American Journal of Transplantation*, **6**, 50–59.

Zakim, D. and Boyer, T.D. (1996) *Hepatology: A Textbook of Liver Disease*, Philadelphia, W.B. Saunders Company.

Zeilinger, K., Sauer, I. M., Pless, G., Strobel, C., Rudzitis, J., Wang, A. G., Nussler, A. K., Grebe, A., Mao, L., Auth, S. H. G., Unger, J., Neuhaus, P. and Gerlach, J. C. (2002) Three-dimensional co-culture of primary human liver cells in bioreactors for *in vitro* drug studies: effects of the initial cell quality on the long-term maintenance of hepatocyte-specific functions. *Atla – Alternatives to Laboratory Animals*, **30**, 525–538.

Zeilinger, K., Holland, G., Sauer, I. M., Efimova, E., Kardassis, D., Obermayer, N., Liu, M., Neuhaus, P. and Gerlach, J. C. (2004) Time course of primary liver cell reorganization in three-dimensional high-density bioreactors for extracorporeal liver support: an immunohistochemical and ultrastructural study. *Tissue Engineering*, **10**, 1113–1124.

20

Kidney tissue engineering

A SAITO, Tokai University, Japan

20.1 Introduction

For a decade regenerative medicine has been under focus and considered as the next generation treatment for diseased tissues or failed organs. The approach utilizes embryonic and bone marrow stem cells to repair diseased tissues or to replace a failed organ with a newly generated one by allowing the stem cells to differentiate. Human embryonic stem cells have not been used clinically in regenerative medicine so far because of ethical hurdles. Autologous bone marrow stem cells and peripheral blood stem cells have been studied for use in regenerative medicine for treating cardiac,[1] neuronal[2,3] and vascular diseases.[4,5] In general, regenerative medicine is expected to be applied as the first step of treatment for organs and tissues in which one cell type plays the main role in the development and functioning of the organs or the tissues.

Generating an organ such as the kidney by using regenerative medicine may be complicated because the basic functional unit of the kidney – a nephron – consists of a glomerulus, the filtration unit containing endothelial, mesangial and epithelial cells; and a tubule, the metabolic and endocrinologic unit that comprises tubular epithelial cells (proximal and distal tubules, loop of Henle and collecting duct). Therefore, an artificial kidney using kidney cells and artificial membranes as the scaffold should be developed as an essential step in kidney regeneration.

This chapter describes the functional evaluation of polymer membranes and tubular epithelial cells as sources of bioartificial tubule devices, and the interaction between them, including active transport of fluids and solutes, scanning electron microscopic findings of the tubular cell-attached artificial membrane, as well as the expression of transporter mRNA of tubular cells on the membranes.

20.2 Present status of kidney regeneration

Regenerative medicine is expected to repair diseased organs and to generate organs using embryonic and bone marrow stem cells and progenitor cells in an adult organ. In the treatment of kidney diseases, bone marrow stem cells were expected to replace affected mesangial cells to improve IgA nephropathy, or to replace damaged tubular epithelial cells in acute tubular necrosis in animal models, despite the hypothesis that bone marrow cells are considered only to be fused with mesangial cells in glomerula or to be fused with tubular epithelial cells in the tubules. Imasawa *et al.*[6] demonstrated that after bone marrow transplantation using bone marrow cells obtained from normal B_6 mice attenuated glomerular lesions in HIGA mice, a murine model of IgA nephropathy. They also demonstrated[7] that after bone marrow transplantation from normal B_6 mice that produced green fluorescence protein (GFP), the GFP-positive cells were confirmed to be located at the site of desmin-positive (mesangial) cells in HIGA mice. Poulsom *et al.*[8] reported that Y-chromosome positive cells were identified at tubular lesions in female mice after bone marrow cells from male mice were transplanted to female mice, and Y-chromosome positive cells were also identified at tubular lesions of renal grafts transplanted from female donors to male recipients in the renal biopsy specimens. Lin *et al.*[9] also confirmed that β-galactosidase-positive cells were located at the site of tubular epithelial cells in the tubular lesions in kidneys with ischaemia-reperfusion injury in female mice after transplantation of haematopoietic bone marrow cells from male Rosa26 mice that express β-galactosidase constitutively, and the male-specific *Sry* gene and Y chromosome were identified in the kidney. Kitamura *et al.*[10] clarified that renal progenitor-like cells obtained from the S3 segment of nephron in rat adult kidneys (rKS56) were differentiated into mature tubular cells defined by aquaporin-1,2 expression in different culture conditions, and these cells responded to parathyroid hormone or vasopressin. Necrotized tubular epithelial cells in the actual site of acute tubular necrosis are presumably replaced by bone marrow stem cells, and/or adult progenitor cells, and bone marrow stem cells might be translocated to the mesangial lesions in the case of proliferative glomerulonephritis including IgA nephropathy. Although it is uncertain whether bone marrow stem cells differentiate into mesangial cells, or only fuse with them, the existence of mesangial cells derived from bone marrow cells appears to ameliorate disease condition of IgA nephropathy.

Embryonic stem cells have the potential to develop a kidney that can be a source for kidney transplantation. However, it is difficult to overcome the ethical hurdles in clinical utilization of human embryonic stem cells. Dekel *et al.*[11] demonstrated that pig or human kidney precursors transplanted in immunosuppressive mouse developed to mouse-size human or porcine kidneys after transplantation of 4-week pig or 8-week human embryos. Although it is difficult to overcome the ethical hurdles, this finding suggests that early human kidney precursors transplanted into renal failure patients, to whom immunosuppressants are administered, develop into adult kidneys in the recipient's abdomen.

Further studies are required to establish kidney regeneration and to ameliorate kidney diseases by treatment with embryonic and bone marrow stem cells. Development of a bioartificial kidney appears to be an essential step to accomplish kidney regeneration.

20.3 Functional limitation of current haemodialysis as an artificial kidney

The kidney functions in maintaining the homeostasis of body fluids by controlling (1) excretion of metabolites, (2) blood volume, (3) electrolyte concentrations, (4) acid–base balance, and (5) metabolic and endocrinologic functions of renal tubules. Patients with chronic kidney diseases[12] in whom the renal function has deteriorated to less than 15% of the norm have to be treated by dialysis or undergo surgical intervention for renal transplantation to prevent the risk of mortality. It has been reported that in 2002 more than 1 million patients underwent dialysis and 200 000 patients underwent transplantation.[13]

In haemodialysis, accumulated metabolites in a patient's blood are removed by diffusion across a semipermeable artificial membrane, and excessive or low densities of electrolytes are normalized by moving to or from dialysate according to Donann's membrane equation. Excessive body fluids are removed under negative pressure added to the dialysate, and the acidic pH of the blood is neutralized by the dialysate containing bicarbonate or sodium acetate. In Japan, patients with renal failure undergo haemodialysis three times per week, and each session lasts for 4 h. Currently, treatment by haemodialysis cannot substitute the normal kidney function because it lasts for only 12 (7.1%) of the 168 h of a week, and some components used in haemodialysis are harmful to patients. Severe complications such as atherosclerosis, dialysis-related amyloidosis and cardiac failure are observed in patients undergoing long-term haemodialysis because of lack of proper kidney function. The current haemodialysis treatment cannot replace the glomerular or tubular function. In order to overcome this problem and prevent dialysis-related complications, intermittent haemodialysis has to be replaced by continuous haemofiltration, and the tubular function has to be included in the new treatment. Hence, development of a bioartificial kidney capable of continuous haemofiltration and having bioartificial tubules is essential, unless functional kidneys are developed from stem cells, or the diseased kidney tissues are healed by bone marrow stem cells or other treatment techniques.

20.4 System configuration for bioartificial kidneys

It has been reported that continuous haemofiltration (10 l/day) facilitated the maintenance of the patient's plasma urea, creatinine, uric acid, inorganic phosphate and β_2-microglobulin at levels lower than those of the same eight patients on conventional haemodialysis.[14–16] At the first step of treatment using bioartificial kidneys, it was intended to develop the system in conjunction with

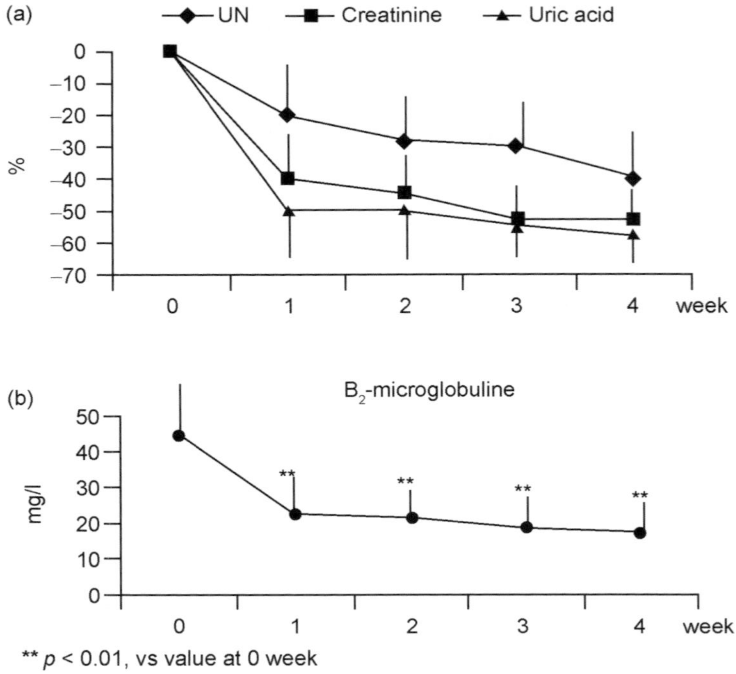

20.1 Changes in plasma urea nitrogen (UN), creatinine (Cr), and uric acid (UA) concentrations for 4 weeks in eight stable haemodialyzed patients treated with haemofiltration with 10 l/day of ultrafiltrate. The zero in the figure shows the predialysis plasma values of these compounds. The concentrations of UN, Cr and UA were maintained at −60 to −40% of the predialysis values. Modified from *Nephrology*, **10** (Suppl.3), Saito, A. Development of bioartificial kidneys, S10–S15, 2003, with permission from Blackwell Publishing.

continuous haemofiltration using 10 l/day of filtrate (7 ml/min) (Fig. 20.1)[15] and a bioartificial tubule device that uses proximal tubular epithelial cells and porous membrane hollow fibre modules to reabsorb 6 l/day (4 ml/min) and discard 4 l/day (3 ml/min).

The basic functional unit of the kidney – a nephron – comprises vascular and tubular portions, and the two function in transporting biological substances via numerous transporters and channels located at the plasma membrane of tubular epithelial and vascular endothelial cells. In a nephron, an efferent capillary that arises from a glomerulus parallels a proximal tubule, Henle's loop, a distal tubule and a collecting duct, and actively transports biologically useful substances between the capillaries and the tubules. A bioartificial kidney should mimic the structure and the function of the nephron. Therefore, a continuous haemofilter was used as a glomerulus, and a tubular cell-attached

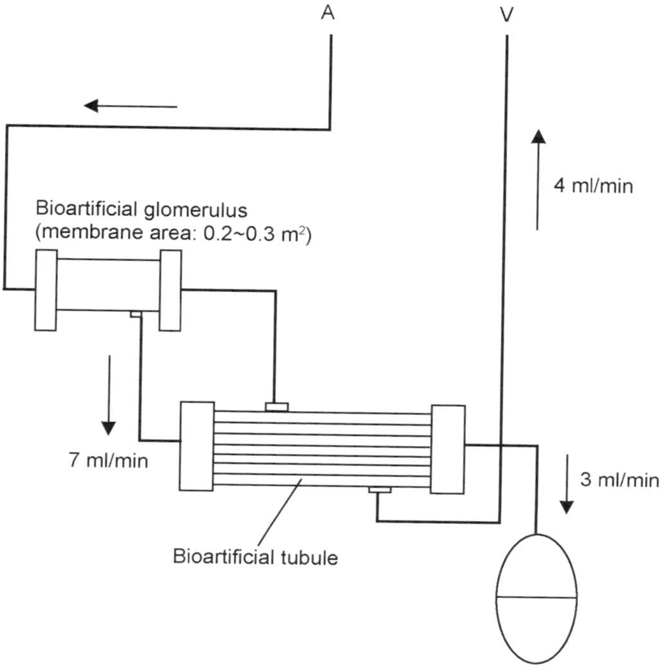

20.2 Flow diagram of a bioartificial kidney which consists of a continuous haemofilter and a bioartificial tubule device using a hollow fibre module and proximal tubular epithelial cells. Ultrafiltrate obtained from a patient with the flow of 7 ml/min is eluted inside the hollow fibres of the bioartificial tubule device. The filtrate is transported to outside the hollow fibres across the cell-attached hollow fibres with the flow of 4 ml/min, and the remaining filtrate is discarded with flow of 3 ml/min. Reprinted from *Artif Organs*, **28**, Saito, A. Research into the development of a wearable bioartificial kidney with a continuous hemofilter and a bioartificial tubule device using tubular epithelial cells, 58–63 (2004), with permission from Blackwell Publishing.

hollow fibre module functions as a bioartificial tubule device. The ultrafiltrate obtained from the patient's blood was eluted inside the hollow fibre capillaries and the blood that went through the haemofilter was eluted outside the tubular epithelial cell-attached capillaries in the bioartificial tubule device (see Fig. 20.2).[16] These two parts are connected by blood and filtrate tubings, and roller pumps control the blood and haemofiltration volumes. Humes *et al.*[17] and Ozgen *et al.*[18] have confirmed that this system functioned as a nephron that could transport water, sodium and glucose and operate like renal tubules *in vitro*. Moreover, Weitzel *et al.*[19] succeeded in treating acute renal failure in patients with multiple organ dysfunction in 2002.

The author[16] intended to develop a wearable type of bioartificial kidney for chronic renal failure patients that could function for prolonged periods and was compact. An artificial kidney that can work for more than a week without using systemic anticoagulation and weighs less than 1 kg is now under development.

20.5 Past and current status of development of bioartificial kidneys

While developing a bioartificial kidney, constructions of both – a bioartificial glomerulus device and a bioartificial tubule device – should be considered. A bioartificial glomerulus in which confluent endothelial cells are attached to polymer membranes has never been successfully developed, because a sufficient amount of ultrafiltrate could not be obtained with the endothelial cell-attached porous membranes. Therefore, researchers working on bioartificial kidneys initially focused on the development of bioartificial tubule devices.

In 1987, Aebischer *et al.*[20,21] first investigated the ability of tubular epithelial cell attachment on the outer surface of semipermeable hollow fibre capillary membranes to function as a bioartificial tubule. They also evaluated the transport of water, phenol red and several biological substances across the Madin–Darby canine kidney (MDCK) and Lewis lung cancer porcine kidney (LLC-PK_1) cell-attached membranes.[22] Humes *et al.*[17] developed renal assist devices (RAD) in which proximal tubular epithelial cells formed confluent monolayers on the inner surfaces of polysulphone hollow fibre membranes. Further, they evaluated glucose transport, bicarbonate production, ammonia excretion, 1,25, hydroxyvitamin D production, etc. by the proximal tubular epithelial cells, (LLC-PK_1 cells and human primary tubular epithelial cells present on the polymer membranes). Humes *et al.* succeeded in scaling up the human proximal tubular epithelial cell-attached membrane area of the RAD to 0.8 m^2, and they treated renal failure dogs by using the RAD for 24 h under an anaesthesia.[23] Survival of the dogs was shown to be prolonged after the RAD treatment. They also treated 10 patients with acute renal failure admitted to the ICU by using continuous haemofiltration and RAD for 24 h; six patients survived for 28 days after the RAD treatments.[24] Although the reason why good prognosis was observed in these patients treated with RAD using tubular epithelial cells has not been completely clarified, plasma levels of G-CSF, interleukin (IL)-6, IL-10, and the ratio of IL-6/IL-10 ratio was confirmed to have decreased after the treatment.[24]

To determine whether the RAD alters patient mortality, a phase II randomized, open label study was carried out in 58 ICU patients with dialysis-dependent acute renal failure in 2005. After 6 h of continuous venous–venous haemofiltration (CVVH), the patients were randomized (2:1) to receive CVVH with or without the RAD for 72 h.[25] The primary end point compared 28-day all-cause mortality in patients receiving conventional CVVH with those receiving CVVH and RAD. Forty patients were randomized to receive RAD therapy. The 28-day mortality rate of patients receiving any duration of RAD therapy was

Table 20.1 Summary of cell type and membrane materials used for each basic research in the development for bioartificial tubule devices

Text of section	First author	Cell type used	Membrane material	Reference number
20.4	Humes HD	LLC-PK_1 cells	Polysulphone	17
	Ozgen N	LLC-PK_1 cells	Polysulphone	18
	Humes HD	human primary tubular epithelial cells	Polysulphone	17, 23, 24
20.5	Aebischer P	MDCK, LLC-PK_1 cells	Polysulphone Polyvinyl chloride acrylic copolymer	20
	Saito A	MDCK cells	Cellulose diacetate	26
	Fujita Y	MDCK cells	Polycarbonate	27
	Terashima M	MDCK, LLC-PK_1 and JTC-12 cells	Polycarbonate	28
20.6.1	Kanai	MDCK, KU-2 cells	Polystyrene	30
	Sato Y	LLC-PK_1 cells	Cellulose acetate, polysulphone	29
20.6.2	Saito A	HK-2 cells	Polysulphone, polyimde, and EVAL	33
20.7	Saito A	MDCK, JTC-12 cells LLC-PK_1 cells	Polysulphone, cellulose acetate polycarbonate	16
	Asano M	JTC-12, MDCK, and BALB3T3 cells	Polycarbonate	34
20.9	Fujita Y	LLC-PK_1 cells	Nucleopore polycarbonate	37

34.3% compared with 55.6% for patients receiving only CVVH. These results suggest that treatment with RAD using human tubular epithelial cells for 24–72 h improves the 28-day mortality rate in ICU patients with acute renal failure.

To prevent complications associated with dialysis in long-term dialysis patients, a bioartificial tubule device that has tubular metabolic function needs to be developed. The bioartificial kidneys for chronic renal failure patients have to function for long durations to lower the treatment costs because the number of chronic renal failure patients is remarkably high, and the treatment duration of these patients is longer than those of acute renal failure patients. The regeneration of peritoneal dialysis effluent by MDCK cell-lined cellulose acetate hollow fibre membrane modules (see Fig. 20.3) under hydraulic pressure,[26] and the electrophysiological capacity of the cell-attached membrane by using the Ussing chamber system[27] were investigated first. The transport ability of water and transepithelial potential, the short circuit current and conductance with MDCK, LLC-PK_1 and JTC-12 cells under conditions with and without osmotic and/or hydraulic pressure and osmotic and hydraulic pressure overload were then investigated.[28] Then, the efficiency of polysulphone and cellulose acetate

Table 20.2 Summary of *in vivo* experiments using animal models in development of bioartificial kidneys

Text of section	First author	Cell type used	Animal model used	Reference number
20.5	Humes HD	Human primary tubular cells	Renal failure dogs	23
	Humes HD	Human primary tunular cells	Acute renal failure patients	17, 24
	Weitzel WF	Human primary tubular cells	Acute renal failure patients	19
	Tumlin J	Human primary tubular cells	Acute renal failure patients	25

membranes was evaluated for use as porous polymer membranes, and the effectiveness of Pronectin-F, collagen type I and IV for use as coating matrices that had to be applied for developing bioartificial tubule devices.[29]

At the first step, bioartificial tubule devices will be utilized in conjunction with intermittent dialysis sessions three times a week. Subsequently, the treatment system might be replaced by a continuous portable type system.

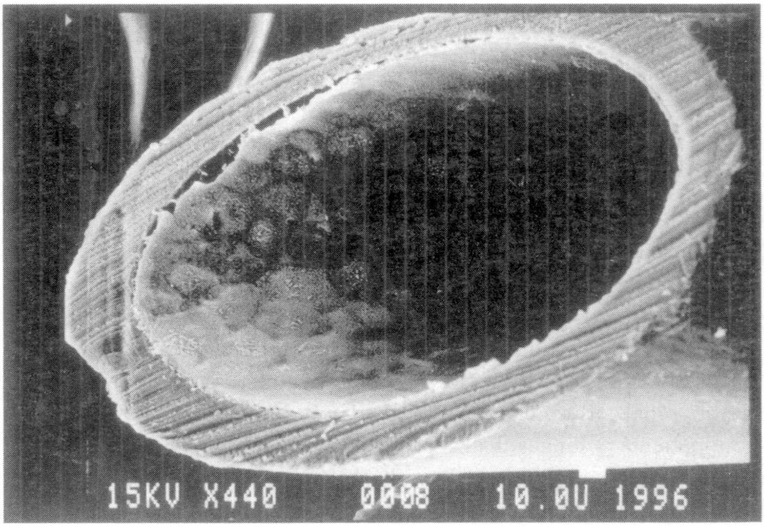

20.3 Cross-sectional finding of scanning electron microscopy of MDCK cell-attached on the inner surface of cellulose acetate hollow fibre (×440). The MDCK cells formed a confluent monolayer on the inner surface of the membrane. Reprinted from *Mater Sci & Eng C*, **6**, Saito, A, Suzuki, H, Bomsztyk, K, Ahmad, S. Regeneration of peritoneal effluent by Madin–Darby canine kidney cells-lined hollow fibers, 221–226 (1998), with permission from Elsevier Science.

Continuous haemofilters and bioartificial tubule devices might be able to function for a longer periods. The long-term function of a bioartificial tubule device was evaluated by using LLC-PK_1 cells and a polysulphone hollow fibre module, and the efficacy of water, sodium and glucose reabsorption across the LLC-PK_1 cell-attached hollow fibre membrane.[18] Water, glucose and sodium reabsorption capacity of the devices were maintained at a certain level for 2 weeks, and it declined thereafter because of excessive growth of tubular epithelial cells on the devices.

20.6 Attachment and proliferation of tubular epithelial cells on polymer membranes

In order to develop a bioartificial tubule device by using renal tubular epithelial cells, knowledge about the functionally differentiated condition of cells is extremely important. Therefore, we first evaluated the characteristics and interactions of some polymer membranes, several tubular epithelial cells and extracellular matrices as the sources of bioartificial tubule device.

20.6.1 Tubular epithelial cell attachment and proliferation on extracellular matrix-coated polymer membranes

Kanai *et al.*[30] investigated the effects of various extracellular matrices (ECM) on MDCK cell- and human renal carcinoma cell lines along with effects of KU-2 cell attachment to surfaces of polymers, tissue culture dishes, 96 well plates (Iwaki Glass, Tokyo, Japan). MDCK cells and KU-2 cells were seeded onto 96 well plates that were precoated with collagen types I and IV, laminin and fibronectin. The degree of cell attachment onto the ECM were measured by the 3-(4,5-dimethyl-2-thiazolyl)-2,5-diphenyl-2H-tetrazolium bromide (MTT) assay. Greater MDCK cell attachment was observed between 60 min and 24 h after cell seeding in 5 μg/ml of laminin, and 60 min and longer or and at 30 min and longer at fibronectin concentrations of 1 and 4 μg/ml or 16 and 32 μg/ml, respectively. On the other hand, better KU-2 cell attachment was observed between 15 and 60 min, or 24 h after seeding in 40 μg/ml laminin. These data suggest that cell types probably differ with regard to the most suitable ECM, the best concentration and the best incubation time.

Sato *et al.*[29] studied the attachment and the proliferation of porcine proximal tubular epithelial cells – LLC-PK_1 – on cellulose diacetate, polysulphone membranes and polystyrene as controls with and without ECM coating in order to develop a bioartificial tubule device by using proximal tubular epithelial cells and porous polymer membranes. Two types of porous filter membranes that were prepared using polysulphone (membrane pore size, 0.22 μm; thickness, 150–200 μm; Millipore, Bedford, MA) or cellulose acetate (Microfilter FR 40; membrane pore size, 0.4 μm; thickness, 40 μm; Fuji Film, Tokyo, Japan) were used in the experimental devices, and a nonporous plate prepared from polystyrene (multiple-well plate for tissue culture) (Iwaki; Asahi Techno Glass, Chiba, Japan) was used as controls in the study. As coating materials, extracellular matrices, Pronectin-F (ProNectin F kit; Iwaki), collagen type I (Cellmatrix-I-P; Nitta Gelatin, Osaka, Japan) or collagen type IV (Cellmatrix-IV; Nitta Gelatin) was used. LLC-PK_1 cells (4×10^4) were seeded onto polysulphone or cellulose acetate membranes (surface area, $4\,cm^2$) with or without the extracellular matrix and cultured in 0.5 ml of DMEM. Twenty-four hours to seven days after cell seeding, the number of cells on the membranes was counted using a cell-counting kit. The changes in number of cells on (b) the polysulphone membrane and (a) the cellulose acetate membrane 24 h to 7 days after cell seeding are shown in Fig. 20.4.[31] Cell proliferation was significantly higher on the polysulphone membrane but not on matrix-coated cellulose acetate membrane within 1 week. Cell proliferation continued was enhanced by coating the membranes with extracellular matrices such as Pronectin F, collagen I and collagen IV. Confluent monolayer formation of cells was observed on matrix-coated polysulphone membrane but not on matrix-coated cellulose acetate membrane within 1 week. Cell number on polystyrene plate was significantly higher than that on the uncoated polysulphone membrane for time periods greater than 24 h ($p < 0.05$).

20.6.2 Tubular epithelial cell attachment and proliferation on polymer membrane without coating

ECM coating is not only a time-consuming and expensive procedure, but is also not a safe method because of the risk of viral contamination if the matrices are obtained from animals such as a pig or a cow. If tubular epithelial cells can attach and proliferate directly on the polymer membranes, the procedure would be safe and economical. The author has studied the interaction between tubular cells and several polymer membranes in order to confirm the characteristics of the membranes that are to be used for direct attachment of tubular epithelial cells.[31] Among them, polyimide, ethylene vinyl alcohol copolymer (EVAL) membranes, and polysulphone membrane as a control, were investigated to evaluate the tubular cell attachment and the proliferation without matrix coating. Attachment and the proliferation of HK-2 cells[32] on non-coated polyamide (Minntech Co., Mineapolis) and ethylene vinyl alcohol copolymer (EVAL; Kurarey Co., Tokyo, Japan) membranes that have hydrophobic and hydrophilic structures on the surfaces were compared with those on polysulphone membrane. HK-2 cells at density of 10^5 cells/ml were seeded in the polysulphone, polyimide and EVAL hollow fibre mini-modules for five each of three different modules, and the modules were incubated for 6 days. Subsequently, the modules were cut and five hollow fibres for each module were examined using scanning electron microscopy. The scanning electron microscopy revealed confluent

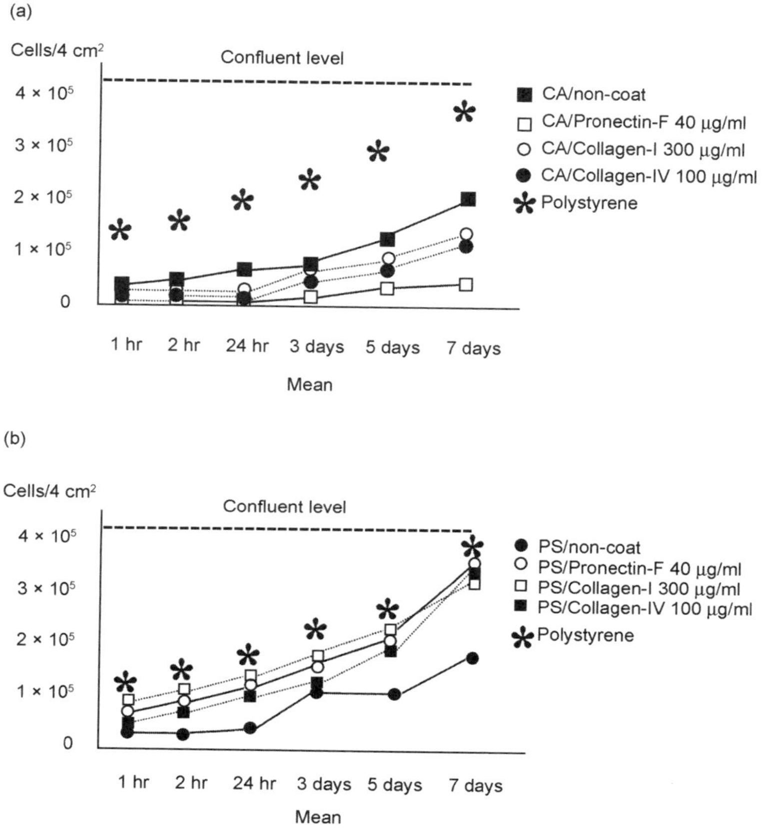

20.4 Attachment and proliferation of LLC-PK_1 cells on cellulose acetate (a) and polysulphone (b) membranes with coating of several extracellular matrices, and polystyrene plates for culture dish as the control were compared for 7 days. The cell proliferation on polysulphone membrane was significantly higher than on the coated cellulose acetate membrane, and was enhanced by coating the membranes with various extracellular matrices. Confluent monolayer formation of the cells was observed on matrix-coated polysulphone membrane but not on matrix-coated cellulose acetate membrane within 1 week. Reprinted from *Therapeutic Apheresis and Dialysis*, **10**, Saito, A, Aung, T, Sekiguchi, K, Sato, Y. Present status and perspective of the development of a bioartificial kidney for chronic renal failure patients, 342–347 (2006), with permission from Blackwell Publishing.

layers of HK-2 on the polyimide and EVAL membranes at 6 days after the seeding, while the cells did not form a confluent layer on the polysulphone membrane (Fig. 20.5).[33] This finding suggests that porous polymer membranes with hydrophobic and hydrophilic structures on the surfaces are suitable for direct attachment of tubular epithelial cells that can be used for the development of bioartificial tubule device. Further investigation is required for the functional evaluation of the treatment with regard to its long-term use.

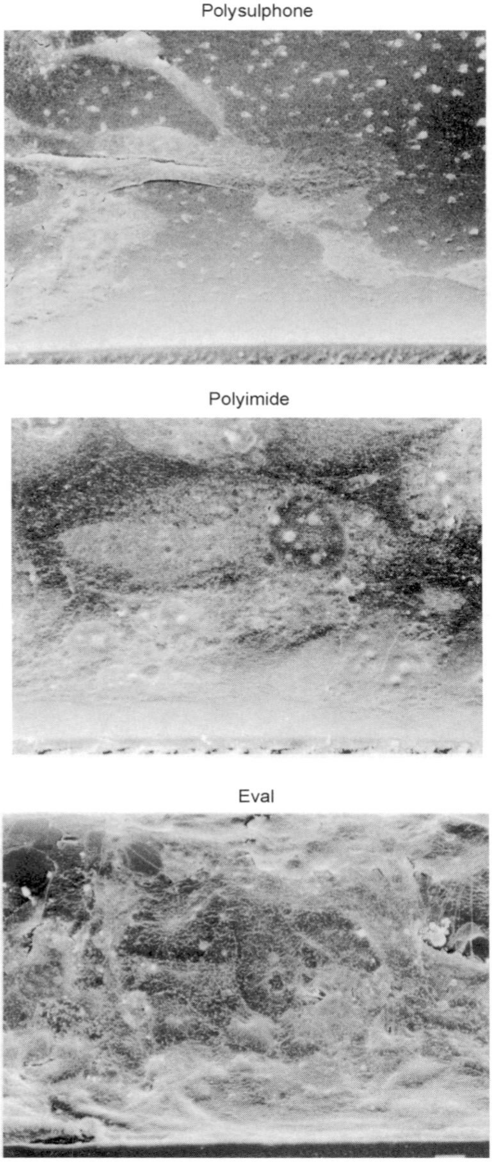

20.5 HK-2 cells, human proximal tubular epithelial cell line, seeded inside of hollow fibre modules of polysulphone, polyimide and ethylene vinyl alcohol copolymer (EVAL) membranes without coating were evaluated the formation of confluent layers 6 days after incubation. As the results, HK-2 cells formed the confluent monolayers on the polyimide and the EVAL membranes while the cells did not form the confluent layer on polysulphone membrane. Reprinted from *Journal of Artificial Organs*, **9** Saito, A. *et al.*, 130–135 (2006), with permission from Springer.

20.7 Function of tubular epithelial cells on polymer membranes

Messenger RNA (mRNA) expression and presence of physiologically active proteins,[29] production of certain metabolites, activation of vitamin D in proximal tubular epithelial cells[17] and incorporation of pentosidine, a moiety of advanced glycation end products (AGEs), into the proximal tubular epithelial cells,[34] etc. on polymer membranes were examined by *in vitro* studies.

Expression of sodium glucose cotransporter-1 (sGLT-1) mRNA at the apical side and facilitated glucose transporter-1 (GLUT-1) mRNA at the basal side of membranes of tubular epithelial cell-layers on polymer membranes were also investigated by Sato *et al.*[29] Expression of sGLT-1 and GLUT-1 mRNA in LLC-PK_1 cell layers on cellulose acetate and polysulphone membranes were examined by reverse transcription-polymerase chain reaction (RT-PCR) 1, 2 and 3 weeks after formation of the confluent monolayers (Fig. 20.6). The sGLT-1 and GLUT-1 mRNA expressions were detected on both cellulose acetate and polysulphone membranes and were maintained for 3 weeks.

The existence of Na^+/K^+ ATPase protein in Madin–Darby canine kidney (MDCK)[27] and JTC-12[16] cell layers on polymer membrane were detected using anti-Na^+/K^+ ATPase antibody by confocal laser scanning microscopy, and Na^+/K^+ ATPase protein was observed to be located at the lateral and basal sides of the plasma membrane in those cells.

Uraemia plasma-containing considerable amounts of pentosidine was added to the apical portion of the medium in which JTC-12, MDCK and BALB3T3 cells (mouse embryonic fibroblasts) were cultured. Subsequently, pentosidine plasma levels were detected by high-performance liquid chromatography (HPLC) using a fluorescence detector at excitation/emission wavelengths of 335/385 nm over time for the culture, and the incorporation of pentosidine into those cells was examined by histochemical staining with antipentosidine rabbit IgG[34]. Pentosidine was incorporated only into the proximal tubular epithelial cells – JTC-12 – 8 h after the addition of uraemic plasma in the apical side of the medium. Pentosidine was not incorporated into any of the cells when it was added in the basal side.

Humes *et al.*[17] demonstrated that 1,25 hydroxy vitamin D production of LLC-PK_1 cell layers was significantly stimulated in the medium containing the parathyroid hormone as compared with that in medium without the parathyroid hormone and was significantly stimulated in the medium without inorganic phosphate as compared with that in the medium with 3.0 mg/dl of inorganic phosphate. Glucose, bicarbonate and ammonia productions of LLC-PK_1 cell layers on the polymer membrane were also confirmed.

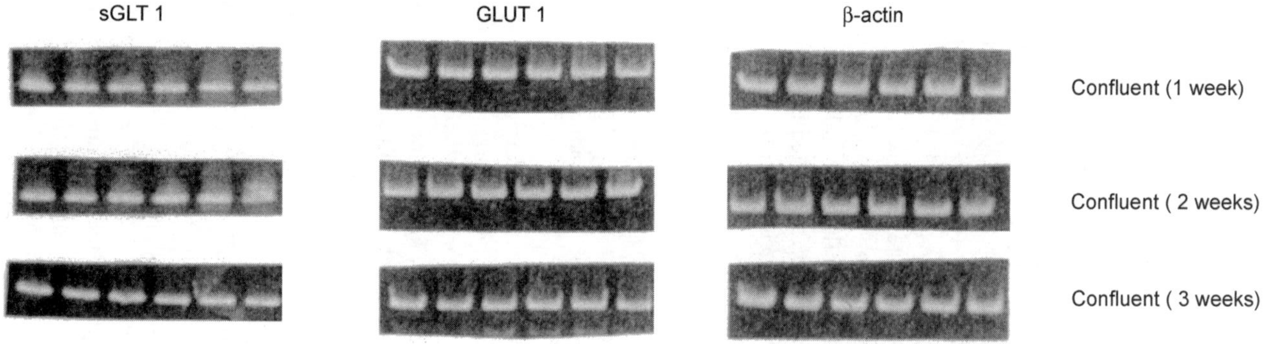

20.6 Expression of sodium-glucose cotransporter-1 (sGLT-1) and facilitated glucose transporter-1 (GLUT-1) mRNA on polysulphone and cellulose acetate membranes was detected by real-time polymerase chain reaction (RT-PCR) during 3 weeks after formation of confluent monolayers. The sGLT-1 and the GLUT-1 mRNA were expressed on both polysulphone and cellulose acetate membranes for 3 weeks after formation of confluent layers.

20.8 Evaluation of a long-term function of LLC-PK_1 cell-attached hollow fibre membrane

Ozgen *et al.*[18] investigated the long-term effectiveness of active transport in proximal tubular epithelial cell-attached polysulphone hollow fibre modules (Nipro Co., Tokyo, Japan) that had an area of 0.4 m^2. Prior to seeding LLC-PK_1 cells, the inner surface of hollow fibres was coated with Pronectin-F. After $10^7\,ml^{-1}$ of LLC-PK_1 cells was seeded in the hollow fibres four times while the module was rotated by 90° in 1 h interval, the modules were incubated for 14 days with O_2 and CO_2 supplementation detecting medium O_2, CO_2 contents and pH (Fig. 20.7). Two types of medium were used, one was eluted inside the hollow fibres containing 50 mg/dl of urea and 5.0 mg/dl of creatinine, and the other was eluted outside the hollow fibres with and without 2.5 g/dl of albumin. The leak rates of urea and creatinine, and transport rates of water, glucose and sodium for 90 min were calculated inside and outside the medium. Further, the concentrations of urea, creatinine, glucose and sodium in both the media were calculated.

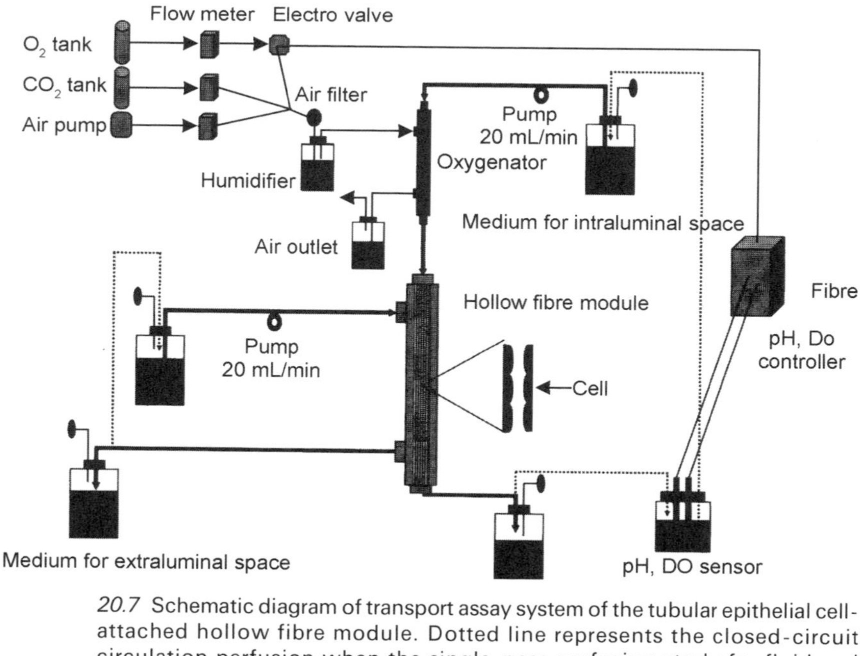

20.7 Schematic diagram of transport assay system of the tubular epithelial cell-attached hollow fibre module. Dotted line represents the closed-circuit circulation perfusion when the single-pass perfusion study for fluid and substances transport was not performed. Reprinted from *Nephrology Dialysis Transplantation*, **19**, Ozgen, N, Terashima, M, Aung, T, Sato, Y, Isoe, C, Kakuta, T, Saito, A. Evaluation of long-term transport ability of a bioartificial tubule device using LLC-PK_1 cells, 2198–2207 (2004), with permission from Oxford University Press.

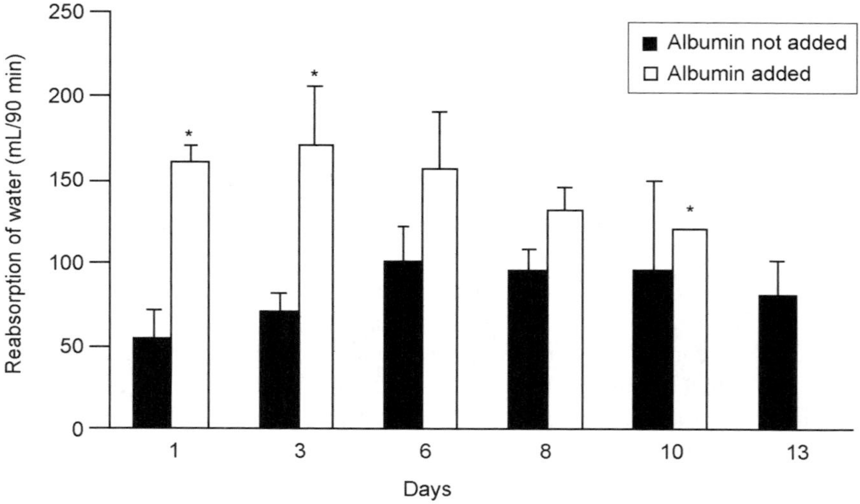

20.8 Reabsorption of water in the tubule device (membrane area: 0.4 m^2) with or without 2.5 g/dl of albumin addition. $^*p < 0.05$ vs. albumin not added. Reprinted from *Nephrology Dialysis Transplantation*, **19**, Ozgen, N, Terashima, M, Aung, T, Sato, Y, Isoe, C, Kakuta, T, Saito, A. Evaluation of long-term transport ability of a bioartificial tubule device using LLC-PK_1 cells, 2198–2207 (2004), with permission from Oxford University Press.

The leak rates of urea and creatinine from the inside the media to the outer environment were ~10%. The transport rates of water and glucose for 90 min are shown in Figs 20.8 and 20.9. The water transport without addition of albumin gradually increased until 8 days after the cell seeding and gradually decreased thereafter. With albumin addition the rate was significantly higher than that without albumin addition 8 days after the cell seeding, and then the transport rate gradually decreased (Fig. 20.8). The sodium transport pattern was similar to the water transport under the conditions with and without albumin addition. The glucose transport was significantly decreased when phlorizin, the inhibitor of sodium/glucose cotransporter (sGLT), was added in the medium inside the hollow fibres (Fig. 20.10). Glucose transport in the cell-attached hollow fibres was considered to be suppressed by the inhibitor of sodium/glucose co-transporter. Scanning electron microscopic findings of the cell-attached hollow fibre at 4, 8 and 13 days are shown in Fig. 20.11. Although LLC-PK_1 cells grew confluently at 4 days and at a high density on a polysulphone hollow fibre membrane, the dense cell layer with uneven cell mass was observed at 13 days after the cell seeding. The tissues comprising kidney tubular epithelial cells are proliferated in a contact-inhibited manner. However, tubular epithelial cells isolated from the kidney tissue seem to lose their original characteristic of contact inhibition. The transport ability of the cell-attached hollow fibres was considered to have decreased significantly 2 weeks after the formation of the confluent layer

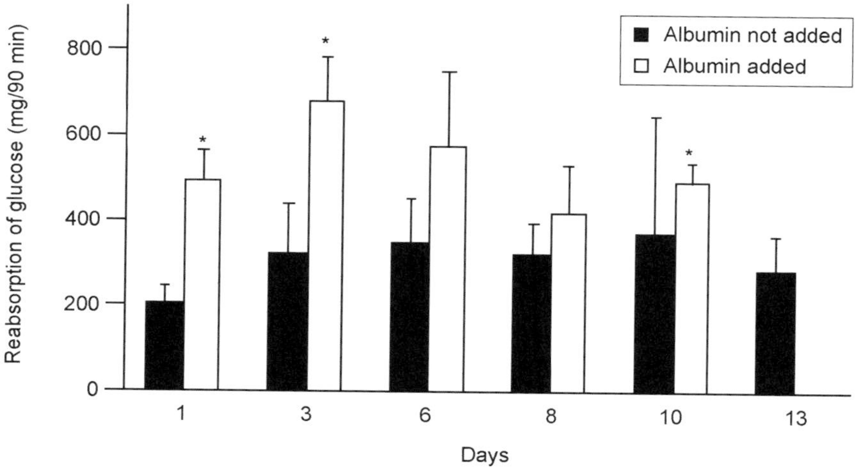

20.9 Reabsorption of glucose in the tubule device (membrane area: 0.4 m^2 with or without 2.5 g/dl of albumin addition. $^*p < 0.05$ vs. albumin added.

due to the multilayer formation. Human primary tubular epithelial cells (RPTEC) were also confirmed to form a multilayer with uneven cell mass. A technique for maintaining a monolayer for long durations must be developed in order to develop bioartificial kidneys for chronic renal failure patients.

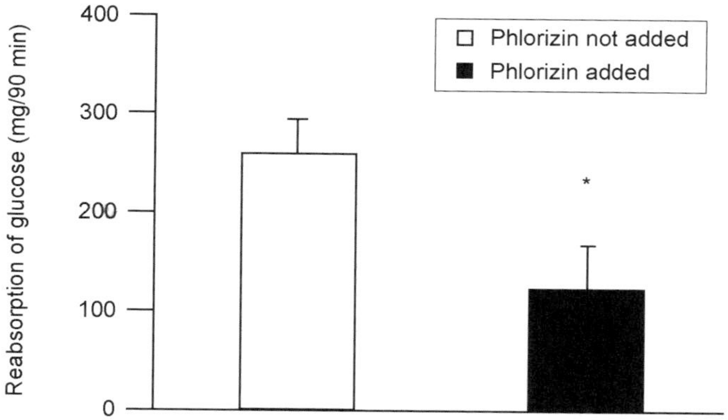

20.10 Comparison of glucose reabsorption amounts from the inner compartment to the outer compartment across the tubular cell-attached hollow fibre modules with or without addition of phlorizin, the sodium dependent glucose transporter inhibitor, in the medium perfusing the inner compartment. Glucose reabsorption was significantly decreased after phlorizin was administrated into the medium perfusing the inner compartment. Reprinted from *Nephrology Dialysis Transplantation*, **19**, Ozgen, N, Terashima, M, Aung, T, Sato, Y, Isoe, C, Kakuta, T, Saito, A. Evaluation of long-term transport ability of a bioartificial tubule device using LLC-PK_1 cells, 2198–2207 (2004), with permission from Oxford University Press.

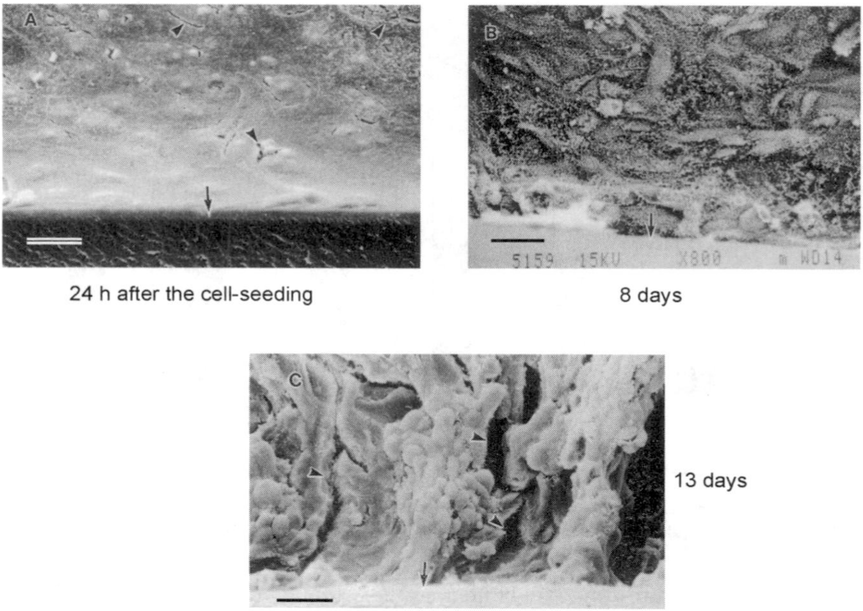

20.11 Scanning electron micrographs of LLC-PK_1 cell-attached hollow fibres retrieved from the modules with 0.4 m^2 of membrane area 4 days, 8 days and 13 days after the cell seeding. Arrows indicate the hollow fibres, Scale bar = 20 μm. Reprinted from *Nephrology Dialysis Transplantation*, **19**, Ozgen, N, Terashima, M, Aung, T, Sato, Y, Isoe, C, Kakuta, T, Saito, A. Evaluation of long-term transport ability of a bioartificial tubule device using LLC-PK_1 cells, 2198–2207 (2004), with permission from Oxford University Press.

20.9 Improvement of the components of a portable bioartificial kidney developed for long-term use

In our previous studies proximal tubular epithelial cell lines such as MDCK cells and LLC-PK_1 cells could never be maintained in confluent monolayers for 2 weeks due to formation of uneven cell mass. Although researchers believed that primary proximal tubular epithelial cells can maintain contact inhibition, and permanently maintain a confluent monolayer, the formation of multilayers of primary human proximal tubular epithelial cells within two weeks of incubation was confirmed (A. Saito, unpublished data). It is considered that tubular epithelial cells isolated from kidney tissues lose their original characteristic of contact inhibition and/or that tubular epithelial cells might change their original characteristics after long-term incubation under unfavourable conditions. We need to develop a new technology by which proximal tubular epithelial cells can maintain a confluent monolayer for a long time by maintaining contact inhibition. Novel porous polymer membranes, or suitable combinations with extracellular matrices, on which tubular epithelial cells can maintain contact inhibition should be identified. Further, appropriate gene transfection into tubular epithelial cells following which the cells can maintain the contact inhibition property should be developed.

For a portable bioartificial kidney, a continuous haemofilter that can function for at least 1 week with minimum anticoagulants is essential; however, continuous haemofilters used currently are required to function only for 24 h under conditions of sufficient systemic anticoagulation. At the first step of generating our bioartificial kidney, an antithrombogenic haemofilter with inner surfaces of hollow fibre capillaries were modified with methacryloyloxyethyl phosphorylcholine (MPC) polymer which was invented by Ueda *et al.*[35,36] This polymer mimics the phospholipid of human cellular membrane can be used for development of artificial glomerulus. It is well known that sufficient glomerular filtration (approximately 150 l/day) is obtained in normal functioning kidneys despite the fact that glomerular basement membrane is covered with endothelial cells. Better glomerular filtration occurs across the glomerular endothelial cells that have larger fenestra than those with other types of endothelial cells. In the near future an endothelial cell-attached haemofilter is expected to be used. A considerable amount of ultrafiltrate might be obtained by using this haemofilter by controlling the fenestra diameter of the endothelial cell membranes and in the absence of any systemic anticoagulation. Heparin-coated blood tubing is used because the entire system would be functional at least for a month.

In order to make a compact bioartificial tubule device, proximal tubular cells can be made functional in which functional protein genes such as a water channel protein and electrolyte pump genes are transfected. The transcellular water transport ability of $LLC\text{-}PK_1$ cells into which rat kidney aquaporin-1 (AQP1) was stably transfected was studied.[31,37] The cells were seeded (1.0×10^5 cells/cm^2) onto Nuclepore polycarbonate membranes at the bottom of Transwell chambers (Corning Life Science, New York). The expression and localization of AQP1 were examined by Western blotting, reverse transcriptase polymer chain reaction (RT-PCR) and immunofluorescence. To measure transcellular water permeation, a simple method was applied, wherein phenol red was used as a cell-impermeant marker of concentration. Rat AQP1-transfected $LLC\text{-}PK_1$ cells had high transcellular osmotic water permeability as compared to wild type $LLC\text{-}PK_1$ cells (Fig. 20.12).[31] The expression of rat AQP1 mRNA and protein bands was confirmed to be stably maintained until a population doubling level of 24. In AQP1-transfected $LLC\text{-}PK_1$ cells, the protein was localized not only on the basolateral side but also on the apical side of the plasma membrane; however, wild-type $LLC\text{-}PK_1$ cells were not stained (Fig. 20.13).[31] It is possible that sufficient AQP1-transfected tubular epithelial cells were supplied for the compaction of the bioartificial renal tubule device.

The treatment with bioartificial tubules requires large amounts of tubular epithelial cells so that it can be widely used for treating renal failure patients worldwide. New technologies for producing a large number of human primary

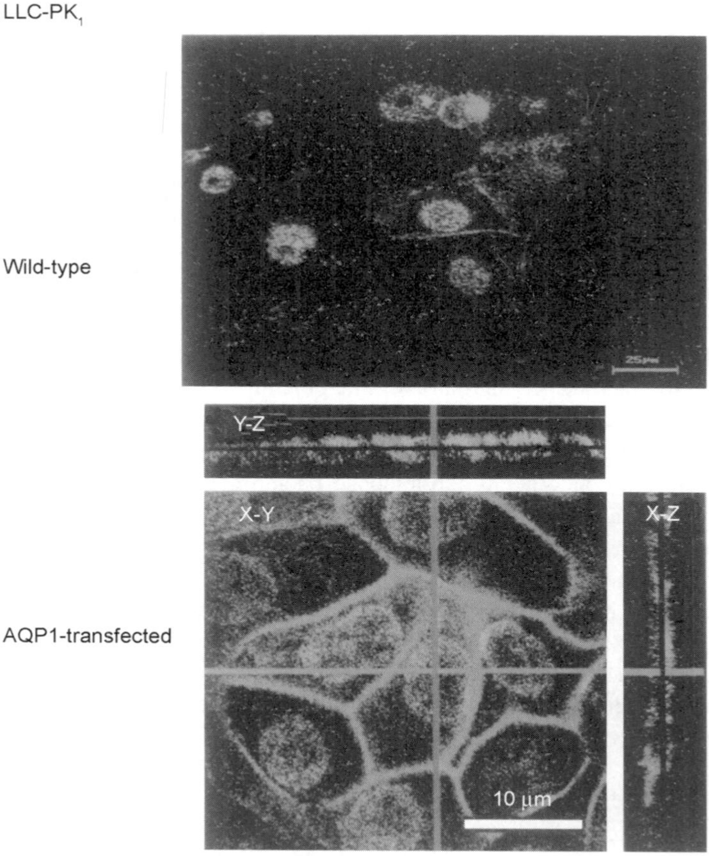

20.12 Immunofluorescence localization of rat aquaporin-1 (AQP1) protein in wild-type (upper column) and AQP1-transfected LLC-PK$_1$ cells using confocal laser scanning microscopy. AQP1 was located mainly at plasma membrane of AQP1-transfected LLC-PK$_1$ cells. Porcine AQP1 was little cross-reacted with rat AQP1 in the wild-type cells. Reprinted from *Therapeutic Apheresis and Dialysis*, **10**, Saito, A, Aung, T, Sekiguchi, K, Sato, Y, Present status and perspective of the development of a bioartificial kidney for chronic renal failure patients, 342–347 (2006), with permission from Blackwell Publishing.

proximal tubular epithelial cells from small number of the cells might be useful for renal failure patients worldwide. Kowolik *et al.*[38] developed new technology for reversing the immortalization of human primary renal proximal tubular epithelial cells (RPTEC: ATCC), using gene cassettes of human telomerase and SV40 T antigen flanked by *lox P* genes. When Cre recombinase is activated after large number of the cells has been obtained, the two gene cassettes are removed.

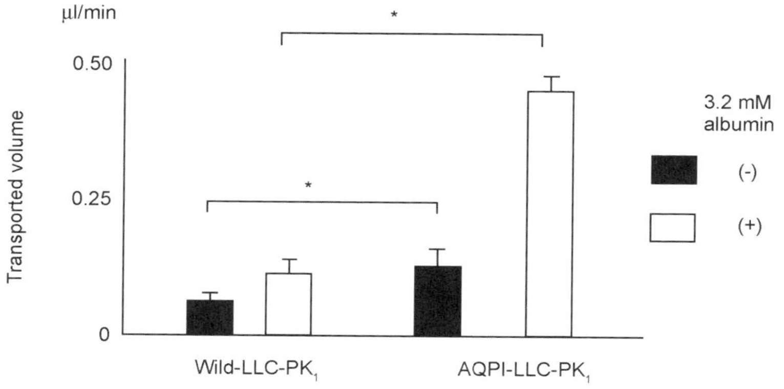

20.13 Comparison of water transport from the apical side to the basal side across the wild-type LLC-PK$_1$ cells and AQP1-transfected LLC-PK$_1$ cells under without or with (open bar) addition of 3.2 mmol/L albumin. Water transport across AQP1-transfected LLC-PK$_1$ was significantly higher than that across wild-type LLC-PK$_1$ cells under with or without addition of albumin. Reprinted from *Therapeutic Apheresis and Dialysis*, **10**, Saito, A, Aung, T, Sekiguchi, K, Sato, Y, Present status and perspective of the development of a bioartificial kidney for chronic renal failure patients, 342–347 (2006), with permission from Blackwell Publishing.

Since this method might be dangerous because of the use of a lenti virus vector, a safer technique needs to be developed.

A portable roller pump has not yet been developed, and it might be difficult for a chronic renal failure patient to use it continuously because the lightest pump weighs 1.9 kg, and most of its components are made of iron. A portable pump using the minimum number of iron parts needs to be developed for patients requiring portable treatment systems.

20.10 Conclusions and future trends

Bioartificial kidneys which comprise an antithrombogenic continuous haemofilter and a bioartificial tubule device with tubular epithelial cells and a hollow fibre module have been studied for two decades. The technique forming confluent monolayers on the semipermeable hollow fibre membrane has been established by Humes *et al.* and the author's group after investigation of the interaction between artificial membranes and several types of tubular epithelial cells. The metabolism and active transport of useful substances in/across the proximal tubular epithelial cell layers also were investigated. Transport of water, sodium and glucose, production of bicarbonate and ammonia, and activation of vitamin D were confirmed to exist in the proximal tubular cells on artificial membranes. Clinical trial evaluating the efficacy of bioartificial tubule devices on the treatment of acute renal failure in patients in ICU revealed the improved mortality rate 28 days after treatment. The bioartificial tubules for treatment of acute renal failure patients with multiple organ dysfunction is considered to be clinically available within several years in the US.

In the development of a bioartificial kidney for maintenance dialysis patients, the techniques for an antithrombogenic continuous haemofilter as an artificial glomerulus and a bioartificial tubule device have never been established. For the treatment of chronic renal failure patients, not only a bioartificial tubule device but also a continuous haemofilter should function longer than is needed for those in acute treatment. Although it is desired that a haemofilter and a tubule device should function for at least 2 weeks in terms of realization of economical and convenient treatment, the bioartificial kidney system has never reached the targeted quality. The antithrombogenic continuous haemofilters, however, in which the inner surface of membrane is modified with haemocompatible compounds such as methacryloyloxyethyl phosphorylcholine, or in which the membrane is covered with autologous endothelial cells, might be developed in a few years. A bioartificial kidney could realize the prevention of long-term complications of maintenance dialysis patients within ten years.

20.11 References

1. Makino S *et al.* (1999), Cardiomyocytes can be generated from marrow stromal cells *in vitro*. *J Clin Invest*, **103**, 697–705.
2. Studer L *et al.* (1998), Transplantation of expanded mesencephalic precursors leads to recovery in parkinsonian rats. *Nat Neurosci*, **1**, 290–295.
3. McDonald JW *et al.* (1999), Transplanted embryonic stem cells survive, differentiate and promote recovery in injured rat spinal cord. *Nat Med*, **5**, 1410–1412.
4. Asahara T *et al.* (1997), Isolation of putative endothelial cells for angiogenesis. *Science*, **275**, 964–967.
5. Kawamoto A *et al.* (2001), Therapeutic potential of *ex vivo* expanded endothelial progenitor cells for myocardial ischemia. *Circulation*, **103**, 634–637.
6. Imasawa T *et al.* (1999), Bone marrow transplantation attenuates murine IgA nephropathy: role of a stem cell disorder. *Kidney Int*, **56**, 1809–1817.
7. Imasawa T *et al.* (2001), The potential of bone marrow-derived cells to differentiate to glomerular mesangial cells. *J Am Soc Nephrol*, **12**, 1401–1409.
8. Poulsom R *et al.* (2001), Bone marrow contributes to renal parenchymal turnover and regeneration. *J Pathol*, **195**, 229–235.
9. Lin F *et al.* (2003), Hematopoietic stem cells contribute to the regeneration of renal tubules after renal ischemia – reperfusion injury in mice. *J Am Soc Nephrol*, **14**, 1188–1199.
10. Kitamura S *et al.* (2005), Establishment and characterization of renal progenitor like cells from S3 segment of nephron in rat adult kidney. *Faseb J*, **19**, 1789–1797.
11. Dekel B *et al.* (2003), Human and porcine early kidney precursors as a new source for transplantation. *Nature Med*, **9**, 53–60.
12. NKF-K/DOQI clinical practice guidelines for chronic kidney disease: evaluation, classification, and stratification. *Am J Kidney Dis* 2002, **39** (Suppl. 1), S14–S266.

13. Moeller S *et al.* (2002), ESRD patients in 2001: global overview of patients, treatment modalities and development trends. *Nephrol Dial Transplant*, **17**, 2071–2076.
14. Saito A *et al.* (1995), Maintaining low concentration of plasma β_2-microglobulin with continuous slow hemofiltration. *Nephrology Dial Transplant*, **10** (Suppl. 3), 52–56.
15. Saito A (2003), Development of bioartificial kidneys. *Nephrology*, **8**, S10–S15.
16. Saito A (2004), Research into the development of a wearable bioartificial kidney with a continuous hemofilter and a bioartificial tubule device using tubular epithelial cells. *Artif Organs*, **28**, 58–63.
17. Humes HD *et al.* (1999), Tissue engineering of a bioartificial renal tubule assist device: In vitro transport and metabolic characteristics. *Kidney Int*, **55**, 2502–2514.
18. Ozgen N *et al.* (2004), Evaluation of long-term transport ability of a bioartificial renal tubule device using LLC-PK_1. *Nephrol Dial Transplant*, **19**, 2198–2207.
19. Weitzel WF *et al.* (2002), Early results with the bioartificial kidney in ICU patients with acute renal failure. *J Am Soc Nephrol*, **13** (Program and Abstract Issue), 642A.
20. Aebischer P *et al.* (1987), The bioartificial kidney: progress toward an ultrafiltration device with renal epithelial cells processing. *Life Support Sys*, **5**, 159–168.
21. Aebischer P *et al.* (1987), Renal epithelial cells grown on semipermeable processor. *Trans Am Soc Artif Intern Organs*, **33**, 96–102.
22. Ip TK *et al.* (1987), Renal epithelial cell-controlled solute transport across permeable membrane as the foundation for bioartificial kidney. *Artif Organs*, **13**, 58–61.
23. Humes HD *et al.* (1999), Replacement of renal function in uremic animals with a tissue-engineered kidney. *Nat Biotechnol*, **17**, 451–455.
24. Humes HD *et al.* (2004), Initial clinical results of the bioartificial kidney containing human cells in ICU patients with acute renal failure. *Kidney Int*, **66**, 1578–1588.
25. Tumlin J *et al.* (2005), Effect of the renal assist device (RAD) on mortality of dialysis-dependent acute renal failure: randomized, open-labeled, multicenter, phase II trial. *J Am Soc Nephrol*, **16** (Abstract Issue), 46A.
26. Saito A *et al.* (1998), Regeneration of peritoneal effluent by Madin–Darby canine kidney cells-lined hollow fibers. *Mater Sci Eng C*, **6**, 221–226.
27. Fujita Y *et al.* (2002), Evaluation of active transport and morphological changes for bioartificial renal tubule device using MDCK cells. *Tissue Eng*, **8**, 13–24.
28. Terashima M *et al.* (2001), Evaluation of water and electrolyte transport of tunular epithelial cells under osmotic and hydraulic pressure for development of bioartificial tubules. *Artif Organs*, **25**, 209–212.
29. Sato Y *et al.* (2005), Evaluation of proliferation and functional differentiation of LLC-PK_1 cells on porous polymer membranes for the development of a bioartificial renal tubule device. *Tissue Eng*, **11**, 1506–1515.
30. Kanai N *et al.* (1999), Effect of extracellular matrix on renal epithelial cell attachment on the polmer substrate. *Artif Organs*, **23**, 114–118.
31. Saito A *et al.* (2006), Present status and perspective of the development of a bioartificial kidney for chronic renal failure patients. *Ther Apher Dial*, **10**, 342–347,
32. Ryan MJ *et al.* (1994), HK-2: An immortalized proximal tubule epithelial cell line from normal adult human kidney. *Kidney Int*, **45**, 48–57.
33. Saito A *et al.* (2006), Present status and perspectives of bioartificial kidneys. *J Artif Organs*, **9**, 130–135.
34. Asano M *et al.* (2002), Renal proximal tubular metabolism of protein-linked pentosidine, an advanced glycation end product. *Nephron*, **91**, 688–694.

35. Ueda H (2005), Asymmetrically functional surface properties on biocompatible phospholipid polymer membrane for bioartificial kidney. Published online 12 December 2005 in Wiley InterScience (www.interscience.wiley.com). DOI:10.1002/jbm.a.30606
36. Ishihara K *et al.* (1994), Hemocompatibility on graft copolymers composed of poly(2-methacryloyloxyethyl phosphorylcholine) side chain and poly(*n*-butyl methacrylate) backbone. *J Biomed Mater Res*, **28**, 225–232.
37. Fujita Y *et al.* (2004), Transcellular water transport and stability of expression in aquaporin1-transfected LLC-PK_1 cells in the development of a portable bioartificial renal tubule device. *Tissue Eng*, **10**, 711–722.
38. Kowolik CM *et al.* (2004), Cre-mediated reversible immortalization of human renal proximal tubular epithelial cells. *Oncogene*, **23**, 5950–5957.

21
Bladder tissue engineering

A M TURNER, University of York, UK, R SUBRAMANIAM and D F M THOMAS, St James's University Hospital, UK and J SOUTHGATE, University of York, UK

Advances in biomaterial technologies and tissue engineering strategies to produce clinically useful quantities of cells for use in tissue and organ replacement surgery have grown alongside an enhanced understanding of urothelial cell biology and the physiology of the urinary bladder. The combination of these factors has begun to show promise for improving the quality of life for patients with intractable urinary incontinence. Passive techniques, whereby materials are placed in the bladder to be infiltrated by cells using the body's own regenerative processes, are being joined by new classes of biomaterial, which combine synthetic and natural properties to provide a biomimetic environment for tissue growth and differentiation. Furthermore, the generation of functional tissue *ex vivo* is on the horizon, with the possibility of 'off-the-shelf' reconstructions becoming closer to reality.

21.1 The bladder – structure and function

The bladder is a complex organ that not only stores urine, but also maintains its electrolyte composition by a combination of limited passive permeability and active ionic transport (Lewis, 2000). By retaining urine at safe, physiological pressures, the bladder protects the kidneys in the upper renal tract from damage (Thomas, 1997). The remarkable capacity and compliance of the bladder is dependent on the structural, biomechanical and biological properties of the smooth muscle wall and the highly specialised urothelial lining, which provides both urinary barrier and mechanosensory functions. In common with all tissue engineering, the successful (i.e. functional) engineering of partial or whole bladder organ constructs requires fit-for-purpose biomaterials based on a comprehensive understanding of bladder structure, biology and physiology. Given that until recently the urinary bladder was considered to be a passive urine storage organ, it is unsurprising that past attempts to reconstruct it with unsuitable materials have resulted in failure.

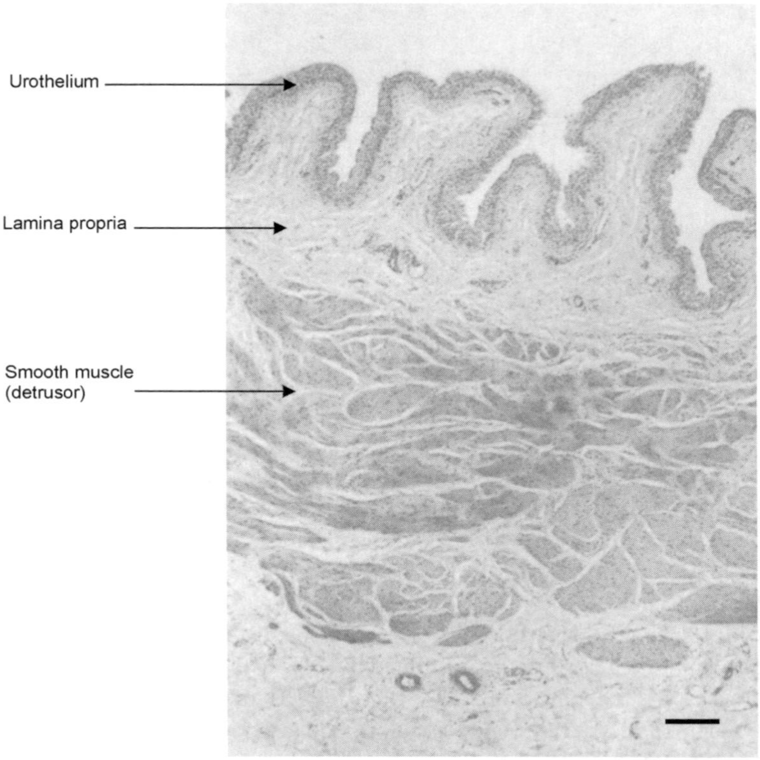

21.1 Structure of human bladder (haematoxylin and eosin staining; scale bar 100 μm).

The mammalian bladder is composed of four distinct layers, with an outer serosal layer surrounding the muscular compartment, which is made up of three loosely arranged layers of smooth muscle bundles (Fig. 21.1). Concentrically within this, the lamina propria is a viscoelastic collagenous connective tissue supporting a variety of cell types, including blood vessels, sensory and motor neurons. A basal lamina separates the lamina propria from the urothelium. The urothelium itself is a transitional epithelium made up of three zones: a single row of basal cells attached to the basement membrane, several layers of intermediate cells and a single, overlying row of superficial 'umbrella' cells that abuts the luminal space. The urothelium is specialised to function as a urinary barrier, with the paracellular barrier maintained by intercellular tight junctions, particularly at the level of the superficial cells (Acharya *et al.*, 2004; Varley *et al.*, 2006) and the transcellular barrier provided by multiple thickened plaques of asymmetric unit membrane (AUM) embedded in the outer leaflet of the apical membrane (Hicks, 1965). The AUM plaques are constituted by the interactions of four uroplakin ('urothelium-plaque') proteins and are a unique feature of urothelium (Wu *et al.*, 1990; Yu *et al.*, 1994). The AUM plaques are formed in the Golgi apparatus and transported to the apical membrane in the form of fusiform vesicles (Tu *et al.*, 2002), thus providing a source of membrane for accommodating changes in surface area and maintaining a low-pressure environment during bladder filling. Such is the relationship between urothelial structure and function, the loss of one component of the AUM can have devastating effects on urothelial structure and transcellular barrier properties (Hu *et al.*, 2000, 2002). Thus, urothelial differentiation antigens not only provide objective markers of urothelial cytodifferentiation, but by virtue of their role in the urothelium, may also be regarded as surrogate markers of urinary barrier function. Unfortunately, expression of these markers is not invariably reported in bladder tissue engineering reports, leading to discrepancy in the interpretation of some studies.

An important feature of the urothelium is its high regenerative capacity. Thus, although the urothelium is regarded as a 'stable' or quiescent tissue with an extremely slow rate of cell turnover, which may be as long as 200 days (Hicks, 1975), it is able to undergo rapid proliferation in response to acute injury. Lavelle and colleagues performed a controlled study of selective urothelial damage in rats, which showed that recovery of transcellular and paracellular components of the urinary barrier occurred within 72 h, with the intermediate cells undergoing rapid maturation to form differentiated umbrella cells (Lavelle *et al.*, 2002). The regenerative and differentiation capacity of urothelium is critical to maintaining the urine-proofing properties of the bladder and has positive implications for developing tissue engineering strategies.

21.2 The clinical need for bladder reconstruction

Any disease process or surgical intervention that renders the bladder unstable, under high pressure, or lacking in capacity or compliance can result in a range of clinical problems ranging from mild to severe chronic urinary incontinence to irreversible kidney damage caused by raised upper tract pressures. Whereas bladder augmentation was historically used for patients with a contracted bladder secondary to tuberculosis, nowadays indications include neuropathic bladder (e.g. secondary to myelomeningocele or spinal cord injury), severe detrusor instability, interstitial cystitis and following radical resection for cancer or trauma. For some patients, conservative measures such as clean intermittent self-catheterisation (CISC) to remove residual volumes of urine and pharmacotherapeutic options to improve bladder stability are enough to satisfactorily improve quality of life. However, for a subset of patients, where intractable incontinence or pain destroys quality of life, or where serious kidney damage is imminent, surgical reconstruction of the bladder is the treatment of choice. As the clinical need for bladder augmentation or reconstruction is the convergent end point of a number of different underlying aetiopathologies, it should also be considered whether the chosen augmentation strategy is likely to affect or compromise the outcome. For example, detrusor myectomy is successful in most patients with idiopathic detrusor instability, but is largely unsuccessful in those with neuropathic bladder secondary to myelomeningocele (Marte *et al.*, 2002; Kumar and Abrams, 2005).

21.3 Concepts and strategies of bladder reconstruction and tissue engineering

The primary aim of reconstruction is to increase the capacity and compliance of the bladder and, in combination with CISC, improve continence and quality of life. The development of sensory self-voiding function is outside current objectives and in all current and proposed bladder reconstruction strategies, it is anticipated that voluntary emptying will be aided by CISC, either via the urethra or, if this is not possible for physical or technical reasons, via an outlet onto the body wall (vesicotomy), such as described by Mitrofanoff (1980).

Non-compliant or hard materials are incompatible with bladder function and natural or synthetic polymers are the most promising candidate materials for bladder wall substitution. Natural materials include vascularised tissue grafts, free tissue grafts, decellularised tissue matrices and purified natural extracellular matrix (ECM) proteins. Synthetic polymers include polyglycolic acid (PGA), polylactic acid (PLA) and polylactic-co-glycolic acid (PLGA) (Kim *et al.*, 2000).

Where acellular or decellularised materials are used to reconstruct or augment the bladder, full tissue integration requires that the material becomes cellularised and is eventually resorbed and replaced. Cellularisation strategies can be categorised as either *passive* or *active* tissue engineering. The passive approach relies upon the regenerative capacity of the bladder to cellularise a biomaterial that has been used to reconstruct the bladder. By contrast, active tissue engineering involves sourcing appropriate autologous stem or somatic cells and seeding them onto the biomaterial *in vitro* to form a live tissue construct. The advantage of the latter approach is that clinically useful numbers of autologous cells can be propagated in the laboratory and used to develop a tissue construct *in vitro*, prior to surgical implantation (Fig. 21.2). A key variable is the extent of development of biomimetic tissues *ex vivo*. The concept of the *internal bioreactor* describes the implantation of the construct at an early stage of its development, under the assumption that the body itself will provide the appropriate environment to nurture and condition the neotissue. Alternatively, an *external bioreactor* may be used to mature the construct *in vitro*, with the aim of delivering tissues that can sustain function *in vivo* from the outset (Korossis *et al.*, 2006). By simulating an appropriate nutrient, biochemical and mechano-stimulatory environment, the hypothesis is that tissues matured *in vitro* will achieve their full functional potential and behave as native tissue equivalents.

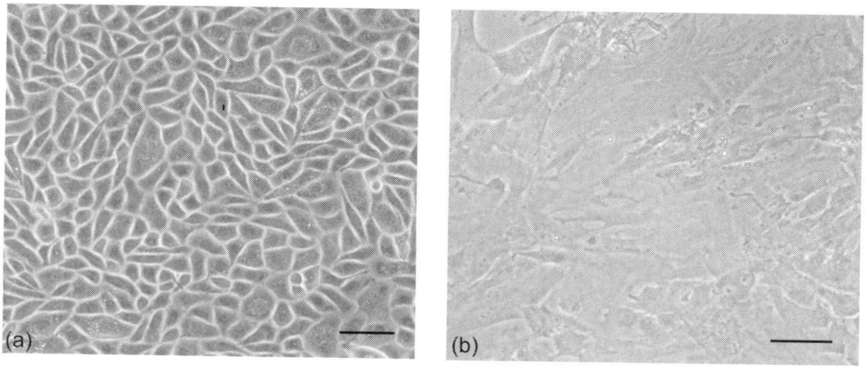

21.2 Phase contrast micrograph showing normal human urothelial (a) and smooth muscle (b) cells in culture (scale bar 50 μm).

21.4 Review of past and current strategies in bladder reconstruction

21.4.1 Vascularised tissue grafts

The use of reconfigured vascularised or pedicled tissue grafts to augment the bladder has a long history. Currently, the most commonly performed procedure for end-stage bladder disease involves replacing or augmenting the bladder with a vascularised segment of the patient's own bowel. This procedure of *enterocystoplasty* involves isolating a full thickness segment of bowel on its vascular pedicle, detubularising it along the antimesenteric border and incorporating it into the bivalved bladder. The procedure brings about improved capacity in severely compromised contracted bladders, with success due to the compliant nature of the host tissue, which is vascularised to ensure survival of the graft. A review of 24 enterocystoplasty series showed that the procedure is successful in as many as 92% of patients with neuropathic bladder (Beier-Holgersen *et al.*, 1994; Greenwell *et al.*, 2001).

Enterocystoplasty was first described in a canine model in 1888 and then in man a year later, but it was not until the mid-twentieth century that the technique became popular for the treatment of the contracted, tuberculous bladder (Tizzoni and Foggi, 1888; Von-Mikulicz, 1889; Couvelaire, 1950). Stomach (gastro-cystoplasty), small intestine (ileocystoplasty) and large intestine (colo-cystoplasty) have all been used as the reconstructing segment, but in the UK, ileocystoplasty is the most commonly performed procedure (Thomas, 1997).

Despite many patients experiencing the benefits of improved continence, improved urodynamic parameters and greater control over voiding, entero-cystoplasty carries with it the potential for a number of serious complications. These are mainly attributable to the fact that bowel mucosa is structurally and physiologically unsuited to exposure to urine and include both early complications associated with all major abdominal surgery and specific, longer-term complications of enterocystoplasty, including spontaneous perforation of the bladder, mucus production by the bowel epithelium, bladder stones, bacteriuria, metabolic disturbances and malignancy (reviewed by Thomas, 1997; Greenwell *et al.*, 2001).

Given that the side effects of enterocystoplasty are related to the interaction of urine with bowel mucosa, the logical progression would be to remove the bowel epithelium to leave the raw muscle surface facing the lumen, so-called seromuscular enterocystoplasty. Experimentally, in rabbit, canine, porcine and bovine surgical models, this approach has resulted in graft fibrosis and shrinkage, attributable to severe inflammation secondary to chemical irritation or infection of the graft, ischaemia or damage to the intestine during dissection (Motley *et al.*, 1990; Salle *et al.*, 1990; Aktug *et al.*, 2001; Fraser *et al.*, 2004; Hafez *et al.*, 2005). The phenomenon was independent of which side of the bowel wall faced the lumen. Severe fibrosis was also observed when a non-seeded vascularised capsule-flap of abdominal wall or gracilis muscle was incorporated into rat bladder (Schoeller *et al.*, 2004). Lima and colleagues, however, in their series of 129 human bladder augmentations using demucosalised intestine, have shown that fibrosis and shrinkage was prevented by the use of a silicon balloon left in the bladder for 2 weeks post-augmentation (Lima *et al.*, 2004). Pedicled omental flaps to repair or augment the bladder (omentocystoplasty) have been largely successful both clinically and in animal models, particularly when used to close defects associated with vesico-vaginal fistula (Kiricuta and Goldstein, 1972).

Although demucosalisation of the bowel prior to incorporation into the bladder has inevitably resulted in graft fibrosis and shrinkage, when urothelium has been allowed to cover the augmenting graft, shrinkage occurred minimally or not at all (Aktug *et al.*, 2001; Schoeller *et al.*, 2004; Hafez *et al.*, 2005). Hafez and colleagues developed an aerosol transfer technique in a porcine model using urothelial and bladder smooth muscle cell suspensions in fibrin glue (Hafez *et al.*, 2005). Autologous urothelial cells with or without smooth muscle cells, isolated at hemicystectomy, were sprayed onto demucosalised colon and then incorporated into the remaining bladder. After 6 weeks, this led to the development of a stratified, multilayered uroplakin-positive urothelium on top of a bladder or colonic smooth muscle submucosa, respectively, and no inflammation was described. Although the procedure did not involve propagation of cells in culture, it feasibly could do, the point of interest being whether the cells would remain capable of developing into a morphologically differentiated tissue *in vivo*. Fraser and colleagues achieved a successful bladder augment in a minipig model by implanting *in vitro*-propagated autologous urothelial cells onto a vascularised, de-epithelialised uterine tissue used as the augmenting segment in a 'composite cystoplasty' (Fraser *et al.*, 2004). The urothelium was transplanted from cell culture to surgical site using a Vicryl™ mesh carrier. Although a stratified urothelium that expressed AUM was identified in the composite utero-cystoplasty, there was evidence of stromal inflammation within both augmenting and native segments. Composite enterocystoplasty was also performed, but incomplete urothelialisation and colonic epithelial cell regrowth occurred in addition to inflammation and contraction of the augmenting segment. As the implanted urothelium was a proliferative, undifferentiated tissue with poor urinary barrier properties, it was hypothesised that implanting a differentiated, functional tissue would encourage the rapid establishment of an effective urinary barrier and provide the stroma with immediate protection from urine-mediated damage. The approach of composite cystoplasty offers advantages over other bladder tissue engineering strategies as it utilises an available vascularised smooth muscle tissue and only requires engineering of the urothelium.

21.4.2 Free tissue grafts

Shortly after the first colocystoplasty was described in 1912, attempts were made to incorporate *free* biological tissue into the bladder. First, Neuhof used a free fascial patch in dogs (Neuhof, 1917) and, since then, split skin grafts, placenta, peritoneum and dural membrane have been used as patches (Draper *et al.*, 1952; Kelami *et al.*, 1970b; Hutschenreiter *et al.*, 1978; Fishman *et al.*, 1987). There were mixed results, with complications often arising as a result of normal functioning of the donor tissue such as hair growth on skin grafts, alongside more general problems, such as graft contraction and stone formation. Nevertheless, the appeal of using free biological tissue has persisted. Stenzl performed detrusor myectomy using free latissimus dorsi (LD) grafts in four dogs (Stenzl *et al.*, 2000). This approach was based on Carpentier's LD cardiac wrap for patients with severe cardiomyopathy, which was the first recorded case of substituting non-skeletal muscle with skeletal muscle (Carpentier and Chachques, 1985). The neurovascular bundles were anastamosed to pelvic equivalents to restore function. Detrusor myectomy was reported to result in a normally functioning bladder, although specific changes in bladder capacity were not reported. The same report indicated that a similar protocol had been performed on two patients with non-functioning bladders, although details of the outcome were limited, and to date, no further reports have appeared in the literature.

21.4.3 Acellular matrices

The decellularisation of an allogeneic or xenogeneic tissue can potentially provide a bio- and tissue-compatible polymeric scaffold or matrix for recellularisation by the recipient's own cells. The two most commonly described preparations for use in bladder reconstruction are porcine small intestinal submucosa (SIS) and bladder acellular matrix graft (BAMG). In their natural states, these ECMs are heavily populated with cells and hence must undergo decellularisation to remove all potentially immunogenic material. Non-crosslinked ECM scaffolds have been described as slow release vehicles of naturally occurring growth factors because, once implanted, they slowly degrade, acting as a scaffold for new ECM proteins produced by ingrowing cells (Kim *et al.*, 2000; Badylak, 2002).

Small intestinal submucosa

Small intestine submucosa has been used as a reconstructive tool in musculoskeletal, vascular and urological specialties with promising results. Its preparation entails the removal of the major cellular components of the bowel wall to leave the collagen- and elastin-rich submucosal layer. When incorporated into a bladder reconstruction, SIS degrades rapidly and completely and the breakdown products enter the circulation and are excreted in the urine (Badylak *et al.*, 1998; Record *et al.*, 2001). In its place, cellular encroachment and infiltration occur rapidly, with the resultant tissue resembling that of the surrounding native organ. Early biocompatibility studies of macerated SIS periureteric injection and bladder patch grafts in pigs demonstrated the potential for smooth muscle and vascular in-growth, combined with a benign immunoreactivity (Knapp *et al.*, 1994). Analysis of SIS patches 11 months after incorporation into rat bladders not only showed replacement by normal bladder tissue, but also vascularisation and re-innervation (Vaught *et al.*, 1996). Furthermore, although of a lower magnitude, appropriate contractile and relaxatory responses were elicited on chemical stimulation of the patch, suggesting expression of neurotransmitter receptors. Similar results were obtained using SIS patches implanted in dogs, which confirmed that the regenerated grafts had similar viscoelastic properties to native bladder, despite having a reduced muscle:collagen ratio (Kropp *et al.*, 1996b). This last fact may explain the decrease in magnitude of contraction observed. In addition, extensive neovascularisation had occurred in the submucosa and it was suggested that afferent nerves had re-innervated the segment (Kropp *et al.*, 1996a, b).

SIS has been shown to be a promising candidate for future clinical studies of bladder reconstruction. However, there is evidence that not all SIS is created equal, with the regenerative potential and incidence of complications being dependent upon the age of the donor pig and the portion of the bowel from which the SIS matrices are derived (Kropp *et al.*, 2004). Furthermore, recent work has cast doubt on the biological safety of commercially available SIS which, *in vitro*, has demonstrated cytotoxic effects on urothelial cells and has been found to retain porcine nuclear residues (Feil *et al.*, 2006).

Bladder acellular matrix graft

For BAMG preparation, the isolation of the bladder submucosa by dissection to leave a cell-depleted tissue, similar to SIS, has been described (Chen *et al.*,

1999). More commonly, split- (urothelium-free) or full-thickness bladder is decellularised (Sutherland *et al.*, 1996; Probst *et al.*, 1997, 2000; Dahms *et al.*, 1998; Piechota *et al.*, 1998; Merguerian *et al.*, 2000; Reddy *et al.*, 2000; Brown *et al.*, 2002; Bolland *et al.*, 2006). Implanting BAMG in the bladders of rats, dogs and pigs has shown regeneration of urothelial and muscle layers, with reinnervation and revascularisation of the graft (Sutherland *et al.*, 1996; Probst *et al.*, 1997, 2000; Piechota *et al.*, 1998; Reddy *et al.*, 2000). As with SIS, appropriate contraction and relaxation has been described in reconstructed bladders, but again at a lower magnitude. One key functional difference, however, relates to the compliance of the material prior to implantation and is attributable to its preparation. As discussed, regenerated SIS had similar viscoelastic properties to native bladder. However, native, non-regenerated SIS segments were 30 times less compliant than either native bladder or regenerated SIS (Kropp *et al.*, 1996b). Conversely, non-regenerated split-thickness and full-thickness BAMG, respectively, exhibited similar biomechanical properties to native bladder from the outset (Dahms *et al.*, 1998; Bolland *et al.*, 2006). Whether this constitutes a disadvantage is moot; the ultimate aim remains to produce a regenerated patch with functionality similar to native bladder tissue.

Potential problems associated with the approach of using decellularised matrices include lithogenesis, graft shrinkage and incomplete and disorganised smooth muscle infiltration. Graft shrinkage due to fibroproliferative change has been shown to increase with time, with up to 48% reduction in graft size (Brown *et al.*, 2002). It should be considered that, although regenerated smooth muscle within the grafts is often disorganised (i.e. does not form bundles) in surgical models, the extent and speed of cell incorporation is very much dependent upon the size of the grafts. In rat models, graft size is small (~0.5 cm^2) whereas it is much greater in larger animals (up to 46 cm^2) and so it is unsurprising that smooth muscle bundles have been reported as scanty at the centres of larger patches (Piechota *et al.*, 1998; Brown *et al.*, 2002). In practical terms, the surface area of bladder augmentations in man is an order of magnitude greater than many described experimentally *in vivo*, and this represents a severe limitation to much bladder tissue engineering research, particularly where rodent models have been used. Lithogenesis too is a particular problem in rat models, with up to 75% and 80% of animals found to have bladder stones in SIS and BAMG reconstructions, respectively (Vaught *et al.*, 1996; Piechota *et al.*, 1998). The problem is not confined to rodents: one group treated pigs with alendronate, an osteoclast inhibitor, to reduce urinary calcium concentrations, following the discovery of microcalcification in the suburothelial zone of BAMG (Reddy *et al.*, 2000), but such treatment does not allow for accurate determination of the risks involved.

Decellularised biomaterials appear to retain biological activity and this may encourage the in-growth of tissue. Furthermore, given that the composition and structure of the ECM is exclusive to individual tissues, there may be advantages in orthotopic-derived matrix; BAMG may be expected to contain more appropriate growth factors for bladder tissue engineering than SIS (Badylak, 2004; Bolland *et al.*, 2006). Infiltration and organisation of smooth muscle bundles in both SIS and BAM grafts may, however, be improved by the incorporation of growth factors and other bioactive substances. BAMG has been shown to be capable of sustained release of exogenous basic fibroblast growth factor and was demonstrated in a dose-dependent manner to significantly reduce graft shrinkage in a rat model of bladder augmentation (Kanematsu *et al.*, 2003).

Natural matrices that undergo chemical or non-chemical crosslinking to enhance the physical attributes and stability of the material are invariably rendered inert and this may ultimately inhibit cellular incorporation (Badylak, 2002). Although some such materials have shown comparable results to SIS and BAMG, further development is necessary to realise the full potential of these biomatrices (Nuininga *et al.*, 2004). One such crosslinked material is Pelvicol™ (Permacol™ in the UK). This decellularised porcine dermis is used clinically in genitor-urinary reconstruction, for example as a corporal patch in Peyronie's disease and as a pubovaginal sling (Santucci and Barber, 2005). However, early *in vitro* and *in vivo* assessments of this non-resorbable material have not been promising. In the laboratory, Kimuli and colleagues reported poor smooth muscle infiltration of the material, possibly as a result of chemical crosslinking (Kimuli *et al.*, 2004). Furthermore, an experimental study of bladder augmentation in rabbits using Pelvicol™ concluded that it was an unsuitable material for the procedure (Ayyildiz *et al.*, 2006).

The limited success of passive tissue engineering using natural tissues and matrices has led to these biomaterials being seeded with urothelial and smooth muscle cells *ex vivo* with the aim of enhancing tissue integration following implantation. Of underlying relevance to this approach is the pioneering work of Baskin and colleagues, who first showed that bladder smooth muscle development from the fetal mesenchyme was dependent upon paracrine interactions with the urothelium (Baskin *et al.*, 1996; DiSandro *et al.*, 1998). The potential for reciprocal interactions between urothelial and smooth muscle compartments during bladder tissue engineering has been investigated both *in vitro* (Fujiyama *et al.*, 1995; Zhang *et al.*, 2000; Ram-Liebig *et al.*, 2004, 2006; Brown *et al.*, 2005) and *in vivo* (Yoo *et al.*, 1998; Master *et al.*, 2003; Zhang *et al.*, 2004). There is some controversy about the precise mechanisms underlying these interactions that is outside the scope of the present review. However, it is clear that heterotypic cell–cell interactions may play an important role in the development of functional tissue-engineered bladders.

21.4.4 Natural extracellular matrix

The ECM has been used extensively as a xenogeneic and allogeneic biomaterial for cells of many types, reflecting its natural evolution as a tissue scaffold.

Collagen, the most abundant protein within the ECM and the major structural protein in the body, is largely responsible for the strength and conformability of natural materials. Collagen has been shown to encourage cell growth, have minimal immunogenicity and can be readily purified and moulded into the desired form, making it an ideal tool for tissue engineering applications (Elbahnasy *et al.*, 1998; Hattori *et al.*, 2006). Purified collagen, however, when used for reconstruction in the urinary tract, has been shown to lose its tensile strength and to be susceptible to tearing during suturing (Elbahnasy *et al.*, 1998). To overcome these problems, collagen has been reinforced with synthetic materials (see below), and with natural biomaterials, including pedicled omental flap (Hattori *et al.*, 2006). The latter investigators employed a porcine *in vivo* model to demonstrate that collagen sponge became vascularised when combined with omentum for 7 days *in vivo* and that only when preconditioned in this way was the collagen sponge able to support passive engineering of the bladder (Hattori *et al.*, 2006). This approach has important implications for other natural or synthetic biomaterials, as it provides a strategy for *in vivo*-preintegration of a scaffold for subsequent use in passive tissue engineering.

21.4.5 Synthetic grafts

The incorporation of synthetic materials alone into the bladder has largely been met with failure, primarily as a result of biological and mechanical incompatibility. Plastics, polyvinyl sponge, polytetrafluoroethylene (Teflon^{TM}) and Japanese paper have all been used to reconstruct the bladder with variable results, but none has been pursued to the present day (Bohne and Urwiller, 1957; Kudish, 1957; Kelami *et al.*, 1970a; Fujita, 1978). Most promising was an experimental bladder reconstruction in rabbits with a 6.25 cm^2 poly(ϵ-benzyloxycarbonyl-L-lysine) membrane (Koiso *et al.*, 1983). By 6 months it was reported that the resorbable membrane was completely replaced with normal urothelium and smooth muscle and there were no recorded complications. Despite such a positive study, no follow-on or clinical studies have been performed to date.

Cell culture techniques have shown that, in the correct environment, urothelial and smooth muscle cells can attach and propagate in treated tissue culture plastic (for example Southgate *et al.*, 1994; Kimuli *et al.*, 2004). This led to the growth of bladder-derived cells on biodegradable synthetic scaffolds (Scriven *et al.*, 2001). Compared with natural materials, production of a synthetic biomaterial should allow full processing control over properties such as strength, biodegradability, microstructure and permeability; however, a fundamental feature of these materials is that they lack the natural signals that regulate cell attachment and growth (Vacanti and Langer, 1999; Danielsson *et al.*, 2006). Atala and colleagues were the first to demonstrate the feasibility of seeding cells onto a purely synthetic matrix for implantation *in vivo* (Oberpenning *et al.*, 1999). Polylactic-co-glycolic acid (PLGA) is a well characterised biomaterial with predictable biodegradability properties, and is widely used as Vicryl^{TM} sutures and meshes within the field of surgery. It is non-toxic and biocompatible with both normal human urothelial and bladder smooth muscle cells (Pariente *et al.*, 2001, 2002; Scriven *et al.*, 2001). These qualities make Vicryl^{TM} an attractive candidate for combination with natural materials to form implantable constructs for bladder reconstruction. Oberpenning used a PGA mesh, moulded into the shape of a bladder and coated with PLGA, and seeded the outer and inner surfaces of the biomaterial with autologous smooth muscle and urothelial cells, respectively (Oberpenning *et al.*, 1999). The constructs were then implanted *in vivo* onto a bladder base remaining after trigone-sparing cystectomy in dogs. Once coated with fibrin glue, the construct was wrapped with omentum and the animals were monitored for up to 11 months. There were no reported complications and mean capacity and compliance of the neobladders were similar to measurements pre-operatively. At 3 months, the polymer had degraded, leaving a vascularised, innervated tissue composed of organised smooth muscle bundles and a stratified urothelium, which was positive with antibodies against AUM. It is unlikely that the same degree of regeneration would have occurred without an omental wrap, such is its ability to induce neovascularisation.

Methods to improve cell attachment and proliferation on synthetic materials have also been explored. One solution is to coat the synthetic material with a biological substance or to use a surface modification procedure prior to seeding to encourage attachment. For example, *in vitro*, cultured smooth muscle cells have been shown to attach and proliferate on a biodegradable polyesterurethane foam pre-treated with fetal bovine serum (Danielsson *et al.*, 2006) and on plasma-coated, electrospun polystyrene (Baker *et al.*, 2006). Alternatively, the material can be combined with one of the aforementioned natural materials to act as a biodegradable scaffold, giving strength and conformability to the structure. It is surprising given the results described by Oberpenning that, when combined with PLGA, collagen hybrid matrices have shown mixed results *in vitro*. In one study, smooth muscle cells were able to proliferate and retain expression of differentiation markers on a gel-based construct, but not on a sponge, the opposite being the case for urothelial cells, which stratified on a sponge but not a gel, although unequivocal markers of differentiation (e.g. AUM) were not shown (Nakanishi *et al.*, 2003). As a biomaterial's properties can have differential effects upon proliferation, migration and differentiation of different cell types, this needs to be taken into consideration when developing the ideal synthetic material. For instance, smooth muscle cells adopted a more natural organisation when grown on electrospun polystyrene scaffolds with fibres aligned as collagen fibres *in vivo* rather than showing a random distribution (Baker *et al.*, 2006). Similarly, urothelial and smooth muscle cells showed improved growth properties on materials where the elastic modulus most closely matched that of the bladder (Rohman *et al.*, 2007).

Meanwhile, Atala and colleagues have made the transition from canine model to clinical trials (Atala *et al.*, 2006). Collagen only and collagen–PGA hybrid scaffolds were seeded with autologous smooth muscle and urothelial cells and implanted into nine patients with severely neuropathic bladders. Three had a collagen-only implant, one had collagen-only implant with an omental wrap and three patients had the collagen–PGA hybrid scaffold with an omental wrap. Two patients were lost to follow-up and one patient with a collagen-only implant underwent conventional augmentation because of progressively rising intra-vesical pressures. For the remaining patients, although followed up annually for up to 5 years, not all were available at each time point and only four had investigations in the fifth year. There were minimal or modest increases in capacity and compliance of the bladders, with the hybrid scaffold achieving the best results. The new bladder tissue was described as having a normal structure, with smooth muscle and stratified urothelium; however, the differentiation status of the urothelium was not reported and so the objective benefit of the approach taken is unclear.

The passive and active tissue engineering approaches described above have all relied on the internal bioreactor to develop a functional, mature bladder tissue. The alternative strategy is to use an external bioreactor to condition an active tissue-engineered construct prior to implantation *in vivo*.

21.5 Cell conditioning in an external bioreactor

21.5.1 Static conditioning

It is widely accepted that cells lose functional and differentiated characteristics when isolated from their host tissue and propagated in cell culture. For all active tissue engineering strategies, the critical question is whether this loss is reversible and whether cultured cells can be induced to differentiate and form functional, biomimetic tissues *in vitro*. Normal human urothelial (NHU) cells in monoculture have been shown to switch from a proliferative to a stratified, differentiated urothelium following manipulation of the growth medium (Cross *et al.*, 2005). The resultant urothelium had functional barrier properties, as assessed by a high transepithelial electrical resistance of $>3000\,\Omega\,cm^2$ (TER) and low diffusive permeability to urea, water and dextran. Recently, the generation of a differentiated porcine urothelium *in vitro* has been described, which had urinary barrier properties and expressed markers of terminal cytodifferentiation, including AUM (Turner *et al.*, 2007). This advance can now be used to progress the approach of composite cystoplasty (described in Section 21.4.1), by transplanting a functional biomimetic urothelium onto the de-epithelialised pedicled smooth muscle bowel segment used to augment the bladder and thus, ensuring immediate implementation of a functional urinary barrier. Advances have also been made in the identification of the molecular pathways involved in inducing urothelial differentiation, with the nuclear hormone receptor peroxisome proliferator activated receptor gamma (PPARγ) identified as a key regulator of urothelial differentiation *in vitro* (Varley *et al.*, 2004a,b, 2006).

21.5.2 Biomechanical conditioning

Further improvement of the biomimetic properties of *in vitro*-generated bladder tissue may be achieved by simulating the physical environment of the bladder (Korossis *et al.*, 2006). So-called 'functional tissue engineering' employs an external bioreactor to condition cells seeded onto a scaffold by controlling nutrition and providing appropriate mechanostimulation. Mechanosensitivity is a requirement in all cells and allows them to respond appropriately to physiological signals, as well as insults such as physical stress and osmotic pressure gradients (Hamill and Martinac, 2001), for example, *in vitro* studies of myofibroblasts showed that proliferation and biosynthetic activity changed with the degree of mechanical stress (Grinnell, 1994). The bladder fills with urine passively and undergoes several fill–void cycles daily; despite the large and often rapid changes in volume, the urinary barrier remains intact. The urothelium manages this by being exquisitely sensitive to stretch, with the superficial cells mobilising AUM-containing cytoplasmic vesicles to open onto the luminal membrane in response to filling, thus maintaining an appropriate surface area (Truschel *et al.*, 2002). Replication of the fill–void cycle would seem to be an appropriate measure when generating a biomimetic tissue *in vivo* and may have significant consequences for tissue functionalisation.

21.6 Future trends

21.6.1 Stem cells

Where the damage to the bladder reflects an underlying irreversible disease process, an autologous biopsy may not provide sufficient healthy cells to propagate clinically useful quantities for active tissue engineering. In such cases, reseeding a biomaterial with pluripotent stem cells may offer a solution. Identification of precursor cells that can give rise to bladder-specific cell types has yet to be achieved, but given the fact that the urothelium is a regenerative self-renewing tissue, the expectation that there are peripheral stem cell populations with urothelial lineage potential is low. Nevertheless, there is early evidence that embryonic stem cells may produce factors that enhance the integration of urothelial and smooth muscle cells into passive scaffolds *in vivo* (Frimberger *et al.*, 2005, 2006) which fits with the observations made by Baskin in the developing bladder (Staack *et al.*, 2005).

21.6.2 Smart biomaterials

At a basic level, Danielsson's polyesterurethane serum-pretreated foam designed to encourage smooth muscle attachment and growth is an example of a smart biomaterial, as it sought to endow a synthetic material with more biological properties (Danielsson *et al.*, 2006). However, the desire to create the perfect biomaterial to support cell growth has encouraged the development of more advanced matrices. Synthetic hydrogel matrices based on poly(ethylene glycol) with integrated vascular endothelial growth factor (VEGF) have demonstrated that angiogenesis can be triggered locally by VEGF activation in response to the release of MMP-2 by human umbilical vein endothelial cells *in vitro* (Zisch *et al.*, 2003). It was suggested that local activation of VEGF in response to remodelling is likely to be similar to the controlled response *in vivo*. The functionalisation of synthetic biomaterials by incorporation of relevant growth factors is an attractive approach assuming that single factors can be shown to entrain appropriate cell behaviours.

21.7 Conclusions

Reconstruction of the urinary bladder is carried out when conservative and medical therapies have failed to alleviate the debilitating symptoms of a small, non-compliant bladder. Enterocystoplasty has provided relief for many patients, but is recognised to come at a price, in terms of various complications and risks resulting from interactions between the bowel epithelial lining and urine. Advances in urothelial cell biology support the strategy of composite cystoplasty, where the bowel epithelium is replaced by an autologous *in vitro*-engineered urothelium. However, active engineering of full bladder wall equivalents is inhibited by the lack of equivalent advances in bladder smooth muscle cell biology and by the need to identify fit-for-purpose synthetic scaffold materials that match the tissue-specific biological and physical requirements of the bladder. Passive bladder engineering requires consideration of the clinical relevance of the model used, particularly in terms of the size of patch; nevertheless, preconditioning of the scaffold to encourage vascularisation and thus cellular infiltration, for instance by pre-integration into the omentum, appears to be a promising approach that may be applicable to natural, synthetic and hybrid polymeric materials alike.

21.8 References

Acharya, P., Beckel, J., Ruiz, W. G., Wang, E., Rojas, R., Birder, L. and Apodaca, G. (2004) Distribution of the tight junction proteins ZO-1, occludin, and claudin-4, -8, and -12 in bladder epithelium. *Am J Physiol Renal Physiol,* **287**, F305–18.

Aktug, T., Ozdemir, T., Agartan, C., Ozer, E., Olguner, M. and Akgur, F. M. (2001) Experimentally prefabricated bladder. *J Urol,* **165**, 2055–8.

Atala, A., Bauer, S. B., Soker, S., Yoo, J. J. and Retik, A. B. (2006) Tissue-engineered autologous bladders for patients needing cystoplasty. *Lancet,* **367**, 1241–6.

Ayyildiz, A., Nuhoglu, B., Huri, E., Ozer, E., Gurdal, M. and Germiyanoglu, C. (2006) Using porcine acellular collagen matrix (Pelvicol) in bladder augmentation: experimental study. *Int Braz J Urol,* **32**, 88–92; discussion 92–3.

Badylak, S. F. (2002) The extracellular matrix as a scaffold for tissue reconstruction. *Semin Cell Dev Biol,* **13**, 377–83.

Badylak, S. F. (2004) Xenogeneic extracellular matrix as a scaffold for tissue reconstruction. *Transpl Immunol,* **12**, 367–77.

Badylak, S. F., Kropp, B., McPherson, T., Liang, H. and Snyder, P. W. (1998) Small intestional submucosa: a rapidly resorbed bioscaffold for augmentation cystoplasty in a dog model. *Tissue Eng,* **4**, 379–87.

Baker, S. C., Atkin, N., Gunning, P. A., Granville, N., Wilson, K., Wilson, D. and Southgate, J. (2006) Characterisation of electrospun polystyrene scaffolds for three-dimensional *in vitro* biological studies. *Biomaterials,* **27**, 3136–46.

Baskin, L. S., Hayward, S. W., Young, P. and Cunha, G. R. (1996) Role of mesenchymal-epithelial interactions in normal bladder development. *J Urol,* **156**, 1820–7.

Beier-Holgersen, R., Kirkeby, L. T. and Nordling, J. (1994) 'Clam' ileocystoplasty. *Scand J Urol Nephrol,* **28**, 55–8.

Bohne, A. W. and Urwiller, K. L. (1957) Experience with urinary bladder regeneration. *J Urol,* **77**, 725–32.

Bolland, F., Korossis, S., Wilshaw, S. P., Ingham, E., Fisher, J., Kearney, J. N. and Southgate, J. (2006) Development and characterisation of a full-thickness acellular porcine bladder matrix for tissue engineering. *Biomaterials*, **28**, 1061–70.

Brown, A. L., Farhat, W., Merguerian, P. A., Wilson, G. J., Khoury, A. E. and Woodhouse, K. A. (2002) 22 week assessment of bladder acellular matrix as a bladder augmentation material in a porcine model. *Biomaterials,* **23**, 2179–90.

Brown, A. L., Brook-Allred, T. T., Waddell, J. E., White, J., Werkmeister, J. A., Ramshaw, J. A., Bagli, D. J. and Woodhouse, K. A. (2005) Bladder acellular matrix as a substrate for studying *in vitro* bladder smooth muscle–urothelial cell interactions. *Biomaterials,* **26**, 529–43.

Carpentier, A. and Chachques, J. C. (1985) Myocardial substitution with a stimulated skeletal muscle: first successful clinical case. *Lancet,* **1**, 1267.

Chen, F., Yoo, J. J. and Atala, A. (1999) Acellular collagen matrix as a possible 'off the shelf' biomaterial for urethral repair. *Urology,* **54**, 407–10.

Couvelaire, R. (1950) The 'little bladder' of genito-urinary tuberculosis; classification, site and variants of bladder–intestine transplants. *J Urol Medicale Chir,* **56**, 381–434.

Cross, W. R., Eardley, I., Leese, H. J. and Southgate, J. (2005) A biomimetic tissue from cultured normal human urothelial cells: analysis of physiological function. *Am J Physiol Renal Physiol,* **289**, F459–68.

Dahms, S. E., Piechota, H. J., Dahiya, R., Lue, T. F. and Tanagho, E. A. (1998) Composition and biomechanical properties of the bladder acellular matrix graft: comparative analysis in rat, pig and human. *Br J Urol,* **82**, 411–19.

Danielsson, C., Ruault, S., Simonet, M., Neuenschwander, P. and Frey, P. (2006) Polyesterurethane foam scaffold for smooth muscle cell tissue engineering. *Biomaterials,* **27**, 1410–15.

Disandro, M. J., Li, Y., Baskin, L. S., Hayward, S. and Cunha, G. (1998) Mesenchymal-epithelial interactions in bladder smooth muscle development: epithelial specificity. *J Urol,* **160**, 1040–6; discussion 1079.

Draper, J. W., Stark, R. B. and Lau, M. W. (1952) Replacement of mucous membrane of urinary bladder with thick-split grafts of skin: experimental observations. *Plast Reconstr Surg,* **10**, 252–9.

Elbahnasy, A. M., Shalhav, A., Hoenig, D. M., Figenshau, R. and Clayman, R. V. (1998) Bladder wall substitution with synthetic and non-intestinal organic materials. *J Urol,* **159**, 628–37.

Feil, G., Christ-Adler, M., Maurer, S., Corvin, S., Rennekampff, H. O., Krug, J., Hennenflotter, J., Kuehs, U., Stenzl, A. and Sievert, K. D. (2006) Investigations of urothelial cells seeded on commercially available small intestine submucosa. *Eur Urol,* **50**, 1330–7.

Fishman, I. J., Flores, F. N., Scott, F. B., Spjut, H. J. and Morrow, B. (1987) Use of fresh placental membranes for bladder reconstruction. *J Urol,* **138**, 1291–4.

Fraser, M., Thomas, D. F., Pitt, E., Harnden, P., Trejdosiewicz, L. K. and Southgate, J. (2004) A surgical model of composite cystoplasty with cultured urothelial cells: a controlled study of gross outcome and urothelial phenotype. *BJU Int,* **93**, 609–16.

Frimberger, D., Morales, N., Shamblott, M., Gearhart, J. D., Gearhart, J. P. and Lakshmanan, Y. (2005) Human embryoid body-derived stem cells in bladder regeneration using rodent model. *Urology,* **65**, 827–32.

Frimberger, D., Morales, N., Gearhart, J. D., Gearhart, J. P. and Lakshmanan, Y. (2006) Human embryoid body-derived stem cells in tissue engineering-enhanced migration in co-culture with bladder smooth muscle and urothelium. *Urology,* **67**, 1298–303.

Fujita, K. (1978) The use of resin-sprayed thin paper for urinary bladder regeneration. *Invest Urol,* **15**, 355–7.

Fujiyama, C., Masaki, Z. and Sugihara, H. (1995) Reconstruction of the urinary bladder mucosa in three-dimensional collagen gel culture: fibroblast-extracellular matrix interactions on the differentiation of transitional epithelial cells. *J Urol,* **153**, 2060–7.

Greenwell, T. J., Venn, S. N. and Mundy, A. R. (2001) Augmentation cystoplasty. *BJU Int,* **88**, 511–25.

Grinnell, F. (1994) Fibroblasts, myofibroblasts, and wound contraction. *J Cell Biol,* **124**, 401–4.

Hafez, A. T., Afshar, K., Bagli, D. J., Bahoric, A., Aitken, K., Smith, C. R. and Khoury, A. E. (2005) Aerosol transfer of bladder urothelial and smooth muscle cells onto demucosalized colonic segments for porcine bladder augmentation *in vivo*: a 6-week experimental study. *J Urol,* **174,** 1663–7; discussion 1667–8.

Hamill, O. P. and Martinac, B. (2001) Molecular basis of mechanotransduction in living cells. *Physiol Rev,* **81**, 685–740.

Hattori, K., Joraku, A., Miyagawa, T., Kawai, K., Oyasu, R. and Akaza, H. (2006) Bladder reconstruction using a collagen patch prefabricated within the omentum. *Int J Urol,* **13**, 529–37.

Hicks, R. M. (1965) The fine structure of the transitional epithelium of rat ureter. *J Cell Biol,* **26**, 25–48.

Hicks, R. M. (1975) The mammalian urinary bladder: an accommodating organ. *Biol Rev Camb Philos Soc,* **50**, 215–46.

Hu, P., Deng, F. M., Liang, F. X., Hu, C. M., Auerbach, A. B., Shapiro, E., Wu, X. R., Kachar, B. and Sun, T. T. (2000) Ablation of uroplakin III gene results in small urothelial plaques, urothelial leakage, and vesicoureteral reflux. *J Cell Biol,* **151**, 961–72.

Hu, P., Meyers, S., Liang, F. X., Deng, F. M., Kachar, B., Zeidel, M. L. and Sun, T. T. (2002) Role of membrane proteins in permeability barrier function: uroplakin ablation elevates urothelial permeability. *Am J Physiol Renal Physiol,* **283**, F1200–7.

Hutschenreiter, G., Rumpelt, H. J., Klippel, K. F. and Hohenfellner, R. (1978) The free peritoneal transplant as substitute for the urinary bladder wall. *Invest Urol,* **15**, 375–9.

Kanematsu, A., Yamamoto, S., Noguchi, T., Ozeki, M., Tabata, Y. and Ogawa, O. (2003) Bladder regeneration by bladder acellular matrix combined with sustained release of exogenous growth factor. *J Urol,* **170**, 1633–8.

Kelami, A., Dustmann, H. O., Ludtke-Handjery, A., Carcamo, V. and Herlld, G. (1970a) Experimental investigations of bladder regeneration using Teflon-felt as a bladder wall substitute. *J Urol,* **104**, 693–8.

Kelami, A., Ludtke-Handjery, A., Korb, G., Rolle, J., Schnell, J. and Danigel, K. H. (1970b) Alloplastic replacement of the urinary bladder wall with lyophilized human dura. *Eur Surg Res,* **2**, 195–202.

Kim, B. S., Baez, C. E. and Atala, A. (2000) Biomaterials for tissue engineering. *World J Urol,* **18**, 2–9.

Kimuli, M., Eardley, I. and Southgate, J. (2004) *In vitro* assessment of decellularized porcine dermis as a matrix for urinary tract reconstruction. *BJU Int,* **94**, 859–66.

Kiricuta, I. and Goldstein, A. M. (1972) The repair of extensive vesicovaginal fistulas with pedicled omentum: a review of 27 cases. *J Urol,* **108**, 724–7.

Knapp, P. M., Lingeman, J. E., Siegel, Y. I., Badylak, S. F. and Demeter, R. J. (1994) Biocompatibility of small-intestinal submucosa in urinary tract as augmentation cystoplasty graft and injectable suspension. *J Endourol,* **8**, 125–30.

Koiso, K., Komai, T. and Niijima, T. (1983) Experimental urinary bladder reconstruction using a synthetic poly(alpha-amino acids) membrane. *Artif Organs,* **7**, 232–7.

Korossis, S., Bolland, F., Ingham, E., Fisher, J., Kearney, J. and Southgate, J. (2006) Review: tissue engineering of the urinary bladder: considering structure–function relationships and the role of mechanotransduction. *Tissue Eng,* **12**, 635–44.

Kropp, B. P., Rippy, M. K., Badylak, S. F., Adams, M. C., Keating, M. A., Rink, R. C. and Thor, K. B. (1996a) Regenerative urinary bladder augmentation using small intestinal submucosa: urodynamic and histopathologic assessment in long-term canine bladder augmentations. *J Urol,* **155**, 2098–104.

Kropp, B. P., Sawyer, B. D., Shannon, H. E., Rippy, M. K., Badylak, S. F., Adams, M. C., Keating, M. A., Rink, R. C. and Thor, K. B. (1996b) Characterization of small intestinal submucosa regenerated canine detrusor: assessment of reinnervation, *in vitro* compliance and contractility. *J Urol,* **156**, 599–607.

Kropp, B. P., Cheng, E. Y., Lin, H. K. and Zhang, Y. (2004) Reliable and reproducible bladder regeneration using unseeded distal small intestinal submucosa. *J Urol,* **172**, 1710–13.

Kudish, H. G. (1957) The use of polyvinyl sponge for experimental cystoplasty. *J Urol,* **78**, 232–5.

Kumar, S. P. and Abrams, P. H. (2005) Detrusor myectomy: long-term results with a minimum follow-up of 2 years. *BJU Int,* **96**, 341–4.

Lavelle, J., Meyers, S., Ramage, R., Bastacky, S., Doty, D., Apodaca, G. and Zeidel, M. L. (2002) Bladder permeability barrier: recovery from selective injury of surface epithelial cells. *Am J Physiol Renal Physiol,* **283**, F242–53.

Lewis, S. A. (2000) Everything you wanted to know about the bladder epithelium but were afraid to ask. *Am J Physiol Renal Physiol,* **278,** F867–74.

Lima, S. V., Araujo, L. A. and Vilar, F. O. (2004) Nonsecretory intestinocystoplasty: a 10-year experience. *J Urol,* **171**, 2636–9; discussion 2639–40.

Marte, A., di Meglio, D., Cotrufo, A. M., di Iorio, G., de Pasquale, M. and Vessella, A. (2002) A long-term follow-up of autoaugmentation in myelodysplastic children. *BJU Int,* **89**, 928–31.

Master, V. A., Wei, G., Liu, W. and Baskin, L. S. (2003) Urothlelium facilitates the recruitment and trans-differentiation of fibroblasts into smooth muscle in acellular matrix. *J Urol,* **170**, 1628–32.

Merguerian, P. A., Reddy, P. P., Barrieras, D. J., Wilson, G. J., Woodhouse, K., Bagli, D. J., McLorie, G. A. and Khoury, A. E. (2000) Acellular bladder matrix allografts in the regeneration of functional bladders: evaluation of large-segment (>24 cm) substitution in a porcine model. *BJU Int,* **85**, 894–8.

Mitrofanoff, P. (1980) Trans-appendicular continent cystostomy in the management of the neurogenic bladder. *Chir Pediatr,* **21**, 297–305.

Motley, R. C., Montgomery, B. T., Zollman, P. E., Holley, K. E. and Kramer, S. A. (1990) Augmentation cystoplasty utilizing de-epithelialized sigmoid colon: a preliminary study. *J Urol,* **143,** 1257–60.

Nakanishi, Y., Chen, G., Komuro, H., Ushida, T., Kaneko, S., Tateishi, T. and Kaneko, M. (2003) Tissue-engineered urinary bladder wall using PLGA mesh–collagen hybrid scaffolds: a comparison study of collagen sponge and gel as a scaffold. *J Pediatr Surg,* **38**, 1781–4.

Neuhof, H. (1917) Fascial transplantation into visceral defects: an experimental and clinical study. *Surg, Gynec Obst,* **25**, 383.

Nuininga, J. E., van Moerkerk, H., Hanssen, A., Hilsbergen, C. A., Oosterwijk-Wakka, J., Oosterwijk, E., de Gier, R. P., Schalken, J. A., van Kuppevelt, T. H. and Feitz, W. F. (2004) A rabbit model to tissue engineer the bladder. *Biomaterials,* **25**, 1657–61.

Oberpenning, F., Meng, J., Yoo, J. J. and Atala, A. (1999) *De novo* reconstitution of a functional mammalian urinary bladder by tissue engineering. *Nat Biotechnol,* **17**, 149–55.

Pariente, J. L., Kim, B. S. and Atala, A. (2001) *In vitro* biocompatibility assessment of naturally derived and synthetic biomaterials using normal human urothelial cells. *J Biomed Mater Res,* **55**, 33–9.

Pariente, J. L., Kim, B. S. and Atala, A. (2002) *In vitro* biocompatibility evaluation of naturally derived and synthetic biomaterials using normal human bladder smooth muscle cells. *J Urol,* **167**, 1867–71.

Piechota, H. J., Dahms, S. E., Nunes, L. S., Dahiya, R., Lue, T. F. and Tanagho, E. A. (1998) *In vitro* functional properties of the rat bladder regenerated by the bladder acellular matrix graft. *J Urol,* **159**, 1717–24.

Probst, M., Dahiya, R., Carrier, S. and Tanagho, E. A. (1997) Reproduction of functional smooth muscle tissue and partial bladder replacement. *Br J Urol,* **79**, 505–15.

Probst, M., Piechota, H. J., Dahiya, R. and Tanagho, E. A. (2000) Homologous bladder augmentation in dog with the bladder acellular matrix graft. *BJU Int,* **85**, 362–71.

Ram-Liebig, G., Meye, A., Hakenberg, O. W., Haase, M., Baretton, G. and Wirth, M. P. (2004) Induction of proliferation and differentiation of cultured urothelial cells on acellular biomaterials. *BJU Int,* **94**, 922–7.

Ram-Liebig, G., Ravens, U., Balana, B., Haase, M., Baretton, G. and Wirth, M. P. (2006) New approaches in the modulation of bladder smooth muscle cells on viable detrusor constructs. *World J Urol,* **24,** 429–37.

Record, R. D., Hillegonds, D., Simmons, C., Tullius, R., Rickey, F. A., Elmore, D. and Badylak, S. F. (2001) *In vivo* degradation of 14C-labeled small intestinal submucosa (SIS) when used for urinary bladder repair. *Biomaterials,* **22**, 2653–9.

Reddy, P. P., Barrieras, D. J., Wilson, G., Bagli, D. J., McLorie, G. A., Khoury, A. E. and Merguerian, P. A. (2000) Regeneration of functional bladder substitutes using large segment acellular matrix allografts in a porcine model. *J Urol,* **164**, 936–41.

Rohman, G., Pettit, J. J., Isaure, I., Cameron, N. R. and Southgate, J. (2007) Influence of the physical properties of two-dimensional polyester substrates on the growth of normal human urothelial and urinary smooth muscle cells *in vitro*. *Biomaterials*, **28**, 2264–74.

Salle, J. L., Fraga, J. C., Lucib, A., Lampertz, M., Jobim, G., Jobim, G. and Putten, A. (1990) Seromuscular enterocystoplasty in dogs. *J Urol,* **144**, 454–6; discussion 460.

Santucci, R. A. and Barber, T. D. (2005) Resorbable extracellular matrix grafts in urologic reconstruction. *Int Braz J Urol,* **31**, 192–203.

Schoeller, T., Neumeister, M. W., Huemer, G. M., Russell, R. C., Lille, S., Otto-Schoeller, A. and Wechselberger, G. (2004) Capsule induction technique in a rat model for bladder wall replacement: an overview. *Biomaterials,* **25**, 1663–73.

Scriven, S. D., Trejdosiewicz, L. K., Thomas, D. F. M. and Southgate, J. (2001) Urothelial cell transplantation using biodegradable snthetic scaffolds. *J Mat Sci Mat Med,* **12**, 991–6.

Southgate, J., Hutton, K. A., Thomas, D. F. and Trejdosiewicz, L. K. (1994) Normal human urothelial cells *in vitro*: proliferation and induction of stratification. *Lab Invest,* **71**, 583–94.

Staack, A., Hayward, S. W., Baskin, L. S. and Cunha, G. R. (2005) Molecular, cellular and developmental biology of urothelium as a basis of bladder regeneration. *Differentiation,* **73**, 121–33.

Stenzl, A., Strasser, H., Klima, G., Eder, I., Frauscher, F., Klocker, H., Bartsch, G. and Ninkovic, M. (2000) Reconstruction of the lower urinary tract using autologous muscle transfer and cell seeding: current status and future perspectives. *World J. Urol.*, **18**, 44–50.

Sutherland, R. S., Baskin, L. S., Hayward, S. W. and Cunha, G. R. (1996) Regeneration of bladder urothelium, smooth muscle, blood vessels and nerves into an acellular tissue matrix. *J Urol,* **156**, 571–7.

Thomas, D. F. (1997) Surgical treatment of urinary incontinence. *Arch Dis Child,* **76**, 377–80.

Tizzoni, G. and Foggi, A. (1888) Die Weiderhestellung der Harnblase. *Centralbl F Chir,* **15**, 921–3.

Truschel, S. T., Wang, E., Ruiz, W. G., Leung, S. M., Rojas, R., Lavelle, J., Zeidel, M., Stoffer, D. and Apodaca, G. (2002) Stretch-regulated exocytosis/endocytosis in bladder umbrella cells. *Mol Biol Cell,* **13**, 830–46.

Tu, L., Sun, T. T. and Kreibich, G. (2002) Specific heterodimer formation is a prerequisite for uroplakins to exit from the endoplasmic reticulum. *Mol Biol Cell,* **13**, 4221–30.

Turner, A. M., Subramaniam, R., Thomas, D. F. M. and Southgate, J. (2007) Generation of a functional, differentiated porcine urothelial tissue *in vitro* for use in composite cystoplasty (submitted).

Vacanti, J. P. and Langer, R. (1999) Tissue engineering: the design and fabrication of living replacement devices for surgical reconstruction and transplantation. *Lancet,* **354 Suppl 1**, SI32–4.

Varley, C. L., Stahlschmidt, J., Lee, W. C., Holder, J., Diggle, C., Selby, P. J., Trejdosiewicz, L. K. and Southgate, J. (2004a) Role of PPARgamma and EGFR signalling in the urothelial terminal differentiation programme. *J Cell Sci,* **117**, 2029–36.

Varley, C. L., Stahlschmidt, J., Smith, B., Stower, M. and Southgate, J. (2004b) Activation of peroxisome proliferator-activated receptor-gamma reverses squamous metaplasia and induces transitional differentiation in normal human urothelial cells. *Am J Pathol,* **164**, 1789–98.

Varley, C. L., Garthwaite, M. A., Cross, W., Hinley, J., Trejdosiewicz, L. K. and Southgate, J. (2006) PPARgamma-regulated tight junction development during human urothelial cytodifferentiation. *J Cell Physiol,* **208**, 407–17.

Vaught, J. D., Kropp, B. P., Sawyer, B. D., Rippy, M. K., Badylak, S. F., Shannon, H. E. and Thor, K. B. (1996) Detrusor regeneration in the rat using porcine small intestinal submucosal grafts: functional innervation and receptor expression. *J Urol,* **155**, 374–8.

Von-Mikulicz, J. (1889) Zur Operation der angebarenen Blaben-Spalte. *Zentralbl Chir,* **20**, 641–3.

Wu, X. R., Manabe, M., Yu, J. and Sun, T. T. (1990) Large scale purification and immunolocalization of bovine uroplakins I, II, and III. Molecular markers of urothelial differentiation. *J Biol Chem,* **265**, 19170–9.

Yoo, J. J., Meng, J., Oberpenning, F. and Atala, A. (1998) Bladder augmentation using allogenic bladder submucosa seeded with cells. *Urology,* **51**, 221–5.

Yu, J., Lin, J. H., Wu, X. R. and Sun, T. T. (1994) Uroplakins Ia and Ib, two major differentiation products of bladder epithelium, belong to a family of four transmembrane domain (4TM) proteins. *J Cell Biol,* **125**, 171–82.

Zhang, Y., Kropp, B. P., Moore, P., Cowan, R., Furness III, P. D., Kolligian, M. E., Frey, P. and Cheng, E. Y. (2000) Coculture of bladder urothelial and smooth muscle cells on small intestinal submucosa: potential applications for tissue engineering technology. *J Urol,* **164,** 928–34; discussion 934–5.

Zhang, Y., Kropp, B. P., Lin, H. K., Cowan, R. and Cheng, E. Y. (2004) Bladder regeneration with cell-seeded small intestinal submucosa. *Tissue Eng,* **10**, 181–7.

Zisch, A. H., Lutolf, M. P., Ehrbar, M., Raeber, G. P., Rizzi, S. C., Davies, N., Schmokel, H., Bezuidenhout, D., Djonov, V., Zilla, P. and Hubbell, J. A. (2003) Cell-demanded release of VEGF from synthetic, biointeractive cell ingrowth matrices for vascularized tissue growth. *Faseb J,* **17**, 2260–2.

22
Nerve bioengineering

P KINGHAM and G TERENGHI, University of Manchester, UK

22.1 Peripheral nerve

22.1.1 Structure

The peripheral nervous system consists of the cranial nerves, spinal nerves, peripheral nerves and the peripheral components of the autonomic nervous system. Peripheral nerves contain a variable mix of fibres, both myelinated and unmyelinated, of three broad types. Motor fibres supply the end plates in skeletal muscle, sensory fibres receive information from viscera, skin, muscle, tendon and joints and autonomic fibres, both sympathetic and parasympathetic, subserve the blood vessels, viscera, sweat glands and arrector pilae muscles.

Individual nerve fibres are supported by a collagen-rich endoneurium and they are grouped in bundles, or fascicles, which are contained within the perineurium. This specialised tissue is composed of flattened perineurial cells alternating with layers of collagen. It functions to maintain the homeostasis of the endoneurial fluid surrounding myelinated and unmyelinated fibres and to provide a barrier to diffusion. Fascicles are embedded in connective tissue which, in its outer layers, condenses to form the epineurium or nerve sheath. A peripheral nerve may contain one or multiple fascicles (Birch *et al.*, 1998). Each peripheral nerve has an extensive blood supply composed of interconnecting epineurial, perineurial and endoneurial plexuses which link with extrinsic regional vessels.

22.1.2 Neuronal cells

The major cellular constituents of the peripheral nervous system are neurons and Schwann cells. The neuronal cell body is generally sited in a ganglion, where numerous neuronal cells are grouped, and from where its processes, dendrites and axons, originate. The axon is a column of neuronal cytoplasm, or axoplasm, enclosed by a cell membrane, or axolemma. Within the axon there is a complex system of axoplasmic reticulum, membranous cisterns, tubes and vesicles, mitochondria, lamellar and multivesicular bodies (Birch *et al.*, 1998). Most important is the axonal cytoskeleton of microtubules, neurofilaments and matrix which provides the apparatus for axoplasmic transport (Hollenbeck, 1989).

The axon carries materials between the cell body and the distal end organs in two forms of transport, fast and slow. Fast axonal transport works in both directions at a rate of 200–400 mm/day. Neurotransmitters synthesised in the cell body are transported anterogradely to the distal end of the axon, while distally uptaken extracellular molecules, such as growth factors, are simultaneously transported retrogradely to the cell body. Slow transport occurs only in an anterograde direction, transporting cytoskeletal components from the cell body to the distal terminus. The rate of slow transport, 1–4 mm/day, is thought to correspond to that of peripheral regeneration following axotomy (Danielsen *et al.*, 1986).

22.1.3 Schwann cells

In the peripheral nervous system, axons are closely associated with Schwann cells (SC). These wrap along the entire length of the larger axons, juxtaposing one another at the nodes of Ranvier and laying down spiral layers of myelin sheath. Each axon–SC unit, or nerve fibre, is contained within a basal lamina. The smaller fibres are arranged in bundles surrounded by similar columns of SC. It is probably the diameter of the axon that determines whether the SCs will lay down a myelin sheath around it (Birch *et al.*, 1998).

SC develop from the neural crest and appear in mature nerves as two different cell types, myelinating and non-myelinating, both derived from the same precursor cell (Jessen and Mirsky 1999), and forming a stable, non-proliferating population. These developmental changes are associated with changes in gene expression and protein synthesis such as up-regulation of myelin proteins P0, myelin basic protein and peripheral myelin protein and down-regulation of NCAM, $p75^{NTR}$ and GFAP. If a mature SC loses contact with axons, it undergoes radical changes in morphology and gene expression that lead to developmental regression or de-differentiation of individual SC and myelin breakdown, followed by proliferation.

22.1.4 Extracellular matrix

The cellular components of the peripheral nervous system are supported by an extracellular matrix (ECM) comprising the SC basal lamina and surrounding extracellular space. The ECM contains a diverse set of macromolecules including laminin-2, collagen types IV and VI, P200, tenascin-C, F-spondin, heparan sulphate and chondroitin sulphate proteoglycans, fibronectin and entactin. During peripheral nerve development SC themselves synthesise and assemble basal lamina ECM and fibril-forming collagens (Chernousov *et al.*, 1998). This synthesis is dependent upon axonal contact (Bunge *et al.*, 1982). Following injury, many ECM molecules are important in promoting axonal growth and regeneration (Ard *et al.*, 1987; Agius and Cochard, 1998). Thus peripheral nerve integrity is maintained by the close coordination and complex interactions of both its cellular and its extracellular components. It is this complexity that makes the tissue engineering assembly of peripheral nerve components so challenging.

22.2 Peripheral nerve injury and regeneration

Following the transection of a peripheral nerve, in the distal nerve stump Wallerian degeneration occurs which is characterised by macrophage recruitment, degradation of both the axonal and myelin components and proliferation of endoneurial fibroblasts and resident SC population, which align within the original basal lamina to form bands of Büngner. In the short term the denervated SC population becomes growth supportive (Gordon *et al.*, 2003) forming a conduit to guide regenerating axons (Son and Thompson, 1995) and expressing neurotrophic factors, such as nerve growth factor, brain-derived neurotrophic factor and ciliary neurotrophic factor, conducive to axonal growth (Heumann *et al.*, 1987; Meyer *et al.*, 1992; Rende *et al.*, 1992).

As early as 3 hours post-axotomy the proximal nerve produces axon sprouts from the node of Ranvier (Torigoe *et al.*, 1996). Regeneration follows after a few days, during which the neuronal cell changes its metabolism from maintenance to regeneration. This latency period varies among species (Hall, 2001), and it is followed by regeneration and axonal branching. This sprouting is gradually modified as a proportion of axons reach their target, while others fail and are pruned back (Brushart, 1993). Regeneration is influenced by neurotropic factors from the distal stump that enhance axonal growth and survival, and guide them towards the correct target, and by microgeometric cues from cell adhesion molecules (Thomas, 1989; Ide, 1996). The effectiveness of this guidance is influenced by the distance between the proximal and distal nerve stumps (Thomas, 1989; Hall, 2001). Ultimately, successful regeneration is dependent upon axons of motor, sensory and autonomic type making appropriate connections first with the distal nerve stump and, finally, with their target organs.

22.3 Peripheral nerve repair

Despite major advances in reconstructive surgery techniques in recent years, results of peripheral nerve repair remain unpredictable. Frequently the end result is impaired function and chronic pain, leading to both disability and a decreased quality of life (Lundborg, 2003). In some cases of peripheral nerve injury, direct end-to-end nerve repair is not feasible due to a gap between the transected nerve ends. This may result from actual nerve tissue loss, retraction of the nerve ends or the necessity to excise crushed or damaged nerve stumps prior to repair.

Repair of a gap injury using an autologous nerve graft was not widely accepted until the 1970s with the advent of microsurgery (Millesi *et al.*, 1976; Matejcik, 2002) and to date it remains the gold standard in bridging a nerve gap (Birch *et al.*, 1998; Battiston *et al.*, 2005). A graft is commonly harvested from the sural nerve in the leg or the medial cutaneous nerve in the forearm of the patient, and it provides a guidance channel to the regrowing axons as it contains a basal lamina scaffold and endogenous Schwann cells. As a natural, non-immunogenic, ready-to-use graft, this method has clear advantages, but functional outcome remains extremely variable (Nunley *et al.*, 1996). Technical modifications such as the vascularised nerve graft have failed to demonstrate significant improvement in outcome (Doi *et al.*, 1992). Also, for the patient, there is additional donor site morbidity, scarring, sensory loss and possible neuroma formation (Wu and Chiu, 1999; Evans, 2001).

Nerve allografting circumvents the problems caused by autogenous nerve harvest and has been used clinically in situations where autografting was not possible (Mackinnon *et al.*, 2001). However, the necessity for immune suppression following this procedure makes it difficult to justify in the majority of cases.

22.4 Bioengineered nerve conduits

In recent years, research has been directed towards the development of an alternative to nerve autograft to bridge the injury gap in the form of a nerve conduit. The ideal conduit should be inert, immunologically compatible, biodegradable and able to support axonal growth and should result in functional recuperation at least comparable, or superior, to that achieved with a nerve graft (Dahlin and Lundborg, 2001; Weber and Dellon, 2001). Attempts to use conduits to repair nerve gap date back to the last century and materials that were used include decalcified bone, rubber tubes, fat and fascial sheaths, blood vessels, and tubes made out of parchment, metal, collagen, tantalum, millipore elastic, polygalactin and polyorthoester (reviewed in Mackinnon and Dellon, 1988) (Table 22.1). More recent reports document the successful use of both biological and synthetic materials (Table 22.2).

22.4.1 Biological conduits

Conduits developed from a biological tissue include acellular muscle grafts, which can be created by either freeze–thawing or heat treatment, destroying cellular component and leaving behind their basal lamina (Whitworth *et al.*, 1995). Muscle grafts have been shown to support regeneration comparable to nerve grafts over a 2 cm rat sciatic nerve gap (Bryan *et al.*, 1993). *In vivo* the graft was penetrated by Schwann cells, fibroblasts, perineural and endothelial cells and axon regeneration was apparent within 3 weeks, while the graft itself

Table 22.1 A selection of synthetic and biological components used for generation of nerve guides

Polymeric materials
Non-resorbable
- Non-porous
  - Ethylene-vinyl acetate copolymer (EVA)
  - Polytetrafluoroethylene (PTFE)
  - Polyethylene (PE)
  - Silicone elastomers (SE)
  - Polyvinyl chloride (PVC)
- Microporous
  - Expanded polytetrafluoroethylene (ePTFE)
  - Millipore (cellulose filter)
- Semipermeable
  - Polyacrylonitrile (PAN)
  - Polyacrylonitrile/polyvinyl chloride (PAN/VC)
  - Polysulphone (PS)
- Piezoelectric
  - Polyvinylidene fluoride (PVOF)
  - PTFE-electret
- Release of trophic factors
  - Ethylene vinyl acetate (EVA)

Bioresorbable
- Polyglycolide (PGA)
- Polylactide (PLLA)
- PGA/PLLA blends
- Polycaprolactone (PCL)
- Polyhydroxybutyrate (PHB)
- Polyhydroxyethyl methacrylate

Biological materials
- Artery/vein
- Muscle
- Decalcified bone
- Collagen/gelatine
- Hyaluronic acid derivatives
- Alginate

Adapted from Valentini, RF and Aebischer, P. Strategies for the engineering of peripheral nervous tissue regeneration. In RP Lanza, R Langer and WL Chick (Eds), *Principles of Tissue Engineering*, Academic Press, RG Landes Company, Austin, 1997, pp. 671–684.

ultimately disappears (Hall, 1997). However, even with the addition of transplanted SC, muscle grafts did not perform as well as autologous nerve graft controls (Keilhoff *et al.*, 2005).

Reasonable clinical outcomes have been reported with the use of vein grafts to bridge sensory nerve lesions in the hand (Risitano *et al.*, 2002). Modifications such as turning the vein inside-out (Wang *et al.*, 1993a) and coating the vein with collagen gel (Wang *et al.*, 1993b) have also had moderate success. However, this technique appears inferior in comparison with muscle grafts

Table 22.2 Human nerve conduit implant studies

Material	Authors	Article reference
Silicone	Merle *et al.*	*Microsurgery* 1989, **10**: 130
	Lundborg *et al.*	*J Hand Surg* 1994, **19B**: 273
	Lundborg *et al.*	*J Hand Surg* 1997, **22A**: 99
	Braga-Silva	*J Hand Surg* 1999, **24B**: 703
	Dahlin *et al.*	*Scand J Plast Reconst Surg* 2001, **35**: 29
	Lundborg *et al.*	*J Hand Surg* 2004, **29B**: 100
PTFE	Stanec *et al.*	*J Reconstr Microsurg* 1998, **14**: 227
	Stanec *et al.*	*Br J Plast Surg* 1998, **51**: 637
	Pogrel *et al.*	*J Oral Maxillofac Surg* 1998, **56**: 319
	Pitta *et al.*	*J Oral Maxillofac Surg* 2001, **59**: 493
PGA	MacKinnon *et al.*	*Plast Reconstr Surg* 1990, **85**: 419
	Crawley *et al.*	*Plast Reconstr Surg* 1992, **90**: 300
	Weber *et al.*	*Plast Reconstr Surg* 2000, **106**: 1036
	Inada *et al.*	*Neurosurgery* 2004, **55**: 640
	Inada *et al.*	*Pain* 2005, **117**: 251
	Navissano *et al.*	*Microsurgery* 2005, **25**: 268
PLCL	Bertleff *et al.*	*J Hand Surg* 2005, **30A**: 513

Adapted from Schlosshauer *et al.* Synthetic nerve guide implants in humans: a comprehensive study. *Neurosurgery* 2006, **59**, 740–748.

(Fansa *et al.*, 2001). Small intestinal mucosa has been proposed as a potential biological nerve conduit (Smith *et al.*, 2004) and, stripped of its mucosal and serosal layers to leave an acellular collagen matrix, it can be fashioned into a roll to bridge a nerve gap (Hadlock *et al.*, 2001). While the results are poor when this conduit is used alone, seeding with cultured SC promotes significant regeneration approaching that achieved by autograft.

Collagen is a biological material that can be shaped into a conduit and collagen tubes obtained from rat tail tendons have been shown to support a moderate nerve regeneration across a 1 cm nerve gap (Brandt *et al.*, 1999). Nerve regeneration was also demonstrated in collagen tubes stabilised by microwave crosslinking, while the absence of crosslinkage resulted in swelling of the tube, obstruction of the lumen and lack of regeneration (Ahmed *et al.*, 2004). Similarly, type one bovine collagen tubes, strengthened by UV irradiation, showed regeneration comparable to that of autografts over 1.5 cm nerve gaps whereas untreated tubes obstructed axonal growth (Itoh *et al.*, 2002). Over longer (2 cm) gaps purified bovine collagen tubes failed to bridge a nerve defect (Yoshii and Oka, 2001).

22.4.2 Synthetic conduits

Silicone conduits have been one of the most extensively used materials, both in experimental models of nerve regeneration and clinically (Fields *et al.*, 1989;

Dahlin *et al.*, 2001; Li *et al.*, 2004). The non-absorbable, inert silicone conduit acts as a biological chamber, allowing the accumulation of growth factors, ECM molecules and SCs which promote nerve regeneration across short nerve gaps (Lundborg *et al.*, 1997). However, a clinical trial using silicone tubes has shown that the conduit can cause symptoms of mild irritation and nerve compression, occasionally necessitating removal (Lundborg *et al.*, 2004). Other non-absorbable materials used for nerve conduits include polysulphone (Aebischer *et al.*, 1989), polyvinyl chloride (Scaravilli, 1984) and Gore-Tex tubing (Pitta *et al.*, 2001). More recently, attention has focused on the increasing availability of biodegradable synthetic materials.

Synthetic polymeric biomaterials are advantageous in their unlimited supply and reproducible properties (Wan *et al.*, 2001). Poly-alpha esters and their copolymers degrade into low molecular weight compounds, such as lactic and glycolic acid, which enter normal metabolic pathways (Wang *et al.*, 1990). A number of polymers from this family have been used to create nerve conduit including poly-L-glycolic acid (Bini *et al.*, 2003), poly-L-lactic acid (Evans, 2001) and poly-L-lactide-co-6-caprolactone (Aldini *et al.*, 1995).

Polyglycolic acid slowly degrades to non-toxic products and it is already used clinically as suture materials and implants (Ginde and Gupta, 1987). Polyglycolic acid nerve conduits showed results comparable with autografting in primate (MacKinnon and Dellon, 1990) and recently, these conduits have been used in a multicentre randomised prospective clinical trial (Battiston *et al.*, 2005). Other synthetic materials such as polygalactin and maxon, similarly approved for use as resorbable surgical sutures, have been investigated experimentally as nerve conduits (Molander *et al.*, 1982; Seckel *et al.*, 1986; Mackinnon and Dellon, 1990). Polyphosphoester, previously designed as a biopolymer for synthetic drug release (Dahiyat *et al.*, 1995) can be formed into conduits which have been shown to support successful regeneration, although their rapid degradation rate represents a limiting factor (Wang *et al.*, 2000).

Biological polymers, such as the polysaccharides, have found uses in a diverse range of fields from food thickening and flavouring to ophthalmic surgery (Byrom, 1987). One class of biopolymers that has attracted particular attention in the creation of medical devices has been polyhydroxyalkanoates or PHA (Zinn *et al.*, 2001). These are bacterial storage compounds constituting multiple different hydroxyalkanoates which have the capacity to confer a range of mechanical and physical properties depending on the type of bacteria and culture environment used (Lee, 1996). The biocompatible, biodegradable, properties of PHA along with the potential to alter their composition to obtain specific mechanical properties and degradation times, makes them attractive tissue engineering materials (Chen and Wu, 2005).

Poly-3-hydroxybutyrate (PHB) was the first PHA to be identified and is the most widely distributed member of the family (Zinn *et al.*, 2001). PHB can be polymerised into resorbable, non-antigenic mats that are strong, flexible and

22.1 (a) PHB polymer sheets showing the flexibility of the material; (b) PHB sheets rolled and sealed to form a conduit suitable for grafting into a nerve injury, as shown (c) in the intra-operative photograph, and (d) in a schematic representation.

easy to handle (Fig. 22.1(a)), making it suitable to use as a wrap-around in direct nerve repair (Hazari *et al.*, 1999a). We have also shown that PHB mats rolled and sealed to create a conduit (Fig 22.1) promote axonal regeneration when bridging both short (Hazari *et al.*, 1999b) and long (Young *et al.*, 2002) nerve gaps. The fibrous composition of the PHB sheets could be orientated longitudinally to the conduit, providing directional and contact guidance to the regrowing axons (Young *et al.*, 2002). PHB conduits become vascularised at an early stage in regeneration which is an essential requisite for the delivery of nutrients to the highly metabolic environment of the regenerating nerve.

22.5 Matrix materials

A major development in the construction of bioengineered nerve conduits has been the progression from the simple concept of tubulisation to the creation of a conduit that more closely mimics the nerve environment. The provision of a three-dimensional matrix supports the regrowing axons and the proliferating Schwann cells, enabling the regeneration through a longer gap length (Hadlock *et al.*, 2000). Within such a matrix there is also the option of incorporating cells and growth factors considered permissive to nerve regeneration, which improve further the function of the conduit (Keilhoff *et al.*, 2005).

The matrix may present different structure, and an amorphous gel has the potential to provide the properties of the ECM such as visco-elasticity, diffusive transport and interstitial flow (Lutolf and Hubbell, 2005). Hydrogels are highly hydrated polymers with water content >30% and composed of hydrophilic polymer chains that are readily degradable (Drury and Mooney, 2003). Their mechanical and structural similarities to ECM make them a preferred choice for tissue engineering. There are both synthetic and naturally occurring hydrogels, the natural ones being particularly applicable for tissue engineering because many are either components of natural ECM or have macromolecular properties similar to it (Drury and Mooney, 2003).

Collagen is the main component of natural ECMs and the most abundant protein in mammalian tissues (Drury and Mooney, 2003) and it has been used as a material for nerve conduits to bridge nerve gaps (Madison *et al.*, 1984, 1985; Labrador *et al.*, 1998). However, collagen and laminin-containing gels (Vitrogel, Zyderm and Matrigel) appeared to impede peripheral nerve regeneration in comparison to saline-filled tubes (Valentini *et al.*, 1987), a finding confirmed in a later study (Guenard *et al.*, 1992). A second ECM component, hyaluronic acid, when added sequentially to a nerve guide via an injection port, shows improvements in axonal count and conduction velocities (Seckel *et al.*, 1995).

Alginate hydrogel, a naturally occurring material extracted from brown algae, has been employed as a matrix for SC transplantation and controlled release of glial growth factor (GGF) within a nerve conduit (Mosahebi *et al.*, 2001; Mohanna *et al.*, 2003, 2005). Combining alginate with fibronectin, a glycoprotein constituent of normal ECM, improved peripheral nerve regeneration *in vivo* (Mosahebi *et al.*, 2003). However, without the additional inclusion of GGF within the conduit, regeneration in alginate-only conduit was found to be poor, possibly because of the slow degradation of the hydrogel, and suggesting that it may cause an obstruction to the regenerating axons (Mohanna *et al.*, 2005).

Bovine collagen conduit filled with a porous collagen–glycosaminoglycan matrix showed good nerve regeneration (Chamberlain *et al.*, 2000). This matrix increased axonal regeneration in both silicone and collagen tubes in comparison with empty controls but only surpassed that of autografts in the collagen tubes. The success of this material may be explained by its structure of axially orientated pore channels. Porous structures offer an alternative luminal scaffold to the gel-based matrix as this conformation may be more permissive to axonal regeneration by allowing space for axonal growth and providing an increased surface area for the adherence of incorporated cells. Porcine collagen can also be processed into a sponge of interconnected pores which is biocompatible and permissive to cell attachment (Hiraoka *et al.*, 2003). When inserted into a PGA conduit, this collagen sponge enabled superior functional recovery compared with autografts experimentally (Nakamura *et al.*, 2004) and it has also shown some success clinically (Inada *et al.*, 2004). Pore orientation may be an important factor in guiding regeneration, and a matrix of longitudinally aligned channels coated with laminin supported SC adherence and nerve regeneration similar to nerve autografts (Hadlock *et al.*, 2000). Synthetic channelled conduits incorporating extracellular matrix molecules and neurotrophic factors have also been investigated (Bender *et al.*, 2004; Yang *et al.*, 2005)

While this approach shows promise, creation of longitudinally aligned channels within a matrix material is a complex and demanding process (Sundback *et al.*, 2003) and it requires a high degree of mechanical stability of the material. An alternative concept is to fill a conduit with longitudinally aligned fibres, offering the same advantages of space and increased surface area. This construction may be able to organise structurally the regeneration and direct the formation of the nerve cable, more closely approximating the bands of Büngner seen in a naturally regenerating nerve. In its simplest form this concept was developed by Lundborg *et al.* (1997) who found that the addition of eight, 250 μm polyamide filaments to a silicone tube allowed the bridgeable gap length in a rat sciatic nerve to be increased from 1 to 1.5 cm. Interestingly the authors observed that this fibrous inner structure clearly enhanced axonal regeneration, but developing axons did not appear to advance in immediate contact with the polyamide filaments.

A subsequent development of this concept was the introduction of synthetic fibres into a silicone conduit using poly-L-Lactide (PLLA) filaments of 40–100 μm in diameter (Ngo *et al.*, 2003). The function of these fibres was shown to be density dependent, and at low densities (16–32 fibres per conduit) they enhance axonal regeneration across a 1 cm sciatic nerve gap compared with empty tubes. Furthermore, the addition of PLLA fibres enhances regeneration when longer gaps (up to 1.8 cm) are used. As with the previous studies the authors noted that macrophages and fibroblasts grew in close association with the PLLA fibres, while SC and axons rarely did (Ngo *et al.*, 2003). The relationship of regrowing axons was studied *in vitro* using bundles of polypropylene filaments coated with various ECM molecules. An increasing pattern of alignment of neurites was more evident when the filaments were of smaller diameter and grouped in bundles, in effect replicating the fascicular structure of the nerve (Wen and Tresco, 2006). Furthermore, SC alignment was found to be dependent on the width, spacing and depth of microgrooves (Hsu *et al.*, 2005) or microcontact printed laminin (Wang *et al.*, 2006).

Natural materials can also be manufactured as fibres, and a collagen conduit filled with collagen fibres supported axonal regeneration across a 1 cm rat sciatic nerve gap (Tong *et al.*, 1994). Coating the fibres with ECM molecules can increase both the speed of regeneration and the number of myelinated axons as shown by the use of longitudinally orientated laminin-coated collagen fibres within a PGA conduit to bridge an 8 cm nerve gap (Matsumoto *et al.*, 2000). Collagen degrades more rapidly than synthetic materials, and in both these studies it was found to be completely resorbed at the time of harvest. This factor may confer an advantage in that the fibres are less likely to cause obstruction as the nerve develops. However it also limits the opportunity to observe the relationship between the fibres and SC, axons and other cells. Other studies also suggest that cells will preferentially grow along longitudinal fibres composed of chitosan, a polysaccharide with properties similar to natural ECM molecules (Yuan *et al.*, 2004). Comparing the growth of cultured SC on 15 μm chitosan fibres or on chitosan membranes, the authors noted both increased cell density and more rapid, directional migration along the fibres. However, an *in vivo* study comparing a collagen matrix arranged either as longitudinally orientated collagen fibres or as a sponge demonstrated superior, although not statistically significant, results with the sponge (Toba *et al.*, 2001).

22.6 Cultured cells and nerve constructions

22.6.1 Schwann cells

While there is clear evidence that both synthetic and natural materials can function as nerve conduits, when they are used alone the results frequently fall short of what can be achieved with an autologous nerve graft. There is a considerable amount of evidence to suggest that the targeted administration of neurotrophic factors to the injured nerve accelerates both motor and sensory nerve regeneration (Derby *et al.*, 1993; Whitworth *et al.*, 1996; Sterne *et al.*, 1997; Bryan *et al.*, 2000; Terris *et al.*, 2001; Fine *et al.*, 2002; Barras *et al.*, 2002; Mohanna *et al.*, 2003, 2005; McKay Hart *et al.*, 2003). This effect can be enhanced by a controlled release of growth factors (Piotrowicz and Shoichet, 2006) in a gradient concentration manner (Moore *et al.*, 2006). However, regenerating axons respond to a multitude of highly coordinated neurotrophic cues which it would be impossible to reproduce simply by the sustained release of a cocktail of growth factors (Terenghi, 1999). An attractive alternative is to introduce into the conduit Schwann cells, which are essential for nerve regeneration. Indeed the ability of SC to align themselves provided directional cues for the regrowing axons (Thompson and Buettner, 2006). This ability to guide regeneration can also be enhanced by the synthesis and secretion by SC of increased amount of neurotrophic factors (Li *et al.*, 2006).

At 6 weeks after transplantation of SC into a multi-lumen synthetic (PLGA) conduit, which was used to bridge a short (7 mm) rat sciatic nerve defect there was a percentage of neural tissue per cross-sectional area statistically similar to autografts (Hadlock *et al.*, 2000). We also showed that retrovirally labelled SCs (Fig. 22.2) can enhance axonal regeneration when transplanted within a PHB conduit, a result which strongly supports the addition of SC to a conduit (Mosahebi *et al.*, 2001). Axonal counts approaching that of autograft controls were demonstrated using a 2 cm synthetic poly-glycolic acid conduit seeded with autologous cultured SC (Cheng and Chen, 2002). In a different conduit model, at 9 months post-implantation, SC induced a functional recovery across a

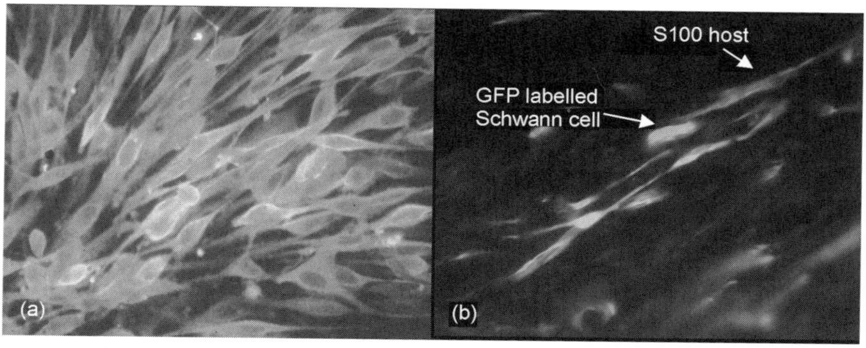

22.2 (a) A culture of Schwann cells showing the typical spindle-shaped morphology. The cells are stained with S100 antibodies, a marker of glial cells. (b) GFP-labelled Schwann cells transplanted into a nerve conduit. The endogenous Schwann cells are stained with S100 antibodies using Cy3 fluorescent reporter label. The transplanted cells are easily recognised, and the co-localisation of GFP and S100 results in yellow staining of the cells.

2 cm nerve gap that was equivalent to autografts, and not seen in empty conduits (Sinis *et al.*, 2005). The addition of SC to acellular vein and muscle grafts improved regeneration in comparison to the same conduits left empty, although control autografts performed superiorly to all experimental alternatives in this study (Fansa and Keilhoff, 2004).

In all these studies, autologous or syngeneic cells were used to avoid immune reactions following SC transplantation. The existing evidence is somewhat conflicting as to how the immune response to transplanted cells may affect regeneration. Genetically marked allogenic SCs can improve axonal regeneration when transplanted in PHB conduits (Mosahebi *et al.*, 2002). Although there was an increased immune reaction in terms of lymphocyte and macrophage count in transplants containing allogenic SCs, axonal in-growth into the conduits was comparable to that observed with syngeneic SC, suggesting that the immune response was not deleterious to regeneration. The study proposes that, despite ultimate rejection of the allogenic cells, the neurotrophic factors they produce may be long lasting and remain functional within the matrix after SC demise. There is some support for this suggestion in a more recent study of allogenic SC transplantation within a spinal cord injury (Hill *et al.*, 2005). While there is progressive death of transplanted cells without immunosuppression, simultaneous infiltration of host SCs occurred in greater numbers than after injury alone. Interestingly, previous studies had suggested that heterologous transplanted SC within a nerve conduit provoke an intense immune reaction (Guenard *et al.*, 1992) and that only autologous SC promote regeneration rates approaching that of standard nerve grafts (Rodriguez *et al.*, 2000).

Extrapolating the experimental evidence into a clinical context, until the absolute risk of allogenic SC transplantation is fully defined, it appears that autologous SC cultured from the injured patient present the only realistic option for the safe use of SC transplants. However, clinically a peripheral nerve biopsy is necessary to obtain autologous SC for culture, with an additional surgical procedure that the concept of a bioengineered graft aspires to avoid. Most importantly the generation of sufficient quantities of SC for transplantation from the patient's peripheral nerve biopsy requires at least 3–6 weeks according to established protocols (Vroemen and Weidner, 2003). This represents a significant delay in managing an acute injury a factor widely recognised to impair overall outcome in peripheral nerve repair (Sulaiman and Gordon, 2000; McKay Hart *et al.*, 2003). A recent study proposes a new method of acute dissociation of SC to generate in a short time a sufficient number of cells to be seeded in a nerve conduit, resulting in good regeneration (Brandt *et al.*, 2005).

22.6.2 Stem cells

An alternative and rapidly developing branch of tissue engineering is stem cell technology. Stem cells are progenitor cells with the capacity both to self-renew and to generate differentiated progeny (Morrison *et al.*, 1997). They can be subdivided into totipotent stem cells – those that can differentiate into all cell types of a particular organism – and multipotent cells that can differentiate into many kinds of cells but not the whole organism (Temple, 2001).

Totipotent or embryonic stem (ES) cells are developed from the inner cell mass of the blastocyst in early embryonic development. Human ES cells were first successfully cultured in 1998 using embryos from therapeutic termination of pregnancies (Shamblott *et al.*, 1998). The applications of these cells in replacement therapies are potentially vast and include the creation of cardiomyocytes, neural and haematopoietic precursors (Keller and Snodgrass, 1999). The isolation and culture of human ES cells are of particular interest to neurobiologists, with the potential to offer an unlimited supply of *in vitro* cultured donor cells for the management of neurodegenerative diseases. ES-derived neural precursors not only morphologically but also functionally integrate into the brain (Wernig *et al.*, 2004). Furthermore, novel techniques are being developed that will extend the differentiation potential of ES cells beyond that of general neuronal precursors into defined neuronal lineages, increasing the potential for targeted replacement of specific cell types (Bibel *et al.*, 2004; Li *et al.*, 2005). However, the scientific and clinical use of embryonic tissues also raises huge ethical dilemmas (Annas *et al.*, 1999; Gilbert, 2004), limiting the current application of embryonic stem cell technology.

Adult multipotent stem cells offer an alternative to the use of fetal tissues. It was observed that the nervous system, previously thought to be incapable of renewal, has within it stem cells with equally exciting potential and particular relevance to nerve regeneration (Davis and Temple, 1994). Neural stem cell isolation and culture was first described by Reynolds and Weiss (1992) and these cells have subsequently been shown to have the capacity to differentiate into mature neurons both *in vitro* (Moe *et al.*, 2005) and *in vivo* (Fricker *et al.*, 1999). Their potential to treat degenerative diseases of the nervous systems has been widely explored (Yandava *et al.*, 1999; Bjorklund and Lindvall, 2000; Akiyama *et al.*, 2002; Uchida *et al.*, 2003) and some improvement in axonal growth has been demonstrated in peripheral nerve regeneration models (Murakami *et al.*, 2003).

Although neuronal progenitor cells have been shown to have benefits in experimental models, clinically, the accessibility of donor tissue is an equally important issue. Such cells are harvested from deep within the brain a procedure that cannot be directly extrapolated to human use. Olfactory ensheathing cells (OEC) from the adult mammalian olfactory bulb may offer a more accessible source of neuronal progenitor cells (Ramon-Cueto *et al.*, 2000). These cells can be harvested from the olfactory mucosa within the nose (Bianco *et al.*, 2004; Murrell *et al.*, 2004) and have been shown experimentally to enhance motor performance in peripheral nerve regeneration (Guntinas-Lichius *et al.*, 2002). Further investigation will offer insight into the potential clinical value of these cells as an autologous donor source for nerve regeneration.

22.6.3 Bone marrow stromal cells

Progenitor cells from an alternative, more clinically attractive source exists in the bone marrow which contains two distinct types of multipotent progenitor cell. The clinical application of haematopoietic stem cells, which replenish blood cell lineages, has been apparent for some time and continues to evolve (Gratwohl *et al.*, 2002). The potential of the second cell population, mesenchymal stem cells or marrow stromal cells, has more recently emerged (Prockop, 1997). The terminology of these cells within the literature is inconsistent. Originally, these cells were referred to as colony-forming unit fibroblastic cells (Owen and Friedenstein, 1988). More recently the abbreviation MSC has been applied to both mesenchymal stem cells (i.e. those that can differentiate into all mesenchymal lineages) and marrow stromal cells (cells arising from the supportive, non-haematopoietic structures of bone marrow) and the two used interchangeably. This creates considerable confusion since the number of true stem cells within the marrow stromal cell population is estimated to be very small (Jiang *et al.*, 2002; Tohill and Terenghi, 2004). For the purposes of this review, the abbreviation MSC will be applied to the heterogeneous population of marrow stromal (non-haematopoietic) cells, of which stem cells probably make up only a small part, and the term stem cell reserved only for those cells shown to be clonogenic and multipotential.

Both haematopoietic and mesenchymal progenitors are clinically advantageous in that they are readily accessible by bone marrow biopsy and their renewable populations represent an abundant source (Fig. 22.3). MSC can be

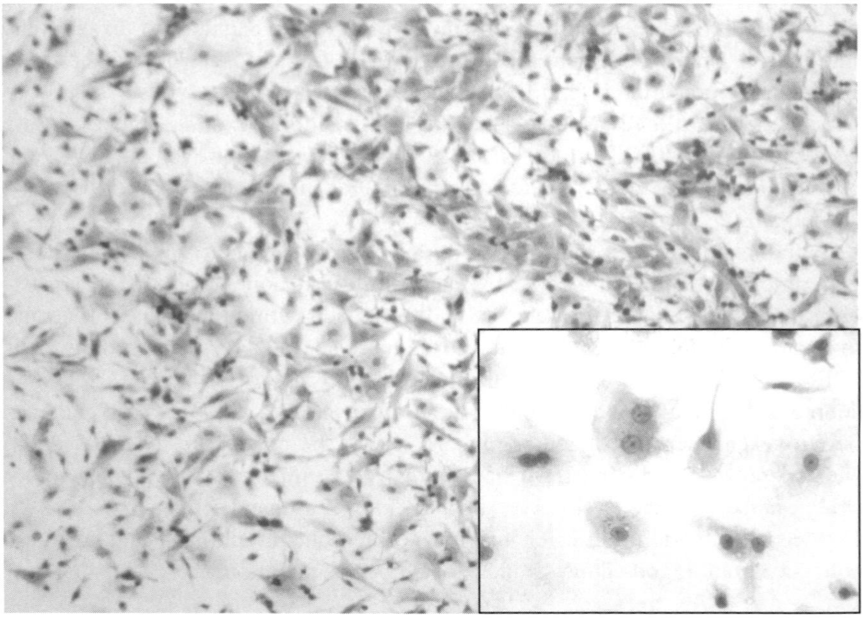

22.3 Culture of undifferentiated bone marrow-derived mesenchymal stem cells 5 days after harvest of the cells. Note the abundant proliferation of the stem cells, which is highlighted in the high magnification inset showing numerous mitotic cell divisions.

separated with relative ease from haematopoietic cells because of their rapid adherence to plastic (Bianco *et al.*, 2001). In culture they will differentiate into all mesenchymal cell lineages including osteoblasts, chondroblasts, adipocytes and myoblasts (Pittenger *et al.*, 1999). This property has been exploited clinically to treat mesenchymal disorders including osteogenesis imperfecta and chemotherapy-induced bone marrow depletion (Gerson, 1999).

While it was originally thought that organ-specific somatic stem cells were capable of only limited differentiation, it is now evident that adult stem cells are able to cross lineage boundaries and differentiate into the diverse cell types previously only thought possible with ES differentiation (Blau *et al.*, 2001; Morrison, 2001; Martin-Rendon and Watt, 2003). A number of different adult stem cell populations including haematopoietic (Orlic *et al.*, 2001; Bonilla *et al.*, 2002; Sigurjonsson *et al.*, 2005), dermal (Joannides *et al.*, 2004; Amoh *et al.*, 2005) and neural (Bjornson *et al.*, 1999) have been shown to display such plasticity. More recently, stromal cells from adipose tissue have been shown to differentiate into other mesenchymal cell lineages and also into neurons and glia (Safford *et al.*, 2004; Sondermann *et al.*, 2005; Izadpanah *et al.*, 2006).

Bone marrow-derived mesenchymal progenitors, in particular, have attracted considerable attention regarding their differentiation potential (Huttmann *et al.*,

2003). *In vitro* work shows differentiation of clonally isolated mesenchymal stem cells into cells displaying characteristics of all three germ layers (Jiang *et al.*, 2002). *In vivo* these cells engraft and differentiate into the haematopoietic system and gastrointestinal tract and also have been shown to differentiate into functional hepatocytes (Schwartz *et al.*, 2002). Additional studies have shown that MSC are capable of differentiating into cells characteristic of myocytes (Ferrari *et al.*, 1998), cardiac muscle (Tomita *et al.*, 1999) and hepatic stem cells (Petersen *et al.*, 1999). More pertinent in the context of nerve regeneration is the ability of these cells to differentiate into both neurons and glia.

In vitro both rodent and human adult MSC have the capacity to differentiate into neurons (Woodbury *et al.*, 2000). Further *in vitro* studies have demonstrated the potential of MSC to express markers of neural progenitors (neurofilament), neurons (neuron-specific nuclear protein, β tubulin III) and glia (glial fibrillary acidic protein) under appropriate culture conditions (Sanchez-Ramos *et al.*, 2000; Kim *et al.*, 2002; Bossolasco *et al.*, 2005). In addition, a selected subset of multipotent bone marrow stem cells can differentiate into cells that display electrophysiological and phenotypic characteristics of neurons following co-culture with astrocytes (Jiang *et al.*, 2003).

These *in vitro* findings have been further supported by *in vivo* work, whereby MSC injected either centrally or peripherally were found months later incorporated throughout the CNS (Azizi *et al.*, 1998; Kopen *et al.*, 1999; Brazelton *et al.*, 2000; Mezey *et al.*, 2000). More recently published evidence suggests that this phenomenon occurs in the human brain following therapeutic bone marrow transplantation (Mezey *et al.*, 2003). Outside of the brain, transplanted MSC have been shown to enhance remyelination of the spinal cord (Sasaki *et al.*, 2001), improve spinal cord conduction velocity (Akiyama *et al.*, 2002) and facilitate both tissue repair and functional recovery following direct injection into a spinal cord injury model (Wu *et al.*, 2003).

22.6.4 MSC in peripheral nerve regeneration

Although the majority of the literature regarding the use of MSC has focused on their application within the central nervous system, studies of MSC applied to peripheral nerve injury have also been reported (Tohill and Terenghi, 2004). The observation that MSC transplanted into the spinal cord showed features characteristic of peripheral rather than central myelin synthesis (Sasaki *et al.*, 2001), is therefore of particular interest as it raised the possibility that these cells may have equal potential in peripheral nerve repair.

When MSC were injected directly into the distal stump of a transected and immediately repaired sciatic nerve, approximately 5% of the cells expressed the Schwann cell marker S100 at 1 month after surgery (Cuevas *et al.*, 2002). A similar technique was also applied to a nerve crush injury. At 3 weeks, the transplanted cells incorporated into the injury site and an unquantified proportion stained positively for the SC markers S100, GFAP and p75^{NTR} (Zhang *et al.*, 2004). These MSC were not characterised prior to injection and it is unclear whether the observed expression of SC markers represents an actual differentiation process occurring *in vivo* and, if this is the case, exactly what proportion of the MSC are able to differentiate. Nonetheless a functional improvement, measured by a walking track test, was noted in the first study, suggesting that the transplanted cells contribute in some way to the regenerative process.

The first study to show the differentiation of MSC along a SC lineage *in vitro* prior to transplantation also identified the culture conditions that can effectively initiate differentiation (Dezawa *et al.*, 2001). Under specific culture conditions, MSCs can be induced to differentiate into cells expressing four SC markers (p75^{NTR}, S-100, GFAP and O4). The differentiated MSC seeded within an artificial nerve conduit bridging a rat sciatic nerve gap promoted nerve regeneration, and the labelled differentiated MSCs co-localised with myelin associated glycoprotein immunostaining, suggesting a direct involvement with myelin formation.

In our studies, a similar protocol confirmed the potential of MSCs to differentiate into cells expressing typical Schwann cell markers (Tohill *et al.*, 2004; Caddick *et al.*, 2006). Following transplantation into a rat sciatic nerve gap model these cells continue to express S-100 and promoted nerve regeneration (Fig. 22.4) which was increased compared with acellular control conduits, and statistically similar to the result seen with SCs (Tohill *et al.*, 2004). This is consistent with *in vitro* functional studies, where co-cultures of differentiated MSC with dissociate dorsal root ganglia neurons promoted neurite extension similar to that observed in SC-neuron co-cultures (Caddick *et al.*, 2006). Interestingly both studies noted that undifferentiated MSC also promoted some moderate regeneration, which supports the findings of Cuevas *et al.* (2002) and Zhang *et al.* (2004), although regeneration was much greater following *in vitro* differentiation of MSC.

More recently, two studies have shown that grafting undifferentiated MSC in nerve conduits resulted in nerve regeneration and motor functional recovery (Chen *et al.*, 2006; Pereira Lopes *et al.*, 2006). Enhanced nerve regeneration was also observed following transplantation of differentiated MSC in various types of artificial nerve conduits (Choi *et al.*, 2005; Zhang *et al.*, 2005; Hou *et al.*, 2006). The demonstration that differentiated MSC are expressing *in vitro* various levels of neurotrophins (Crigler *et al.*, 2006), and they produce myelination of axon (Keilhoff *et al.*, 2006) similar to SC is further supporting evidence of the potential of these cells for application in nerve bioengineering.

Scepticism exists in the literature regarding the true plasticity of MSC. Cell fusion *in vivo* and *in vitro* has been reported to be responsible for some, if not all differentiation observed (Terada *et al.*, 2002; Alvarez-Dolado *et al.*, 2003; Wang *et al.*, 2003). Other reports have failed to reproduce the findings so far described

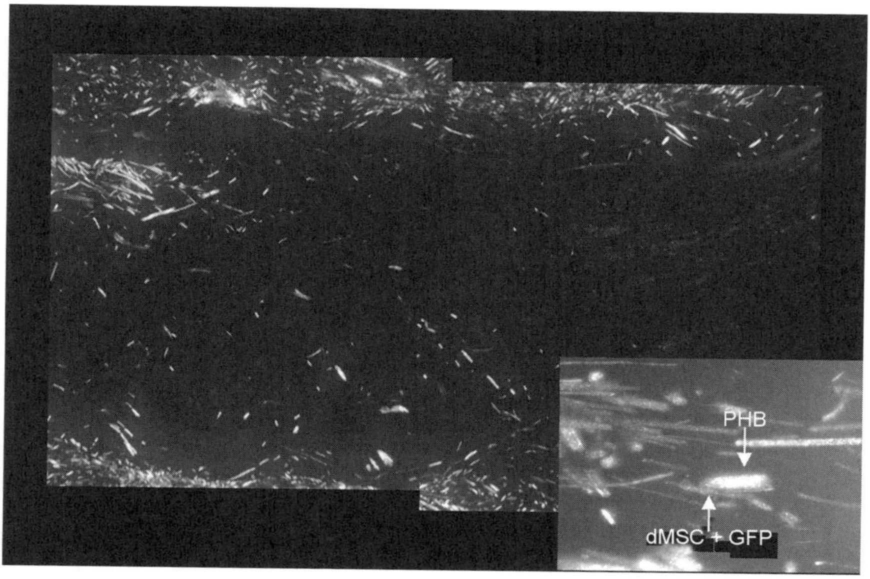

22.4 Axonal regeneration into a PHB conduit. The axon are stained with a Pan-Neurofilament antiserum (Cy3 fluorescent reporter). The PHB polymer fibres forming the nerve conduit interweave with the regenerating axons. The inset shows a higher magnification of GFP-labelled differentiated stem cells (dMSC + GFP) in close apposition with the PHB fibres, which offer contact guidance to both glial cells and axons.

(Castro *et al.*, 2002; Wehner *et al.*, 2003) or argued that positive immunoreactivity for neuronal markers may be an artefact of cellular shrinkage (Lu *et al.*, 2004). This is possibly due to technical differences between studies and difficulties in their interpretation (reviewed in Pauwelyn and Verfaillie, 2006).

22.6.5 Immunological characteristics of MSC

There is some evidence to suggest that MSC may have a degree of immune privilege when transplanted as allografts or xenografts (Azizi *et al.*, 1998). It is widely considered that PNS is not an immune privileged site when transplanted with allogenic tissue (Guenard *et al.*, 1992; Mosahebi *et al.*, 2002). Thus any findings that support the safe transplantation of MSC between HLA incompatible individuals are clearly worthy of consideration if one is to consider using these cells to create a ready to use nerve graft.

Liechty *et al.* (2000) examined immunogenicity by injecting human MSC into fetal sheep at gestation periods before and after the development of immunological competence. They showed engraftment and differentiation of the transplanted cells in wide range of tissues. More importantly, even after the development of immune competence there was no evidence of rejection, although long-term engraftment was limited to particular tissue types. The authors suggest that this is a result of both local immune suppression by human MSC and limited immunogenicity due to the fact that they do not express MHC class II HLAs.

More recent *in vitro* experimental work supports these findings (Di Nicola *et al.*, 2002; Jorgensen *et al.*, 2003; Tse *et al.*, 2003). It appears from these studies that the observed immune privilege of MSC occurs by two separate mechanisms. Firstly, MSC do not express MHC class II (although it is present intracellularly) or T-cell co-stimulatory molecules on their cell surface and are thus poorly recognised by T cells. Secondly MSC actively suppress T-cell proliferation by a mechanism involving the production of soluble factors. The exact process by which this occurs remains unclear, although it has been demonstrated that T-cell apoptosis is not involved (Di Nicola *et al.*, 2002).

These data come from studies on undifferentiated MSC. However there also exists evidence to suggest that, following *in vitro* differentiation, the immunological properties so far described are retained (Le Blanc *et al.*, 2003). While this study looked only at MSC differentiated to mesenchymal lineages, it is tempting to consider that such properties may also be retained following differentiation into other cell types. The immunological properties of MSC differentiated towards a glial or neuronal lineage remain to be investigated, but some progress has been made in understanding the outcome of allogeneic stem cell transplantation (Barrett, 2006), and the option of transplanting a cell which escapes immune rejection remains clinically appealing.

22.7 Conclusions

This review highlights the potential of bioengineered conduits for peripheral nerve repair, also as an alternative to current autologous nerve graft repair. Many materials have been tested experimentally and some have also been used in clinical studies as empty conduits. At present, it is difficult to decide what is the most promising material, and possibly a better answer might be given when a combination of nerve conduits and transplanted cells has been applied clinically. Although many ethical and practical problems are still to be resolved before this can happen, the fast pace of this type of research would indicate that the clinical application of a bioengineered nerve constructs will be in the near future.

22.8 References

Aebischer P, Guenard V and Brace S (1989), 'Peripheral nerve regeneration through blinded-ended semipermeable guidance channels; effect of the molecular weight cutoff', *J Neurosci*, **9**, 3590–3595.

Agius E and Cochard P (1998), 'Comparison of neurite outgrowth induced by intact and injured sciatic nerves; a confocal and functional analysis', *J Neurosci*, **18**, 328–338.

Ahmed M R, Venkateshwarlu U and Jayakumar R (2004), 'Multilayered peptide incorporated collagen tubules for peripheral nerve repair', *Biomaterials* , **25**, 2585–2594.

Akiyama Y, Radtke C, Honmou O and Kocsis J D (2002), 'Remyelination of the spinal cord following intravenous delivery of bone marrow cells', *Glia*, **39**, 229–236.

Aldini N N, Perego G, Cella G D, Maltarello M C and Fini M (1995), 'Effectiveness of a bioabsorbable conduit in the repair of peripheral nerves', *Biomaterials*, **17**, 959–962.

Alvarez-Dolado M, Pardal R, Garcia-Verdugo J M, Fike J R, Lee H O, Pfeffer K, Lois C, Morrison S J and Alvarez-Buylla A (2003), 'Fusion of bone-marrow-derived cells with Purkinje neurons, cardiomyocytes and hepatocytes', *Nature*, **425**, 968–973.

Amoh Y, Li L, Campillo R, Kawahara K, Katsuoka K, Penman S and Hoffman R M (2005), 'Implanted hair follicle stem cells form Schwann cells that support repair of severed peripheral nerves', *Proc Nat Acad Sci, USA*, **102**, 17734–17738.

Annas G J, Caplan A and Elias S (1999), 'Stem cell politics, ethics and medical progress', *Nature Med*, **5**, 1339–1341.

Ard M D, Bunge R P and Bunge M B (1987), 'Comparison of the Schwann cell surface and Schwann cell extracellular matrix as promoters of neurite growth', *J Neurocytol*, **16**, 539–555.

Azizi S A, Stokes D, Augelli B J, DeGirolamo C and Prockop D J (1998), 'Engraftment and migration of human bone marrow stromal cells implanted in the brains of albino rats – similarities to astrocyte grafts', *Proc Nat Acad Sci, USA*, **95**, 3908–3913.

Barras F M, Pasche P, Bouche N, Aebischer P and Zurn A D (2002), 'Glial cell line-derived neurotrophic factor released by synthetic guidance channels promotes facial nerve regeneration in the rat', *J Neurosci Res*, **70**, 746–755.

Barrett J (2006), 'Improving outcome of allogeneic stem cell transplantation by immunomodulation of the early post-transplant environment', *Curr Opin Immunol*, **18**, 592–598.

Battiston B, Gueno S, Ferrero M and Tos P (2005), 'Nerve repair by means of tubulization; literature review and personal clinical experience comparing biological and synthetic conduits for sensory nerve repair', *Microsurgery*, **25**, 258–267.

Bender M D, Bennett J M, Waddell R L Doctor J S and Marra K G (2004), 'Multi-channelled biodegradable polymer/CultiSpher composite nerve guides', *Biomaterials*, **25**, 1269–1278.

Bianco J I, Perry C, Harkin D G, Mackay-Sim A and Feron F (2004), 'Neurotrophin 3 promotes purification and proliferation of olfactory ensheathing cells from human nose', *Glia*, **45**, 111–123.

Bianco P, Riminucci M, Gronthas S and Gehron Robey P (2001), 'Bone marrow stromal stem cells: nature, biology and potential applications', *Stem Cells*, **19**, 180–193.

Bibel M, Richter J, Schrenk K, Tucker K L, Staiger V, Korte M, Goetz M and Barde Y-A (2004), 'Differentiation of mouse embryonic stems cells into a defined neuronal lineage', *Nature Neurosci*, **7**, 1003–1009.

Bini T B, Gao S, Xu X, Wang S, Ramakrishna S and Leong K W (2003), 'Peripheral nerve regeneration by microbraided poly-L-lactide-co-glycolide biodegradable polymer fibres', *J Biomed Mat Res*, **68A**, 286–295.

Birch R, Bonney G and Parry C B W (1998), *Surgical Disorders of the Peripheral Nerves*, London, Churchill Livingstone.

Bjorklund A and Lindvall O (2000), 'Cell replacement therapies for central nervous system disorders', *Nature Neurosci*, **3**, 537–544.

Bjornson C R R, Rietze R L, Reynolds B A, Magli C M and Vescovi A L (1999), 'Turning brain into blood: an haematopoietic fate adopted by adult neural stem cells *in vivo*', *Science*, **283**, 534–537.

Blau H M, Brazelton T R and Weimann J M (2001), 'The evolving concept of a stem cell: entity or function?', *Cell*, **105**, 829–841.

Bonilla S, Alarcon P, Villaverde R, Aparicio P, Silva A and Martinez S (2002), 'Haematopoietic progenitor cells from adult bone marrow differentiate into cells that express oligodendroglial antigens in the neonatal mouse brain', *Eur J Neurosci*, **15**, 575–585.

Bossolasco P, Cova L, Calzarossa C, Rimoldi S G, Borsotti C, Lambertenghi Deliliers G, Silani V, Soligo D and Polli E (2005), 'Neuro-glial differentiation of human bone marrow stem cells *in vitro*', *Exp Neurol*, **193**, 312–325.

Brandt J, Dahlin L B and Lundborg G (1999), 'Autologous tendons used as grafts for bridging peripheral nerve defects', *J Hand Surg*, **24B**, 284–290.

Brandt J, Nilsson A, Kanje M, Lundborg G and Dahlin L B (2005), 'Acutely-dissociated Schwann cells used in tendon autografts for bridging nerve defects in rats: a new principle for tissue engineering in nerve reconstruction', *Scand J Plast Reconstr Surg Hand Surg*, **39**, 321–325.

Brazelton T, Rossi F M V, Keshet G I and Blau H M (2000), 'From marrow to brain: expression of neuronal phenotypes in adult mice', *Science*, **290**, 1775–1779.

Brushart T M E (1993), 'Motor axons preferentially reinnervate motor pathways', *J Neurosci*, **13**, 2730–2738.

Bryan D J, Miller R A D C P and Wang K K (1993), 'Immunocytochemstry of skeletal muscle basal lamina grafts in nerve regeneration', *Plast Reconstr Surg*, **92**, 927–940.

Bryan D J, Holway A H, Wang K K, Silva A E, Trantolo D J, Wise D and Summerhayes I C (2000), 'Influence of glial growth factor and Schwann cells in a bioresorbable guidance channel on peripheral nerve regeneration', *Tissue Eng*, **6**, 129–138.

Bunge M B, Williams A K and Wood P M (1982), 'Neuron-Schwann cell interaction in basal lamina formation', *Develop Biol*, **92**, 449–460.

Byrom D (1987), 'Polymer synthesis by microorganisms: technology and economics', *Tibtech*, **5**, 246–250.

Caddick J, Kingham P J, Gardiner N J, Wiberg M and Terenghi G (2006), 'Phenotypic and functional characteristics of mesenchymal stem cells differentiated along a Schwann cell lineage', *Glia*, **54**, 840–849.

Castro R F, Jackson K A, Goodell M A, Robertson C S, Liu H and Shine H D (2002), 'Failure of bone marrow cells to transdifferentiate into neural cells *in vivo*', *Science*, **297**, 1299–1301.

Chamberlain L J, Yannas I V, Hsu H-P, Strichartz G R and Spector M (2000), 'Near-terminus axonal structure and function following rat sciatic nerve regeneration through a collagen–GAG matrix in a ten-millimeter gap', *J Neurosci Res*, **60**, 666–677.

Chen G-Q and Wu Q (2005), 'The application of polyhydroxalkanoates as tissue engineering materials', *Biomaterials*, **26**, 6565–6578.

Chen X, Wang XD, Chen G, Lin WW, Yao J and Gu XS (2006), 'Study of *in vivo* differentiation of rat bone marrow stromal cells into Schwann cell-like cells', *Microsurgery*, **26**(2), 111–115.

Cheng B and Chen Z (2002), 'Fabricating autologous tissue to engineer artificial nerve', *Microsurgery*, **22**, 133–127.

Chernousov M A, Stahl R C and Carey D J (1998), 'Schwann cells use a novel collagen-dependent mechanism for fibronectin fibril assembly', *J Cell Sci*, **111**, 2763–2777.

Choi B H, Zhu S J, Kim B Y, Huh J Y, Lee J H and Jung J H (2005), 'Transplantation of cultured bone marrow stromal cells to improve peripheral nerve regeneration', *Int J Oral Maxillofac Surg*, **34**, 537–542.

Crigler L, Robey R C, Asawachaicharn A, Gaupp D and Phinney D G (2006), 'Human mesenchymal stem cell subpopulations express a variety of neuro-regulatory molecules and promote neuronal cell survival and neuritogenesis', *Exp Neurol*, **198**, 54–64.

Cuevas P F C, Dujovny M, Garcia-Gomez I, Cuevas B, Gonzalez-Corrochano R and Diaz-Gonzalez D (2002), 'Peripheral nerve regeneration by bone marrow stromal cells', *Neurol Res*, **24**, 634–638.

Dahiyat B I, Richards M and Leong K W (1995), 'Controlled release from polyphosphoester matrices', *J Control Release*, **33**, 13–21.

Dahlin L and Lundborg G (2001), 'Use of tubes in peripheral nerve repair', *Neurosurg Clin North Am*, **12**, 341–352.

Dahlin L B, Anagnostaki L and Lundborg G (2001), 'Tissue response to silicone tubes used to repair human median and ulnar nerves', *Scand J Plast Reconstr Surg Hand Surg*, **35**, 29–34.

Danielsen N, Lundborg G and Frizell M (1986), 'Nerve repair and axonal transport: outgrowth delay and regeneration rate after transection and repair of rabbit hypoglossal nerve', *Brain Res*, **376**, 125–132.

Davis A and Temple S (1994), 'A self renewing multipotential stem cell in embryonic rat cerebral cortex', *Nature*, **17**, 263–266.

Derby A, Engleman V W, Frierdich G E, Neises G, Rapp S R and Roufa D G (1993), 'Nerve growth factor facilitates regeneration across nerve gaps: morphological and behavioural studies in rat sciatic nerve', *Exp Neurol*, **119**, 176–191.

Dezawa M, Takahashi I, Esaki M, Takano M and Sawada H (2001), 'Sciatic nerve regeneration in rats induced by transplantation of *in vitro* differentiated bone-marrow stromal cells', *Eur J Neurosci*, **14**, 1771–1776.

Di Nicola M, Stella C, Magni M, Milanesi M, Longoni P D, Matteucci P, Grisanti S and Gianni A M (2002), 'Human bone marrow stromal cells suppress T-lymphocyte proliferation induced by cellular or nonspecific mitogenic stimuli', *Blood*, **99**, 3838–3843.

Doi K, Tamaru K, Sakai K, Kuwata N, Kurafuji Y and Kawai S (1992), 'A comparison of vascularized and conventional sural nerve grafts', *J Hand Surg*, **17A**, 670–676.

Drury J L and Mooney D J (2003), 'Hydrogels for tissue engineering: scaffold design variables and applications', *Biomaterials*, **24**, 4337–4351.

Evans G R D (2001), 'Peripheral nerve injury: a review and approach to tissue engineered constructs', *Anat Rec*, **263**, 396–404.

Fansa H and Keilhoff G (2004), 'Comparison of different biogenic matrices seeded with cultured Schwann cells for bridging peripheral nerve defects', *Neurol Res*, **26**, 167–173.

Fansa H, Keilhoff G, Wolf G and Schneider W (2001), 'Tissue engineering of peripheral nerves: a comparison of venous and acellular muscle grafts with cultured Schwann cells', *Plast Reconstr Surg*, **107**, 485–494.

Ferrari G, Cusella-De Angelis G, Coletta M, Paolucci E, Stornaiuolo A, Cossu G and Mavilio F (1998), 'Muscle regeneration by bone marrow-derived myogenic progenitors', *Science*, **279**, 1528–1530.

Fields R D, Le Beau J M, Longo F M and Ellisman M H (1989), 'Nerve regeneration through artificial tubular implants', *Prog Neurobiol*, **33**, 87–134.

Fine E G, Decosterd I, Papaloizos M, Zurn A D and Aebischer P (2002), 'GDNF and

NGF released by synthetic guidance channels support sciatic nerve regeneration across a long gap', *Eur J Neurosci*, **15**, 589–601.

Fricker R A, Carpenter M K, Winkler C, Greco C, Gates M A and Bjorklund A (1999), 'Site-specific migration and neuronal differentiation of human neural progenitor cells after transplantation in the adult rat brain', *J Neurosci*, **19**, 5990–6005.

Gerson S L (1999), 'Mesenchymal stem cells: no longer second class marrow citizens', *Nature Med*, **5**, 309–313.

Gilbert D M (2004), 'The future of human embryonic stem cell research: addressing ethical conflict with responsible scientific research', *Med Sci Monit*, **10**, RA99–103.

Ginde R M and Gupta R K (1987), '*In vitro* chemical degradation of poly-glycolic acid pellets and fibres', *J Appl Polym Sci*, **33**, 2411–2429.

Gordon T, Sulaiman O and Boyd J G (2003), 'Experimental strategies to promote functional recovery after peripheral nerve injuries', *J Periph Nerv Syst*, **8**, 236–250.

Gratwohl A, Baldomero H, Horisbrger B, Schmid C, Passweg J and Urbano-Ispizua A (2002), 'Current trends in haematopoietic stem cell transplantation in Europe', *Blood*, **100**, 2374–2386.

Guenard V, Kleitman N, Morrissey T K, Bunge R P and Aebischer P (1992), 'Synergic Schwann cells derived from adult nerves seeded in semipermeable guidance channels enhance peripheral nerve regeneration', *J Neurosci*, **12**, 3310–3320.

Guntinas-Lichius O, Wewetzer K, Tomov T L, Azzolin N, Kazemi S, Streppel M, Neiss W F and Angelov D N (2002), 'Transplantation of olfactory mucosa minimizes axonal branching and promotes the recovery of the vibrissae motor performance after facial nerve repair in rats', *J Neurosci*, **22**, 7121–7131.

Hadlock T, Sundback C, Hunter D, Cheney M and Vacanti J (2000), 'A polymer foam conduit seeded with Schwann cells promotes guided peripheral nerve regeneration', *Tissue Eng*, **6**, 119–127.

Hadlock T A, Sundback C A, Hunter D A, Vacanti J P and Cheney M L (2001), 'A new artificial nerve graft containing rolled Schwann cell monolayers', *Microsurgery*, **21**, 96–101.

Hall S (1997), 'Axonal regeneration through acellular muscle grafts', *J Anat*, **190**, 57–71.

Hall S (2001), 'Nerve repair: a neurobiologist's view', *J Hand Surg*, **26B**, 129–136.

Hazari A, Johansson-Ruden G, Junemo-Bostrom K, Ljungberg C, Terenghi G, Green C and Wiberg M (1999a), 'A new resorbable wrap-around implant as an alternative nerve repair technique', *J Hand Surg*, **24B**, 291–295.

Hazari A, Wiberg M, Johansson-Ruden G, Green C and Terenghi G (1999b), 'A resorbable nerve conduit as an alternative to nerve autograft in nerve gap repair', *Br J Plast Surg*, **52**, 653–657.

Heumann R, Korsching S, Bandtlow C and Thoenen H (1987), 'Changes of nerve growth factor synthesis in non-neuronal cells in response to sciatic nerve transection', *J Cell Biol*, **104**, 1623–1631.

Hill C E, Moon L D F, Wood P M and Bunge M B (2005), 'Labelled Schwann cell transplantation: cell loss, host Schwann cell replacement and strategies to enhance survival', *Glia*, **53**, 338–343.

Hiraoka Y, Kimura Y, Ueda H and Tabata Y (2003), 'Fabrication and biocompatibility of collagen sponge reinforced with poly-glycolic acid fiber', *Tissue Eng*, **9**, 1101–1112.

Hollenbeck P J (1989), 'The transport and assembly of the axonal cytoskeleton', *J Cell Biol*, **108**, 223–227.

Hou S-Y, Zhang H-Y, Quan D-P, Liu X-L and Zhu J-K (2006), 'Tissue-engineered peripheral nerve grafting by differentiated bone marrow stromal cells', *Neuroscience*, **140**, 101–110.

Hsu S H, Chen C Y, Lu P S, Lai C S and Chen C J (2005), 'Oriented Schwann cell growth on microgrooved surfaces', *Biotechnol Bioeng*, **92**, 579–588.

Huttmann A, Li C L and Duhrsen U (2003), 'Bone marrow-derived stem cells and plasticity', *Ann Hematol*, **82**, 599–604.

Ide C (1996), 'Peripheral nerve regeneration', *Neurosci Res*, **25**, 101–121.

Inada Y, Morimoto S, Takakura Y and Nakamura T (2004), 'Regeneration of peripheral nerve gaps with a polyglycolic acid–collagen tube', *Neurosurgery*, **55**, 640–648.

Itoh S, Takakuda K, Kawabata S, Aso Y, Kasai K, Itoh H and Shinomiya K (2002), 'Evaluation of cross-linking procedures of collagen tubes used in peripheral nerve repair', *Biomaterials*, **23**, 4475–4481.

Izadpanah R, Trygg C, Patel B, Kriedt C, Dufour J, Gimble J M and Bunnell B A (2006), 'Biological properties of mesenchymal stem cells derived from bone marrow and adipose tissue, *J Cell Biochem*, **99**, 1285–1297.

Jessen K R and Mirsky R (1999), 'Schwann cells and their precursors emerge as major regulators of nerve developments', *Trends Neurosci*, **22**, 402–410.

Jiang Y, Jahagirdar B N, Reinhardt R L, Schwartz R E, Keene C D, Ortiz-Gonzalez X R, Reyes M, Lenvik T, Blackstad M, Du J, Aldrich S, Lisburg A, Low W, Largaespada D A and Verfaillie C M (2002), 'Pluripotency of mesenchymal stem cells derived from adult marrow', *Nature*, **418**, 41–49.

Jiang Y, Henderson D, Blackstad M, Chen A, Miller R F and Verfaillie C M (2003), 'Neuroectodermal differentiation from mouse multipotent adult progenitor cells', *Proc Nat Acad Sci, USA*, **100**, 11854–11860.

Joannides A, Gaughwin P, Schwiening C, Majed H, Sterling J, Compston A and Chandran S (2004), 'Efficient generation of neural precursors from adult human skin: astrocytes promote neurogenesis from skin-derived stem cells', *Lancet*, **364**, 172–178.

Jorgensen C, Djouad F, Apparailly F and Noel D (2003), 'Engineering mesenchymal stem cells for immunotherapy', *Gene Ther*, **10**, 928–931.

Keilhoff G, Pratsch F, Wolf G and Fansa H (2005), 'Bridging extra large defects of peripheral nerves: possibilities and limitations of alternative biological grafts from acellular muscle and Schwann cells', *Tissue Eng*, **11**, 1004–1014.

Keilhoff G, Stang F, Goihl A, Wolf G and Fansa H (2006), 'Transdifferentiated mesenchymal stem cells as an alternative therapy in supporting nerve regeneration and myelination', *Cell Mol Neurobiol*, **26**, 1235–1252.

Keller G and Snodgrass H R (1999), 'Human embryonic stem cells: the future is now', *Nature Med*, **5**, 151–152.

Kim B J, Seo J H, Bubien J K and Oh Y S (2002), 'Differentiation of adult bone marrow stem cells into neuroprogenitor cells *in vitro*', *Neuro Rep*, **13**, 1185–1188.

Kopen G C, Prockop D J and Phinney D G (1999), 'Marrow stromal cells migrate throughout forebrain and cerebellum, and they differentiate into astrocytes after injection into neonatal mouse brains', *Proc Nat Acad Sci USA*, **96**, 10711–10716.

Labrador R O, Buti M and Navarro X (1998), 'Influence of collagen and laminin gels concentration on nerve regeneration after resection and tube repair'. *Exp Neurol*, **149**, 243–252.

Le Blanc K, Tammik C, Rosendahl K, Zetterberg E and Ringden O (2003), 'HLA expression and immunologic properties of differentiated and undifferentiated mesenchymal stem cells', *Exp Hematol*, **31**, 890–896.

Lee S Y (1996), 'Bacterial polyhydroxyalkanoates', *Biotech Bioeng*, **49**, 1–14.

Li J, Yan J G, Ai X, Hu S, Gu Y D, Matloub H S and Sangeer J R (2004), 'Ultrastructural analysis of peripheral nerve regeneration within a nerve conduit', *J Reconstr Microsurg*, **20**, 565–569.

Li Q, Ping P, Jiang H and Liu K (2006), 'Nerve conduit filled with GDNF gene-modified Schwann cells enhances regeneration of the peripheral nerve', *Microsurgery*, **26**, 116–121.

Li X-J, Du Z-W, Zarnowska E D, Pankratz M, Hansen L O, Pearce R A and Zhang S-C (2005), 'Specification of motoneurons from human embryonic stems cells', *Nature Biotech*, **23**, 215–221.

Liechty K W, MacKenzie T C, Shaaban A, Radu A, Moseley A B, Deans R, Marshak D R and Flake A W (2000), 'Human mesenchymal stem cells engraft and demonstrate site-specific differentiation after *in utero* transplantation in sheep', *Nature Med*, **6**, 1282–1286.

Lu P, Blesch A and Tuszynski M H (2004), 'Induction of bone marrow stromal cells to neurons: differentiation, transdifferentiation, or artefact', *J Neurosci Res*, **77**, 174–191.

Lundborg G (2003), 'Nerve injury and repair – a challenge to the plastic brain', *J Peripher Nerv Syst*, **8**, 209–226.

Lundborg G, Dahlin L, Dohi D, Kanje M and Terada N (1997), 'A new type of bioartificial nerve graft for bridging extended defects in nerves', *J Hand Surg*, **22B**, 299–303.

Lundborg G, Rosen B, Dahlin L, Holmberg J and Rosen I (2004), 'Tubular repair of the median or ulnar nerve in the human forearm: a 5-year follow-up', *J Hand Surg*, **29B**, 100–107.

Lutolf M P and Hubbell J A (2005), 'Synthetic biomaterials as instructive extracellular microenvironments for morphogenesis in tissue engineering', *Nature Biotech*, **23**, 47–55.

Mackinnon S E and Dellon A L (1988), *Surgery of the Peripheral Nerve*, New York, Thieme Medical Publishers Inc.

Mackinnon S E and Dellon A L (1990), 'A study of nerve regeneration across synthetic (Maxon) and biologic (collagen) nerve conduits for nerve gaps up to 5cm in the primate', *J Microsurg*, **6**, 117–121.

Mackinnon S E, Doolabh V B, Novak C B and Trulock E P (2001), 'Clinical outcome following nerve allograft transplantation', *Plast Reconstr Surg*, **107**, 1419–1429.

Madison R, Sidman R L, Nyilas Em, Chiu T-H and Greatorex D (1984), 'Non-toxic nerve guide tubes support neovascular growth in transected rat optic nerve', *Exp Neurol*, **86**, 448–461.

Madison R, da Silva C F, Dikkes P, Chiu T H and Sidman R L (1985), 'Increased rate of peripheral nerve regeneration using bioresorbable nerve guides and a laminin-containing gel', *Exp Neurol*, **88**, 767–772.

Martin-Rendon E and Watt S M (2003), 'Stem cell plasticity', *Br J Haematol*, **122**, 877–891.

Matejcik V (2002), 'Peripheral nerve reconstruction by autograft', *Injury*, **33**, 627–631.

Matsumoto K, Ohnishi K, Kiyotani T, Sekine T, Ueda H, Nakamura T, Endo K and Shimizu Y (2000), 'Peripheral nerve regeneration across an 80 mm gap bridged by a polyglycolic acid (PGA)–collagen tube filled with laminin-coated fibres: a histological and electrophysiological evaluation of regenerated nerves', *Brain Res*, **868**, 325–328.

McKay Hart A, Wiberg M and Terenghi G (2003), 'Exogenous leukemia inhibitory factor enhances nerve regeneration after late secondary repair using a bioartificial nerve conduit', *Br J Plast Surg*, **56**, 444–450.

Meyer M, Matsuoka I, Wetmore C, Olson L and Thoenen H (1992), 'Enhanced synthesis of brain-derived neurotrophic factor in the lesioned peripheral nerve: different mechanisms are responsible for the regulation of BDNF and NGF mRNA', *J Cell Biol*, **119**, 45–54.

Mezey E, Chandross K, Harta G, Maki R A and McKercher S R (2000), 'Turning blood into brain: Cells bearing neuronal antigens generated *in vivo* from bone marrow', *Science*, **290**, 1779–1782.

Mezey E, Key S, Vogelsang G, Szalayova I, Lange G D and Crain B (2003), 'Transplanted bone marrow generates new neurons in human brains', *Proc Nat Acad Sci USA*, **100**, 1364–1369.

Millesi H, Berger A and Meissl G (1976), 'Further experiences with interfascicular grafting on the median, ulnar and radial nerves', *J Bone Joint Surg Am*, **58**, 227–230.

Moe M C, Varghese M, Danilov A I, Westerlund U, Ramm-Petteersen J, Brundin L, Svensson M, Berg-Johnsen J and Langmoen I A (2005), 'Multipotent progenitor cells from the adult human brain: neurophysiological differentiation to mature neurons', *Brain*, **128**, 2189–2199.

Mohanna P-N, Young R C, Wiberg M and Terenghi G (2003), 'A composite poly-hydroxybutyrate-glial growth factor conduit for long nerve gap repairs', *J Anat*, **203**, 553–585.

Mohanna P-N, Terenghi G and Wiberg M (2005), 'Composite PHB-GGF conduit for long nerve gap repair: a long term evaluation', *Scand J Plast Reconstr Surg Hand Surg*, **39**, 129–137.

Molander H Olsson Y, Engkvist O, Bowald S and Erikson I (1982), 'Regeneration of peripheral nerve through a polygalactin tube', *Muscle Nerve*, **5**, 54–57.

Moore K, Macsween M and Shoichet M (2006), 'Immobilized concentration gradients of neurotrophic factors guide neurite outgrowth of primary neurons in macroporous scaffolds', *Tissue Eng*, **12**(2), 267–278.

Morrison S J (2001), 'Stem cell potential: can anything make anything?' *Curr Biol*, **11**, R7–R9.

Morrison S J, Shah N M and Anderson D J (1997), 'Regulatory mechanisms in stem cell biology', *Cell*, **88**, 287–298.

Mosahebi A, Woodward B, Wiberg M, Martin R and Terenghi G (2001), 'Retroviral labelling of Schwann cells: *in vitro* characterisation and *in vivo* transplantation to improve peripheral nerve regeneration', *Glia*, **34**, 8–17.

Mosahebi A, Fuller P, Wiberg M and Terenghi G (2002), 'Effect of allogeneic Schwann cell transplantation on peripheral nerve regeneration', *Exp Neurol*, **173**, 213–223.

Mosahebi A, Wiberg M and Terenghi G (2003), 'Addition of fibronectin to alginate matrix improves peripheral nerve regeneration in tissue-engineered conduits', *Tissue Eng*, **9**, 209–218.

Murakami T, Fujimoto Y, Yasunaga Y, Ishida O, Tanaka N, Ikuta Y and Ochi M (2003), 'Transplanted neuronal progenitor cells in a peripheral nerve gap promote nerve repair', *Brain Res*, **974**, 17–24.

Murrell W, Feron F, Wetzig A, Cameron N, Splatt K, Bellette B, Bianco J, Perry C, Lee G and Mackay-Sim A (2004), 'Multipotent stem cells from adult olfactory mucosa', *Develop Dynam*, **233**, 496–515.

Nakamura T, Inada Y, Fukuda S, Yoshitani M, Nakada A, Itoi S-I, Kanemaru S-I, Endo K and Shimizu Y (2004), 'Experimental study on the regeneration of peripheral nerve gaps through a polyglycolic acid–collagen (PGA–collagen) tube', *Brain Res*, **1027**, 18–29.

Ngo T-T B, Waggoner P J, Romero A A, Nelson K D, Eberhart R C and Smith G M (2003), 'Poly-L-lactide microfilaments enhance peripheral nerve regeneration across extended nerve lesions', *J Neurosci Res*, **72**, 227–238.

Nunley J A, Saies A D, Sandow M J and Urbaniak J R (1996), 'Results of interfascicular nerve grafting for radial nerve lesions', *Microsurgery*, **17**, 431–437.

Orlic D, Kajstura J, Chimenti S, Jakoniuk I, Anderson S M, Li B, Pickel J, McKay R, Nadal-Ginard B, Bodine D M, Leri A and Anversa P (2001), 'Bone marrow cells regenerate infarcted myocardium', *Nature*, **410**, 701–705.

Owen M and Friedenstein A (1988), 'Stromal stem cells: marrow-derived osteogenic precursor', *Ciba Foundation Symposium*, 42–60.

Pauwelyn K A and Verfaillie C M (2006), 'Transplantation of undifferentiated, bone marrow-derived stem cells', *Curr Top Dev Biol*, **74**, 201–251.

Pereira Lopes F R, de Moura Campos L C, Correa J D Jr, Balduino A, Lora S, Langone F, Borojevic R and Blanco Martinez A M (2006), 'Bone marrow stromal cells and resorbable collagen guidance tubes enhance sciatic nerve regeneration in mice', *Exp Neur*, **198**, 457–468.

Petersen B E, Bowen W C, Patrene K D, Mars W M, Sullivan A K, Murase N, Boggs S S, Greenburger J S and Goff J P (1999), 'Bone marrow as a potential source of hepatic oval cells', *Science*, **284**, 1168–1170.

Piotrowicz A and Shoichet M S (2006), 'Nerve guidance channels as drug delivery vehicles', *Biomaterials*, **27**, 2018–2027.

Pitta M C, Wolford L M, Mehra P and Hopkin J (2001), 'Use of Gore-Tex tubing as conduit for inferior alveolar and lingual nerve repair: experience with 6 cases', *J Oral Maxillofac Surg*, **59**, 493–496.

Pittenger M F, Mackay A M, Beck S, Jaiswal R K, Douglas R, Mosca J, Moorman M A, Simonetti D W, Craig S and Marshak D R (1999), 'Multilineage potential of adult human mesenchymal stem cells', *Science*, **284**, 143–147.

Prockop D J (1997), 'Marrow stromal cells as stem cells for nonhematopoietic tissues', *Science*, **276**, 71–74.

Ramon-Cueto A, Cordero M I, Santos-Benito F F and Avila J (2000), 'Functional recovery of paraplegic rats and motor axon regeneration in their spinal cords by olfactory ensheathing glia', *Neuron*, **24**, 425–435.

Rende M, Muir D, Ruoslahti E, Hagg T, Varon S and Manthorpe M (1992), 'Immunolocalization of ciliary neuronotrophic factor in adult rat sciatic nerve', *Glia*, **5**, 25–32.

Reynolds B A and Weiss S (1992), 'Generation of neurons and astrocytes from isolated cells of the adult mammalian central nervous system', *Science*, **255**, 1707–1710.

Risitano G, Cavallaro G, Merrino T, Coppolino S and Ruggeri F (2002), 'Clinical results and thoughts on sensory nerve repair by autologous vein graft in emergency hand reconstruction', *Chir Main*, **21**, 194–197.

Rodriguez F J, Verdu E, Ceballos D and Navarro X (2000), 'Nerve guides seeded with autologous Schwann cells improve nerve regeneration', *Exp Neurol*, **161**, 571–584.

Safford K, Safford S, Gimble J, Shetty A and Rice H (2004), 'Characterization of neuronal/glial differentiation of murine adipose-derived adult stromal cells', *Exp Neurol*, **187**, 319–328.

Sanchez-Ramos J, Song S, Cardozo-Pelaez F, Hazzi C, Stedefort T, Willing A, Freeman T B, Saporta S, Janssen W, Patel N, Cooper D R and Sanberg P R (2000), 'Adult bone marrow stromal cells differentiate into neural cells *in vitro*', *Exp Neurol*, **164**, 247–256.

Sasaki M, Honmou O, Akiyama Y, Uede T, Hashi K and Kocsis J D (2001),

'Transplantation of an acutely isolated bone marrow fraction repairs demyelinated adult rat spinal cord axons', *Glia*, **35**, 26–34.

Scaravilli F (1984), 'The influence of distal environment on peripheral nerve regeneration across a gap', *J Neurocytol*, **13**, 1027–1041.

Schwartz R E, Reyes M, Koodie L, Jiang Y, Blackstad M, Lund T, Lenvik T, Johnson S, Hu W-S and Verfaillie C M (2002), 'Multipotent adult progenitor cells from bone marrow differentiate into functional hepatocyte-like cells', *J Clin Invest*, **109**, 1291–1302.

Seckel B R, Ryan S E, Gagne R G, Chiu T H and Watkins E (1986), 'Target specific nerve regeneration through a nerve guide in the rat', *Plastic and Reconstructive Surgery*, **78**, 793–800.

Seckel B R, Jones D, Hekimian K J, Wang K-K, Chakalis D P and Costas P D (1995), 'Hyaluronic acid through a new injectable nerve guide delivery system enhances peripheral nerve regeneration in the rat', *J Neurosci Res*, **40**, 3138–3324.

Shamblott M J, Axelman J, Wang S, Bugg E M, Littlefield J W, Donovan P J, Blumenthal P D, Huggins G R and Gearhart J D (1998), 'Derivation of pluripotent stem cells from cultured human primordial germ cells', *Proc Nat Acad Sci USA*, **95**, 13726–13731.

Sigurjonsson O E, Perreault M-C, Egeland T and Glover J C (2005), 'Adult human haematopoietic stem cells produce neurons efficiently in the regenerating chicken embryo spinal cord', *Proc Nat Acad Sci USA*, **102**, 5227–5232.

Sinis N E S H, Schulte-Eversum C, Schlosshauer B, Doser M, Dietz K, Rosner H, Muller H W and Haerle M (2005), 'Nerve regeneration across a 2-cm gap in the rat median nerve using a resorbable nerve conduit filled with Schwann cells', *J Neurosurg*, **103**, 1067–1076.

Smith R M, Wiedl C, Chubb P and Greene C H (2004), 'Rose of small intestine submucosa (SIS) as a nerve conduit: preliminary report', *J Investig Surg*, **17**, 339–344.

Son Y-J and Thompson W J (1995), 'Schwann cell processes guide regeneration of peripheral axons', *Neuron*, **14**, 125–132.

Sondermann Freitas C S and Dalmau S R (2006), 'Multiple sources of non-embryonic multipotent stem cells: processed lipoaspirates and dermis as promising alternatives to bone-marrow-derived cell therapies', *Cell Tissue Res*, **325**, 403–411.

Sterne G D, Coulton G R, Brown R A, Green C J and Terenghi G (1997), 'Neurotrophin-3-enhanced nerve regeneration selectively improves recovery of muscle fibers expressing myosin heavy chains 2b', *J Cell Biol*, **139**, 709–715.

Sulaiman O A R and Gordon T (2000), 'Effects of short- and long-term Schwann cell denervation on peripheral nerve regeneration, myelination and size', *Glia*, **32**, 234–246.

Sundback C, Hadlock T, Cheney M and Vacanti J (2003), 'Manufacture of porous polymer nerve conduits by a novel low-pressure injection moulding process', *Biomaterials*, **24**, 819–830.

Temple S (2001), 'Stem cell plasticity – building the brain of our dreams', *Nature Rev*, **2**, 513–520.

Terada N, Hamazaki T, Oka M, Hoki M, Mastalerz D M, Nakano Y, Meyer E M, Morel L, Petersen B E and Scott E W (2002), 'Bone marrow cells adopt the phenotype of other cells by spontaneous cell fusion', *Nature*, **416**, 542–545.

Terenghi G (1999), 'Peripheral nerve regeneration and neurotrophic factors', *J Anat*, **194**, 1–14.

Terris D J, Toft K M, Moir M, Lum J and Wang M (2001), 'Brain-derived neurotrophic factor-enriched collagen tubule as a substitute for autologous nerve grafts', *Arch Otolaryngol*, **127**, 294–298.

Thomas P K (1989), 'Focal nerve injury: guidance factors during axonal regeneration', *Muscle Nerve*, **12**, 796–802.

Thompson D M and Buettner H M (2006), 'Neurite outgrowth is directed by Schwann cell alignment in the absence of other guidance cues', *Ann Biomed Eng*, **34**, 161–168.

Toba T, Nakamura T, Shimizu Y, Matsumoto K, Ohnishi K, Fukuda S, Yoshitani M, Ueda H, Hori Y and Endo K (2001), 'Regeneration of canine peroneal nerve with the use of a polyglycolic acid-collagen tube filled with laminin-soaked collagen sponge: a comparative study of collagen sponge and collagen fibres as filling materials for nerve conduits', *J Biomed Mat Res*, **58**, 622–630.

Tohill M and Terenghi G (2004), 'Stem-cell plasticity and therapy for injuries of the peripheral nervous system', *Biotechnol Appl Biochem*, **40**, 17–24.

Tohill M, Mantovani C, Wiberg M and Terenghi G (2004), 'Rat bone marrow mesenchymal stem cells express glial markers and stimulate nerve regeneration', *Neurosci Lett*, **362**, 200–203.

Tomita S, Li R-K, Weisel R D, Mickle D A G, Kim E-J, Sakai T and Jia Z-Q (1999), 'Autologous transplantation of bone marrow cells improves damaged heart function', *Circulation*, **100**, II247–II256.

Tong X-J, Hirai K-I, Shimada H, Mizutani Y, Isumi T, Toda N and Yu P (1994), 'Sciatic nerve regeneration navigated by laminin–fibronectin double coated biodegradable collagen grafts in rats', *Brain Res*, **663**, 155–162.

Torigoe K, Tanaka H-F, Takahashi A, Awaya A and Hashimoto K (1996), 'Basic behaviour of migratory Schwann cells in peripheral nerve regeneration', *Exp Neurol*, **137**, 301–308.

Tse W T, Pendleton J D, Beyer W M, Egalka M C and Guinan E C (2003), 'Suppression of allogeneic T-cell proliferation by human marrow stromal cells: Implications in transplantation', *Transplantation*, **75**, 389–397.

Uchida K, Okano H, Hayashi T, Mine Y, Yoshikuni T, Nomura T and Kawase T (2003), 'Grafted swine neuroepithelial stem cell scan form myelinated axons and both efferent and afferent synapses with xenogenic rat neurons', *J Neurosci Res*, **72**, 661–669.

Valentini R F, Aebischer P, Winn S R and Galletti P M (1987), 'Collagen-and laminin-containing gels impede peripheral nerve regeneration through semipermeable nerve guidance channels', *Exp Neurol*, **98**, 350–356.

Vroemen M and Weidner N (2003), 'Purification of Schwann cells by selection of p75 low affinity nerve growth factor receptor expressing cells from adult peripheral nerve', *J Neurosci Meth*, **124**, 135–143.

Wan A C A, Mao H-Q, Wang S, Leong K W, Ong L K L L and Yu H (2001), 'Fabrication of poly-phosphoester nerve guides by immersion precipitation and the control of porosity', *Biomaterials*, **22**, 1147–1156.

Wang D Y, Huang Y C, Chiang H, Wo A. M and Huang Y Y (2006), 'Microcontact printing of laminin on oxygen plasma activated substrates for the alignment and growth of Schwann cells', *J Biomed Mater Res B Appl Biomater*, **80**, 447–453.

Wang H T, Palmer H, Linhardt R J, Flanagan D R and Schmitt E (1990), 'Degradation of poly-ester microspheres', *Biomaterials*, **11**, 679–685.

Wang K K, Costas P D, Bryan D J, Jones D S and Seckel B R (1993a), 'Inside-out vein graft promotes improved nerve regeneration in rats', *Microsurgery*, **14**, 608–618.

Wang K K, Costas P D, Jones D S, Miller R A and Seckel B R (1993b), 'Sleeve insertion and collagen coating improve nerve regeneration through vein conduits', *J Microsurg*, **9**, 39–48.

Wang S, Wan A C A, Xu X, Gao S, Mao H-Q, Leong K W and Yu H (2000), 'A new nerve guide conduit material composed of a biodegradable poly-phosphoester', *Biomaterials*, **22**, 1157–1169.

Wang X, Willenbring H, Akkari Y, Torimaru Y, Foster M, Al-Dhalimy M, Lagasse E, Finegold M, Olson S and Grompe M (2003), 'Cell fusion is the principal source of bone-marrow-derived hepatocytes', *Nature*, **422**, 897–901.

Weber R A and Dellon A L (2001), 'Synthetic nerve conduits: indications and technique', *Sem Neurosurg*, **12**, 341–352.

Wehner T, Bontert M, Eyupoglu I, Prass K, Prinz M, Klett F F, Heinze M, Bechmann I, Nitsch R, Kirchhoff F, Kettenmann H, Dirnagl U and Priller J (2003), 'Bone marrow-derived cells expressing green fluorescent protein under the control of the glial fibrillary acidic protein promoter do not differentiate into astrocytes *in vitro* and *in vivo*', *J Neurosci*, **23**, 5004–5011.

Wen X and Tresco P A (2006), 'Effect of filament diameter and extracellular matrix molecule precoating on neurite outgrowth and Schwann cell behavior on multifilament entubulation bridging device *in vitro*', *J Biomed Mat Res*, **76**, 626–637.

Wernig M, Benninger F, Schmandt T, Rade M, Tucker K L, Bussow H, Beck H and Brustle O (2004), 'Functional integration of embryonic stem cell-derived neurons *in vivo*', *J Neurosci*, **24**, 5258–5268.

Whitworth I H, Dore C, Hall S M, Green C and Terenghi G (1995), 'Different muscle graft denaturing methods and their use for nerve repair', *Br J Plast Surg*, **48**, 493–499.

Whitworth I H, Brown R A, Dore C J, Anand P, Green C J and Terenghi G (1996), 'Nerve growth factor enhances nerve regeneration through fibronectin grafts', *J Hand Surg*, **21B**, 514–522.

Woodbury D, Schwarz E J, Prockop D J and Black I B (2000), 'Adult rat and human bone marrow stromal cells differentiate into neurons', *J Neurosci Res*, **61**, 364–370.

Wu J and Chiu D T (1999), 'Painful neuromas: a review of treatment modalities', *Ann Plast Surg*, **43**, 661–667.

Wu S, Suzuki Y, Ejiri Y, Noda T, Bai H, Kitada M, Kataoka K, Ohta M, Chou H and Ide C (2003), 'Bone marrow stromal cells enhance differentiation of cocultured neurosphere cells and promote regeneration of injured spinal cord', *J Neurosci Res*, **72**, 343–351.

Yandava B D, Billinghurst L L and Snyder E Y (1999), 'Global cell replacement is feasible via neural stem cell transplantation: Evidence from the dysmyelinated *shiverer* mouse brain', *Proc Nat Acad Sci USA*, **96**, 7029–7036.

Yang Y, de Laporte L, Rives C B, Jang J-H, Lin W-C, Shull K R and Shea L D (2005), 'Neurotrophin releasing single and multiple lumen nerve conduits', *J Control Release*, **104**, 433–446.

Yoshii S and Oka M (2001), 'Collagen filaments as a scaffold for nerve regeneration', *J Biomed Mat Res*, **56**, 400–405.

Young R C, Wiberg M and Terenghi G (2002), 'Poly-3-hydroxybutyrate (PHB): a resorbable conduit for long-gap repair in peripheral nerves', *Br J Plast Surg*, **55**, 235–240.

Yuan Y, Zhang P, Yang Y, Wang X and Gu X (2004), 'The interaction of Schwann cells with chitosan membranes and fibers *in vitro*,' *Biomaterials*, **25**, 4273–4278.

Zhang P, He X, Liu K, Zhao F, Fu Z, Zhang D, Zhang Q and Jiang B (2004), 'Bone marrow stromal cells differentiated into functional Schwann cells in injured rats sciatic nerve', *Artif Cells Blood Substit Immobil Biotechnol*, **32**, 509–518.

Zhang P, He X, Zhao F, Zhang D, Fu Z and Jiang B (2005), 'Bridging small-gap peripheral nerve defects using biodegradable chitin conduits with cultured Schwann and bone marrow stromal cells in rats', *J Reconstr Microsurg*, **21**, 565–572.

Zinn M, Whitholt B and Egli T (2001), 'Occurrence, synthesis and medical application of bacterial polyhydroxyalkanoate', *Adv Drug Deliv Rev*, **53**, 5–21.

23
Lung tissue engineering

A E BISHOP and H J RIPPON, Imperial College London, UK

23.1 Introduction

The lungs are essential organs for life and, not surprisingly, respiratory diseases are a leading cause of death worldwide. In the UK, respiratory diseases, predominantly those related to smoking, kill one in five people and the costs to the NHS are enormous; £6.6 billion in 2004 (source: British Thoracic Society, www.brit-thoracic.org.uk). However, despite this severe impact on society, funding for research in this area is disproportionately low and the pathogenetic mechanisms of many lung diseases remain poorly understood.

The complex structure of the lung makes it a particularly difficult target for any tissue engineering strategy and the construction of lung tissue for human implantation remains a remote target. However, some progress is being made towards achieving targeted lung repair and the creation of a biohybrid device that can augment the function of damaged lung. Some of the most pivotal advances have been made in the area of stem cell biology, which has moved to the forefront of medical research in the past few years. Being the basis of natural pathways for tissue maintenance and repair, stem cells represent a key target for mediating repair *in vivo*. Targeted activation of endogenous stem cell pools can augment the body's innate regenerative capability. In cases where this approach is not sufficient, stem cells can provide an abundant, renewable source for the generation of pulmonary cells *in vitro* that can then be delivered to the lung, alone or on scaffolds, for the repair of even more extensive tissue damage. A further application of pulmonary cells grown *in vitro* from stem cells is the construction of *in vitro* screening systems for pharmaceutical or toxicological applications. This chapter describes the progress being made towards producing lung constructs.

23.2 Lung structure

The mammalian lung is made up of a series of branching airways, each smaller than the last, at the terminus of which are the small alveolar sacs which act as the gas exchange unit. There are about three hundred million alveoli in the adult human lung, giving the surface area of between 60 and 80 m^2 that is needed for human respiration. To replicate such a complex structure in the laboratory is an enormous challenge for tissue engineers. The pulmonary tree contains several distinct anatomical regions, lined throughout by a continuous confluent layer of different types of epithelial cells that forms the interface with the air space. Under the top epithelial cell layer lies the vasculature lined by endothelial cells that also form a continuum, running from the main pulmonary artery, through the capillary network of the distal lung, to the pulmonary vein and then to the left atrium of the heart. In between the epithelial and endothelial layers lies the pulmonary interstitium which comprises a range of connective tissue cell types providing a scaffold for cell layers within the lung. The following summarizes the structure of these three main components of the lung.

The proximal airways, the trachea and major bronchi, are lined by pseudostratified epithelium containing ciliated and mucous secretory (or goblet) cells on the luminal surface. Neuroendocrine cells are also scattered in the epithelium (Lauwreyns and Cokelaere, 1973; Cutz and Orange, 1977). Ciliated epithelial cells extend into the smaller airways but here, rather than goblet cells, the bronchioles possess the cuboidal, non-ciliated Clara cells. The alveoli are lined by two distinct but related types of epithelial phenotypes; the flattened squamous (type I) and cuboidal (type II) pneumocytes. The type I pneumocyte is the cell type across which gas exchange occurs. Type II pneumocytes secrete pulmonary surfactant to lower surface tension and prevent airway collapse, and can differentiate to form type I pneumocytes.

The vasculature of the lung is different from that of the systemic circulation in that the vessels are low resistance. Throughout the pulmonary vasculature the endothelial cell layer is continuous, bathed on one side by blood and on the other by interstitial fluid. Unlike the epithelium, pulmonary endothelial cells share most characteristics regardless of their position in the pulmonary tree, although there are some small differences related to the function of each region. Venous endothelial cells are thinner and polygonal, and have fewer organelles, while alveolar endothelial cells have an avesicular zone, the basal lamina of which fuses with that of type I pneumocytes to form the air–blood barrier.

The interstitium is the space between the epithelium lining the air space, the vascular endothelium and the pleural mesothelium. The cells of the interstitium form a spectrum of types from smooth muscle to fibroblasts. Immune cells, in the form of macrophages and lymphocytes, mainly T cells, are present in the interstitial fluid. A web of elastin and collagen fibres, interwoven with fibronectin fibrils and proteoglycan molecules, permeates the interstitial space and forms the matrix that supports the structural integrity of the lung while allowing its plastic deformation during respiration.

23.3 Sources of cells for lung tissue engineering

Primary cells collected from body fluids, tissue explant outgrowth or tissue disaggregation can be used for tissue engineering but are usually difficult to access, have a limited lifespan and do not maintain their phenotype for long *in vitro*. They have been used successfully, however, in the construction of *in vitro* models of lung. For tissue engineering though, it is proposed that stem cells provide a virtually inexhaustible cell source. Large-scale production of cells for tissue engineering could be achieved if undifferentiated stem cells were expanded in culture and driven in batches to differentiate into the relevant target cell types. Consequently, current research is focused on developing means to expand stem cells, promote their efficient differentiation to specific lineages, to purify the required populations and to manipulate the resulting cells into a form suitable for implantation (see Polak and Bishop, 2006, for review).

For each tissue engineering strategy using stem cells, the initial step is the selection of the most appropriate cell type on which to base it. Stem cells can be isolated from embryonic, fetal or adult tissue, but stem cells do not all have the same differentiation potential and the range of cell types to which they can differentiate varies according to their origin and the signals to which they are exposed. As mentioned previously, the structural complexity and cellular heterogeneity of the lung renders the study of lung-specific stem cell biology difficult and, therefore, the isolation of sufficient numbers of pulmonary stem cells for potential therapeutic applications is simply unrealistic at this time. For this reason, attempts have been made to produce pulmonary epithelium from non-lung stem cells. Most lung tissue engineering to date has exploited embryonic stem cells, in view of their relative availability, known provenance, pluripotency and proliferative capacity (see Rippon and Bishop, 2004, for review). Preliminary work derived alveolar airway epithelium, specifically type II pneumocytes, from murine embryonic stem cells using a commercial medium that maintains differentiated characteristics in primary lung epithelial cell cultures (Ali *et al*, 2002; Rippon *et al.*, 2004). These cells fulfil the type II cell phenotype by expressing surfactant proteins (Fig. 23.1), possessing microvilli and lamellar bodies and by differentiating *in vitro* to type I pneumocytes. A more recent protocol that has been developed provides a three-step, largely serum-free strategy for the generation of distal lung epithelial progenitor cells. These cells appear to parallel those present in the early branching lung at approximately E10-11 of murine embryonic life which differentiate further during branching morphogenesis and postnatal alveolarization to form Clara cells and type I and type II pneumocytes (Rippon *et al.*, 2006).

Lung epithelial cell yields of more than a few per cent of the differentiated embryonic stem cell population have been difficult to obtain by the manipulation of cell culture medium alone, an observation that also holds true for other cell lineages derived from the endodermal germ layer. Consequently,

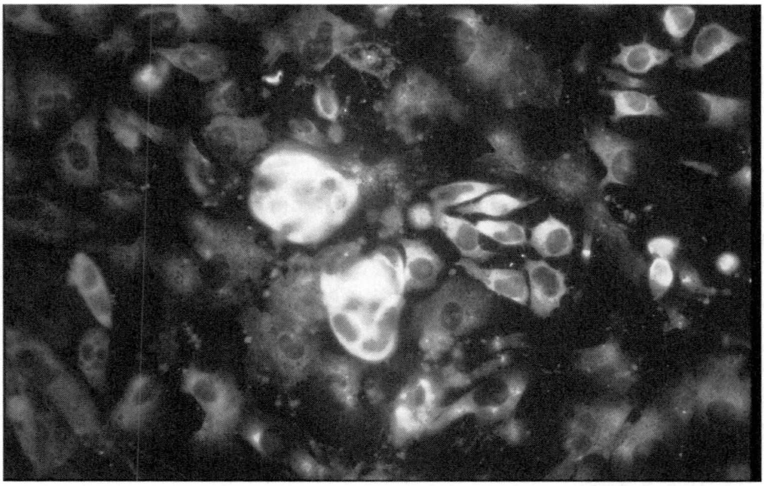

23.1 A heterogeneous population of murine embryonic stem cells (ESC) grown in medium containing a mix of growth factors that maintain differentiated characteristics in primary lung epithelial cell cultures. Type II pneumocytes that had differentiated from the ESC were identified by immunofluorescent staining for their specific marker, surfactant protein C.

other means have been tested for their ability to drive formation of pulmonary epithelium from embryonic stem cells. Thus, it has been reported that recapitulation of the natural environment encountered by differentiating pulmonary epithelium in the embryo, by combining medium supplementation with cell culture at an air–liquid interface, can induce the formation of fully differentiated tracheobronchial airway epithelium from murine embryonic stem cells (Coraux *et al.*, 2005). The inductive power of lung mesenchyme on embryonic stem cells has also been investigated, in view of the critical role that underlying mesenchymal cells play in regulating lung epithelial differentiation *in vivo*. Early differentiating embryonic stem cell spheroids were wrapped in microdissected lung mesenchyme taken from E11.5 murine embryos. After only 5 days, mesenchyme and embryonic stem cells had coalesced and small channels could be seen that were lined by cells expressing markers of alveolar epithelium (Van Vranken *et al.*, 2005; Fig. 23.2). A similar approach of organotypic regeneration has recently been employed (Denham *et al.*, 2006), to show that the inductive properties of murine lung mesenchyme persist for only a short developmental window around embryonic day 11.5, the stage at which distal lung epithelial differentiation is initiated *in vivo*. Clearly, this latter approach is too limited to provide cells or tissue for implantation but it provides an *in vitro* model of distal lung development for investigation of mesodermal–endodermal interactions.

Another approach that has been used involves a method originally used to convert fibroblasts into T cells using T-cell extracts (Håkelien *et al.*, 2002) that

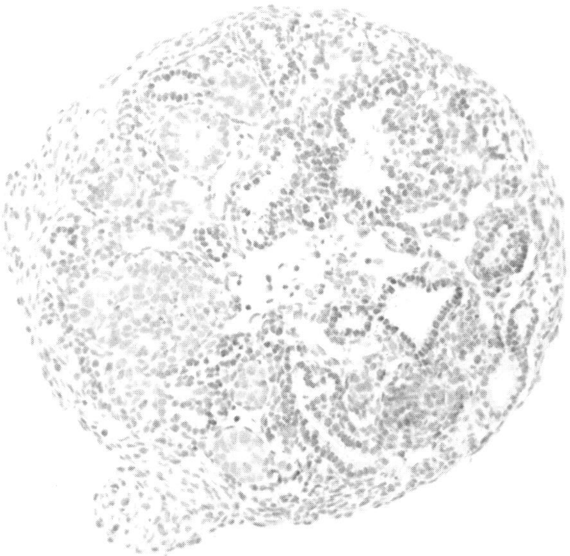

23.2 Murine embryonic lung mesenchyme wrapped around an embryoid body and cultured for 5 days. After this time, the two layers had coalesced and small channels could be seen lined by cells immunostained for the marker of immature distal lung alveolar epithelium, thyroid transcription-1.

has been adapted and applied to murine embryonic stem cells. In this method the membranes of stem cells are permeabilized and exposed to crude lysates of a type II pneumocyte cell line. Following membrane re-sealing and culture in basic, unsupplemented medium, the cells were found to differentiate to type II pneumocytes with an efficiency of approximately 10%, well in excess of the efficiency achieved with most other methods (Qin *et al.*, 2005).

Technology developed with murine cell lines is being rapidly translated to human embryonic and somatic stem cells in order to progress to clinic. For embryonic stem cells, human lines were isolated in 1998 and it is only within the last few years that they have become readily available to laboratories lacking the highly specialized expertise and facilities to isolate their own human embryonic stem cell lines. Recently, type II pneumocytes have been successfully differentiated from human embryonic stem cells (Samadikuchaksaraei *et al.*, 2006), a major step in the development of stem cell-based treatments for lung disease. Derivation of pulmonary epithelium *in vitro* from non-pulmonary somatic stem cells has not yet been reported in the literature.

23.4 Lung tissue constructs

Lung tissue has been constructed for *in vitro* use, mainly for drug delivery research and toxicological testing. For example, EpiAirway™ (www.Matteck.com)

comprises normal human tracheal and bronchial epithelial cells cultured to form a pseudo-stratified structure that resembles epithelial tissue of airways. Similar models have been made of bronchus specifically for the *in vitro* study of the pathogenesis of asthma (Chakir *et al.*, 2001; Paquette *et al.*, 2003, 2004). Attempts have been made to engineer trachea for implantation (Kojima and Vacanti, 2004) and the first successful human application has been reported (Omori *et al.*, 2005). In addition to direct implantation of an engineered construct, there has been early work on the engineering of trachea by aerosol delivery of cells (Roberts *et al.*, 2005). An atomizer has been developed, but not yet tested in animals, that successfully sprays mammalian chondrocytes and tracheal epithelial cells with more than 70% viability and no effect on growth rate.

The generation in the laboratory of the complex gas-exchange component of the lung for human implantation, however, remains an elusive goal. Pneumocytes attach fairly well to a range of polymer or bioactive scaffolds (Table 23.1), particularly if an appropriate extracellular matrix protein is coated on the scaffold surface, e.g. laminin (Tan *et al.*, 2003; Fig. 23.3). Attempts to create distal lung

Table 23.1 Culture of pulmonary epithelial cells on biological and non-biological 3-dimensional scaffolds

Cells	Scaffold/polymer	Results	Reference
Dissociated fetal rat lung	Collagen-glycosaminoglycan (GAG)	Cell-mediated contraction of scaffold and formation of small 3D cellular structures	Chen *et al.* (2005)
Dissociated fetal rat lung	Collagen/gelatin sponges	Cells spontaneously formed alveolar-like structures containing mainly type II pneumocytes	Douglas and Chapple (1976); Douglas and Teel (1976); Douglas *et al.*, (1976, 1983)
Primary neonatal rat type II pneumocytes	Collagen gel	Cells aggregated into alveolar-like structures containing differentiated type II pneumocytes	Saito *et al.* (1985)
Primary rat type II pneumocytes	Collagen gel	Cells aggregated and branched into cystic structures containing both type II and type I pneumocyte-like cells	Sugihara *et al.* (1993)
Primary adult rat type II pneumocytes	Matrigel (Engelbreth–Holm–Swarm tumour basement membrane) and laminin	Cells cultured on matrigel formed differentiated, polarised aggregates. Cells on laminin transdifferentiated to type I pneumocyte-like cells	Shannon *et al.* (1990)

Table 23.1 Continued

Cells	Scaffold/polymer	Results	Reference
Subpopulation of primary type II pneumocytes	Matrigel (Engelbreth–Holm–Swarm tumour basement membrane)	Cells acquired some differentiated type II pneumocyte characteristics and aggregated	Kalina *et al.* (1993)
Fetal rabbit distal lung epithelial cells	Matrigel (Engelbreth–Holm–Swarm tumour basement membrane)	Cells in thick gels associated into spherical, polarized epithelial structures	Blau *et al.* (1988), Chinoy *et al.* (1994)
Mixed population from embryonic murine lung	Matrigel (Engelbreth–Holm–Swarm tumour basement membrane)	Cells formed branching 3D structures expressing surfactant protein C	Mondrinos *et al.* (2006)
Mixed population from embryonic murine lung	Poly-lactic-co-glycolytic acid (PLGA) foams	Cells grew into scaffold but differentiated epithelium was not maintained, even in presence of tissue-specific growth factors	Mondrinos *et al.* (2006)
Mixed population from embryonic murine lung	Poly-L-lactic acid (PLLA) nanofibres	Cells grew into scaffold but differentiated epithelium was not maintained, even in presence of tissue-specific growth factors	Mondrinos *et al.* (2006)
MLE-12 cell line	Poly-DL-lactic acid (PDLLA) disks and foams	Scaffold was non-toxic and supported cell growth	Lin *et al.* (2006)
MLE-12 cell line	Laminin-coated 58S sol–gel foams	Coated scaffold supported cell migration and growth	Tan *et al.* (2003)

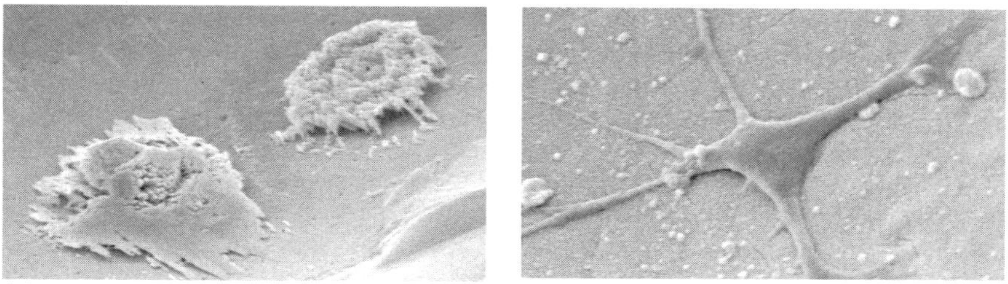

(a) (b)

23.3 Scanning electron micrographs showing pneumocytes grown on poly-DL-lactic acid: (a) without coating and (b) coated with the extracellular matrix protein laminin. The beneficial effect of laminin can be seen clearly as the pneumocyte has spread and attached closely to the polymer surface.

began 30 years ago when dispersed rat fetal pneumocytes grown on a collagen sponge aggregated around a lumen to form something that crudely resembled alveoli (Douglas and Chapple, 1976; Douglas and Teel, 1976; Douglas *et al.*, 1976). These 'alveolar-like structures', which formed within 2 days of culture and could be maintained for up to 6 weeks, were found to consist mainly of type II pneumocytes that secreted surfactant (Douglas *et al.*, 1983). Similar apparent reconstitution of alveolar structures *in vitro* has been reported for neonatal (Saito *et al.*, 1985) and adult rat lung cells grown on collagen gels (Sugihara *et al.*, 1993) or basement membrane (Shannon *et al.*, 1990; Kalina *et al.*, 1993) and for fetal rabbit lung cells grown on basement membrane (Blau *et al.*, 1988; Chinoy *et al.*, 1994). Recent developments include growth of dissociated rat lung cells on a collagen-glycosaminoglycan (GAG) tissue engineering scaffold (Chen *et al.*, 2005). Alveolar-like structures were observed in the scaffold and the construct exhibited cell-mediated contraction suggesting the formation of elastic units, as occurs *in vivo*. This work also indicated the potential usefulness for lung construction of collagen–GAG, a scaffold that is already well established in tissue engineering, e.g. for skin (Yannas *et al.*, 1989). Another group has reported the growth of mixed embryonic murine lung cells (epithelium, endothelium and mesenchyme) in 3D on Matrigel and the synthetic polymers poly-lactic-co-glycotic acid (PLGA) and poly-L-lactic acid (PLLA) formed into either foams or nanofibrous scaffolds (Mondrinos *et al.*, 2006). In this system, it was possible to upregulate the branching and sacculation of alveolar-like tissue when cells were cultured in Matrigel by adding bFGF, FGF-7 and FGF-10 to the culture medium.

Artificial lungs, in the form of ECMOs (extracorporeal membrane oxygenators) or ECLS (life support systems), have been used for some time in clinic but

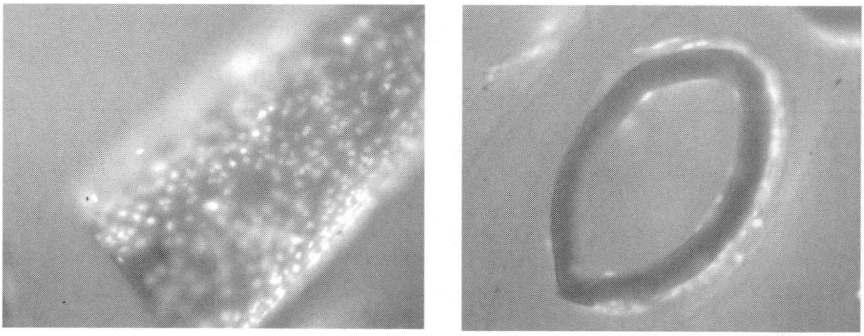

(a) (b)

23.4 A single polymethylpentene (PMP) fibre from the gas-exchange module of the Novalung interventional lung-assist device with pneumocytes (with fluorescently stained nuclei) growing on the surface: (a) side view and (b) cross-section. These PMP fibres are woven to form mats that are stacked in the device.

these are known to be relatively inefficient and are entirely mechanical. These systems are used, for example, in cardiopulmonary bypass during heart surgery and also, to a lesser extent, to support patients in respiratory failure. Work is in progress to improve these machines, for example by increasing efficiency and/or making them intrathoracic, with a view to creating systems that can provide extended respiratory support (see Federspiel and Henchir, 2004, for review). A potential means to achieve this would be to make the devices into biohybrids; a major step on the long, rocky road to making lung tissue for implantation could be the incorporation of epithelial and/or endothelial cells in extra-corporeal gas-exchange devices. For this latter possibility, our team at Imperial College London is collaborating with a company, Novalung™ (http://www.novalung.com), to try to increase the efficiency and lifespan of their interventional lung-assist device (Fischer *et al.*, 2006) by incorporating stem cell-derived lung phenotypes, thereby generating a biohybrid system (Fig. 23.4).

23.5 Conclusions

Lung diseases are extremely prevalent worldwide, yet many remain effectively incurable without a transplant and the numbers of donor organs continue to dwindle. The complexity and cellular heterogeneity of the lung make it a singularly difficult target for tissue engineering and we are a long way from constructing functional lung tissue for implantation. The latest research suggests that controlled and targeted lung repair could be effected by at least three methods: by activating the body's existing repair pathways and stem cell pools, by implanting healthy cells or tissue grown from stem cells in the laboratory or a combination of elements of both strategies. The major expansion of the field of stem cell biology and its applications in medicine in the past decade has provided tissue engineers with a new, abundant and very plastic source with which to try to derive lung tissue. A major step towards translating these efforts into clinical use could be using stem cell-derived lung cells outside the body, as part of extracorporeal gas-exchange devices. However, although exciting advances have been made in the area of lung tissue engineering in recent years, the time-frame for transferring any of the approaches being tested into clinic remains unknown.

23.6 References

Ali NN, Edgar AJ, Samadikuchaksaraei A, Timson CM, Romanska HM, Polak JM, Bishop AM (2002), 'Derivation of type II alveolar epithelial cells from murine embryonic stem cells', *Tissue Eng*, **8**, 541–550.

Blau H, Guzowski DE, Siddiqu ZA, Scarpelli EM, Bienkowski RS (1988), 'Fetal type 2 pneumocytes form alveolar-like structures and maintain long-term differentiation on extracellular matrix', *J Cell Physiol*, **136**, 203–214.

Chakir J, Page N, Hamid Q, Laviolette M, Boulet LP, Rouabhia M (2001), 'Bronchial muscosa produced by tissue engineering: a new tool to study cellular interactions in asthma', *J Allergy Clin Immunol*, **107**, 36–40.

Chen P, Marsilio E, Goldstein RH, Yannas IV, Spector M (2005), 'Formation of lung alveolar-like structures in collagen–glycosaminoglycan scaffolds *in vitro*', *Tissue Eng*, **11**, 1436–1448.

Chinoy MR, Antonio-Santiago MT, Scarpelli EM (1994), 'Maturation of undifferentiated lung epithelial cells into type II cells *in vitro*: a temporal process that parallels cell differentiation *in vivo*', *Anat Rec*, **240**, 545–554.

Coraux C, Nawrocki-Raby B, Hinnrasky J, Kileztky C, Gaillard D, Dani C, Puchelle E (2005), 'Embryonic stem cells generate airway epithelial tissue', *Am J Respir Cell Mol Biol*, **32**, 87–92.

Cutz E, Orange RP (1977), 'Mast cells and endocrine (APUD) cells of the lung', in Lichtenstein LM and Atisten KF, *Asthma: Physiology, Immunopharmacology and Treatment*, Academic Press, New York, 51–76.

Denham M, Cole TJ, Mollard R (2006), 'Embryonic stem cells form glandular structures and express surfactant protein-C following culture with dissociated fetal respiratory tissue', *Am J Physiol Lung Cell Mol Physiol*, **290**, L1210–1215.

Douglas WH, Chapple PJ (1976), 'Characterization of monolayer cultures of type II alveolar pneumonocytes that produce pulmonary surfactant *in vitro*', *Dev Biol Stand*, **37**, 71–76.

Douglas WH, Teel RW (1976), 'An organotypic *in vitro* model system for studying pulmonary surfactant production by type II alveolar pneumonocytes', *Am Rev Respir Dis*, **113**, 17–23.

Douglas WH, Moorman GW, Teel RW (1976), 'The formation of histotypic structures from monodisperse fetal rat lung cells cultured on a three-dimensional substrate', *In Vitro*, **12**, 373–381.

Douglas WH, Sommers-Smith SK, Johnston JM (1983), 'Phosphatidate phosphohydrolase activity as a marker for surfactant synthesis in organotypic cultures of type II alveolar pneumonocytes', *J Cell Sci*, **60**, 199–207.

Federspiel WJ, Henchir, KA (2004), 'Artificial lungs: basic principles and current applications' in Wnek GE, Bowlin GL, *Encyclopedia of Biomaterials and Biomedical Engineering*, Marcel Dekker Inc., New York, 910–921.

Fischer S, Simon AR, Welte T, Hoeper MM, Meyer A, Tessmann R, Gohrbandt B, Gottlieb J, Haverich A, Strueber M (2006), 'Bridge to lung transplantation with the novel pumpless interventional lung assist device NovaLung', *J Thorac Cardiovasc Surg*, **131**, 719–723

Håkelien AM, Landsverk HB, Robl JM, Skalhegg BS, Collas P (2002), 'Reprogramming fibroblasts to express T-cell functions using cell extracts', *Nat Biotechnol*, **20**, 460–466.

Kalina M, Riklis S, Blau H (1993), 'Pulmonary epithelial cell proliferation in primary culture of alveolar type II cells', *Exp Lung Res*, **19**, 153–175.

Kojima K, Vacanti CA (2004), 'Generation of a tissue-engineered tracheal equivalent', *Biotechnol Appl Biochem*, **39**, 257–262.

Lauweryns JM, Cokelaere M (1973), 'Hypoxia-sensitive neuro-epithelial bodies: intrapulmonary secretory neuroreceptors modulated by the CNS', *Z Zellforsch Milrosk Anat*, **145**, 521–540.

Lin YM, Boccaccini AR, Polak JM, Bishop AE, Maquet V (2006), 'Biocompatibility of poly-DL-lactic acid (PDLLA) for lung tissue engineering', *J Biomater Appl*, **21**(2), 109–118.

Mondrinos MJ, Koutzaki S, Jiwanmall E, Li M, Dechadarevian JP, Lelkes PI, Finck CM

(2006), 'Engineering three-dimensional pulmonary tissue constructs', *Tissue Eng*, **12** (4), 717–728.

Omori K, Nakamura T, Kanemaru S, Asato R, Tanaka S, Magrufov A, Ito J, Shimizu T (2005), 'Regenerative medicine of the trachea: the first human case', *Ann Ontol Rhinol Laryngol*, **114**, 429–433.

Paquette JS, Tremblay P, Bernier V, Auger FA, Laviolette M, Germain L, Boutet M, Boulet LP, Goulet F (2003), 'Production of tissue-engineered three-dimensional human bronchial models', *In Vitro Cell Dev Biol Anim*, **39**, 213–220.

Paquette JS, Moulin V, Tremblay P, Bernier V, Boutet M, Laviolette M, Auger FA, Boulet LP, Goulet F (2004), 'Tissue-engineered human asthmatic bronchial equivalents', *Eur Cell Mater*, **10**, 1–11.

Polak JM, Bishop AE (2006), 'Stem cells and tissue engineering: past, present and future', *Ann New York Acad Sci*, **1068**, 352–366.

Qin MD, Tai GP, Collas P, Polak JM, Bishop AE (2005), 'Cell extract-derived differentiation of embryonic stem cells', *Stem Cells*, **23**, 712–718.

Rippon HJ, Bishop AE (2004), 'Embryonic stem cells', *Cell Prolif*, **27**, 23–34.

Rippon HJ, Ali NN, Polak JM, Bishop AE (2004), 'Initial observations on the effect of medium composition on the differentiation of murine embryonic stem cells to alveolar type II cells', *Cloning & Stem Cells*, **6**, 49–56.

Rippon HJ, Polak JM, Qin M, Bishop AE (2006), 'Derivation of distal lung epithelial progenitors from murine embryonic stem cells using a novel 3-step differentiation protocol', *Stem Cells*, **24**, 1389–1398.

Roberts A, Wyslouzil BE, Bonassar L (2005), 'Aerosol delivery of mammalian cells for tissue engineering', *Biotechnol Bioeng*, **91**, 801–807.

Saito K, Lwebuga-Mukasa J, Barrett C, Light D, Warshaw JB (1985), 'Characteristics of primary isolates of alveolar type II cells from neonatal rats', *Exp Lung Res,* 8, 213–225.

Samadikuchaksaraei A, Cohen S, Isaac K, Rippon HJ, Polak JM, Bielby RC, Bishop AE (2006), 'Derivation of type II pneumocytes from human embryonic stem cells', *Tissue Eng*, **12**, 867–875.

Shannon JM, Emrie PA, Fisher JH, Kuroki Y, Jennings SD, Mason RJ (1990), 'Effect of a reconstituted basement membrane on expression of surfactant apoproteins in cultured adult rat alveolar type II cells', *Am J Cell Resp Cell Mol Biol*, **2**, 183–192.

Sugihara H, Toda S, Miyabara S, Fujiyama C, Yonemitsu N (1993), 'Reconstruction of alveolus-like structure from alveolar type II epithelial cells in three-dimensional collagen gel matrix culture', *Am J Pathol*, **142**, 783–792.

Tan A, Romanska HM, Lenza R, Jones J, Hench LL, Polak JM, Bishop AE (2003), 'The effect of 58S bioactive sol–gel-derived foams on the growth of murine lung epithelial cells', *Key Eng Mat*, **240–242**, 719–724.

Van Vranken B, Romanska HM, Polak JM, Rippon HJ, Shannon JM, Bishop AE (2005), 'Co-culture of embryonic stem cells with pulmonary mesenchyme: a micro-environment that promotes differentiation of pulmonary epithelium', *Tissue Eng*, **11**, 1177–1187.

Yannas IV, Lee E, Orgill DP, Skrabut EM, Murphy GF (1989), 'Synthesis and characterization of a model extracellular matrix that induces partial regeneration of adult mammalian skin', *Proc Nat Acad Sci USA*, **86**, 933–937.

24
Intestine tissue engineering

D A J LLOYD and S M GABE, St Mark's Hospital, UK

24.1 Introduction

Short bowel syndrome is characterised by gastrointestinal fluid loss, malabsorption and progressive malnutrition; if fluids, electrolytes and nutrients are not replaced, patients become dehydrated and malnourished. This may ultimately be fatal. The commonest cause of SBS is loss of functional absorptive intestinal surface usually due to extensive small bowel resection. Patient survival is dependent on long-term parenteral nutrition, with its associated morbidity and mortality. Surgical treatment to lengthen the bowel has been described in children but is of limited value in adults. Small intestinal transplantation can be considered but it requires aggressive immunosupression and survival rates are not yet as high as other solid organ grafts.

A novel approach to the treatment of patients with intestinal failure due to short bowel syndrome is the replacement of lost small intestine with bioengineered tissue. Human trials of tissue-engineered bladder[1] have given a tantalising insight into the therapeutic possibilities of this evolving technology. Potentially, the resultant increase in absorptive capacity that might be achieved via tissue engineering of the small intestine could allow a patient with short bowel syndrome to be weaned from parenteral nutrition. However, the structural and functional complexities of the small intestine are considerable and, partly as a result of this, tissue engineering of the small intestine is still in its infancy. This chapter will review progress made to date in the field of small intestinal tissue engineering, will highlight the limitations of the current models and techniques, and will explore possible future research directions.

24.2 Approaches to tissue engineering of the small intestine

Conceptually, there are two possible approaches to tissue engineering of the small intestine. The first approach is to try to replicate the anatomical structure of the small intestine with the assumption that this will produce functional replacement tissue. However, the small intestine is anatomically complex with its morphology reflecting not only its function as an absorptive surface but also its role as a barrier against the external environment (see Fig. 24.1). The production of innervated muscle layers, vascular and lymphatic networks and appropriate lymphoid tissue in addition to a functional mucosal surface is a considerable challenge. The second approach is to concentrate primarily on producing a functional absorptive surface without necessarily reproducing the exact anatomical structure of the small intestine. While ultimately this is likely to be inferior to the production of functional replica tissue, which would be considered to be the gold standard of tissue engineering, it is likely to yield greater therapeutic possibilities in the short to medium term.

The majority of studies to date have concentrated on the production of a tissue engineered layer of small intestinal mucosa, often referred to as neomucosa or neointestine.[2–13] Generation of an intact neomucosal layer is understandably considered to be vital for the manufacture of functional replacement intestinal tissue. However, it must be remembered that the function of the small intestine is also dependent on an adequate vascular supply and lymphatic drainage as well as coordinated peristalsis dependent on correctly innervated muscular layers.

Two main techniques have been used in order to produce small intestinal neomucosa in animal models. The simplest method has been to achieve

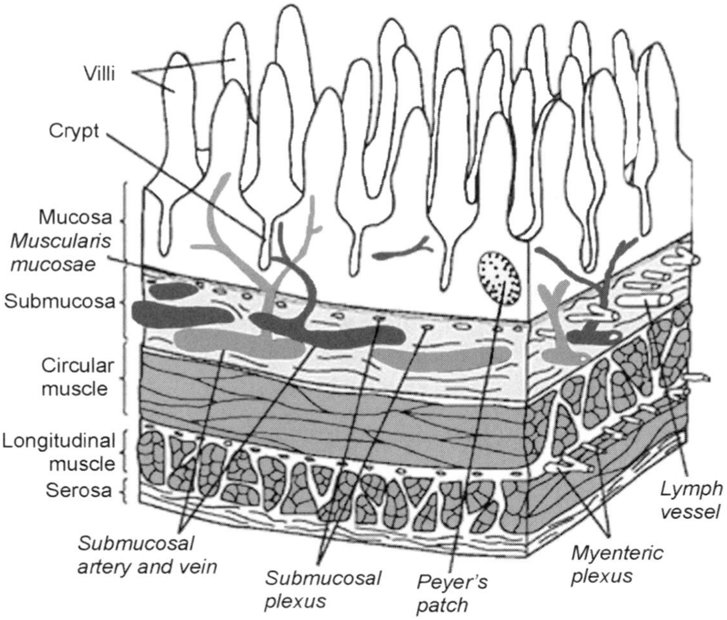

24.1 Schematic diagram illustrating the structural complexity and various component layers of the small intestinal wall.

intestinal lengthening by interposition of a tubular artificial scaffold between segments of healthy small intestine. The scaffold then forms a framework for in-growth of mucosa from the healthy intestinal tissue.[2, 14–16] The alternative technique has been to attempt to transplant intestinal stem cell populations, harvested from neonatal animals, onto either denuded bowel or artificial scaffolds.[3–5,9–11,13,17–25] Both of these techniques, along with their relative strengths and weaknesses, will be discussed in more detail below.

24.3 Artificial scaffolds

Availability of a suitable scaffold material is vital to the tissue engineering of any organ. A successful scaffold needs to have a range of physical, chemical and biological properties that are tailored to the tissue that it is to support. It must also be biocompatible and not elicit a significant foreign body reaction. Scaffold materials can be either natural or synthetic and a summary of materials used as scaffolds for small intestinal tissue engineering is shown in Table 24.1. Natural scaffolds include a range of decellularised biological matrices, such as small intestine submucosa (SIS) and acellular dermal matrix, and processed biologically derived materials, such as collagen sponge. Synthetic scaffolds have been created from a range of different bioresorbable polymers and co-polymers including polyglycolic acid (PGA), polylactide (PLA) and poly(D,L-lactide-co-glycolide) (PLGA) using a range of different engineering techniques.[26] Cross-sectional images of examples of natural and synthetic scaffolds are shown in Figs 24.2 and Fig. 24.3. As will be discussed, in the setting of small intestinal tissue engineering natural materials have been used principally as scaffolds for intestinal lengthening procedures.[2,14–16] Conversely, experiments attempting to create neointestine by transplanting stem cell cultures have employed synthetic scaffolds.[3,4,13,18]

The 3D structure of the scaffold material is of vital importance as it not only will influence the mechanical properties of the structure but will also affect cell migration and adhesion. The scaffolds used for tissue engineering, both natural and synthetic, are porous structures with interconnected pore networks. The larger pores allow cellular infiltration and migration and penetration by blood vessels and lymphatics. The smaller pores allow diffusion of oxygen and nutrients into and waste products out of the structure. An advantage of synthetic scaffolds is that it is possible to manipulate the exact physical properties of the material. By altering pore size and pore density it is possible to change both the physical and the biological characteristics of the synthetic scaffold. It is also easier to control the overall shape of synthetic compared with natural scaffolds.

Another important property of the scaffolds used for tissue engineering is the ability to promote cell adhesion, migration and proliferation. This is heavily influenced by the surface properties of the scaffold material. In general, cell

Table 24.1 Materials used as scaffolds for small intestinal tissue engineering

Type	Structure	Features/properties	Experimental design	References
Natural	Small intestine submucosa (SIS)	Manufacture by mechanical removal of mucosa and muscular layers from porcine small intestine followed by osmotic lysis of remaining cells. Resulting membrane ~80–100 μm thick. Greater success with multi-layered sheets.	Intestinal patch and intestinal lengthening	17, 32, 33
Natural	Surgisis (Cook Biotech Inc.)	Commercially available SIS-type scaffold material derived from porcine small intestine.	Intestinal lengthening	4
Natural	Collagen sponge	Collagen extracted from porcine skin (70–80% type I collagen, 20–30% type III atelocollagen). Fibres whipped, freeze-moulded and freeze-dried. Stabilised by heating to cross-link collagen fibres and subsequent basal coating with polyglycolic acid (PGA).	Intestinal lengthening	18, 30, 31
Natural	Acellular dermal matrix (ADM)	Commercially available ADM (AlloDerm, Cell Life) sutured to created a tubular structure.	Intestinal lengthening	16
Natural	Acellular gastric wall	Gastric tissue decellularised by detergent-enzymatic treatment and sutured to create a tubular structure.	Intestinal lengthening	34
Synthetic	Polyglycolic acid (PGA)	Sheets of non-woven PGA (15 μm fibre diameter and 250 μm average pore diameter) wrapped into a tubular structure and stabilised by coating with 5% polylactic acid (PLA). Improved cellular adhesion demonstrated after coating with type 1 collagen.	Stem cell transplantation	5, 7, 13, 20–24, 49, 50, 52, 58
Synthetic	Poly(D,L-lactide-co-glycolide) (PLGA)	PLGA foam manufactured by thermally induced phase separation. Radially oriented interconnected pores with large size distribution (50–300 μm). Rolled into tubular structure and opposing edges joined by dissolving in chloroform and opposing.	Stem cell transplantation	15, 27

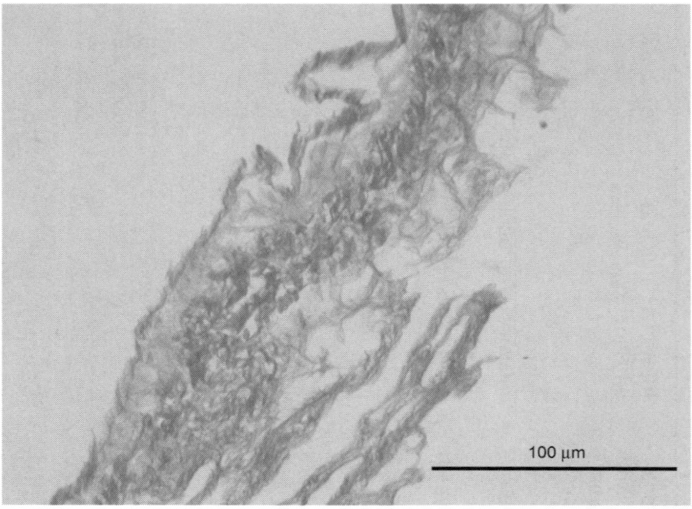

24.2 Light microscopy of decellularised section of rat colon processed in preparation for intestinal lengthening procedure. Section is stained with Mallory's elastin to demonstrate the lattice of collagen and elastin fibres.

adhesion is enhanced by a rough rather than a smooth fibre surface. This is seen in natural scaffolds such as SIS due to the high collagen content. However, the majority of synthetic polymers used to create scaffolds are hydrophobic with a relatively smooth fibre surface. In order to promote cellular interactions, a number of surface modification techniques have been employed, including surface coating, chemical modification and plasma treatment.[27] Collagen coating of PGA scaffolds has been shown to improve adhesion of intestinal epithelial cells[3] although it is not known how such modified synthetic scaffolds would compare with natural scaffold materials.

Creating the ideal scaffold for small intestinal tissue engineering is considerably more difficult than for other hard and soft tissues such as bone or cartilage owing to the increased structural complexity. The ultimate goal is to produce flexible multilayered tubular neointestinal tissue rather than a uniform solid structure such as bone. The production of tubular scaffolds is technically more challenging as the structure must have enough strength to remain patent when initially implanted but must degrade over time to allow replacement by intestinal tissue and subsequent tissue expansion. The time taken for the biodegradable scaffold to break down is critical; if it is too rapid then the lumen may collapse but if it is too slow then the growth of the neointestine may be impeded. The degradation properties of synthetic scaffolds can be modulated by altering the chemical composition and pore structure. While it is possible to alter the physical properties of natural scaffold materials, for example by crosslinking collagen fibres,[16] the overall effects are more modest than for synthetic

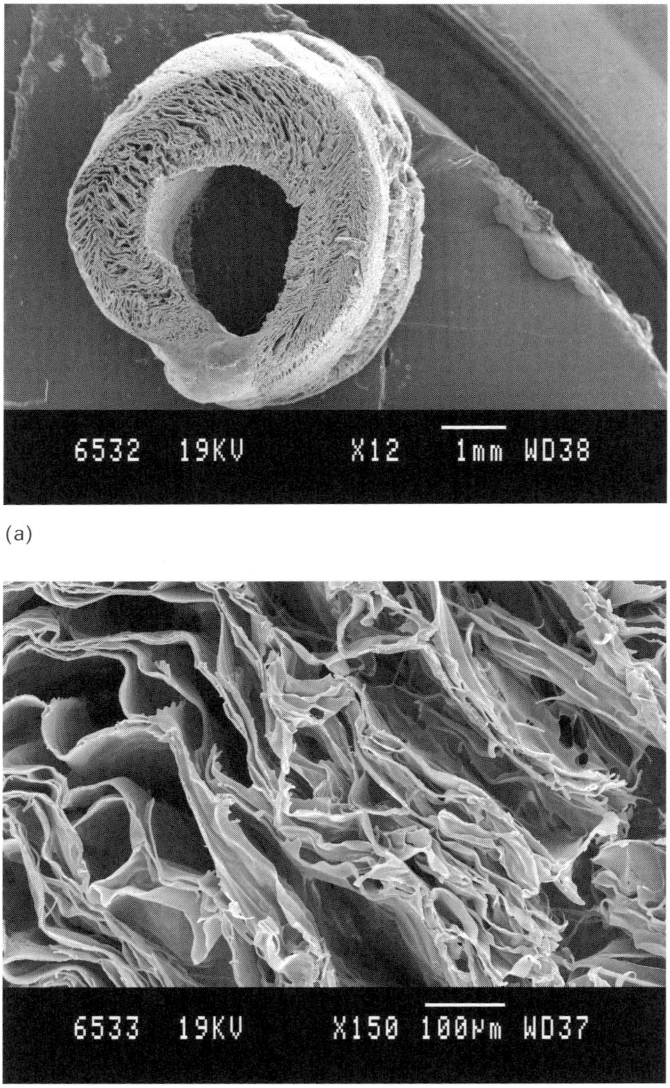

(b)

24.3 Scanning electron microscopy of PLGA foam scaffold showing (a) tubular structure with chloroform join and (b) cross-section structure with interconnecting pores of varying sizes.

materials. To date, the use of natural scaffold materials has been limited to intestine lengthening procedures[2,14–16] where the limited strength of the scaffold materials has been augmented by initially supporting the structure on silicone tube stents.[16,28,29]

24.4 Intestinal lengthening using artificial scaffolds

A number of groups have attempted to tissue engineer small intestine by interposing artificial tubular scaffolds between sections of healthy intestinal tissue.[2,14–16] Various different materials have been used to produce these artificial scaffolds, including freeze moulded collagen fibres,[16,28,29] small intestine submucosa[2,15,30,31] and acellular matrices derived from gastric wall[32] and dermis.[14] The basic concept is to employ the natural regenerative potential of the small intestine to populate an artificial scaffold designed to promote the growth of cells in from adjoining healthy intestinal tissue.

Experiments attempting to lengthen the bowel using tubular scaffolds have developed from earlier studies that demonstrated that small bowel defects could be successfully patched using autologous serosa, abdominal wall muscle or peritoneum.[6–8,12,33] These experiments demonstrated in-growth of mucosa along the margins of the patch-small intestine anastamosis[7] with formation of neomucosa. Histologically, the morphology of this neomucosa was similar to that of the surrounding small intestine although the central areas of the patch remained only partially covered with immature mucosa at 36 weeks.[7] When serosal patches were used, the resulting neomucosa had aminopeptidase, maltase and lactase activities similar to that of the surrounding native intestine although activities of alkaline phosphatase and sucrase were reduced.[6,12] The functional potential of this neointestinal patch was further supported by experiments in a porcine model of short bowel syndrome which demonstrated increased weight gain after patching with colonic serosa.[6]

24.4.1 Experimental models using tubular scaffolds

Attempts to create tubular scaffolds using colonic serosa were of limited success, with few animals surviving the procedure and only partial coverage of the scaffolds with neomucosa.[34] Recent experimental models using artificial tubular scaffolds have been more successful. Several studies from Japan have employed collagen sponge soaked in autologous blood to produce a tubular scaffold for intestinal lengthening in a canine model.[16,28,29] Collagen sponges were formed by enzyme extraction of collagen from porcine skin followed by whipping and freeze moulding to produce flat, sponge-like structures. These were then stabilised by inducing crosslinking between the collagen strands by heating and then further reinforced by application of polyglycolic acid.[16] In the initial experiments in beagle dogs, a 5cm section of defunctioned jejunum was resected and replaced with a silicone tube stent. This silicone tube was then wrapped in collagen sponge which had been soaked in autologous blood which was in turn wrapped with omentum.[16] These initial experiments demonstrated the growth of neomucosa inwards from the healthy jejunum, but not the development of a muscular or serosal layer.[16] In further experiments using the same basic model the collagen sponge was seeded with mesenchymal stem cells derived from bone marrow in an attempt to produce a muscle layer.[28] However, these experiments were essentially unsuccessful with only a very thin muscle layer developing below the mucosal surface.[28] More recent experiments from the same group have succeeded in creating an intact smooth muscle layer after 12 weeks by seeding the collagen sponge with autologous smooth muscle cells derived from stomach wall, although only in ileal patch grafts.[29]

There has been considerable interest in the use of SIS in the tissue engineering of a range of tissues including urinary tract, tendon and blood vessels. SIS is primarily an acellular collagen-based matrix; it is manufactured from small intestine by mechanical removal of the mucosa and muscular layers followed by osmotic lysis of any remaining cells. Unlike other scaffold materials that have been used for intestinal tissue engineering, SIS has been shown to retain growth factors with properties similar to FGF and TGF-β.[35] Initial experiments showed that porcine SIS could be used as a patch to repair relatively large defects in canine small bowel and that there was migration of cells into the patch which at 3 months resulted in a mucosal layer, disorganised smooth muscle layers and a serosal layer similar to native small intestine.[15] However, attempts to interpose a segment of tubular SIS between divided small intestinal loops resulted in anastamotic leakage in all animals.[15] Further experiments in rodents have demonstrated that 2 cm long tubular SIS scaffolds could be successfully interposed into defunctioned small intestine.[30,31] By 12 weeks the entire lumen was covered with neomucosa and by 24 weeks there were also distinct smooth muscle and serosal layers.[30] More recently, similar experiments using Surgisis, a porcine-derived matrix very similar to SIS, have demonstrated well-organised layers of mucosa, smooth muscle and serosa 24 weeks after interposition of 3 cm scaffolds into defunctioned ileal loops in rodent models.[2]

24.4.2 Limitations

The use of artificial scaffolds to lengthen the small intestine by interposition alone is appealing in its simplicity. The procedure is relatively straightforward and there is no requirement for exogenous biological tissue. Perhaps most importantly, the most recent studies have demonstrated that these techniques can yield well-organised tissue with distinct mucosal, muscle and serosal layers. However, there are drawbacks. In order to prevent anastamotic leakage it is necessary to defunction the loop of bowel into which the tubular scaffold is interposed. In the studies where the interposed scaffold was not taken out of the flow of luminal contents, all animals died of peritonitis.[14,15] In patients who already have compromised intestinal function due to short bowel syndrome, the temporary loss of further functional intestine is clearly undesirable. The exact length of time that the scaffold-containing loop of intestine needs to be defunctioned for is uncertain; only a single study has reported successful reanastamosis, and this after 8 weeks.[29] It seems very likely that the time required for adequate engraftment of the scaffold will depend on the length of the implant, given that growth of intestinal tissue into the scaffold appears to occur only from the anastamoses. As such it may be difficult and potentially very slow to achieve significant lengthening of the small intestine using interposition of artificial scaffolds alone. The possibilities would either be to interpose multiple short scaffolds which would be technically very difficult and would also increase the risk of anastamotic breakdown, or would be to interpose one or two longer scaffolds which would engraft much more slowly and hence would require a section of small intestine to be defunctioned for a considerable length of time. Neither of these options is ideal in an individual who already has short bowel syndrome, especially in an adult who is likely to have had a number of previous operations.

24.5 Transplantation of intestinal stem cell cultures

An alternative to implanting acellular scaffolds to lengthen the small intestine is the creation of neointestinal tissue prior to anastamosis. Native intestinal mucosa has an impressive capacity for replication and regeneration both under normal physiological conditions and following injury. This regenerative capacity is dependent on the activity of intestinal epithelial stem cells. Advances in the ability to identify and culture intestinal stem cell populations have made it possible to transplant stem cells onto artificial scaffolds in order to create neointestine.

24.5.1 Intestinal stem cells

Intestinal stem cells are found towards the base of the epithelial crypts. The majority of stem cell divisions are believed to result in a single daughter cell and a single stem cell which retains the original template DNA. These daughter cells then undergo further divisions to produce a population of transit-amplifying (TA) cells. These TA cells are rapidly proliferating and divide and further differentiate to produce the different epithelial cell lines. The concept that a single intestinal stem cell can give rise to all intestinal epithelial cell lines, known as the Unitarian hypothesis, is supported by a significant body of evidence. Stem cell daughter cells and initial TA cells retain their clonogenicity and are able to revert back to stem cells if the crypt is damaged and existing stem cells are lost. However, as the TA cells divide further they lose their capacity for clonal expansion. Enterocytes, goblet cells and enteroendocrine cells undergo further differentiation as they migrate upwards towards the tip of the villus. They then are either shed into the intestinal lumen or undergo apoptosis.

While the hierarchical pattern of cell proliferation and differentiation from crypt to villus is firmly established, the exact number of the stems cells in each crypt is less clear. This is historically due to the lack of reliable molecular stem cell markers although the discovery of Musashi-1 and Hes1 as putative stem cell markers is encouraging. It is believed that there are four to six stem cells in each crypt which are located in a specific stem cell compartment also known as the 'stem cell niche'. This niche comprises the intestinal epithelial stem cell, neighbouring proliferating cells and adjacent mesenchymal cells such as the pericryptal fibroblasts and intestinal subepithelial myofibroblasts. These mesenchymal cells are believed to play an important role in the maintenance of the stem cell population and the control and regulation of proliferation via the secretion of various peptides. There are complex signalling pathways between the different components of the stem cell niche, and understanding of this signalling is increasing rapidly.

24.5.2 Isolation and culture of intestinal stem cells

A necessary initial step in the creation of intestinal neomucosa by seeding artificial scaffolds with intestinal tissue has been the development of techniques to produce primary cultures of intestinal epithelium. In the early 1990s Evans and colleagues described a method by which disaggregated intestinal tissue, termed intestinal organoids, were derived from neonatal rat small intestine by partial digestion using a mixture of collagenase and dispase.[36] As the tryptic activity of the enzymatic solution is low there is not complete dissociation of the epithelial cells. The resulting intestinal organoids are cellular aggregates consisting of polarised intestinal epithelium surrounding a core of mesenchymal cells[36] (see Fig. 24.4). It is believed that they contain intestinal stem cells, other progenitor cells and epithelium along with the mesenchymal stroma.[36] *In vitro* studies demonstrated that the intestinal organoids could be maintained in cell culture. Interestingly, more extensive dissociation by prolonged enzymatic action to yield single cell suspensions appeared to inhibit cell proliferation.[36] This would appear to support the importance of maintaining the stem cell niche.

Subsequent experiments demonstrated that suspensions of organoid units transplanted into subcutaneous pockets in adult rodents could develop into small, short, tubular structures which consisted of a central mucin filled lumen surrounded by a circumferential epithelial layer.[10] As with the *in vitro* experiments, implantation of single cell suspensions did not result in the formation of cysts containing neomucosa.[37] When intestinal organoids derived from 5–8-day-old rats were implanted into nude mice, 39% developed into neointestinal cysts; when intestinal organoids were implanted into inbred rats the success rate was 84%. The maximum length of these structures was 5 mm in the nude mice and 8 mm in the rats. As early as 2 weeks after organoid unit transplantation, the epithelial layer had formed crypts and villi and was histologically similar to

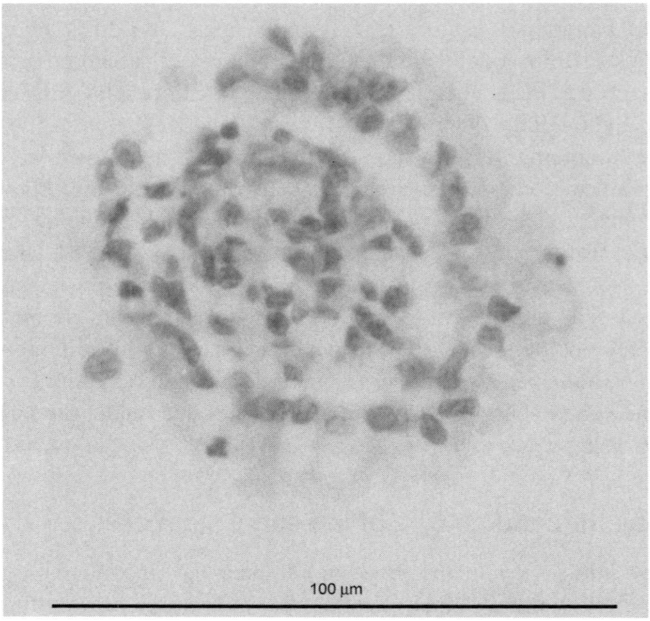

24.4 Light microscopy of an intestinal organoid embedded in agar gel, sectioned, and stained with haematoxylin and eosin. An epithelial layer is seen surrounding a core of mesenchymal cells.

small intestinal mucosa. This neomucosa was shown to contain all epithelial cell lineages including Paneth cells, which were not identified in 6-day-old neonatal small intestine.[10] This is of particular relevance as it suggests development of these cells from pluripotent stem cells in the transplanted intestinal organoids rather than simply from multiplication of more differentiated cells already in the transplanted tissue. Lactase, sucrase, aminopeptidase and alkaline phosphatase activity was also demonstrated in the neomucosa as was sodium-dependent glucose transport,[9] suggesting that the neointestine had functional potential. In addition to the neomucosal components, smooth muscle-like cells were identified adjacent to the neomucosa although they had not developed into discrete muscle layers.[10]

While the majority of subsequent studies have used rodent tissue exclusively, it is noteworthy that Sattar and colleagues demonstrated that intestinal organoids produced from human fetal small intestine could be successfully implanted subcutaneously into SCID mice to produce cysts of neomucosa similar to that described in rat models.[38] Tissue was obtained following terminations of pregnancy between 12 and 20 weeks. Successful development of subcutaneously implanted intestinal organoids was achieved in a similar proportion of experiments as in rodent models and, similarly, the resultant neomucosa contained all mucosal cell lines at 50 days after implantation.[38]

24.5.3 Implantation onto scaffolds

Experiments by Campbell's group, following on from their work on the isolation of intestinal organoids, showed that small intestinal mucosa could be regenerated by seeding intestinal organoids onto ascending colon after mucosectomy.[17] Fourteen days after implantation onto loops of defunctioned mucosectomised colon, neomucosa had developed in 76% of animals.[17] This neomucosa was histologically similar to small intestine and contained enterocytes, goblet cells, Paneth cells and enteroendocrine cells.[17] No regeneration was seen on control loops of mucosectomised colon, confirming that regeneration was due to proliferation of the implanted intestinal organoids rather than to incomplete mucosectomy.

More recently, Avansino and colleagues have developed a model in which intestinal organoids derived from the distal ileum of neonatal mice and rats were implanted onto segments of mucosectomised jejunum which had been defunctioned and tied off to prevent loss of implanted organoids.[23] After 2 weeks, neointestine containing all four intestinal epithelial cell lineages had developed on the mucosectomised jejunum.[23] Initial experiments suggested that there was an optimal dose of intestinal organoids for maximal engraftment.[39] The percentage of the mucosal surface covered with neomucosa (as opposed to native jejunal mucosa) was determined by immunohistochemical staining using antibodies against the ileal bile acid transporter (IBAT) and confirmed in further experiments using intestinal organoids harvested from green fluorescent protein (GFP) positive animals.[23] Disappointingly, the maximal covering by neomucosa was only 18% and varied considerable depending on the methods used for mucosectomy.[23] However, further experiments suggest that these segments of tissue-engineered ileum had functional potential. After resection of the native terminal ileum in a rat model, anastamosis of tissue-engineered ileum in continuity with native intestine was shown to significantly attenuate bile acid loss compared to animals that did not have neointestine inserted; total bile acid loss in the rats with tissue-engineered ileum was similar to animals with an intact native ileum although taurocholate uptake was less than in normal animals.[24]

The implantation of intestinal organoids onto artificial scaffolds has been pioneered by a research group in the USA headed by Joseph Vacanti. This group has employed 1 cm long tubular scaffolds created from polyglycolic acid (PGA) fibre meshes stabilised by spraying with poly(L-lactic acid) (PLLA) and a 50/50 copolymer of poly(D,L-lactic-co-glycolic acid) (PLGA).[40] In the initial experiments intestinal organoids were seeded onto the polymer scaffold 2 h prior to implantation into the omentum of adult rats; intestinal organoids engrafted on 16 out of 19 scaffolds with the formation of small cyst-like structures with a maximum length of 3.6 mm.[18] Histological analysis of cysts harvested between 2 and 8 weeks confirmed the presence of neomucosa containing columnar epithelium, goblet cells and Paneth cells.[18]

Further experiments by the Vacanti group demonstrated that intestinal organoid engraftment could be enhanced by coating the scaffolds with type I collagen with cysts forming from 93% of collagen coated scaffolds versus 64% of non-coated scaffolds.[3] Collagen coated scaffolds also resulted in considerably larger cysts at 6 weeks with a maximal length of 30 mm documented.[3] Histology again demonstrated columnar epithelium and goblet cells in the mucosal layer and smooth muscle-like cells in the submucosa.[3] Immunohistochemistry detected cells staining positively for sucrase on the apical epithelial surface of the neomucosa and Ussing chamber experiments demonstrated a potential difference across the mucosa, although this was significantly less than across normal ileal mucosa.[3]

Studies employing immunohistochemical staining for CD34, a vascular endothelial marker, have demonstrated vascularisation of the tissue-engineered intestine.[41] Attempts were made to compare capillary density and tissue concentrations of vascular endothelial growth factor (VEGF) and basic fibroblast growth factor (bFGF) with native small intestine taken from juvenile and adult rats.[41] Such comparisons are difficult to interpret given the morphological differences between the tissue engineered and native intestines. However, the authors concluded that although VEGF and bFGF were present in the neo-intestine, the relatively low concentrations suggested that there were other growth factors involved in angiogenesis.[41] A separate study analysing similar tissue also reported the presence of lymphatic vessels in the neointestinal cysts, although, again, the pattern of lymphangiogenesis appeared different from that seen in native small intestine.[42]

Our own group has developed a similar model for small intestinal tissue engineering in which intestinal organoids are implanted onto PLGA scaffolds.[13] Sheets of PLGA foam are made by a thermally induced phase separation process. This results in a porous structure with radially oriented interconnected pores ranging in size from 50 to 300 μm. The PLGA foam is rolled into a tube and the opposing edges dissolved with chloroform an then pressed together to join.[43] These PLGA scaffolds have been implanted subcutaneously into rats. Unlike experiments using PGA scaffolds, the PLGA scaffolds are left *in situ* for several weeks to allow them to become cellularised prior to implantation of intestinal organoids. This has facilitated successful development of intestinal neomucosa with fewer implanted organoids than previously described[13] (see Fig. 24.5). Further experiments have demonstrated that the resultant neomucosa contains replicating and differentiated cells and remains viable at 12 weeks after intestinal organoid implantation.[25]

The Vacanti investigators have developed their model further and have successfully anastamosed neointestine onto native small intestine. Neointestinal cysts were opened longitudinally 3 weeks after initial scaffold implantation and were joined to native jejunum via a side-to-side anastamosis.[5,21] Examination after a further 7 weeks revealed significantly greater villus number and height as

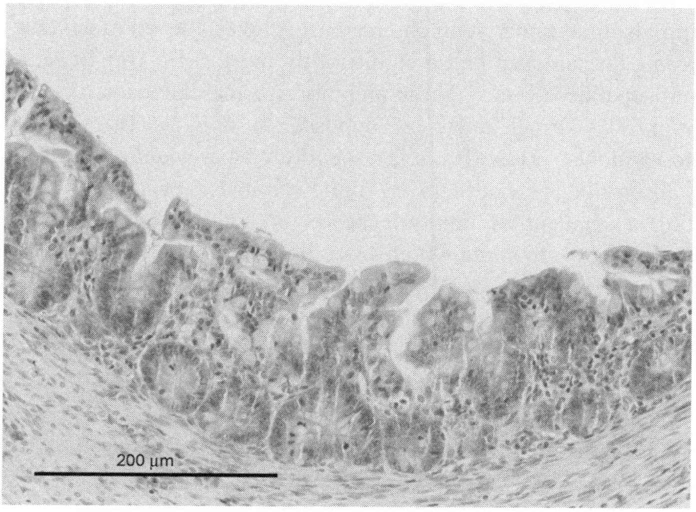

24.5 Haematoxylin and eosin staining of PLGA scaffold 4 weeks after organoid implantation. The appearance of the mucosa and submucosa lining the luminal surface of the scaffold is similar to that of normal small intestinal tissue. There is columnar epithelium containing mucin producing goblet cells which has developed into crypts and villi in some areas. The submucosa contains macrophages, fibroblasts and other inflammatory cells.

well as greater surface area of the neomucosa that had been anastamosed compared with that which had not.[5,21] There was also significantly greater expression of the glucose transporter SGLT1 in the anastamosed neomucosa.[11] Neomucosal morphology was maintained to 36 weeks.[22] A single study has also demonstrated the feasibility of end-to-end anastamosis with overall patency rates of 78% at 10 weeks.[20]

More detailed immunohistochemical studies demonstrated cellular proliferation rates in the mucosal layer of anastamosed neointestine to be similar to those seen in native small intestine; proliferation rates were significantly lower in non-anastamosed neointestine.[11] Apoptotic rates were similar in the anastamosed and non-anastamosed neointestine.[11] Interestingly, these studies also demonstrated immune cell subsets in the neointestine that had been anastamosed to native intestine with population densities that were similar to native jejunum.[44] The immune cell populations appeared to develop with time and not to develop in the neointestine that had not been anastamosed to native small intestine, leading the authors to conclude that the development of the neomucosal immune system was dependent both on exposure to luminal content and to the duration of exposure.[44]

The possible therapeutic potential of the neointestinal cysts created by the Vacanti group has been glimpsed in a couple of studies where neointestinal cysts were anastamosed to native jejunum after massive enterectomy.[19,22] Compared with animals that underwent enterectomy alone, the animals that had had neointestine anastamosed had a significantly more rapid weight recovery after initial post-operative loss.[19] These animals also maintained serum B_{12} concentrations in the normal range as opposed to animals that had undergone enterectomy alone.[19] However, it is not entirely clear whether these effects were due to absorption of nutrients by the neomucosa or due to the effect of implanting a segment of immotile bowel on gut transit. Although the neointestine was shown to contain both smooth muscle-like cells staining positively for smooth muscle actin and neuronal cells staining positively for S100, gut transit times were significantly longer in the rats that had had neointestine anastamosed to native jejunum (1825 minutes $\pm$ 753 compared with 982 $\pm$ 300).[19] This may have resulted in prolonged exposure to, and hence improved absorption by, the remnant native small intestine. Interestingly, myenteric denervation of a segment of ileum has been shown to significantly attenuate weight loss after 80% enterectomy in a rat model,[45] suggesting that delaying gut transit has a significant effect on intestinal absorptive capacity.

24.5.4 Limitations

The possibility of creating segments of functional neointestine from neonatal tissue over a relatively short course is exciting. The techniques described above have demonstrated tissue can be created without having to defunction loops of small bowel and that this neointestine can then be successfully anastamosed with native small intestine. Similar techniques have also been used to create tissue-engineered oesophageal, gastric and colonic mucosa.[46–49] However, there are some significant limitations at present. The principal problem is the source of donor tissue used to produce the intestinal organoids coupled with the low yield of the tissue engineering process.

Experiments to date have obtained intestinal organoids from either neonatal or fetal small intestine, with studies reporting a yield of about 40 000 intestinal organoids from the small intestine of a single neonatal rat.[3] Fetal and neonatal tissue is not ideal for the harvesting of intestinal organoids in a clinical scenario, and ultimately an autologous source of tissue for implantation will need to be found in order to avoid the requirement for long-term immunosuppression. The most recent studies by the Vacanti group have implanted up to 100 000 intestinal organoids per 1 cm long biosynthetic scaffold in order to create a single cyst of neomucosa[50] which clearly demonstrates the inefficiency of the tissue engineering process. Implantation onto denuded native jejunum would appear to require fewer intestinal organoids with Avansino and colleagues reporting optimal results with the implantation of 10 000 intestinal organoids per 3 cm jejunal segment[39] although it is noteworthy that the overall percentage of successful implantations was considerably lower than that reported by the Vacanti group.[50]

In addition to the limitations relating to sourcing intestinal organoids and the yield of the process, the neointestine produced by transplanting intestinal stem cell clusters appears less well developed than that produced by interposition of artificial scaffolds into healthy small intestine. Prolonged follow-up of neointestine created by seeding of synthetic scaffolds with intestinal organoids revealed a well-developed mucosal layer;[22] however, even after anastamosis with native small intestine there were not well defined mucosal, muscular and serosal layers as seen in some experiments using biological scaffolds interposed between healthy loops of native bowel.[2] It remains unclear whether or not further remodelling of the muscular and serosal layers would occur beyond the 36 weeks follow-up period.

24.6 Growth factors

Growth and regeneration of the small intestine is under the control of a wide range of growth factors and cytokines. Attempts have been made to try to speed and enhance the development of tissue-engineered small intestine by manipulating these growth factors. Studies of neointestinal development on serosal patches in rabbits demonstrated that infusion of urogastrone resulted in more rapid in-growth of neointestine from the surrounding native bowel.[51] This effect was shown to be dose dependent[52] and prolonged administration resulted in an increase in the amount of neomucosa produced.[53]

Using the same model it was shown that 50% enterectomy resulted in greater ingrowth of neomucosa over the serosal patch.[54] This is unsurprising given the adaptive response seen in individuals after massive small bowel resection and it is interesting to note that the effects of urogastrone are not synergistic with those of massive small bowel resection.[55] Experiments using synthetic scaffolds seeded with intestinal organoids also demonstrated increased neointestinal cyst length and diameter in animals that had undergone small bowel resection compared with controls.[56] As with native small intestine, contact with luminal contents has also shown to be an important stimulus to neomucosal growth and development. Studies using tubular scaffolds to lengthen the small intestine and those employing transplantation of intestinal organoids onto synthetic scaffolds have consistently demonstrated increased neomucosal growth and maturation when in continuity with the native small bowel.[22]

More recently it has been shown that GLP-2 has a stimulatory effect on neointestinal development. Parenteral administration of GLP-2 to rodents implanted with scaffolds seeded with intestinal organoids resulted in significantly greater villus height and crypt depth in the resulting neomucosa, along with increased crypt proliferation and reduced apoptosis. In addition, there was also evidence of apical transporter up-regulation.[50] These effects are similar to those seen in native small intestine after GLP-2 administration. Taken together, the findings above suggest that the response of neointestine to proliferative stimuli may be very similar to that of native small intestine.

24.7 Future trends

The progress made in the field of small intestinal tissue engineering is exciting but, as has been highlighted above, there are significant limitations to the techniques that have been developed. Intestinal lengthening procedures have produced the most morphologically correct neointestine with well-demarcated mucosal, muscle and serosal layers, but the process is slow and requires defunctioning of a proportion of the small intestine. Theoretically, it may be possible to combine the technique with transplantation of intestinal stem cells in order to speed the generation of neomucosa and allow earlier reanastamosis, and implantation of multiple cell lines may augment the regeneration of both mucosal and muscle layers.

Transplantation of intestinal stem cells is limited by a lack of suitable donor tissue and the low yield of the process. In order to avoid the problems of tissue rejection and the need for long-term immunosuppression an autologous source of donor tissue is necessary. It is unlikely that sufficient intestinal stem cells could be harvested from native small intestine given that attempts to expand harvested intestinal tissue *in vitro* have been unsuccessful to date. However, it may be possible to transplant haematopoietic stem cells harvested from either bone marrow or from the peripheral circulation, and induce transdifferentiation into intestinal stem cells. Studies suggest that bone marrow-derived mesenchymal cells can differentiate into pericryptal myofibroblasts in both mice and humans. A recent study has shown that in IL-10 knockout mice undergoing bone marrow transplantation, 30% of colonic subepithelial myofibroblasts are of bone marrow origin in normal mucosa, increasing to 45% in inflamed mucosa 3 months after transplantation.[57] Transdifferentiated epithelial cells have also been reported in intestinal mucosa several years after bone marrow transplantation.[58] Of note, bone marrow cells have been used successfully in the tissue engineering of vascular tissue.[59]

It is likely that it is possible to improve the yield of intestinal tissue engineering by modulating the scaffold properties in order to accelerate and augment neointestinal growth and development. As mentioned above, intestinal organoid engraftment onto denuded jejunum appears more successful than engraftment onto synthetic scaffolds as suggested by the lower numbers of intestinal organoids required. This probably reflects the optimal surface characteristics and pre-existing vascularisation of the denuded jejunum. Hybrid scaffolds combining synthetic and biological materials may maintain the flexibility of synthetic compounds while simultaneously gaining the superior surface characteristics of biologically derived tissue. Pre-implantation will allow vascularisation of the scaffolds prior to transplantation of intestinal organoids or other stem cell populations, and the yield of pre-implanted PLGA scaffolds compares favourably with PGA scaffolds that were not pre-implanted.[13] It may also be possible to improve vascularisation by coating scaffolds either with vascular growth factors such as VEGF[60] or with compounds such as bioactive glass, which stimulate endogenous growth factor release and promote blood vessel growth.[61]

24.8 Conclusions

Tissue engineering of the small intestine offers a novel treatment for patients with short bowel syndrome which avoids the potential complications of long-term parenteral nutrition and intestinal transplantation. Small intestinal tissue engineering is particularly challenging due to the considerable anatomical and functional complexity of the gastrointestinal tract. Current techniques are based on either lengthening the remaining intestine using biologically derived scaffolds or implantation of intestinal stem cell populations, derived from neonatal tissue, onto synthetic scaffolds. Both processes have advantages but also disadvantages which would prevent clinical application at present. However, the fields of material engineering and cell biology are evolving rapidly and it seems likely that intestinal tissue engineering may become a viable therapeutic option in the not too distant future.

24.9 References

1. Atala A, Bauer SB, Soker S, Yoo JJ, Retik AB. Tissue-engineered autologous bladders for patients needing cystoplasty. *Lancet* 2006; **367**: 1241–1246.
2. Ansaloni L, Bonasoni P, Cambrini P, Catena F, De CA, Gagliardi S, Gazzotti F, Peruzzi S, Santini D, Taffurelli M. Experimental evaluation of Surgisis as scaffold for neointestine regeneration in a rat model. *Transplant Proc* 2006; **38**: 1844–1848.
3. Choi RS, Riegler M, Pothoulakis C, Kim BS, Mooney D, Vacanti M, Vacanti JP. Studies of brush border enzymes, basement membrane components, and electrophysiology of tissue-engineered neointestine. *J Pediatr Surg* 1998; **33**: 991–996.
4. De Faria W, Tryphonopoulos P, Kleiner G, Santiago S, Gandia C, Ruiz P, Tzakis A. Study of the development and evolution of neointestine in a rat model. *Transplant Proc* 2004; **36**: 375–376.
5. Kim SS, Kaihara S, Benvenuto MS, Choi RS, Kim BS, Mooney DJ, Vacanti JP. Effects of anastomosis of tissue-engineered neointestine to native small bowel. *J Surg Res* 1999; **87**: 6–13.
6. Binnington HB, Tumbleson ME, Ternberg JL. Use of jejunal neomucosa in the treatment of the short gut syndrome in pigs. *J Pediatr Surg* 1975; **10**: 617–621.
7. Binnington HB, Sumner H, Lesker P, Alpers DA, Ternberg JL. Functional characteristics of surgically induced jejunal neomucosa. *Surgery* 1974; **75**: 805–810.
8. Lillemoe KD, Berry WR, Harmon JW, Tai YH, Weichbrod RH, Cogen MA. Use of vascularized abdominal wall pedicle flaps to grow small bowel neomucosa. *Surgery* 1982; **91**: 293–300.
9. Tait IS, Penny JI, Campbell FC. Does neomucosa induced by small bowel stem cell transplantation have adequate function? *Am J Surg* 1995; **169**: 120–125.
10. Tait IS, Flint N, Campbell FC, Evans GS. Generation of neomucosa *in vivo* by

transplantation of dissociated rat postnatal small intestinal epithelium. *Differentiation* 1994; **56**: 91–100.
11. Tavakkolizadeh A, Berger UV, Stephen AE, Kim BS, Mooney D, Hediger MA, Ashley SW, Vacanti JP, Whang EE. Tissue-engineered neomucosa: morphology, enterocyte dynamics, and SGLT1 expression topography. *Transplantation* 2003; **75**: 181–185.
12. Thompson JS, Vanderhoof JA, Antonson DL, Newland JR, Hodgson PE. Comparison of techniques for growing small bowel neomucosa. *J Surg Res* 1984; **36**: 401–406.
13. Lloyd DA, Ansari TI, Gundabolu P, Shurey S, Maquet V, Sibbons PD, Boccaccini AR, Gabe SM. A pilot study investigating a novel subcutaneously implanted pre-cellularised scaffold for tissue engineering of intestinal mucosa. *Eur Cell Mater* 2006; **11**: 27–33.
14. Pahari MP, Raman A, Bloomenthal A, Costa MA, Bradley SP, Banner B, Rastellini C, Cicalese L. A novel approach for intestinal elongation using acellular dermal matrix: an experimental study in rats. *Transplant Proc* 2006; **38**: 1849–1850.
15. Chen MK, Badylak SF. Small bowel tissue engineering using small intestinal submucosa as a scaffold. *J Surg Res* 2001; **99**: 352–358.
16. Hori Y, Nakamura T, Matsumoto K, Kurokawa Y, Satomi S, Shimizu Y. Tissue engineering of the small intestine by acellular collagen sponge scaffold grafting. *Int J Artif Organs* 2001; **24**: 50–54.
17. Tait IS, Evans GS, Flint N, Campbell FC. Colonic mucosal replacement by syngeneic small intestinal stem cell transplantation. *Am J Surg* 1994; **167**: 67–72.
18. Choi RS, Vacanti JP. Preliminary studies of tissue-engineered intestine using isolated epithelial organoid units on tubular synthetic biodegradable scaffolds. *Transplant Proc* 1997; **29**: 848–851.
19. Grikscheit TC, Siddique A, Ochoa ER, Srinivasan A, Alsberg E, Hodin RA, Vacanti JP. Tissue-engineered small intestine improves recovery after massive small bowel resection. *Ann Surg* 2004; **240**: 748–754.
20. Kaihara S, Kim S, Benvenuto M, Kim BS, Mooney DJ, Tanaka K, Vacanti JP. End-to-end anastomosis between tissue-engineered intestine and native small bowel. *Tissue Eng* 1999; **5**: 339–346.
21. Kaihara S, Kim SS, Benvenuto M, Choi R, Kim BS, Mooney D, Tanaka K, Vacanti JP. Successful anastomosis between tissue-engineered intestine and native small bowel. *Transplantation* 1999; **67**: 241–245.
22. Kaihara S, Kim SS, Kim BS, Mooney D, Tanaka K, Vacanti JP. Long-term follow-up of tissue-engineered intestine after anastomosis to native small bowel. *Transplantation* 2000; **69**: 1927–1932.
23. Avansino JR, Chen DC, Hoagland VD, Woolman JD, Stelzner M. Orthotopic transplantation of intestinal mucosal organoids in rodents. *Surgery* 2006; **140**: 423–434.
24. Avansino JR, Chen DC, Hoagland VD, Woolman JD, Haigh WG, Stelzner M. Treatment of bile acid malabsorption using ileal stem cell transplantation. *J Am Coll Surg* 2005; **201**: 710–720.
25. Lloyd DA, Ansari T, Shurey S, Maquet V, Sibbons PD, Boccaccini AR, Gabe SM. Prolonged maintenance of neointestine using subcutaneously implanted tubular scaffolds in a rat model. *Transplant Proc* 2006; **38**: 3097–3099.
26. Maquet V, Jerome R. Design of macroporous biodegradable polymer scaffold for cell transplantation. *Mat Sci Forum* 1997; **250**: 15–42.
27. Wang S, Cui W, Bei J. Bulk and surface modifications of polylactide. *Anal Bioanal Chem* 2005; **381**: 547–556.

28. Hori Y, Nakamura T, Kimura D, Kaino K, Kurokawa Y, Satomi S, Shimizu Y. Experimental study on tissue engineering of the small intestine by mesenchymal stem cell seeding. *J Surg Res* 2002; **102**: 156–160.
29. Nakase Y, Hagiwara A, Nakamura T, Kin S, Nakashima S, Yoshikawa T, Fukuda K, Kuriu Y, Miyagawa K, Sakakura C, Otsuji E, Shimizu Y, Ikada Y, Yamagishi H. Tissue engineering of small intestinal tissue using collagen sponge scaffolds seeded with smooth muscle cells. *Tissue Eng* 2006; **12**: 403–412.
30. Wang ZQ, Watanabe Y, Toki A. Experimental assessment of small intestinal submucosa as a small bowel graft in a rat model. *J Pediatr Surg* 2003; **38**: 1596–1601.
31. Wang ZQ, Watanabe Y, Noda T, Yoshida A, Oyama T, Toki A. Morphologic evaluation of regenerated small bowel by small intestinal submucosa. *J Pediatr Surg* 2005; **40**: 1898–1902.
32. Parnigotto PP, Marzaro M, Artusi T, Perrino G, Conconi MT. Short bowel syndrome: experimental approach to increase intestinal surface in rats by gastric homologous acellular matrix. *J Pediatr Surg* 2000; **35**: 1304–1308.
33. Erez I, Rode H, Cywes S. Enteroperitoneal anastomosis for short bowel syndrome. *Harefuah* 1992; **123**: 5–8, 72.
34. Thompson JS. Neomucosal growth in serosa lined intestinal tunnels. *J Surg Res* 1990; **49**: 1–7.
35. Voytik-Harbin SL, Brightman AO, Kraine MR, Waisner B, Badylak SF. Identification of extractable growth factors from small intestinal submucosa. *J Cell Biochem* 1997; **67**: 478–491.
36. Evans GS, Flint N, Somers AS, Eyden B, Potten CS. The development of a method for the preparation of rat intestinal epithelial cell primary cultures. *J Cell Sci* 1992; **101 (Pt 1)**: 219–231.
37. Patel HR, Tait IS, Evans GS, Campbell FC. Influence of cell interactions in a novel model of postnatal mucosal regeneration. *Gut* 1996; **38**: 679–686.
38. Sattar A, Robson SC, Patel HR, Angus B, Campbell FC. Expression of growth regulatory genes in a SCID mouse-human model of intestinal epithelial regeneration. *J Pathol* 1999; **187**: 229–236.
39. Avansino JR, Chen DC, Woolman JD, Hoagland VD, Stelzner M. Engraftment of mucosal stem cells into murine jejunum is dependent on optimal dose of cells. *J Surg Res* 2006; **132**: 74–79.
40. Mooney DJ, Mazzoni CL, Breuer C, McNamara K, Hern D, Vacanti JP, Langer R. Stabilized polyglycolic acid fibre-based tubes for tissue engineering. *Biomaterials* 1996; **17**: 115–124.
41. Gardner-Thorpe J, Grikscheit TC, Ito H, Perez A, Ashley SW, Vacanti JP, Whang EE. Angiogenesis in tissue-engineered small intestine. *Tissue Eng* 2003; **9**: 1255–1261.
42. Duxbury MS, Grikscheit TC, Gardner-Thorpe J, Rocha FG, Ito H, Perez A, Ashley SW, Vacanti JP, Whang EE. Lymphangiogenesis in tissue-engineered small intestine. *Transplantation* 2004; **77**: 1162–1166.
43. Day RM, Boccaccini AR, Maquet V, Shurey S, Forbes A, Gabe SM, Jerome R. *In vivo* characterisation of a novel bioresorbable poly(lactide-co-glycolide) tubular foam scaffold for tissue engineering applications. *J Mater Sci Mater Med* 2004; **15**: 729–734.
44. Perez A, Grikscheit TC, Blumberg RS, Ashley SW, Vacanti JP, Whang EE. Tissue-engineered small intestine: ontogeny of the immune system. *Transplantation* 2002; **74**: 619–623.

45. Garcia SB, Kawasaky MC, Silva JC, Garcia-Rodrigues AC, Borelli-Bovo TJ, Iglesias AC, Zucoloto S. Intrinsic myenteric denervation: a new model to increase the intestinal absorptive surface in short-bowel syndrome. *J Surg Res* 1999; **85**: 200–203.
46. Grikscheit T, Ochoa ER, Srinivasan A, Gaissert H, Vacanti JP. Tissue-engineered esophagus: experimental substitution by onlay patch or interposition. *J Thorac Cardiovasc Surg* 2003; **126**: 537–544.
47. Grikscheit TC, Ogilvie JB, Ochoa ER, Alsberg E, Mooney D, Vacanti JP. Tissue-engineered colon exhibits function *in vivo*. *Surgery* 2002; **132**: 200–204.
48. Grikscheit TC, Ochoa ER, Ramsanahie A, Alsberg E, Mooney D, Whang EE, Vacanti JP. Tissue-engineered large intestine resembles native colon with appropriate *in vitro* physiology and architecture. *Ann Surg* 2003; **238**: 35–41.
49. Maemura T, Ogawa K, Shin M, Mochizuki H, Vacanti JP. Assessment of tissue-engineered stomach derived from isolated epithelium organoid units. *Transplant Proc* 2004; **36**: 1595–1599.
50. Ramsanahie A, Duxbury MS, Grikscheit TC, Perez A, Rhoads DB, Gardner-Thorpe J, Ogilvie J, Ashley SW, Vacanti JP, Whang EE. Effect of GLP-2 on mucosal morphology and SGLT1 expression in tissue-engineered neointestine. *Am J Physiol Gastrointest Liver Physiol* 2003; **285**: G1345–G1352.
51. Thompson JS, Sharp JG, Saxena SK, McCullagh KG. Stimulation of neomucosal growth by systemic urogastrone. *J Surg Res* 1987; **42**: 402–410.
52. Thompson JS, Saxena SK, Sharp JG. Effect of urogastrone on intestinal regeneration is dose-dependent. *Cell Tissue Kinet* 1988; **21**: 183–191.
53. Thompson JS, Saxena SK, Sharp JG. Effect of the duration of infusion of urogastrone on intestinal regeneration in rabbits. *Cell Tissue Kinet* 1989; **22**: 303–309.
54. Bragg LE, Thompson JS. The influence of intestinal resection on the growth of intestinal neomucosa. *J Surg Res* 1989; **46**: 306–310.
55. Thompson JS, Bragg LE, Saxena SK. The effect of intestinal resection and urogastrone on intestinal regeneration. *Arch Surg* 1990; **125**: 1617–1621.
56. Kim SS, Kaihara S, Benvenuto MS, Choi RS, Kim BS, Mooney DJ, Taylor GA, Vacanti JP. Regenerative signals for intestinal epithelial organoid units transplanted on biodegradable polymer scaffolds for tissue engineering of small intestine. *Transplantation* 1999; **67**: 227–233.
57. Bamba S, Otto WR, Lee CY, Brittan M, Preston SL, Wright NA. The contribution of bone marrow to colonic subepithelial myofibroblasts in interleukin-10 knockout mice. *Gut* 2005; **54**: A17.
58. Okamoto R, Yajima T, Yamazaki M, Kanai T, Mukai M, Okamoto S, Ikeda Y, Hibi T, Inazawa J, Watanabe M. Damaged epithelia regenerated by bone marrow-derived cells in the human gastrointestinal tract. *Nat Med* 2002; **8**: 1011–1017.
59. Hibino N, Shin'oka T, Matsumura G, Ikada Y, Kurosawa H. The tissue-engineered vascular graft using bone marrow without culture. *J Thorac Cardiovasc Surg* 2005; **129**: 1064–1070.
60. Murphy WL, Peters MC, Kohn DH, Mooney DJ. Sustained release of vascular endothelial growth factor from mineralized poly(lactide-co-glycolide) scaffolds for tissue engineering. *Biomaterials* 2000; **21**: 2521–2527.
61. Day RM, Maquet V, Boccaccini AR, Jerome R, Forbes A. *In vitro* and *in vivo* analysis of macroporous biodegradable poly(D,L-lactide-co-glycolide) scaffolds containing bioactive glass. *J Biomed Mater Res A* 2005; **75**: 778–787.

25

Micromechanics of hydroxyapatite-based biomaterials and tissue engineering scaffolds

A FRITSCH and L DORMIEUX, Ecole Nationale des Ponts et Chaussées (LMSGC-ENPC), France, C HELLMICH, Vienna University of Technology (TU Wien), Austria and J SANAHUJA, Lafarge Research Center, France

25.1 Introduction

Hydroxyapatite (HA, with chemical formula $Ca_{10}(PO_4)_6(OH)_2$ in its pure ('stoichiometric') form) biomaterials production has been a major field in biomaterials science and biomechanical engineering due to its excellent biocompatibility, and since its chemical composition, structure and mechanical properties are similar to bone mineral (Hench and Jones, 2005). With the aim of mimicking the bone mineral and its important biological and mechanical properties within bone tissues, HA is widely used for biomedical applications: they encompass coating of orthopaedic and dental implants (Dorozhkin and Epple, 2002), artificial hard tissue replacement implants in orthopaedics, maxillofacial and dental implant surgery (Charrière *et al.*, 2001), either in a pure state (Frame *et al.*, 1981; Mastrogiacomo *et al.*, 2006) or as composite, with ceramic, metallic or polymer inclusions as reinforcing component (Verma *et al.*, 2006).

Different processing methods for HA powders lead to formation of materials close to biological HA (so-called calcium-deficient HA synthesized through biomineralization; Kikuchi *et al.*, 2004), but never identical to it. The majority of these methods fall into two categories: (i) wet methods such as precipitation processes and (ii) solid state reactions (Suchanek and Yoshimura, 1998). In a precipitation process (Jarcho *et al.*, 1976; Akao *et al.*, 1981), the exact stoichiometric quantities of calcium- and phosphate-containing solutions are mixed at $pH > 9$ and at temperatures below 100 °C, followed by boiling for several days under a CO_2-free atmosphere, filtration and drying. Blade, needle or rod-shaped crystals of nanometre size can be produced. The crystal's crystallinity and Ca/P ratio depend strongly on the preparation technique and are standardly lower than that of stoichiometric HA, i.e. possibly closer to that biologically generated HA. Solid state reactions between different non-HA calcium phosphates and CaO, $Ca(OH)_2$ or $CaCO_3$ deliver a stoichiometric and well-crystallized HA product, but require high temperatures (above 1200 °C) and long heat treatment times. There are many alternative methods (Suchanek and Yoshimura, 1998) to prepare HA powders besides the two aforementioned, such as the sol–gel method, the flux method, electrocrystallization, thermal deposition.

The processing of conventional porous HA ceramics typically involves the following steps (Hench and Jones, 2005). A preliminary component ('green body') is formed out of the starting material, which can be in the form of a powder, a slurry or a colloidal suspension. The former are pressed to produce a powder compact, and the latter are shaped by a slip or tape casting process. During sintering of the green body (at temperatures above 1000 °C) its density and strength increase. Finally, the dense compact can be machined to obtain the desired shape.

Typical examples for powder-based production of porous HA biomaterials were produced by the following researchers (see also Table 25.1):

- Peelen *et al.* (1978) mixed commercially available HA powders with a 10% hydrogen peroxide solution, poured it into a mould, and controlled the porosity of HA ceramics by a variation of the sintering temperature (Tables 25.1 and 25.2).
- Akao *et al.* (1981) precipitated HA powder, which was compacted and sintered at different temperatures (Tables 25.1–25.3).
- De With *et al.* (1981) compacted and sintered isostatically pressed HA powder (Tables 25.1 and 25.3).
- Shareef *et al.* (1993) produced mixtures with different weight ratios of commercially available fine and coarse powders. Ring-shaped samples were formed by uniaxial pressing and then sintered (Tables 25.1 and 25.2).
- Arita *et al.* (1995) used mixing of starting powders and a casting process to produce green bodies made of HA before sintering (Tables 25.1 and 25.3).
- Martin and Brown (1995) prepared calcium-deficient HA formed in aqueous solutions at physiological temperature. The authors realized two different liquid-to-solid weight ratios, resulting in two different porosities (Tables 25.1 and 25.2).
- Liu (1998) prepared HA powder by mixing of starting powders. Water and polyvinyl butyral powder were added to HA before casting the slurry and sintering the green bodies (Tables 25.1–25.3).
- Charrière *et al.* (2001) mixed commercially available powders in an aqueous solution and used a casting process to obtain HA cement (Tables 25.1 and 25.3).

The mechanical and microstructural properties, i.e. stiffness/strength and porosity, of these materials (see Tables 25.2 and 25.3) will be used as to validate the theoretical developments described in this chapter. We will go beyond statistical correlations between porosity and stiffness/strength or empirical structure–property relationships (Rao and Boehm, 1974; Driessen *et al.*, 1982;

Table 25.1 Hydroxyapatite-based porous biomaterials used for model validation: survey on processing, pore size, and mechanical characterization methods

Literature reference	Source material(s)	Processing steps	Shape/size of samples	Typical pore size	Mechanical characterization method
Peelen *et al.* (1978)	Commercially available HA powder	Mixing of HA powder with 10% hydrogen peroxide solution, poured into mould, sintering	Cylindrical ($d = 1$ cm, $h = 1$–1.5 cm)	1–200 μm	Uniaxial, quasi-static compressive test (Table 25.3)
Akao *et al.* (1981)	$Ca(OH)_2$, H_3PO_4	Mixing of starting powders to precipitate HA powder, mixed with water and cornstarch, compaction, sintering	Bars ($5 \times 5 \times 10\,cm^3$)	$\sim 1\ \mu$m (pore size $\approx$ grain size, see also Figs 2–4 of the reference)	Uniaxial, quasi-static compressive test (Tables 25.2 and 25.3)
De With *et al.* (1981)	Commercially available HA powder	Mixing of HA powder with water, compaction, sintering	Cylindrical ($d = 5$ mm, $h = 15$ mm)	1–5 μm (see Figs 2 and 7 of the reference)	Ultrasonic pulse-echo technique (Table 25.3)
Shareef *et al.* (1993)	Commercially available fine and coarse HA powders	Mixing of HA powders, compaction, sintering	Ring-shaped (inner diameter 34 mm)	1 μm	Quasi-static tensile test (Stanford ring bursting test, Table 25.2)
Arita *et al.* (1995)	$CaHPO_4$, $CaCO_3$	Mixing of starting powders with water, tape casting, sintering	Discs ($d = 2.54$ cm)	$\sim$1 μm (see Fig. 6 of the reference)	Resonance frequency method (Table 25.3)

Table 25.1 Continued

Literature reference	Source material(s)	Processing steps	Shape/size of samples	Typical pore size	Mechanical characterization method
Martin and Brown (1995)	$CaHPO_4$, $Ca_4(PO_4)_2O$	Mixing of starting powders with water, precipitation, compaction at low temperature	Cylinders ($d \sim 6.40$ mm, $h = 5.09$–6.39 mm)	~1–2 μm	Uniaxial, quasi-static compressive test (Table 25.2)
Liu (1998)	$Ca(OH)_2$, H_3PO_4	Mixing of starting powders in solution, mixing of HA powder with water and polyvinyl butyral powder in a slurry, slip casting, sintering	Bars ($5 \times 8 \times 50$ mm^3)	2–200 μm	Quasi-static tensile test (three-point bending; Tables 25.2 and 25.3)
Charrière *et al.* (2001)	$CaHPO_4$, $CaCO_3$	Mixing of starting powders with polyacrylic acid solution in suspension, poured into mould, slip casting	Hollow cylinders ($d = 18$ mm, $h = 40$ mm)	~ 1 μm	Uniaxial, quasi-static compressive test (Table 25.3)

Table 25.2 Experimental compressive strength $\Sigma_{exp}^{ult,c}$, bending strength $\Sigma_{exp}^{ult,b}$, and tensile strength $\Sigma_{exp}^{ult,t}$ of hydroxyapatite biomaterials, as function of porosity ϕ

Peelen *et al.* (1978)		Akao *et al.* (1981)		Shareef *et al.* (1993)		Martin and Brown (1995)		Liu (1998)	
ϕ (%)	$\Sigma_{exp}^{ult,c}$ (MPa)	ϕ (%)	$\Sigma_{exp}^{ult,c}$ (MPa)	ϕ (%)	$\Sigma_{exp}^{ult,t}$ (MPa)	ϕ (%)	$\Sigma_{exp}^{ult,c}$ (MPa)	ϕ (%)	$\Sigma_{exp}^{ult,b}$ (MPa)
36	160	2.8	509	12.2	37.1	0.27	172.5	20.2	25.5
48	114	3.9	465	16.1	32.8	0.39	119.0	26.8	20.0
60	69	9.1	415	20.6	31.8			29.0	16.8
65	45	19.4	308	24.8	24.2			32.6	13.9
70	30			27.3	23.6			39.6	14.4
				29.2	20.0			42.8	11.1
								50.9	7.2
								54.5	8.0

Table 25.3 Experimental Young's modulus E_{exp} and Poisson's ratio ν_{exp} of hydroxyapatite biomaterials, as function of porosity ϕ

Akao *et al.* (1981)		De With *et al.* (1981)			Arita *et al.* (1995)		Liu (1998)		Charrière *et al.* (2001)	
ϕ (%)	E_{exp} (GPa)	ϕ (%)	E_{exp} (GPa)	ν_{exp} (1)	ϕ (%)	E_{exp} (GPa)	ϕ (%)	E_{exp} (GPa)	ϕ (%)	E_{exp} (GPa)
2.8	88	3	112	0.275	6	88	8	93	44	13.5
3.9	85	6	103	0.272	28	41	17	78		
9.1	80	9	93	0.265	33	32	21	66		
19.4	44	17	78	0.253	35	29	32	44		
		22	67	0.242	50	14	44	22		
		27	54	0.238	52	10	54	18		

Katz and Harper, 1990), to describe the microstructures governing the overall mechanical behaviour. The chapter is organized as follows.

First, we will present the fundamentals of continuum micromechanics (Section 25.2). This mathematical theory allows for explanation of mechanical interactions at the microstructural observation scales of the HA biomaterials. In this way, macroscopic mechanical properties can be linked to microstructural characteristics ('structure–property relationships'). Then, we develop a continuum micromechanical concept for elasticity and strength of mono-porosity biomaterials made of hydroxyapatite (Section 25.3). Afterwards, we strictly verify the validity of these micromechanics models for numerous different porous HA biomaterials, on the basis of physically and statistically independent morphological and mechanical tests (Section 25.4). Before concluding, we will show how to extend the developed theories to HA biomaterials with a hierarchical structure, i.e. with a double-porosity (Section 25.5). Such biomaterials have recently been introduced for tissue engineering purposes. Our models will show the role of micro- and macroporosity on the stiffness and strength properties of highly porous tissue engineering scaffolds.

25.2 Fundamentals of continuum micromechanics

25.2.1 Representative volume element and phase properties

In continuum micromechanics (Hill, 1963; Suquet, 1997; Zaoui, 2002), a material is understood as a macro-homogeneous, but micro-heterogeneous body filling a representative volume element (RVE) with characteristic length l, $l \gg d$, d standing for the characteristic length of inhomogeneities within the RVE, and $l \ll L$, L standing for the characteristic lengths of geometry or loading of a structure built up by the material defined on the RVE (Fig. 25.1). In general, the microstructure within each RVE is so complicated that it cannot be described in complete detail. Therefore, N_r quasi-homogeneous subdomains with known physical quantities are reasonably chosen. They are called material phases (Fig. 25.1a).

Quantitative phase properties are volume fractions f_r of phases $r = 1, \ldots, N_r$, (average) elastic or strength properties of phases. As regards phase elasticity, the fourth-order stiffness tensor $\mathbb{C}_r$ relates the (average 'microscopic') second-order strain tensor in phase r, $\boldsymbol{\varepsilon}_r$, to the (average 'microscopic') second-order stress tensor in phase r, $\boldsymbol{\sigma}_r$,

$$\boldsymbol{\sigma}_r = \mathbb{C}_r : \boldsymbol{\varepsilon}_r \qquad [25.1]$$

As regards phase strength, brittle failure can be associated to the boundary of an elastic domain $f_r(\boldsymbol{\sigma}) < 0$,

$$f_r(\boldsymbol{\sigma}) = 0 \qquad [25.2]$$

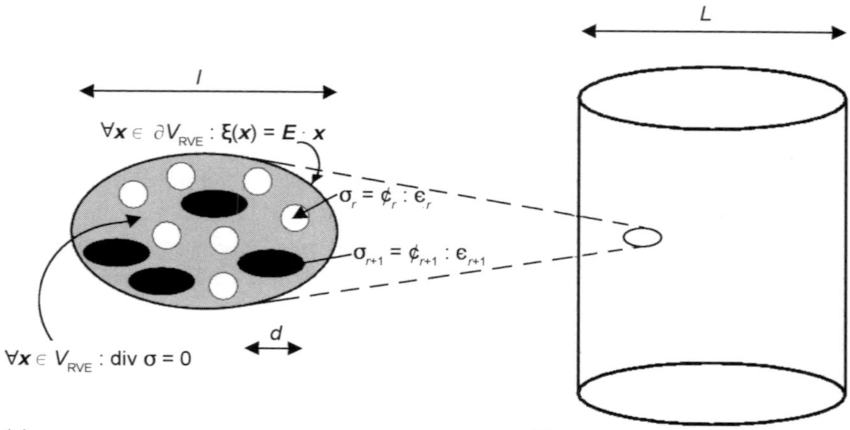

(a) (b)

25.1 (a) Loading of a representative volume element, built up by phases *r* with stiffness $\mathbb{C}_r$ and strength properties $f_r(\boldsymbol{\sigma}) = 0$, according to continuum micromechanics (Hashin, 1983; Zaoui, 2002): Displacements ξ, related to a constant (homogenized) strain $\boldsymbol{E}$, are imposed at the boundary of the RVE; (b) structure built up of material defined on RVE (a).

defined in the space of microstresses $\boldsymbol{\sigma}(\boldsymbol{x})$, $\boldsymbol{x}$ being the position vector for locations within or at the boundary of the RVE.

Also the spatial arrangement of the phases needs to be specified. In this respect, two cases are of particular interest: (i) one or several inclusion phases with different shapes are embedded in a contiguous 'matrix' phase (as in a reinforced composite material), or (ii) mutual contact of all disorderly arranged phases (as in a polycrystal).

25.2.2 Averaging – homogenization

The central goal of continuum micromechanics is to estimate the mechanical properties (such as elasticity or strength) of the material defined on the RVE (the macro-homogeneous, but micro-heterogeneous, medium) from the aforementioned phase properties. This procedure is referred to as homogenization or one homogenization step. Therefore, homogeneous ('macroscopic') strains $\boldsymbol{E}$ are imposed onto the RVE, in terms of displacements at its boundary ∂V:

$$\forall \boldsymbol{x} \in \partial V : \xi(\boldsymbol{x}) = \boldsymbol{E} \cdot \boldsymbol{x} \tag{25.3}$$

As a consequence, the resulting kinematically compatible microstrains $\boldsymbol{\varepsilon}(\boldsymbol{x})$ throughout the RVE with volume V_{RVE} fulfil the average condition (Hashin, 1983):

$$\boldsymbol{E} = \langle \boldsymbol{\varepsilon} \rangle = \frac{1}{V_{\text{RVE}}} \int_{V_{\text{RVE}}} \boldsymbol{\varepsilon} \, \mathrm{d}V = \sum_r f_r \boldsymbol{\varepsilon}_r \tag{25.4}$$

providing a link between 'micro' and 'macro' strains. Analogously, homogenized ('macroscopic') stresses $\boldsymbol{\Sigma}$ are defined as the spatial average over the RVE, of the microstresses $\boldsymbol{\sigma}(\boldsymbol{x})$:

$$\boldsymbol{\Sigma} = \langle \boldsymbol{\sigma} \rangle = \frac{1}{V_{\mathrm{RVE}}} \int_{V_{\mathrm{RVE}}} \boldsymbol{\sigma} \, \mathrm{d}V = \sum_r f_r \boldsymbol{\sigma}_r \qquad [25.5]$$

Homogenized ('macroscopic') stresses and strains, $\boldsymbol{\Sigma}$ and $\boldsymbol{E}$, are related by the homogenized ('macroscopic') stiffness tensor $\mathbb{C}$:

$$\boldsymbol{\Sigma} = \mathbb{C} : \boldsymbol{E} \qquad [25.6]$$

which needs to be linked to the stiffnesses $\mathbb{c}_r$, the shape and the spatial arrangement of the phases (Section 25.2.1). This link is based on the linear relation between the homogenized ('macroscopic') strain $\boldsymbol{E}$ and the average ('microscopic') strain $\boldsymbol{\varepsilon}_r$, resulting from the superposition principle valid for linear elasticity [25.1] (Hill, 1963). This relation is expressed in terms of the fourth-order concentration tensors $\mathbb{A}_r$ of each of the phases r.

$$\boldsymbol{\varepsilon}_r = \mathbb{A}_r : \boldsymbol{E} \qquad [25.7]$$

Insertion of Eq. [25.7] into Eq. [25.1] and averaging over all phases according to Eq. [25.5] leads to

$$\boldsymbol{\Sigma} = \sum_r f_r \mathbb{c}_r : \mathbb{A}_r : \boldsymbol{E} \qquad [25.8]$$

From Eq. [25.8] and Eq. [25.6] we can identify the sought relation between the phase stiffness tensors $\mathbb{c}_r$ and the overall homogenized stiffness $\mathbb{C}$ of the RVE,

$$\mathbb{C} = \sum_r f_r \mathbb{c}_r : \mathbb{A}_r \qquad [25.9]$$

The concentration tensors $\mathbb{A}_r$ can be suitably estimated from Eshelby's 1957 matrix-inclusion problem, according to Zaoui (2002) and Benveniste (1987):

$$\mathbb{A}_r^{est} = [\mathbb{I} + \mathbb{P}_r^0 : (\mathbb{c}_r - \mathbb{C}^0)]^{-1} : \left\{ \sum_r f_s [\mathbb{I} + \mathbb{P}_s^0 : (\mathbb{c}_s - \mathbb{C}^0)]^{-1} \right\}^{-1} \qquad [25.10]$$

where $\mathbb{I}$, $I_{ijkl} = 1/2(\delta_{ik}\delta_{jl} + \delta_{il}\delta_{kj})$, is the fourth-order unity tensor, δ_{ij} (Kronecker delta) are the components of second-order identity tensor $\mathbf{1}$, and the fourth-order Hill tensor $\mathbb{P}_r^0$ accounts for the shape of phase r, represented as an ellipsoidal inclusion embedded in a fictitious matrix of stiffness $\mathbb{C}^0$. For isotropic matrices (which is the case considered throughout this chapter), $\mathbb{P}_r^0$ is accessible via the Eshelby tensor:

$$\mathbb{S}_r^{esh,0} = \mathbb{P}_r^0 : \mathbb{C}^{0,-1} \qquad [25.11]$$

documented in Eshelby (1957); see also Section 25.3.

Back-substitution of Eq. [25.10] into Eq. [25.9] delivers the sought estimate for the homogenized ('macroscopic') stiffness tensor, $\mathbb{C}^{est}$, as

$$\mathbb{C}^{est} = \sum_r f_r \mathbb{c}_r : [\mathbb{I} + \mathbb{P}_r^0 : (\mathbb{c}_r - \mathbb{C}^0)]^{-1} : \left\{ \sum_r f_s [\mathbb{I} + \mathbb{P}_s^0 : (\mathbb{c}_s - \mathbb{C}^0)]^{-1} \right\}^{-1} \qquad [25.12]$$

Choice of 'matrix stiffness' $\mathbb{C}^0$ determines which type of interactions between the phases is considered: For $\mathbb{C}^0$ coinciding with one of the phase stiffnesses (Mori–Tanaka scheme; Mori and Tanaka, 1973), a composite material is represented (contiguous matrix with inclusions); for $\mathbb{C}^0 = \mathbb{C}^{est}$ (self-consistent scheme; Hershey, 1954; Kröner, 1958), a dispersed arrangement of the phases is considered (typical for polycrystals).

As long as the average phase strains $\boldsymbol{\varepsilon}_r$ are relevant for brittle phase failure, resulting in overall failure of the RVE, the concentration relation [25.7] allows for translation of the brittle failure criterion of the weakest phase $r = w$ into a macroscopic ('homogenized') brittle failure criterion, according to [25.2], [25.1], [25.7] and [25.6],

$$\begin{aligned} f_w(\boldsymbol{\sigma}) = 0 &= f_w(\mathbb{c}_w : \boldsymbol{\varepsilon}_w) = f_w(\mathbb{c}_w : \mathbb{A}_w : \boldsymbol{E}) \\ &= f_w(\mathbb{c}_w : \mathbb{A}_w : \mathbb{C}^{-1} : \boldsymbol{\Sigma}) = \mathscr{F}(\boldsymbol{\Sigma}) \end{aligned} \qquad [25.13]$$

In case strain heterogeneities within the phases become relevant for failure, one has to resort to a different strategy, alluded to in Section 25.5.

Fourth-order tensor operations such as the ones occurring in [25.1] and [25.6]–[25.12] can be suitably evaluated in a vector/matrix-based software, through a compressed vector/matrix notation with normalized tensorial basis, often referred to as the Kelvin or the Mandel notation, see e.g. Cowin (2003) for details.

25.3 Micromechanical representation of mono-porosity biomaterials made of hydroxyapatite – stiffness and strength estimates

In the line of the concept presented in Section 25.2, we envision monoporosity biomaterials made of HA as porous polycrystals consisting of HA needles (Shareef *et al.*, 1993, Fig. 7; Liu, 1998, Fig. 2) with stiffness $\mathbb{c}_{HA}$ and volume fraction $(1 - \phi)$, being oriented in all space directions, and of spherical (empty) pores with vanishing stiffness and volume fraction ϕ (porosity) (see Figs 25.2 and 25.3).

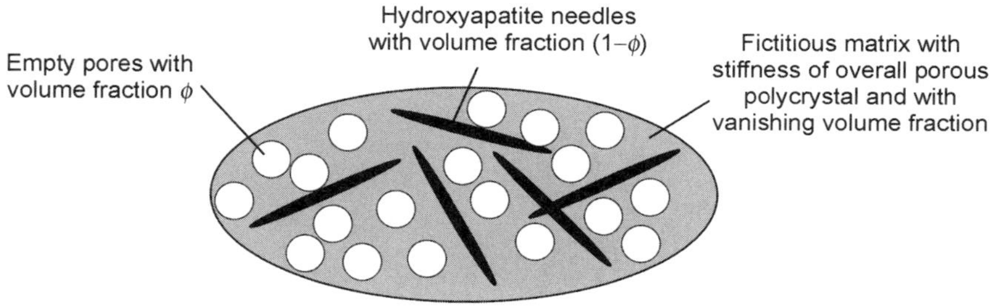

25.2 RVE of polycrystal representing monoporosity biomaterial made of hydroxyapatite: uniform orientation distribution of cylindrical (needle-like) inclusions and spherical (empty) pores.

25.3.1 Stiffness estimate

In a reference frame ($\boldsymbol{e}_1$, $\boldsymbol{e}_2$, $\boldsymbol{e}_3$), the HA needle orientation vector $\boldsymbol{N} = \boldsymbol{e}_r$ is given by Euler angles ϑ and φ (see Fig. 25.3). Specification of [25.12] for $\mathbb{C}^0 = \mathbb{C}^{est} = \mathbb{C}_{poly}$ (self-consistent scheme) and for an infinite number of solid phases related to orientation directions $\boldsymbol{N} = \boldsymbol{e}_r(\vartheta, \varphi)$, which are uniformly distributed in space ($\varphi \in [0, 2\pi]$; $\vartheta \in [0, \pi]$), yields the homogenized stiffness of the porous hydroxyapatite biomaterial depicted in Fig. 25.2 (Fritsch *et al.*, 2006):

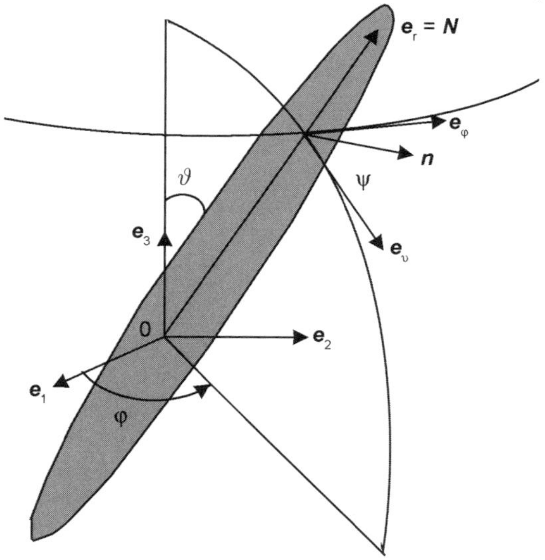

25.3 Cylindrical (needle-like) HA inclusions oriented along vector $\boldsymbol{N}$ and inclined by angles ϑ and φ with respect to the reference frame ($\boldsymbol{e}_1$, $\boldsymbol{e}_2$, $\boldsymbol{e}_3$); local base frame $\boldsymbol{e}_r$, $\boldsymbol{e}_\vartheta$, $\boldsymbol{e}_\varphi$ is attached to the needle.

$$\mathbb{C}_{poly} = (1-\phi)\mathbb{c}_{HA} : \langle[\mathbb{I} + \mathbb{P}_{cyl}^{poly} : (\mathbb{c}_{HA} - \mathbb{C}_{poly})]^{-1}\rangle : \left\{(1-\phi)\langle[\mathbb{I} + \mathbb{P}_{cyl}^{poly} :(\mathbb{c}_{HA} - \mathbb{C}_{poly})]^{-1}\rangle + \phi(\mathbb{I} - \mathbb{P}_{sph}^{poly} : \mathbb{C}_{poly})^{-1}\right\}^{-1} \quad [25.14]$$

with the angular average

$$\langle[\mathbb{I} + \mathbb{P}_{cyl}^{poly} : (\mathbb{c}_{HA} - \mathbb{C}_{poly})]^{-1}\rangle = \int_{\varphi=0}^{2\pi}\int_{\vartheta=0}^{\pi}[\mathbb{I} + \mathbb{P}_{cyl}^{poly}(\vartheta,\varphi) : (\mathbb{c}_{HA} - \mathbb{C}_{poly})]^{-1}\frac{\sin\vartheta\,\mathrm{d}\vartheta\,\mathrm{d}\varphi}{4\pi} \quad [25.15]$$

$\mathbb{P}_{sph}^{poly}$ and $\mathbb{P}_{cyl}^{poly}$ are the fourth-order Hill tensors for spherical and cylindrical inclusions, respectively, in an isotropic matrix with stiffness $\mathbb{C}_{poly} = 3k_{poly}\mathbb{J} + 2\mu_{poly}\mathbb{K}$; $\mathbb{J}$, $J = \frac{1}{3}\delta_{ij}\delta_{kl}$, is the volumetric part of the fourth-order unity tensor $\mathbb{I}$, and $\mathbb{K} = \mathbb{I} - \mathbb{J}$ is its deviatoric part. The Hill tensors are related to Eshelby tensors via Eq. [25.11]. The Eshelby tensor $\mathbb{S}_{sph}^{esh}$ corresponding to spherical inclusions (pores in Fig. 25.2) reads as

$$\mathbb{S}_{sph}^{esh} = \frac{3k_{poly}}{3k_{poly} + 4\mu_{poly}}\mathbb{J} + \frac{6(k_{poly} + 2\mu_{poly})}{5(3k_{poly} + 4\mu_{poly})}\mathbb{K} \quad [25.16]$$

In the base frame $(\boldsymbol{e}_\vartheta, \boldsymbol{e}_\varphi, \boldsymbol{e}_r)$, $(1 = \vartheta,\ 2 = \varphi,\ 3 = r$, see Fig. 25.3 for Euler angles φ and ϑ), attached to individual solid needles, the non-zero components of the Eshelby tensor $\mathbb{S}_{sph}^{esh}$ corresponding to cylindrical inclusions read as

$$S_{cyl,1111}^{esh} = S_{cyl,2222}^{esh} = \frac{5-4\nu_{poly}}{8(1-\nu_{poly})};\ S_{cyl,1122}^{esh} = S_{cyl,2211}^{esh} = \frac{-1+4\nu_{poly}}{8(1-\nu_{poly})};$$

$$S_{cyl,1133}^{esh} = S_{cyl,2233}^{esh} = \frac{\nu_{poly}}{2(1-\nu_{poly})};$$

$$S_{cyl,2323}^{esh} = S_{cyl,3232}^{esh} = S_{cyl,3223}^{esh} = S_{cyl,2332}^{esh} = S_{cyl,3131}^{esh} = S_{cyl,1313}^{esh} = S_{cyl,1331}^{esh} = S_{cyl,3113}^{esh} = \tfrac{1}{4};$$

$$S_{cyl,1212}^{esh} = S_{cyl,2121}^{esh} = S_{cyl,2112}^{esh} = S_{cyl,1221}^{esh} = \frac{3-4\nu_{poly}}{8(1-\nu_{poly})} \quad [25.17]$$

with ν_{poly} as Poisson's ratio of the polycrystal,

$$\nu_{poly} = \frac{3k_{poly} - 2\mu_{poly}}{6k_{poly} + 2\mu_{poly}} \quad [25.18]$$

Following standard tensor calculus (Salençon, 2001), the tensor components of $\mathbb{P}_{cyl}^{poly}(\varphi,\vartheta) = \mathbb{S}_{cyl}^{esh}(\varphi,\vartheta) : \mathbb{C}_{poly}$, being related to differently orientated inclusions, are transformed into one, single base frame $(\boldsymbol{e}_1, \boldsymbol{e}_2, \boldsymbol{e}_3)$, in order to evaluate the integrals in [25.14] and [25.15].

25.3.2 Strength estimate

Strength of the porous polycrystal made up of HA needles (see Fig. 25.2 for its RVE) is related to brittle failure of the most unfavourably stressed single needle. Therefore, the macroscopic stress (and strain) state needs to be related to corresponding stress and strain states in the individual needles. Accordingly, we specify the concentration relations [25.7] and [25.10] for the monoporosity biomaterial defined through [25.14]–[25.18], resulting in:

$$\boldsymbol{\varepsilon}_{HA}(\varphi,\vartheta) = [\mathbb{I} + \mathbb{P}^{poly}_{cyl}(\varphi,\vartheta) : (\mathbb{c}_{HA} - \mathbb{C}_{poly})]^{-1} : \Big\{(1-\phi)\langle \mathbb{I} + \mathbb{P}^{poly}_{cyl}(\varphi,\vartheta) : (\mathbb{c}_{HA} - \mathbb{C}_{poly})]^{-1}\rangle + \phi(\mathbb{I} - \mathbb{P}^{poly}_{sph} : \mathbb{C}_{poly})^{-1}\Big\}^{-1} : \boldsymbol{E} \tag{25.19}$$

When employing phase elasticity [25.1] to HA, and overall elasticity [25.6] to the monoporous biomaterial according to [25.14], the concentration relation [25.19] can be recast in terms of stresses:

$$\boldsymbol{\sigma}_{HA}(\varphi,\vartheta) = \mathbb{c}_{HA} : \Big\{[\mathbb{I} + \mathbb{P}^{poly}_{cyl}(\varphi,\vartheta) : (\mathbb{c}_{HA} - \mathbb{C}_{poly})]^{-1} : \Big\{(1-\phi)\langle[\mathbb{I} + \mathbb{P}^{poly}_{cyl}(\varphi,\vartheta) : (\mathbb{c}_{HA} - \mathbb{C}_{poly})]^{-1}\rangle + \phi(\mathbb{I} - \mathbb{P}^{poly}_{sph} : \mathbb{C}_{poly})^{-1}\Big\}^{-1}\Big\} : \mathbb{C}^{-1}_{poly} : \boldsymbol{\Sigma} = \mathbb{B}_{HA}(\varphi,\vartheta) : \boldsymbol{\Sigma} \tag{25.20}$$

with $\mathbb{B}_{HA}(\varphi,\vartheta)$ as the so-called stress concentration factor of needle with orientation $\boldsymbol{N}(\varphi,\vartheta)$. We consider that needle failure is governed by the normal stress $\sigma_{HA,NN}(\varphi,\vartheta)$ in needle direction and by the shear stress in planes orthogonal to the needle direction, $\sigma_{HA,Nn}(\varphi,\vartheta;\psi)$ (see Fig. 25.3):

$$\sigma_{HA,NN}(\varphi,\vartheta) = \boldsymbol{N}\cdot\boldsymbol{\sigma}_{HA}(\varphi,\vartheta)\cdot\boldsymbol{N} \tag{25.21}$$

$$\sigma_{HA,Nn}(\varphi,\vartheta,;\psi) = \boldsymbol{n}(\psi)\cdot\boldsymbol{\sigma}_{HA}(\varphi,\vartheta)\cdot\boldsymbol{N} \tag{25.22}$$

depending on the direction $\boldsymbol{n}$ orthogonal to $\boldsymbol{N}$, specified through angle ψ (Fig. 25.3):

$$\boldsymbol{n} = \cos\psi\,\boldsymbol{e}_\vartheta + \sin\psi\,\boldsymbol{e}_\varphi \tag{25.23}$$

More specifically, the failure criterion for the single needle considers interaction between tensile strength $\sigma^{ult,t}_{HA}$ and shear strength $\sigma^{ult,s}_{HA}$, and it reads as

$$\vartheta = 0..\pi, \psi = 0..2\pi : f_{HA}(\boldsymbol{\sigma}) = \max_{\vartheta}(\beta \max_{\psi}|\sigma_{HA,Nn}| + \sigma_{HA,NN}) - \sigma^{ult,t}_{HA} = 0 \tag{25.24}$$

with $\beta = \sigma^{ult,t}_{HA}/\sigma^{ult,s}_{HA}$ being the ratio between uniaxial tensile strength $\sigma^{ult,t}_{HA}$, and the shear strength $\sigma^{ult,s}_{HA}$ of pure hydroxyapatite. Use of [25.20] to [25.23] in [25.24] yields a macroscopic failure criterion in the format of [25.13]:

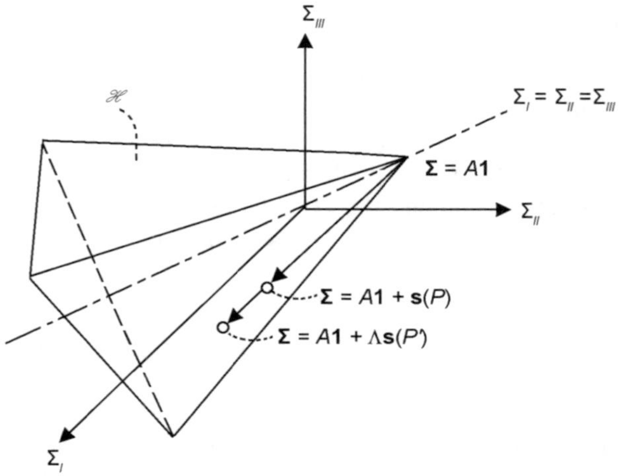

25.4 Homothetic elastic domain $\mathscr{H}$ of monoporosity biomaterial in principal stress space (Σ_I, Σ_{II} and Σ_{III} are the principal stresses of macroscopic stress tensor $\mathbf{\Sigma}$, $\mathbf{\Sigma} = \Sigma_I e_I \otimes e_I + \Sigma_{II} e_{II} \otimes e_{II} + \Sigma_{III} e_{III} \otimes e_{III}$).

$$\mathscr{F}(\mathbf{\Sigma}) = \max_{\vartheta}\left\{\beta \max_{\psi} |\boldsymbol{N} \cdot \mathbb{B}_{HA}(\varphi, \vartheta) : \mathbf{\Sigma} \cdot \boldsymbol{n}(\psi)| + \boldsymbol{N} \cdot \mathbb{B}_{HA}(\varphi, \vartheta) : \mathbf{\Sigma} \cdot \boldsymbol{N}\right\} - \sigma_{HA}^{ult,t} = 0 \qquad [25.25]$$

and a corresponding elastic domain,

$$\mathscr{F}(\mathbf{\Sigma}) < 0 \qquad [25.26]$$

with $\mathbb{B}_{HA}$ according to [25.20]. We can show that, in the principal macroscopic stress space, the shape of failure criterion [25.25] for monoporous biomaterials made of hydroxyapatite is homothetic ('cone'-type), see Fig. 25.4 and Section 25.7. We also will evaluate the criterion [25.25] for uniaxial macroscopic stress states $\mathbf{\Sigma} = \pm\Sigma_{ref}\boldsymbol{e}_3 \otimes \boldsymbol{e}_3$: Insertion of these stress states into [25.20]–[25.24] yields an equation for Σ_{ref}, the corresponding results $\Sigma_{poly}^{ult,t}$ and $\Sigma_{poly}^{ult,c}$ being model predictions of macroscopic uniaxial strengths as functions of ('microscopic') needle strength and porosity (see Figs 25.9 and 25.10, and Section 25.4 for further details).

25.4 Model validation

25.4.1 Strategy for model validation through independent test data

In the line of Popper, who stated that a theory – as long as it has not been falsified – will be 'the more satisfactory the greater the severity of independent tests it survives', cited from Mayr (1997, p. 49), the verification of the micromechanical representation of HA biomaterials ([25.14]–[25.18] for elasticity and

Table 25.4 'Universal' (biomaterial-independent) isotropic phase properties of pure hydroxyapatite needles

Young's modulus E_{HA} Poisson's ratio ν_{HA}	114 GPa 0.27	from Katz and Ukraincik (1971)
Uniaxial tensile strength $\sigma_{HA}^{ult,t}$ Uniaxial shear strength $\sigma_{HA}^{ult,s}$	52.2 GPa 80.3 GPa	from Shareef *et al.* (1993) and Akao *et al.* (1981); see Section 25.4.2 for details

[25.19]–[25.26] for strength) will rest on two independent experimental sets, as it has been successfully done for other material classes such as bone (Hellmich and Ulm, 2002; Hellmich *et al.*, 2004; Fritsch and Hellmich, 2007) or wood (Hofstetter *et al.*, 2005, 2006). Biomaterial-specific macroscopic (homogenized) stiffnesses $\mathbb{C}_{poly}$ (Young's moduli E_{poly} and Poisson's ratios ν_{poly}), and uniaxial (tensile and compressive) strengths ($\Sigma_{poly}^{ult,t}$ and $\Sigma_{poly}^{ult,c}$), predicted by the micromechanics model [25.14]–[25.26] on the basis of biomaterial-independent ('universal') elastic and strength properties of pure hydroxyapatite (experimental set I, Table 25.4) for biomaterial-specific porosities ϕ (experimental set IIa, Tables 25.2 and 25.3), are compared to corresponding biomaterial-specific experimentally determined moduli E_{exp} and Poisson's ratios ν_{exp} (experimental set IIb-1, Table 25.3) and uniaxial tensile/compressive strength values (experimental set IIb-2, Table 25.2). Since we avoided introduction of micromorphological features which cannot be experimentally quantified (such as the precise crystal shape), all material parameters are directly related to well-defined experiments.

25.4.2 'Universal' mechanical properties of (biomaterial-independent) hydroxyapatite – experimental set I

Experiments with an ultrasonic interferometer coupled with a solid media pressure apparatus (Katz and Ukraincik, 1971; Gilmore and Katz, 1982) reveal the isotropic elastic constants for dense HA powder ($\phi = 0$), Young's modulus $E_{HA} = 114$ GPa, and Poisson's ratio $\nu_{HA} = 0.27$ (equivalent to bulk modulus $k_{HA} = E_{HA}/3/(1 - 2\nu_{HA}) = 82.6$ GPa and shear modulus $\mu_{HA} = E_{HA}/2/(1 + \nu_{HA}) = 44.9$ GPa).

The authors are not aware of direct strength tests on pure HA (with $\phi = 0$). Therefore, we consider one uniaxial tensile test, $\Sigma_{exp}^{ult,t} = 37.1$ MPa, and one uniaxial compressive test, $\Sigma_{exp}^{ult,c} = 509$ MPa, on fairly dense samples (with $\phi = 12.2\%$ and $\phi = 2.8\%$, respectively), conducted by Shareef *et al.* (1993) and Akao *et al.* (1981), respectively (see Table 25.2). From these two tests, we back-calculate, via evaluation of [25.20]–[25.25] for $\mathbf{\Sigma} = \Sigma_{exp}^{ult,t} \boldsymbol{e}_3 \otimes \boldsymbol{e}_3$ and $\mathbf{\Sigma} = -\Sigma_{exp}^{ult,c} \boldsymbol{e}_3 \otimes \boldsymbol{e}_3$, the 'universal' tensile and shear strength of pure hydroxyapatite, $\sigma_{HA}^{ult,t}$ and $\sigma_{HA}^{ult,s}$ (Table 25.4).

25.4.3 Biomaterial-specific porosities – experimental set IIa

Porosity of HA biomaterials is standardly calculated from mass M and volume V of well-defined samples on the basis of the mass density of pure hydroxyapatite, $\rho_{HA} = 3.16\,\mathrm{g/cm^3}$:

$$\phi = \frac{M}{V\,\rho_{HA}} \qquad [25.27]$$

Corresponding porosity values have been reported by Peelen *et al.* (1978), Akao *et al.* (1981), De With *et al.* (1981), Shareef *et al.* (1993), Arita *et al.* (1995), Martin and Brown (1995), Liu (1998) and Charrière *et al.* (2001); see Tables 25.2 and 25.3.

25.4.4 Biomaterial-specific elasticity experiments on hydroxyapatite biomaterials (experimental set IIb-1)

Elastic properties of HA biomaterials were determined through uniaxial quasi-static mechanical tests (Akao *et al.*, 1981; Charrière *et al.*, 2001), but also through ultrasonic techniques (De With *et al.*, 1981; Liu, 1998), or resonance frequency tests (Arita *et al.* 1995).

In uniaxial quasi-static experiments, an increasing (vertical) force is applied to the tested sample, and the deformation of the sample is simultaneously measured. Division of the force by the cross-sectional area of the specimen yields the (quasi-homogeneous) uniaxial stress in the sample, while its strain state is provided through strain gauges or back-calculated from the relative displacements of the top and the bottom of the specimen. The gradient of the corresponding stress–strain curve gives access to Young's modulus. Respective experimental results are documented for cuboidal specimens (Akao *et al.*, 1981) and hollow cylindrical specimens (Charrière *et al.*, 2001); see Tables 25.1 and 25.3 as well as Fig. 25.7.

A typical ultrasonic device consists of a pulser-receiver (Fig. 25.5(a)), an oscilloscope, and several ultrasonic transducers (Fig. 25.5(b)). The pulser unit emits an electrical square-pulse of typically up to 400 V, with typical frequencies ranging from the kHz to the MHz range. The piezoelectric elements inside the ultrasonic transducers transform such electrical signals into mechanical signals (when operating in the sending mode, transferring, via a coupling medium, the mechanical signals to one side of the specimen under investigation), or they transform mechanical signals back to electrical signals (when receiving mechanical signals from the opposite side of the specimen under investigation). The oscilloscope gives access to the time of flight of the ultrasonic wave through the specimen, t_s, which provides, together with the travel distance through the specimen, l_s, the phase velocity of the wave as:

$$v_s = \frac{l_s}{t_s} \text{ (transmission-through configuration)} \qquad [25.28]$$

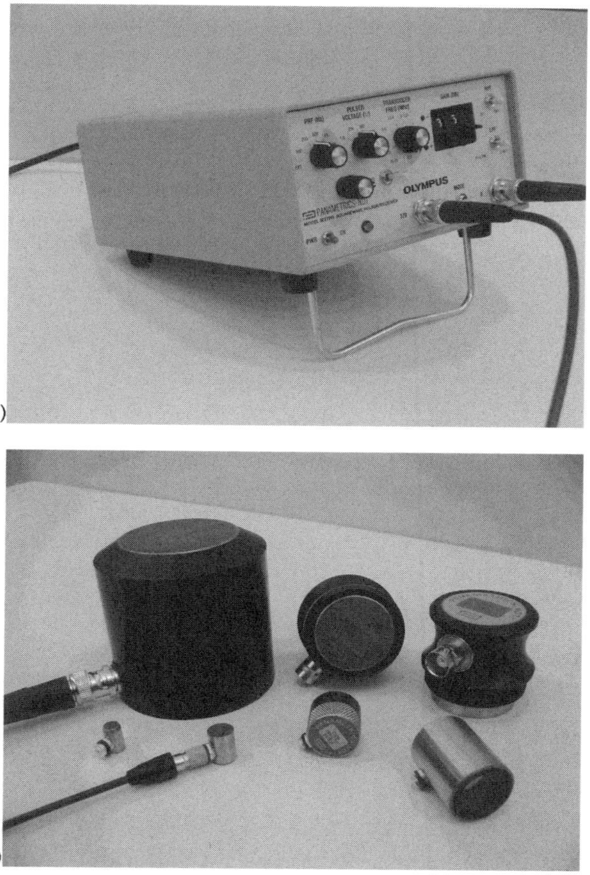

(a)

(b)

25.5 Ultrasonic measurement equipment (Müllner *et al.*, 2007) used at the Institute for Mechanics of Materials and Structures, Vienna University of Technology, Vienna, Austria.

In the pulse-echo configuration, only one transducer is used, which acts both as transmitter and receiver, and the wave propagation velocity follows to be:

$$v_s = \frac{2l_s}{t_s} \text{ (pulse-echo configuration)} \quad [25.29]$$

Frequency f and wave velocity v_s give access to the wavelength λ, through

$$\lambda = \frac{v_s}{f} \quad [25.30]$$

If the wavelength is considerably smaller than the diameter of the specimen, a (compressional) bulk wave, i.e. a laterally constrained wave, propagates with velocity v_L in a quasi-infinite medium. On the other hand, if the wavelength is considerably larger than the diameter of the specimen, a bar wave propagates with velocity v_{bar}, i.e. the specimen acts as one-dimensional bar without lateral constraints (Ashman *et al.*, 1984). In contrast, shear waves propagate identically in quasi-infinite media and bar-like structures (Ashman *et al.*, 1987).

As regards bulk waves, combination of the conservation law of linear momentum, of the generalized Hooke's law, of the linearized strain tensor, and of the general plane wave solution for the displacements inside an infinite solid medium yields the elasticity tensor components C_{1111} and C_{1212} as functions of the material mass density ρ and the longitudinal and transverse bulk wave propagation velocities v_L and v_T, reading for isotropic materials as (Carcione, 2001):

$$C_{1111} = \rho v_L^2 \quad \text{and} \quad C_{1212} = G = \rho v_T^2 \qquad [25.31]$$

with G as the shear modulus.

Combination of [25.31] with the definitions of the engineering constants Young's modulus E and Poisson's ratio ν, yields the latter as functions of the wave velocities, in the form

$$E = \frac{v_T^2(3v_L^2 - 4v_T^2)}{v_L^2 - v_T^2} \qquad [25.32]$$

and

$$\nu = \frac{E}{2G} - 1 = \frac{\frac{v_L^2}{2} - v_T^2}{v_L^2 - v_T^2} \qquad [25.33]$$

respectively.

In the case of bar wave propagation (Kolsky, 1953), the measured bar wave velocity v_{bar} gives direct access to the Young's modulus,

$$E = \rho\, v_{bar}^2 \qquad [25.34]$$

Since the wavelength is a measure for the loading of the structure ($\lambda \approx D$ in Fig. 25.1), the mechanical properties [25.31]–[25.34] are related to an RVE with characteristic length $l \ll \lambda$; see Fig. 25.6.

Respective experimental results are documented for bar-shaped specimens (Liu, 1998) and cylindrical samples (De With *et al.*, 1981), see Tables 25.1 and 25.3 as well as Figs 25.7 and 25.8.

In resonance frequency tests, beam type specimens are excited in the flexural vibration mode, and the corresponding free vibration gives access to the fundamental resonance frequency ω. The latter gives access, via beam theory, to Young's modulus of the sample:

$$E = \frac{12\omega^2 \rho_s l^4}{\pi^4 b^2} \qquad [25.35]$$

with ρ_s as the mass density, l_s as the length of specimen, and b_s as the side length

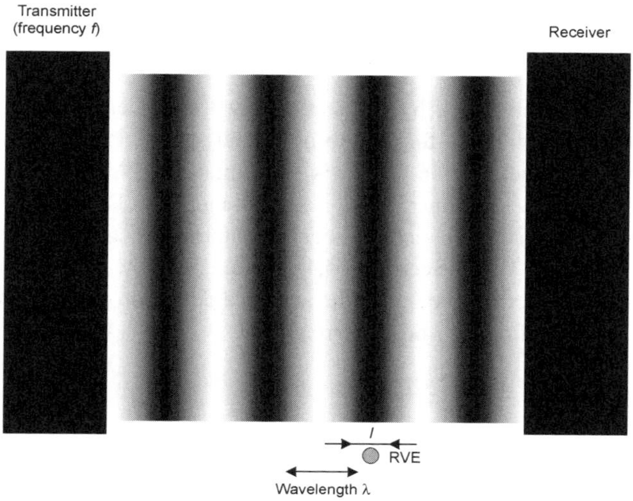

25.6 Schematic, grey-scale based illustration of stress magnitude in specimens tested ultrasonically with frequency f: characterization of material element (RVE) with characteristic length l separated by scale from the wavelength λ; see also Fritsch and Hellmich (2007).

of the square cross-section of the specimen. Respective experimental results are documented for disc-shaped samples (Arita *et al.*, 1995); see Tables 25.1 and 25.3 as well as Fig. 25.7.

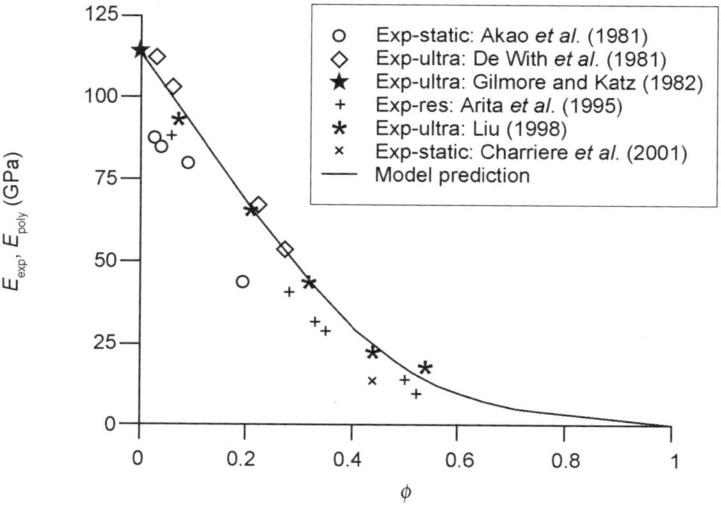

25.7 Comparison between model predictions (E_{poly}) [25.14]–[25.19] and experiments (E_{exp}) for Young's modulus of different monoporosity biomaterials made of hydroxyapatite, as function of porosity ϕ; ultra = ultrasonic tests, res = resonance frequency tests, static = quasi-static uniaxial tests.

25.4.5 Comparison between biomaterial-specific stiffness predictions and corresponding experiments

The stiffness values predicted by the homogenization scheme [25.14]–[25.18] (see Section 25.3 and Fig. 25.2) for biomaterial-specific porosities (Section 25.4.3, experimental set IIa) on the basis of biomaterial-independent ('universal') stiffness of HA (Section 25.4.2, experimental set I) are compared to corresponding experimentally determined biomaterial-specific stiffness values from experimental set IIb-1 (Section 5.4.4). To quantify the model's predictive capabilities we consider the mean and the standard deviation of the relative error between stiffness predictions and experiments,

$$\bar{e} = \frac{1}{n}\sum e_i = \frac{1}{n}\sum \frac{q_{poly} - q_{exp}}{q_{exp}} \qquad [25.36]$$

$$e_S = \left[\frac{1}{n-1}\sum (e_i - \bar{e})^2\right]^{1/2} \qquad [25.37]$$

where q has to be replaced by the quantity in question, E or ν, and with summation over n values q_{exp} per experimental protocol.

Insertion of biomaterial-specific porosities (Table 25.3) into Eq. [25.14] delivers, together with [25.15]–[25.18], biomaterial-specific stiffness estimates for the effective Young's modulus E_{poly} and the effective Poisson's ratio ν_{poly}. These stiffness predictions are compared to corresponding experimental stiffness values (Figs 25.7 and 25.8). The satisfactory agreement between model

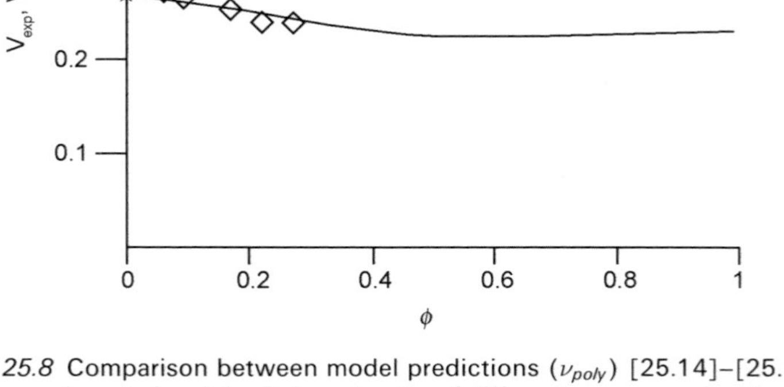

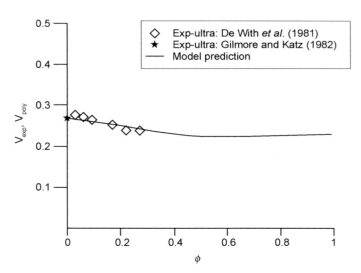

25.8 Comparison between model predictions (ν_{poly}) [25.14]–[25.19] and experiments (ν_{exp}) for Poisson's ratio of different monoporosity biomaterials made of hydroxyapatite, as function of porosity ϕ; ultra = ultrasonic tests.

predictions and experiments is quantified by prediction errors of $16 \pm 25\%$ (mean value $\pm$ standard deviation according to [25.36] and [25.37]) for Young's modulus, and of $-0.4 \pm 2.3\%$ for Poisson's ratio.

25.4.6 Biomaterial-specific strength experiments on hydroxyapatite biomaterials (experimental set IIb-2)

In uniaxial compressive quasi-static tests, a sharp decrease of stress after a stress peak in the stress–strain diagram (Akao *et al.*, 1981; Martin and Brown, 1995; see Section 25.4.4 for the derivation of stress–strain curves) indicates brittle material failure, as observed for all biomaterials described herein, and the aforementioned stress peak is referred to as the ultimate stress or uniaxial strength $\Sigma_{exp}^{ult,c}$. Respective experimental results are documented for cylindrical samples (Peelen *et al.*, 1978) and bars (Akao *et al.*, 1981); see Tables 25.1 and 25.2 as well as Fig. 25.10.

In three-point bending tests, a force F_s is applied to the centre of a beam specimen, and the maximum normal stress $\Sigma^{ult} = \Sigma^{ult} \boldsymbol{e}_3 \otimes \boldsymbol{e}_3$ in the bar-type sample is calculated according to beam theory,

$$\Sigma_{exp}^{ult,b} = \frac{3F_s l_s}{2b_s h_s^2} \qquad [25.38]$$

with l_s, b_s and h_s as the length, width and height of the specimen with rectangular cross-section, respectively. Respective experimental results are documented by Liu (1998); see Tables 25.1 and 25.2 as well as Fig. 25.9.

In the Stanford ring bursting test, ring-shaped specimens are pressurized internally, in order to generate a tensile hoop stress in the ring. The pressure is

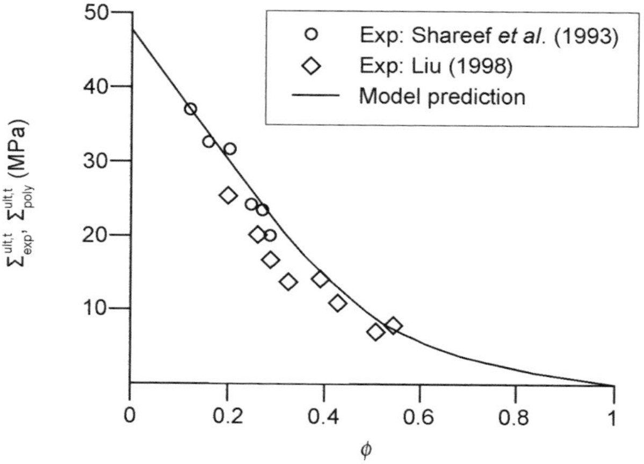

25.9 Comparison between model predictions [25.14]–[25.26] and experiments for tensile strength of different monoporosity biomaterials made of hydroxyapatite, as function of porosity ϕ.

increased until the sample fails. The tensile stress in the ring is calculated according to:

$$\Sigma_{exp}^{ult,t} = \frac{r_s p_i}{d_s} \qquad [25.39]$$

with r_s as the inner diameter of the ring, p_i as the internal pressure, and d_s as the wall thickness of the ring. Respective experimental results are documented by Shareef *et al.* (1993); see Tables 25.1 and 25.2 as well as Fig. 25.9.

25.4.7 Comparison between biomaterial-specific strength predictions and corresponding experiments

The strength values predicted by the homogenization scheme [25.19]–[25.26] (see Section 25.3 and Fig. 25.2) for biomaterial-specific porosities (Section 25.4.3, experimental set IIa) on the basis of biomaterial-independent ('universal') uniaxial tensile and compressive strengths of HA (Section 25.4.2, experimental set I) are compared with corresponding experimentally determined biomaterial-specific uniaxial tensile and compressive strength values from experimental set IIb-2 (Section 25.4.6).

Insertion of biomaterial-specific porosities (Table 25.2) into Eqs. [25.14]–[25.25] delivers together with $\sigma_{HA}^{ult,t}$ and $\sigma_{HA}^{ult,s}$ (Table 25.4) biomaterial-specific strength estimates for uniaxial tensile strength ($\Sigma_{poly}^{ult,t}$) and uniaxial compressive strength ($\Sigma_{poly}^{ult,c}$). These strength predictions are compared to corresponding experimental strength values (Figs 25.9 and 25.10). The satisfactory agreement between model predictions and experiments is quantified by prediction errors of $14 \pm 15\%$ (mean value $\pm$ standard deviation according to [25.36] and [25.37]

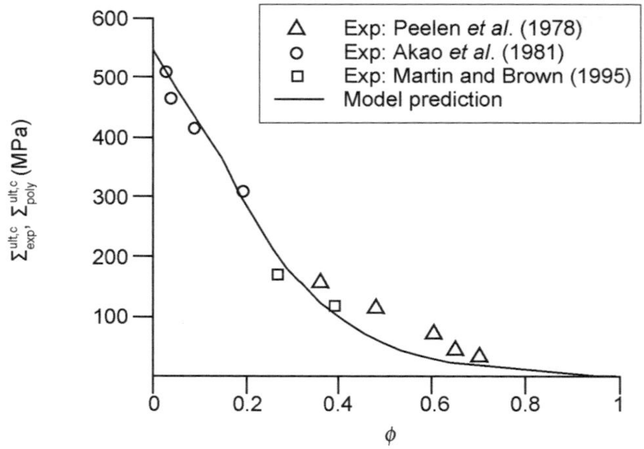

25.10 Comparison between model predictions [25.14]–[25.26] and experiments for compressive strength of different monoporosity biomaterials made of hydroxyapatite, as function of porosity ϕ.

with $q_{poly} = \Sigma_{poly}^{ult,t}$ or $\Sigma_{poly}^{ult,c}$, and with $q_{exp} = \Sigma_{exp}^{ult,t}$ or $\Sigma_{exp}^{ult,c}$) for uniaxial tensile strength, and of $-21 \pm 28\%$ for uniaxial compressive strength.

25.5 Continuum micromechanics model for 'hierarchical' hydroxyapatite biomaterials with two pore spaces used for tissue engineering

Classical HA materials, referred to throughout this chapter so far, exhibit one pore space with pore size in the micrometre range typically. These pores are often too small as to allow for activities of biological cell seeded within them – which, however, is a prerequisite for the biomaterial to act as tissue engineering scaffold, i.e. to provide initial stiffness and strength to large bone defect, before being resorbed and replaced by new bone. Therefore, cutting edge production technologies (e.g. foam methods, Mastrogiacomo *et al.*, 2006) allow for formation of large, 100–500 μm sized spherical pores (so-called macroporosity), as to host biological cells, e.g. bone marrow stromal cells. Still, the solid material between the spherical pores contains a porosity, with a typical pore size ranging from the sub-micrometre level to at most several micrometres. Such a double-porosity has important mechanical implications, which we tackle in this final section: we describe the introduction of more than one RVE for the representation of a hierarchically organized material (Section 25.5.1), which is followed by the elasticity (25.5.2) and the strength prediction (25.5.3) of such materials.

25.5.1 Multistep homogenization – micromechanical representation of double-porosity tissue engineering scaffolds made of hydroxyapatite

In a hierarchically organized material, a multistep homogenization approach is necessary in order to upscale the mechanical properties. If a single phase of an RVE, described in Section 25.2, exhibits a heterogeneous microstructure itself, its mechanical behaviour can be estimated by introduction of an RVE within this phase, with dimensions $l_2 \leq d$, comprising again smaller phases with characteristic length $d_2 \ll l_2$, and so on. This leads to a multistep homogenization scheme; see Fig. 25.11.

New-generation tissue engineering scaffolds made of hydroxyapatite will be modelled in the framework of a two-step homogenization scheme (Fig. 25.12). The first homogenization step (Fig. 25.12a) refers to several hundreds of micrometres characteristic length where micrometre-sized needle-shaped HA crystals and spherical (empty) micropores form a porous polycrystal. This first homogenization step is identical to the one described for elasticity and strength in Section 25.3, inclusive of the material properties of Table 25.4. The second homogenization step refers to an RVE of several millimetres characteristic length

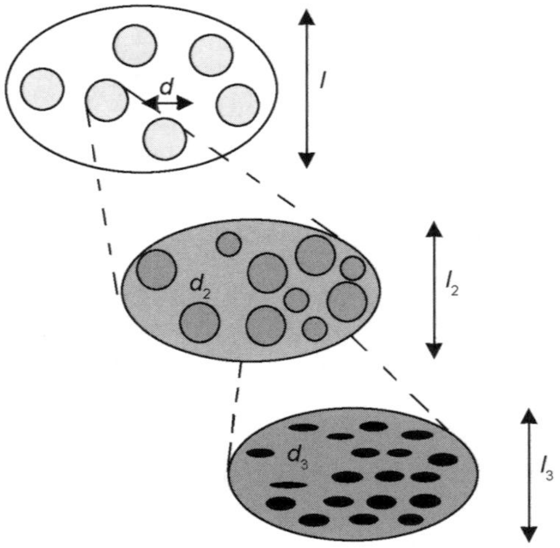

25.11 Multistep homogenization.

(Fig. 25.12b) where spherical macropores (with volume fraction Φ) are embedded into a contiguous solid phase (filling volume V_S, $V_S = V_{PORO}(1 - \Phi)$) which is built up by the (micro)porous polycrystal material from homogenization step 1.

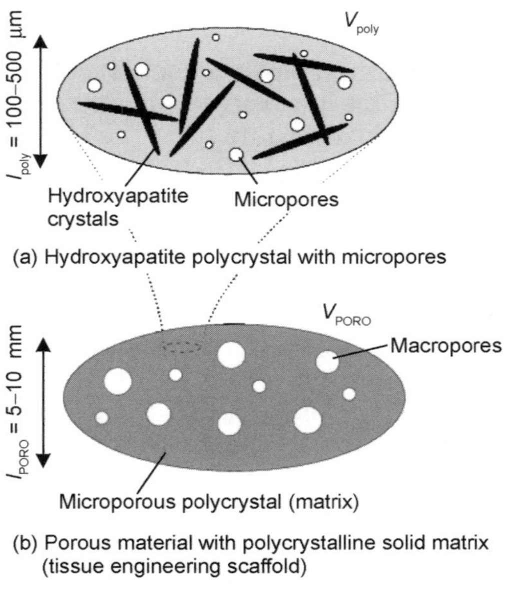

25.12 Micromechanical representation of a double-porosity biomaterial made up of hydroxyapatite, by means of a two-step homogenization procedure.

25.5.2 Elastic properties

The Mori–Tanaka homogenization scheme has been proven as suitable tool to upscale the elastic properties of the solid phase (k_{poly} and μ_{poly} defined through [25.14]–[25.15]) to the stiffness of such a porous material, see e.g. Dormieux (2005), Dormieux *et al.* (2006):

$$\mathbb{C}_{PORO} = (1-\Phi)\mathbb{C}_{poly} : ((1-\Phi)\mathbb{I} + \Phi(\mathbb{I} - \mathbb{S}^{esh}_{sph})^{-1})^{-1} \qquad [25.40]$$

with the Eshelby tensor $\mathbb{S}^{esh}_{sph}$ for spherical inclusions according to [25.16] so that

$$k_{PORO} = \frac{4k_{poly}\mu_{poly}(1-\Phi)}{3k_{poly}\Phi + 4\mu_{poly}} \qquad [25.41]$$

$$\mu_{PORO} = \mu_{poly}\frac{(1-\Phi)(9k_{poly} + 8\mu_{poly})}{9k_{poly}(1+\frac{2}{3}\Phi) + 8\mu_{poly}(1+\frac{3}{2}\Phi)} \qquad [25.42]$$

See Figs 25.13 and 25.14 for the elastic properties of double-porosity tissue-engineering scaffolds made of hydroxyapatite. According to [25.6], $\mathbb{C}_{PORO} = 3k_{PORO}\mathbb{J} + 2\mu_{PORO}\mathbb{K}$ links the stresses $\boldsymbol{\Sigma}_{PORO}$ related to an RVE of biomaterial with macropores (V_{PORO}) to the corresponding strain $\boldsymbol{E}_{PORO}$.

25.5.3 Strength properties

We approximate the homothetic criterion [25.25] (see also Section 25.7) of the polycrystalline solid matrix by means of a Drucker–Prager criterion (cone in the principal stress space), reading as:

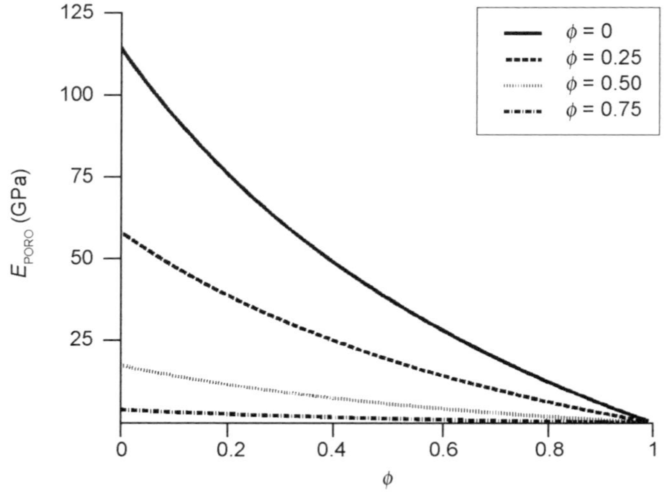

25.13 Model predictions for Young's modulus of a double-porosity tissue engineering scaffold made of hydroxyapatite, with microporosities ϕ and macroporosities Φ.

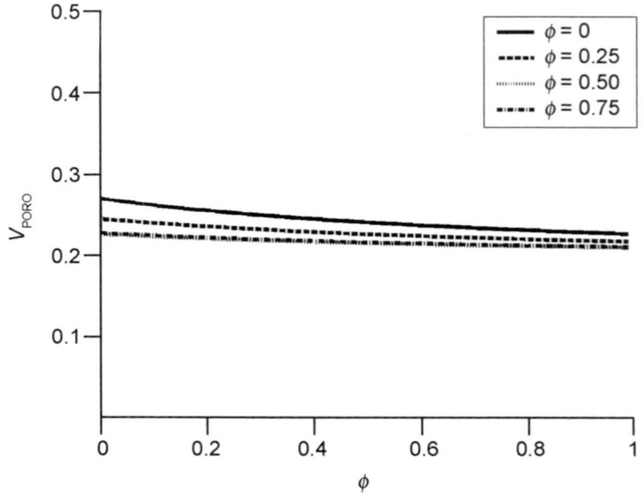

25.14 Model predictions for Poisson's ratio of a double-porosity tissue engineering scaffold made of hydroxyapatite, with microporosities ϕ and macroporosities Φ.

$$\alpha(\sigma_{poly,m} - h) + \sigma_{poly,d} \leq 0 \qquad [25.43]$$

In [25.43] $\sigma_{poly,m}$ and $\sigma_{poly,d}$ are the mean stress and equivalent deviatoric stress of the second-order stress tensor $\boldsymbol{\sigma}_{poly}$ (identical to $\boldsymbol{\Sigma}_{poly}$ in Section 25.3, but now in the role of a 'microscopic' phase stress):

$$\sigma_{poly,m} = \frac{1}{3} \operatorname{tr} \boldsymbol{\sigma}_{poly},$$

$$\sigma_{poly,d} = \sqrt{\frac{1}{2} \boldsymbol{\sigma}_{poly,d} : \boldsymbol{\sigma}_{poly,d}}$$

with

$$\boldsymbol{\sigma}_{poly,d} = \boldsymbol{\sigma}_{poly} - \sigma_{poly,m}\mathbf{1} \qquad [25.44]$$

and the friction angle α and the cohesion h are determined such that the microporosity-strength relations ($\sigma_{poly}^{ult,c}(\phi) \equiv \Sigma_{poly}^{ult,c}(\phi)$ and $\sigma_{poly}^{ult,t}(\phi) \equiv \Sigma_{poly}^{ult,t}(\phi)$) predicted by the homothetic criterion [25.25] (Figs 25.9 and 25.10) are exactly reproduced. Hence, they read as

$$\alpha(\phi) = \sqrt{3}\frac{\sigma_{poly}^{ult,c}(\phi) - \sigma_{poly}^{ult,t}(\phi)}{\sigma_{poly}^{ult,c}(\phi) + \sigma_{poly}^{ult,t}(\phi)} \qquad [25.45]$$

$$h(\phi) = \frac{2}{3}\frac{\sigma_{poly}^{ult,c}(\phi)\sigma_{poly}^{ult,t}(\phi)}{\sigma_{poly}^{ult,c}(\phi) - \sigma_{poly}^{ult,t}(\phi)} \qquad [25.46]$$

Again, we have used lower case letters for polycrystalline strength, as they refer to the 'microscopic' level of the second homogenization step.

We consider brittle failure of the overall porous medium if the polycrystal failure criterion [25.43] together with [25.44]–[25.46] is reached in highly stressed regions of the polycrystalline matrix. The corresponding ('micro'-) heterogeneity within the solid matrix has recently been shown (Dormieux *et al.*, 2002) to be reasonably considerable through so-called (homogeneous) *effective* ('micro'-) stresses, such as the square root of the spatial average over the solid material phase, of the squares of equivalent deviatoric ('micro'-)stresses:

$$\sqrt{\langle \sigma_d^2 \rangle_S} = \sqrt{\frac{1}{V_S} \int_{V_S} \frac{1}{2} \boldsymbol{\sigma}_d(\boldsymbol{x}) : \boldsymbol{\sigma}_d(\boldsymbol{x}) \, \mathrm{d}V} \qquad [25.47]$$

$$\text{with } \boldsymbol{\sigma}_d = \boldsymbol{\sigma} - \frac{1}{3} \operatorname{tr} \boldsymbol{\sigma}(\boldsymbol{x}) \, \mathbf{1} \qquad [25.48]$$

with V_S as the volume of the solid matrix between the macropores in RVE of Fig. 25.12(b).

Energy considerations (Dormieux *et al.*, 2002; Dormieux, 2005, p. 132) allow for relation of the effective deviatoric stresses [25.47], used to approximate $\sigma_{poly,d}$ in [25.43], to the macroscopic stresses $\boldsymbol{\Sigma}_{PORO}$ acting on the RVE of tissue engineering scaffold (Fig. 25.12b):

$$\sigma_{poly,d}^2 \approx \langle \sigma_d^2 \rangle_S = \left[-\frac{\partial}{\partial \mu_{poly}} \left(\frac{1}{k_{PORO}} \right) \Sigma_{PORO,m}^2 - \frac{\partial}{\partial \mu_{poly}} \left(\frac{1}{\mu_{PORO}} \right) \Sigma_{PORO,d}^2 \right] \frac{\mu_{poly}^2}{1 - \Phi} \qquad [25.49]$$

The effective mean stress level in the solid matrix is chosen as the stress average over the solid phase:

$$\sigma_{poly,m} \approx \langle \sigma_m \rangle_S = \frac{1}{V_S} \int_{V_S} \frac{1}{3} \operatorname{tr} \boldsymbol{\sigma}(\boldsymbol{x}) \, \mathrm{d}V = \frac{\Sigma_{PORO,m}}{1 - \Phi} \qquad [25.50]$$

Use of Eqs. [25.50] and [25.49], together with [25.41]–[25.48], in [25.43] yields a brittle failure criterion at the scale of the porous material with polycrystalline microporous matrix in the solid phase:

$$\begin{aligned} \mathscr{F}(\boldsymbol{\Sigma}_{PORO}) = {} & \left[\frac{3\Phi}{4} - \alpha(\phi)^2 \right] \Sigma_{PORO,m}^2 \\ & + \left[\frac{2\Phi(23 - 50\nu_{poly}(\phi) + 35\nu_{poly}^2(\phi))}{(-7 + 5\nu_{poly}(\phi))^2} + 1 \right] \Sigma_{PORO,d}^2 \\ & + 2[\alpha(\phi)]^2 h(\phi)(1 - \Phi)\Sigma_{PORO,m} - [\alpha(\phi)]^2 [h(\phi)]^2 (1 - \Phi)^2 = 0 \end{aligned} \qquad [25.51]$$

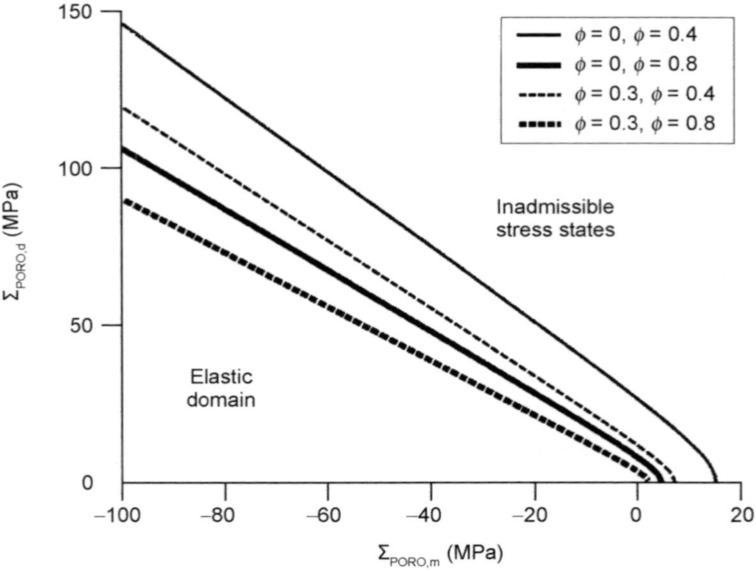

25.15 Elastic domain of a double-porosity tissue engineering scaffold made of hydroxyapatite, with microporosities ϕ and macroporosities Φ, Eq. [25.52]. Abscissa and ordinate correspond to the mean and the equivalent deviatoric stress, respectively.

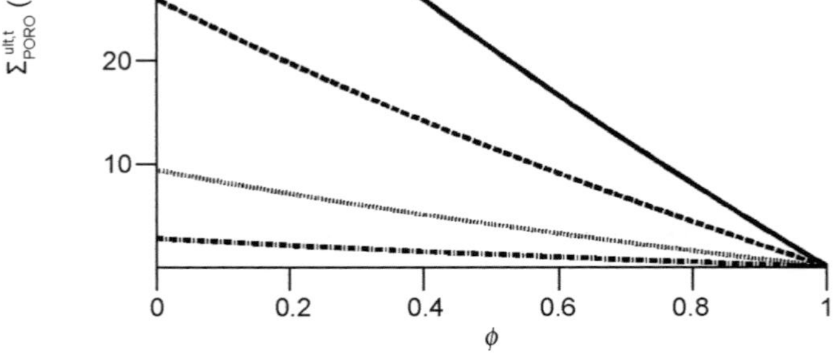

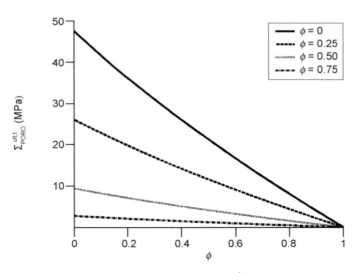

25.16 Uniaxial tensile strength prediction $\Sigma_{PORO}^{ult,t}$ of a double-porosity tissue engineering scaffold made of hydroxyapatite, with microporosities ϕ and macroporosities φ, Eq. [25.51] with $\Sigma_{PORO,m} = \Sigma_{PORO}^{ult,t}/3$, $\Sigma_{PORO,d} = \Sigma_{PORO}^{ult,t}/\sqrt{3}$.

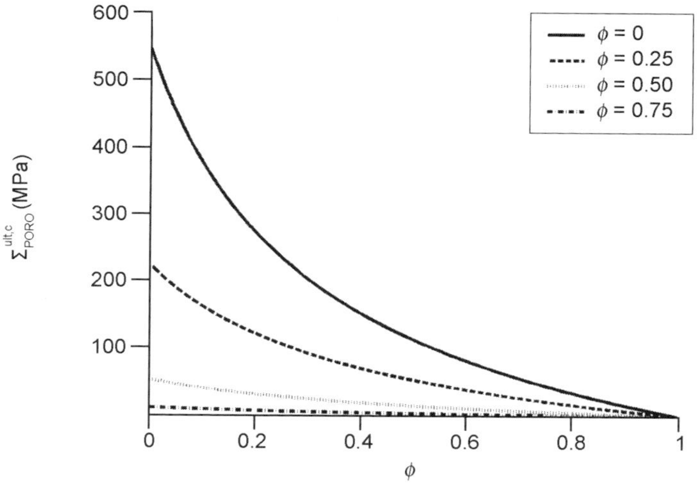

25.17 Uniaxial compressive strength prediction $\Sigma_{PORO}^{ult,t}$ of a double-porosity tissue engineering scaffold made of hydroxyapatite, with microporosities ϕ and macroporosities φ, Eq. [25.51] with $\Sigma_{PORO,m} = \Sigma_{PORO}^{ult,c}/3$, $\Sigma_{PORO,d} = \Sigma_{PORO}^{ult,c}/\sqrt{3}$.

The corresponding elastic domain $\mathscr{F}(\mathbf{\Sigma}_{PORO}) < 0$ depends on the micro/macroporosity contributions (Fig. 25.15), as do uniaxial tensile and compressive strengths of the double-porosity tissue engineering scaffold material (Figs 25.16 and 25.17).

25.6 Conclusions and future trends

We have developed a continuum micromechanical concept for elasticity and strength of mono-porosity biomaterials made of hydroxyapatite, which was verified with independent experimental sets. Extension to double-porosity tissue engineering scaffolds has been shown. We are optimistic that – once suitable data comprising morphological information as well as stiffness and strength are available – a thorough experimental validation will also reveal satisfying predictive capabilities of the double-porosity models.

Such models have the potential to considerably improve biomaterial design. Nowadays the latter is largely done in a trial-and-error procedure. Based on a number of mechanical and/or acoustical tests, new material design parameters are *guessed*. On the other hand, with well-validated micromechanics models, the mechanical implications of changes in the microstructure can be predicted so that minimization of material failure risk allows for optimization of key design parameters, such as porosities or geometries of microstructures. Hence, we believe that micromechanical theories can considerably speed up the future improvement of tissue engineering scaffolds.

25.7 Appendix: Homothetic ('cone-type') shape of failure criterion for hydroxyapatite biomaterials – Drucker–Prager approximation

In the macroscopic principal stress space (coordinates Σ_I, Σ_{II} and Σ_{III} the principal stresses of the macroscopic stress tensor $\boldsymbol{\Sigma} = \Sigma_I \boldsymbol{e}_I \otimes \boldsymbol{e}_I + \Sigma_{II} \boldsymbol{e}_{II} \otimes \boldsymbol{e}_{II} + \Sigma_{III} \boldsymbol{e}_{III} \otimes \boldsymbol{e}_{III}$, $\Sigma_{ij} = 0$ for $i \neq j$, see Eq. [25.5]), the domain of admissible stresses has a homothetic ('cone or pyramid-type') shape. This means that if a point P (related to stress state $\boldsymbol{\Sigma} = \mathbf{1}A + \boldsymbol{s}$) lies within the domain, also all the points P' lying on a ray from the apex A through P (related to stress states $\boldsymbol{\Sigma} = \mathbf{1}A + \Lambda \boldsymbol{s}$) are part of the domain (Fig. 25.4). $\Lambda > 0$ describes the dilation of stress states with respect to apex A.

To prove this, we first have to show that failure criterion [25.25] has a finite hydrostatic tensile limit representing the apex. Evaluation of concentration relation [25.20] for any hydrostatic macroscopic stress state, $\boldsymbol{\Sigma} = H\mathbf{1}$, yields vanishing needle shear stress (see [25.22]),

$$\sigma_{HA,Nn} = \boldsymbol{N} \cdot (\mathbb{B}(\boldsymbol{N}) : H\mathbf{1}) \cdot \boldsymbol{n} = 0 \qquad [25.52]$$

and non-zero needle normal stress (see [25.21]),

$$\sigma_{HA,NN} = \boldsymbol{N} \cdot (\mathbb{B}(\boldsymbol{N}) : H\mathbf{1}) \cdot \boldsymbol{N} \leq \sigma_{HA}^{ult,t} \qquad [25.53]$$

Use of [25.52] and [25.53] in failure criterion [25.24] and in corresponding elastic domain $f_{HA}(\boldsymbol{\sigma}) < 0$ yields:

$$H \leq (1-\phi)\sigma_{HA}^{ult,t} \frac{k_{HA}(\mu_{HA} + 3(k_{poly} + \mu_{poly}))}{(3k_{HA} - 2k_{poly})\mu_{HA} + 3k_{HA}(k_{poly} + \mu_{poly})} = A \qquad [25.54]$$

Considering $k_{HA} > k_{poly}$, we can assure that the fraction term in [25.54] is positive (and smaller than 1) so that a finite hydrostatic limit $A > 0$ according to [25.54] represents the apex in Fig. 25.4.

Secondly, we have to show that all macroscopic stress states (Σ_I, Σ_{II}, Σ_{III}) belonging to the elastic domain (inclusive its boundary) make up a homothety in the macroscopic principal stress space. At the apex, $\boldsymbol{\Sigma} = A\mathbf{1}$, criterion [25.25] reads

$$\sigma_{NN} = \boldsymbol{N} \cdot (\mathbb{B}(\boldsymbol{N}) : A\mathbf{1}) \cdot \boldsymbol{N} = \sigma_{HA}^{ult,t} \qquad [25.55]$$

We characterize a stress state lying in the elastic domain or on its boundary (i.e. a stress state being part of the homothety) by addition of a tensor $\boldsymbol{s}$ to the hydrostatic stress state of the apex, $\boldsymbol{\Sigma} = A\mathbf{1} + \boldsymbol{s} \in \mathscr{H}$ (see Fig. 25.4) so that the criteria [25.25] and [25.26] now read as

$$\max_{\vartheta}(\beta \max_{\psi} |\boldsymbol{N} \cdot (\mathbb{B}(\boldsymbol{N}) : (A\mathbf{1} + \boldsymbol{s})) \cdot \boldsymbol{n}| + (\boldsymbol{N} \cdot (\mathbb{B}(\boldsymbol{N}) : (A\mathbf{1} + \boldsymbol{s})) \cdot \boldsymbol{N})) \leq \sigma_{HA}^{ult,t} \qquad [25.56]$$

It has to be shown that any other stress state given by the dilation $\mathbf{\Sigma} = A\mathbf{1} + \Lambda \boldsymbol{s}$, $\Lambda > 0$, is also an element of the homothety $\mathscr{H}$, i.e. also fulfils criteria [25.25] and [25.26]. Insertion of [25.52] specified for $H = A$, and of [25.55] into [25.56] yields

$$\max_{\vartheta}(\beta \max_{\psi} |\boldsymbol{N} \cdot (\mathbb{B}(\boldsymbol{N}) : \boldsymbol{s}) \cdot \boldsymbol{n}| + (\boldsymbol{N} \cdot (\mathbb{B}(\boldsymbol{N}) : \boldsymbol{s}) \cdot \boldsymbol{N})) \leq 0 \qquad [25.57]$$

Multiplying [25.57] by $\Lambda > 0$ yields

$$\max_{\vartheta}(\beta \max_{\psi} |\boldsymbol{N} \cdot (\mathbb{B}(\boldsymbol{N}) : \Lambda\boldsymbol{s}) \cdot \boldsymbol{n}| + (\boldsymbol{N} \cdot (\mathbb{B}(\boldsymbol{N}) : \Lambda\boldsymbol{s}) \cdot \boldsymbol{N})) \leq 0 \qquad [25.58]$$

Using [25.55] and insertion of [25.52] specified for $A = H$, and of [25.55] into [25.58] finally yields

$$\max_{\vartheta}(\beta \max_{\psi} |\boldsymbol{N} \cdot (\mathbb{B}(\boldsymbol{N}) : (A\mathbf{1} + \Lambda\boldsymbol{s})) \cdot \boldsymbol{n}| + (\boldsymbol{N} \cdot (\mathbb{B}(\boldsymbol{N}) : (A\mathbf{1} + \Lambda\boldsymbol{s})) \cdot \boldsymbol{N})) \leq \sigma_{HA}^{ult,t} \qquad [25.59]$$

which is equivalent to saying that $\mathbf{\Sigma} = A\mathbf{1} + \Lambda\boldsymbol{s} \in \mathscr{H}$, i.e. that $\mathbf{\Sigma} = A\mathbf{1} + \Lambda\boldsymbol{s}$ fulfils criteria [25.25] and [25.26]. Hence, the domain of macroscopic admissible stress states obeys the properties of a homothety with apex A in the principal stress space (Fig. 25.4).

25.8 Nomenclature

A	macroscopic hydrostatic stress state related to the apex of failure criteria for monoporosity biomaterials
$\mathbb{A}_r$	fourth-order strain concentration tensor of phase r
$\mathbb{A}_r^{est}$	estimate of fourth-order strain concentration tensor of phase r
$\mathbb{B}_{HA}$	fourth-order stress concentration tensor for single HA crystals
b_s	side length of HA biomaterial sample
$\mathbb{C}$	fourth-order homogenized stiffness tensor
$\mathbb{C}^{est}$	estimate of fourth-order homogenized stiffness tensor
$\mathbb{C}^0$	fourth-order stiffness tensor of infinite matrix surrounding an ellipsoidal inclusion
$\mathbb{C}_{poly}$	fourth-order homogenized stiffness tensor of monoporosity biomaterial made of HA
$\mathbb{C}_{PORO}$	fourth-order homogenized stiffness tensor of a double-porosity tissue engineering scaffold made of HA
$\mathbb{c}_{HA}$	fourth-order stiffness tensor of single HA crystals within the RVE V_{poly}
$\mathbb{c}_r$	fourth-order stiffness tensor of phase r
d, d_2, d_3	characteristic length of inhomogeneities within an RVE
d_s	wall thickness of a ring-shaped HA biomaterial sample
$\boldsymbol{E}$	second-order 'macroscopic' strain tensor

E_{exp}	experimental Young's modulus of monoporosity biomaterial made of HA
E_{HA}	Young's modulus of single HA crystals within the RVE V_{poly}
E_{poly}	homogenized Young's modulus of monoporosity biomaterial made of HA
$\boldsymbol{E}_{PORO}$	second-order macroscopic strain tensor (related to RVE V_{PORO} of a double-porosity tissue engineering scaffold made of HA)
$\bar{e}$	mean of relative error between predictions and experiments
e_S	standard deviation of relative error between predictions and experiments
$\boldsymbol{e}_1, \boldsymbol{e}_2, \boldsymbol{e}_3$	unit base vectors of Cartesian reference base frame
$\boldsymbol{e}_\vartheta, \boldsymbol{e}_\varphi, \boldsymbol{e}_r$	unit base vectors of Cartesian local base frame of a single crystal
$\mathscr{F}(\boldsymbol{\Sigma})$	boundary of elastic domain in space of macrostresses
F_s	force applied to the centre of a beam in a three-point bending test
f	ultrasonic excitation frequency
f_r	volume fraction of phase r
$f_r(\boldsymbol{\sigma})$	boundary of elastic domain of phase r in space of microstresses
G	shear modulus
H	hydrostatic stress
$\mathscr{H}$	homothety
h	cohesion of solid phase within RVE V_{PORO} obeying to a Drucker–Prager criterion
h_s	height of a HA biomaterial sample
$\mathbb{I}$	fourth-order identity tensor
$\mathbb{J}$	volumetric part of fourth-order identity tensor $\mathbb{I}$
$\mathbb{K}$	deviatoric part of fourth-order identity tensor $\mathbb{I}$
k_{HA}	bulk modulus of single HA crystals within the RVE V_{poly}
k_{poly}	homogenized bulk modulus of monoporosity biomaterial made of HA
k_{PORO}	homogenized bulk modulus of a double-porosity tissue engineering scaffold made of HA
L	characteristic length of a structure containing an RVE
l_s	length of a HA biomaterial sample
l, l_2, l_3	characteristic lengths of RVEs
M	mass of a HA biomaterial sample
$\boldsymbol{N}$	orientation vector aligned with longitudinal axis of needle
N_r	number of phases within an RVE
$\boldsymbol{n}$	orientation vector perpendicular to $\boldsymbol{N}$
$\mathbb{P}_r^0$	fourth-order Hill tensor characterizing the interaction between the phase r and the matrix $\mathbb{C}^0$
$\mathbb{P}_{cyl}^{poly}$	fourth-order Hill tensor for cylindrical inclusion in matrix with stiffness $\mathbb{C}_{poly}$
$\mathbb{P}_{sph}^{poly}$	fourth-order Hill tensor for spherical inclusion in matrix with stiffness $\mathbb{C}_{poly}$

p_i	internal pressure applied to a ring-shaped HA biomaterial sample during a Stanford ring bursting test
RVE	representative volume element
r, s	index for phases
r_s	inner diameter of a ring-shaped HA biomaterial sample
$\mathbb{S}_r^{esh,0}$	fourth-order Eshelby tensor for phase r embedded in matrix $\mathbb{C}^0$
$\mathbb{S}_{cyl}^{esh}$	fourth-order Eshelby tensor for cylindrical inclusion embedded in isotropic matrix with stiffness $\mathbb{C}_{poly}$
$\mathbb{S}_{sph}^{esh}$	fourth-order Eshelby tensor for spherical inclusion embedded in isotropic matrix with stiffness $\mathbb{C}_{poly}$
$\boldsymbol{s}$	second-order stress tensor characterizing a stress state lying on the surface ('cone') limiting the elastic domain
t_s	transition time of an ultrasonic wave through a specimen
tr	trace of a second-order tensor
V	volume of a HA biomaterial sample
V_{poly}	volume of an RVE of monoporosity biomaterial made of HA
V_S	volume of the matrix between the macropores in RVE of a porous material with polycrystalline solid matrix
V_{PORO}	volume of an RVE of a double-porosity tissue engineering scaffold made of HA
V_{RVE}	volume of an RVE
v_{bar}	ultrasonic wave propagation velocity of a bar wave
v_L	ultrasonic wave propagation velocity of a bulk wave in longitudinal direction
ν_s	ultrasonic wave propagation velocity within a HA biomaterial sample
v_T	ultrasonic wave propagation velocity of a bulk wave in transversal direction
w	index denoting weakest phase
$\boldsymbol{x}$	position vector within an RVE
α	friction angle of solid phase within RVE V_{PORO} obeying to a Drucker–Prager criterion
β	ratio between uniaxial tensile strength and shear strength of pure HA
δ_{ij}	Kronecker delta (components of second-order identity tensor $\mathbf{1}$)
$\boldsymbol{\varepsilon}_{HA}$	second-order strain tensor field within single HA crystals
$\boldsymbol{\varepsilon}_r$	second-order strain tensor field of phase r
ϑ	latitudinal cordinate of spherical coordinate system
Λ	dilation factor for a stress state lying on the surface ('cone') limiting the elastic domain
λ	ultrasonic wave length
μ_{HA}	shear modulus of single HA crystals within the RVE V_{poly}
μ_{poly}	homogenized shear modulus of monoporosity biomaterial made of HA

μ_{PORO}	homogenized shear modulus of a double-porosity tissue engineering scaffold made of HA
ν_{exp}	experimental Poisson's ratio of monoporosity biomaterial made of HA
ν_{HA}	Poisson's ratio of single HA crystals within the RVE V_{poly}
ν_{poly}	homogenized Poisson's ratio of monoporosity biomaterial made of HA
ξ	displacements within an RVE and at its boundary
ρ	material mass density
ρ_{HA}	mass density of pure HA
ρ_s	mass density of an HA biomaterial sample
$\mathbf{\Sigma}$	second-order 'macroscopic' stress tensor
Σ_{I}, Σ_{II}, Σ_{III}	principal stresses of macroscopic stress tensor $\mathbf{\Sigma}$
$\Sigma_{poly}^{ult,t}$	predicted uniaxial tensile strength of monoporosity biomaterial made of HA
$\Sigma_{poly}^{ult,c}$	predicted uniaxial compressive strength of monoporosity biomaterial made of HA
$\Sigma_{exp}^{ult,t}$	experimental uniaxial tensile strength of monoporosity biomaterial made of HA
$\Sigma_{exp}^{ult,c}$	experimental uniaxial compressive strength of monoporosity biomaterial made of HA
$\mathbf{\Sigma}_{PORO}$	second-order macroscopic stress tensor (related to RVE V_{PORO} of a double-porosity tissue engineering scaffold made of HA)
$\Sigma_{PORO,m}$	'macroscopic' mean stress (related to RVE V_{PORO} of a double-porosity tissue engineering scaffold made of HA)
$\Sigma_{PORO,d}$	'macroscopic' equivalent deviatoric stress (related to RVE V_{PORO} of a double-porosity tissue engineering scaffold made of HA)
$\Sigma_{PORO}^{ult,t}$	predicted uniaxial tensile strength of a double-porosity tissue engineering scaffold made of HA
$\Sigma_{PORO}^{ult,c}$	predicted uniaxial compressive strength of a double-porosity tissue engineering scaffold made of HA
Σ_{ref}	component of uniaxial stress tensor $\mathbf{\Sigma}$ imposed on boundary of monoporosity biomaterial made of HA
$\boldsymbol{\sigma}_{HA}(\varphi,\vartheta)$	second-order stress tensor field within single HA crystals
$\sigma_{HA,NN}(\varphi,\vartheta)$	normal component of stress tensor $\boldsymbol{\sigma}(\varphi,\vartheta)$ in needle direction
$\sigma_{HA,Nn}(\varphi,\vartheta)$	shear component of stress tensor $\boldsymbol{\sigma}(\varphi,\vartheta)$ in planes orthogonal to the needle direction
$\sigma_{HA}^{ult,t}$	uniaxial tensile strength of pure HA
$\sigma_{HA}^{ult,s}$	shear strength of pure HA
$\boldsymbol{\sigma}_{poly}$	second-order 'macroscopic' stress tensor (related to RVE V_{poly} of monoporosity biomaterial made of HA)
$\sigma_{poly,m}$	'macroscopic' mean stress (related to RVE V_{poly} of monoporosity biomaterial made of HA)

$\sigma_{poly,d}$	'macroscopic' equivalent deviatoric stress (related to RVE V_{poly} of monoporosity biomaterial made of HA)
$\boldsymbol{\sigma}_r$	second-order stress tensor field of phase r
φ	longitudinal coordinate of spherical coordinate system
ϕ	volume fraction of micropores within RVE V_{poly}
Φ	volume fraction of macropores within RVE V_{PORO}
ψ	longitudinal coordinate of vector $\boldsymbol{n}$
ω	fundamental flexural resonant frequency
∂V	boundary of an RVE
$\mathbf{1}$	second-order identity tensor
$\langle(.)\rangle_V = 1/V \int_V (.)V$	average of quantity (.) over volume V
.	first-order tensor contraction
:	second-order tensor contraction
$\otimes$	dyadic product of tensors

25.9 References

Akao M, Aoki H, Kato K (1981), 'Mechanical properties of sintered hydroxyapatite for prosthetic applications', *Journal of Materials Science*, **16**, 809–812.

Arita I, Wilkinson D, Mondragón M, Castaño V (1995), 'Chemistry and sintering behaviour of thin hydroxyapatite ceramics with controlled porosity', *Biomaterials*, **16**, 403–408.

Ashman R B, Cowin S C, Van Buskirk W C, Rice J C (1984), 'A continuous wave technique for the measurement of the elastic properties of cortical bone', *Journal of Biomechanics*, **17**, 349–361.

Ashman R B, Corin J D, Turner C H (1987), 'Elastic properties of cancellous bone – Measurement by an ultrasonic technique', *Journal of Biomechanics*, **20**, 979–986.

Benveniste Y (1987), 'A new approach to the application of Mori-Tanaka's theory in composite materials', *Mechanics of Materials*, **6**, 147–157.

Carcione J M (2001), *Wave Fields in Real Media: Wave Propagation in Anisotropic, Anelastic and Porous Media, Handbook of Geophysical Exploration*, Vol. 31, Oxford, Pergamon.

Charrière E, Terrazzoni S, Pittet C, Mordasini P, Dutoit M, Lemaître J, Zysset P (2001), 'Mechanical characterization of brushite and hydroxyapatite cements', *Biomaterials*, **22** (21), 2937–2945.

Cowin S C (2003), 'A recasting of anisotropic poroelasticity in matrices of tensor components', *Transport in Porous Media*, **50**, 35–56.

De With G, van Dijk H, Hattu N, Prijs K (1981), 'Preparation, microstructure and mechanical properties of dense polycrystalline hydroxy apatite', *Journal of Materials Science*, **16**, 1592–1598.

Dormieux L (2005), 'Poroelasticity and strength of fully or partially saturated porous materials', in Dormieux L, Ulm F J (Eds), *Applied Micromechanics of Porous Media, CISM Vol. 480*, New York, Springer, pp. 109–152.

Dormieux L, Molinari A, Kondo D (2002), 'Micromechanical approach to the behavior of poroelastic materials', *Journal of the Mechanics and Physics of Solids*, **50**, 2203–2231.

Dormieux L, Kondo D, Ulm F-J (2006), *Microporomechanics*, Chichester, Wiley.

Dorozhkin S V, Epple M (2002), 'Biological and medical significance of calcium phosphates', *Angewandte Chemie International Edition*, **41**, 3130–3146.

Driessen A, Klein C, de Groot K (1982), 'Preparation and some properties of sintered β-whitlockite', Biomaterials, **3**, 113–116.

Eshelby J D (1957), 'The determination of the elastic field of an ellipsoidal inclusion, and related problems', *Proceedings of the Royal Society London, Series A*, **241**, 376–396.

Frame J W, Browne R M, Brady C L (1981), 'Hydroxyapatite as a bone substitute in the jaws', *Biomaterials*, **2** (1), 19–22.

Fritsch A, Dormieux L, Hellmich C (2006), 'Porous polycrystals built up by uniformly and axisymmetrically oriented needles: homogenization of elastic properties', *Comptes Rendus Mecanique*, **334** (3), 151–157.

Fritsch A, Hellmich C (2007), ' "Universal" microstructural patterns in cortical and trabecular, extracellular and extravascular bone materials: micromechanics-based prediction of anisotropic elasticity', *Journal of Theoretical Biology*, **244** (4), 597–620.

Gilmore R, Katz J L (1982), 'Elastic properties of apatites', *Journal of Materials Science*, **17**, 1131–1141.

Hashin Z (1983), 'Analysis of composite materials: a survey', *Journal of Applied Mechanics*, **29**, 143–150.

Hellmich C, Ulm F-J (2002), 'A micromechanical model for the ultrastructural stiffness of mineralized tissues', *Journal of Engineering Mechanics (ASCE)*, **128** (8), 898–908.

Hellmich C, Barthélémy J-F, Dormieux L (2004), 'Mineral–collagen interactions in elasticity of bone ultrastructure – a continuum micromechanics approach', *European Journal of Mechanics A – Solids*, **23**, 783–810.

Hench L L, Jones R J (Eds.) (2005), *Biomaterials, Artifical Organs and Tissue Engineering*, Cambridge, Woodhead Publishing.

Hershey A V (1954), 'The elasticity of an isotropic aggregate of anisotropic cubic crystals', *Journal of Applied Mechanics (ASME)*, **21**, 236–240.

Hill R (1963), 'Elastic properties of reinforced solids: some theoretical principles', *Journal of the Mechanics and Physics of Solids*, **11** (5), 357–362.

Hofstetter K, Hellmich C, Eberhardsteiner J (2005), 'Development and experimental validation of a continuum micromechanics model for the elasticity of wood', *European Journal of Mechanics A – Solids*, **24** (6), 1030–1053.

Hofstetter K, Hellmich C, Eberhardsteiner J (2006), 'The influence of the microfibril angle on wood stiffness: a continuum micromechanics approach', *Computer Assisted Mechanics and Engineering Sciences*, **13**, 523–536.

Jarcho M, Bolen C, Thomas M, Bobick J, Kay J, Doremus R (1976), 'Hydroxylapatite synthesis and characterisation in dense polycrystalline form', *Journal of Materials Science*, **11**, 2027–2035.

Katz J L, Harper R (1990), 'Calcium phosphates and apatites', in Williams D (Ed.), *Concise Encyclopedia of Medical and Dental Materials*, Oxford, Pergamon Press, pp. 87–95.

Katz J L, Ukraincik K (1971), 'On the anisotropic elastic properties of hydroxyapatite', *Journal of Biomechanics*, **4** (3), 221–227.

Kikuchi M, Ikoma T, Itoh S, Matsumoto H N, Koyama Y, Takakuda K, Shinomiya K, Tanaka J (2004), 'Biomimetic synthesis of bone-like nanocomposites using the self-organization mechanism of hydroxyapatite and collagen', *Composites Sciences and Technology*, **64** (6), 819–825.

Kolsky H (1953), *Stress Waves in Solids*, Oxford, Clarendon Press.

Kröner E (1958), 'Computation of the elastic constants of polycrystals from constants of single crystals', *Zeitschrift für Physik*, **151**, 504–518.

Liu D-M (1998), 'Preparation and characterisation of porous hydroxyapatite bioceramic via a slip-casting route', *Ceramics International*, **24**, 441–446.

Martin R I and Brown P W (1995), 'Mechanical properties of hydroxyapatite formed at physiological temperature', *Journal of Materials Science: Materials in Medicine*, **6**, 138–143.

Mastrogiacomo M, Scaglione S, Martinetti R, Dolcini L, Beltrame F, Cancedda R, Quarto R (2006), 'Role of scaffold internal structure on *in vivo* bone formation in macroporous calcium phosphate bioceramics', *Biomaterials*, **27** (17), 3230–3237.

Mayr E (1997), *This is Biology – The Science of the Living World*, Cambridge, MA, Harvard University Press.

Mori T, Tanaka K (1973), 'Average stress in matrix and average elastic energy of materials with misfitting inclusions', *Acta Metallurgica*, **21** (5), 571–574.

Müllner H W, Fritsch A, Kohlhauser C, Reihsner R, Hellmich C, Godlinski D, Rota A, Slesinski R, Eberhardsteiner J (2007), 'Acoustical and poromechanical characterization of titanium scaffolds for biomedical applications', *Strain*, accepted for publication.

Peelen J, Rejda B, de Groot K (1978), 'Preparation and properties of sintered hydroxylapatite', *Ceramurgia International*, **4** (2), 71–74.

Rao W, Boehm R (1974), 'A study of sintered apatites', *Journal of Dental Research*, **53** (6), 1351–1355.

Salençon J (2001), *Handbook of Continuum Mechanics: General Concepts, Thermoelasticity*, Berlin, Springer.

Shareef M, Messer P, Noort R (1993), 'Fabrication, characterization and fracture study of a machinable hydroxyapatite ceramic', *Biomaterials*, **14** (1), 69–75.

Suchanek W, Yoshimura M (1998), 'Processing and properties of hydroxyapatite-based biomaterials for use as hard tissue replacement implants', *Journal of Materials Research*, **13** (1), 94–117.

Suquet P (Ed.) (1997), *Continuum Micromechanics*, New York, Springer.

Verma D, Katti K, Katti D (2006), 'Bioactivity in *in situ* hydroxyapatite-polycaprolactone composites', *Journal of Biomedical Materials Research Part A*, **78A** (4), 772–780.

Zaoui A (2002), 'Continuum micromechanics: survey', *Journal of Engineering Mechanics (ASCE)*, **128** (8), 808–816.

26

Cartilage tissue engineering

J E GOUGH, University of Manchester, UK

26.1 Introduction

Cartilage has been a major focus of tissue engineering and materials for potential cartilage repair have been investigated for decades (Bray and Merrill, 1973; Klompmaker *et al.*, 1991). Although cartilage appears to be a simple tissue, containing only one cell type, it remains a major challenge for tissue engineers. This is due to its limited capacity for self-repair and its biomechanics (high resistance to compression, tension and shearing with some resilience and elasticity). These features contribute to the difficulty of integration of tissue-engineered constructs with the host tissue. Many researchers focus on an osteochondral approach to cartilage repair to improve integration with the subchondral bone. Although there are several types of cartilage found within the body (hyaline, fibrocartilage, elastic cartilage) this chapter will focus mainly on articular hyaline cartilage and will cover the structure of cartilage, the need for cartilage repair, cell sources, growth factors, effects of loading and some of the materials and scaffolds being developed for cartilage tissue engineering.

26.2 Structure, cellularity and extracellular matrix

Cartilage is generally an avascular tissue which has limited capacity for self-repair. Nutrition and transfer of metabolites is therefore via diffusion through the matrix and cells generally exist in a low-oxygen environment. Cartilage consists of chondrocytes embedded in a collagen and proteoglycan-rich extracellular matrix (ECM), with a high matrix : cell ratio. Chondrocytes are found within lacunae, either singly or in groups. The matrix directly surrounding the cells is called the pericellular matrix (PCM) and the cells together with this PCM have been termed the *chondron*. The PCM has been found to have a slightly different matrix composition from the rest of the ECM (Guilak *et al.*, 2006).

Chondrocytes have a prominent endoplasmic reticulum and Golgi apparatus due to their high level of macromolecule synthesis and secretion. Cartilage structure is not homogeneous and has three major zones, the superficial zone

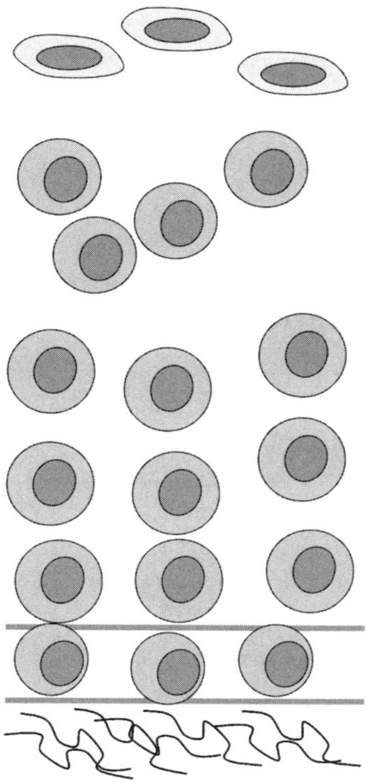

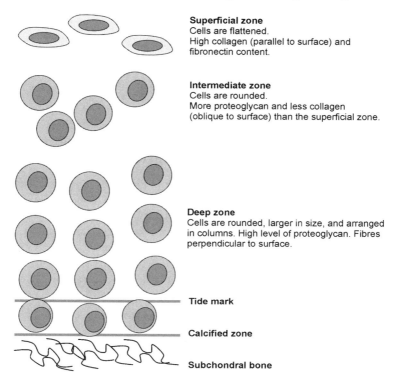

26.1 Cellularity of cartilage.

(surface) containing flattened cells, the intermediate zone containing rounded but randomly orientated cells and the deep zone containing rounded cells that appear to be organised in parallel columns, as shown in Fig. 26.1. Below this deep zone, a calcified zone is found adjacent to the subchondral bone. The composition of the ECM within these zones also varies. Apart from mature chondrocytes, a population of cartilage progenitor cells has been isolated from the surface zone of articular cartilage (Dowthwaite *et al.*, 2004).

The major macromolecules found within the matrix are collagen type II (90–95% of the total collagen content), proteoglycans such as aggrecan, hyaluronan and the sulphated glycosaminoglycans (GAGs) chondroitin sulphate and keratan sulphate. Other collagen types are also found such as types II, VI, IX, X and XI along with other proteoglycans such as decorin, biglycan and perlecan (Guilak *et al.*, 2006).

26.3 The need for cartilage repair

The most common disease affecting cartilage is osteoarthritis (OA) and sports injuries are common causes of trauma and damage. OA occurs as a result of wear and tear of the cartilage within joints, causing inflammation and is often associated with the ageing process. Symptoms of OA include chronic pain, stiffness and loss of mobility and it affects approximately 8 million people in the UK. Damage to the articular cartilage of the knee commonly occurs as a result of a sports injury and often occurs as partial or full thickness tears to the medial and lateral surfaces of the femur and tibia.

26.4 Current treatments including autologous chondrocyte transplantation

Total joint replacement, such as replacing the knee, can be used to treat advanced cases of OA that have not responded to non-surgical procedures (including physiotherapy and anti-inflammatory drugs). However this procedure is limited to more elderly patients due to the lifetime of the implant.

For damage to articular cartilage, debridement can be used to improve the contour of the damaged area by removing debris. This gives immediate short-term relief for small defects, but does not induce the formation of repair tissue unless the subchondral bone is penetrated, which allows mesenchymal cells to migrate to the damaged area and allow repair tissue to form. This procedure led to the development of ways to penetrate the subchondral bone, such as microfracture surgery and abrasion arthroplasty. Microfracture surgery involves the formation of holes into the subchondral bone, and abrasion arthroplasty uses an arthroscopic burring device to remove tissue below the damaged area and into the subchondral bone. Both of these techniques provide only short-term relief and the repair tissue that forms tends to be fibrocartilage, rich in collagen type I and of lower mechanical compressive strength, therefore resembling scar tissue.

26.4.1 Autologous chondrocyte transplantation (ACT)

ACT is available for lesions of articular cartilage in young healthy adults with no evidence of osteoarthritis. Brittberg *et al.* (1994) originally performed this procedure in 16 patients, summarised in Fig. 26.2. Biopsies of cartilage were taken of 300–500 mg and chondrocytes were isolated enzymatically and grown in the patient's own serum. Approximately 2.6–5 million cells in 50–100 μl were injected into the cartilage lesion which was then covered with a periosteal flap taken from the patient's proximal medial tibia. Genzyme Corporation supply autologous chondrocytes as an FDA-approved product known as Carticel® which is a 12 million autologous cell pellet. BioTissue Technologies now offer an ACT service using a biodegradable polymer fleece with a resorbable gel component which delivers the cells into the pores of the fleece (Kaps *et al.*, 2004; Barnewitz *et al.*, 2006), therefore adopting the tissue engineering approach to cartilage repair. Genzyme is currently conducting preclinical studies of their next-generation Carticel, which may require only minimally invasive or

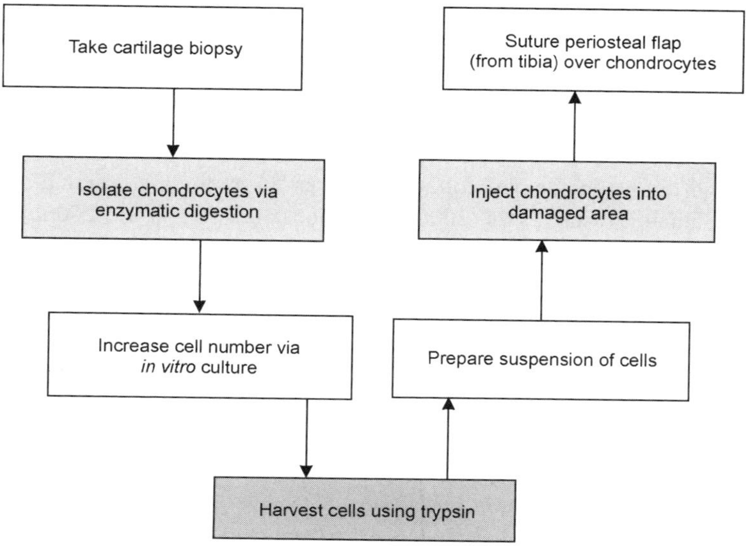

26.2 Method of autologous chondrocyte transplantation, based on Brittberg *et al.* (1994).

arthroscopic surgery, which uses a fleece matrix as a support for the autologous cells (Carticel II®).

26.5 Cell source

For cartilage repair strategies using a tissue engineering approach, the choice of cell type is important from a research point of view, and essential from a clinical point of view. For research purposes the choice of cell type must be appropriate: if animal cells other than human cells are used then there must be confidence that the results gained reflect those of human cells. It is generally considered that the patient's own cells will be used for clinical tissue engineering. There are therefore two choices of cell types: the patient's cartilage cells can be used for expansion and seeding into the tissue construct, or the patient's own stem cells could be used. Unlike osteoblasts, chondrocytes are not highly motile cells and so are unlikely to migrate from the native tissue to populate a scaffold. As a cell source for cartilage tissue engineering, human mesenchymal stem cells from bone marrow have received much research attention due to their capacity to differentiate into chondrocytes. Other sources of stem cells for cartilage have also been investigated such as adipose tissue (Awad *et al.*, 2004; Betre *et al.*, 2006). Human adipose-derived adult stem cells (hADAS) have been cultured in an elastin-like polypeptide (ELP, properties discussed below). hADAS were cultured with or without TGFβ, dexamethasone and in low oxygen tension (5%) (Wang *et al.*, 2005). Regardless of culture conditions, cells were found to increase their hyaline-like matrix synthesis (collagen II and GAG) and decrease proliferation. The effect of low oxygen tension is discussed later in this chapter. Agarose, alginate (gels) and gelatin (porous scaffold) have also been used to study cartilage-like tissue formation by hADAS cells (Awad *et al.*, 2004). TGFβ1 caused increased protein and proteoglycan synthesis with specifically chondroitin sulphate and collagen II observed by immunostaining.

Human (Hwang *et al.*, 2006a) and mouse (Hwang *et al.*, 2006b) embryonic stem cells have also been investigated for potential cartilage repair strategies. Poly(ethylene glycol)-diacrylate (PEGDA) hydrogels modified with RGD (arginine, glycine, aspartate) in medium containing TGFβ1 were used to culture human ES cells. RGD modification resulted in an increase in Safranin O staining and collagens I, II and X. Mouse ES cells were cultured in PEGDA hydrogels with and without addition of glucosamine over 21 days. A dose-dependent effect was observed with 2 mM glucosamine causing increased aggrecan mRNA expression, collagen types I, II and X immunostaining and GAG production demonstrated by Safranin O staining.

Various *in vivo* animal models have been, and are constantly used to attempt to evaluate scaffolds and strategies for cartilage repair. Many of these *in vivo* studies have been inconclusive or have demonstrated poor results, mainly failing due to poor integration with the native cartilage tissue or by forming fibrous cartilage or scar tissue. A variety of non-human animal cells have also been used in *in vitro* models. Giannoni *et al.* (2005) have studied the inter-species variability in the response of chondrocytes cultured *in vitro* during expansion. Chondrocytes isolated from sheep, dog or human cartilage were expanded for 1, 6 or 12 cell duplications. Cell phenotype was monitored and redifferentiation capacity was analysed using micromass pellet assay and an *in vivo* assay of ectopic cartilage formation in a mouse model. Sheep chondrocytes were found to maintain their phenotype better than that of human chondrocytes, with dog chondrocytes being least able to maintain their phenotype. After expansion, sheep chondrocytes were observed to spontaneously reform hyaline-like cartilage whereas human chondrocytes redifferentiated only under stimulation with chondrogenic inducers (dexamethasone and TGFβ1) and dog chondrocytes lost their capacity to redifferentiate even in the presence of chondrogenic inducers. This study and the varying success of *in vivo* models highlights the need for care in cell sourcing and animal model choice for cartilage tissue engineering.

26.6 Materials

A wide range of materials, both natural and synthetic, have been investigated for potential cartilage repair, some of which are summarised in Tables 26.1–26.4 (these summary tables are not intended to be exhaustive lists but contain examples of some of the materials investigated with some appropriate

Table 26.1 Synthetic materials used for cartilage tissue engineering

Material	Structure	Additional molecules	Cell type	References
PLA	Porous scaffold		Sheep chondrocytes	Gugala and Gogolewski (2000)
PGA	Non-woven scaffold		Bovine chondrocytes	Ma and Langer (1999)
PLGA	Porous scaffold		Rabbit MSCs/rabbit cartilage defect	Uematsu *et al.* (2005)
PLGA	Microspheres		Bovine chondrocytes, rabbit chondrocytes/athymic mouse model	Mercier *et al.* (2005), Kang *et al.* (2005)
Polyglactin/ polydioxanone			Human chondrocytes/nude mice, horse chondrocytes/horse defect	Kaps *et al.* (2004), Barnewitz *et al.* (2006)
Poly(hydroxybutyrate-co-hydroxyhexanoate)	Porous scaffold		Rabbit chondrocytes	Deng *et al.* (2002, 2003)
Photocrosslinked poly(ethylene oxide)	Hydrogel		Bovine chondrocytes	Bryant and Anseth (2001)
PCL	Nanofibrous scaffold	TGF-β1	Human MSCs	Li *et al.* (2005)
PEMA/THFMA	Foamed scaffold		Bovine chondrocytes	Barry *et al.* (2004)
p(NiPAAm-co-AAc)	Hydrogel	TGF-β3, ascorbate, dexamethasone	Rabbit chondrocytes	Na *et al.* (2006)
poly (NIPAAm-co-BMA)/PEG	Thermo-reversible copolymer hydrogel	IGF-1, TGF-β1	Bovine chondrocytes	Yasuda *et al.* (2006)
PEG	Hydrogel, photo-polymerisable hydrogel	Glucosamine, TGF-β	Murine ES cell, goat MSCs	Hwang *et al.* (2006a,b), Williams *et al.* (2003)
PEG fumarate gelatin		TGF-β1	Bovine chondrocytes	Park *et al.* (2004, 2005)
PEGT/PBT	Porous scaffold, fibrous scaffold, porous scaffold		Bovine chondrocytes/nude mice, bovine and human chondrocytes, bovine chondrocytes	Malda *et al.* (2005), Woodfield *et al.* (2004), Mahmood *et al.* (2005)

Table 26.2 Natural materials used for cartilage tissue engineering

Material	Structure	Additional molecules	Cell type	References
Alginate	Hydrogel		Bovine chondrocytes	Yoon *et al.* (2007), Stevens *et al.* (2004)
Alginate/fibrin		None, TGF-β1	Rabbit chondrocytes/rabbit defect, Human chondrocytes	Perka *et al.* (2000a,b)
Chitosan, hyalruonan	Fibrous sweets		Rabbit chondrocytes	Yamane *et al.* (2005)
Cellulose	Non-woven fabric		Bovine chondrocytes	Muller *et al.* (2006)
Collagen Type II	Porous scaffolf Electrospun scaffold		Bovine chondrocytes, Human chondrocyte cell line	Pieper *et al.* (2002) Shields *et al.* (2004)
Collagen I	Porous sponge, porous scaffold, Porous scaffold	 +/− Chondroitin sulphate bFGF	Bovine chondrocytes, Bovine chondrocytes Rat chondrocytes/nude mice (subcutaneous)	Mizuno *et al.* (2001) van Susante *et al.* (2001) Fujisato *et al.* (1996)
Collagen I/II	Porous scaffold	+/− Chondroitin sulphate	Rabbit defect, canine chondrocytes	Buma *et al.* (2003) Nehrer *et al.* (1997) Lee *et al.* (2000)
Collagen I/II/III	Porous scaffold		Ovine MSCs, sheep defect	Dorotka *et al.* (2005)
Collagen I, hyaluronan	Porous sponge		Bovine chondrocytes	Alleman *et al.* (2001)
Gelatin, chondroitin, hyaluronan	Porous scaffold		Porcine chondrocytes	Chang *et al.* (2003)
Hyaluronan	Non-woven mesh Non-woven mesh Non-woven mesh Non-woven mesh	 TGF-β1, bFGF TGF-β1 TGF-β1, EGF, bFGF	Rabbit chondrocytes/rabbit defect Human chondrocytes Human MSCs Chick embryo chondrocytes	Grigolo *et al.* (2001) Grigolo *et al.* (2002) Lisignoli *et al.* (2005) Girotto *et al.* (2003)
Fibrin	Hydrogel, injectable gel		Bovine chondrocytes, *in vivo* mouse and swine	Eyrich *et al.* (2007) Peretti *et al.* (2006)
Silk (*Bombyx mori*)	Porous scaffold	TGF-β1. bFGF	Human chondrocytes, human MSCs	Wang *et al.* (2006) Hofmann *et al.* (2006)

Table 26.3 Natural and synthetic blends for cartilage tissue engineering

Natural	Synthetic	Structure	Additional molecules	Cell type	References
Alginate	PLA	Gel within porous PLA	TGF-β1	Human MSCs	Caterson *et al.* (2001), Wayne *et al.* (2005)
Collagen	PEG				Taguchi *et al.* (2005)
Hyaluronan	PLGA			Chondrocytes	Yoo *et al.* (2005)
Chondroitin, Hyaluronate	PLGA, gelatin			Rabbit MSCs	Fan *et al.* (2006)
Fibrin glue	PCL		TGF-β1	*In vivo* rabbit model	Huang *et al.* (2002)
RGD-modified	Poly(ethylene glycol)-diacrylate	Hydrogels	+/− hyaluronic acid, collagen I	Human embryonic stem cells	Hwang *et al.* (2006a)
Gelatin	pNIPAAm	*In situ* formable scaffold		*In vivo*	Ibusuki *et al.* (2003)

Table 26.4 Extracellular matrix mimics for cartilage tissue engineering

Material	Structure	Cell type	References
Elastin-like polypeptide	Hydrogel, injectable enzyme crosslinked gel	Human adipose stem cells, porcine chondrocytes	Betre *et al.* (2002, 2006), McHale *et al.* (2005)
Lys-Leu-Asp 12	Hydrogel	Bovine chondrocytes	Kisiday *et al.* (2002, 2004)
A range of Fmoc-dipeptides	Hydrogels	Bovine chondrocytes	Jayawarna *et al.* (2006)
Collagen mimetic peptide	Hydrogel	Bovine chondrocytes	Lee *et al.* (2000)
PEG-based with MMP sensitive peptides	Hydrogel	Bovine chondrocytes	Park *et al.* (2004)

references). Owing to the wide range of materials investigated, only examples that have been made into 3D scaffolds have been included.

26.6.1 Synthetic materials

Many synthetic materials, mostly polymers, have been investigated to act as templates for cartilage regeneration and examples are shown in Table 26.1. These include the commonly investigated FDA-approved polymers poly(lactic acid) (PLA), poly(glycolic acid) (PGA) and polycaprolactone (PCL) and have been formed into either porous, fibrous or hydrogel scaffolds. Properties of many of these polymers have been reviewed by Hutmacher (2000).

26.6.2 Natural materials

In recent years there has been a push to investigate natural materials for tissue engineering, especially those natural polymers that are present in the body and some examples are shown in Table 26.2. Much research has focused on the biomolecules found in cartilage such as collagen. In particular, collagen types I and II have been investigated with the conclusion made that collagen type II tends to elicit a more favourable chondrocyte response and matrix formation. This is not surprising since collagen type II is the major collagen in cartilage (Nehrer *et al.*, 1997; Veilleux *et al.*, 2004).

26.6.3 Extracellular matrix mimics

The ideal material for tissue engineering any tissue or organ of the body will have properties that mimic those of the native tissue. Instead of using bulk synthetic or natural materials, there is a recent drive towards the use of materials that more closely mimic ECM by engineering specific fibre sizes, porosity and hydration, gelation and mechanical properties. Lee *et al.* (2006) have reported a collagen mimetic peptide hydrogel (with PEG) for cartilage tissue engineering. The collagen mimetic peptide Pro-Hyp-Gly forms a triple helical conformation, resembling that of natural collagen. A photopolymerisable derivative was copolymerised with poly(ethylene oxide) diacrylate to create a novel hydrogel. Encapsulated chondrocytes were able to synthesise a GAG and collagen-rich ECM. A genetically engineered elastin-like polypeptide (ELP) has also been developed using recombinant DNA techniques, composed of a pentapeptide repeat of Val-Pro-Gly-Xaa-Gly (where Xaa is any amino acid except Pro) (Betre *et al.*, 2002). This ELP forms an aqueous solution which when raised above its transition temperature (35°C), precipitates and forms a gelatinous structure. Porcine chondrocytes were encapsulated during gelation and were observed to maintain their rounded morphology and produce a GAG and collagen containing ECM.

The ECM is constantly remodelled by the cells within it and therefore for materials to truly mimic it, such materials must be susceptible to breakdown by cell-derived proteases and remodelling. Park *et al.* (2004) have incorporated matrix metalloproteinase (MMP) sensitive linkers within PEG hydrogels which resulted in diffuse matrix production by bovine chondrocytes, and increase in the gene expression levels of aggrecan and collagen II.

Recently the area of self-assembly has gained much interest in the tissue engineering field, including that of cartilage tissue engineering. Self-assembly is a process that occurs spontaneously in nature during, for example, the assembly of proteins and lipid bilayers. Recently this process has been exploited to form self-assembling peptide hydrogels from simple amino acids (Holmes *et al.*, 2000; Holmes, 2002; Silva *et al.*, 2004). The range in properties of the amino acids can be exploited to design rationally designed scaffolds for tissue repair, developing a range of scaffolds with specific properties. Kisiday *et al.* (2002, 2004) reported on a self-assembling peptide hydrogel for potential cartilage repair. The hydrogel was composed of lysine (K), leucine (L) and aspartic acid (D) and had the sequence AcN-KLDLKLDLKLDL-CNH_2 (referred to as KLD-12) which contains alternating hydrophobic and hydrophilic residues that promotes β-sheet formation. Bovine chondrocytes were encapsulated within the gels and were found to maintain the rounded chondrocyte morphology and to synthesise proteoglycans. GAG and collagen II staining revealed presence in the gel, especially within the pericellular matrix region surrounding the cells. ECM synthesis was accompanied by an increase in mechanical properties over time where equilibrium modulus and dynamic stiffness reached approximately 1/5 to 1/3 of native cartilage tissue by the 28th day (see Fig. 26.3).

Jayawarna *et al.* (2006) have developed a novel self-assembling dipeptide hydrogel for potential cartilage repair, which supports morphology and

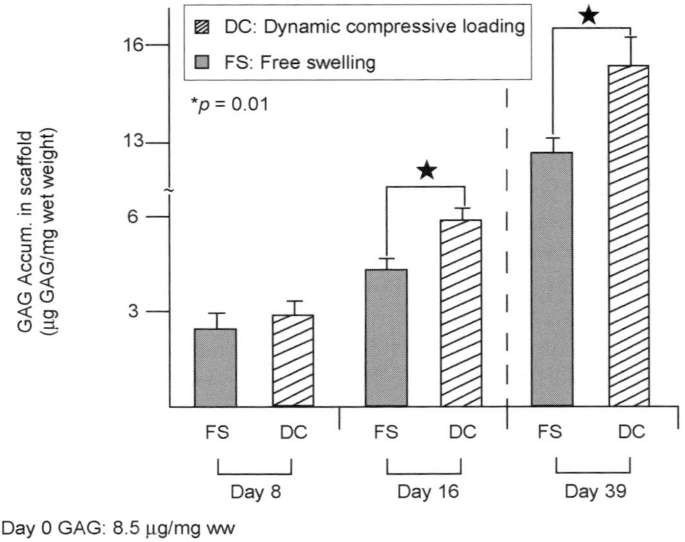

26.3 Chondrocyte biosynthesis in loaded self-assembled peptide scaffolds. GAG accumulation in chondrocyte-seeded peptide hydrogels subjected to alternate day loading or maintained in free-swelling conditions. By day 16 and onwards, GAG accumulation in compressed gels was significantly greater than the unloaded samples. Figure taken with permission from *Journal of Biomechanics*, **37**, Kisiday J.D. *et al.*, Effects of dynamic compressive loading on chondrocyte biosynthesis in self-assembling peptide scaffolds, pp 595–604, Copyright (2004), with permission from Elsevier.

proliferation of encapsulated bovine chondrocytes (as shown in Fig. 26.4). Dipeptides were linked to Fmoc (fluorenylmethoxycarbonyl) which has been used extensively in the self-assembly process due to interactions between π electrons of the aromatic fluorenyl rings which causes π stacking. Seven Fmoc-dipeptides were examined containing combinations of glycine (Gly), alanine (Ala), leucine (Leu), lysine (Lys) and phenylalanine (Phe). These gels had varying hydrophobicity and fibre diameter, which ranged from approximately 20 to 70 nm determined by cryo-scanning electron microscopy. Three gels were chosen for cell studies based on their ability to form stable gels at physiological pH. Chondrocytes were found to increase in number over 7 days and maintain the rounded chondrocyte morphology when encapsulated within Phe-Phe, Gly-Gly/Phe-Phe and Lys/Phe-Phe.

26.7 Growth factors and oxygen

26.7.1 Growth factors

The inclusion of growth factors into tissue engineering scaffolds or culture medium is a huge area of research due to their ubiquitous roles within the body

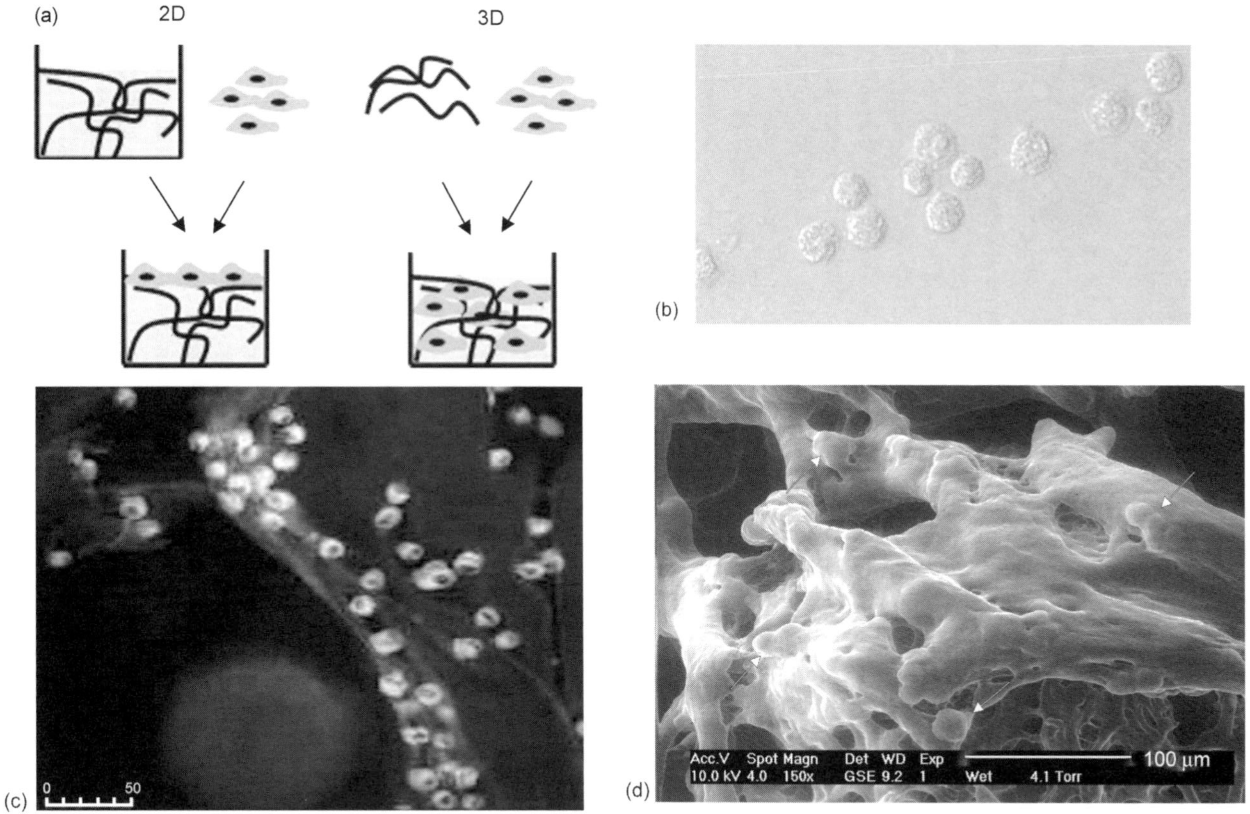

26.4 Dipeptide hydrogels for chondrocyte culture: (a) diagram showing 2D and 3D cell culture; (b) light micrograph demonstrating retention of rounded chondrocyte morphology on Phe-Phe gel surface; (c) two-photon fluorescence micrograph showing cell colonisation within the 3D Phe-Phe hydrogel structure; (d) ESEM image showing Gly-Gly/Phe-Phe gel structure and presence of chondrocytes (arrows). Scale bar is 50 μm. Image taken with permission from Jayawarna *et al.* (2006).

in controlling cell proliferation, differentiation and survival. Many growth factors have been analysed with respect to cartilage tissue engineering including bone morphogenetic proteins (BMPs), some of which are known to affect the chondrocyte phenotype, insulin-like growth factor 1 (IGF-1) and transforming growth factor-β1 (TGF-β1),

Blunk *et al.* (2002) investigated the effects of IGF-1, TGF-β1 and the cytokine interleukin 4 (IL-4) on bovine articular chondrocytes cultured over 4 weeks on polyglycolic acid (PGA) scaffolds (non-woven mesh of 13 μm diameter fibres). Scaffold wet weight and ECM components were increased by addition of IGF-1 (10–300 ng/ml, caused increase in GAG), IL-4 and TGF-β1 (30 ng/ml, caused increase in collagen). In a follow-on study, Gooch *et al.* (2002) investigated the effects of BMPs 2, 12 and 13 on bovine articular chondrocytes cultured over 4 weeks on the same PGA scaffolds. Addition of 100 ng/ml of the BMPs caused an increase in mass of the scaffolds and an increase in GAG content but not total collagen. Interestingly BMP-2 was found to promote chondrocyte hypertrophy, suggesting it affects chondrocyte maturation.

In 2% agarose gels and after a pre-culture period of 2 days, Mauck *et al.* (2003) found an increase in proteoglycan and collagen synthesis by bovine chondrocytes and an increase in the equilibrium aggregate modulus of the gels on addition of 10 ng/ml TGF-β1 and 100 ng/ml IGF-1. Additional loading of these gels had further effects as discussed below. Miyanishi *et al.* (2006a) investigated the effects of 10 ng/ml TGF-β3 on hMSCs over 14 days. This resulted in an over 10-fold increase in expression of SOX9, type II collagen and aggrecan.

26.7.2 Oxygen

Since cartilage is generally avascular, chondrocytes exist in a low oxygen tension environment. Research has shown that matrix synthesis and gene expression can be altered by changing the oxygen tension. Generally ‘low oxygen tension’ refers to 5% rather than the usual 20% O_2. *h*ADAS cells were suspended in alginate beads and cultured in control or chondrogenic media (supplemented with ascorbic acid, dexamethasone, ITS, TGF-β1), in either low oxygen (5%) or atmospheric oxygen tension (20%) for up to 14 days. Chondrogenic conditions and low oxygen tension inhibited the proliferation of *h*ADAS cells, but caused an increase in the rate of protein, total collagen synthesis and GAG synthesis. Immunohistochemistry showed presence of collagen type II and chondroitin-4-sulphate (Wang *et al.*, 2005). Culture in low oxygen has been found to cause redifferentiation (increase in GAGs and collagen II) of dedifferentiated chondrocytes compared to atmospheric oxygen tension (Murphy and Sambanis, 2001; Domm *et al.*, 2004). Similar results have been found by Kurz *et al.* (2004), who demonstrated that chondrocytes cultured on collagen VIII membranes in low oxygen showed an increase in collagen II expression and proteoglycan synthesis and a decrease in collagen I. This response was found to be further enhanced by culture in alginate.

26.8 Loading

Cartilage is a connective tissue under a variety of mechanical forces, both in the adult and during development. To obtain cartilage-like tissue that mimics native cartilage tissue with regard to extracellular matrix composition and arrangement, the application of appropriate mechanical forces is thought to be essential. As chondrocytes are known to respond to mechanical forces, the application of these forces during tissue engineering strategies for cartilage have been investigated.

Kisiday *et al.* (2004) have investigated the effects of dynamic compressive loading of chondrocytes in their KLD-12 self-assembled peptide scaffolds. Alternate day loading (compared with continuous loading) was found to stimulate proteoglycan synthesis (Fig. 26.3), which also resulted in an increase in the equilibrium and dynamic compressive stiffness of the gels. Sinusoidal dynamic compression was applied with 2.5% strain amplitude superimposed on 5% static offset strain at a frequency of 1.0 Hz which was performed non-continuously in cycles of 30 m–1 h compression followed by 30 m–7 h free swelling. Although proteoglycan synthesis was increased, interestingly the alternate day loading did not have an effect on total collagen synthesis, therefore suggesting that the conditions preferentially upregulated synthesis of molecules that are involved in compression resistance (proteoglycans).

It is likely that many extracellular stimuli will have synergistic effects on the chondrocyte phenotype as cells within the body are under the influence of a plethora of factors, including cytokines, growth factors, mechanical forces and fluid flow. Mauck *et al.* (2003) studied the effects of TGF-β1 (10 ng/ml) or IGF-1 (100 ng/ml) in conjunction with dynamic deformational loading on bovine chondrocytes cultured in 2% agarose gels, after a pre-culture period of 2 days. Loading (10% peak to peak strain at 1 Hz frequency, 3 h/day, 5 days/week, 1 h loading, 1 h rest) was found to cause an increase in proteoglycan and collagen content and in the equilibrium aggregate modulus. In conjunction with addition of TGF-β1 and IGF-1, proteoglycan and collagen was found to further increase, as did the equilibrium modulus. Loading also caused collagen II to be distributed throughout the gel rather than restricted to the pericellular area in the unloaded control.

Miyanishi *et al.* (2006a) investigated the effects of intermittent hydrostatic pressure (IHP) (10 MPa, 1 Hz, 4 h per day) on adult human MSCs over 3, 7 and 14 days culture. The effects were tested in conjunction with TGF-β3 (10 ng/ml). Expression of *Sox-9*, *aggrecan* and *collagen II* genes were determined using quantitative real-time polymerase chain reaction (PCR). Addition of TGF-β3 caused an increase in expression of these three genes as mentioned above but additional IHP caused a further increase. Increases were also noted in IHP-only treated samples. Increased deposition of collagen II and aggrecan-rich ECM was also noted in the individually treated samples and those treated with both TGF-β3 and IHP. In a further study Miyanishi *et al.* (2006b) investigated the effects of varying the IHP (0.1, 1 and 10 MPa) under the same conditions. Expression of *Sox-9* and *aggrecan* were found to increase with 0.1 MPa, whereas collagen required 10 MPa for a maximal increase. Sulphated GAG content of the ECM was found to increase with 1 and 10 MPa whereas collagen deposition required 10 MPa to be increased. This study (and others) confirms the complexity and synergistic effects of extracellular factors on chondrocyte responses.

26.9 Osteochondral defects

Owing to the difficulty in integrating tissue engineered cartilage constructs with the host tissue, many researchers focus on an osteochondral approach to cartilage repair to improve integration with the subchondral bone. These generally consist of one scaffold, often termed a biphasic or hybrid scaffold, with properties applicable to both osteoblasts and chondrocytes. Commonly the chondrocyte phase is a polymer (such as PLA, PLGA, or a natural material such as hyaluronan, collagen) with additional glass or ceramic phase (such as calcium phosphate, hydroxyapatite) for osteoblast culture.

Examples that have shown some success under the experimental conditions reported include D,L-PLGA/L-PLA with L-PLGA/tricalcium phosphate (TCP) (Sherwood *et al.*, 2002), injectable calcium phosphate and a hyaluronan derivative (Gao *et al.*, 2002), PLA/hydroxyapatite (Schek *et al.*, 2004), collagen gel with resorbable β-tricalcium phosphate (Tanaka *et al.*, 2005) and PCL reinforced with TCP (Shao *et al.*, 2006).

26.10 Conclusions and future trends

There are many materials that have been investigated for cartilage tissue engineering, including both natural and synthetic materials that have demonstrated chondrocyte or stem cell viability and matrix formation. There are many methods available for the production of porous 3D scaffolds for colonisation by cells and subsequent tissue formation. Although at first glance cartilage appears to be a relatively simple tissue containing only one cell type, it is clear that it is a complex tissue with high resistance to compression. The arrangement of cells and ECM components is complex and therefore will be difficult to accurately mimic in a tissue-engineered construct. This raises the question that is common for many tissues of the body, do we really need to go into such detail for tissue engineering? If a more simple engineered construct is implanted, will the normal body conditions direct the eventual detailed cellularity and matrix organisation? This may well occur as tissues of the body are constantly being remodelled depending upon the changing loads to which they are subjected. The effects of loading on the chondrocyte phenotype are clearly important; however, there is great discrepancy in the literature, demonstrating the sensitivity of the cells to respond to certain mechanical forces and the culture environment in which they are in.

Cartilage is itself a hydrogel so the ideal scaffold or template for cartilage engineering is also likely to be a hydrogel. ECM mimics are possibly the most relevant current materials. The ECM is constantly being remodelled and therefore implantation of a matrix mimic that has the potential to be remodelled along with the existing tissue is likely to be the future of cartilage engineering. The main drawback of this approach, however, is matching the biomechanical properties of the ECM mimics and the native cartilage. Ultimately natural materials or ECM mimics that resemble biomolecules found in cartilage are likely to produce the desired cellular effect but not necessarily the desired biomechanical effect.

26.11 References

Allemann, M., Mizuno, S., Eid, K., Yates, K. E., Zaleske, D., Glowacki, J., 2001. Effects of hyaluronan on engineered articular cartilage extracellular matrix gene expression in 3-dimensional collagen scaffolds. *Journal of Biomedical Materials Research*, **55**, 13–19.

Awad, H. A., Quinn Wickham, M., Leddy, H. A., Gimble, J. M., Guilak, F., 2004. Chondrogenic differentiation of adipose-derived adult stem cells in agarose, alginate, and gelatin scaffolds. *Biomaterials*, **25**, 3211–3222.

Barnewitz, D., Endres, M., Krüger, I., Becker, A., Zimmermann, J., Wilke, I., Ringe, J., Sittinger, M., Kaps, C., 2006. Treatment of articular cartilage defects in horses with polymer-based cartilage tissue engineering grafts. *Biomaterials*, **27**, 2882–2889.

Barry, J. J. A., Gidda, H. S., Scotchford, C. A., Howdle, S. M., 2004. Porous methacrylate scaffolds: supercritical fluid fabrication and *in vitro* chondrocyte responses. *Biomaterials*, **25**, 3559–3568.

Betre, H., Setton L. A., Meyer, D. E., Chilkoti, A., 2002. Characterization of a genetically engineered elastin-like polypeptide for cartilaginous tissue repair. *Biomacromolecules*, **3**, 910–916.

Betre, H., Ong, S. R., Guilak, F., Chilkoti, A., Fermor, B., Setton, L. A., 2006. Chondrocytic differentiation of human adipose-derived adult stem cells in elastin-like polypeptide. *Biomaterials*, **27** (1), 91–99.

Blunk, T., Sieminski, A. L., Gooch, K. J., Courter, D. L., Hollander, A. P., Nahir, M., Langer, R., Vunjak-Novakovic, G., Freed, J. E., 2002. Differential effects of growth factors on tissue-engineered cartilage. *Tissue Engineering*, **8** (1), 73–84.

Bray, J. C., Merrill, E. W., 1973. Poly(vinyl alcohol) hydrogels for synthetic articular cartilage. *Journal of Biomedical Materials Research*, **7** (5), 431–443.

Brittberg, M., Lindahl, A., Nilsson, A., Ohlsson, C., Isaksson, O., Peterson, L., 1994. Treatment of deep cartilage defects in the knee with autologous chondrocyte transplantation. *New England Journal of Medicine*, **331** (14), 889–895.

Bryant, S. J., Anseth, K. S., 2001. The effects of scaffold thickness on tissue engineered cartilage in photocrosslinked poly(ethylene oxide) hydrogels. *Biomaterials*, **22**, 619–626.

Buma, P., Pieper, J. S., van Tienen, T., van Susante, J. L. C., van der Kraan, P. M., Veerkamp, J. H., van den Berg, W. B., Veth, R. P. H., van Kuppevelt, T. H., 2003. Cross-linked type I and type II collagenous matrices for the repair of full-thickness articular cartilage defects – a study in rabbits. *Biomaterials*, **24**, 3255–3263.

Caterson, E. J., Nesti, L. J., Li, W. J., Danielson, K. G., Albert, T. J., Vaccaro, A. R., Tuan, R. S., 2001. Three-dimensional cartilage formation by bone marrow-derived cells seeded ion polylactide/alginate amalgam. *Journal of Biomedical Materials Research*, **57**, 394–403.

Chang, C. H., Liu, H. C., Lin, C. C., Chou, C. H., Lin, F. H., 2003. Gelatin–chondroitin–hyaluronan tri copolymer scaffold for cartilage tissue engineering. *Biomaterials*, **24**, 4853–4858.

Deng, Y., Zhao, K., Zhang, X .F., Hu, P., Chen, G. Q., 2002. Study on the three-dimensional proliferation of rabbit articular cartilage-derived chondrocytes on polyhydroxyalkanoate scaffolds. *Biomaterials*, **23**, 4049–4056.

Deng, Y., Lin, X. S., Zheng, Z., Deng, J. G., Chen, J. C., Ma, H., Chen, G. Q., 2003. Poly(hydroxybutyrate-co-hydroxyhexanoate) promoted production of extracellular matrix of articular cartilage chondrocytes *in vitro*. *Biomaterials*, **24**, 4273–4281.

Domm, C., Schunke, M., Steinhagen, J., Freitag, S., Kurz, B., 2004. Influence of various alginate brands on the redifferentiation of dedifferentiated bovine articular chondrocytes in alginate bead culture under high and low oxygen tension. *Tissue Engineering*, **10**, 1796–1805.

Dorotka, R., Windberger, U., Macfelda, K., Bindreiter, U., Toma, C., Nehrer, S., 2005. Repair of articular cartilage defects treated by microfracture and a three-dimensional collagen matrix. *Biomaterials*, **26**, 3617–3629.

Dowthwaite. G. P., Bishop, J. C., Redman, S. N., Khan, I. M., Rooney, P., Evans D. J. R., Haughton, L., Bayram, Z., Boyer, S., Thomson, B., Wolfe, M. S., Archer, C. W., 2004. The surface of articular cartilage contains a progenitor cell population. *Journal of Cell Science*, **117** (6), 889–897.

Eyrich, D., Brandl, F., Appel, B., Wiese, H., Maier, G., Wenzel, M., Staudenmaier, R., Goepferich, A., Blunk, T., 2007. Long-term stable fibrin gels for cartilage engineering. *Biomaterials*, **28**, 55–65.

Fan, H. B., Hu, Y. Y., Zhang, C. L., Li, X. S., Lv, R. Qin, L., Zhu, R., 2006. Cartilage regeneration using mesenchymal stem cells and a PLGA-gelatin/chondroitin/hyaluronate hybrid scaffold. *Biomaterials*, **27**, 4573–4580.

Fujisato, T., Sajiki, T., Liu, Q., Ikada, Y., 1996. Effect of basic fibroblast growth factor on cartilage regeneration in chondrocyte-seeded collagen sponge scaffold. *Biomaterials*, **17**, 155–162.

Gao, J., Dennis, J. E., Solchaga, L. A., Goldberg, V. M., Caplan, A. I., 2002. Repair of osteochondral defect with tissue-engineered two-phase composite material of injectable calcium phosphate and hyaluronan sponge. *Tissue Engineering*, **8**, 827–837.

Giannoni, P., Crovace, A., Malpeli, M., Maggi, E., Arbico, R., Cancedda, R., Dozin, B., 2005. Species variability in the differentiation potential of *in vitro*-expanded articular chondrocytes restricts predictive studies on cartilage repair using animal models. *Tissue Engineering*, **11**, 237–248.

Girotto, D., Urbani, S., Brun, P., Renier, D., Barbucci, R., Abatangelo, G., 2003. Tissue-specific gene expression in chondrocytes grown on three-dimensional hyaluronic acid scaffolds. *Biomaterials*, **24**, 3265–3275.

Gooch, K. J., Blunk, T., Courter, D. L., Sieminski, A. L., Vunjak-Novakovic, G., Freed, L. E., 2002. Bone morphogenetic proteins-2, -12, and -13 modulate *in vitro* development of engineered cartilage. *Tissue Engineering*, **8** (1), 591–601.

Grigolo, B., Roseti, L., Fiorini, M., Fini, M., Giavaresi, G., Aldini, N. N., Giardino, R., Facchini, A., 2001. Transplantation of chondrocytes seeded on a hyaluronan derivative (Hyaff (R)-11) into cartilage defects in rabbits. *Biomaterials*, **22**, 2417–2424.

Grigolo, B., Lisignoli, G., Piacentini, A., Fiorini, M., Gobbi, P., Mazzotti, G., Duca, M., Pavesio, A., Facchini, A., 2002. Evidence for redifferentiation of human chondrocytes grown on a hyaluronan-based biomaterial (HYAFF (R) 11): molecular, immunohistochemical and ultrastructural analysis. *Biomaterials*, **23**, 1187–1195.

Gugala, Z., Gogolewski, S., 2000. *In vitro* growth and activity of primary chondrocytes on a resorbable polylactide three-dimensional scaffold. *Journal of Biomedical Materials Research*, **49**, 183–191.

Guilak, F., Alexopoulos, L. G., Upton, M. L., Youn, I., Choi, J. B., Cao, L., Setton, L. A., Haider, M. A., 2006. The pericellular matrix as a transducer of biomechanical and biochemical signals in articular cartilage. *Annals of the New York Academy of Science*, **1068**, 498–512.

Hofmann, S., Knecht, S., Langer, R., Kaplan, D. L., Vunjak-Novakovic, G., Merkle, H.,P., Meinel, L., 2006. Cartilage-like tissue engineering using silk scaffolds and mesenchymal stem cells. *Tissue Engineering*, **12**, 2729–2738.

Holmes, T. C., 2002. Novel peptide-based biomaterial scaffolds for tissue engineering. *Trends in Biotechnology*, **20** (1), 16–21.

Holmes, T. C., de Lacalle, S., Su, X., Liu, G. S., Rich, A., Zhang, S. G., 2000. Extensive neurite outgrowth and active synapse formation on self-assembling peptide scaffolds. *Proceedings of the National Academy of Sciences*, **97** (12), 6728–6733.

Huang, Q., Hutmacher, D. W., Lee, E. H., 2002. *In vivo* mesenchymal cell recruitment by a scaffold loaded with transforming growth factor beta 1 and the potential for *in situ* chondrogenesis. *Tissue Engineering*, **8**, 469–482.

Hutmacher, D. W., 2000.Scaffolds in tissue engineering bone and cartilage. *Biomaterials*, **21**, Sp. Iss., 2529–2543.

Hwang, N.S., Varghese, S., Zhang, Z., Elisseeff, J., 2006a. Chondrogenic differentiation of human embryonic stem cell-derived cells in arginine-glycine-aspartate modified hydrogels. *Tissue Engineering*, **12**, 2695–2706.

Hwang, N. S., Varghese, S., Theprungsirikul, P., Canver, A., Elisseeff, J., 2006b. Enhanced chondrogenic differentiation of murine embryonic stem cells in hydrogels with glucosamine. *Biomaterials*, **27** (36), 6015–6023.

Ibusuki, S., Iwamoto, Y., Matsuda, T., 2003. System-engineered cartilage using poly(N-isopropylacrylamide)-grafted gelatin as *in situ*-formable scaffold: *in vivo* performance. *Tissue Engineering*, **9**, 1133–1142.

Jayawarna, V., Ali, M., Jowitt, T. A., Miller, A. F., Saiani, A., Gough, J. E., Ulijn, R.V., 2006. Nano-structured hydrogels for 3D cell culture through self-assembly of Fmoc-dipeptides. *Advanced Materials*, **18** (5), 611–614.

Kang, S. W., Jeon, O., Kim, B. S., 2005. Poly(lactic-co-glycolic acid) microspheres as an injectable scaffold for cartilage tissue engineering. *Tissue Engineering*, **11**, 438–447.

Kaps, C., Fuchs, S., Endres, M., Vetterlein, S., Krenn, V., Perka, C., Sittinger, M., 2004. Molecular characterization of tissue-engineered human articular chondrocyte transplants based on resorbable polymer fleeces. *Orthopade*, **33**, 76–84.

Kisiday, J., Jin, M., Kurz, B., Hung, H., Semino, C., Zhang, S., Grodzinsky, A. J., 2002. Self-assembling peptide hydrogel fosters chondrocyte extracellular matrix production and cell division: implications for cartilage tissue repair. *Proceedings of the National Academy of Sciences*, **99** (15), 9996–10001.

Kisiday, J. D., Jin, M., DiMicco, M. A., Kurz, B., Grodzinsky, A. J., 2004. Effects of dynamic compressive loading on chondrocyte biosynthesisin self-assembling peptide scaffolds. *Journal of Biomechanics*, **37**, 595–604.

Klompmaker, J., Jansen, H. W. B., Veth, R. P. H., DeGroot, J. H., Nijenhuis A. J., Pennings, A. J., 1991. Porous polymer implant for repair of meniscal lesions – a preliminary study in dogs. *Biomaterials*, **12** (9), 810–816.

Kurz, B., Domm, C., Jin, M. S., Sellckau, R., Schunke, M., 2004. Tissue engineering of articular cartilage under the influence of collagen I/III membranes and low oxygen tension. *Tissue Engineering*, **10**, 1277–1286.

Lee, C. R., Breinan, H. A., Nehrer, S., Spector, M., 2000. Articular cartilage chondrocytes in type I and type II collagen-GAG matrices exhibit contractile behavior *in vitro*. *Tissue Engineering*, **6**, 555–565.

Lee, H. J., Lee, J.-S., Chansakul, T., Yu, C., Elisseeff, J. H., Yu, S. M., 2006. Collagen mimetic peptide-conjugated photopolymerizable PEG hydrogel. *Biomaterials*, **27**, 5268–5276.

Li, W. J., Tuli, R., Okafor, C., Derfoul, A., Danielson, K. G., Hall, D. J., Tuan, R. S., 2005. A three-dimensional nanofibrous scaffold for cartilage tissue engineering using human mesenchymal stem cells. *Biomaterials*, **26**, 599–609.

Lisignoli, G., Cristino, S., Piacentini, A., Toneguzzi, S., Grassi, F., Cavallo, C., Zini, N., Solimando, L., Maraldi, N. M., Facchini, A., 2005. Cellular and molecular events during chondrogenesis of human mesenchymal stromal cells grown in a three-dimensional hyaluronan based scaffold. *Biomaterials*, **26**, 5677–5686.

Ma, P. X., Langer, R., 1999. Morphology and mechanical function of long-term *in vitro* engineered cartilage. *Journal of Biomedical Materials Research*, **44**, 217–221.

Mahmood, T. A., Shastri, V. P., van Blitterswijk, C. A., Langer, R., Riesle, J., 2005. Tissue engineering of bovine articular cartilage within porous poly(ether ester) copolymer scaffolds with different structures. *Tissue Engineering*, **11**, 1244–1253.

Malda, J., Woodfield, T. B. F., van der Vloodt, F., Wilson, C., Martens, D. E., Tramper, J., van Blitterswijk, C. A., Riesle, J., 2005. The effect of PEGT/PBT scaffold architecture on the composition of tissue engineered cartilage. *Biomaterials*, **26**, 63–72.

Mauck, R. L., Nicoll, S. B., Seyhan, S. L., Ateshian, G. A., Hung, C. T., 2003. Synergistic action of growth factors and dynamic loading for articular cartilage tissue engineering. *Tissue Engineering*, **9**, 597–611.

McHale, M. K., Setton, L. A., Chilkoti, A., 2005. Synthesis and *in vitro* evaluation of enzymatically cross-linked elastin-like polypeptide gels for cartilaginous tissue repair. *Tissue Engineering*, **11**, 1768–1779.

Mercier, N. R., Costantino, H. R., Tracy, M. A., Bonassar, L. J., 2005. Poly(lactide-co-glycolide) microspheres as a moldable scaffold for cartilage tissue engineering. *Biomaterials*, **26**, 1945–1952.

Miyanishi, K., Trindade, M. C. D., Lindsey, D .P., Beaupre, G. S., Carter, D. R., Goodman, S. B., Schurman, D. J., Smith, R. L., 2006a. Effects of hydrostatic pressure and transforming growth factor-beta 3 on adult human mesenchymal stem cell chondrogenesis *in vitro*. *Tissue Engineering*, **12**, 1419–1428.

Miyanishi, K., Trindade, M. C. D. Lindsey, D. P., Beaupre, G. S., Carter, D. R., Goodman, S. B., Schurman, D. J., Smith, R. L., 2006b. Dose- and time-dependent effects of cyclic hydrostatic pressure on transforming growth factor-beta 3-induced chondrogenesis by adult human mesenchymal stem cells *in vitro*. *Tissue Engineering*, **12**, 2253–2262.

Mizuno, S., Allemann, F., Glowacki, J., 2001. Effects of medium perfusion on matrix production by bovine chondrocytes in three-dimensional collagen sponges. *Journal of Biomedical Materials Research*, **56**, 368–375.

Muller, F. A., Muller, L., Hofmann, I., Greil, P., Wenzel, M. M., Staudenmaier, R., 2006. Cellulose-based scaffold materials for cartilage tissue engineering. *Biomaterials*, **27** (21), 3955–3963.

Murphy, C. L., Sambanis, A., 2001. Effect of oxygen tension and alginate encapsulation on restoration of the differentiated phenotype of passaged chondrocytes. *Tissue Engineering*, **7**, 791–803.

Na, K., Park, J. H., Kim, S. W., Sun, B. K., Woo, D. G., Chung, H. M., Park, K. H., 2006. Delivery of dexamethasone, ascorbate, and growth factor (TGF beta-3) in thermo-reversible hydrogel constructs embedded with rabbit chondrocytes. *Biomaterials*, **27**, 5851–5957.

Nehrer, S., Breinan, H .A., Ramappa, A., Young, G., Shortkroff, S., Louie, L. K., Sledge, C. B., Yannas, I. V., Spector, M., 1997. Matrix collagen type and pore size influence behaviour of seeded canine chondrocytes. *Biomaterials*, **18**, 769–776.

Park, H., Temenoff, J. S., Holland, T. A., Tabata, Y., Mikos, A. G., 2005. Delivery of TGF-beta 1 and chondrocytes via injectable, biodegradable hydrogels for cartilage tissue engineering applications. *Biomaterials*, **26**, 7095–7103.

Park, Y., Lutolf, M. P., Hubbell, J. A., Hunziker, E. B., Wong, M., 2004. Bovine primary chondrocyte culture in synthetic matrix metalloproteinase-sensitive poly(ethylene glycol)-based hydrogels as a scaffold for cartilage repair. *Tissue Engineering*, **10**, 151–522.

Peretti, G. M., Xu, J. W., Bonassar, L. J., Kirchhoff, C. H., Yaremchuk, M. J., Randolph, M. A., 2006. Review of injectable cartilage engineering using fibrin gel in mice and swine models. *Tissue Engineering*, **12**, 1151–1168.

Perka, C., Schultz, O., Spitzer, R. S., Lindenhayn, K., 2000a. The influence of transforming growth factor beta 1 on mesenchymal cell repair of full-thickness cartilage defects. *Journal of Biomedical Materials Research*, **52**, 543–552.

Perka, C., Spitzer, R. S., Lindenhayn, K., Sittinger, M., Schultz, O., 2000b. Matrix-mixed culture: new methodology for chondrocyte culture and preparation of cartilage transplants. *Journal of Biomedical Materials Research*, **49**, 305–311.

Pieper, J. S., van der Kraan, P. M., Hafmans, T., Kamp, J., Buma, P., van Susante, J. L. C., van den Berg, W. B., Veerkamp, J. H., van Kuppevelt, T. H., 2002. Crosslinked type II collagen matrices: preparation, characterization, and potential for cartilage engineering. *Biomaterials*, **23**, 3183–3192.

Schek, R. M., Taboas, J. M., Segvich, S. J., Hollister, S. J., Krebsbach, P. H., 2004. Engineered osteochondral grafts using biphasic composite solid free-form fabricated scaffolds. *Tissue Engineering*, **10**, 1376–1385.

Shao, X., Goh, J. C. H., Hutmacher, D. W., Lee, E. H., Zigang, G., 2006. Repair of large articular osteochondral defects using hybrid scaffolds and bone marrow-derived mesenchymal stem cells in a rabbit model. *Tissue Engineering*, **12**, 1539–1551.

Sherwood, J. K., Riley, S.L., Palazzolo, R., Brown, S. C., Monkhouse, D. C., Coates, M., Griffith, L. G., Landeen, L. K., Ratcliffe, A., 2002. A three-dimensional osteochondral composite scaffold for articular cartilage repair. *Biomaterials*, **23**, 4739–4751.

Shields, K. J., Beckman, M. J., Bowlin, G. L., Wayne, J. S., 2004. Mechanical properties and cellular proliferation of electrospun collagen type II. *Tissue Engineering*, **10**, 1510–1517.

Silva, G. A., Czeisler, C., Niece, K. L., Beniash, E., Harrington, D.A., Kessler, J. A., Stupp, S. I., 2004. Selective differentiation of neural progenitor cells by high-epitope density nanofibers. *Science*, **303** (5662), 1352–1355.

Stevens, M. M., Qanadilo, H. F., Langer, R., Shastri, V. P., 2004. A rapid-curing alginate gel system: utility in periosteum-derived cartilage tissue engineering. *Biomaterials*, **25**, 887–894.

van Susante, J. L. C., Pieper, J., Buma, P., van Kuppevelt, T. H., van Beuningen, H., van der Kraan, P. M., Veerkamp, J. H., van den Berg, W. B., Veth, R. P. H., 2001. Linkage of chondroitin-sulfate to type I collagen scaffolds stimulates the bioactivity of seeded chondrocytes *in vitro*. *Biomaterials*, **22**, 2359–2369.

Taguchi, T., Xu, L. M., Kobayashi, H., Taniguchi, A., Kataoka, K., Tanaka, J., 2005. Encapsulation of chondrocytes in injectable alkali-treated collagen gels prepared using poly(ethylene glycol)-based 4-armed star polymer. *Biomaterials*, **26**, 1247–1252.

Tanaka, T., Komaki, H., Chazano, M., Fujii, K., 2005. Use of a biphasic graft constructed with chondrocytes overlying a β-tricalcium phosphate block in the treatment of rabbit osteochondral defects. *Tissue Engineering*, **11**, 331–339.

Uematsu, K., Hattori, K., Ishimoto,Y., Yamauchi, J., Habata, T., Takakura, Y., Ohgushi, H., Fukuchi, T., Sato, M., 2005. Cartilage regeneration using mesenchymal stem cells and a three-dimensional poly-lactic-glycolic acid (PLGA) scaffold. *Biomaterials*, **26**, 4273–4279.

Veilleux, N. H., Yannas, I. V., Spector, M., 2004. Effect of passage number and collagen type on the proliferative, biosynthetic, and contractile activity of adult canine articular chondrocytes in type I and II collagen-glycosaminoglycan matrices *in vitro*. *Tissue Engineering*, **10**, 119–127.

Wang, D. W., Fermor, B., Gimble, J. M., Awad, H. A., Guilak, F., 2005. Influence of oxygen on the proliferation and metabolism of adipose derived adult stem cells. *Journal of Cellular Physiology*, **204**, 184–191.

Wang, Y. Z., Blasioli, D. J., Kim, H. J., Kim, H. S., Kaplan, D. L., 2006. Cartilage tissue engineering with silk scaffolds and human articular chondrocytes. *Biomaterials*, **27**, 4434–4442.

Wayne, J. S., McDowell, C. L., Shields, K. J., Tuan, R. S., 2005. *In vivo* response of polylactic acid-alginate scaffolds and bone marrow-derived cells for cartilage tissue engineering. *Tissue Engineering*, **11**, 953–963.

Williams, C. G., Kim, T. K., Taboas, A., Malik, A., Manson, P., Elisseeff, J., 2003. *In vitro* chondrogenesis of bone marrow-derived mesenchymal stem cells in a photopolymerizing hydrogel. *Tissue Engineering*, **9**, 679–688.

Woodfield, T. B. F., Malda, J., de Wijn, J., Peters, F., Riesle, J., van Blitterswijk, C. A., 2004. Design of porous scaffolds for cartilage tissue engineering using a three-dimensional fiber-deposition technique. *Biomaterials*, **25**, 4149–4161.

Yamane, S., Iwasaki, N., Majima, T., Funakoshi, T., Masuko, T., Harada, K., Minami, A., Monde, K., Nishimura, S., 2005. Feasibility of chitosan-based hyaluronic acid hybrid biomaterial for a novel scaffold in cartilage tissue engineering. *Biomaterials*, **26**, 611–619.

Yasuda, A., Kojima, K., Tinsley, K. W., Yoshioka, H., Mori, Y., Vacanti, C. A., 2006. *In vitro* culture of chondrocytes in a novel thermoreversible gelation polymer scaffold containing growth factors. *Tissue Engineering*, **12**, 1237–1245.

Yoo, H. S., Lee, E. A., Yoon, J. J., Park, T. G., 2005. Hyaluronic acid modified biodegradable scaffolds for cartilage tissue engineering. *Biomaterials*, **26**, 1925–1933.

Yoon, D. M., Hawkins, E. C., Francke-Carroll, S., Fisher, J. P., 2007. Effect of construct properties on encapsulated chondrocyte expression of insulin-like growth factor-1. *Biomaterials*, **28**, 299–306.

Index